D0746579

No Time to be Brief

No Time to be Brief

A scientific biography of Wolfgang Pauli

Charles P. Enz
University of Geneva

OXFORD
UNIVERSITY PRESS

OXFORD

UNIVERSITY PRESS

Great Clarendon Street. Oxford ox2 6dp

Oxford University Press is a department of the University of Oxford.
It furthers the University's objective of excellence in research. scholarship.
and education by publishing worldwide in

Oxford New York

Auckland Cape Town Dar es Salaam Hong Kong Karachi
Kuala Lumpur Madrid Melbourne Mexico City Nairobi
New Delhi Shanghai Taipei Toronto

With offices in

Argentina Austria Brazil Chile Czech Republic France Greece
Guatemala Hungary Italy Japan Poland Portugal Singapore
South Korea Switzerland Thailand Turkey Ukraine Vietnam

Oxford is a registered trade mark of Oxford University Press
in the UK and in certain other countries

Published in the United States
by Oxford University Press Inc.. New York

© Oxford University Press 2002

The moral rights of the author have been asserted
Database right Oxford University Press (maker)

First published 2006
First published in paperback 2010

All rights reserved. No part of this publication may be reproduced.
stored in a retrieval system. or transmitted. in any form or by any means.
without the prior permission in writing of Oxford University Press.
or as expressly permitted by law. or under terms agreed with the appropriate
reprographics rights organization. Enquiries concerning reproduction
outside the scope of the above should be sent to the Rights Department.
Oxford University Press. at the address above
You must not circulate this book in any other binding or cover
and you must impose the same condition on any acquirer

British Library Cataloguing in Publication Data
Data available

Library of Congress Cataloging in Publication Data
Data available

Typeset by the author
Printed in Great Britain
by the
MPG Books Group. Bodmin and King's Lynn

ISBN 978-0-19-958815-2(Pbk.) 978-0-19-856479-9(Hbk.)

1 3 5 7 9 10 8 6 4 2

Preface

It was my fate to be Wolfgang Pauli's assistant at the time of his death in December 1958. When Pauli's widow, Franca, was considering a biography of her late husband, she considered that I was the obvious candidate. On 24 October 1983 she wrote to me in German: 'Herewith I request Professor Charles P. Enz, Geneva, to write the biography of his former teacher of theoretical physics, Professor Wolfgang Pauli.' This was a challenge particularly because I had a personal knowledge of Pauli's impatience with vague expression.

Although I began collecting material for this task at once, my preoccupation with teaching and research forced a pace upon me which was clearly too slow for Mrs Pauli. When after 2 1/2 years, she did not see a biography forthcoming, she became impatient and decided that there whould never be a biography written by me. There was still no manuscript when Franca Pauli passed away on 11 July 1987. It is only since my retirement in 1995 that I have been able to devote all my time to this endeavour. It saddens me that I could not present Mrs Pauli with my work, which I therefore dedicate to her memory.

There was no question that this biography should be written in any language but English. Since much of Pauli's enormous correspondence, edited in several volumes by Karl von Meyenn, is written in German, I chose to include numerous passages from such letters in English translation.

In writing this book I enjoyed the assistance of several people. First, Dr Beat Glaus, the former director of the science history collection at the main library of ETH (Eidgenössische Technische Hochschule), the Federal Institute of Technology in Zurich. His unassuming, friendly, and expert advice and support were essential to the outcome of this enterprise. Also essential was the critical counsel I received from Dr Vincent Frank-Steiner (Basle), a nephew of Paul Rosbaud, the editor of many works by and friend of Pauli. On the origins of Pauli's family I received help from Professor Gerhard Oberkofler (Innsbruck), Professor Hans Warhanek (Vienna), Dr František Smutný (Prague), and Dr Daniel Pascheles (Zurich).

I am particularly indebted to Professor Markus Fierz for his many letters sharing with me his deep knowledge of Pauli's thinking. I also benefited from commentaries by Professor Karl von Meyenn, Dr Herbert van Erkelens, and Professor Wolf Beiglböck, and many letters from Bernhard Lötscher. I also acknowledge the cooperation of Anja Tschörtner and Sönke Adlung at Oxford University Press.

For very essential support on the computer I thank Dr Andreas Malaspinas, the expert at the Department of Theoretical Physics, University of Geneva. My thanks also go to the head of this department, Professor Jean-Pierre Eckmann, and his successor, Professor Markus Büttiker, for allowing me to use the department's facilities. I thank the staff of the Physics Library, Joceline Favre, Claire-Lise Held, and Christina Sironi-Windahl, for their friendly help in procuring documents. Margaret E. Schmid, the former secretary of Scherrer and Pauli, I thank for many comments and photographs. Finally, I have to thank my wife Ilse for her patience.

Geneva   C.P.E.
October 2001

Contents

1
Pauli's family

The ancestors

With me everything is more complicated than with you, the ancestors as well as the religion. The former are Jewish as well as Christian (among the Christians a somewhat decadent, now extinct Austrian noble family as well as very healthy and robust Czechs also). The Jewish ancestors were in Prague for a long time, certainly a very characteristic city (the 'Golem' by Meyrink had always fascinated me) where, however, I never was . . . . [1a]

This autobiographical note of 1950, contained in a letter from Pauli to Aniela Jaffé (1903–91), the secretary of Carl Gustav Jung (1875–1961), conveys a first glimpse of Wolfgang Pauli's roots that are to be found in nineteenth-century Vienna, the capital of the Austrian Empire, as well as in Prague, the capital of the kingdom of Bohemia which in 1866 came under the Austrian crown. Bohemia's population was predominantly Czech, with a small but influential German minority and Jews in between. Three of Pauli's grandparents belonged to the small, isolated, but dominant German-speaking population in Prague. Culturally, an important fraction of this minority were Jews who had moved out of the ghetto in the 1850s.

Located between the Old Town Square (Altstädter Ring) and the Moldavia (Vltava) river, the Prague ghetto contained some five synagogues, the oldest dating from the thirteenth century, and a Jewish cemetery which dates back to the thirteenth century and is the oldest Jewish cemetery still in existence in Europe. The ghetto was created in the 13th century in order to separate the Jews from the Christian population and was surrounded by a wall. The latter was torn down in 1848 when in all the big cities of the Austrian Empire namely Vienna, Prague, Budapest, and Milan, a revolution erupted which forced Count Clemens Lothar von Metternich, the dominant figure of the post-Napoleonic Restoration, to abdicate.

The Ghetto was then called 'Josefov' or 'Josef's Town' in honour of Emperor Josef II (1741–90), the son of Empress Maria Theresia (1717–80), called the 'Volkskaiser', who had introduced equal rights for the Jews. But it was only at the turn of the century that the slum-like ghetto was renovated, which induced Franz Kafka (1883–1924) to observe that the unhealthy old Jewish town (Judenstadt) was much more real in his mind than the hygienic new town. Among the group of young German writers in Prague, to which belonged Rainer Maria Rilke (1875–1926), Franz Werfel (1890–1945), and Gustav Meyrink (1868–1932) of the *Golem*, Kafka was one of the few who spoke Czech fluently. The Germans generally considered Czech to be the language of servants. But humiliated Czechs reacted by attacking the houses of Germans or Jews at night.

Although Pauli was never in Prague, this city, as we shall see, contained his most important roots. During the reign (1347–78), of the Bohemian king Charles IV, Prague was the flourishing imperial capital where in 1348 the university named after Charles was founded, the oldest in German lands. Prague was then the largest city in Europe after Paris. Towards the end of the reign of Emperor Rudolf II (1552–1612), Johannes Kepler (1571–1630) moved from Graz to Prague in 1600 to work with the famous astronomer Tycho Brahe (1546–1601). The latter, however, died and Kepler succeeded him as imperial mathematician and discovered his famous three laws. In 1604 the Swiss watchmaker and mathematician Jost Bürgi (1552–1632) also entered the service of Emperor Rudolf II and became friendly with Kepler. The latter encouraged Bürgi to work out his mathematical findings— Bürgi had discovered logarithms long before John Napier (1550–1617) (see Ref. [1b]). Also in Prague the High Rabbi Löw (1520-1609) was said to have created in 1580, by his erudition in the Jewish tradition of the Cabbala, a golem who could act as his servant. In Hebrew *Golem* means a formless mass, but among the Jewish mystics it is a human figure of clay that pious masters were able to animate [2a,b]. More than a quarter of a millenium after Kepler, in 1867, another scientist moved from Graz to Prague, Ernst Mach (1838–1916) who, for 28 years, was the dominant figure in experimental and physiological physics and the philosophy of science at Charles University. During Mach's tenure in 1882 the university was divided into the German Ferdinand University and the Czech Charles University.

Wolf Pascheles, the grandfather of Pauli's father, was an editor and a noted author of Jewish literature in Prague; in fact, he is listed in most of the Jewish encyclopedias [2b-e]. He was born in Bohemia on 11 May 1814, the son of poor parents, and died in Prague on 22 November 1857 [3,4a]. Speculation has it [4c] that the Pascheles family originated in Spain where they had the name Pasquales. The likely reason for the migration to the East was the Inquisition. This also suggests that many other names ending in 'es' are of Spanish origin (see Ref. [5], p. 8). According to Wolf Pascheles' own handwritten notes, 'I was five years old when I came into the "teaching house for poor children" (*Armenkinderlehrhaus*). My mother died when I was one-and-a-half years old.' At age 14 'in the year 1828 I wrote a small popular prayer book for Israelite women and selichots [liturgical poems recited as prayers of repentance and/or forgiveness [3]], as the first epidemic of cholera broke out in Prague ... '. In the year 1837 he acquired the bookseller's right, and the following year he married Sara Taubeles, the daughter of a 'respected Talmudist'. At that time he took his poor old father into his business. 'In the year 1846 he conceived the plan to collect the popular tales of the Israelites and to edit them together with biographies of famous Israelites, Jewish stories, etc. under the title "Sippurim"' (translated from Ref. [4a]).

In 1852 Wolf Pascheles continued to publish his *Sippurim* in 17 more issues. For a copy he presented to Emperor Franz Josef he received the golden medal *Viribus unitis*. In the same year he also founded the *Pascheles illustrierter Volkskalender* which appeared in 18 annual issues [2e]. This was the origin of the Publishing House of Hebrew Literature which his son Jacob finally established in Prague's Old Town Square (Altstädter Grosser Ring) [3].

Wolf's son Jacob W. Pascheles, Pauli's grandfather, was born in Prague on 2 January 1839. He married Helen Utitz (Ref. [5], p. 102) of the influential Utitz family [5] whose descendants owned 'one of the largest leather factories on the Continent, the firm Utitz Brothers in Prague' (translated from Ref. [4b]; these were Helen's brothers, see also Ref. [5], pp. 114-15). Helen Utitz was born in Prague in February 1841 (Ref. [5], p. 89). Jacob and Helen Pascheles-Utitz had seven children ([4c]; see also Ref. [5], p. 117), of which, however, only five are listed in the Utitz family tree of Ref. [5]: Oskar, Isidor, Wolf, Olga, and Julius (two daughters are missing there).

The family lived at No. 7 of Prague's Old Town Square (see Fig. 2 of Ref. [3]) while the bookshop was at No. 11, next to the Kinsky Palace where the German Old Town High School (*Deutsches Staatsgymnasium Altstadt*) was located. Jacob Pascheles 'was steeped in all aspects of Jewish culture.' (Ref. [5], p. 107). He was an elder of the congregation of Prague's Gipsy's Synagogue (*Zigeuner Synagoge*) and in this function assisted at the confirmation (bar mitzvah) of Franz Kafka [3] on 13 June 1896 (see the printed invitation by his father Hermann Kafka on p. 28 of Ref. [6a]). Hermann Kafka had a garment store on the ground floor of the Kinsky Palace (see his card and the picture of the Kinsky Palace on pp. 24–5 of Ref. [6a]). For a time the Kafka family also lived in the Old Town Square but moved to Celetná Street (*Zeltner Gasse*) in 1896 [3] (see the city map on pp. 52–3 of Ref. [6a]). Jacob's brother Samuel W. Pascheles also lived on this street [3]. Jacob Pascheles died in Prague on 23 November 1897 and was buried in the Old Jewish Cemetery on Meiselgasse, two blocks down from the Old Town Square (see the city map on pp. 52–3 of Ref. [6a]) where his tombstone still stands (see Fig. 1 of Ref. [3]). The inscription on the tombstone, in Hebrew and German, reads: 'Jacob W. Pascheles, editorial bookseller (*Verlagsbuchhändler*), principal (*Vorsteher*) of the Israelite burial fraternity (*Beerdigungsbruderschaft*) and of the Gipsy Synagogue in Prague, born 2 January 1839, died 23 November 1897'. His wife Helen outlived him by 30 years; she died in 1927 [4b].

Friedrich Schütz, the father of Pauli's mother, was born in Prague on 24 April 1845 [6b, 7a,b]. He received his education in the schools of the Piarists and later in the Commercial Academy of Prague [2e]. In high school he developed an active interest in literature and theatre, writing dramatic poetry. And he engaged in a lifelong career as a writer and journalist. In 1869 he moved to Vienna, after several of his plays had already been performed; his first theatre piece, 'Gegenseitig', performed at the Carl Theatre in Prague, had been a complete success [7b]. In Vienna in 1873 he joined the influential liberal

Viennese newspaper the *Neue Freie Presse* [7b], which 'was almost exclusively dependent on Jewish money and talent' (Ref. [5], p. 112), as a collaborator on the political section and as a serial contributor and theatre chronicler, and also in 1873 became a member of the editorial board. In 1894 he was sent to Russia to report on the coronation of Czar Nicholas II. This journey gave rise in 1897 to a book on Russia, *Das heutige Russland* [6b]. Friedrich Schütz wrote well over a dozen theatre pieces, mostly comedies with political content, which were produced on the first German-speaking stages. Several pieces are listed in the literature surveys published between 1868 and 1900: 'William the Conqueror' (a comedy in two acts), 'Systematic' (a comedy in two acts), 'Old Girls' (a comedy in one act), and a military prank [8a].

In 1875 Friedrich Schütz married Bertha Dillner, with whom he had two daughters: Bertha, the mother of Wolfgang Pauli [7a,b], and Rika, who was seven years younger [8b]. His last years were devoted to writing his recollections of political life. This work, *Werden und Wirken des Bürgerministeriums*, which remained unfinished, was edited by his daughter Bertha in 1909 [6b]. He died in Vienna after a long illness on 22 December 1908 [7c]. Contrary to what could be concluded from the quotation introducing this section, Friedrich Schütz was Jewish but non-practising in later years. He figures in Ref. [2e] among the 'Jewish men and women'. As follows from Pauli's letter of 29 May 1940 to Frank Aydelotte, the director of the Institute for Advanced Study in Princeton, according to German law Pauli was 75 per cent Jewish [7d].

Bertha Schütz-Dillner was born in Vienna on 10 November 1847, the daughter of Otto Dillner von Dillnersdorf, *Polizei-Oberkommissär* (Police Staff Officer) [6b] of noble descent. She sang in concerts as a child and entered the then new opera school in Vienna. From there she immediately went on the stage at the Kärntnertor Theatre in 1865 and, one year later, joined the opera houses in Cologne and in Prague where, between 1866 and 1872, she reached the summit of her musical career. It was only in 1873 that she was able to join the Court Opera in her home town, and for 12 years she held numerous soprano and mezzo-soprano parts and also gave guest performances in Berlin, Stuttgart, Dresden, and Weimar. At the end of 1884 she retired from the Court Opera where she was made an honorary member [6b,7e].

Bertha Schütz-Dillner had acquired an unusually large repertoire, excluding only vehement or frivolous characters, and had sung the principal parts of a great many operas during her career. She was a woman of noble appearance and warm human contact and had an exceptionally harmonious marriage and family life [7e]. Young 'Wolfi' or 'Wolferl', as Wolfgang Pauli was then called, loved his grandmother who sang and played the piano for him. Much later, in 1953, she appears as the 'distinguished lady' ('*vornehme Dame*') in Pauli's fantasy *The Piano Lesson* [9] where she blends in with 'the girl that lives in Küsnacht', namely Pauli's later friend Marie-Louise von Franz (1915–98) whose father, the 'captain' (Hauptmann) in the fantasy, had also been an imperial officer in Vienna. Bertha Schütz-Dillner died in Vienna on 27 June 1916 [7e,8b].

The parents

On 4 November 1955 my father died from heart failure at an advanced age.
This also results in a considerable change in the unconscious and I suspect
that for me this means a change (Wandlung) of the shadow. Indeed, with
me the shadow was projected for a long time upon the father, and I had to
learn gradually to distinguish the dream figure (Traumfigur) of the shadow
from the real father. Correspondingly, the binding of the luminous anima to
the shadow or the devil (. . .) previously often appeared projected upon the
'evil stepmother' (his much younger second wife whom my father now leaves
behind) and upon my father. [11a]

This retrospect, contained in a long letter from Pauli to Jung in 1956 with the
title 'Expressions (*Aussagen*) of the Psyche', shows the complex relationship of
the grown-up Pauli with his father. The reasons for this complexity, according
to Pauli's widow Franca, were twofold: first, that Pauli's father let himself
be baptized; and second, that he remarried [11b]. In Pauli's childhood the
family lived in grandfather Friedrich Schütz's house at Anton-Frank-Gasse
18 in Währing, Vienna's 18th district. Anton-Frank-Gasse is a short street off
Gymnasiumstrasse which continues northwards into the 19th district of Döbling
and whose name derives from the Döbling Gymnasium (kaiserlich königliches
Staatsgymnasium, Imperial State High School) which young Pauli attended. To
the north-west, at Hartäckerstrasse, is the Döbling cemetery where the Schütz
family tomb is located.

This part of Vienna is some 5 km to the north-west of the inner city,
the old town built around Saint Stephen's Cathedral. The inner city is neatly
separated from the newer districts by the monumental Ringstrasse, where in
the era of liberalism, the Town Hall (Rathaus), the Parliament (Reichsrat), the
Imperial Court Opera (kaiserlich königliche Hofoper), the Hofburg Theatre,
the Museums of Art and of Natural History, and the university, founded in
1365, were built between 1870 and 1880. The Liberals had come to power in
1860 but, gradually, new social and political groups claimed access to power:
Christian-Socialists, Pangermanists, and Slavic nationalists. At the turn of the
century Vienna was in turmoil and liberalism lost hold of government. In 1897
the anti-semitic leader of the Christian-Socialist Party, Lueger, '*der schöne*
Karl' (beautiful Charles), became mayor of the city of Vienna where one year
later chaotic events greeted the 50th anniversary of the reign of Emperor Franz
Josef (1830–1916). For a more extensive account of the history of Vienna see,
e.g. Ref. [13a].

This was the time of the upheaval of the young against their fathers, as exemplified by Sigmund Freud (1856–1939) in his self-analysis *The Interpretation*
of Dreams, published in 1900. For this *fin de siècle* modernity these chaotic
times provided a fertile brew producing a new culture, irreverent towards the
baroque tradition and based on science, psychology, instincts, and aesthetics.
Among the most prominent representatives were: in art, Gustav Klimt (1862–
1918), the leader of the Secession movement; in literature, Arthur Schnitzler
(1862–1931), Hugo von Hofmannsthal (1874-1929), and the cultural critic

Fig. 1.1 The Pauli house at Anton-Frank-Gasse 16, Vienna (CERN)

Karl Kraus (1874-1936); and in music, Gustav Mahler (1860–1911), the director of the Court Opera, and Arnold Schönberg (1874–1951). Theodor Herzl (1860–1904), the founder of Zionism, and, somewhat later, Oskar Kokoschka (1881–1980), the co-founder of expressionism, and Erwin Schrödinger (1887–1961), the founder of wave mechanics, were also products of this fertile brew. For a masterly exposition of the politics and culture of *fin de siècle* Vienna see Ref. [13b].

Pauli's father Wolfgang Josef Pascheles, called Wolf like his grandfather, was born in Prague on 11 September 1869. He attended the German Old Town High School in Prague from 1879 to 1887 where, for the first four years, he was a classmate of Ernst Mach's oldest son Ludwig who was born on 10 November 1868 [3]. Seven years after Wolf finished at high school Franz Kafka started at the same school. In 1887 Wolf Pascheles and Ludwig Mach met again when they both registered at the Medical Faculty of the German Carl-Ferdinand University in Prague where Ernst Mach was professor of experimental physics from 1867 to 1895. It is very likely that Wolf and Ludwig attended the 'Practicum of

Physics for Medical Students' that Mach taught there [3].

According to his handwritten curriculum vitae [14a] of 1897, Wolf Pascheles received his doctor's degree from Carl-Ferdinand University on 24 April 1893. He emphasizes that 'during the whole University period the undersigned frequented the physical University institute and there occupied himself also in the purely physical direction' (translated from Ref. [14a]). This shows not only that he had direct contact with Ernst Mach, quite independent of son Ludwig, but also that from the beginning, physics played an essential role in his scientific career. Indeed, the curriculum vitae closes with an acknowledgement to his teachers Ernst Mach and the physiologists Ewald Hering, a colleague of Mach in Prague (see Ref. [13a]), and Franz Hofmeister, professor in Strasbourg, who 'exerted a decisive influence on the direction and method of work of the undersigned'. In fact, Ernst Mach, already in Prague, submitted to the Imperial Academy of Sciences in Vienna short communications concerning the young Pascheles, 'for whom he remained a fatherly friend to the end' [15b]. As will become evident in the course of this chapter, Ernst Mach also had a decisive influence on Wolfgang Pauli's destiny.

Wolf Pascheles' curriculum vitae also informs us that 'On 1 January 1893, ... on invitation of his present head [Hermann Nothnagel], the undersigned left his native town' to work as a medical doctor at the Imperial (kaiserlich königliches) Rudolf Hospital in Vienna, where he rose from 'external praeparand' to 'stationary departmental assistant' [14a]. He also worked at the Vienna Policlinic [15c]. In 1898 he became a Privatdozent for internal medicine at the University of Vienna [14b]; he then lived at Gluckgasse in the inner city, near the Old Imperial Castle (Alte Hofburg) [8c].

In 1895, when he was already 57 years old, Ernst Mach accepted the third professorship of philosophy at the University of Vienna which included the commission of a course (Lehrauftrag) on philosophy, in particular the history and theory of the inductive sciences [16a]. In the same year his son Ludwig obtained his doctorate in Prague [3]. Mach's move to Vienna may have had several reasons. On the one hand, the working conditions in Mach's laboratory in Prague had deteriorated, and son Ludwig saw himself constrained to take over most of the responsibilities. But perhaps more seriously was the suicide the year before of Mach's other son, Heinrich, who had just successfully completed his studies at Göttingen. Mach, who had four sons and one daughter, never quite recovered from this loss [16b]. In Vienna Mach, who did not believe in the reality of atoms, became an antagonist of Ludwig Boltzmann (1844-1906), his colleague in theoretical physics, and an unconditional proponent of the atomic foundations of thermodynamics (see e.g. Ref. [13a]).

Wolf Pascheles' move to Vienna appears as a complete break with his past; it looks as if he had only waited for his father to die. On 28 July 1898 he obtained official permission to change his family name to Pauli, on 7 March 1899 he quit the Israelite cult community, and on 2 May 1899 married Bertha Camilla Schütz, after having been baptized a Catholic [8c,d]. The reason for this break, no doubt, was the rising anti-Semitism which represented a real danger to his ambition to pursue a university career. Indeed, Freud, who like Pascheles had started his career in Vienna at a hospital in 1886, remained there for 10 years with scant opportunity for research and no prestige. It was only in 1902, and not without making use of influential connections, that he at last was nominated university professor [13b]. 'For Wolf and the ten per cent of Jews who converted during the last decades of the nineteenth century, changing religion had little to do with their theological beliefs. It was a necessary step to further their careers. They had not been "good Jews", and they had no intention of becoming "good Christians". All Wolf wanted was to be a good professor of chemical medicine' (Ref. [5], p. 121).

And, indeed, it seemed to work, because in 1907 he received the title of 'ausserordentlicher Universitätsprofessor'. Although this position was purely honorary, he had become director of the physico-chemical department of the Biological Experimental Institute the year before [14b]. In 1920 he was nominated 'ausserordentlicher Professor ad personam' for biophysical chemistry at the Medical Faculty [14c]. And in 1922 he became 'Ordinarius' (full professor) and director of the new Institute for Medical Colloid Chemistry at the university which, located at Währingerstrasse, two blocks down Gymnasiumstrasse from his home at Anton-Frank-Gasse, had been created for him [15a]. In fact, he had become internationally known through his many important publications which include three books: *Physical Chemistry in the Service of Medicine* (New York, 1907), *Kolloidchemie der Eiweisskörper* (Dresden–Leipzig, 1920), which was translated into English (London, 1922) and had an extended second edition (1933), and *Elektrochemie der Kolloide*, with E. Valko (Berlin, 1929) [15a].

It is not clear why Wolf Pascheles chose Pauli as his new family name. While preservation of the initial was the rule in legal name changes, Jewish converts often tried to preserve several letters. Smutný has speculated [3] that the name Pauli might be reminiscent of the house where the Pascheles had lived in Prague. Indeed, the house at No. 7 on the Old Town Square was formerly a convent of Paulans (Minims of St Francis of Paola, originally known as Hermits of St. Francis of Assisi [3]). Interestingly, one of Wolf's brothers, Isidor Pascheles, also changed his name to Pauli. The son of Isidor, Felix Pauli, is mentioned as 'my cousin Felix Pauli' and as 'Felix Pauli, c/o O. Pascheles, Zürich 6, Ottikerstrasse 14' in the letters [535] of 11 November, [536] of 17 November, and [537] of 26 November 1938 from Wolfgang Pauli to P. A. M. Dirac, respectively. In these letters Pauli thanks Dirac for helping his cousin who had fled to Switzerland from the Nazis and hoped to emigrate to the United States via England [17]. In the second letter the initial O. stands of course for Oskar, the other brother of Wolf, who in the Utitz family tree is designated as the founder of the 'Swiss family' and who is the grandfather of Daniel Pascheles of Ref. [4]. To Oskar 'fell the highest honor, namely to alone support my mother till the end of her life in the year 1927' (translated from Ref. [4b]).

The name change appears explicitly in Wolf's publications. Indeed, there is a paper 'The physical changes of state of the proteins' (*Die physikalischen Zustandsänderungen der Eiweisskörper*), *Archiv für die gesamte Physiologie*

Fig. 1.2 Pauli's father Wolf Pascheles-Pauli

78, 315 (1899), where in the references on p. 345 three papers are quoted under 1) with the remark 'These papers are still published under my former name *Pascheles*' (*Diese Arbeiten sind noch unter meinem früheren Namen* Pascheles *veröffentlicht*). These papers already belong to research on the important functional connection between the physico-chemical behaviour of proteins and their molecular constitution [15a], the unravelling of which made the name of Wolfgang Pauli famous. This was also the reason why the younger Pauli signed his publications 'Wolfgang Pauli, jr.' until he assumed the professorship of theoretical physics at the Federal Institute of Technology, ETH (Eidgenössische Technische Hochschule), in Zurich on 1 April 1928.

Pauli's mother Bertha Camilla Pauli-Schütz, called Maria, was born in Vienna on 29 November 1878. Seven years later, on 27 July 1885, her sister Rika was born and later became Mrs Rika Adam but died very young on 3 July 1916 [8b]. Bertha Pauli was 'one of the few truly strong personalities among Austrian women' (translated from Ref. [18]). Strongly motivated by beauty and righteousness she had an extensive knowledge of dramatic literature and had written a book on the French Revolution [18]. Even after Wolfgang's birth she attended the classical section [18] of the well-known high school for girls of Dr Eugenie Schwarzwald and passed the final exams (*Matura*) at age 27 [19]. At that time it was practically impossible for girls to attend a public high school, as is exemplified by the Döbling High School that her son Wolfgang attended 13 years later, which admitted girls starting in the school year 1922–3 (see Ref. [20], p. 16).

In 1911 Bertha and Wolf Pauli left the Catholic Church for no known reason and became evangelical [6b]. Bertha was a feminist and a pacifist and, like her father, was a collaborator on the *Neue Freie Presse* where she wrote theatre critiques and historical essays. In 1912 she spoke to a large assembly on the necessity of maintaining peace in the face of the crisis in the Balkans. After the war when son Wolfgang had left home to pursue his studies, Bertha Pauli-Schütz, the socialist, took part in the election campaign of 1919 and in articles in the *Arbeiterzeitung*, the workers' daily, urged women to vote for the Social-Democratic Party [6b]. In addition she wrote beautiful and knowledgeable essays on Mary Antoinette, Lady Byron, and other women [18]. But, gradually, resignation took hold of her, and 'the sense of incapability ever stayed with her, this perhaps the deepest cause of her early death. For . . . she strove after too much, and unsatisfied ambition turned into restlessness, restlessness transformed into existential anxiety' (translated from Ref. [18]). On 15 November 1927 she died from 'Hearth-shaped pneumonia. Poisoning by narcotic poison' (*Herdförmige Lungenentzündung. Vergiftung durch narkotisches Gift*) [8e].

Some time after his wife's death Wolfgang's father married the young sculptor Maria Rottler who was the same age as his son [21a]. This new life continued until almost the end of 1938 when Wolfgang's father felt threatened by the Nazis who had annexed Austria on 13 March 1938. He fled to Zurich while Maria Rottler spent the war years in Vienna. On 20 November 1938 son Wolfgang had written to the president of the School Council of ETH in Zurich, Professor Arthur Rohn: 'I am sorry that today I have to bother you with a personal matter. It concerns the possibility for my father to obtain an entrance permit to stay in Switzerland. The situation in Austria has now come to a point where, according to my latest information, there exists a personal threat for my father based on race. He has been forbidden to write abroad and to enter the University and it is uncertain how long he will be allowed to stay in his apartment' (translated from Ref. [21b], document II.12). The president responded the following day saying that he saw no difficulty in obtaining an entrance permit and suggested that Wolfgang's son contact the Cantonal authorities (Ref. [21b], II.13).

Leaving all his laboratory equipment behind, Wolfgang's father received generous hospitality in Zurich from Professor Paul Karrer, Nobel Laureate and director of the Institute of Organic Chemistry at the University of Zurich. Here he wrote 10 more publications which were based on the experimental investigations he had conducted in his institute in Vienna before his flight [24].

After the war he enjoyed in Zurich a most happy marriage with his second wife [15b]. And his 80th birthday became the occasion for numerous colleagues, pupils, and friends to gather there and honour a scientist many thought worthy of the Nobel Prize (he had won other prestigious prizes) and a man of unlimited kindness [15a,b]. He died in Zurich on 4 November 1955, one year after the University of Vienna had bestowed an honorary doctorate upon him [24]. He had been a member of the German Academy of Natural Scientists 'Leopoldina' since 1904 and a corresponding member of the Austrian (since 1937) and the Bavarian (since 1950) Academies of Sciences [15c].

Wolfgang and his sister Hertha

Among my books there is a rather dusty container, in which there is a silver cup in fin-de-siècle *style, and in this cup there lies a card. . . . This cup, of course, is a christening cup, and on the card is inscribed in old-fashioned flowery type: 'Dr. E. Mach, Professor an der Universität Wien'. It so happened that my father was very friendly with his family, and at that time was intellectually entirely under his influence. He (Mach) had thus graciously agreed to assume the role of my godfather. He evidently was a stronger personality than the catholic priest, and the result seems to be that in this way I am baptized antimetaphysical instead of catholic. In any case, the card remains in the cup and in spite of my larger spiritual transformations (*Wandlungen) *in later years it still remains a label which I myself carry, namely: 'of antimetaphysical descent'. Indeed, Mach considered metaphysics, somewhat simplifying, to be the origin of all evil in the world–that is, psychologically speaking, as the devil himself–and that cup with the card in it remained a symbol for the* aqua permanens *which exorcizes the evil metaphysical spirits.* [25]

Three hundred years after Johannes Kepler moved to Prague, discovering his famous laws there, the year of Max Planck's discovery of the quantum of action h and Sigmund Freud's publication of *The Interpretation of Dreams* witnessed the birth of Wolfgang Ernst Friedrich Pauli on 25 April 1900. All three of these left a mark on Pauli's destiny. His second name was chosen in honour of Ernst Mach who 'had thus graciously agreed to assume the role of my godfather' [25]. In view of Mach's well-known anti-clerical attitude [16a] his acceptance of this role appears as a great favour for his protégé, Wolfgang the father, who, much later, characterized his acquaintance with Ernst Mach as 'the most important event of my spiritual development' [27a]. The third name was chosen in honour of grandfather Friedrich Schütz whom young Wolfi liked to accompany on walks to the inner city and visits to antique shops [19]. Already typical was the reaction of the 4-year-old on such a walk correcting his aunt when she explained to him 'now we are walking over the Danube Canal'. 'No, Aunt Erna', Wolfi replied, 'this is the Wien Canal which flows into the Danube Canal' [27b].

Wolfi had all the children's diseases and always felt bored (*'es war ihm immer fad'*; this is a uniquely Viennese expression). His childhood was uneventful

Fig. 1.3 At the age of 20 months, with mother Bertha Camilla Pauli-Schütz, 1901 (CERN)

until his seventh year when on 4 September 1906 his sister Hertha Ernestina (another reference to Ernst Mach!) was born [28]. Her birth was a great shock to Wolfi who, while adoring his mother and grandmother, was jealous and felt slighted [29]. This shock may be seen as Pauli's encounter with his anima, i.e. the unconscious female side of his psyche (see Ref [12], p. 228). Indeed, 30 years later Pauli associates his sister's birth with the birth of his anima which, he argues, has something to do with the number 7. He writes: 'In the seventh year of my life was the birth of my sister. *The 7, therefore, is a hint to the birth of the anima* (who indeed also appeared in later dreams)' (translated from letter [16], Ref. [12], p. 20).

In his 10th year, son Wolfgang started high school at the Döbling Gymnasium. Quite early his inclination, stimulated by his father, was directed towards the natural sciences. At age 14 Wolfgang had already assimilated calculus [30]. Godfather Ernst Mach advised on the choice of scientific readings and dedicated to 'Meinem lieben Patenkind [godchild] Wolf in freundlichem Gedenken.

München-Vaterstetten, 17. Oct. 1913' his famous work on mechanics [31a] whose 'tendency', according to the preface, 'is an antimetaphysical one' [31b]. In a letter dated 12 January 1914 from Haar near Munich (Vaterstetten near Haar, see below), addressing Pauli's father as 'Hochgeehrter Herr College', Mach also recommended the professor of mathematics at the university, Wilhelm Wirtinger, as a consultant [31c,19]. Wilhelm Wirtinger (1865–1945), considered the greatest mathematician in Austria, had a distinguished career, rewarded by several prizes and honorary doctorates. His main research topic was the theory of functions. Having himself published his first paper at the age of 20, he must have felt sympathetic towards the young Pauli [32].

In the letter quoted at the beginning of this section Pauli paints a fascinating picture of his godfather Mach who lived at Hofstattgasse 3, the second-next parallel street south of Anton-Frank-Gasse: 'He was an accomplished master of experiments, and his apartment was chock-full of prisms, spectroscopes, stroboscopes, electrifying machines, etc. When I visited him he always showed me a nice experiment, realized in order to partly eliminate and partly support, by correcting the thinking which is always untrustworthy and causes delusions and errors. Always presuming his own psychology to be generally valid, he recommended everybody to use that inferior auxiliary function [in C.G. Jung's sense] as "economically" as possible (economy of thinking). His own thinking closely followed the sensual perceptions, tools and appliances' (translated from letter [60], Ref. [12], p. 105).

At the time when young Pauli met his godfather, Mach was no longer in full command of all his faculties. In 1897 he had suffered a stroke that paralysed the right side of his body, and in 1901 his health forced him to take early retirement. But his considerable popularity is reflected by the fact that he was immediately offered a title of nobility, and this in spite of his socialist views. Mach, however, declined and accepted instead life membership in the Upper House of Parliament. In 1912 his health deteriorated even more, but through the devotion of his wife and particularly his son Ludwig, he was able to continue his scientific activities. Since his childhood Ludwig had served his father in complete devotion. He had not only sacrificed his medical career but later also renounced the position of US representative of the Zeiss optical works in Jena in order to be able to assist his father in his optical research. Ludwig, 'who became wealthy through his patents.' (Ref. [16b], p. 272), bought a property in the small village of Vaterstetten near the town of Haar, a suburb east of Munich, and urged his father to join him there. In the Spring of 1913 Ernst Mach moved to Vaterstetten where he died on 19 February 1916 [16a,b]. It was in Vaterstetten that the Pauli father and son visited godfather Mach for the last time in summer 1914 [30].

Young Wolfgang was not a model pupil, as his grades of the eight years spent at the Döbling Gymnasium show (Ref. [20], p. 49). He excelled in mathematics and physics and also in philosophy, but in Latin and Greek his grades remained just 'satisfactory' over the years. Forty years after receiving his diploma (*Reifezeugnis*) on 2 July 1918 [33a] Pauli wrote to his former

Fig. 1.4 As a high school student, about 1916 (CERN)

German teacher Alois Hornung: 'That the mention "with distinction" is written on it [the diploma] somewhat surprises me even today. For, my talents were specifically in mathematics and physics while otherwise I was a mediocre pupil. You will be surprised when I now say: I am glad to have attended the humanistic highschool and have learned Latin as well as Greek . . . but later I became interested in scientific texts from the 17th century [see Essay 21 in Ref. [23]], as well as in the Greek philosophers (*Naturphilosophen*) [see Essay 16 in Ref. [23]]. . . . I also have the best remembrance of your lessons in German literature' [33b].

Gradually the pacifist and socialist convictions of Pauli's mother also began to influence him. As Pauli's classmate Eric Hula, who later became a professor of economics at New York University, observed much later: 'After the eruption of World War I, a passionate interest for politics awoke in him [Pauli] which certainly was also nourished by his socialistically oriented and literarily active mother. The longer lasted the war the keener became his opposition against it and, generally, against the whole "establishment"' [33c].

The quintessence of this establishment of course was the Imperial Court. On the death of Emperor Franz Josef on 21 November 1916 the young Charles I (1887–1922) succeeded to the throne. His reign, however, lasted only until the end of the war. On 11 November 1918 he had to abdicate, and four months later he and his young wife Zita of Bourbon-Parma (1892–1989) had to seek asylum in Switzerland. But during the short period of her reign Empress Zita gained the sympathy of the people by her efforts to alleviate the hardships caused by the war during the bad winter of 1916–17 [34]. On her birthday on 9 May mass was celebrated in the churches. For this occasion Wolfgang devised a special kind of joke. He asked the family's female cook to write a letter of vigorous protest to the school director complaining that Pauli's entire class was planning demonstratively to read his mother's daily, the *Arbeiterzeitung*, during the anniversary mass [35]. In devising jokes in class Pauli came always first: he was great at imitating the professors, and it was he who invented the nickname '*das U-Boot*' (the submarine) for the tiny Professor Hornung because of his propensity for surfacing unexpectedly in the middle of a group of pupils (see the first article of Ref. [35]).

Pauli's class has gone down in local history as the 'class of the geniuses'. It counted among its 27 boys two later Nobel Prize winners, besides Pauli the winner of the Chemistry Prize in 1938, Richard Kuhn; then two famous actors, Hans Thimig of the well-known Viennese family of actors, and Gustav Diessl who later married the great singer Maria Cebotari, but also the conductor Herbert Winkler who became a leading figure in the musical life of Japan; and three university professors, two directors of medical schools, a politician, and several industrialists [35]. The graduation picture, however, only shows 17 boys, one of them in uniform; many had already joined the army. Immediately after graduation Richard Kuhn had also to report for service in the Imperial Signal Corps while Pauli remained the only one to be exempt—because of faint-heartedness (*Herzschwäche*). None of his classmates was surprised [35]. But Pauli did not die from heart failure.

Why then did Pauli in 1958 only half-jokingly remark on Walter Thirring's (b. 1927) acceptance of the chair of theoretical physics at the University of Vienna—which his father Hans had occupied before—by using the expression '*geistige Einöde*' (spiritual desert) for his home town [36a]? The fact is that, although he could not hide his imperial Austrian roots, Pauli harboured rather mixed feelings towards his country and his home town which had long lost its central position in politics as well as in science. So when Valentin Telegdi (b. 1922) was introduced to him in 1946, Pauli asked him: 'When did you leave Vienna?' He answered: '1938, Herr Professor.' And Pauli replied: 'Me already in 1918. I always had a good intuition' (translated from Ref. [36b], p. 118).

In 1949, in a letter to his friend Hans Thirring (1888–1976), he considered it 'the supreme luck . . . that may happen to an Austrian if he is able to leave this country—if possible for good (the age of 21 years might just not yet be too late for this; I myself was 18 at the time)' (translated from Ref. [36c]).

Indeed, after his graduation Pauli himself chose to study in Munich and not in Vienna. The reason is understandable. On 7 October 1915 Fritz Hasenöhrl (1874–1915), Ludwig Boltzmann's successor to the chair of theoretical physics at the Institute of Physics at the University of Vienna, was killed by a grenade on the Italian front in South Tirol. After Arnold Sommerfeld (1868–1951) had declined to succeed Hasenöhrl in 1917 [37] and the other leading candidate, the Pole Marian von Smoluchowski (1872–1917), had died unexpectedly the same year (Ref. [38], p. 99), a period of intrigue and arguments, complicated by political and racial considerations, began. It lasted until the retirement in 1920 of Franz Exner (1849–1926), the professor of experimental physics and director of the IInd Institute of Physics. The main contenders for Exner's chair then were Gustav Jäger (1865–1938) and Felix Ehrenhaft (1879–1952). Finally, Jäger got the chair but 'a new Institute (III) was created for Ehrenhaft. The Viennese physicists had thus managed to achieve mediocrity in all their professorships' (Ref. [38], pp. 129-130; see also Ref. [13a], pp. 252–4).

After losing his chair in the Nazi annexation of Austria in 1938, Ehrenhaft first went to England and later emigrated to the United States and became a US citizen. But in 1947 he returned to Vienna and took over the direction of the Ist Institute of Physics. Before emigrating Ehrenhaft had maintained that he had observed particles with fractional elementary charge, arguing with Robert Andrews Millikan (1868–1953) in the United States (Ref. [13a], pp. 83–6) who, in 1923, received the Nobel Prize 'for his work on the elementary charge of electricity and on the photo-electric effect'. In the United States, Ehrenhaft had announced that he had observed magnetic monopoles. Pauli's judgement on Ehrenhaft—which is also a judgement on Vienna—is expressed in the following joke: '*Ja der* Ehrenhaft, he is the exact opposite of Hitler. One should never have let Ehrenhaft get out of Vienna and Hitler never get in!' (translated from Ref. [36b], p. 118).

At the time of his graduation young Pauli, miraculously, had already become an expert in the general theory of relativity, submitting his first paper from Vienna on 22 September 1918. But at the University of Vienna this domain of research was represented only by the young generation of Erwin Schrödinger and Hans Thirring. In 1921 the latter obtained the associate professorship offered the year before to Schrödinger, becoming full professor only in 1927. In turn, Schrödinger, who had admired both Hasenöhrl and Smoluchowski, left his private dozentship in 1920 for Jena where Max Wien, the cousin of the more famous Willy Wien, had offered him a better paid assistantship (Ref. [38], pp. 130–2; Ref. [13a], p. 228). In Munich, on the other hand, Sommerfeld was the uncontested leader of German theoretical physics.

The Pauli-Schütz family was still complete, except for the death of the Schütz grandparents, when in 1916 grandmother Helen Pascheles, who had lost her son Julius in the war (Refs. [4b,5]), came for a visit 'to find consolation in the company of Wolf's children. During this visit they were told, for the first time, that their name was really Pascheles and that their father was Jewish' [39]. When Hertha grew up, Wolfgang had already left home on his studies, so that brother and sister were never very close. However, their relationship became quite affectionate, as an undated airmail letter from Hertha shows, in which she greets her brother with truly Austrian charm: '*Love von uns Beiden. Ich umarme Dich. Immer Deine Hertha* (I embrace you. Always yours)' [27b]. Later, when mother Bertha died and daughter Hertha left home to start her career as an actress, the family home at Anton-Frank-Gasse was left deserted. When father Wolfgang remarried he chose with his young wife a new residence in the 9th District, closer to the centre but still near his institute (Ref. [40a,b], p. 17).

Fig. 1.5 Pauli's sister Hertha, 1933 (CERN)

Hertha left the Gymnasium to study at the Academy of Dramatic Arts in Vienna. At the age of 19 she began her career as an actress at the Lobe Theatre in Breslau (today the Polish city of Wrocław) [15c] where in 1929 she married her actor colleague Carl Behr. In her first role she charmed her theatre critics, and Max Reinhardt (1873-1943) brought her to the famous German Theatre in Berlin [15c,40c] which he directed from 1905 to 1932. Reinhardt, together with Hugo von Hofmannsthal, also created the Salzburg Festivals. On emigrating to the United States he founded a theatre school in Hollywood. In Berlin Hertha collaborated on radio and film projects and became a friend of Ödön Horváth [40d]. In 1933, when the Nazis introduced their first prohibitions, Hertha returned to Vienna where she wrote several books, her first novel *Toni ein Frauenleben für Ferdinand Raimund* appearing in 1936 [41a]. In Vienna Hertha belonged to the famous circles of artists, together with the poet Walter Mehring, who was charmed by her beauty. And together with Carl (Carli) Frucht she founded the literary agency *Österreichische Korrespondenz* [15c,40c]. But on Sunday 13 March 1938, the day Austria was annexed by the Nazis, Hertha fled to Paris via Zurich where she missed brother Wolfgang who was in Cambridge for invited lectures (see page 311; Hertha in Ref. [40a,b], p. 34 erroneously writes that Pauli was in Oxford). Shortly before, Hertha had published a biography of the pacifist Baroness Bertha von Suttner (Vienna, 1937) who had won the Nobel Peace Prize of 1905 and of whom Hertha painted an affectionate portrait bearing the features of her own late mother Bertha [15c,40c]. It was this book that made Hertha's name known. She was visited by the New York publisher Blanche Knopf who wanted to take the manuscript with her. But Hertha also became known in Germany where the book was promptly prohibited (Refs [40a,b], p. 10 and [15c,40c]).

In Paris, where Hertha again met Carl Frucht and Walter Mehring as well as many other friends who, like her, had fled Austria or Germany, she worked as a publisher's representative until the German occupation [41b]. But then Hertha had to flee again. On 11 June 1940 she left Paris with Walter Mehring and some other friends and headed south. On a tortuous route, facing possible death by air attacks and later seizure by the Vichy police or the Gestapo, she reached Marseilles in July where, while waiting in despair and hope for a rescuing US visa, she again met many companions of a similar fate like Franz and Alma Werfel. Aided by the 'most daring underground fighter of World War II', the American Varian Fry (Ref. [40a,b], p. 229), she crossed, alone, the Pyrenees into Spain. At last she reached Lisbon from where the ship *Nea Hellas* left at midnight on 3 September 1940 for New York (Ref. [40a,b], p. 258). A few weeks earlier, but independently, brother Wolfgang and his wife Franca followed the same route to the United States, leaving Geneva on 31 July by train via Barcelona to Lisbon [42]. And after some time Mehring and Frucht also arrived in New York.

The passionate story of this flight, sweetened by the delicate narration of her romance with Gilbert in southern France during her Paris years, is recounted in Hertha's beautiful autobiographical novel *Break of Time* [40a,b] which 'may be counted among the valid literary testimonies of German emigrants' destinies during the Third Reich. From the distance of thirty years the author looked back: "This narration shall build a bridge that unites the Today with the Yesterday— for my friends and myself. A bridge over the cleft of time and beyond, made of thoughts, memories, images"' (translated from Ref. [41a]).

In 1941 Hertha arrived in Hollywood with the aid of the International Rescue Committee, where she worked briefly on the screenwriters' team with Metro-Goldwyn-Mayer. By 1943 she was writing for the *Reader's Digest*. And she published many more books, both in German and in English, in particular *Alfred Nobel, dynamite king, architect of peace* (Fischer, New York, 1942),

Silent night, the story of a song (Knopf, New York, 1946) and, finally, her Suttner biography, *Das Genie eines liebenden Herzens* (1955; translation: *Cry of the heart, the story of Bertha von Suttner* (Washburn, New York, 1957)), as well as, *The Secret of Sarajevo, the story of Franz Ferdinand and Sophie* (Appleton-Century-Crofts, New York, 1965; in German: *Das Geheimnis von Sarajevo*, 1966) [41a,c]. The first of these books was translated by another emigrant, Ernst Basch from Munich, who adopted in the United States the name E. B. Ashton. This collaboration gradually turned into a lifelong partnership. In 1951 the two settled in a small farmhouse at 102 Woodhull Road in Huntington, Long Island, near New York [40c,41c]. In 1952 Hertha became a US citizen. She died at the hospital in Bay Shore, Long Island, on 9 February 1973 [41b]. The urn containing her ashes as well as those of her husband, who died on 20 February 1983, were entombed in the Schütz family grave in Döbling Cemetery [8c,43]. Hertha was a member of the PEN Club and in 1967 was awarded the Silver Medal of Honour of the Republic of Austria [41a,b].

Notes and references

[1] (a) Translated from: W. Pauli, letter [1172] of 28 November 1950 to Aniela Jaffé, in: K. von Meyenn (ed.), *Wolfgang Pauli. Scientific Correspondence with Bohr, Einstein, Heisenberg, a.o., Volume IV, Part I: 1950-1952* (Springer, Berlin and Heidelberg, 1996), p. 198; (b) H. Lutstorf and M. Walter, *Jost Bürgi's 'Progress Tabulen' (Logarithmen) nachgerechnet und kommentiert* (Schriftenreihe der ETH-Bibliothek, Nr. 28, Zurich, 1992).

[2] (a) G. Scholem, *Zur Kabbala und ihrer Symbolik* (Rhein Verlag, 8th edn 1960; Suhrkamp, Frankfurt/Main, 1995), Section V; (b) *Jewish Encyclopedia* (Funk and Wagnalls, New York and London, 1905); (c) *Encyclopaedia Judaica* (Keter, Jerusalem, 1971); (d) *Jüdisches Lexikon* (Jüdischer Verlag, Berlin, 1928); (e) S. Wininger (ed.) *Grosse Jüdische National-Biographie* (Buchdruckerei 'Arta', Czernowitz, 1936).

[3] F. Smutný, 'Ernst Mach and Wolfgang Pauli's Ancestors in Prague', *Gesnerus* **46**, 183 (1989).

[4] (a) *Nekrolog* for Wolf Pascheles by his son Jacob (in German, with Hebrew quotations); (b) *Rede des lieben Vaters anlässlich seines 80. Geburtstags, 6. Dezember 1956, Restaurant Belvoir in Zürich*. (c) Letter dated 3 July 1995 from Dr Daniel Pascheles, Zurich, to Charles P. Enz. I thank Dr Pascheles for copies of the quoted documents.

[5] Gerda Hoffer, *The Utitz Legacy. A Personalized History of Central European Jewry* (Posner and Sons, Jerusalem, 1988); German translation: *Nathan Ben Simon und seine Kinder. Eine europäisch-jüdische Familiengeschichte* (Deutscher Taschenbuch Verlag, Munich, 1996).

[6] (a) K. Wagenbach, *Franz Kafka* (Rowohlt Taschenbuch Verlag, Reinbek-Hamburg, 1984); (b) *Österreichisches biographisches Lexikon 1815-1950* (Verlag der Österreichischen Akademie der Wissenschaften, Vienna, 1999), XI. Band, pp. 300–2.

[7] (a) C. von Wurzbach (ed.), *Biographisches Lexikon des Kaiserthums Oesterreich* (Druck und Verlag der k. k. Hof- und Staatsdruckerei, Vienna, 1876); (b) *Das geistige Wien. Künstler- und Schriftsteller-Lexikon* (C. Daberkow's Verlag, Vienna, 1893), Erster Band; (c) *Wiener Abendpost* (Supplement to *Wiener Zeitung*), Vienna, 23 December 1908; (d) K. von Meyenn (ed.), *Wolfgang Pauli. Scientific Correspondence with Bohr, Einstein, Heisenberg, a.o., Volume III: 1940–1949* (Springer, Berlin and Heidelberg, 1993), p. xxviii, and *Volume IV: 1950-1952* (Berlin and Heidelberg, 1996), p. 201; (e) *Neue Freie Presse*, Vienna, 28 June 1916. I thank Professor Gerhard Oberkofler, Archiv, Universität Innsbruck, for copies of the documents (c) and (e).

[8] (a) W. Heinsius (ed.), *Allgemeines Bücher-Lexikon* (Brockhaus, Leipzig, 1878, 1890), Vol. 15 (1868–74), Vol. 18 (1885–8); H. Weise (ed.), *Hinrichs' Fünfjahres-Katalog* (Hinrichs, Leipzig, 1901), Vol. 10 (1896-1900); (b) *Inscription on Family tombstone*, Döbling Cemetery (Wien 19, Hartäckerstrasse 65); (c) *Wiener Adressbuch Lehmann*; (d) *Traubuch der Pfarre Währing* (Wien 18, Maynollogasse 3); (e) *Totenbeschau-Befund* Nr. 25552, Vienna, 17 November 1927. (f) Information contained in a letter dated 19 June 1986 from Professor Hans Warhanek, Institut für Experimentalphysik der Universität Wien, to Charles P. Enz. I thank Professor Warhanek and his wife for the information contained in (c), (d) and (e).

[9] W. Pauli, *Die Klavierstunde. Eine aktive Phantasie über das Unbewusste. Frl. Dr. Marie-Louise v. Franz in Freundschaft gewidmet*, in Ref. [10], pp. 317–30.

[10] H. Atmanspacher, H. Primas, and E. Wertenschlag-Birkhäuser (eds), *Der Pauli-Jung-Dialog und seine Bedeutung für die moderne Wissenschaft* (Springer, Berlin and Heidelberg, 1995).

[11] (a) Translated from: W. Pauli, letter [69] of 23 October 1956 to C.G. Jung, in Ref. [12], p. 150 (see also p. 91). (b) Notes of my phone conversation with Franca Pauli on 30 July 1984.

[12] C. A. Meier (ed., with the cooperation of C. P. Enz and M. Fierz), *Wolfgang Pauli und C.G. Jung. Ein Briefwechsel 1932-1958* (Springer, Berlin and Heidelberg, 1992).

[13] (a) J. Mehra and H. Rechenberg, *The Historical Development of Quantum Theory. Volume 5. Erwin Schrödinger and the Rise of Wave Mechanics* (Springer, New York, 1987), Chapter I; (b) C. E. Schorske, *Fin-de-siècle Vienna* (Knopf, New York, 1979).

[14] *Akten Wolf Pascheles-Pauli*: (a) Curriculum vitae, handwritten, dated 'Wien, 10. Juni 1897' and signed 'Dr. W. Pascheles'; (b) *K. K. Minis-*

terium für Kultus und Unterricht, report Nr. 20346, 26 April 1913; (c) *Österreichische Präsidentschaftskanzlei*, report Z. 3765/Pr., 3 September 1920. Allg. Verwaltungsarchiv Wien (Akten des Ministeriums für Cultus u. Unterricht). I thank Professor Gerhard Oberkofler, Archiv, Universität Innsbruck, for copies of these documents.

[15] (a) E. I. Valko, 'Professor Wolfgang Pauli zum achzigsten Geburtstag', *Österreichische Chemikerzeitung* **50**, Heft 9 (1949); (b) A. Chwala, 'Zum achzigsten Geburtstag Wolfgang Paulis', in: *Festschrift Wolfgang Pauli, Monatshefte für Chemie und verwandte Teile anderer Wissenschaften*, Österr. Akad. Wiss., Math.-Naturwiss. Kl. (Springer, Vienna, 1950); (c) W. Killy and R. Vierhaus (eds), *Deutsche Biographische Enzyklopädie (DBE)* (Saur, Munich, 1996).

[16] (a) K. D. Heller, *Ernst Mach. Wegbereiter der modernen Physik. Mit ausgewählten Kapiteln aus seinem Werk* (Springer, Vienna, 1964); (b) J. T. Blackmore, *Ernst Mach. His Work, Life, and Influence* (University of California Press, Berkeley, 1972).

[17] K. von Meyenn (ed.), *Wolfgang Pauli. Scientific Correspondence with Bohr, Einstein, Heisenberg, a.o., Volume II: 1930-1939* (Springer, Berlin and Heidelberg, 1985).

[18] 'Zur Erinnerung an Berta Pauli. Von einem Freunde', in: *Neue Freie Presse*, Vienna, 17 November 1927. I thank Professor Gerhard Oberkofler, Archiv, Universität Innsbruck, for a copy of this document.

[19] Karl von Meyenn, *Wolfgang Pauli (1900-1958) 'Antimetaphysisch statt katholisch getauft'*, in Ref. [20], p. 47.

[20] *100 Jahre Gymnasium Gymnasiumstrasse (1885-1985)* (Gisteldruck, Vienna).

[21] (a) *Note*: In one of our numerous conversations Pauli's widow Franca told me (notes of a phone call, 12 July 1984) that Pauli's father had divorced his first wife Bertha. However, document [8d] clearly states under '*Stand*' (situation): '*verh.*' (*verheiratet*, married). Therefore, in Refs. [22a], p. 23 and [22b], p. 18 the situation was described as 'separation' (*Trennung*). However, reality may have been more subtle: a possible cause for the 'sense of incapability . . . and unsatisfied ambition . . .' may have been the mere thought that a much younger and artistically more successful woman might have intruded on the life or thoughts of her husband. See also the quotation at the beginning of this section. Document [8e] also states under '*Glaubensbekenntnis*' (confession of faith): '*evangl. A.B.*' (evangelical '*Augsburger Bekenntnis*', see e.g. *Österreich Lexikon* (Österreichischer Bundesverlag, Vienna and Munich, 1966), Vol. 1, p. 292), and gives as the address Anton-Frank-Gasse 18. (b) C. P. Enz, B. Glaus, and G. Oberkofler (eds.), *Wolfgang Pauli und sein Wirken an der ETH* (vdf Hochschulverlag ETH, Zurich, 1997).

[22] (a) C. P. Enz, 'Rationales und Irrationales im Leben Wolfgang Paulis', in Ref. [10], p. 21; (b) C. P. Enz, 'Wolfgang Pauli (1900-1958). A Biographical Introduction', in Ref. [23], p. 13.

[23] C. P. Enz and K. von Meyenn (eds), *Wolfgang Pauli. Writings on Physics and Philosophy* (Springer, Berlin and Heidelberg, 1994).

[24] P. Karrer, 'Wolfgang Josef Pauli †', in: *Neue Zürcher Zeitung*, 22 November 1955.

[25] Translation of an excerpt of: W. Pauli, letter [60] of 31 March 1953 to C. G. Jung, in Ref. [12], pp. 104–5. This excerpt is exhibited, together with the silver cup and Mach's card, in the Pauli Room at CERN (European Organization for Nuclear Research), Geneva. (The name CERN derives from the original French designation, Conseil Européen pour la Recherche Nucléaire. CERN is the custodian of the Pauli estate and rights, and keeper of the *Pauli Archive*.) The translation is taken from Ref. [26], pp. 766 and 787.

[26] C. P. Enz, 'W. Pauli's Scientific Work', in: J. Mehra (ed.), *The Physicist's Conception of Nature* (Reidel, Dordrecht, 1973), pp. 766–99.

[27] (a) Letter from Wolfgang father to Ernst Mach of 17 February 1913, quoted in Ref. [19]. (b) Undated airmail letter from Hertha to Wolfgang, PLC Bi 120, *Pauli Archive*.

[28] *Note*: Inexplicably, most references to Hertha Pauli give her year of birth as 1909, see e.g. Ref. [8b], Ref. [12], footnote *a*, p. 21 in Ref. [15c], and Ref. [41c] below. But 1906 agrees with Pauli's own information in letter [16] of Ref. [12] quoted in the text. I thank Dr K. von Meyenn for his letter of 15 April 2000 insisting that 1906 is correct.

[29] Notes of my conversations with Franca Pauli on 8 April 1984.

[30] K. von Meyenn, 'Einleitende Bemerkungen zur Neuausgabe', in: *Wolfgang Pauli. Physik und Erkenntnistheorie* (Vieweg, Braunschweig, 1984), pp. ix–x. See also Ref. [19].

[31] (a) E. Mach, *Die Mechanik in ihrer Entwicklung historisch kritisch dargestellt* (Brockhaus, Leipzig, 7. verbesserte und vermehrte Auflage, 1912), English translation, containing additions and alterations up to the ninth (final) edition, by T. J. McCormack, *The Science of Mechanics, a critical and historical account of its development* (Open Court Publications, La Salle, IL, and London, 1942). (b) The German original with the dated dedication is in the Pauli Room at CERN, Geneva. (c) Letter PLC Bi 137, *Pauli Archive*.

[32] 'Wilhelm Wirtinger', in: *Almanach für das Jahr 1945, Akademie der Wissenschaften in Wien* (Wien, 95. Jahrgang, 1947), p. 336. I thank Professor Gerhard Oberkofler, Archiv, Universität Innsbruck, for a copy of this document.

[33] (a) Document PLC Bi 42, *Pauli Archive*. (b) Translated from: W. Pauli, letter of 22 June 1958 to Alois Hornung, PLC Bi 128, *Pauli Archive*, quoted in Ref. [19]. (c) Translated from: Erich Hula, letter of 6 May

1968 to Pauli, PLC Bi 145, *Pauli Archive*, quoted in Ref. [19]. See also: A. Hermann, *Die Funktion und Bedeutung von Briefen*, in: A. Hermann, K. v. Meyenn, and V. F. Weisskopf (eds), *Wolfgang Pauli. Scientific Correspondence with Bohr, Einstein, Heisenberg, a.o., Vol. I: 1919–1929* (Springer, New York, 1979), p. xxiv.

[34] G. Brook-Shepherd, *The Last Empress* (Harper-Collins, London, 1991).

[35] Ref. [26], p. 767. The *Hamburger Abendblatt* of 3 March 1962 carries the story on p. 52 under the title 'Die Klasse der Genies'. See also *Neue Illustrierte Wochenschau* (Vienna) of 29 May 1955 which, under the same title, reports in particular on a conference by Richard Kuhn at the University of Vienna. Copies of both papers, as well as the group photograph of Ref. [20], p. 37, which they also carry, are in the file PLC Bi at the *Pauli Archive*. Pauli's widow Franca confirmed to me that Pauli had told her this story.

[36] (a) My recollection; see also Ref. [26], p. 767; (b) V. L. Telegdi, 'Pauli-Anekdoten', in: C. P. Enz und K. von Meyenn (eds), *Wolfgang Pauli. Das Gewissen der Physik* (Vieweg, Braunschweig, 1988); (c) W. Pauli, letter to Hans Thirring of 23 February 1949. Nachlass Hans Thirring (Zentralbibliothek für Physik, Vienna).

[37] F. Sauter (ed.), *Arnold Sommerfeld. Gesammelte Schriften* (Vieweg, Braunschweig, 1968), Vol. 4, p. 681.

[38] W. Moore, *Schrödinger. Life and Thought* (Cambridge University Press, Cambridge, 1989).

[39] Related by Hertha Pauli in Zurich after World War II, see Ref. [5], p. 128. *Note*: This information contrasts with the description given in: J. Mehra and H. Rechenberg, *The Historical Development of Quantum Theory. Volume 1. The Quantum Theory of Planck, Einstein, Bohr and Sommerfeld: Its Foundation and the Rise of Its Difficulties. 1900-1925* (Springer, New York, 1982), Part 2, footnote 633, p. 377. In this footnote Mehra states that Pauli 'grew up without knowing anything about his family's Jewish background' and quotes an interview with Paul Ewald (AHQP Interview, 8 May 1962, p. 15) claiming that, on Ewald's observation, 'there was no doubt about his [Pauli's] being a Jew'. Pauli replied, 'I? No. Nobody has ever told me that and I don't believe I am.'

[40] (a) H. Pauli, *Der Riss der Zeit geht durch mein Herz* (Zsolnay, Vienna, 1970); (b) paperback edition of [40a] (Ullstein-Buch, Nr. 30243, Frankfurt/M and Berlin, 1990); (c) A. Stoltenberg, *Nachwort*, in [40b]; (d) H. Caplan and B. Rosenblatt (eds), *International Biographical Dictionary of Central European Emigrés 1933-1945, Volume II: The Arts, Sciences, and Literature* (Saur, Munich, 1983).

[41] (a) 'Hertha Pauli zum Gedenken', in: *Neue Zürcher Zeitung*, 15 February 1973, p. 25; (b) *Hertha Pauli, 63, wrote juveniles*, obituary, Niels Bohr Library, American Center for Physics, One Physics Ellipse, College Park,

Maryland; (c) file *Pauli Hertha Ernestina 1909-*, Public Library, New York.

[42] W. Pauli, letter [600] of 3 September 1940 to M. Fierz, p. 35, and letter [605] of 15 October 1940 to H. Bhabha, p. 44, in: K. von Meyenn (ed.), *Wolfgang Pauli. Scientific Correspondence with Bohr, Einstein, Heisenberg, a.o., Volume III: 1940- -1949* (Springer, Berlin and Heidelberg, 1993).

[43] *Note*: Ref. [8f] records the recollection of an official of the cemetery administration saying that the urn of E. B. Ashton was brought by a grandson living in Germany. There was a note saying that a Mr Tim Unger, 1000 Berlin 37, Schrockstrasse 24, had ordered the interment of the urn. Hertha Pauli had no children, see Ref [40a,b], p. 148. A photograph of the tomb is contained in the file PLC Bi 127 at the *Pauli Archive*.

2
The prodigy of relativity theory

The *Encyklopädie Article*

... he [Pauli] was not only familiar with the most subtle arguments in the Theory of Relativity through his own research work, but was also fully conversant with the literature of the subject
... the article aims at including all of the more valuable contributions to the theory which have appeared up to the end of 1920. Beyond this, the author's own opinions are to be found in many places throughout the article. [1]

No one studying this mature, grandly conceived work would believe that the author is a man of twenty-one. [4]

Both this preface by Pauli's teacher Arnold Sommerfeld (1868–1951) and the review on Pauli's *Encyklopädie Article* by Albert Einstein (1879–1955) himself confirm the prodigious faculties of the young Pauli. Also Lise Meitner (1878–1968), a long-standing friend of Pauli's, wrote in a letter to Pauli's widow Franca on 22 June 1959: 'I often thought of and also have told it that in the fall of 1921 I have met Sommerfeld in Lund, and that he told me he had such a gifted student that the latter could not learn anything any more from him, but because of the university laws valid in Germany he had to sit through (*absitzen*) 6 semesters in order to make his doctorate. Therefore he, Sommerfeld, had set this student on an encyclopedia article. When I asked for the name, Sommerfeld mentioned Wolfgang Pauli who was already known to me!' (translated from a letter by Lise Meitner from Stockholm, PLC, *Pauli Archive*).

How did Pauli during his last years at the Döbling Gymnasium manage not only to understand the theory of relativity in all its technical details but also to read the highly specialized literature used in his first publication 'On the energy components of the gravitational field' [6]? Part of the answer, most likely, is that during his high school years Pauli had private lessons with the theoretical physicist Dr Hans Adolf Bauer (1891–1953) [7a] who in 1920 became honorary dozent and in 1936 *ausserordentlicher* (associate) professor of theoretical physics at the Institute of Technology (*Technische Hochschule*) of Vienna [7b]. Bauer, like Schrödinger and Hans Thirring, then worked on problems of general relativity. Indeed, all three of them published papers in 1918 which are explicitly mentioned in Sections 59–61 of Pauli's *Encyklopädie*

Fig. 2.1 Pauli as a student, about 1919 (CERN)

Article [2, 3]. Bauer had worked out the gravitational field of spheres consisting of a compressible liquid (Section 59) while Thirring calculated the effect of a rotating sun on planetary motion which became known as the Lense–Thirring effect (Section 60). Bauer and Schrödinger, on the other hand, had analysed the properties of the energy components of the gravitational field in some special cases (Section 61). Since the latter problem is closely related to Pauli's first publication [6], it is clear that he was directly influenced by Bauer's and Schrödinger's works.

So Pauli must have been a frequent visitor to the new building of the University's Physics Department which had been completed in spring 1913, in time to host, in September, the important 85th Meeting of the German Natural Scientists and Physicians attended by over 7000 participants. It was at this conference that Einstein gave his first comprehensive public lecture on the preliminary theory of general relativity that he had worked out in collaboration with his friend Marcel Grossmann (1878–1936) who was professor of mathematics at ETH [8]. (This report is quoted in footnote 274 of Refs [2, 3].) 'The site faces Boltzmanngasse, next to the Institute for Radium Research and about a kilometre from the main university building on the Ring. The elegant five-storey limestone structure has a classically austere style. It includes a large lecture hall, extensive laboratories and workshops, many offices and seminar rooms, and one of the best physics libraries in Europe' (Ref. [9a], p. 70; see also Ref. [9b]). Seventy years later a meeting of former pupils and friends of Wolfgang Pauli was held in the same building to commemorate the 25th anniversary of Pauli's death [10].

In a radio address on 19 December 1958 commemorating Pauli's death, Hans Thirring recalled that around 1915 or 1916 a younger colleague from Pauli's school had told him: 'Imagine . . . in the 5th grade we have a pupil who is phenomenally gifted in mathematics and physics, to the point that he promises to become a new Gauss or Boltzmann.' And Thirring continued: 'On the occasion of a vacation by Pauli in Vienna I also had the opportunity to meet the young man personally. . . . He came to me, not to ask questions, but with a concrete proposition to continue some related [with Thirring's mentioned work], quite tricky calculations which I had not quite dared to attack, remarking that this problem should be approached at once. We sat down, each of us separately at a table, and went on calculating until our heads fumed. We did arrive at the same result but with the difference that he proceeded with much more elegance and skill so that he finished much earlier than me. Rarely has the ingenious superiority of a colleague become so evident to me as at that encounter with the then 20 year old student Pauli' (translated from Ref. [11a], pp. 687-8; see also Ref. [11b], p. 13).

These 'related, quite tricky calculations' concerned the relativity of the centrifugal and Coriolis accelerations (which for the present discussion we abbreviate as c and C, respectively) that Pauli discusses in detail towards the end of Section 60 of Refs [2, 3]. Considering a point particle moving inside a rotating spherical shell of mass M and radius a the effect, according to Pauli, is $(m/a)(c + C)$ where $m = kM/c^2$ is the gravitational radius of the shell (k = gravitational constant, c = velocity of light) [12a]. And Pauli concludes: 'Since this ratio [m/a] is quite minute for all masses obtainable in practice, there is no hope of verifying experimentally this decisive and important result; and we can understand the reason for the negative result of Newton's primitive experiment with the rotating pail, and of the improved experiment by B. and T. Friedländer [*Absolute and relative Bewegung*, Berlin, 1896], which attempted to demonstrate the presence of centrifugal forces inside a heavy rotating flywheel' (Ref. [2a], p. 175). Einstein had emphasized this conclusion in his report to the Vienna meeting of 1913 mentioned above where he calls the effect the 'hypothesis of the relativity of inertia' (Ref. [8], pp. 1261-2). At the relativity congress of 1955 in Berne, Pauli observed on this problem: 'However, the G-field [the g_{ik}, see below], which according to Einstein is just the ether in a new form, retains its conceptual independence over against matter. Einstein has repeatedly stated that he would find it more satisfactory if the G-field were to vanish identically in the absence of matter. He called this fundamental principle "Mach's principle" in honour of Ernst Mach, who paved the way for later thought on the general theory of relativity by his critique of absolute space' (Ref. [12b], pp. 109-10).

After this first glimpse into the *Encyklopädie Article* let us now look more closely at this 'grandly conceived work' [4]. It is organized in five parts: *Part I.*

The foundations of the special theory of relativity; Part II. Mathematical tools; Part III. Special theory of relativity. Further elaborations; Part IV. General theory of relativity; Part V. Theories on the nature of charged elementary particles. While, according to the notice at the end of Ref. [3], the article was completed in December 1920, at least as in the quoted literature, this disposition essentially already figures in a letter from Pauli of 7 May 1920 to the editor of the astronomical part (volume VI) of the *Encyklopädie*, Samuel Oppenheim (1884–1928) (see Ref. [13b], p. 1913), in which a common notation for the different volumes is proposed [13c]. The only difference with this early proposal is that the part on general relativity is divided into three sections covering the foundations, the field equations, and questions of cosmology, respectively. The relatively small space that the section on general relativity occupies in the final form of the article is quite striking. The justification may be seen by the fact that, by giving in *Part II* a quite exhaustive discussion of the mathematical tools concerning both the special and the general theory of relativity, Pauli was able to concentrate the discussion in *Part III* and *Part IV* on the purely physical content of the theories. This emphasis on the mathematical tools is commented upon very favourably in the review that Erwin Madelung (1881–1972) from the Department of theoretical physics in Kiel wrote on the special edition in book form [3]. He concluded: 'The book is to be valued as an outstanding achievement' (translated from Ref. [13d]).

In the original (Ref. [3]) a footnote to the general title expresses Pauli's 'warm gratitude [*wärmsten Dank*] to Geheimrat Klein, for the great interest he has shown in this article, for his active help in proof-reading, and for his valuable advice on many occasions' [2]. Felix Klein (1849–1925) was a dominant figure and a reformer of mathematics teaching in Göttingen, 'the place of mathematical high culture' (Ref. [15a], p. 675). In 1894 he had started the gigantic project of the *Encyklopädie der Mathematischen Wissenschaften* which he directed for 30 years. According to Klein's plan it should cover not only mathematics but also geodesy, astronomy, and theoretical physics. While Klein himself edited volume IV on mechanics he asked his favourite pupil and friend Arnold Sommerfeld to be the editor of volume V on physics. Sommerfeld had first asked Einstein himself to write the part on relativity theory. However, when Einstein declined, he planned to undertake this task himself, and he asked Pauli to assist him. However, Pauli's first draft impressed Sommerfeld so much that he let him do the whole article by himself, which, in view of the illustrous names of the other contributors, was an act of extraordinary confidence in his student (Ref. [14], p. 13). Sommerfeld, who at the time was president of the German Physical Society, also proposed Pauli for membership. Pauli became a member on 14 November 1919 [15b].

There exist four letters that Klein addressed to Pauli during the last stages of completion of the manuscript, all written between March and May 1921 (Ref. [14], letters [10]–[13]). They are of great interest because of the precision of Klein's advice, the objectivity of his critique, and his superior judgement. One therefore understands at once Pauli's 'warm gratitude' towards Klein and also Sommerfeld's enthusiastic description in his autobiographical sketch that he wrote shortly before his death: 'Overwhelming was the impression that I received ... from F. Klein's grand personality. ... I have always considered Klein as my true teacher .... Of decisive influence for my later teaching activity was the example of his high-flying elocution' (translated from Ref. [15a], p. 675). Pauli took into account Klein's advice expressed in these letters mainly in the many footnotes (they are numbered from 1 to 373). In the first letter, dated 8 March 1921, Klein acknowledges receipt of the manuscript. But then, emphasizing Henri Poincaré's (1854–1912) priority before Einstein in recognizing the group properties of the Lorentz transformations of special relativity, on the one hand, and the competition between Einstein and Hilbert concerning general relativity, on the other, Klein advises Pauli to pay great attention to the historical facts. Pauli took notice in Section 1 (*Part I*) and particularly in the footnotes 11 and 14 where Poincaré is given due credit. And in Section 23 (*Part II*) and Section 57 (*Part IV*) concerning one of the central problems of the *Encyklopädie Article*, namely the variational principles in general relativity (see the next section), the essential contibutions of the main actors Hendrik A. Lorentz (1853–1929), David Hilbert (1862–1943), Albert Einstein, Hermann Weyl (1885–1955), and Felix Klein are quoted in footnotes 100 to 104, respectively, which are repeated in footnote 317.

Of particular interest is footnote 315 which gives credit to Hilbert for first having recognized the important but subtle fact that because of the possibility in general relativity of choosing arbitrary coordinates, the general solution of the field equations must contain four arbitrary functions, so that there can be only four equations less than there are unknowns, namely the 10 components of the metric tensor g_{ik} which defines the invariant line element ds by $ds^2 = g_{ik}dx^i dx^k$ (as usual, repeated indices are summed from 1 to 4). But this arbitrariness does not invalidate causality because all possible solutions describe the same physical situation. On the other hand, as recalled by Klein in his fourth letter [13], dated 8 May 1921, Einstein, in his Göttingen lectures of June/July 1915—and until his breakthrough of 25 November 1915—had not realized this subtle fact, believing instead that the field equations were invariant only under linear transformations (see e.g. Section 14c of Ref. [16]).

Interestingly, in footnote 315 Pauli also mentions Mach's early work on energy conservation [17a] in an edition of 1877 where 'E. Mach had already, on the basis of relativistic considerations, arrived at the result that the number of equations expressing the physical laws should in fact be less than the number of unknowns'. The passage reads: 'If the number [of equations among the natural phenomena, abbreviated here as P] were larger than or equal to the number of the P, then all the P, of course, would be overdetermined or, at least, completely determined. The fact of the variability of Nature, therefore, proves the number of equations to be less than the number of the P' (translated from Ref. [17b], pp. 25–6). In footnote 277 finally, following Klein's advice in the fourth letter, Pauli credits Hilbert for having 'At the same time as Einstein, and independently ... formulated the generally covariant field equations', but criticizes Hilbert's axiomatic presentation. Pauli then adds his own, physical criticism, namely that Hilbert's derivation is based only on the particular matter system of Mie's theory (see the next section).

But in spite of Klein's favourable opinion: 'Your presentation is already much more to my taste' [namely, as compared with von Laue's 'unbecoming' book] (translated from Ref [14], letter [11]); 'it was not all that perfect', as Pauli himself observed in a conversation with Jagdish Mehra in Spring 1958: 'I had not even mentioned the Bianchi identities!' [18]. And, indeed, on looking up this item in the subject index of Ref. [2a] one is referred to *Note 7* of the *Supplementary Notes* to the reedition in English of 1958. In *Notes 6* and *7* it is shown that 'It is logically possible to abandon any derivation of the "geodesic components" Γ^l_{ik} from the metric of the space defined by the tensor g_{ik}' (Ref. [2a], p. 208). The geodesic components Γ^i_{lk} define the parallel displacement of a contravariant vector (upper index) a^i along dx^k to $a^i + \bar{d}a^i$ where $\bar{d}a^i = -\Gamma^i_{lk}a^l dx^k$. When a^i is displaced along a closed parallelogram formed by dx^j and dx^k ($j \neq k$) the result in general deviates from a^i. This deviation $\delta^2 a^i$ defines the curvature tensor R^i_{ljk} by $\delta^2 a^i = R^i_{ljk}a^l dx^j dx^k$ (see Ref. [2a], Sections 14 and 16).

While the Γ^l_{ik} are defined in *Note 6* without imposing the symmetry $\Gamma^l_{ik} = \Gamma^l_{ki}$, this symmetry is explicitly used in *Note 7* in order to develop a general affine (as opposed to metric) tensor calculus which leads to a generalized curvature tensor R^h_{ikl}. Even in this generalized affine framework (see the next section) the *Bianchi identities* exist; they state that the cyclic sum (klm) of the covariant derivative $R^h_{ikl;m}$ of the Riemann tensor vanishes,

$$R^h_{ikl;m} + R^h_{ilm;k} + R^h_{imk;l} = 0. \tag{2.1}$$

The covariant derivative of a vector a^i along dx^k is defined by the difference between the increment $da^i = (\partial a^i/\partial x^k)dx^k$ and that of parallel displacement $\bar{d}a^i$, i.e. by $a^i_{;k}dx^k = \delta a^i = da^i - \bar{d}a^i = (\partial a^i/\partial x^k + \Gamma^i_{lk}a^l)dx^k$. For a covariant vector (lower index) $b_i = g_{ik}b^k$ the derivative $b_{i;k}$ is defined by the fact that a scalar has no parallel displacement, so that $(a^i b_i)_{;k} = \partial(a^i b_i)/\partial x^k$ and by the product rule $(a^i b_i)_{;k} = a^i_{;k}b_i + a^i b_{i;k}$. These definitions generalize to any tensor (see Ref. [2a], Section 20).

The most important result derived in *Note 7* are the contracted Bianchi identities

$$G^{k}_{i\ ;k} = 0, \tag{2.2}$$

where $i = 1, 2, 3$ are the spatial directions and $i = 4$ is the time direction. Here $G_{ik} = R_{ik} - (1/2)g_{ik}R$ is called the Einstein tensor in Refs [2, 3], $R_{ik} = R_{ki} = R^h_{ihk}$ is the contracted curvature or Ricci tensor, $R = g^{ik}R_{ik}$ the curvature invariant, and g^{ik} is the inverse of the metric tensor $g_{ik} = g_{ki}$ (as usual, the raising or lowering of an index is obtained by applying g^{ik} or g_{ik}, respectively). As shown in *Note 7*, Eqs (2.2) follow from (2.1) by first applying the contraction $h = k$ and then contracting with g^{im}, making use of the identities $g^{im}_{;k} = 0$ and $g^{im}R^k_{ilm} = R^k_l$, the latter being a consequence of the symmetry $R_{hilm} = R_{mlih}$. Equations (2.2) are therefore only valid in a metric (Riemannian) space.

This last remark indicates that because of the physical importance of the contracted Bianchi identities (2.2) the more general affine connection cannot be physically significant. It is surprising that in Section (b) of *Note 7* Pauli does not mention the originator of this generalization of relativity theory, namely Eddington [20a] whose book *The Mathematical Theory of Relativity* of 1923 [20b] is cited only in *Note 23*, on p. 225 of Ref. [2a]. In a long letter to Arthur Stanley Eddington (1882–1944) written in Copenhagen on 20 September 1923, which contains his deepest convictions on the state of physics at the time, Pauli writes: 'Nobody is able to establish empirically an affine connection between vectors in neighboring points without previously having determined the line element [the g_{ik}]. For this reason and in contrast to you and Einstein, I consider the invention by the mathematicians, that it is possible to found a geometry on an affine connection, as momentarily irrelevant for physics' (translated from Ref. [14], p. 118).

The importance of the contracted Bianchi identities (2.2) lies in their allowing for the four arbitrary functions in the general solution of Einstein's field equations

$$G_{ik} = -\kappa T_{ik} \qquad (2.3)$$

alluded to above in relation to footnote 315. In Eq. (2.3) κ is Einstein's gravitational constant ($8\pi/c^4$ times Newton's constant k, see Ref. [2a], Eq. (409), p. 163) and T_{ik} the material energy–momentum tensor. As emphasized by Pais (Ref. [16], pp. 222, 256, and 275), in November 1915 Einstein (and also Hilbert) still did not know the relations (2.2) and therefore was not aware that the energy–momentum conservation laws

$$T_{i\;;k}^{\;k} = 0 \qquad (2.4)$$

are a consequence of Eqs (2.2) and (2.3).

In the *Encyklopädie Article* the identity (2.2) does occur, although in disguised form, as Eq. (182a) (Ref. [2a], p. 69 where, however, $\mathbf{G}_{rs}$ should read $\mathbf{G}^{rs}$ as in Ref. [3]). This equation follows from (2.2) by multiplication with $\sqrt{-g}$ and identifying $\mathbf{G}_i^{\;k} = \sqrt{-g}G_i^{\;k}$ where $g = \det|g_{ik}|$. The second term in Eq. (182a) which is due to the definition (152) of the covariant derivative is obtained by using the relations $\Gamma_{r,ki} + \Gamma_{k,ri} = (g_{rk})_i$ and $\sqrt{-g}\Gamma^r_{ri} = (\sqrt{-g})_i$ where $(...)_i$ here stands for partial derivation by x^i. This shows that Pauli's judgement of his own youthful work, namely that it 'was not all that perfect', is rather too severe: the essential identity (2.2) was there! Pais notes: 'The name Bianchi [Luigi Bianchi (1856–1928), a former pupil of Klein] appears neither in any of the five editions of Weyl's *Raum, Zeit, Materie* (the fifth edition appeared in 1923) nor in Pauli's review article of 1921' (Ref. [16], p. 276).

The derivation of the contracted Bianchi identities (182a) in Refs [2, 3] is unnecessarily heavy and also incomplete in so far as the derivation of the key relation (179) is relegated to Weyl's *Raum, Zeit, Materie* (footnote 105).

However, the complicated variational expression (181) from which Eq. (182a) immediately follows is so general that it also yields a less obvious but historically important and fundamental result, namely Eq. (184). Using the field equations (2.3) to replace G_i^k by the material energy–momentum tensor and defining $t_i^k = -U_i^k/\kappa$, this equation reads

$$\frac{\partial}{\partial x^k}[\sqrt{-g}(T_i^{\ k} + t_i^{\ k})] = 0. \tag{2.5}$$

Einstein derived this alternative form of the local energy–momentum conservation law (2.4) for the first time in his major review paper of 1916 [21], still not knowing the identities (2.2). While in this paper Einstein calculated the quantity $t_i^{\ k}$ only for the special class of coordinate systems satisfying $g = -1$, he came back to the derivation of Eq. (2.5) in his follow-up paper 'Hamiltonian Principle and General Relativity' [23] where, however, he did not bother to work out explicitly the very complicated expression of $t_i^{\ k}$.

Now Eq. (2.5) is interesting because it can be derived using only the properties of the gravitational field, without explicit knowledge of the matter field. And it is fundamental because it allows one to obtain an energy–momentum conservation law in integral form, which is not possible with the covariant forms (2.4) or (2.2) because of the additional term in Eq. (182a). By integrating (2.5) over a finite spatial volume V such that, for $k = 1, 2, 3, T_i^{\ k} + t_i^{\ k} = 0$ on the boundary of V, one finds that the *total energy–momentum*

$$J_i = \int_V \sqrt{-g}(T_i^{\ 4} + t_i^{\ 4})d^3x = \text{const.} \tag{2.6}$$

In other words, the energy and momentum of a closed system are conserved. Einstein called the $t_i^{\ k}$ the 'energy–momentum components of the gravitational field' [21, 23]. However, as discussed by Pauli in Section 61 of Refs [2, 3] where the law (2.6) figures as Eq. (447), this interpretation is not without problems. For, although Eq. (2.5) is perfectly covariant, the $t_i^{\ k}$ do not form a general tensor and may be made to vanish at any arbitrary space – time point. But even worse properties of the $t_i^{\ k}$ were exhibited by Schrödinger and Bauer in some special cases; their works, which were mentioned at the beginning of this section, are quoted in footnotes 348 and 349, respectively.

This clearly shows the importance of knowing the general form of the $t_i^{\ k}$ which Einstein had not calculated. This was the task that the 18-year-old pupil Pauli set himself for his first publication [6]. The austere style of this paper puts some strain on the reader's goodwill, in that the latter has a hard time finding out what the main result is—namely, Eq. (7). After announcing the purpose of the paper in two sentences, Pauli gives the definition of $t_i^{\ k}$ in Eqs (1) to (4), which are the same, respectively, as Eqs (183) and (178) of Refs [2, 3] for $U_i^{\ k}$ and G, called G^* here (the geodesic components Γ^l_{ik} are written in the old Christoffel notation $\{_i{}^k_l\}$). The hard part is to derive the very first formula, namely Eq. (5), for which the reader is directed to the same reference of Weyl given in footnote 105 of Refs [2, 3]. Armed with this result one quickly arrives at Eq. (7). And with the help of the identity

$$\sqrt{-g}\left[2\frac{\partial g^{mk}}{\partial x^i} - g^{mk}g_{lj}\frac{\partial g_{lj}}{\partial x^i}\right] = 2\frac{\partial(\sqrt{-g}g^{mk})}{\partial x^i}$$

it is easy to verify that Eq. (7) is the same as Eq. (185) of Refs [2, 3] where in footnote 106 Pauli's paper [6] is quoted for the general case and Einstein's paper [21] for the case $g = -1$. In the remainder of the paper Pauli applies this result first to Einstein's case $g = -1$ and then to the situation of a weak gravitational field, for which the g_{ik} deviate only little from the free values 1 for $i = k$ and 0 for $i \neq k$.

Fig. 2.2 Standing, with Einstein, autumn 1926 (photo by P. Ehrenfest, CERN)

In writing the *Supplementary Notes* 35 years later (see the preface of Ref. [2]) Pauli remedied the unsatisfactory presentation of the derivation of the

Fig. 2.3 Sitting, with Einstein, autumn 1926 (photo by P. Ehrenfest, CERN)

contracted Bianchi identities (182a) in *Note 8*. The method used had been developed by Attilio Palatini (1889–1949) (see Ref. [13b], pp. 1939–40) back in 1919 in a paper that was known to Pauli when he wrote the *Encyklopädie Article* and which is mentioned in footnote 105. Now the history of the re-edition of this work including *Supplementary Notes* is an interesting story in itself which I cannot completely suppress here, although it belongs to a much later chapter of this book. As we shall see, after writing the *Encyklopädie Article* Pauli was no longer very active in the domain of relativity theory, except for a series of papers on the five-dimensional formulation in the early 1930s and a paper with Einstein in 1943. His activity had shifted to quantum theory.

Nevertheless, Pauli always retained a keen interest in relativity. And eventually, the question of re-editing the *Encyklopädie Article* came up. There exists an unpublished preface written by Pauli exactly 30 years after the publication of the *Encyklopädie Article*, on 30 July 1951, where one reads: 'I am now asked by a publisher to write a preface for a new edition' [24]. This publisher was Paul Rosbaud (1895–1963) who 'at the end of 1951 . . . accepted [the publisher Robert] Maxwell's invitation to become Scientific Director of a new venture called Pergamon Press' [25]. In the context of the present section the following comment in this preface is worth noting: 'I [Pauli] regret that the cyclic identities of Bianchi for the curvature tensor are not used in this book . . . [follows Eq. (2.1) above]. When doubly contracted, these identities lead in the most direct way to Eq. (182) of sect. 23' [24].

At the same time as this preface, Pauli decided to lecture in his *Spezialvorlesung* (special course) of two hours per week at ETH on special and general relativity. This course, which took place in the winter term of 1951/2, thus looks like a 'warming-up' exercise for the task of writing supplementary notes for a re-edition of the *Encyklopädie Article*. However, questions of copyright transfer by Teubner delayed the project [26]. In 1953 Pauli's motivation towards relativity grew considerably because in the summer of 1955 a large international conference 'Fünfzig Jahre Relativitätstheorie' was to be held in Berne, over which Pauli was going to preside [27]. So he decided to give in the summer of 1953 a *Spezialvorlesung* on problems of general relativity.

Working at that time as an assistant at ETH I regularly attended Pauli's *Spezialvorlesung*. Comparing my notes from the 1951/2 course with the *Supplementary Notes* I may remark that Pauli, among many other things, considered in the part on the special theory the experiments by Ives and Stilwell on the second-order Doppler effect mentioned in *Note 5*. And in the part on the general theory he emphasized the possibility of defining the Γ^l_{ik} in an affine space, as done in *Note 6*. He then developed the Ricci–Levi-Cività calculus and gave an elegant derivation of the Bianchi identities contained in *Note 7(c)*.

On the other hand, at the end of the course on problems of general relativity Pauli asked me if I would write out my notes for him, which shows that he attached some importance to this course. Based on my notes of 1953 I prepared a package of 111 handwritten pages [28] which I gave to Pauli without keeping a copy (much later CERN made me one).

Returning to *Note 8*, the key relation (180) of Refs [2, 3] is proved there and on pp. 17 – 19 of Ref. [28] by using Palatini's idea that, although the Γ^l_{ik} do not form a tensor, any difference of two sets of geodesic components does, as follows from the transformation property of the Γ^l_{ik} (see, e.g., Eq. (1.13) of Ref. [28]). In *Note 8*, in addition, the 40 functions Γ^l_{ik} are varied independently from the g_{ik}, as done by Palatini. However, in view of the quantum nature of matter Pauli expresses doubts there about the physical justification of such independent variations of the Γ^l_{ik} because it is not obvious that 'the matter part of the action integral does not contain the Γ^r_{ik} explicitly'.

In this section the *Encyklopädie Article* has been examined mainly in relation to Pauli's first publication [6]. The next section will discuss it in connection with his second and third papers which, as we shall see, expressed views that Pauli held throughout his whole life. Later it will also be seen, in relation to the five-dimensional formulation of general relativity theory, that the important *Supplementary Note* 23 concerning this problem is based very much on the corresponding chapter of the *Spezialvorlesung* of 1953 [28].

Pauli and Hermann Weyl

Dear Mr. Pauli.
I am extremely pleased to be able to welcome you as collaborator. However,

it is almost inconceivable to me how you could possibly have succeeded at so young an age to get hold of all the means of knowledge and to acquire the liberty of thought that is needed to assimilate the theory of relativity. [29]

With this welcome into the community of experts in relativity theory a personal relationship between Hermann Weyl and Wolfgang Pauli began that was marked by many periods of social and professional contact in Zurich and in Princeton, New Jersey. The initiation must have come in a letter from Pauli sometime prior to 10 May 1919, the date of Weyl's letter quoted above, in which Pauli asked for information on the third edition of Weyl's celebrated book *Space, Time, Matter* [30] quoted many times in the *Encyklopädie Article* [2, 3], and informed him about his own work on Weyl's theory. This was Pauli's second paper which he submitted on 4 June 1919 from Sommerfeld's institute in Munich [31]. A footnote in this paper (Ref. [31], p. 466) shows Pauli being uncharacteristically respectful towards Weyl when he points out an error in the sign of some of Weyl's formulae with the words 'I should like to express, with due respect, the opinion that a small oversight has occurred in Weyl's paper' (translation from Ref. [32], p. 767). Soon, however, the biting criticism for which Pauli became known and often feared also shows in his relation towards Weyl. Already in the third paper, his second on the subject of Weyl's theory, submitted from Munich on 3 November 1919 [33], Pauli raises at the end a 'physical–conceptual doubt' which is of fundamental importance. He says:

In Weyl's theory we continually operate with the field strength in the interior of the electron. However, for the physicist the latter is only defined as force acting on a test body, and since there are no test bodies smaller than the electron itself, the notion of electric field strength in a mathematical point seems to be an empty fiction, void of content. One would rather like to maintain that in physics only quantities be introduced which are observable in principle. Should we with the continuum theories for the field in the interior of the electron be on a wrong track at all? (Translated from Ref. [33], p. 749)

This criticism from the 19-year-old Pauli reflects the positivist attitude of Einstein much more than that of his godfather Mach whose emphasis on the direct physiological role of the senses in perception made him reject the reality of the atomic realm. Indeed, Einstein says in his obituary for Mach: 'Concepts . . . only make sense insofar as the objects to which they refer can be exhibited, as also the points of view according to which these concepts are associated to these objects (analysis of the concepts)' (translated from Ref. [34], p. 102). Here Einstein's 'objects', as Pauli's 'quantities' in the above quotation, already have the meaning of the 'observables' which 10 years after this obituary became so important in the development of quantum mechanics. On the other hand, Einstein acknowledged Mach's influence on his theory of relativity: 'The quoted lines [from *The Science of Mechanics*] show that Mach has clearly recognized the weak sides of classical mechanics and that he was not far from demanding a general theory of relativity, and this already almost half a century ago!' (translated from Ref. [34], p. 103).

Pauli repeated the above criticism contained in his third paper [33] in a discussion at the 86th Meeting of the German Natural Scientists and Physicians at Nauheim from 18 to 26 September 1920 [35], after Weyl had presented his theory. At this conference, which was the first scientific meeting Pauli attended, he met many of the scientists for the first time, most prominently Hermann Weyl himself. There he also, very likely, was introduced to Einstein by Lise Meitner (see Ref. [11b], note I.23, p. 92). Pauli formulated his criticism as a somewhat provocative question addressed to Einstein: 'I therefore want to ask Professor Einstein whether he agrees with the view that one may expect the solution of the problem of matter only from a modification of our conception of space (and perhaps also of time) and of the electric field in the sense of atomism, or if he considers the mentioned doubts as not relevant and is of the opinion that one should hold on to the foundations of the continuum theories' (translated from Ref. [35], p. 650). Einstein evaded the question and talked in the fashion of his obituary for Mach of associating concepts to experiences.

In the *Encyklopädie Article* Pauli's criticism appears first in his discussion of the theory of the electron by Gustav Mie (1868–1957) in Section 64 (Ref. [2a], p. 191). In 1912 Mie was the first to set up a theory that could describe electrically charged 'elementary particles' (the electron and the hydrogen nucleus). By this was meant a spherically symmetrical charge distribution inside the radius of the particle, maintained in equilibrium by electrical forces—the only ones, apart from gravitation, known at the time. In Mie's theory the repulsive Coulomb forces are compensated by electrical forces resulting from a modification of standard electrodynamics in the interior of the particle. This is achieved by deriving electrodynamics from a Lagrangian density L, called the world function, which is a function of the six electromagnetic field components F_{ik} satisfying $F_{ik} = -F_{ki}$ and the four electromagnetic potentials ϕ_i. Now equilibrium is described quite generally by a static potential given by $\phi_i = 0$ ($i = 1, 2, 3$), $\phi_4 = V/e$ (e = elementary electric charge). But while in standard electrodynamics potentials ϕ_i and ϕ_i + const. are equivalent since, according to Eq. (2.9) below, they determine the same fields F_{ik}, in Mie's theory ϕ_i and ϕ_i + const. give rise to different functionals L. From this Pauli concludes: 'A material particle will therefore not be able to exist in a constant external potential field. This, to us, seems to constitute a very weighty argument against Mie's theory' (Ref. [2a], p. 192).

There remained two theories of charged elementary particles to be discussed in *Part V* of the *Encyklopädie Article*: the 'extremely profound theory' of Weyl presented in Section 65 (see below) and Einstein's theory which 'assumes that the material particles are held together solely by gravitational forces', in Section 66. Concerning the latter Pauli concluded: 'According to the theory of Einstein ... every static spherically symmetrical distribution of electricity is in equilibrium [last sentence in italics]. However satisfactory the foundations of this theory may be, it, too, is not capable of providing an answer to the problem of the structure of matter' (Ref. [2a], p. 205). This most urgent problem in Pauli's view, therefore, was nowhere near a solution. And, as we shall see, it was not furthered by Weyl's theory either. Therefore Pauli closes the *Encyklopädie*

Article with the visionary observation that 'new elements which are foreign to the continuum concept of the field will have to be added to the basic structure of the theories developed so far, before one can arrive at a satisfactory solution of the problem of matter' (Ref. [2a], p. 206). Two years later, in the letter to Eddington quoted in the last section, Pauli emphasizes again 'that the notion of field only makes sense if we are able to indicate a reaction which in principle allows us to measure the field strength in every space–time point if we so desire' (translated from Ref. [14], p. 117). In later years, when Pauli's attention is focused on quantum field theory, this problem reappears and with no less urgency as the duality between the field and its test body.

But now let us return to Pauli's relationship with Weyl and follow it through the years. In January 1928, shortly before he joined Weyl as professor at ETH in Zurich where Weyl had been appointed in 1913, Pauli in a letter from Hamburg vividly recalls the time 'when for the first time I wrote you a letter. Without falling into elegiac considerations I wish to express the hope that in the meantime you may find me to have become yet somewhat more grown-up' (translated from letter [181] of 29 January 1928, Ref. [14], pp. 427–8). Five months later he writes, referring to a professorship of theoretical physics that Weyl held at Princeton University in the United States in 1928–9, before leaving Zurich with regret to become, in 1930, the successor of Hilbert in Göttingen: 'the conclusion appears unrefutable that, at least for some time, you want to be judged not according to your successes in the domain of pure mathematics but based on your loyal but unlucky love towards physics' (translated from letter [227] of 1 July 1929, Ref. [14], p. 505). In the next paragraph of this letter Pauli alludes to the 'lyrical character' of Weyl's papers which, I think, is impossible to ignore in reading Weyl. Indeed, at the centenary celebration of Hermann Weyl, his younger son Michael quoted Weyl as saying: 'I am almost more concerned with expression and form than with cognition itself' and: 'I love the sort of poetry that has a strong human appeal, be it tender or stormy, from Goethe's lyrics to Werfel's way to thrusting himself onto your bosom. Also, the sort of listening with held breath to the quiet voices of objects, tenderly caressing their down, which you find in Rilke's poetry, can give me joy' (Ref. [36], pp. 97–8).

Pauli could not suppress his sarcastic but all too honest response to this poetic nature of Weyl. 'Pauli esteemed Weyl very much, but they didn't fit together. Weyl had very polished manners, he somewhat irritated Pauli [*ging Pauli etwas auf die Nerven*]' [37]. Towards the end of this letter Pauli writes: 'How do you like it over there and how do the Americans like what in your case I have designated as make-up [*Schminke*]?' (translated from letter [227], p. 506, see also Ref. [38]). Weyl, understandably, did not appreciate this last remark, as is reflected in Pauli's apologies in the following letter two months later: 'I am very sorry that the "malignancies" [*Bosheiten*] of my last letter have annoyed you; to produce this effect was not at all their purpose, but they were meant to stand quite by themselves' (translated from letter [235] of 26 August 1929, Ref. [14], p. 518); Pauli found them '*so schön* [so beautiful]'. Like the nickname 'das U-Boot' that the high school student gave to Professor Hornung (see p. 15), this is just another example of Pauli's urge to invent funny labels for people he liked and appreciated. Later examples are 'der Chorknabe des Papstes' (the Pope's choir boy), 'die Karikatur des schaffenden Gottes' (the caricature of the creating God) [38] (see also Ref. [39a]).

Hermann Weyl was born in Elmshorn, Prussia, on 9 November 1885; he was evangelical, the son of a bank director. He studied mathematics in Göttingen and Munich. In Göttingen, where he received much stimulation from Hermann Minkowski (1864–1909) and particularly from Hilbert, he obtained his doctorate in 1908 and in 1910 became a Privatdozent. In autumn 1913 he married Helen Joseph. At the same time Weyl began his 17-year period as professor at the Swiss Federal Institute of Technology, ETH in Zurich, which in his own judgement was the most fruitful of his life and which left engendered a strong emotional attachment to Zurich and the ETH.

However, this period also caused him some sufferings in deciding to renounce numerous tempting calls from other universities, in particular to succeed Felix Klein in Göttingen in 1920 and a call from Columbia University in New York in 1927. But he could not resist the temptation of becoming Hilbert's successor. The rise of the Nazis and the fact that his wife was Jewish, however, influenced Weyl to accept, after some hesitation, an invitation in 1932 from the Institute for Advanced Study in Princeton where he remained until his retirement in 1951. In 1948 his wife Helen died, and in 1950 he married Ellen Bär, the widow of the experimental physicist at the University of Zurich, Richard Bär (1892–1940), from the Zurich banking family of Julius Bär. During his retirement Weyl divided his time between Princeton and Zurich where he lived in the house of the Bär family at Bergstrasse and where he died following a stroke on 8 December 1955, four weeks after celebrating his 70th birthday. For more details on Hermann Weyl's work see Refs [39b,c].

Weyl publicly admitted the seriousness of Pauli's criticism in his encomium at a dinner in honour of Pauli's Nobel Prize held at the Institute for Advanced Study on 10 December 1945. He said: 'Perhaps I am among the first with whom he established scientific contacts, for the first papers he published dealt with a unified field theory of gravitation and electromagnetism which I had propounded in 1918. He dealt with it in a truly Paulinean fashion—namely, he dealt it a pernicious blow' [40].

The idea of Weyl's theory is to generalize Riemannian geometry, not by giving up the metric altogether as in the affine connection mentioned in the last section, but by making the length of a vector dependent on the path along which it is carried around. This is a generalization of Riemannian geometry where only the direction of a vector is path dependent but not its length, while in Euclidean geometry neither direction nor length is path dependent. The key notion allowing vectors to be 'carried around' is parallel displacement. It says (see the previous section) that, in an infinitesimal displacement dx^i, a vector ξ^i suffers an infinitesimal change given by Eq. (64) of Section 14 of the Encyklopädie Article, i.e. by

$$d\xi^i = -\Gamma^i_{rs}\xi^r dx^s. \tag{2.7}$$

This affine connection does not contain the metric (the g_{ik}) at all, and there is no a priori reason for the geodesic components to satisfy the symmetry $\Gamma^i_{rs} = \Gamma^i_{sr}$. This general non-symmetrical affine connection is mentioned in *Note 6* and discussed, with other non-symmetrical theories, in *Note 23* of Ref. [2]. However, as observed by Eddington in Section 92 of his book [20b], in order to have a geometry, the 'parallelogram law' must hold. This means that, choosing for two given infinitesimal displacements dx^i_1 and dx^i_2 the vectors $\xi^i_1 = dx^i_2$ and $\xi^i_2 = dx^i_1$, parallel displacement generates two new vectors $\xi^i_1 + d\xi^i_1$ and $\xi^i_2 + d\xi^i_2$ which, when added to ξ^i_2 and ξ^i_1, respectively, must yield a closed 'parallelogram'. As shown by Weyl—who calls this law 'commutativity' in Ref [41a]—the symmetry $\Gamma^i_{rs} = \Gamma^i_{sr}$ then immediately follows from Eq. (2.7).

In Weyl's theory an infinitesimal displacement dx^i induces a change of an area a by $da = -a\phi_i dx^i$ where $\phi_i(x)$ is a given vector field. Since length ceases to have an absolute meaning, an area a may be 'gauged' by an arbitrary positive function $\lambda(x)$ to $\bar{a} = \lambda a$. From this follows Weyl's *gauge transformation*

$$\bar{g}_{ik} = \lambda g_{ik}; \quad \bar{\phi}_i = \phi_i - \frac{1}{\lambda}\frac{\partial\lambda}{\partial x^i} \tag{2.8}$$

(note the misprint in the first of Eqs (477) in Ref. [2a]). This also implies a modification of the geodesic components which now satisfy the generalized relation $\Gamma_{r,ki} + \Gamma_{k,ri} = (g_{rk})_i + g_{rk}\phi_i$. It follows immediately that the quantities

$$F_{ik} = e\left[\frac{\partial\phi_k}{\partial x^i} - \frac{\partial\phi_i}{\partial x^k}\right], \tag{2.9}$$

where e is a constant, are invariant under the gauge transformation (2.8), while $F_{ik} = 0$ determines the Riemannian case. From this Weyl concluded that the F_{ik} may be identified with the electromagnetic field, in which case the constant e in Eq. (2.9) has the dimension of electric charge. In the *Encyklopädie Article* Pauli writes:'Just as in Einstein's theory the gravitational effects are closely linked with the behaviour of measuring rods and clocks, such that they follow from it unambiguously, so the same holds in Weyl's theory for electromagnetic effects. In this sense, both gravitation and electricity appear as a consequence of the world metric in this theory.' But in a new paragraph he states: 'This point of view had to [be] modified subsequently by Weyl. For, the basic assumptions of the theory in its original form lead to deductions which seem to be in contradiction with experiment. This was stressed by Einstein [quoted in footnote 386]' (Ref. [2a], p. 195). The argument developed by Pauli is that, by integrating the above relation $da/a = -\phi_i dx^i$ over a closed path, $a = a_0 \exp(-\int \phi_i dx^i)$, a clock (atom) of frequency f_0 put during a time t into a region where there is a static electric potential V, as described by $\phi_i = 0$ ($i = 1, 2, 3$), $\phi_4 = V/e$, will be detuned to $f = f_0 \exp(Vct/2e)$ (c = velocity of light, e = any electric charge).

In Weyl's theory the analogue of the four contracted Bianchi identities (2.2) of Einstein's theory are five identities that the 14 fields g_{ik} and ϕ_i must satisfy, the new law added to energy–momentum conservation being conservation of charge. This means: 'In Weyl's theory the charge conservation law is formally on exactly the same footing as the energy conservation law' (Ref. [2a], p. 200). But this formal analogy also holds, although not 'on exactly the same footing', in the more realistic case when electromagnetism and gravitation are described by independent theories. It was this analogy that 10 years later motivated Pauli to hold fast to energy conservation when Bohr was ready to abandon it in the face of its apparent violation in β-decay because, Pauli argued, violation of charge conservation has never been observed. The only issue for Pauli then was to postulate a new, very elusive particle, the neutrino (see page 213).

To see the emergence of the conservation laws we must look more closely at the variational approach to relativity theory as developed in the fundamental works of Lorentz, Hilbert, Einstein, Weyl, and Klein quoted in footnotes 100 to 104 of the *Encyklopädie Article* and mentioned in the last section. In this approach the content of the theory is expressed by the condition that the action integral

$$A = \int W \sqrt{-g} d^4 x, \qquad (2.10)$$

taken over an arbitrary finite space–time region, be stationary with respect to a variation of the fields g_{ik} and, in Weyl's theory, also of the ϕ_i. Here the action density W is a function of all the fields and may be taken as the sum of a gravitational part W_g and a material part W_m. Since at the time under consideration the only known non-gravitational forces were electromagnetic, it was natural to take for W_m the Lagrangian density of Maxwell's theory, $W_m = (1/4) F_{ik} F^{ik}$. As for W_g, it can be argued that in Einstein's theory the curvature invariant R is, up to a constant factor, the only choice. This choice, however, does not work in Weyl's theory because the integrand in (2.10) must be scale invariant. But according to Eq. (2.8), g scales with λ^4 and R with λ^{-1}. In his first paper Weyl opted for $W_g = R_{hikl} R^{hikl}$ as 'the simplest and most natural choice' [41a]. However, other possible choices of W_g are R^2 and $R_{ik} R^{ik}$ or any linear combination of the three invariants mentioned. Although Weyl refers to the two additional invariants at the end of his next paper, he considers them as 'artificial constructs' [41b].

Pauli, on the other hand, examines in his second paper [31] the general linear combination

$$W = c_1 R_{hikl} R^{hikl} + c_2 R_{ik} R^{ik} + c_3 R^2 + \frac{1}{4} F_{ik} F^{ik} \qquad (2.11)$$

with respect to three problems: the observable effects, the form of the field equations, and the stability of the electron. Section 58 of the *Encyklopädie Article* gives a masterly presentation of the two observable effects, namely the perihelion advance of the planet Mars and the deflection of starlight by the sun, and emphasizes the excellent numerical agreement of Einstein's theory with observation. In his second paper [31] Pauli observes that for $c_1 = 0$ in (2.11) Weyl's theory gives the same result for these effects as Einstein's. The reason is that, for $c_1 = 0$ and $F_{ik} = 0$, the action integral (2.10) is always stationary if $R_{ik} = 0$. But these are just the field equations (2.3) in the absence of matter, i.e. if $T_{ik} = 0$, which is the case in the region outside the sun that determines the gravitational field responsible for both observable effects.

However, in the case $c_1 \neq 0$ it is not at all evident that Weyl's theory should yield the observable effects correctly since in this case the field equations resulting from Eqs (2.10) and (2.11) are quite different from Einstein's, even when $F_{ik} = 0$. These general field equations are derived in an extremely lengthy calculation in Pauli's second paper, the result being contained in Eqs (27), (28), and (49) of Ref. [31]. The observable effects, however, are not calculated in this paper. It is Pauli's third publication [33] that addresses this problem for the case of Weyl's preferred choice, $W = R_{hikl}R^{hikl}$. There Pauli shows that, to second order in the ratio m/r where $m = kM/c^2$ is the gravitational radius of the sun, Einstein's expression for the g_{ik} is obtained, so that the observable effects come out identical to Einstein's.

Concerning the third problem, Pauli notes in his second paper: 'For the second time [footnote 1: "The first attempt in this direction is the theory of Mie"], therefore, the possibility seems to be at hand to explain and to understand out of a deeper foundation the existence of the electron' (translated from Ref. [31], p. 461). As Weyl had already showed for his preferred choice of the action density W in Ref. [41a] and Pauli in his second paper [31] generalized to the full expression (2.11), the static case implies that the curvature invariant R must be constant. But in his physical conclusions Pauli goes beyond Weyl when he says: 'If anywhere in the universe there is matter, that also means an electric charge, then R cannot be zero. . . . One very likely would have expected that sufficiently far away from all matter Euclidean geometry would hold. But we see that this is not the case; for, R does not vanish even there, which has a physical meaning only if the world is closed. Thus we are forced in the most natural way to the requirement already inferred by Einstein from the fact that the fixed stars do not escape to infinity but persist in static equilibrium occupying a finite portion of space. But while Einstein's theory of gravitation was compelled to take refuge in a new hypothesis to explain this, namely the cosmological λ-term, here the same emerges completely without constraint and automatically' (translated from Ref. [31], pp. 461–2).

All this sounds quite sympathetic towards Weyl's theory. In fact, two years later Pauli, in a short review of the fourth edition of Weyl's book, praises Weyl's modified point of view concerning the problem of matter 'in which the electrical elementary particles, in contrast to Mie's perception, appear as true singularities of the field, and which also allows us to discern the limits of field physics' (translated from Ref. [42a]). These limits are expressed in Pauli's question quoted at the beginning of this section: 'Should we with the continuum theories for the field in the interior of the electron be on a wrong track at all?' Einstein had immediately grasped the seriousness of this question when he wrote to Max Born on 27 January 1920: 'Pauli's objection is directed not only against Weyl but against any other continuum theory, also one that treats the electron as a singularity' (translated from Ref. [42b], p. 43).

However, the important negative conclusion that Pauli draws in his second paper is this: 'The differential equations for the static field are symmetrical with respect to positive and negative electricity' (translated from Ref [31], p. 462). It is this conclusion that led Pauli to close the section on Weyl's theory in the *Encyklopädie Article* with the statement: 'Summarizing, we can say that Weyl's theory has not succeeded in getting any nearer to solving the problem of the structure of matter' (Ref. [2a], p. 202). Here, however, *Supplementary Note 21* provides the necessary historical correction: 'After the properties of negatons and positons [electrons and positrons] had turned out to [be] exactly symmetrical, the negative antiproton was found experimentally. . . . All arguments in the text, which are based "on the asymmetry between the two kinds of electricity" have therefore to be discarded' (Ref. [2a], p. 223).

Thus the 'pernicious blow' that Weyl mentioned in his *encomium* for Pauli's Nobel Prize quoted above [40], in reality did not come from Pauli's but from Einstein's objection, namely that in Weyl's theory measuring rods and clocks are history dependent. In his allocution for Weyl's 70th birthday Pauli recalled this episode as follows: "That which was common to us indeed began already with my early works on your extension of the Riemannian geometry by non-integrability of the length. Einstein's critique, well-known to you, belonged to my first impressions of Einstein at the physicists' meeting in Bad Nauheim 1919 [1920, see Ref. [35]]. Referring to the sharpness and well-defined nature of the spectral lines he called this extension of geometry from the physical point of view a luxury generalization' (translated from Ref. [42c]).

The mentioned difficulty, expressed in Weyl's gauge transformation (2.8), encountered an unexpected issue 10 years later. Indeed, it turned out that it is not the gravitational field $g_{ik}(x)$ but the complex quantum mechanical wavefunction $\psi(x)$ that is 'gauged' such that Eqs (2.8) are to be replaced by

$$\psi' = e^{if}\psi; \; \phi_i' = \phi_i + \frac{\hbar c}{e}\frac{\partial f}{\partial x^i} \qquad (2.12)$$

($2\pi\hbar$ = Planck's constant, e = elementary charge). It was Fritz London (1900–54) who modified the above integrated gauge relation $a = a_0 \exp(-\alpha \int \phi_i dx^i)$, where in Weyl's case $\alpha = 1$, by choosing an imaginary value $\alpha = -ie/\hbar c$ which to him was somehow suggested by the way the vector potential is introduced in quantum mechanics [43a]. The correct invariance (2.12), however, was first introduced into quantum mechanics—in a relativistically covariant way—by Vladimir Fock (1898–1974) [43b]. Here the second equation (2.12) is the same as the second equation (2.8) if we identify $\log \lambda = -(\hbar c/e)f$, which is the reason why the name 'gauge transformation' had been maintained by Weyl [43c] for Eqs (2.12), although in absolute value, the wavefunction ψ is invariant. This historical turn is commented upon by Pauli in *Note 22* of the *Encyklopädie Article* where Eqs (1) are the above Eqs (2.12) (see, however,

[44]), and it is acknowledged by Weyl in an addendum of 1955 to the reprinted paper [41a] in Ref. [45], p. 192. An interesting account of Weyl's theory and the history of gauge theory is contained in Section II of the Centenary Lecture for Hermann Weyl delivered by Chen Ning Yang (b. 1922) at ETH in 1985 [46].

Notes and references

[1] Preface by A. Sommerfeld to the German special edition in book form, in Ref. [2a], p. xi.

[2] (a) W. Pauli, *Theory of Relativity*, translated from Ref. [3] by G. Field. *With Supplementary Notes by the Author* (Pergamon Press, London, 1958; paper-back edition: Dover, New York, 1981). (b) Wolfgang Pauli, *Teoria della Relatività*, traduzione di Paolo Gulmanelli (Editore Boringhieri, Torino, 1958); (c) Wolfgang Pauli, *Relativitätstheorie, Ergänzende Anmerkungen* aus dem englischen Original übersetzt von Klaus Schilling (editore Boringhieri, Torino, 1963).

[3] W. Pauli jr., 'Relativitätstheorie', in: *Encyklopädie der mathematischen Wissenschaften* (Teubner, Leipzig, 1921), Vol. V/19, pp. 539–775; reprinted in book form (Teubner, Leipzig, 1921). Reprinted with the *Supplementary Notes* of Ref. [2a] in Ref. [5], Vol. 1, pp. 1–263.

[4] A. Einstein, *Naturwissenschaften* **10**, 184 (1922), translated in the preface to Ref. [5], Vol. 1, p. x.

[5] R. Kronig and V. F. Weisskopf (eds), *Collected Scientific Papers by Wolfgang Pauli. In Two Volumes* (Wiley Interscience, New York, 1964).

[6] W. Pauli jr., 'Über die Energiekomponenten des Gravitationsfeldes', Physikalische Zeitschrift **20**, 25 (1919). Submitted from Vienna on 22 September 1918. Reprinted in Ref. [5], Vol. 2, pp. 10–12.

[7] (a) P. Urban, 'Wolfgang Pauli †', *Acta Physica Austriaca* **12**, 217 (1959); (b) H. Sequenz (ed.), *150 Jahre Technische Hochschule in Wien 1815–1965* (Wien, 1965), p. 167. I thank Professor Gerhard Oberkofler, Archiv, Universität Innsbruck, for a copy of this document.

[8] '85. Versammlung deutscher Naturforscher und Ärzte in Wien vom 21. bis 28. September 1913', *Physikalische Zeitschrift* **14**, 1073–266 (1913).

[9] (a) W. Moore, *Schrödinger. Life and Thought* (Cambridge University Press, Cambridge, 1989); (b) J. Mehra and H. Rechenberg, *The Historical Development of Quantum Theory*, Vol. 5: *Erwin Schrödinger and the Rise of Wave Mechanics* (Springer, New York, 1987), Section I.4.

[10] C. P. Enz und K. von Meyenn (eds) *Wolfgang Pauli. Das Gewissen der Physik* (Vieweg, Braunschweig, 1988).

[11] (a) K. von Meyenn (ed., with the cooperation of A. Hermann and V. F. Weisskopf), *Wolfgang Pauli. Scientific Correspondence with Bohr, Einstein, Heisenberg, a.o., Volume II: 1930–1939* (Springer, Berlin and

Heidelberg, 1985); (b) S. Richter, *Wolfgang Pauli. Die Jahre 1918–1930. Skizzen zu einer wissenschaftlichen Biographie* (Verlag Sauerländer, Aarau, 1979).

[12] (a) *Note*: On p. 688 of Ref. [11a] this result is erroneously quoted as the sum of the centrifugal and Coriolis accelerations (with factor 1!), 'and that the additional terms practically remain unmeasurably small'. This error might perhaps be due to an inaccurate translation of the relevant passage in Ref. [2], p. 175: 'but these two terms would have to be multiplied', instead of 'are multiplied'. Anyway, the conclusion drawn from this erroneous quotation, that 'Mach's Principle', namely the equivalence of the rotating and the fixed coordinate systems, holds, is of course not true, as Pauli and Einstein [8] had already emphasized. (b) W. Pauli, 'The Theory of Relativity and Science', in Ref. [13a], pp. 107–11.

[13] (a) C. P. Enz and K. von Meyenn (eds), *Wolfgang Pauli. Writings on Physics and Philosophy* (Springer, Berlin and Heidelberg, 1994); (b) *J. C. Poggendorff's biographisch-literarisches Handwörterbuch. Band VI: 1923 bis 1931*; (c) W. Pauli, letter [5] to Samuel Oppenheim of 7 May 1920, in Ref. [14]; (d) E. Madelung, 'W. Pauli jun., Relativitätstheorie (Sonderabdruck aus der Enzyklopädie der mathematischen Wissenschaften)', *Physikalische Zeitschrift* **23**, 192 (1922).

[14] A. Hermann, K. von Meyenn, and V. F. Weisskopf (eds), *Wolfgang Pauli. Scientific Correspondence with Bohr, Einstein, Heisenberg, a.o., Volume I: 1919-1929* (Springer, New York, 1979).

[15] (a) A. Sommerfeld, 'Autobiographische Skizze', in: F. Sauter (ed.), *Arnold Sommerfeld. Gesammelte Schriften* (Vieweg, Braunschweig, 1968), Vol. 4, pp. 673–9; (b) *Verhandlungen der Deutschen Physikalischen Gesellschaft* **21**, 795 (1919).

[16] A. Pais, *'Subtle is the Lord ...'. The Science and the Life of Albert Einstein* (Oxford University Press, Oxford, 1982).

[17] (a) E. Mach, *Die Geschichte und die Wurzel des Satzes von der Erhaltung der Arbeit* (Vortrag vom 15. November 1871, Gesellschaft der Wissenschaft; J. G. Clave, Prag, 1872); English translation by P. E. B. Jourdain, *History and Root of the Principle of the Conservation of Energy* (Open Court Publications, Chicago, 1911); (b) K. D. Heller, *Ernst Mach. Wegbereiter der modernen Physik. Mit ausgewählten Kapiteln aus seinem Werk* (Springer, Vienna, 1964).

[18] J. Mehra, 'Einstein, Hilbert, and the Theory of Gravitation', in Ref. [19], footnote 242, p. 170.

[19] J. Mehra (ed.), *The Physicist's Conception of Nature* (Reidel, Dordrecht, 1973).

[20] (a) A. S. Eddington, 'A Generalisation of Weyl's Theory of the Electromagnetic and Gravitational Fields', *Proceedings of the Royal Society* A

99, 104 (1921); (b) A. S. Eddington, *The Mathematical Theory of Relativity* (Cambridge University Press, Cambridge, 1923; 3rd edn Chelsea, New York, 1975).

[21] A. Einstein, 'Die Grundlage der allgemeinen Relativitätstheorie', *Annalen der Physik* **49**, 769 (1916); reproduced in book form (J. A. Barth, Leipzig, 1916). Reproduced as document 30 in Ref. [22].

[22] A. J. Knox, M. J. Klein, and R. Schulmann (eds), *The Collected Papers of Albert Einstein, Volume 6: The Berlin Years 1914-1917* (Princeton University Press, Princeton, NJ, 1996).

[23] A. Einstein, 'Hamiltonsches Prinzip und allgemeine Relativitätstheorie', *Sitzungsberichte der Berliner Akademie* **42**, 1111 (1916). Reproduced as document 41 in Ref. [22].

[24] Unpublished preface, handwritten original, dated 30 July 1951, and typed copy, signed and dated 31 July 1951, in Box 9, *Pauli Archive*.

[25] C. P. Enz, Preface, in Ref. [13a], p. 3.

[26] *Note*: The publishing house B. G. Teubner was established in Leipzig since 1811. But the Soviet occupation after World War II rendered its publishing activity increasingly difficult so that, sometime after 1952, Teubner was established in Stuttgart. See letter from Teubner to Pauli, dated Stuttgart, 1 June 1955. Item 9/553, Box 9, *Pauli Archive*. For more details on this correspondence see the comment after letter [2244] in: K. von Meyenn (ed.), *Wolfgang Pauli. Scientific Correspondence with Bohr, Einstein, Heisenberg, a.o., Volume IV, Part III: 1955-1956* (Springer, Berlin and Heidelberg, 2001).

[27] A. Mercier and M. Kervaire (eds), 'Jubilee of Relativity Theory', *Helvetica Physica Acta*, Supplement **4** (1956), 286 pp.

[28] 'Probleme der Allgemeinen Relativitätstheorie, Vorlesung gehalten an der Eidg. Techn. Hochschule, Zürich, 1953, von Prof. Dr. W. Pauli. Ausgearbeitet von Ch. Enz' (handwritten, 111 pp.), *Pauli Archive*. Pauli's personal notes that he used as preparation for this course are contained in Box 9, *Pauli Archive*.

[29] Translated from: W. Pauli, letter [1] of 10 May 1919 to H, Weyl, in Ref. [14].

[30] H. Weyl, *Raum, Zeit, Materie; Vorlesungen über allgemeine Relativitätstheorie* (Springer, Berlin, 3. umgearbeitete Auflage, 1919); English translation by H. L. Brose, *Space, Time, Matter* (Dover, New York, 1952).

[31] W. Pauli jr., 'Zur Theorie der Gravitation und der Elektrizität von Hermann Weyl', *Physikalische Zeitschrift* **20**, 457 (1919). Reprinted in Ref. [5], Vol. 2, pp. 13–23.

[32] C. P. Enz, 'W. Pauli's Scientific Work', in Ref. [19], pp 766–99.

[33] W. Pauli jr., 'Merkurperihelbewegung und Strahlenablenkung in Weyls Gravitationstheorie', *Verhandlungen der Deutschen Physikalischen Gesellschaft* **21**, 742 (1919). Reprinted in Ref. [5], Vol. 2, pp. 1–9.

[34] A. Einstein, 'Ernst Mach', *Physikalische Zeitschrift* **17**, 101 (1916). Reprinted in Ref. [17b], pp. 151–7 and in Ref. [22], pp. 278–81.

[35] 'Vorträge und Diskussionen von der 86. Naturforscherversammlung in Nauheim vom 10.–25. September 1920', *Physikalische Zeitschrift* **21**, 649-720 (1920).

[36] K. Chandrasekharan (ed.), *Hermann Weyl 1885–1985. Centenary Lectures delivered by C. N. Yang, R. Penrose, A. Borel at ETH Zürich* (ETH Zürich and Springer, Berlin and Heidelberg, 1986).

[37] Notes of my conversation with Franca Pauli on 8 April 1984.

[38] J. Mehra, Preface, in: J. Mehra and H. Rechenberg, *The Historical Development of Quantum Theory. Volume 1. The Quantum Theory of Planck, Einstein, Bohr and Sommerfeld: Its Foundation and the Rise of Its Difficulties: 1900-1925* (Springer, New York, 1982), Part 1, footnotes 16-18, p. xxiii.

[39] (a) V. L. Telegdi, 'Pauli-Anekdoten', in Ref. [10], pp. 115–20; (b) G. Frei and U. Stammbach (eds), *Hermann Weyl und die Mathematik an der ETH Zürich 1913-1930* (Birkhäuser, Basle, 1992); (c) D. Speiser, 'Hermann Weyl 1885–1955', *Physikalische Blätter* **42**, 39–44 (1986).

[40] H. Weyl, Encomium, *Science* **103**, 216 (1946).

[41] (a) H. Weyl, 'Gravitation und Elektrizität, *Sitzungsberichte der Preussischen Akademie der Wissenschaften 1918*, p. 465. Reprinted with discussion remarks by Einstein and Weyl in Ref. [45], pp. 179–92. (b) H. Weyl, 'Eine neue Erweiterung der Relativitätstheorie', *Annalen der Physik* **59**, 101 (1919).

[42] (a) W. Pauli jr., 'H. Weyl, Raum, Zeit, Materie. 4. Auflage', *Physikalische Zeitschrift* **23**, 256 (1922); (b) Albert Einstein, Hedwig und Max Born, *Briefwechsel 1916–1955, kommentiert von Max Born* (Nymphenburger Verlagshandlung, München, 1969). (c) Translated from: W. Pauli, Allocution [2183] on the occasion of the 70th birthday of Hermann Weyl on 9 November 1955, in: K. von Meyenn (ed.), *Wolfgang Pauli. Scientific Correspondence with Bohr, Einstein, Heisenberg, a.o., Volume IV, Part III: 1955-1956* (Springer, Berlin and Heidelberg, 2001).

[43] (a) F. London, 'Quantenmechanische Deutung der Theorie von Weyl', *Zeitschrift für Physik*, **42**, 375 (1927); (b) V. Fock, 'Über die invariante Form der Wellen- und der Bewegungsgleichungen für einen geladenen Massenpunkt', *Zeitschrift für Physik* **39**, 226 (1927); (c) H. Weyl, 'Elektron und Gravitation. I.', *Zeitschrift für Physik* **56**, 330 (1929).

[44] The first equation (1) of Note 22 (Ref. [2a], p. 223) contains a serious error: $-i$ should be $+$ (without the imaginary i!).

[45] B. Eckmann *et al.* (eds), *Selecta Hermann Weyl. Published on the occasion of his seventieth birthday by the Federal Institute of Technology in Zurich and the Institute for Advanced Study in Princeton* (Birkhäuser, Basle and Stuttgart, 1956).

[46] C. N. Yang, 'Hermann Weyl's Contribution to Physics', in Ref. [36], p. 7.

3
Student in Munich

In Sommerfeld's current

In the autumn of 1906, on the proposition of Röntgen, I received a call to Munich as the indirect successor of Boltzmann. . . . In Munich I had the occasion to give courses for the first time on the various domains of theoretical physics and special lectures on questions of current interest. From the beginning I tended, and I have not shied from any effort, to found in Munich, through an activity of seminars and colloquia, a nursery of theoretical physics. [1]

This is how Sommerfeld describes the establishment of his school in his *Autobiographical Sketch* [1]. Pauli went to this school, matriculating at the Ludwig–Maximilian University in Munich on 4 October 1918 [3] and taking up residence at Theresienstrasse 66 (see Ref. [4], p. 536), two blocks to the south-west of the university's main building. The reputation of a prodigy of relativity had already preceded him, as is evident from Sommerfeld's astonished remarks (see Ref. [4], p. 1). 'From the Viennese intelligentsia I got a downright amazing specimen around me in the young Pauli, son of a Viennese medical chemist. A freshman! His talents even surpass those of Debye by a large margin' wrote Sommerfeld in a letter to his Austrian colleague Josef von Geitler on 14 January 1919 (translated from Ref. [5a], p. 29, see also Ref. [5b], Section 11(g)).

But Munich greeted Pauli with chaos. The dreadful 'rape-seed winter' of 1916–17 had been as bad in Munich as in Vienna (see page 16). And hopes of a breakthrough by the German Army on the western front dimmed after the failure of the large offensive in March 1918. In July the Allies started a counter-offensive. In October their troups threatened to advance towards Bavaria and the Austrian Empire collapsed. On 7 November there were peace demonstrations in Munich, the socialist leader Kurt Eisner proclaimed the (first) 'Räterepublik' (Soviet Republic) and the Bavarian king Ludwig III was dismissed. Two days later similar events took place in Berlin, and the abdication of Emperor William II (1859–1941) and the proclamation of the republic ended the 47 years of the Second German Empire.

Chaos again reigned in Munich when Eisner was assassinated on 21 February 1919 and a second 'Räterepublik' was proclaimed on 13 April. The university was closed, newspapers were prohibited, food and weapons were seized.

Things were made even worse by a spell of extreme cold and the lack of coal; half a metre of snow fell on 1 April. During the Easter week from 14 to 22 April the population reacted in its misery with a total strike against the 'Red Terror'. But from Berlin a forceful army was sent against Bavaria. On 30 April 1919 Munich was surrounded, and by the evening of 2 May the 'White Army' was in command of the city. However, for several days Munich was subject to a 'White Terror' of revenge (see Chapter 4, 'Red and White', in Ref. [7a]).

At the national level the Weimar Constitution was adopted in Weimar by the National Assembly and signed by the Imperial (Reichs) President Friedrich Ebert on 11 August 1919. This was the beginning of the Weimar Democratic Republic which lasted until 1933 when it was abolished by Hitler's National Socialists. Although it had a stabilizing effect, this reunification was not the end of the misery nor of the instability. This is eloquently described in the following passage in a letter from Einstein to Paul Ehrenfest (1880–1933) written on 7 April 1920, just a few weeks after the 'Kapp–Putsch', the rightist–nationalist revolution led by Wolfgang Kapp in Berlin, which spread to other parts of Germany: 'Externally calm has returned again here [Berlin]. But there are gaps of oppositions of marked acuteness. Arbitrariness of the sabre and exasperation against it. In town, want and hunger are dreadful. Children mortality is horrible. Where in reality we are heading to, nobody knows. The state has sunk into an extreme swoon. Next to it the main powers quarrel: Sabre, money, extreme socialist unions' (translated from Ref. [5a], p. 27).

Einstein openly pleaded for the Weimar Republic and for international understanding, and this in the face of violent anti-Semitic agitation against him and criticism of his relativity theory supported by several leading experimentalists. In this criticism Philipp Lenard's (1862–1947) prestige of the Nobel Prize won in 1905 'for his work on cathode rays' stood against Einstein's rising international reputation. Pauli, in his *Encyklopädie Article* (see pages 26f), mentions Lenard's criticism only in footnote 287a, saying explicitly that he 'cannot agree with these objections'. And Weyl reports from the Nauheim meeting of September 1920, where Pauli had met Einstein and Weyl for the first time (see page 38): 'The last and most dramatic part [of the meeting], the general discussion on relativity theory, developed essentially into a duel between Einstein and Lenard. Planck, as chairman, ruled with dexterity, rigor and impartiality; it was not to a small part his merit that this "Nauheim relativity conversation" proceeded with dignity. In it opposing epistemological perceptions of principle concerning science clashed against each other' (translated from Ref. [7b], p. 14).

The Versailles Peace Treaty signed on 28 June 1919 demanded drastic territorial sacrifices from Germany and prohibited reunification with Austria. Equally painful were the economic demands, in particular the high reparation payments. After a murderous war this treaty expressed the bitterness of the Allies against what they considered German guilt for the war. But right at the outset, in October 1914, German intellectuals had protested against this accusation in the *Manifesto of the 93* (see Ref. [5b], p. 234). Towards the end of the war German science and culture were indeed systematically isolated (see Ref. [5a], pp. 31–2), so that Planck himself defiantly wrote in 1918: 'When the enemies have taken away defence (*Wehr*) and might from our fatherland, when in the innermost serious crises have erupted and perhaps even more serious ones are imminent, there is one thing no external and internal enemy has yet taken from us: this is the position that German science holds in the world. But to hold this position and, if necessary, to defend it by all means, this is what our [Prussian] Academy, as the noblest scientific institution of the state, is called for among the first' (translated from Ref. [5a], p. 25). More than three years later Sommerfeld wrote in a letter of 12 February 1922 to Paul Epstein: 'The enormous Allied lie of German guilt for the war will continue to reign for still some more years, but not eternally' (translated from Ref. [5a], p. 30).

What was Pauli's reaction to all this? There are no letters to his family in Vienna, in which such testimony could perhaps be found. Such letters must have been destroyed either when his father Wolfgang fled to Zurich in 1939 (see pages 11–12) or when Pauli and his wife Franca left Zurich for the United States in 1940 (see pages 339–40). But Pauli was singularly uninterested in the daily happenings of local political and economical events. He considered reading newspapers a waste of time. This is vividly illustrated by the following episode recalled by Margaret Schmid, the much-esteemed secretary of the physics department at ETH in the mid-1950s. It was one day around 1957 when Pauli called Ms Schmid to come to his office two storeys above, asking her to help him in a money transfer from Germany. As usual, she had advice to offer: 'Herr Professor, you surely have read in the newspaper last week that money transfers with Germany have been liberated.' Quite unexpectedly, Pauli became nervous and started pacing his office from desk to wall and back. Finally he said: 'You disappoint me. I never would have thought that you, of all people, would expect me to read newpapers or listen to radio.' Excusing herself she promised Pauli she would clear up the question, and Pauli was happy again: 'I knew you could help me!' (Ref. [8a], p. 285).

There is also a recollection of Pauli from the early 1920s in an unpublished allocution he had addressed 'An Hermann Weyl zum 9. Nov. 1955', for Weyl's 70th birthday on 9 November (Weyl died one month later on 8 December 1955). There Pauli writes: 'After Nauheim [the meeting of the German natural scientists and physicians, see pages 38 and 50] there was the following year a common railway ride from Munich to the physicists' and mathematicians' meeting in Jena [September 1921], where also Edgar Meyer [1879–1960] appeared. You both came from Zurich which at that time was unknown to me, and I remember well the cheerful mood on the train. The war was over, with Sommerfeld I had come to ride the just current (*Fahrwasser*); why should I as a young man then still have cared for the political and economical circumstances in Germany and Austria?' (translated from letter [2183], Ref. [7c], p. 401; this quotation also appears in Ref. [4], p. 33; see also Ref. [7b], note I.28, p. 93). During Eisner's 'Räterepublik', on Tuesday, 10 December 1918, when Pauli was in his first semester, he gave a talk entitled 'Remarks on General Relativity' at Sommerfeld's seminar (see Ref. [7b], note I.6, p. 91). On 4 June 1919 when Munich was still occupied by the 'White Army', he submitted his second paper on relativity theory (see pages 42–4).

But Pauli was not alone in taking scant notice of the tumultuous events around him. Sommerfeld, for one, succeeded in publishing two of his most important papers during the height of the 'Red Terror' in April 1919 (published 1919 and 1920, see Ref. [1], p. 678). Sommerfeld's institute and apartment, both located in the university building, were well shielded from the outside world—except for the students, of course. That the latter could be a nuisance became evident when Einstein was invited by Sommerfeld and several of his colleagues to give a talk in Munich in November 1921. The extremist representatives of the students refused to give assurances to the rector that no disruptions would occur. Einstein learned of it through the press in time to cancel his invitation. Apart from the lecture room and his own office Sommerfeld's institute consisted of a laboratory in the cellar, a seminar room and a small library. For his Ph.D. students he was the model of a hard-working scientist who always found time for their problems. The most important weekly event for them was the seminar in which they had to report on current research papers and which was the most efficient means of selection (see the chapter on *Sommerfeld's* seminar in Ref. [7a]).

Sommerfeld also was an accomplished teacher. The main course of four hours per week covered all of classical physics in a cycle of three years. It included a problem session of one hour which was supervised by his assistant. The first assistant had been Peter Debye (1884–1966) who had accompanied Sommerfeld to Munich in 1906 (see Ref. [1], p. 677) and who later became a professor at Cornell University in Ithaca, New York, and in 1936 won the Nobel Prize in Chemistry. During Pauli's time the assistant was Peter Paul Ewald (1888–1985) who had obtained his doctorate with Sommerfeld in 1912 and, after one year with Hilbert in Göttingen, became a Privatdozent (lecturer) in Munich in 1914, an associate professor in Stuttgart in 1921 (see Ref. [8b], footnote 447, p. 281) and later a professor in Belfast (Northern Ireland) and in Brooklyn (New York). Ewald's successor as Sommerfeld's assistant was Pauli's friend Gregor Wentzel (1898–1978) who obtained his doctorate with Sommerfeld in 1921 and, the following year, became a Privatdozent in Munich (see Ref. [8b], footnote 609, p. 356). After his exam in 1921 (see below) Pauli himself held a deputy assistantship. Karl Ferdinand Herzfeld (1892–1978) and Wilhelm Lenz (1888–1957), Pauli's future boss in Hamburg, were Privatedozenten, and Pauli's friend Werner Heisenberg (1901–76) and Otto Laporte (1903–71) were students.

Sommerfeld's three-year cycle consisted of the following courses: I. Mechanik, II. Mechanik der deformierbaren Medien, III. Elektrodynamik, IV. Optik, V. Thermodynamik und Statistik, VI. Partielle Differentialgleichungen der Physik. Eventually, all these courses appeared in print as separate volumes of the *Vorlesungen über theoretische Physik*, the first two during the war in 1942 and 1944, respectively, volume VI in 1947, volume III in 1948, and volume IV in 1950. In a review of volume IV in *Optics*, Pauli praises the 'senior master' (Altmeister) for a 'conceivably unconventional book' [8c]. These five volumes were edited by Sommerfeld himself, based on the notes of former students. Unfortunately, they were printed on very bad-quality paper, particularly the first two, which, however, did not prevent them from being in high demand—I succeeded in getting copies from Germany as a student after the war. The missing volume V appeared in 1952, one year after Sommerfeld died, edited by F. Bopp and J. Meixner. This celebrated lecture series saw many editions and translations into English and Russian (see Ref. [2], pp. 685–8). In his *Autobiographical Sketch* Sommerfeld himself comments on the creation of this lecture series as follows: 'During the war years I decided on persuasion to edit my old course-lectures in amplified form. ... Without this work I could not possibly have endured the political convulsions of the war period' (Ref. [1], p. 679).

In addition to the cycle Sommerfeld gave a *Spezialvorlesung* (special course) of two hours weekly on current research topics. In distinction to Sommerfeld, Pauli, when he came to ETH in 1928, established a two-year cycle whose content, however, except from being more concentrated, was not much different from Sommerfeld's (compare Ref. [9a]). He adopted essentially the same teaching schedule (see Ref. [8a], pp. 431–4), with one important difference, however. With the exception of one hour of *Spezialvorlesung* before lunch all of Pauli's teaching took place in the afternoons, whereas some of Sommerfeld's lectures started at 9 a.m. Now during his Munich years Pauli had developed the habit of sampling the night-life of Schwabing, the elegant, cultural northern district of the city of Munich, returning late to spend much of the night working. Therefore he used to drop in only at the end of Sommerfeld's morning lecture just to check the blackboard for what the topic had been, and Sommerfeld tolerated this irregular behaviour (see Ref. [9b], p. 45).

Officially, though, Pauli registered for most of the courses of Sommerfeld's three-year cycle. According to his *Kollegienbuch* [3], the booklet in which each student had to list the courses and seminars they were attending, he had registered only for the problem session of 'Thermodynamik und Statistik' in his first term (winter 1918/19). But in his fourth semester he took a two-hour introductory course on statistical mechanics with Privatdozent Herzfeld. During the second term, in summer 1919, Pauli registered not only for the 'Partielle Differentialgleichungen der Physik' but also for the *Spezialvorlesung* on 'Relativitätstheorie'. In the winter of 1919/20 he registered for 'Optik' and also took the *Spezialvorlesung* on the dynamics of the crystal lattices given by Privatdozent Peter Paul Ewald (b. 1888). In summer 1920 Pauli registered for 'Elektrodynamik', and in the last two terms he again only took the *Spezialvorlesung* which was devoted to the theory of X-rays in winter 1920/21 and to electro- and magneto-optics in the summer of 1921. The two missing courses of the cycle were thus mechanics and mechanics of deformable media, topics Pauli had learned from the book that his godfather Mach had dedicated to him when he was just 13 years of age (see page 14).

It seems that in his first semester Pauli had not yet developed his nightly life-style since, according to his own insertions in his *Kollegienbuch* [3], two of the compulsory courses took place every morning from Monday to Friday, namely 'Unorganische Experimentalchemie' with Professor Richard Willstätter from 9 to 10 a.m. and Professor Willy Wien's course 'Experimentalphysik I' (introduction, heat, electricity) given by Dr Graetz from 10 to 11 a.m. In his last semester in summer 1921 Pauli also took Wien's exacting laboratory course of eight hours 'Physikalisches Prakticum', together with 'fellow sufferer' Otto Laporte (Ref. [8b], p. 380). This course was of vital importance because Wien and Sommerfeld were the examiners of the main field of physics in Pauli's oral doctor's examination called *examen rigorosum*. This exam took place on 25 July 1921, and Pauli got the best grade of 1 in all the disciplines (Ref. [4], p. 32). Two years later Werner Heisenberg failed Wien's examination, receiving the low grade of 5 which resulted in a poor average of 3 (Ref. [7a], p. 152).

In the 1880s Arnold Sommerfeld, Willy Wien, and his cousin Max Wien (see page 17), and also Hermann Minkowski, all simultaneously attended the *Altstädter Gymnasium* (Old Town High School) of Königsberg in Prussia (Ref. [1], p. 673), which after World War II became Kaliningrad in the USSR. A lecture by Hermann Minkowski (1864–1909) on space and time in 1908 inspired Sommerfeld in his first *Spezialvorlesung* on relativity (Ref. [1], p. 677). Wilhelm Wien (1864–1928), who had won the Nobel Prize in 1911 'for his discovery regarding the laws governing the radiation of heat', became the successor of Röntgen in 1920 to the chair of experimental physics in Munich. Wilhelm Röntgen (1845–1923) had been the first recipient of the Nobel Prize in Physics in 1901 for his discovery of X-rays.

Sommerfeld used to engage his students in the research activity of his institute right from the start. Thus, in parallel to his work on Weyl's theory of relativity (see page 37) Pauli had already worked in 1919 on problems of quantum theory, to which he had been initiated in his first semester in a course of one hour per week by Sommerfeld on atomic structure and spectral lines, the subject of his famous book whose first edition appeared that very year [10a]. This book gave a detailed account of the enormous spectroscopic material and its interpretation in terms of the old quantum theory developed by Bohr and Sommerfeld, and in which Sommerfeld made great use of his favourite mathematical method of integration in the complex plane. Here 'old' is used in comparison with the new quantum mechanics created by Heisenberg in 1925 and the wave mechanics proposed by Schrödinger in 1926 (see e.g. Ref. [10c]). Because of the rapid development of the field of atomic physics Sommerfeld had to recast and expand his book twice even before the new era. In addition to the four editions of this time there were four more editions after 1925, as well as translations into English, Russian, and French (see Ref. [2], pp. 685–8).

The preface to the first edition of *Atombau und Spektrallinien* [10a], dated September 1919, is noteworthy for its poetic turn of phrase: 'What we are nowadays hearing of the language of spectra is the true music of the spheres within the atom, chords of integral relationships, an order and harmony that becomes ever more perfect in spite of the manifold variety. The theory of spectral lines will bear the name of Bohr for all time. But yet another name will be permanently associated with it, that of Planck. All integral laws of spectral lines and of atomic theory spring originally from the quantum theory. It is the mysterious *organon* on which Nature plays her music of the spectra, and according to the rhythm of which she regulates the structure of the atoms and nuclei' (translation taken from Ref. [11a], p. 64). That, one would think, was language to provoke sarcastic comments from Pauli (see page 39). However, the only remark which is on record stems from one of Sommerfeld's lectures when Pauli turned to Heisenberg, observing 'Doesn't he look like an old Hussar colonel?' (translated from Ref. [9b], p. 42).

To thank Sommerfeld for sending him a copy of the fourth edition of *Atombau und Spektrallinien* [10b] Pauli wrote to Sommerfeld on 6 December 1924 from Hamburg: 'With it you have caused me great joy. It was probably the first time that I have not helped you at all with the new edition, but perhaps I will be able to do so again in the future' (translated from letter [72], Ref. [4], p. 182). This serviceability was as much a trait of Pauli's character as his sarcasm which Sommerfeld seems to have been spared. Pauli, like all of Sommerfeld's disciples, had a deep veneration for him. Thus, for instance, on introducing his fourth paper which is a critique of some published false results on special relativity, Pauli writes: 'As reporter (*Referent*) on the theory of relativity in the mathematical Encyclopedia, on the special advice of my highly revered teacher (*meinem hochverehrten Lehrer*), Herrn Professor Sommerfeld, I have decided to undertake the verification' [12]. Pauli used the same expression '*meinem hochverehrten Lehrer*' in the acknowledgement of Sommerfeld's 'incitation' (*Anregung*) in his Ph.D. work (see Ref. [14] and the following section).

Sommerfeld, in fact, was the only person with whom Pauli had a lifelong pupil–master relationship. Even as an established professor in Zurich, Pauli behaved rather submissively in front of Sommerfeld, bowing in respect and responding with '*Ja, Herr Professor*', '*Nein, Herr Professor*', which amused a few people in the known (see Ref. [11b], p. 85). In his essay quoted above Pauli writes: '[Sommerfeld] has given inspiration to an ever growing circle of disciples in Munich. This circle of disciples, dispersed over many countries on both sides of the Atlantic, among whom I gratefully count myself, takes care that the intellectual tradition which Sommerfeld passed on to us will be transmitted to academic youth and thereby to posterity. This tradition goes back to Sommerfeld's teacher Felix Klein, and through him also to Riemann. . . . It is not only with whole numbers that Sommerfeld's pupils will always feel at home, but also in the complex plane the use of which he, Sommerfeld, was so fond of in evaluating phase integrals and in discussing solutions of partial differential equations' (Ref. [11a], p. 68).

Perhaps the most thoughtful expression of Pauli's feelings towards his teacher is the one contained in his letter written on Sommerfeld's 70th birthday on 5 December 1938. Referring to a paper on one of Sommerfeld's favourite subjects that Pauli had discreetly delayed in order for it to appear at this event

[15a], Pauli observes that, as a reaction to this delay, he 'only could notice that *severe frown* with you that had always instilled me so much awe, since the year 1918, in which year I indeed saw you for the first time'. Pauli comments: 'A deeper secret, of course, is why it was just you that succeeded to instill awe in me—this, no doubt, many others would have liked to learn more precisely from you, in particular, all my later bosses, also Herr Bohr.' Furthermore: 'And then I also remember many integer numbers which, in those cheerful student days, I considered with less love than you, and even not without mockery. Therefore, you also have taken revenge on me, and the sequence 2, 8, 18, 32... even returned for so long, until I succeeded at last in fixing it by prohibitions where it belongs [the exclusion principle].' In conclusion Pauli observes 'that, for me, the years 1918 to 21 in Munich, which you have given me, are far from being exhausted!' (translated from letter [537a], Ref. [15b]). Understandably, Sommerfeld was moved (see letter [538], Ref. [15b]).

However, Pauli was very independent in his research in Munich. Apart from the two instances where he gives credit to his 'highly revered teacher' mentioned above, one cannot find any acknowledgements to Sommerfeld in Pauli's publications. He himself always generously shared his ideas with colleagues in discussions—he was most active in Sommerfeld's seminars—and in his enormous production of letters. All the more sarcastic was his reaction when he found out that some of his ideas had been published, without acknowledging him, by somebody else, even if he was Sommerfeld himself: 'Should I at one time be too lazy to publish a matter myself . . . I will communicate it to you [Sommerfeld] by letter. You then surely will sooner or later publish it in one form or another. . . . This is indeed a very pleasant method of publishing, a service that already once Herr Landé has rendered me so well' (translated from letter [70] from Pauli to Sommerfeld in November 1924, Ref. [4], p. 173; see letter [46] from Pauli to Landé on 23 September 1923). Alfred Landé (1888–1975) was one of the leading scientists in the problem of the Zeeman effect, i.e. the splitting of spectral lines in a magnetic field, with whom Pauli corresponded from 1919 (see Ref. [4]). After serving as a Privatdozent in Frankfurt from 1920 to 1922 Landé became an associate professor in Tübingen. In 1931 he emigrated to the United States where he became a professor of physics in Columbus, Ohio, until his retirement in 1960 (see Ref. [8b], footnote 661, p. 399).

First encounter with magnetism

Many thanks for your letter. I am glad to hear that you have turned to poor magnetism, neglected for quite some time, and thereupon announce beautiful forthcoming results, and I am also glad that we have come to agree about it since, clearly, it would be nonsense to work in parallel and out of coordination at a time when the working domain of model businesses (Modellangelegenheiten) surely is large enough to make, on the contrary, the utmost economical division of labor an urgent requirement. [15c]

With these words Schrödinger greeted Pauli's arrival from Jena on 12 July 1920 in his typical flowery style in the first of his letters to Pauli recorded in Ref. [4]. More specifically, Schrödinger continues: 'Very curious I am to see your diamagnetic note. For, it seems to me that, e.g. with helium, the Bohr orbits are quite insufficient to explain the relatively strong diamagnetism ($\kappa = -11 \times 10^{-9}$), even if the latter were not partially eclipsed by an orientational effect, though obviously very weak here' [15c]. The note on diamagnetism, submitted by Pauli from Munich on 18 June 1920, investigates monatomic gases, excluding molecular gases and non-gaseous states because of interatomic couplings [16]. Pauli follows the method of quadrupole moments that Debye had developed in order to explain the van der Waals cohesive forces acting between the molecules or atoms of a gas [17].

Pauli's calculation is based on the old Bohr–Sommerfeld quantum theory of stationary elliptical orbits of the electrons around the nuclei, as described in Sommerfeld's celebrated book *Atomic Structure and Spectral Lines* [10a] mentioned in the previous section. According to this theory the charge e of the electron gives rise, in its motion around the nucleus, to an elementary Ampère current which, for a circular orbit, is $J = e\omega/2\pi$ where ω is the angular velocity. If r is the radius of the circle and m the electron's mass, its angular momentum is $L = mr^2\omega$. This results in a magnetic moment vector of (see e.g. Ref. [9a], *Volume 1. Electrodynamics*, pp. 73–4)

$$\vec{\mu} = \frac{e}{2mc}\mathbf{L}. \tag{3.1}$$

Here $\mathbf{L}$ is the angular momentum vector whose direction is perpendicular to the plane of the electron's orbit and whose length, according to the old quantum theory, is quantized in multiples of $\hbar = h/2\pi$ (h = Planck's constant),

$$|\mathbf{L}| = k\hbar; \ k = 1, 2, ..., n. \tag{3.2}$$

k and $n = 1, 2, ...$ are called the azimuthal and the principal quantum number, respectively.

From Eqs (3.1) and (3.2) it then follows that the minimum magnetic moment of an electron bound in an atom is one Bohr magneton $\mu_B = e\hbar/2mc$. Here the new quantum theory has produced an important modification. By introducing the orbital angular momentum quantum number $l = k - 1 = 0, 1, 2, ..., n - 1$ the right-hand side of Eq. (3.2) is replaced by $\sqrt{l(l + 1)}\hbar$. Thus in the new theory the minimum magnetic moment is zero. With the above calculation for a circular orbit one can see from Eq. (3.1) that the magnetic moment $|\mu|$ may be expressed in terms of the electric quadrupole moment er^2. The difference with Debye's theory is that in expressing the diamagnetism in terms of the quadrupole moments of the atoms with respect to the axis of the applied magnetic field an additional difficulty arises from the fact that the electron orbits in general are elliptical and not perpendicular to the field direction. This introduces a time dependence of the quadrupole moments which, therefore, must be averaged. A further average involves the direction of $\mathbf{L}$ relative to the field direction.

Introducing, for the individual electron, two sets of Cartesian coordinates x, y, z and ξ, η, ζ with the z-axis in the field direction and the ζ-axis perpendicular to the plane of the orbit, ξ and η being along the principal axes of the ellipse, the quadrupole moment of this electron is $\Theta_\perp = e(x^2 + y^2) = e(R^2 - z^2)$, where R is the distance of the electron from the nucleus. Denoting $\alpha_1, \alpha_2, \alpha_3$ as the projections on the field direction of the unit vectors in the ξ-, η-, and ζ-directions, respectively, it follows that $z = \xi\alpha_1 + \eta\alpha_2 + \zeta\alpha_3$. And since directional averaging yields $\langle \alpha_i \alpha_j \rangle = (1/3)\delta_{ij}$, one finds, summing over all the electrons of the atom, $\overline{\sum \Theta_\perp} = (2/3)\overline{\sum eR^2}$, where the bar means averaging over time. According to Larmor's theorem (see e.g. Ref. [10a], p. 429) a quadrupole moment $\Theta_\perp$ gives rise to a diamagnetic susceptibility per mole $(eL/4mc^2)\Theta_\perp$ where $L = 6.03 \times 10^{23}$ mol^{-1} is Avogadro's number. This results in the famous atomic or Langevin–Pauli diamagnetism per mole

$$\chi_{LP} = \frac{e^2 L}{6mc^2} \overline{\sum R^2}. \tag{3.3}$$

As Pauli emphasizes in a footnote, this result is strictly temperature independent. A detailed historical account of this and many other results may be found in the well-known book by van Vleck [18a], on which Pauli wrote a very favorable review in 1933 [18b].

Making use of measured values for helium and argon Pauli calculates the electric quadrupole moments of these atoms. However, from the size of He and Ar he estimates that the quadrupole moment and hence the susceptibility should be at least 10 times smaller for He and 13 to 40 times smaller for Ar, in qualitative agreement with Schrödinger's statement mentioned above [15c]. For comparison, modern susceptibility values are -1.88×10^{-6} cgs for He and -19.6×10^{-6} cgs for Ar [19], which are about 24 and 12 times smaller, respectively, than the values used by Pauli (see also Ref. [20], footnote 76). This discussion shows Pauli's concern for experimental data which was quite typical of his way of doing physics. It was in this concern that the influence of his godfather, Ernst Mach, perhaps manifested itself most clearly, much more so, it seems to me, than in Mach's positivist thinking (see page 14).

In an introductory paragraph to the paper presented above [16], Pauli examines the possible reasons for the absence of paramagnetism in monatomic gases, stating that 'Either the atom in its natural (unexcited) state possesses no magnetic moment or the atom does have one, but in switching on the external magnetic field, the Maxwell–Boltzmann distribution of the directions of the magnetic axes of the atoms does not set in but the probabilities of these directions stay unchanged so that in the whole no magnetization results' (translated from Ref. [16]). And in a footnote he refers to the diamagnetic helium as a case where the second possibility might be realized. These statements invite a number of historical comments.

The first remark is that classical statistical mechanics indeed gives no paramagnetism since the magnetic moments must be thought to be generated by 'elementary Ampère currents' perpendicular to these moments. In other words, 'the probabilities of these directions stay unchanged'. In his course on statistical mechanics Pauli gives a very compact proof of this result (see Ref. [9a], *Volume 4. Statistical Mechanics*, pp. 50–1 and Figure 13.2). In the older literature this classical result goes under the name of 'van Leeuwen's theorem' and is interpreted as exact cancellation of the paramagnetic and diamagnetic contributions or, for free electrons, as effect due to the walls of the container (see Ref. [18a], Sections 24–6). Therefore, as Pauli emphasizes in the above reference ([9a], Vol. 4), permanent magnetic moments can only be understood in quantum theory. In the old Bohr–Sommerfeld theory they were the result of the 'elementary Ampère currents' of the stationary electronic orbits, as discussed above. However, in a statistical mechanics calculation one was not allowed to enquire about this origin of the moments (see the remark in Ref. [9a], Vol. 4, p. 50). On the other hand, in the new quantum theory where the stationary states carry no such currents, it is still possible to justify Eq. (3.1) (see e.g. Ref. [10c], pp. 111–12).

The second remark concerns helium. This atom, with a nuclear charge $+2e$ and two electrons circling it, is the simplest three-body problem in quantum theory. Its diamagnetism mentioned in the footnote of Pauli's paper [16] gave rise to competing models by Bohr in 1913 (see Ref. [10a], *Zusatz* 12, p. 511) and by many others after him (see Ref. [8b], Section IV.2 and Ref. [10c], p. 623). Sommerfeld, in Chapter 3, Section 6 of Ref. [10b], only discusses the model by Edwin Crawford Kemble (1889–1984) of 1921 and his own of 1916. Kemble, like Bohr, assumed that both electrons had circular orbits with quantum numbers $n = k = 1$ but that their angular momenta L_1 and L_2, instead of being parallel, formed an angle of 120°, so that the resultant $J = L_1 + L_2$ had length $1\hbar$, i.e. an inner quantum number (see Ref. [10b], pp. 342 and 579) of $j = 1$ and, hence, a magnetic moment of one Bohr magneton. This model therefore corresponds to the second possibility in the above quotation from Ref. [16] and, in order to obtain diamagnetism, one must invoke the 'elementary Ampère currents', contrary to the rules of the game in the old quantum theory.

In turn, Sommerfeld's model respected these rules by requiring $j = 0$. However, the choice of circular orbits with $n = k = 1$ then implies that $L_1 + L_2 = 0$, which means that the two orbits are identical but that the electrons circulate in opposite directions. But this is unacceptable because the electrons would bump into each other. Therefore Sommerfeld advanced sophisticated arguments to get $k = 1/2$ for both electrons which now results in elliptical but coplanar orbits and opposite circulation directions. To avoid collisions Sommerfeld assumed that one electron is in aphelion when the other passes through perihelion (see Fig. 41, p. 203 of Ref. [10b]). This description of the helium problem gives an idea of the typical difficulties encountered in the old quantum theory. In fact, the more rigorous formulation of this theory presented in the following section makes it evident that two-electron problems like helium were impossible to treat adequately. And, as will be discussed in the next chapter, helium became one of the stumbling blocks of the old theory.

At the Nauheim meeting in September 1920, where Pauli took part in the discussion after Weyl's talk (see page 38), he also presented his second paper on magnetism [21]. This time he turned to paramagnetic gases whose atoms possess a permanent magnetic moment μ produced by a non-zero total angular momentum $\mathbf{J}$ resulting from all the electron orbits in the atom. While the Langevin–Pauli diamagnetism (3.3) contains no specific quantum feature—$\hbar$ does not appear—paramagnetism is significantly modified by quantum theory. This is what Pauli strove to show in this paper [21] which is of historical relevance because there Pauli proposed as the unit for measuring magnetic moments what he called the Bohr magneton μ_B, which does contain $\hbar$ (see above). In so doing Pauli wanted to show that an empirical unit introduced by Pierre Weiss (1865–1940) in 1911, called the Weiss magneton, had no physical foundation. Weiss, who as a professor at ETH in Zurich from 1902 to 1919 had made his name with extensive magnetic precision measurements, wanted to show that all magnetic moments were integer multiples of the Weiss magneton which is 4.97 times smaller than μ_B (see footnote 3, p. 228 of Ref. [18a]).

With this paper [21] Pauli somewhat stole the show from Sommerfeld since it was the latter who in 1916 had introduced the fundamental idea of directional quantization in a paper on the Zeeman effect [22a,c]. In the section on the theory of the magneton in the fourth edition of his *Atombau und Spektrallinien* (Ref. [10b], Chapter 8, Section 4) Sommerfeld honoured Pauli's paper with just two sentences (pp. 641–2; see, however, Ref. [22b] where full credit is given to Pauli). Perhaps he felt that 'Sommerfeld magneton' would have been an appropriate name. Directional quantization is the requirement that the cosine of the angle θ between μ and the magnetic field direction only takes discrete values $\cos\theta = m/j$ with $m = \pm 1, \pm 2, ..., \pm j$ where, as before, $|\mathbf{J}| = j\hbar$ (note that in Ref. [21] Pauli's notation for j, m is n, k). According to Bohr, the value $m = 0$ was to be excluded, a rule that Schrödinger questioned in the letter quoted above [15c] and which has no basis in the new quantum theory. Now the Langevin-Debye paramagnetism per mole is [18a]

$$\chi_{LD} = \frac{\mu^2 L}{kT} \langle \cos^2 \theta \rangle, \tag{3.4}$$

where k is Boltzmann's constant and T the absolute temperature. (Pauli uses the notation R for k/L, which is confusing since usually R designates the gas constant kL.) Classically, the average is over the continuum of angles θ whereas directional quantization with Bohr's rule gives $\langle \cos^2 \theta \rangle = (j+1)(2j+1)/6j^2$. While the true atomic magneton number is just j, the measured value is what Pauli called the 'fictive magneton number' $p = j\sqrt{3\langle \cos^2 \theta \rangle}$.

Unfortunately, no data from paramagnetic atomic gases were available to Pauli, so he turned to the paramagnetic molecular gases NO and O_2. But these cases give rise, in quantum theory, to additional complications because there is a new vector coming into play, namely the figure axis $\mathbf{f}$ (called '*Figurenachse*') joining the two atoms. Giving reasons for rejecting the case when $\mathbf{J}$ is along $\mathbf{f}$, Pauli assumes that $\mathbf{J}$ is perpendicular to $\mathbf{f}$ and further supposes that thermal agitation does not lead to an angular momentum along $\mathbf{f}$. Then, making use of an argument related to the figure in Schrödinger's letter of Ref. [15c] (alluded to in footnote 3 on p. 617 of Ref. [21]), Pauli concludes that thermal rotation cannot contribute to the directional quantization. This molecular problem then reduces to the atomic case. Therefore, with values $j = 1$ for NO and $j = 2$ for O_2, Pauli finds $p_{NO} = \sqrt{3}$ and $p_{O_2} = \sqrt{15/2}$, which compare well with the experimental numbers 1.8 and 2.8, respectively. The latter, which are taken from work by Weiss and Piccard, are in surprisingly good agreement with modern values (1.82 and 2.84) given in Ref. [19]. Much later Pauli came back to this paper [21] in the last section of his long review article on the old quantum theory published in 1926 (Ref. [23], Ziff. 48). There, instead of Bohr's rule mentioned above, he used the prescription of the new theory which leads to a correction factor of $2j/(2j + 1)$ and hence to $\langle \cos^2 \theta \rangle = (j + 1)/3j$. In this review he also mentioned Weiss's argument for using the Weiss magneton.

Directional quantization also plays a crucial role when the quantization axis is defined by an applied electric field **F** instead of a magnetic field. This is what Pauli demonstrates in a third paper belonging to a collection of works on gases in external fields [24]. He submitted this paper from Munich on 30 July 1921 to the new journal *Zeitschrift für Physik* which had been founded the year before as a complement to the *Verhandlungen der Deutschen Physikalischen Gesellschaft* by Karl Scheel (1866–1936), who also became its first editor. Considering rigid diatomic molecules with an electric dipole moment μ, the polarizability α is again given by the Langevin–Debye formula (3.4). However, Pauli explicitly assumed (Ref. [24], footnote 1, p. 321) that these molecules possess no electronic angular momentum, so that only thermal rotation gives rise to one, which is called $\mathbf{K}/2\pi$, and $J/2\pi$ is its projection onto the direction of the electric field **F**. On the other hand, θ is the angle between the figure axis **f** of the molecule and **F**.

According to the laws of mechanics, **K** is a constant of the motion between collisions of the molecule; and, as Schrödinger notes in the letter of Ref. [15c], **K** is perpendicular to **f**. Therefore θ again has a time dependence which, as in the diamagnetic case, must be averaged before taking the statistical average. But for this purpose it is more convenient to calculate, instead of the polarizability α, the molar polarization

$$P = \alpha F = L\mu \langle \overline{\cos \theta} \rangle \tag{3.5}$$

where, as before, the bar means the time average over the thermal rotation with fixed **K** and the angle brackets indicate the statistical average over the directions of **K**. Now, surprisingly, Pauli finds that, classically and for small fields ($\mu F \ll kT$), $\langle \overline{\cos \theta} \rangle = 0$ because in this case $\langle J^2/\mathbf{K}^2 \rangle = 1/3$. But this means that classically the entire polarization is due to the few molecules for which **K** is perpendicular to **F** and so small that it is perturbed by the field **F**, so that these molecules oscillate around **F** instead of rotating with constant **K**, exactly as in a spherical pendulum.

For the quantum calculation Pauli introduces directional quantization in the form $|\mathbf{K}| = 2\pi\hbar j$ and $J = 2\pi\hbar m$ with $m = \pm 1, \pm 2, ..., \pm j$ and $j = 1, 2, ...$

(Pauli uses the notation m, n instead of j, m.) The result of evaluating the double sums on j and m is, for small fields,

$$\overline{\langle \cos \theta \rangle} = C \frac{\mu F}{kT} \tag{3.6}$$

with $C = 1.5367$. Comparing this with the Langevin–Debye formula (3.4) with the conventional value $\langle \cos^2 \theta \rangle = 1/3$, Pauli finds the ratio of quantum to 'classical' moment $\mu_{qu}/\mu_{cl} = \sqrt{3C} = 2.1471$. It is interesting to note that in the evaluation of the infinite sums over j (his m), Pauli refers to Leonhard Euler's work, a part of which he had studied in highschool at age 13 (see Ref. [4], p. 536).

However, Pauli later questioned the validity of his assumption of a vanishing electronic angular momentum in the dipolar molecules considered by him in Ref. [24], namely the halogen hydrogens, HCl etc. Indeed, HCl possesses one outer electron sitting on the Cl⁻ ion which therefore behaves like the rare gas argon. But Bohr, in his fundamental analysis of the constitution of the periodic table of the elements (see page 88), whose beginnings are contemporaneous with Pauli's paper [24], argued that the rare gases and hence also the halogen ions should have $j = 1$. Bohr's idea was (see Ref. [10b], pp. 188 and 204) that, e.g. in argon, the orbits with strong excentricity (low $l = k - 1$) penetrate into the closed shells where the electrons do not contribute to the angular momentum. Pauli considered this new situation in a paper he wrote in Copenhagen together with Bohr's assistant, Hendrik Kramers (1894–1952) [25], the essential conclusion of which being 'that the orbital plane of the outer electron [in such a molecule] no longer rotates rigidly with the molecule ion' (translated from Ref. [25], p. 367).

But this was still not the last word in applying directional quantization to the Langevin–Debye formula (3.4). With the advent of the new quantum theory in 1925 several authors studied the quantum mechanical problem of the spherical pendulum, also called the rotator (see e.g. Ref. [18a], Section 37, in particular footnote 21). One of them was Lucy Mensing who had obtained her doctorate with Wilhelm Lenz in Hamburg in 1926 and later became an assistant in Tübingen (Ref. [4], p. 328). She had used the Heisenberg matrix method in a solution of the rotator and later also applied her results to problems with external fields, i.e. the Stark effect and the Zeeman effect (Ref. [4], p. 142). Pauli engaged her in a recalculation of Eq. (3.5) above according to the new theory. Making use of the modified notation $\mu \cos \theta = \alpha(j, m) F$, so that, according to Eqs (3.5) and (3.6),

$$\overline{\alpha} = L\langle \alpha(j, m) \rangle = LC \frac{\mu^2}{kT}, \tag{3.7}$$

Mensing and Pauli found the conventional value $C = 1/3$!

This paper was submitted by Mensing from Hamburg on 25 June 1926; Pauli was in Zurich attending the first of the *Vortragswochen* (weeks of lectures) organized every few years by the physics professors of ETH (see Ref. [8a], p.

18) which took place from 21 to 26 June 1926 (see Ref. [4], pp. 326, 334, 539). But before returning to the original value $1/3$, the constant C had witnessed even more outlandish values. Indeed, as mentioned in Mensing and Pauli's paper (see also the table in Ref. [18a], p. 107), Linus Pauling in the United States had published in the *Physical Review* of 1926 a value $C = 4.570$ which he had obtained by applying the new theory to obtain $\mathbf{K}^2/(2\pi\hbar)^2 = j(j+1)$, as explained after Eq. (3.2), but still using the classical evaluation of $\overline{\cos\theta}$. Thus Mensing and Pauli's paper brought a long and confusing development to a close and helped establish faith in the new quantum theory. However, before arriving at this point Pauli still had to contribute his share to the old theory through his Ph.D. work, which is the subject of the next section.

Doctor Pauli junior

Wolfgang had set himself a more difficult problem. He wanted to examine if, in a complicated system for which one was just barely capable of doing the calculations, Bohr's theory and the Bohr–Sommerfeld quantum conditions led to the experimentally correct result. For, in our Munich discussions doubts had come to us whether the hitherto obtained successes of the theory were not limited to simple systems and whether a failure might not occur already in the more complicated system to be studied by Wolfgang. [26]

This recollection of Heisenberg sets the stage beautifully for the present discussion which is concerned with Pauli's doctoral thesis. Independent and proud but good natured he chose, albeit on Sommerfeld's urging, the most difficult subject. Spoiled by having been an infant prodigy, knowing that his knowledge of mathematics and classical mechanics was superior to that of his co-students or even of the Privatdozenten, he wanted to know what was behind 'the music of the spheres' that Sommerfeld had enthused about in the preface of his *Atombau und Spektrallinien*. And this is how his choice came about. As mentioned on page 54, Pauli had taken Sommerfeld's *Spezialvorlesung* of one hour on 'Atombau und Spektrallinien' in his first term during the bad winter of 1918/19. In his *Kollegienbuch* he had marked the time of the lectures: Mondays from 6 to 7 p.m., a most favourable time, indeed. Not surprisingly, in the first edition of Sommerfeld's famous book, the preface to which is dated 2 September 1919, Pauli's criticism is already acknowledged: 'Also the preceding calculations point (according to a personal communication of Herrn W. Pauli) to a contradiction with the experience, which speaks against the model' (translated from Ref. [10a], p. 510).

The model referred to in this quotation concerns the hydrogen molecule ion H_2^+ which is formed by two protons and one electron, and is described on p. 287 and in *Zusatz* 11 on p. 508 of the first edition of Sommerfeld's book [10a]. The two hydrogen nuclei (protons) define the figure axis $\mathbf{f}$ and the electron moves around $\mathbf{f}$ in a circle of radius a in the median plane which is at a distance c from both nuclei (I use Pauli's notation of Ref. [14] instead of Sommerfeld's b). Equilibrium of the nuclei implies that the projection on $\mathbf{f}$ of the Coulomb attraction between one nucleus and the electron must equal the repulsion between the two nuclei. This determines the ratio $(a/c)^2 = 16^{1/3} - 1$. To ensure dynamical equilibrium of the electron the centrifugal force $ma\omega^2$, where m is the electron's mass and ω its angular velocity, must compensate the attraction by the two nuclei. This leads to $ma^3\omega^2 = (a/c)^3 e^2/2$.

A third condition follows from quantum theory. In the ground state, which has the quantum number $k = 1$, the electron's angular momentum is, according to Eq. (3.2), $ma^2\omega = \hbar$. Combining the three conditions one finds for the kinetic energy of the electron

$$E_{kin} = \frac{m}{2}(a\omega)^2 = \frac{me^4}{8\hbar^2}\left(\frac{a}{c}\right)^6 = \frac{R_\infty}{4}(16^{1/3} - 1)^3. \qquad (3.8)$$

Here $R_\infty = me^4/2\hbar^2 = 13.606$ eV is the Rydberg constant (expressed in electron volts) which is the ionization energy of the hydrogen atom for infinite nuclear mass. In Ref. [10a] Sommerfeld writes $R_\infty = Nh$ where $h = 2\pi\hbar$ is Planck's constant. The numerical constant in Eq. (3.8) is $(16^{1/3} - 1)^3/4 = 0.8777$. The other important quantity is the distance $2c$ between the nuclei in the H_2^+ ion. One deduces from the three conditions

$$c = \frac{2\hbar^2}{me^2}\left(\frac{c}{a}\right)^4 = \frac{2}{(16^{1/3} - 1)^2}a_B, \qquad (3.9)$$

or $c = 0.8658a_B$. Here $a_B = \hbar^2/(me^2) = 0.5292$Å is the Bohr radius, which is the radius of the ground state orbit in the hydrogen atom, taken for infinite nuclear mass, and which Pauli and Sommerfeld call a_1. (The corresponding value for H_2 follows from Eq. (13) on p. 285 of Ref. [10a]: $c = a/\sqrt{3} = 0.55a_B$.)

Now in *Zusatz* 5 of Ref. [10a] Sommerfeld proves for a stable system subject to arbitrary Coulomb forces that in a time average (designated by a bar)

$$2\overline{E_{kin}} = -\overline{E_{pot}}. \qquad (3.10)$$

Thus the total energy of H_2^+ is $W_+ = -E_{kin} = -0.8777R_\infty$. (The corresponding value for H_2 is, according to Eq. (14) on p. 286 of Ref. [10a], $W = -2.20R_\infty$.) This shows that H_2^+ is unstable against dissociation into a hydrogen atom H and a proton at rest since the energy of H is by definition $W_1 = -R_\infty$. However, as Sommerfeld discusses on p. 510 of Ref. [10a], ionization experiments on the hydrogen molecule H_2 indicate that $W_+ < W_1$. Therefore the above model cannot be right. And, indeed, Pauli proves in his thesis that 'the circular orbits in the median plane, for which the nuclei are in equilibrium, are unstable with respect to an impulse on the electron perpendicular to the median plane' (translated from Ref. [14], p. 220).

As preliminaries to his analysis Pauli takes up, in Section 1, the discussion of Sommerfeld's condition $W_1 < W_+$ and the empirical evidence for it, concluding 'that electron collisions do not always proceed inelastically, even when it would be expected on purely energetic ground' (translated from Ref. [14], p. 181). This surprising statement is clarified in Section 3 since, indeed, in a radiation-less transition between two quantum states the energy change ΔE of a colliding electron can only assume discrete (positive or negative) values. However, in an extension of Bohr's correspondence principle to such collisions one may say that, since for large quantum numbers ΔE becomes essentially continuous, in this limit classical mechanics becomes valid again. For the number of collisions to be calculated in this limit it is, however, necessary to average over the initial conditions but keep the energy of the colliding electron fixed. Pauli then calls this the mechanical correspondence principle. The average contained in its formulation presages the probabilistic nature of the new quantum theory! As another preliminary Pauli states in Section 2 the stability condition which requires that any possible infinitesimal variation of the initial positions and velocities of all the nuclei and electrons gives rise to only infinitesimal variations of the quantized electronic orbits. Pauli calls quantum states satisfying this adiabatic hypothesis, in agreement with Sommerfeld, dynamically stable.

Zusatz 8 of Ref. [10a] and also *Zusatz* 7 of Ref. [10b] describe the basis of the old quantum theory, namely the Hamiltonian formalism of classical mechanics which has been used in celestial mechanics since the nineteenth century—both Sommerfeld and Pauli quote Charlier's treatise [27a] on this subject. In the Hamiltonian formalism a mechanical system is characterized by a Hamiltonian function H of $2f$ canonically conjugate variables, the coordinates $q_1, q_2, ..., q_f$ and the momenta $p_1, p_2, ..., p_f$, where f is the number of degrees of freedom. H determines the dynamics of the system through the $2f$ canonical equations of motion

$$\frac{dq_k}{dt} = \frac{\partial H}{\partial p_k}; \frac{dp_k}{dt} = -\frac{\partial H}{\partial q_k}; \ k = 1, 2, ..., f \tag{3.11}$$

which are invariant under general canonical transformations of the variables.

In order for the Bohr–Sommerfeld quantization prescription to work it is necessary to assume that the atom or molecule under consideration can be described as a conditionally or multiply periodic system. This means that there exists a canonical transformation to so-called action and angle variables $J_1, J_2, ..., J_f$ and $w_1, w_2, ..., w_f$, respectively, such that the Hamiltonian function does not depend on the angles w_k. Then it follows from Eqs (3.11) that the J_k are constants of the motion and, therefore, may be quantized,

$$J_k = \int p_k dq_k = 2\pi n_k \hbar; \ k = 1, 2, ..., f, \tag{3.12}$$

where the n_k are positive integers or zero and the integral is over a period. This condition shows that the 'action variables' must have the dimension of an action. The angle variables, therefore, are dimensionless, and their time derivatives $dw_k/dt = \nu_k$ have the dimension of frequencies which, according to Eqs (3.11), are again functions of the J and hence are also constants, so that by integration $w_k = \nu_k t + \text{const}$. Thus the dynamical problem separates.

One then shows (see the mentioned *Zusätze* in Refs [10a,b]) that the original coordinates q_k are periodic functions of the w with period 1 and hence may be developed into f-fold Fourier series. And since the w are linear functions of time the q_k are then conditionally or multiply periodic. Since the ratios of the ν_k are in general irrational, the trajectory $q_1(t), q_2(t), ..., q_f(t)$ in the course of time densely fills a whole region in the f-dimensional coordinate space.

This is the general situation encountered with one electron where separation works. Hence it becomes evident that problems involving more than one electron are impossible to treat adequately in the old quantum theory. Therefore, the H_2^+ ion is the most complicated problem that the old theory could handle. It has $f = 9$ degrees of freedom, of which three define the position of the centre of mass and two the direction **f** of the figure axis joining the two nuclei. The remaining four degrees of freedom are the three coordinates of the electron's position and the nuclear coordinate c introduced above. Since the nuclei are almost 2000 times heavier than the electron their motions are mere 'tremors' (Pauli, on p. 224 of Ref. [14], calls them '*Zittern der Kerne*') with respect to the centre of mass and to the figure axis (note that in the old theory even in atoms the nucleus shows 'tremors' with respect to the center of mass!). Quoting Charlier [27a] and the even much earlier lectures by Jacobi [27b] Pauli notes that the two-centre problem with fixed nuclei becomes conditionally periodic in elliptical coordinates and hence may be solved by separation of the three variables $\lambda = (r_1 + r_2)/2c$, $\mu = (r_1 - r_2)/2c$, and the angle φ around the z-axis which is along **f**. Here r_1 and r_2 are the distances of the electron from the two nuclei defining the foci of the rotation ellipsoids $\lambda = $ const. and hyperboloids $\pm\mu = $ const. In these coordinates the median plane is defined by $\mu = 0$. Denoting p_λ, p_μ, and p_φ as the canonically conjugate momenta, there are three quantum conditions (3.12) for $k = 1, 2, 3$.

In Section 6 of Ref [14] Pauli gives a classification of the possible electronic orbits, first distinguishing between planar orbits, for which $\varphi = $ const., and spatial orbits depending on φ. In the first class, a detailed discussion of which had been given by Charlier in the first volume of Ref. [27a], Pauli distinguishes four types of orbits: oscillations in the median plane called '*Pendelbahnen*', satellite orbits around one of the nuclei filling the region between an ellipse and one branch of an hyperbola, orbits circling both nuclei filling the region inside an ellipse, called '*Lemniskatenbahnen*' and planetary orbits filling the region between two ellipses.

The analysis of the class of spatial orbits, which is more involved, leads to the following three types: unsymmetrical orbits which stay on one side of the median plane and fill a ring-shaped region limited by two ellipsoids and two hyperboloids, symmetrical orbits—with respect to the median plane—filling a tube-shaped region limited by two ellipsoids and one hyperboloid, and, finally, orbits in the median plane filling a plane ring limited by two circles. The last degenerate into Sommerfeld's circular orbit discussed above when the two limiting circles coincide. Section 7 is devoted to the equilibrium orbits which are determined by the three quantum conditions (3.12) and the equilibrium condition for the nuclei, namely that in a time average the force on each nucleus vanishes. Pauli shows very elegantly (Ref. [14], p. 198) that this condition is equivalent to Sommerfeld's simpler relation (3.10). Together with the stability condition mentioned above this eliminates not only all planar orbits but also Sommerfeld's circular orbit. Pauli also recovers in his Eq. (67) Sommerfeld's relations (3.8) and (3.9) above, and the value $W_+ = -0.8777R_\infty$ is the lowest of all the energy values Pauli was able calculate.

The most important part of the thesis is the calculation of the energy and size of the ground state of the symmetrical class of orbits in Section 10— the unsymmetrical class was too complicated for him to calculate—and the presentation of the main results in Section 11. The quantum numbers of this ground state are $n_1 = 0, n_2 = n_3 = 1$, the first meaning that the two limiting ellipsoids coincide. Equations (94) of Ref. [14] gives the numerical values: $W_+ = -0.5175R_\infty$ (Pauli uses the notation W for W_+) and $c = 2.7679a_B$. It must have been a rather unpleasant surprise for Pauli that these and many other results, even some concerning the unsymmetrical class of orbits, were independently obtained in a doctoral dissertation at the University of Utrecht in 1922 [28]. The author responsible, K. F. Niessen, was able to show, in particular, that for $n_1 + n_2 + n_3 \leq 6$ there are no stable orbits in the unsymmetrical class so that the possibility for this class to contain any physical states was practically excluded. Pauli reviewed this work, together with his own, in Ref. [29].

It is of considerable interest to compare the results of Pauli and Niessen with what follows from the new quantum theory. Shortly after its creation the new theory was applied to the helium atom and to the hydrogen molecule. Because these problems concerned two electrons, both had been unaccessible to the old theory. But even in the new theory they required some additional pioneering effort. In the case of He there was the notion of the exchange interaction introduced by Heisenberg [30a] and in the case of H_2 the notion of the covalent bond introduced by Heitler and London [30b], both being consequences of Pauli's exclusion principle (see e.g. Chapter 9 in Ref. [10c] or Ref. [31]). The H_2^+ ion, on the other hand, does not suffer from these complications and, actually, is the simplest molecular system.

In the winter term of 1928–29 a young Hungarian graduate student came from Sommerfeld's institute to Leipzig to work with Heisenberg who the year before had become, at age 25, the professor of theoretical physics there. This student was Edward Teller (b. 1908), often called the 'father of the American hydrogen bomb' (see Ref. [32a], p. 7). Having studied chemistry at Karlsruhe he was interested in molecules, just like Heisenberg at that time. The latter handed him two contradicting recent papers on the ground state of the hydrogen molecule ion, one by a Dane and one by an American. Teller had to find out who was right and also to look into the excited states. With the aid of a mechanical calculator Teller started an extensive numerical evaluation which in February 1930 became his Ph.D. thesis (see Ref. [32a], pp. 14 and 30) and also gave rise to a publication [32b]. The introduction to this paper is of interest here because it gives the history of the problem, including the references, starting with Pauli and Niessen and identifying the 'Dane' as Ø. Burrau and the 'American' most probably as the Briton Sir Alan Herries Wilson (b. 1906) who later became known for his fundamental work on metals and semiconductors. A further investigation of the H_2^+ problem which Pauli found 'extremely useful' was carried out by Georg Cecil Jaffé (1880–1965) who had just lost his professorship at Giessen in the wake of the elimination of Jews from state positions by the National Socialists in 1933 (see Refs [32c] and [15b], pp. 232–3). The natural method in the case of the hydrogen molecule ion is that of the molecular orbitals (see Ref. [33a], pp. 359–71) which, as Sommerfeld observed, could even be used as a basis for the H_2 calculation (see Ref. [10c], p. 625).

One of the experts on both the H_2 and the H_2^+ problems, Owen Willans Richardson (1879–1959), a close friend of Niels Bohr, who won the Nobel Prize of 1928 'for his work on the thermionic phenomenon and especially for the discovery of the law named after him', presented a nice review of both problems at the 200th Anniversary Celebration for Luigi Galvani in Bologna in October 1937 [33b] where Pauli also participated. Richardson's values are the following (see also Table 39, pp. 530–5 of Ref. [33a]). The dissociation energy for $H_2 \rightarrow H + H$ is $D_0 = 4.476$ eV; it is to be compared with the 3.6 eV which Pauli took from the experimental work of J. Franck. The total dissociation energy of H_2 is then $D_0 + 2R_\infty = 2.3290R_\infty$ which is to be compared with Sommerfeld's $-W = 2.20R_\infty$. The dissociation energy for $H_2^+ \rightarrow H + H^+$ is $D_0^+ = 0.1946R_\infty$. Hence the total dissociation energy of H_2^+ is $D_0^+ + R_\infty = 1.1946R_\infty$ which is to be compared with Pauli's $J_{H_2^+} = -W = 0.5175R_\infty$ and with Sommerfeld's $-W_+ = 0.88R_\infty$. The ionization energy $H_2 \rightarrow H_2^+ + e$ is then $I_0 = D_0 + R_\infty - D_0^+ = 15.434$ eV which is to be compared with Pauli's $J_{H_2} = 23.7$ eV. Finally, the equilibrium distance $H^+ - H^+$ is $1.4015a_B$ for H_2, which is to be compared with Sommerfeld's $2c = 1.10a_B$, and is $2.0200a_B$ for H_2^+, which is to be compared with Pauli's $2c = 5.5358a_B$ and with Sommerfeld's $2c = 1.7316a_B$. A further feature introduced by the new quantum theory was the zero-point energy of the vibrational ground state of the two nuclei in H_2 and in H_2^+ which has to be added to the ground state energy and which Richardson calls G_0 and G_0^+, respectively, in Fig. 1 of Ref. [33b] (see also Fig. 47 on p. 642 of Ref. [10c]).

The main result obtained for H_2^+ with the new quantum theory is thus the positive dissociation energy D_0^+ which was negative for Sommerfeld's unstable orbit, namely $-0.12R_\infty$, and even $-0.4828R_\infty$ for Pauli's symmetrical ground state! Correspondingly, the size of Pauli's ground state was much too large while Sommerfeld's orbit was too small. This failure of Pauli's efforts, together with the challenge coming from an unknown co-student from Holland, were likely to bother Pauli. According to his friend Gregor Wentzel, he was deeply discouraged towards the end of his Ph.D. work (Ref. [7b], note II.24, p. 95). Sommerfeld, in the minutes to the *examen rigorosum*, praised Pauli's mechanical correspondence principle as 'one of the general considerations which could gain considerable importance' and expressed the hope that 'sooner or later the results [concerning H_2^+] of the work may be verfiable by observation' [34] (see also Ref [4], note e, p. 71 and Ref. [7b], note II.29, p. 95). And he mentions Pauli's and Niessen's works in a footnote in the fourth edition of his *Atombau und Spektrallinien*, observing that the symmetric spatial ground state orbit 'gives a typical picture for the possibility of non-polar binding in general' (translated from Ref. [10b], p. 176).

In the first volume of his *Lectures on Atom Mechanics* [35a,b] Born devotes Section 39 to the problem of the two centres. Describing Pauli's two classes of planar and spatial orbits first, Born discusses both Pauli's and Niessen's results in the last two paragraphs. He closes the section with the pessimistic observation that the resulting energy values 'cannot be made to agree with the measurements of the ionization and excitation voltages. For this reason we abstain from entering more closely into this model of H_2^+. Where the reason for the failure of the theory must be sought is quite unclear at present. But we shall see in the following that the treatment of the atomic problems with the aid of classical mechanics leads to wrong results as soon as several electrons are present, that is when a three-body problem is at hand. Maybe here also the artificial transformation of the many-body problem into a one-body problem on the basis of the small ratio of the electron to the nuclear mass is not permissible' (translated from Ref. [35a], p. 281). The only other place in the literature where the Pauli–Niessen model is mentioned seems to be in James Franck and Pascual Jordan's book *Excitations of Quantum Jumps by Collisions*, and this only to warn experimenters that the model 'yields wrong values of the ionization energy' [36]. Pauli himself never quoted his Ph.D. work.

The sad fact is that with the two-centre problem Pauli was positioned on the wrong front. As the quotation from Born's *Lectures* [35a] shows, the decisive battle was fought on the front of the one-centre problem, the atom. And it is in reporting from this front that some years later Pauli wrote his long review on the old quantum theory in Geiger and Scheel's *Handbuch der Physik* [23] mentioned in the last section. But in so doing he was fighting a rearguard action, and not in the battle to conquer the new quantum theory, which he later described in a review of fundamental importance [37], again published in Geiger and Scheel's *Handbuch*. He used to call the first review his 'Old Testament', implying that salvation could come only from the 'New Testament' (see Refs. [38a], p. 423 and [38b]). Much later, in a conversation with Jagdish Mehra in Berkeley, California, in the spring of 1958, the year of his death, Pauli pondered: 'When I was young I believed I was the best "formalist" of my time. I thought I was a revolutionary. When the big problems would come I would solve them and write about them. The big problems came and passed by, others solved them and wrote about them. I yet was a classicist and not a revolutionary' (translated from Ref. [8b], p. xxiv; see also Ref. [5], p. 314, where a similar statement by Pauli, reported by Res Jost, is mentioned).

This, however, is a much too pessimistic view. As we shall see in the next chapter, Pauli also fought on the atomic front, although without publishing anything, and, above all, he discovered the exclusion principle and introduced nuclear spin—both discussed in the 'Old Testament'—before the dawn of the new theory. Concerning the quantization of multiply periodic systems, a fundamental criticism, however, remained. It has been raised in recent years by one of Pauli's own diploma students, Martin Gutzwiller (*1925), a friend from my own student days at ETH (see Ref. [8a], pp. 434–5). In discussing the hydrogen molecule ion in Section 3.4 of his celebrated book [40a] Gutzwiller writes on p. 37: 'Pauli worked exactly in the no-person's land between classical and quantum mechanics. . . . His quantization conditions were too restrictive.'

The point is that Pauli and the other experts actually should have known better since Einstein, in a completely ignored paper [41] in 1917, made the important observation that Bohr–Sommerfeld quantization was not so much a problem of separation of variables but rather of establishing what today is called the invariant tori, the absence of which 'prevents any use of the quantization rules and removes the system from being "quantized" in this way' (Ref. [40a], p. 208). And Gutzwiller continues: 'Such total neglect of an incisive comment on a "hot subject" by the world's best-known physicist is almost beyond comprehension, in particular, since many close colleagues wrote large and learned reviews of the whole topic in the early 1920's' (Ref. [40a], p. 208; see also Ref. [40b], note 4). Among the 'learned reviews' mentioned by Gutzwiller are Refs [10a,b], [35], and [23]. This seems to be one of the first instances where 'mathematical physics' proved the intuition of the leading physicists to be wrong.

Notes and references

[1] Translated from: A. Sommerfeld, *Autobiographische Skizze*, in Ref. [2], pp. 673–9.

[2] F. Sauter (ed.), *Arnold Sommerfeld. Gesammelte Schriften* (Vieweg, Braunschweig, 1968), Vol. IV.

[3] *Pauli's Kollegienbuch der Universität München*, PLC-Bi 56, *Pauli Archive*.

[4] A. Hermann, K. von Meyenn, and V. F. Weisskopf (eds), *Wolfgang Pauli. Scientific Correspondence with Bohr, Einstein, Heisenberg, a.o., Volume I: 1919-1929* (Springer, New York, 1979).

[5] (a) K. von Meyenn, 'Ist die Quantentheorie milieubedingt?', in Ref. [6]; (b) A. Pais, *Inward Bound. Of Matter and Forces in the Physical World* (Oxford University Press, Oxford, 1986).

[6] K. von Meyenn (ed.), *Quantenmechanik und Weimarer Republik* (Vieweg, Braunschweig, 1994).

[7] (a) D. C. Cassidy, *Uncertainty. The Life and Science of Werner Heisenberg* (Freeman, New York, 1992); German translation by A. and G. Kleinert, *Werner Heisenberg: Leben und Werk* (Spektrum Akademischer Verlag, Heidelberg, 1995); (b) S. Richter, *Wolfgang Pauli. Die Jahre 1918-1930. Skizzen zu einer wissenschaftlichen Biographie* (Sauerländer,

Aarau, 1979); (c) K. von Meyenn (ed.), *Wolfgang Pauli. Scientific Correspondence with Bohr, Einstein, Heisenberg, a.o., Volume IV, Part III: 1955–1956* (Springer, Berlin and Heidelberg, 2001).

[8] (a) C. P. Enz, B. Glaus, and G. Oberkofler (eds), *Wolfgang Pauli und sein Wirken an der ETH Zürich* (vdf Hochschulverlag ETH, Zurich, 1997); (b) J. Mehra and H. Rechenberg, *The Historical Development of Quantum Theory. Volume 1. The Quantum Theory of Planck, Einstein, Bohr and Sommerfeld: Its Foundation and the Rise of Its Difficulties 1900–1925* (Springer, New York, 1982); (c) W. Pauli, 'Vorlesungen über theoretische Physik. Bd. IV: Optik. Von A. Sommerfeld', *Zeitschrift für angewandte Mathematik und Physik (ZAMP)* **2**, 215 (1951). Reprinted in Ref. [13], Vol. 2, p. 1407.

[9] (a) C. P. Enz (ed.), *Pauli Lectures on Physics*, translated by H. R. Lewis and S. Margulies (MIT Press, Cambridge, MA, 1973): *Volume 1. Electrodynamics, Volume 2. Optics and the Theory of Electrons, Volume 3. Thermodynamics and the Kinetic Theory of Gases, Volume 4. Statistical Mechanics, Volume 5. Wave Mechanics, Volume 6. Selected Topics in Field Quantization*; (b) W. Heisenberg, *Der Teil und das Ganze. Gespräche im Umkreis der Atomphysik* (Piper, Munich, 1969); English translation by A. J. Pomerans, *Physics and Beyond* (Allen & Unwin, London, 1971).

[10] A. Sommerfeld, *Atombau und Spektrallinien* (a) (Vieweg, Braunschweig, 1919); (b) (Vieweg, Braunschweig, Vierte umgearbeitete Auflage, 1924); (c) *Wellenmechanischer Ergänzungsband* (Vieweg, Braunschweig, 3. umgearbeitete und erweiterte Auflage, 1951).

[11] (a) W. Pauli, 'Sommerfeld's Contributions to Quantum Theory', in: C. P. Enz and K. von Meyenn (eds), *Wolfgang Pauli. Writings on Physics and Philosophy* (Springer, Berlin and Heidelberg, 1994), pp. 59–68; (b) V. F. Weisskopf, *The Joy of Insight* (Basic Books, New York, 1991).

[12] W. Pauli jr., 'Die Ausbreitung des Lichtes in bewegten Medien. (Bemerkungen über zwei Arbeiten von Herrn K. Uller.)', *Mathematische Annalen* **82**, 113 (1920). Reprinted in Ref. [13], Vol. 2, pp. 29–35.

[13] R. Kronig and V. F. Weisskopf (eds), *Collected Scientific Papers by Wolfgang Pauli. In Two Volumes* (Wiley Interscience, New York, 1964).

[14] W. Pauli jr., 'Über das Modell des Wasserstoffmolekülions', *Annalen der Physik* (4) **68**, 177 (1922). Reprinted in Ref. [13], Vol. 2, pp. 70–133.

[15] (a) W. Pauli, 'On Asymptotic Series for Functions in the Theory of Diffraction of Light', *Physical Review* **54**, 924 (1938). Reprinted in Ref. [13], Vol. 2, pp. 835–42. (b) K. von Meyenn (ed., with the cooperation of A. Hermann and V. F. Weisskopf), *Wolfgang Pauli. Scientific Correspondence with Bohr, Einstein, Heisenberg a.o., Volume II: 1930–1939* (Springer, Berlin and Heidelberg, 1985). (c) Translated from: E. Schrödinger, letter [8] of 12 July 1920 to Pauli, in Ref. [4].

[16] W. Pauli jr., 'Theoretische Bemerkungen über den Diamagnetismus einatomiger Gase', *Zeitschrift für Physik* **2**, 201 (1920). Reprinted in Ref. [13], Vol. 2, pp. 24–8.

[17] P. Debye, 'Die van der Waals'schen Kohäsionskräfte', *Physikalische Zeitschrift* **21**, 178 (1920).

[18] (a) J. H. Van Vleck, *The Theory of Electric and Magnetic Suscepti-bilities* (Oxford University Press, Oxford, 1932, 1952); (b) W. Pauli, 'Van Vleck, J.H. The theory of electric and magnetic susceptibilities', *Naturwissenschaften* **21**, 239 (1933). Reprinted in Ref. [13], Vol. 2, pp. 1400–01.

[19] C. D. Hodgman *et al.* (eds), *Handbook of Chemistry and Physics* (The Chemical Rubber Co., Cleveland, Ohio, 1962).

[20] C. P. Enz, 'W. Pauli's Scientific Work', in: J. Mehra (ed.), *The Physicist's Conception of Nature* (Reidel, Dordrecht, 1973).

[21] W. Pauli jr., 'Quantentheorie und Magneton', *Physikalische Zeitschrift* **21**, 615 (1920). Reprined in Ref. [13], Vol. 2, pp. 36–8.

[22] (a) A. Sommerfeld, 'Zur Theorie des Zeeman-Effekts der Wasserstoff-linien, mit einem Anhang über den Stark-Effekt', *Physikalische Zeitschrift* **17**, 491 (1916); (b) A. Sommerfeld, 'Spektroskopische Magnetonen-zahlen', *Physikalische Zeitschrift* **24**, 360 (1923); (c) P. Debye, 'Quanten-hypothese und Zeeman-Effekt', *Physikalische Zeitschrift* **17**, 507 (1916).

[23] W. Pauli, 'Quantentheorie', in: H. Geiger and K. Scheel (eds), *Handbuch der Physik* (Springer, Berlin, 1926), Vol. 23, pp. 1–278. Reprinted in Ref. [13], Vol. 1, pp. 271–548.

[24] W. Pauli, 'Zur Theorie der Dielektrizitätskonstante zweiatomiger Dipol-gase', *Zeitschrift für Physik* **6**, 319 (1921). Reprinted in Ref. [13], Vol. 2, pp. 39–47.

[25] H. A. Kramers und W. Pauli jr., 'Zur Theorie der Bandenspektren', *Zeitschrift für Physik* **13**, 351 (1923). Reprinted in Ref. [13], Vol. 2, pp. 134–50.

[26] Translated from Ref. [9b], pp. 55–6.

[27] (a) C. L. Charlier, *Die Mechanik des Himmels. Vorlesungen* (Veit, Leipzig, 1902, 1907), Vols 1 and 2; (b) A. Clebsch (ed.), *C.G.J. Jacobi, Vorlesun-gen über Dynamik, gehalten an der Universität Königsberg im Win-tersemester 1842–1843* (G. Reimer, Berlin, 1866).

[28] K. F. Niessen, 'Zur Quantentheorie des Wasserstoffmolekülions', *An-nalen der Physik* (4) **70**, 129 (1923).

[29] W. Pauli, review of Refs [14] and [28] in *Physikalische Berichte* **4**, 642 (1923). Reprinted in: C. P. Enz and K. von Meyenn (eds), *Wolfgang Pauli. Das Gewissen der Physik* (Vieweg, Braunschweig, 1988), pp. 161–5.

[30] (a) W. Heisenberg, 'Über die Spektra von Atomsystemen mit zwei Elek-tronen', *Zeitschrift für Physik* **39**, 499 (1926). Reprinted in: W. Blum, H.-P. Dürr, and H. Rechenberg (eds), *Werner Heisenberg Collected Works.*

Series A/Part I. Original Scientific Papers (Springer, Berlin and Heidelberg, 1985), pp. 531–50. (b) W. Heitler and F. London, 'Wechselwirkung neutraler Atome und homöopolare Bindung nach der Quantenmechanik', *Zeitschrift für Physik* **44**, 455 (1927).

[31] W. Heitler, *Elementary Wave Mechanics with Applications to Quantum Chemistry* (Clarendon Press, Oxford, 1956), 2nd edn.

[32] (a) B. Geyer (ed.), *Interview von Konrad Lindner: Ein Atomphysiker erzählt. Edward Teller zwischen Leipzig und Livermore* (Universität Leipzig, Leipzig, 1998); (b) E. Teller, 'Über das Wasserstoffmolekülion', *Zeitschrift für Physik* **61**, 458 (1930); (c) G. Jaffé, 'Zur Theorie des Wasserstoffmolekülions', *Zeitschrift für Physik* **87**, 535 (1934).

[33] (a) G. Herzberg, *Molecular Spectra and Molecular Structure. I. Spectra of Diatomic Molecules* (Van Nostrand, New York, 1950, 2nd edn); (b) O. W. Richardson, 'The Test of Wave Mechanics in Molecular Spectra and some Recent Developments in the Spectrum of (H_2)', *Nuovo Cimento* **15**, 232 (1938).

[34] *Protokoll über das Examen rigorosum*, 25. Juli 1921 (handwritten; copy), PLC Bi 95/1, *Pauli Archive*.

[35] (a) M. Born, *Vorlesungen über Atommechanik* (Julius Springer, Berlin, 1925), Erster Band. (b) *Note*: The second volume of Ref. [35a] appeared in 1930 under the title *Elementare Quantenmechanik* with Pascual Jordan as co-author. It treats the new mechanics in the matrix form of Born, Heisenberg, and Jordan. Pauli has written reviews on both volumes in *Die Naturwissenschaften* **13**, 487 (1925) and **18**, 602 (1930), reprinted in Ref. [13], Vol. 2, pp. 1392 and 1397, respectively. While Pauli's review of Volume 1 is quite benevolent, expressing the hope that it 'would accelerate the formation of a simpler and more unified theory of the atoms with more than one electron', Volume 2 is the target of some sarcastic remarks concerning the admittedly one-sided presentation.

[36] J. Franck und P. Jordan, *Anregungen von Quantensprüngen durch Stösse* (Julius Springer, Berlin, 1926), p. 265.

[37] W. Pauli, 'Die allgemeinen Prinzipien der Wellenmechanik', in: H. Geiger and K. Scheel (eds), *Handbuch der Physik* (Springer, Berlin, 1933), 2nd edn, Vol. 24, Part 1, pp. 83–272.

[38] (a) M. Fierz, 'Pauli Wolfgang', in: C. C. Gillispie (ed., with the cooperation of S. G. Navashin and W. Piso), *Dictionary of Scientific Biography* (Scribner's Sons, New York, 1974), Vol. 10, pp. 422–5; (b) C. P. Enz, 'Pauli and the Development of Quantum Theory', Preface in: W. Pauli, *General Principles of Quantum Mechanics*, translated from Refs [39] and [37] by P. Achuthan and K. Venkatesan (Springer, Berlin and Heidelberg, 1980).

[39] W. Pauli, 'Die allgemeinen Prinzipien der Wellenmechanik', in: S. Flügge (ed.), *Handbuch der Physik* (Springer, Berlin, 1958), Vol. 5, Part 1, pp. 1–168.

[40] M. C. Gutzwiller, (a) *Chaos in Classical and Quantum Mechanics* (Springer, New York, 1991); (b) 'The Quantization of a Classical Ergodic System', *Physica* **5**D, 183 (1982).

[41] A. Einstein, 'Zum Quantensatz von Sommerfeld und Epstein', *Verhandlungen der Deutschen Physikalischen Gesellschaft* **19**, 82 (1917).

4
Scientific collaborator

Born's assistant in Göttingen

Little Pauli is very stimulating; such a good assistant I will never have again. Unfortunately, in summer he wants to join Lenz in Hamburg. [1]

Commenting later on this letter to Einstein, Born, however, observed that 'the report on the "little Pauli" is not complete. I remember that he liked to sleep late, and he missed the lecture at eleven o'clock more than once. We then sent him our maid at half-past-ten to be sure that he was up. Without question, he was a first-rate genius, but my apprehension "such a good assistant I will never have again" was yet unjustified. His successor Heisenberg was as intelligent and more conscientious at that: him we did not have to have waken or otherwise remind him of his duties' [1]. Born's impression of Pauli being 'little' may have been psychological since he characterized Pauli's behaviour as 'childlike' (*kindlich*) (see Ref. [2], p. 88). In fact, Pauli was not exceptionally small, 166 cm according to his Swiss Military Certificate [3a] which he got in 1950, after having become a Swiss citizen the year before. And he was only a little smaller than Born, as may be seen from a picture in which Born pulls Pauli by his ear as a symbolic punishment for his sleeping so late (see Ref. [4], p. 35).

After obtaining his promotion certificate on 25 July 1921 [3b] and the certificate of leave from the university on 13 October [3c] Pauli left Munich for the small town of Göttingen. Where he spent his nights in Göttingen so as to sleep so late is not on record, however. Anyway, he only stayed one winter, taking the opportunity offered by Wilhelm Lenz to move to the big harbour city of Hamburg in April 1922 (see Ref. [2], p. 102). Of course, in physics and particularly in mathematics, the city of Göttingen was second to none in Germany. Founded in 1734, its university was host to the most famous mathematicians of their time: Carl Friedrich Gauss (1777–1855) and Bernhard Riemann (1826–1866). And in physics Wilhelm Weber (1804–1891) had developed electromagnetic theory before Maxwell and with Gauss had built the first needle telegraph. For a detailed description of the Göttingen tradition see Ref. [5a], Section III.1.

When Pauli arrived in Göttingen, Max Born (1882–1970) had been established as a professor there for just one semester. Born had spent his youth in Breslau which after World War II became the Polish town of Wrocław and

Fig. 4.1 Pauli with Max Born, Hamburg 1925 (CERN)

where at the end of the communist era the square in front of the physics building has been renamed after him. Born first came to Göttingen as a graduate student in 1904 where he came to know David Hilbert and Hermann Minkowski and where, two years later, he got his Ph.D. with a paper on elasticity theory for which he received a prize. In 1915 he left Göttingen for Berlin where, as an associate professor, he spent the war years on military duty. In 1919 he became full professor at the University of Frankfurt am Main, and in the spring of 1921 he came back to Göttingen, bringing his 'private assistant' Emmerich (Imre) Brody with him (see Ref. [2], pp. 82 and 86).

Pauli, like Brody and like Heisenberg after him, were paid from private funds whereas Friedrich Hund (1896–1997), after finishing his doctorate under Born's direction in 1922, became Born's regular assistant financed by the university, until he became a Privatdozent in 1925. These private funds were provided by the New York financier Henry Goldman, the co-founder of Goldman and Sachs & Co., who had been annoyed by the harsh conditions imposed on Germany after the war (see Ref. [5b], pp. 157–8). Born expressed his gratitude towards Henry Goldman by dedicating his *Vorlesungen über Atommechanik* to him [6a]. Max Born and James Franck (1882–1964), who with Gustav Hertz had received the Nobel Prize of 1925 'for their discovery of the laws governing the impact of an electron upon an atom', became the successors of Peter Debye (1884–1966). Debye had left Göttingen in 1920 to become professor of experimental physics and director of the Physics Institute at ETH in Zürich. Born was director of the Theoretical Institute and Franck director of the Second Experimental Institute while the First Experimental Institute was directed by

Robert Pohl (1884–1976) who had been appointed to a third chair of physics in 1920. The respective collaborators of Born, Franck, and Pohl were soon called '*die Bornierten*' (the narrow-minded), '*die Frankierten*' (those paid for), and '*die Polierten*' (the polished ones). For a detailed account of the Born–Franck era in Göttingen see Ref. [5a], Section III.2.

Alas, these relatively good times did not last. In 1933, when the National Socialists came to power, Born was dismissed and Franck resigned. Born left Göttingen with his family on 9 May for South Tirol where in the coming weeks his residence at Wolkenstein (Selva Gardena) became a meeting place for many friends: Pauli and his sister Hertha, Born's Göttingen colleague Hermann Weyl, Schrödinger and his wife Anny, F. A. Lindemann who later became Lord Cherwell, and the pianist Arthur Schnabel (see Ref. [6b], Part 2, Section 4: *Wolkenstein* and Ref. [6c], Chapter 7, *Tirolean Adventures*; see also letter [320] from Pauli to Heisenberg of 22 July 1933 in Ref. [6d]). In July Born obtained a position at Cambridge University and in 1936 he accepted a call to the University of Edinburgh, retiring in 1953 to Bad Pyrmont in Germany where, one year later, he received the Nobel Prize 'for his fundamental research in quantum mechanics, especially for his statistical interpretation of the wavefunction'. Franck, on the other hand, went to Bohr's institute in Copenhagen in the autumn of 1933 and two years later emigrated to the United States where in 1938 he became a professor at the University of Chicago, retiring to Göttingen in 1964. For more details on the disastrous effects of the rise of Nazism see Ref. [5b], Chapter 15: A New Regime and Ref. [6d], 'Das Jahr 1933', particularly pp. 146 and 167.

Pauli's thesis work continued to occupy him in Göttingen where he worked out the improved and extended version of 64 pages discussed above on pages 63–9. It was only on 4 March 1922 that he submitted this paper for publication [7]. In Munich Pauli's relationship with Heisenberg had developed into a close friendship which now gave rise to a lively exchange of letters. Like Pauli, Werner Heisenberg had grown up in an intellectual environment; his father had become a professor of Byzantine studies at the University of Munich in 1909. But quite contrary to Pauli, Heisenberg's passion were hiking in the Bavarian mountains accompanied by discussions around a campfire with his friends from the German youth movement. Professionally, though, he followed Pauli's path for several years, beginning his studies with Sommerfeld two years after Pauli, following him as Born's private assistant in Göttingen in the winter of 1922/3, and again succeeding Pauli for seven months as Bohr's collaborator in Copenhagen where he returned in 1926. But his discovery of quantum mechanics in June 1925 boosted his career to the highest level; it was crowned by the Nobel Prize of 1932. Already in the autumn of 1927 he received a call as professor of theoretical physics to the University of Leipzig, half a year before Pauli was nominated at ETH in Zürich. Then, however, life became bleak for Heisenberg also, when the conflict between his idealistic and patriotic views of the German cultural heritage and real life under the Nazis induced him to hold fast to his position in Germany (see Ref. [5b]). The friendship between Pauli and Heisenberg, however, survived World War II and, as we shall see, bloomed again shortly before Pauli's death.

In this early correspondence with Pauli, from which, unfortunately, Pauli's letters are missing, Heisenberg explains his new theory of the anomalous Zeeman effect of atoms with one and two outer electrons (see the following section). The Zeeman effect is the splitting of spectral lines in an applied magnetic field, discovered by Pieter Zeeman (1985–43) and explained in its normal form by Henrik A. Lorentz. In 1902, they shared the second Nobel Prize 'for their researches into the influence of magnetism upon radiation phenomena'. Heisenberg first discusses the 'doublet atoms' characterized by a 'Zeeman doublet', but then also speculates about 'triplet atoms' (letters [16], [17], [18] of 19 and 25 November and of 17 December 1921 in Ref. [9]). Heisenberg submitted the work described in these letters on 17 December 1921; it became his first publication [10a]. As discussed in the first of these letters, the main new idea was to divide the time-averaged total angular momentum between the outer electrons and the atomic core. For the ground state of doublet atoms, which have one outer electron, this meant a half $\hbar$ each, thereby introducing half quantum numbers. Of course, this idea was not entirely new since, as discussed on page 59, Sommerfeld had used it in his model of the helium atom; and Sommerfeld's original motivation had also been to split an integer number of quanta $\hbar$ between the electron and nucleus (see Ref. [10b], pp. 199–200).

In an elaboration of these ideas given in this third letter, Heisenberg points to the existence of an inner magnetic field due to the outer electron which gives rise to directional quantization of the atomic core even in the absence of an applied magnetic field. It is this complication that Sommerfeld, quoting 'Heisenberg's magneto-optic model' [10a], held as responsible for the complexity of the spectra, saying 'that the complex structure (*Komplexstruktur*) is a consequence of an inner-atomic magnetic field . . . originating from the electronic revolutions and precessions of the atomic structure' (translated from Ref. [10b], p. 662). This magnetic origin of the complex structure in the multiplicity of the spectra was contrary to the view held by Bohr who 'sees the multiplicities as essentially electric deviations from the central symmetry' (translated from letter [21] from Heisenberg to Pauli of 6 March 1922, Ref. [9], p. 56). Although Heisenberg's model was too restrictive, because the generalization to triplet systems was already unsatisfactory, it stimulated Pauli's own work on the anomalous Zeeman effect, as we shall see. On the other hand, Sommerfeld, who had discussed Heisenberg's paper [10a] at length in the previous edition of his *Atombau und Spektrallinien*, was led to drop it in the fourth edition (see the footnote on p. 629 of Ref. [10b]).

This correspondence between Heisenberg and Pauli was interrupted by the Christmas vacation. From a postcard to Walther Gerlach in Frankfurt am Main, dated 1 January 1922 (Ref. [9], p. 54), we learn that Pauli spent the vacation with his family in Vienna and that on his way back to Göttingen he wished to visit Gerlach and also Erwin Madelung who had succeeded Born there. Pauli's main interest in this visit was the fact that Gerlach was preparing an experiment conceived by Otto Stern to detect directional quantization in an atomic beam. On 17 February 1922 Pauli sent Gerlach a second postcard (Ref. [9], p. 55) to congratulate him on the successful experiment [11a]. Otto Stern (1888–1969) had obtained his Ph.D. in physical chemistry at the University of Breslau in 1912. But, disposing of private means, he joined Einstein in Prague where the latter had a chair at Karl-Ferdinand University from April 1911, and followed Einstein to the ETH in Zurich in August 1912 where he became a Privatdozent in 1913. The next year Stern moved to Frankfurt where he had developed the technique of molecular beams since 1919. However, in September 1921 he went to Rostock as associate professor and the following year he became professor of physical chemistry at the University of Hamburg. He left Germany in 1933 and went to the Carnegie Institute of Technology in Pittsburgh, Pennsylvania, from where he retired in 1945, dividing his time between Berkeley and Zurich. Stern was awarded the Nobel Prize in 1943 'for his contributions to the development of the molecular ray method and his discovery of the magnetic moment of the proton' (see Ref. [11b], p. 486).

In Göttingen Pauli collaborated with Born on the formulation of a general perturbation theory in view of its application to the helium atom and to the cases of atoms in external electric and magnetic fields. In the letter to Einstein of 21 October 1921 Born comments on this endeavour as follows: 'With Pauli I am about to do quantum calculations on atoms using the approximation procedure that Brody and I have developed recently in Zeitschrift für Physik at the example of a system of oscillators' (translated from Ref. [2], p. 88). What finally came out of this collaboration was a paper submitted to *Zeitschrift für Physik* on 29 May 1922 when Pauli had already left Göttingen for Hamburg [12]. As reviewed in the introduction, this approximation method, well known from the astronomical calculations of planetary orbits, had been introduced into atomic physics by Bohr in 1918 and elaborated upon by his assistant Kramers in 1920 and by Born and Brody in 1921. Born and Pauli developed the general formalism in the canonical framework of Eqs (3.11) above.

Since in the canonical formalism the perturbation is described by an inter-action energy term characterized by a small parameter λ, the physical quantities are obtained as power series in λ. The difficult problem is to prove the con-vergence of these series for arbitrary times. In Section 3 of Ref. [12] Born and Pauli give a detailed discussion of this problem, referring extensively to the fundamental work of Henri Poincaré on astronomical perturbation theory [13a]. Poincaré, Pauli once observed, 'was a favourite author of mine in my young years' (translated from Ref. [13b]). Pauli also wrote a highly knowl-edgeable review on this subject [13c]. The main conclusion of Born and Pauli's paper [12] is that in a Fourier development of the perturbation the existence of commensurabilities with large integers τ_k among the frequencies ν_k (see pages 65–6) will prevent the integration of the equations of motion by separation of the variables. When applied to the problem of crossed electric and magnetic fields this implies diffuse spectral lines in agreement with Bohr's point of view. On the other hand, Paul S. Epstein, using a particular perturbation procedure called the Delaunay method, claimed that the problem of the atom with one electron in crossed fields continues to be multiply periodic (see page 65) and hence that the spectral lines remain sharp.

In order to understand these last statements one has to consider Bohr's fundamental condition for the emission of a spectral line of frequency ν,

$$h\nu = E_i - E_f. \tag{4.1}$$

Here h is Planck's constant, and E_i and E_f are the energies, also called spectral terms, of the initial and final quantum states of the system, namely an atom or a molecule emitting the radiation. For a multiply periodic system these spectral terms are obtained as discrete solutions of the equations of motion subject to the quantum conditions. The discreteness of the spectral terms then implies that the resulting spectral lines are sharp.

While Born and Pauli's paper [12] exerted a considerable stimulation on the subsequent calculations of spectral terms, particularly for the helium atom, their criticism of Epstein's work was much too general to be conclusive. In fact, it turned out that the case of the crossed fields was sufficiently simple for the Delaunay method to be successful. Pauli learned about this turn of the problem in a seminar that Epstein gave in Hamburg in July 1923. And he immediately informed Bohr in a postcard to Bohr's experimental collaborator Hans M. Hansen (1886–1956), sent from the resort town of Bansin on the Baltic coast where Pauli spent some time after visiting Hamburg from Copenhagen: 'Please could you tell Prof. Bohr the following: In Hamburg I met Epstein who is on a visit to Germany, and he gave a rather long talk on the case of the Balmer lines in crossed electric and magnetic fields. The result is that, contrary to our hitherto held opinion, mechanics here also leads to harmonic solutions; the mechanical equations for the secular perturbation are integrable. Therefore, sharp components are to be expected. I have verified it and found that Epstein is right. I write this in order that Bohr does not write something wrong on this case in the Guthrie Lecture' (translated from Ref. [9], postcard [38] of 6 July 1923). Bohr had delivered the Seventh Guthrie Lecture on 'The effect of electric and magnetic fields on spectral lines' to the Royal Society the year before on 24 March 1922 (see Ref. [9], footnote 9, p. 101).

Paul S. Epstein (1883–1966) was born in Warsaw, studied in Moscow and did his Ph.D. with Sommerfeld in Munich. After two years as Privatdozent at the University of Zurich he emigrated to the United States in 1921 and became a professor at the California Institute of Technology in Pasadena. His work on atomic dynamics in crossed electric and magnetic fields [14] had an important result in late 1923 when Oskar Klein, making use of a method that Bohr had used to derive the Stark effect, (see Ref. [15]), proved in an elementary way that crossed fields gave rise to a doubly periodic motion composed of independent precessions around each of the fields [17a]. The Stark effect which is the splitting of spectral lines in an applied electric field, had been discovered by Johannes Stark (1874–1957) who had received the Nobel Prize in 1919 'for his discovery of the Doppler effect in canal rays and of the splitting of spectral lines in electric fields' but who later became, with Philipp Lenard (see page 50), the main proponent of the 'Deutsche Physik' of the Nazi era.

Klein's method was cast into a particularly elegant form by Wilhelm Lenz who also derived the quantized energy of the perturbed motion [17b]. But, unfortunately, this last result still contained an error in the assignment of the quantum numbers. The final word on this controversial problem is contained in the most compact and elegant formulation given by Pauli in his 'Quantentheorie' (Ref. [18], Ziff. 25, Eqs (134), (135)). However, hidden in Lenz's paper is a true little gem which, of course, did not escape Pauli's attention. Indeed, in a letter to Kramers from Hamburg, dated 19 December 1923, he says in a footnote: 'Lenz now has found a very beautiful derivation of Epstein's formula for which not even the canonical equations are used' (translated from Ref. [9], p. 136). This gem is what Lenz and Pauli called the '*Achsenvektor*' but which today is known as Lenz (or Lenz-Runge) vector

$$\mathbf{a} = \frac{1}{Ze^2}(\mathbf{L} \times \dot{\mathbf{r}}) + \frac{\mathbf{r}}{|\mathbf{r}|} \tag{4.2}$$

where $\mathbf{L}$ is the angular momentum vector and $\mathbf{r}$ the position vector of the electron with its origin in the nucleus which has the charge Ze.

As will be discussed in the next chapter, this vector, which is a constant of the motion only in the case of the non-relativistic Kepler problem of a $(1/r)$-potential, played a decisive role in Pauli's spectacular solution of the hydrogen atom in the framework of Heisenberg's new matrix mechanics [19a]. In this paper Pauli used the trick of adding external electric and magnetic fields in order to lift the degeneracy of the problem. The same idea also appears in a paper written the year before on the intensities of the so-called combination lines that occur in the Stark effect [19b]. While in the field-free case the spectral lines emitted according to Bohr's condition (4.1) satisfy the selection rule $\Delta k = k_f - k_i = \pm 1$ of the azimutal quantum number k, combination lines result from the selection rule $\Delta k = 0, \pm 2$. Pauli discussed this problem in Ziff. 42 of his 'Quantentheorie' [18].

Born and Pauli tried to apply the general formalism of perturbation theory to the problem of the helium atom. Born reports on this project in a letter to Einstein dated 30 April 1922: 'Pauli, unfortunately, has gone to be with Lenz in Hamburg. We recently have started a common paper, a continuation of the one published with Brody on the quantization of anharmonic oscillators. The approximation method developed there may be applied to all systems for which the "unperturbed" system is conditionally periodic and the perturbation function (*Störungsfunktion*, not '*Strömungsfunktion*'!) may be developed in powers of a parameter. Thereby the case that the unperturbed system is degenerate equally goes along and just leads to Bohr's method of the "secular perturbations". Incidentally, we now really have understood part of Bohr's reflections. Also we have started to calculate the ortho-helium (2 coplanar electrons) and could at once confirm Bohr's old claim that the inner electron circulates rapidly on an ellipse whose large axis always points towards the slowly circulating electron.

Pauli has taken the work along to Hamburg and wants to finish it there since I have no time because of the Encyclopedia article [20].' (translated from Ref. [2], p. 102).

Four months later, after having met Bohr in Göttingen in June (see pages 86-7), Pauli reported to Bohr also on a calculation concerning the ground state of helium: 'In the meantime I have occupied myself very thoroughly with your model for the ground state (*Normalzustand*) of the He-atom, and I have started a large calculation on it. It is not completed yet, but I hope very much that the question whether this model yields the correct ionization potential may be decided. I am looking forward very much to discussing this matter also with Herrn Dr. Kramers. It is, of course, not at all my intention to interfere with him in a publication, and I have attacked this problem of He in the ground state only because I am already very curious of the result' (translated from letter [24] from Pauli to Bohr from Hamburg, dated 5 September 1922, Ref. [9], p. 62). Neither of the two announced papers by Born and Pauli and by Pauli alone have been published. As mentioned in the just quoted letter from Pauli to Bohr, Kramers was working on the ground state of helium which belongs to the term system of para-helium[21a]. Born, assisted by Heisenberg, who during the winter term of 1922/3 held Pauli's position, then turned to the excited states of helium [21b]. He comments on this change of private assistant in a letter to Einstein dated 7 April 1923, as follows: 'In winter I had Heisenberg with me (since Sommerfeld was in America [spending the winter term in Madison]); he is at least as gifted as Pauli but personally nicer and more pleasing. Also he plays the piano very well' (translated from Ref. [2], p. 109).

The distinction between *para* and *ortho* states had been arrived at on purely empirical grounds. Born, in his *Vorlesungen*, observes: 'Why there are two term systems, a singlet system (para-helium) to which belongs the ground state and a doublet system (ortho-helium), and why the two do not combine cannot be treated from the point of view of this book' (translated from Ref. [6a], p. 220). This enigma again could only be solved by the *New Quantum Theory*. Para and ortho states of two-electron systems are, respectively, symmetric and anti-symmetric in the spatial variables but, because of the Pauli exclusion principle and of the spin $1/2$ of the electron, their spin states are anti-symmetric (spin 0) and symmetric (spin 1), respectively (see e.g. Ref. [22a,b]). Thus the ground state is para while each excited state splits into a doublet in which the ortho state is lower than the para state by twice the exchange energy in this state (see Fig. 23, p. 93 of Ref. [22b] and Fig 98, p. 490 of Ref. [10b]).

In his *Vorlesungen* [6a] Born devotes Sections 48 and 49 to the ground state and the excited states of helium, respectively, treating the Coulomb repulsion between the two electrons as the perturbation. The quantum numbers of the ground state are $n = k = 1$ for both electrons. Two models are then possible, the planar model of Bohr which has $j = 2$ and the spatial model of Bohr and Kemble for which $j = 1$ (see page 59). In the planar model the two electrons are always at opposite points of a diameter of the common circle (radius r). Hence the perturbation $\lambda H_1 = e^2/2r$ is a constant and must be added to the potential energy $V(r) = 2Ze^2/r$, where $Z = 2$ for helium. Since $\lambda H_1/V(r) = 1/4Z$ one sees that the perturbation parameter may be taken as $\lambda = 1/Z$. The problem then reduces to two one-electron problems with half the total potential energy each, i.e. Z^*e^2/r, where $Z^* = Z - (1/4)$. The total ground state energy is then $W = -2Z^{*2}R_\infty$, where $R_\infty = me^4/2\hbar^2 = 13.606$ eV is the Rydberg constant (see page 64). This leads to an ionization energy of

$$W_{ion} = (2Z^{*2} - Z^2)R_\infty = \frac{17}{8}R_\infty = 28.91 \text{ eV} \qquad (4.3)$$

which is to be compared with the experimental value 24.6 eV obtained by Franck from electron scattering (see Ref. [6a], Section 48, Eqs (13) and (14)). The agreement, therefore, is not satisfactory. But in addition, the perturbation H_1 gives rise to a phase relation resulting in an instability of the motion.

The spatial model was examined in great detail by Kramers (Ref. [21a], see also Section 12.I in Ref. [23]). Here the two electrons determine a diameter whose centre always lies on the axis defined by the total angular momentum vector **J** (see Fig. 40 in Section 48 of Ref. [6a]). Therefore H_1 only depends on one angle, so that the average $\lambda\overline{H_1}$ over the unperturbed cycle is easily determined. Calculating the total energy to second order in $\lambda = 1/Z$, Kramers obtained [21a]

$$W = -Z^2 \left(2 - 1.373\frac{1}{Z} + 0.271\frac{1}{Z^2} \right) R_\infty = -5.53R_\infty \qquad (4.4)$$

which is to be compared with the experimental value of $-5.807R_\infty$. The result (4.4) is almost 4 eV too small, and this motion also is unstable. Thus none of the models for the ground state of helium is satisfactory. There remain the excited states.

Highly excited (para or ortho) states may be formed by adding a loosely bound electron to the He^+ ion. This suggests taking the inverse average radius of the added electron as a perturbation parameter or, equivalently, $\lambda = 1/k$ where k is the associated azimutal quantum number (see page 57). Thanking Herrn W. Pauli for valuable advice, the authors of Ref. [21b], i.e. essentially Heisenberg, indulge in lengthy and complicated calculations. However, the results (Eq. (23) and the table in Section 49 of Ref [6a]) are unconvincing and do not warrant a closer look at this problem since the ground state is of much more fundamental importance. So let us close this section with Pauli's opinion given in his 'Quantentheorie'—which contains no explicit calculation for two-electron atoms: 'The simplest case of the occurence of equivalent electrons is the ground state (*Normalzustand*) of the helium atom, in which both electrons have to be viewed as having one quantum. In spite of numerous attempts [here Refs [21a] and [10b], Chapter 3, Section 6, are quoted] already in this case it has not been possible to find an atomic model which satisfies all requirements and, in particular, which agrees with the empirical value of the fundamental term of the helium spectrum. And in view of the mentioned difficulty concerning the periodicity property of the orbits [namely, exchanges of angular momentum between equivalent orbits], it is likely that hardly any model for the ground state of the helium atom based on classical mechanics exists at all' (translated from Ref. [18], Ziff. 27, pp. 166–7).

Under the spell of Niels Bohr

The 70th anniversary of Niels Bohr's birthday reminds me of a long and still continuing common pilgrimage since the year 1922, in which so many stations are involved. [24]

This is how Pauli saw his relationship with the great man 33 years after their first encounter in Göttingen in June 1922. The early letters from Pauli to Bohr show rare emotions of tenderness which were the first manifestations of a deep friendship. From his fifth letter (Ref. [9], letter [39]), written on his vacation on the Baltic coast in July 1923 (see the previous section) until his acceptance of the professorship at ETH in December 1927, Pauli invariably (with one exception, see below) addresses Bohr as '*Sehr verehrter lieber* (very revered dear) *Herr Professor*'. Bohr, in his first letter of the year 1928 (Ref. [9], letter [177]) starts calling Pauli '*Du*' both in Danish and in German, instead of the formal '*De*' and '*Sie*', respectively. Pauli only hesitantly followed suit. Indeed, in a postscript to letter [179] Pauli writes: 'I see, I should perhaps have written each time '*Du*' instead of '*Sie*'; please have patience, gradually I will get used to it all right' (translated from Ref. [9], p. 426). At about the same time the letters from Heisenberg to Pauli (no replies from Pauli are recorded in this period) switch from '*Sie*' to '*Du*'. This subtlety of many European languages is often a sensitive measure of the closeness of a relationship.

In spite of the 14 years of difference in age the friendship between Bohr and Pauli was indeed truly reciprocal, as witnessed already in Bohr's first letter to Pauli, written in touchingly faulty German: 'It was a great joy for me to get the opportunity, during my visit to Göttingen, to know you. I am delighted that you come to Copenhagen and hope very much that yourself also you will have pleasure' (translated from letter [22] from Bohr to Pauli dated 3 July 1922, Ref. [9], p. 60). And 36 years later Bohr wrote in the *Foreword* of the Memorial Volume for Pauli: 'When Pauli, after working with Born in Göttingen, came to Copenhagen in 1922, he at once became, with his acutely critical and untiringly searching mind, a great source of stimulation to our group. Especially he endeared himself to us by his intellectual honesty, expressed with candour and humour in scientific discussions as well as in all other human relations' (Ref. [25a], pp. 1–2). In turn, Pauli was profoundly impressed. In his first recorded letter from Copenhagen he wrote to Rudolf Ladenburg (1882–1952) in Berlin on 14 November 1922: 'I feel very well here and learn Bohr's physics. The latter is of an entirely different order of magnitude from the rest of physics' (translated from letter [29a], Ref. [25b], p. 736).

In Copenhagen Niels Bohr (1885–1962) 'grew up in a home which was not only intellectual, but which was also among the most distinguished in outlook and humanity' (Ref. [25c], p. 15). Five years before earning his Ph.D. from the University of Copenhagen in 1911 Bohr had already won a gold medal for his first research work. The decisive period leading to Bohr's three fundamental papers of 1913 on the quantum dynamics of the planetary atomic model were the four months he spent with Ernest Rutherford (1871–1937) in Manchester in 1912. In order to explain his group's scattering experiments with α-particles, Rutherford had assumed such a planetary model consisting of a nucleus of positive charge Ze surrounded by Z electrons, an idea which goes back to Wilhelm Weber (see the previous section). In Manchester Bohr had the decisive idea that in order to have stable orbits of the electrons around the nucleus he had to assume an equation of the form of Planck's quantum condition $E = h\nu$. At the end of July Bohr returned to Copenhagen where he got married and in September 1912 he became assistant at the University of Copenhagen (see U. Hoyer, '1. The Rutherford Memorandum (1912)' in Ref. [16a], pp. 103–5 and Ref. [26a], pp. 45-50).

Under Rutherford's direction from 1907 to 1919 the physics laboratory in Manchester was second in England only to the Cavendish Laboratory in Cambridge, directed by Joseph John Thomson (1856–1940) who had received the Nobel Prize in 1906 'for his theoretical and experimental investigations on the conduction of electricity by gases' and was the main discoverer of the electron. Rutherford grew up near Nelson in New Zealand. In 1895 he came to the Cavendish on a studentship and was invited by Thomson to collaborate on the newly discovered X-rays. From 1898 to 1907 he was professor at McGill University in Montreal, Canada, where in 1902 he discovered with his collaborator Frederick Soddy (1877–1956) the fundamental exponential decay law defining the half-life of radioactive substances.

It was in his Manchester period that in 1911 Rutherford concluded from the experiments of scattering α-particles on metal foils, done by his collaborator from Germany, Hans Wilhelm Geiger (1882–1945), and his student Ernest Marsden (1889–1970), that the atom was a great void of 10^{-8}cm diameter with a heavy core of only 10^{-13}cm diameter at the centre. It was also in this Manchester period that Rutherford was awarded the Nobel Prize in Chemistry of 1908 'for his investigations of the elements, and the chemistry of radioactive substances'.

In 1919 Rutherford became J. J. Thomson's successor as director of the Cavendish. With him came his student James Chadwick (1891–1974) who was to be his closest collaborator. In the same year Rutherford discovered the artificial disintegration of stable elements on bombardment with α-particles. Chadwick, on the other hand, in 1932 discovered the neutron, whose likely properties Rutherford had predicted in his Bakerian Lecture to the Royal Society in 1920. Chadwick was awarded the Nobel Prize in 1935 for his discovery. Rutherford received a large number of honours: in 1922 the Copley Medal, the highest award given by the Royal Society, and in 1925 the Order of Merit from the hands of King George V. He was made a knight in 1914 and a peer (Baron Rutherford of Nelson) in 1931 (see Refs [26b,c]).

While Bohr's work in Manchester only concerned the ground state he subsequently postulated that, more generally, all the stationary states of the electron are determined by Planck's quantum condition. Realizing that the hydrogen atom must have $Z = 1$, he concluded that, for circular orbits,

$$E_{kin} = \frac{n}{2}\hbar\omega \qquad (4.5)$$

where $E_{kin} = m(r\omega)^2/2$ is the kinetic energy, ω the angular velocity and r the radius of the orbit. It is easy to see that, for $k = n$ which means circular orbits, Eq. (4.5) is the same as Eq. (3.2) (see page 57). Since here Eq. (3.10) expresses the equilibrium condition $2E_{kin} = -E_{pot} = e^2/r$, Eq. (4.5) yields for the velocity, $r\omega = e^2/n\hbar$. Hence the total energy $E_{kin} + E_{pot}$ takes the quantized form

$$E_n = -\frac{R_\infty}{n^2} \qquad (4.6)$$

where $R_\infty = me^4/2\hbar^2$. It was only early in 1913 that Bohr learned about the empirical formula for the frequency of the emitted spectral lines of the mathematics teacher from Basle, Johann Jakob Balmer (1825–98). Using Eq. (4.6) for the spectral terms in (4.1) Bohr obtained Balmer's formula. Thus the theory of the hydrogen atom, described in the first paper, was achieved. In the second paper Bohr elaborated on the case of atoms, and discussed molecules in the third [27a,b]. For more details on the history of the atomic models and Bohr's fundamental papers see Ref. [5a], Section II.2, Refs. [28a,b] and Chapters 7 and 8 of Ref. [28c].

When Bohr accepted the invitation to give the well-publicized Wolfskehl Lectures [29a] in Göttingen in June 1922 he had already reached the peak of his fame. He had obtained a professorship at the University of Copenhagen in 1916, although with a very poor endowment. But on 3 March 1921 a new institute of theoretical physics was dedicated at Blegdamsvej (Bleachers' Way) 15, some 2 km to the north of the centre of town marked by the Tivoli amusement park and the railway station, in a building which is still in use today (see Ref. [16b], p. 26 where a historic photograph is reproduced). Bohr and his family of three sons—three more were to come—occupied the upper floors while on the ground floor offices and laboratories were installed, as well as the famous lecture room appearing on many historic photographs and which has survived essentially unscathed. Only six months after the Wolfskehl Lectures, on 11 December 1922, Bohr was honoured with the Nobel Prize 'for his investigations of the structure of atoms, and of the radiation emanating from them' (see Sections 9c and 12a,b of Ref. [28c]).

It is not surprising, therefore, that these seven lectures which took place from 12 to 14 and from 19 to 22 June, were a major event in physics and became known as the 'Bohr Festspiele', attended by everybody able to get there: from Göttingen the mathematicians Felix Klein, David Hilbert, Carl Runge, and

Richard Courant and the physicists Max Born, James Franck, and Robert Pohl with their collaborators; from Munich Arnold Sommerfeld and his student Werner Heisenberg for whom Sommerfeld even paid the travelling expenses; from Hamburg Wolfgang Pauli; from Frankfurt Walther Gerlach, Alfred Landé, and Erwin Madelung; and from Leiden Paul Ehrenfest. Bohr himself was accompanied by his collaborators Oskar Klein and Wilhelm Oseen, all in all about 100 participants. On his return to Copenhagen Bohr was accompanied by Pauli as far as Hamburg where Wilhelm Lenz, who had been unable to attend the lectures, came to greet Bohr at the railway station (see Bohr's first letter to Pauli mentioned above and Ref. [9], p. 58).

But, although Bohr had prepared these lectures very carefully [16c], Friedrich Hund, Born's 'official' assistant, later recalled that 'Bohr did not speak very clearly, and we junior people were not allowed to sit in the front rows reserved for important guests' (Ref. [5a], p. 345). Richard Courant (1888–1972) who was a good friend of both Niels Bohr and his younger brother Harald, a well-known mathematician, recalls: 'Niels was invited to give the famous "Gastvortraege" where the unforgettable Ehrenfest very actively participated, and where the most fruitful contact between Niels and Sommerfeld's young meteoric student Heisenberg was established in a flash by remarks Heisenberg made in the discussion.' And, speaking of a private lecture Bohr once gave him and brother Harald, Courant continues: 'It was a typical Bohr lecture such as we all have experienced so often, excitingly inspiring, though neither acoustically nor otherwise completely understandable. When we interrupted Niels to ask him for some clarification, he angrily protested: "Of course you cannot understand what I try to say now; this may perhaps become understandable, but only after you have heard the story as a whole and have understood the end"' (Ref. [29b], pp. 302–3).

In his recollections Heisenberg paints an idyllic picture of the event: 'In the year 1922 early summer had adorned Göttingen, the pleasant small town of the villas and gardens on the slope of Hainberg, with innumerable blooming bushes, roses and flowerbeds so that already the external splendor justified the designation we later gave these days: The "Bohr Festivals" of Göttingen' (translated from Ref. [30], p. 58). Then Heisenberg describes Bohr's lecturing style much in the way of Courant's characterization quoted above, and he continues: 'Of course, we had learnt Bohr's theory from Sommerfeld so that we knew what it was about. But what was said in Bohr's tongue sounded differently from Sommerfeld's. ... After each lecture there was a discussion, and at the end of the third lecture I dared to advance a critical remark' (translated from Ref. [30], p. 59).

Heisenberg's critical remark referred to Kramers' work on the influence of an electric field on the fine structure of the hydrogen lines [31]. In this paper Kramers addressed the ambitious problem of calculating the Stark effect by perturbation theory for electric fields so small that the splitting of the hydrogen lines became comparable with their fine structure, which is a relativistic effect proportional to c^{-2} first calculated by Sommerfeld in 1915 [32]. Since the measurement of the Stark effect in such small fields is extremely difficult, Heisenberg made the clever remark that such a situation could be realized by dispersion experiments with long-wavelength radiation, since an infinite wavelength corresponds to a weak static electric field. But then he apparently argued that classical dispersion theory would involve an orbital angular frequency ω as in Eq. (4.5) above and not a frequency ν determined by Bohr's quantum condition (4.1), while Kramers of course used the latter and, hence, could not be right. Unfortunately, it seems impossible to reconstruct the exact meaning of Heisenberg's criticism (see Ref. [23], pp. 128–9). In his recollections Heisenberg uses the form of a Platonic dialogue to describe the discussions with Bohr that followed his remark in the afternoon during a long walk over the Hainberg, on which Bohr had invited him (Ref. [30], pp. 59–65; see also Ref. [5b], pp. 128–30).

The editor's introduction to Volume 4 of Bohr's *Collected Works* (Ref. [16c], p. 23) mentions Rudolph Minkowsky, a nephew of Hermann Minkowski, the well-known inventor of the four-dimensional space–time (see Ref. [5a], footnote 946, p. 621), 'and another person', namely Erich Hückel (see Ref. [5a], p. 345), who were asked to prepare notes of the seven lectures. These notes, supplemented by figures and tables supplied by Bohr, are reproduced in Ref. [16c], pp. 341–419. 'The first three lectures dealt with the fundamental principles of the quantum theory of line spectra, as applied to hydrogen and to the influence of external fields upon the hydrogen spectrum. The fourth lecture treated the general features of the spectra of elements with several electrons. In the second half of this lecture the spectrum of helium was discussed. The fifth lecture dealt with the structure of the elements in the second period of the periodic system, and with the structure of the sodium atom [33a]. The sixth lecture dealt with the structure of all the remaining elements, in particular with the explanation of the family of rare earths and the platinum family. It was stated that the unknown element with atomic number 72 should not be a rare earth but should have chemical properties similar to those of zirconium [33b]. The last lecture was devoted to a discussion of X-ray spectra, and level diagrams of krypton, xenon, and niton (radon), taken from recent work of Coster, were shown' (Ref. [16c], p. 23). For more details on these lectures see Ref. [5a], Section III.4.

No discussion remarks by Pauli are recorded in the 'Bohr Festivals', although he obviously participated in the 'endless discussion rounds in appartments, coffee houses and on walking tours' (Ref. [5b], p. 128). However, in closing his first letter to Bohr, sent from Hamburg on 7 July 1922, Pauli writes: 'Finally I wish to thank you very much for the great kindness with which you informed me in Göttingen about the most diverse questions, which for me was of inestimable benefit' (translated from letter [23], Ref. [9], p. 62). Pauli himself later recalled: 'A new phase of my scientific life began when I met Niels Bohr personally for the first time. This was in 1922, when he gave a series of guest lectures at Göttingen in which he reported on his theoretical investigations on the periodic system of elements. . . . During these meetings in Göttingen Bohr came to me one day, accompanied by his assistant, Oskar Klein (now professor in Stockholm), and asked me whether I could come to him in Copenhagen for a year. He needed a collaborator in the editing of his works, which he wanted to publish in German. I was much surprised, and after considering a little while I answered with that certainty of which only a young man is capable: "I hardly think that the scientific demands which you will make on me will cause me any difficulty, but the learning of a foreign tongue like Danish far exceeds my abilities." The result was a hearty burst of laughter from Bohr and Klein, and I went to Copenhagen in the fall of 1922, where both of my contentions were shown to be wrong' (Ref. [34], pp. 213–14).

After the 'Bohr Festivals' not only Pauli's encounter with Bohr but also his acquaintance with Paul Ehrenfest (1889–1933) and with Oskar Klein (1894–1977) grew into lasting friendships. In Ehrenfest particularly, Pauli discovered a kindred critical mind full of pungent wit. This special relationship is beautifully illustrated by the following well-known anecdote reported by Oskar Klein: 'From this time [the 'Bohr Festivals'] originates the first story about their "war of wit" (*Witzkrieg*), which story is related with the fact that many years earlier Ehrenfest, together with his wife Tatjana, had written a very original and thorough but also very controversial article on statistical mechanics for the *Enzyklopädie der Mathematischen Wissenschaften*. At the occasion referred to, Ehrenfest stood somewhat offside of Pauli, fixing him in his mocking manner, and then said: "Herr Pauli, your Encyclopedia article pleases me more than yourself". To which Pauli answered in utter calm: "How strange, with you just the reverse occurs to me." Later, during Ehrenfest's last years [before his suicide] he and Pauli became good friends, and the "war of wit" was continued in a cosy form' (translated from Ref. [35a]; see also Ref. [35b], p. 31 and Ref. [35c], p. 115). Also well known is Ehrenfest's mocking lamentation calling Pauli 'God's whip' (*die Geissel Gottes*) which Pauli subsequently used to sign his letters to Ehrenfest. In a postscript to letter [125] to Kramers on 8 March 1926 Pauli writes: 'This title Ehrenfest has given me. I am very proud of it!' (translated from Ref. [9], p. 307). Later Ehrenfest addressed Pauli also as 'Lieber fürchterlicher (dreadful) Pauli' and as 'Sanct Pauli' (letters [152] and [210] dated 24 January 1927 and 26 September 1928, Ref. [9], p. 371 and p. 475, respectively).

On 1 April 1922 Pauli became Wilhelm Lenz's assistant at the University of Hamburg where his future friend Erich Hecke (1887–1947) had held the chair of mathematics since the foundation of the university in 1919, while Wilhelm Lenz had been appointed to the chair of theoretical physics in 1921. Before leaving for Copenhagen Pauli attended the meeting of the German Natural Scientists and Physicians which took place in Leipzig from 17 to 24 September. In order to mark the 100th anniversary of the society Einstein had been invited to present the main address, but due to anti-Semitic threats he withdrew. The political and economic crises persisted unrelentingly; with almost 100 million marks to the dollar in January 1923 inflation was at its peak. Thus, leaving for neutral Denmark offered some relief, although neither inflation nor social unrest had spared this country (see Ref. [28c], Section 9c).

In Copenhagen Pauli at last met Hendrik A. Kramers with whom he had already shared much of his research activity. Not surprisingly, therefore, a collaboration immediately ensued which led to a common paper on the theory of band spectra discussed on page 62. But the two also had much in common outside of physics. Both Kramers and Pauli were exceptionally well read in literature and philosophy, and both were prolific letter writers, both wrote fantasies, and both were music lovers. But while Pauli never practised music, Kramers played the piano and was an excellent cellist. Born in the port city of Rotterdam in 1894, Kramers enrolled in physics and mathematics at the University of Leiden in 1912 where he took the famous 'Monday morning' lectures of H. A. Lorentz , quite aware that he had the same initials as the great master. Kramers took the main courses with Ehrenfest, but was unable to share his total commitment to physics. This gradually led to an estrangement between Ehrenfest and Kramers (see Ref. [23], Chapter 10). Thus in the summer of 1916, in the middle of the war, Kramers went on his own to Copenhagen where he attended a student meeting and introduced himself to Bohr in a postcard asking to meet him—and it worked! He became Bohr's student and first collaborator (Ref. [23], Chapter 7). In 1919 he went to Leiden, together with Bohr, and submitted his Ph.D. thesis (Ref. [23], Chapter 11), and in 1926 he accepted a call to the University of Utrecht as professor of theoretical physics (Ref. [23], Section 15.I; see also Ref. [5a], Section III.4 and Ref. [35d], pp. 150–7).

Besides working on his own problems (see the following section), in Copenhagen Pauli offered the same service towards Bohr as earlier in Munich towards Sommerfeld whom he had helped with the editing of the first three editions of his *Atombau und Spektrallinien* (see pages 54–5). It was Pauli who translated Bohr's Nobel Lecture into German. Indeed, in a letter sent to Bohr from his summer vacation on the Baltic coast, Pauli writes: 'Today I received an information from [editor Arnold] Berliner . . . concerning the German translation of your Nobel Lecture which is supposed to appear in "Naturwissenschaften". . . . I answered him today that I cannot settle anything without you' (translated from letter [39] of 16 July 1923, Ref. [9], p. 102). In the same letter Pauli sends 'compliments also to your spouse. I promise her that I will not be *too* assiduous with you' This quotation testifies to the close relationship between Bohr's family and his collaborators: 'The children got to know the visiting scientists. Memories of our childhood are linked with many "uncles", among them Uncle Kramers, Uncle Klein, Uncle Hevesy, and Uncle Heisenberg' (Ref. [28c], p. 251). In the Niels Bohr Archive in Copenhagen there is a photograph of Pauli with Bohr's sons Christian (born 1916) and Hans (born 1918) dating from 1922 to 1923. In 1926 the Bohr family moved to new quarters in 'the villa' on Blegdamsvej 17 which today is occupied by the Niels Bohr Archive. And in 1932 Bohr was selected to occupy the Residence of Honour in the Carlsberg breweries, less than 2 km to the west of the centre of town where he and his wife lived until his death in 1962 (see Sections 12e and 15e of Ref. [28c]).

In the 1930s the institute at Blegdamsvej had grown into a world centre of theoretical physics, and Bohr kept abreast of the developments in quantum electrodynamics and in nuclear physics. But his interest was also directed towards the problems of the living; at a congress in Copenhagen in 1932 Bohr gave the famous lecture on 'Light and Life'. Later, in 1936, he concentrated his attention increasingly on nuclear reactions. In the middle of the war in 1943 Bohr was brought to England via Sweden, in order to participate there and later in Los Alamos in the United States on the development of the atom bomb. However, his contacts with Roosevelt and Churchill were a failure. After the war Bohr acted in favour of complete openness concerning research in atomic weapons. He addressed an open letter to the United Nations in 1950 and was awarded the first Atoms for Peace Prize in 1957. He obtained also honorary degrees from over 30 universities. Towards the end of his life he increasingly devoted his time to philosophical questions related to *atomic physics and human knowledge* (see e.g. Ref. [28c]).

Struggling with the anomalous Zeeman effect

I toiled for a very long time with the anomalous Zeeman effect whereby I often got on a wrong track, and I examined and rejected again a great many hypotheses. But it would just not agree! This, for once, has turned profoundly crooked! I was in complete despair for some time. [36]

As in the previous year, Pauli spent the Christmas vacation of 1922 with his family in Vienna where he visited the physics department, and on the way back to Copenhagen he also saw Heisenberg in Göttingen (see Ref. [9], pp. 77–8). Then came Pauli's year on the anomalous Zeeman effect. It was his second encounter with magnetism, and it was much tougher but also more fundamental than his first. For, although the explanation by the old quantum theory of the Stark effect of an electric field on the spectral lines had been relatively successful (see page 80), the essence of the Zeeman effect due to a magnetic field H was that it was anomalous in most cases. Besides the helium atom it was in fact the anomalous Zeeman effect that most drastically showed the shortcomings of the old theory. For historical details see also Ref. [5a], Section IV.4.

As it turned out, the culprit behind these anomalies was the spin of the electron, i.e. its 'inner' angular momentum which measures a half $\hbar$, and not an integer value as in Eq. (3.2). Within six months, between December 1924 and June 1925, the seeds of the three cornerstones of modern quantum theory were sown: the exclusion principle, the electron spin, and quantum mechanics. But to get there it was necessary (because of the angular momentum of the atomic core, see below) to abandon assigning each (outer) electron of an atom a single stationary state characterized by the quantum numbers n and k (see page 57) and to describe the resulting complex structure of the spectra (see Ref. [18], p. 209), in particular the anomalous Zeeman effect, in terms of empirical rules. The inherent intricacies of this approach made Pauli unhappy,

Fig. 4.2 Writing, April 1924 (CERN)

as he later recalled: 'A colleague who met me strolling rather aimlessly in the beautiful streets of Copenhagen said to me in a friendly manner, "You look very unhappy"; whereupon I answered fiercely, "How can one look happy when he is thinking about the anomalous Zeeman effect?"' (Ref. [34], p. 214).

Phenomenologically, the Zeeman splittings $\Delta\nu$ of the spectral terms (see page 78) may be written simply as

$$\Delta\nu = mg\Delta\nu_0. \qquad (4.7)$$

Here $\Delta\nu_0 = \mu_B H/\hbar$ is the normal splitting derived by Lorentz, expressed in units of the Bohr magneton $\mu_B = e\hbar/2mc$ (see page 60); g is the spectroscopic splitting factor introduced by Alfred Landé (b. 1888) [37a] and m the directional quantum number which varies in unit steps between $-j$ and $+j$. However, for the inner quantum number j of page 59 half-integer values now have to be admitted. In terms of Eq. (4.7) the first empirical regularities observed at the turn of the twentieth century by Thomas Preston (1860–1900) and by Carl Runge (1856–1927) may be expressed as properties of g: Preston's rule states that g is independent of the atomic charge Z and of the principal quantum number n, and Runge's rule states that g is a rational number.

The spectroscopic splitting factor introduced in Ref. [37a], which was Landé's first contribution to the anomalous Zeeman effect, turned out to be the appropriate description; it was taken over into the new quantum theory (see Ref. [38], Chapter XVI). Its physical meaning follows from the fact that $h\Delta\nu = \vec{\mu} \cdot \mathbf{H}$ is the energy of the atomic magnetic moment $\vec{\mu}$ in the field $\mathbf{H}$ so that, according to (4.7),

$$g = \frac{\mu_\|}{m\mu_B} \tag{4.8}$$

where $\mu_\|$ is the moment parallel to the field. Hence g represents the ratio of the magnetic moment in units of μ_B to the angular momentum in units of $\hbar$ (since projection on the field direction does not alter this ratio). And the anomaly consists in deviations from the value 1. In his first paper Landé had already noticed that for an s-term, for which $l = k - 1 = 0$, one had to put $g = 2$ (see Ref. [37a], p. 239, in particular footnote 1). Here the standard notation s ('sharp'), p ('principal'), d ('diffuse'), b ('Bergmann', later called f for 'fundamental'), ..., for the spectral series with $l = 0, 1, 2, 3, \ldots$, respectively, is used (in the new theory l was recognized as orbital angular momentum, see page 57). The origin of this *magneto-mechanical anomaly* of the spectroscopic splitting factor again turned out to reside in the electron spin, for which the value 2 must be assigned to g (see Ref. [38], Eq. (1), p. 149).

The observed complex structure could be ordered into multiplets satisfying Rydberg's alternation rule (*Wechselsatz*) [39a] which states that in spectra of atoms with an even number of outer or valence electrons only odd multiplets occur, while in spectra of atoms with an odd number of valence electrons there are only even multiplets (see Ref. [18], Ziff. 45). The best-known examples were the alkali and alkaline earth metals which have one and two valence electrons, respectively. In a magnetic field the spectroscopic terms corresponding to these multiplets split according to Eq. (4.7). For doublets and triplets the associated values mg are listed in Landé's first paper (Table 2 of Ref. [37a] where j is called n). In order to obtain the Zeeman patterns one needs a selection rule. While several such rules were discussed by Landé [37a], the most important is the one proposed by Sommerfeld, namely

$$\Delta j = 0 \text{ or } \pm 1 \tag{4.9}$$

($j = \Delta j = 0$ excluded) for the difference between the j-values of the initial and final states [39b]. This rule is still valid in the new quantum theory (see Ref. [38], Eq. (6), p. 60).

Heisenberg then succeeded in constructing a mechanical model which could account for many features of the observed doublet and triplet patterns (see page 78) [10a]. In this vector model the energy is due to the magnetic moments $\vec{\mu}_r$ which in the case of the doublet are those of the valence electron ($r = 1$) and of the atomic core ($r = 2$). In the absence of an external field this energy reduces to the interaction energy which is proportional to the product $\vec{\mu}_1 \cdot \vec{\mu}_2$. In the triplet case Heisenberg assumed the inner valence electron to be so strongly coupled to the core that, magnetically, the two form a unit. This interaction energy may be described by an inner magnetic field $\mathbf{H}_i$ which the (outer) valence electron produces at the core and which follows from the Biot–Savart law (see e.g. Ref. [40], p. 71). In an external field $\mathbf{H}$ the total magnetic energy therefore may be written

$$\Delta E = \vec{\mu}_1 \cdot \mathbf{H} + \vec{\mu}_2 \cdot (\mathbf{H} + \mathbf{H}_i). \tag{4.10}$$

Heisenberg assigned an inner quantum number j of $n - 1/2$ to the valence electron and $1/2$ to the core of a doublet atom while for a triplet atom his assignment was $n - 1/2$, $-1/2$, and 0 for the two valence electrons and the core, respectively, where $n = 1, 2, \ldots$ is the principal quantum number occurring in Eq. (4.6). In order to be able to use these values in Eq. (4.10) this equation must be time averaged over the precessions of $\vec{\mu}_1$ and $\vec{\mu}_2$. In weak fields $H \ll H_i$ this precession is around the resultant $\vec{\mu}_1 + \vec{\mu}_2$. In this way Heisenberg was able to reproduce Landé's results [37a]. In the general case the direction of the core moment $\vec{\mu}_2$ relative to $\mathbf{H}$ and to $\vec{\mu}_1$ is determined by an extremum condition. This yields more general formulae depending explicitly on the field H.

In the case of a strong field where 'strong' means that $H/H_i \sim \Delta\nu_0/\Delta\nu > 1$, the complex structure and, according to Eq. (4.7), also the g-values are modified. This situation is known as the Paschen–Back effect. It is this case to which Pauli turned his attention in his first paper on the subject [41]. This paper is extensively based on the latest results which Landé had obtained for a large class of multiplets in the weak-field case [42a]. In Table 3 of this reference Landé lists the g-values as functions of three quantum numbers, J, K and R, which, expressed in the modern notation j, l, and s, are $J = j + 1/2$, $K = l + 1/2$, and $R = s + 1/2$, respectively. Landé's table is reproduced in Pauli's 'Quantentheorie' (Ref. [18], p. 228), where the modern notation is already adopted. The quantum number s (not to be confounded with 's-state', see above) designates the total spin of all the valence electrons, i.e. $s = 0$ for singlets, $s = 1/2$ for doublets, $s = 1$ for triplets, etc. But note that this quantum number was associated with the core (*Rumpf*)! Landé called $2R = 2s + 1$ the 'core quantum number' (*Rumpfquantenzahl*). It determines the permanent multiplicity of the multiplet which is attained for $l > s$. The g-values in Table 3 are calculated according to formula (8) of Ref. [42a], which in modern notation is

$$g(j; l, s) = 1 + \frac{j(j + 1) + s(s + 1) - l(l + 1)}{2j(j + 1)} = \frac{3}{2} + \frac{(s - l)(s + l + 1)}{2j(j + 1)} \tag{4.11}$$

where the second expression exhibits the symmetry relative to the value $s = l$. It is quite remarkable that this formula agrees exactly with the modern expression (at least if spin–orbit interaction is negligible, see Ref. [38], Eq. (2), p. 381).

In the J, K-plane of Table 3 of Ref. [42a] only the squares satisfying Landé's Eq. (4), namely

$$|l - s| \leq j \leq l + s \qquad (4.12)$$

are occupied. Taking the sum of the values (4.11) over the interval (4.12) one obtains the remarkable result

$$\sum_{j=|l-s|}^{l+s} g(j; l, s) = \begin{cases} 2s + 1; & l > s \\ (3/2)(2l + 1); & l = s \\ 2(2l + 1); & l < s \end{cases} \qquad (4.13)$$

valid for $m \leq |l - s|$ (see *Note* [43]). To obtain this result it suffices to observe that the Runge denominator in Eq. (4.11) may be written as $1/j(j + 1) = 1/j + 1/(j + 1)$. In particular, Equation (4.13) says that the average g-value of a multiplet with permanent multiplicity, $l > s$, is 1 (see Ref. [42a], footnote 1, p. 193). Eq. (4.13) also represents the sum, for fixed m, of the energies (4.7) in units $h\Delta\nu_0$. But this is the 'principle' that Heisenberg had used to determine the permitted values of j (his $n^* = n - 1/2$; see Eq. (10) of Ref. [10a]). However, Heisenberg tried to justify this principle using an argument of statistical conservation of angular momentum in the emission of the radiation which neither Landé (see the mentioned footnote) nor Pauli (see Ref. [41], footnote 1, p. 163) could accept.

Of course, Landé did not derive Eq. (4.11) just by manipulating numbers. Like Heisenberg he used the vector model (see Fig. 1 in Ref. [42a]). For the angular momenta it says that, in units of $\hbar$, $\mathbf{j} = \mathbf{l} + \mathbf{s}$. And, designating the associated magnetic moments expressed in Bohr magnetons as $\vec{\mu}_j$, $\vec{\mu}_l$, and $\vec{\mu}_s$, respectively, where the last two are Heisenberg's $\vec{\mu}_1$ and $\vec{\mu}_2$, it follows from the above definition of g that $\vec{\mu}_j = g\mathbf{j}$, whereas $\vec{\mu}_l = \mathbf{l}$ and $\vec{\mu}_s = 2\mathbf{s}$. In the absence of an external field $\vec{\mu}_l$ and $\vec{\mu}_s$ precess around $\vec{\mu}_j$. If $\mathbf{e}$ is the unit vector in the direction of $\mathbf{j}$ then $(\mathbf{l} + \mathbf{s}) \cdot \mathbf{e} = |\mathbf{j}|$ and $(\mathbf{l} + 2\mathbf{s}) \cdot \mathbf{e} = g|\mathbf{j}|$, so that $\mathbf{s} \cdot \mathbf{e} = (g - 1)|\mathbf{j}|$. Since in the triangle formed by the vectors $\mathbf{j}$, $\mathbf{l}$, $\mathbf{s}$ the relation $\mathbf{s}^2 - (\mathbf{s} \cdot \mathbf{e})^2 = \mathbf{l}^2 - (|\mathbf{j}| - \mathbf{s} \cdot \mathbf{e})^2$ holds, one finds $\mathbf{s} \cdot \mathbf{e} = (\mathbf{j}^2 - \mathbf{l}^2 + \mathbf{s}^2)/2|\mathbf{j}|$, from which Eq. (4.11) follows, provided that one identifies the lengths of the vectors as (see page 57)

$$|\mathbf{j}| = \sqrt{j(j + 1)} \qquad (4.14)$$

etc., which is an example of the 'piercing' (*Durchbrechung*) of (classical) mechanics expected mainly by Bohr (see e.g. Ref. [42b], p. 121). This is the calculation that led Landé in Ref. [42a] to his equations (18') and (21').

Forgoing any mention of the vector model, Pauli observed in his first paper [41] that Sommerfeld's selection rule (4.9) fixes j only up to a constant. He had hoped in vain to use this freedom to solve the puzzle of the normalization (4.14). To Sommerfeld he wrote 'however, the difference [between several expressions of g] *cannot* be made to disappear by a change of the undetermined additive constant in the normalization of i, k and j' (translated from letter [40] of 19 July 1923, Ref. [9], p. 105). As compared with Landé's J, K, and R Pauli's choice was $j_{Pauli} = J + 1/2 = j + 1$, $k = K + 1/2 = l + 1$, and $i = R = s + 1/2$. The bases of Pauli's analysis are the Tables 1 to 4 which give the values of

mg in strong-fields as functions of m and l for doublets ($s = 1/2$), triplets ($s = 1$), quartets ($s = 3/2$), and quintets ($s = 2$), respectively. While Tables 1 and 2 agreed with known results, Tables 3 and 4, which are reproduced in his 'Quantentheorie' (Ref. [18], p. 235), were predictions. But these mg-values also are the result of implicitly applying the vector model. In the above letter to Sommerfeld, Pauli admits that 'I never would have arrived at the representation of the term values at strong-fields given there if I had not been guided by model considerations' (translated from Ref. [9], p. 105). The vector model is explicitly discussed in Pauli's 'Quantentheorie' [18] at the end of Ziff. 39.

However, in a very strong magnetic field $H \gg H_i$, the interaction term in Eq. (4.10) may be neglected, so that the magnetic moments $\vec{\mu}_l$ and $\vec{\mu}_s$ no longer precess around their resultant but separately around the applied field **H** (see Ref. [42b], p. 122). This means that the projection of l and s on the field direction are directional quantum numbers m_l and m_s (called m_1 and μ, respectively, in Ref. [41]) that vary in unit steps in the respective intervals (see Eq. (12) of Ref. [42b])

$$-l \leq m_l \leq l; \; -s \leq m_s \leq s. \tag{4.15}$$

Thus the projections of l + s = j and of l + 2s = gj on **H** yield Eqs (3) and (4) of Ref. [41], respectively, i.e. $m_l + m_s = m$ and $m_l + 2m_s = gm$, so that (see Eq. (16) of Ref. [42b])

$$g(m_l, m_s; l, s) = \frac{m_l + 2m_s}{m_l + m_s}. \tag{4.16}$$

The entries in Pauli's Tables 1 to 4 are now obtained as follows: Draw for each pair of values l and s the table (matrix) labelled by m_l and m_s satisfying (4.15) and put as entries the values $m_l + 2m_s$. Then reading the matrix parallel to its main diagonal yields the line $k = l + 1$ of Table $2s$ in Ref. [41]. Taking the sum of the values (4.16) over the intervals (4.15) at constant $m \leq |l - s|$ (see *Note* [43]), one recovers the result (4.13), so that

$$\sum_j g(j; l, s) = \sum_{m_l, m_s}^{m_l + m_s = m} g(m_l, m_s; l, s). \tag{4.17}$$

This is the remarkable permanence of the g-sums (see Eq. (5) in Ref. [42b]) which lies behind Heisenberg's 'principle' expressed by Eq. (10) of Ref. [10a]. It already shows the amazing intuition of the 20-year-old Heisenberg in his first publication.

Pauli's decision to devote his attention in his first paper [41] on the anomalous Zeeman effect to the strong-field case is now seen to be motivated by the fact that the projection on the field direction does not involve at all the ambiguity contained in the definition (4.14) of the length of the vectors l, s, and j, but only directional quantization. As he admits himself in the above letter to Sommerfeld, Pauli was of course fully aware of the vector model and discussed it in a (lost) letter to Landé. But his sense of rigour did not allow him to use it in a publication. Its use by Landé in Fig. 1 and Eqs (18′), (21′) of Ref. [42a] and in Section 5 of Ref. [42b] incited Pauli to write to Landé somewhat ironically: 'You have rendered me a very great service with the publication of my communication by letter to you, for I could hardly have tranquillized my conscience if I had published these unfinished (*unabgeschlossenen*) model considerations in a work of my own. On the other hand, I like the idea that now they are published anyhow without that I have to assume full responsibility for them' (translated from letter [46] from Pauli to Landé, Copenhagen, dated 23 September 1923, Ref. [9], p. 120, also mentioned on page 56). The truly remarkable idea Pauli had in his first paper [41] was to derive the zero-field expression (4.11) of the g-factor from the strong-field expression (4.16) by making use of Heisenberg's 'principle' of the permanence of the g-sums. Thereby he was able to avoid the 'piercing of classical mechanics' expressed by Eq. (4.14) in favour of the empirically supported invariance property (4.17) and thus to put Eq. (4.11) on more solid ground. Although Pauli did the calculation only numerically for the special case of Table 4 he claimed that 'it is possible to carry through the calculation also algebraically instead of numerically' (translated from Ref. [41], p. 164). That this claim is indeed true is shown in the rather technical *Note* [43].

The fact that Pauli did not bother to publish this algebraic derivation may be seen as another expression of his pessimistic state of mind. Even during his summer vacation at the seaside resort of Bansing on the Baltic he laments in the letter to Sommerfeld quoted above about his 'unfortunate (*unglückseliges*) work on the anomalous Zeeman effect' (namely, Ref. [41]). That this summer of 1923 had not been a very happy one for Pauli is most dramatically expressed in the letter that he addressed to Bohr from Copenhagen on 12 September: 'I felt very offended about your inciting Kramers to travel with you to England. For, I have looked forward for a long time to divert myself a little with him, which was very necessary for me after my recent scientific failures. . . . Although I cannot really consider all this as very considerate towards me, I still would not take it into consideration if I could convince myself that you really gained a significant benefit from Kramers' accompaniment. . . . But it offends me very much, and I perceive it as an unmerited injustice against me that you have pondered so little the benefit and the necessity of the decision to travel with Kramers—a decision that grieves me so much because the presence of my friend at this moment would psychically mean very much to me. I have always striven honestly to helping you as best I could, and in spite of it all it will be a great joy for me if I may do so again some time in the future. Now, however, I am so sad about the impending departure of Kramers and I feel so unwell that I must ask you to excuse me Saturday evening. I would only spoil everybody's evening.' And he closes by saying 'Don't think badly of your very sad Pauli who until now has so much honoured and loved you' (translated from letter [43], Ref. [9], pp. 111–12).

Bohr left with Kramers for England anyway, from where he sailed alone to the United States on 19 September; it was Bohr's first visit there (see Ref. [16b], pp. 44–5). Three weeks after writing the above letter to Bohr, on 4 October Pauli left Copenhagen to return to Hamburg. Earlier, on Monday 1 October, Hans M. Hansen (1886–1956) had reported to Bohr: 'Tonight Pauli talks to the Physical Society; on Thursday he is leaving, and I have the impression that he is now again quite happy so that his stay is ending in every respect in harmony' (translated from Ref. [9], note d, p. 112). Had Pauli been unhappy because he could not do better than that other prodigy, the one-year younger Heisenberg? I think *Note* [43] suggests that this hypothesis is unjustified. Indeed, I consider Pauli's idea described in *Note* [43] to be of an historical importance not generally appreciated. See also Ref. [38] where in the *Problem* on p. 390 it is stated that, for any values of m (including those considered in *Note* [43]), Eq. (4.17) 'is known as Pauli's *g-permanence rule*' Pauli in Ziff. 40 of his 'Quantentheorie' [18] calls it a sum rule (*Summenregel*) and gives credit to Heisenberg for it.

Nevertheless, Pauli returned to the problem shortly after returning to Hamburg, taking up the vector model—he calls it the '*Ersatzmodell*' (substitution model)—more resolutely. The problem he set himself was the association of the terms of the complex structure in strong and weak external fields [44] or, more specifically, 'the association of the values of m_r [$= m_s$] or m_k [$= m_l$] at strong-fields to the values of j [$= j + 1$] at weak fields at given r [$= s + 1$], k [$= l + 1$] and m' (translated from Ref. [44], p. 372). The specific goal was to derive from a very general form of the '*Ersatzmodell*' his assignments (I) or (II), namely Eqs (8) of Ref. [42b] which Landé had obtained with the help of his rather unconvincing minimum principle (13). Denoting the angular momenta (in units of $\hbar$) of the core and of the outer electron as r ($= s + 1$) and k ($= l + 1$), respectively, and introducing action variables p_r and p_k for the components along the field direction and their associated angle variables $2\pi\varphi_r$ and $2\pi\varphi_k$ (see page 65), Pauli wrote the three terms of Heisenberg's equation (4.10) above in the most general form as

$$H = \lambda[f_1(p_r) + f_2(p_k)] + f_3(\gamma). \tag{4.18}$$

Here λ measures the field strength and γ is the angle between the vectors r and k. In the strong-field case, $\lambda \gg 1$, where f_3 is negligible, it follows immediately from the canonical equations (3.11) applied to the Hamiltonian (4.18) that $\lambda f_1'(p_r) = d\varphi_r/dt = h\omega_r$ and $\lambda f_2'(p_k) = d\varphi_k/dt = h\omega_k$ (in Ref. [44] there is no λ in these formulae), p_r and p_k being constants of the motion. ω_r and ω_k are therefore the frequencies of the precessions around the field direction. In the case of weak fields, $\lambda \to 0$, Eq. (4.18) implies that $\gamma = $ const., so that the vectors r and k precess around their resultant j with frequency $\omega_j = \partial H/h\partial j$. Under the assumption that for any value of λ the precession patterns (the signs of $\omega_r - \omega_k$ and of ω_j) stay the same, a careful discussion of the closed path of the integral $\int p_k d\varphi$ where $\varphi = \varphi_k - \varphi_r$ then yields the assignments (I) or (II).

In Copenhagen Pauli had found some solace not so much by strolling in the beautiful streets as by thinking about problems other than the anomalous

Zeeman effect. One problem in particular caught his interest during this time: the thermal equilibrium between radiation and free electrons [45]. It was motivated by the tremendous boost that the photon idea advanced by Einstein in 1905 received in 1923 when Arthur Holly Compton (1892–1962) experimentally confirmed the relativistic kinematics for the scattering between a photon and a free electron at rest derived independently by himself and by Peter Debye [46a,b]. This experiment [46a], which was received with enormous interest by the international physics community won Compton half of the Nobel Prize of 1927 'for his discovery of the effect named after him'. On the other hand, in 1917 Einstein had derived Planck's spectral distribution law for radiation in equilibrium at absolute temperature T, the so-called black-body radiation,

$$\rho_\nu = \frac{\alpha \nu^3}{e^{h\nu/kT} - 1}, \qquad (4.19)$$

from simple assumptions about elementary (atomic or molecular) emission and absorption processes [47]. In Eq. (4.19) $\alpha = 8\pi h/c^3$ and k is Boltzmann's constant. According to Einstein an atom (molecule) spontaneously emits radiation of frequency ν satisfying Eq. (4.1) in a transition from a state i of higher excitation to a state f of lower excitation with a probability per unit time A_f^i. On the other hand, the presence of radiation gives rise to induced emission and absorption with probabilities per unit time $B_f^i \rho_\nu$ and $B_i^f \rho_\nu$, respectively. Assuming for the atomic density a Maxwell–Boltzmann distribution $n_i \propto \exp(-E_i/kT)$, equilibrium is assured by the condition that the number of processes per unit time is the same in both directions,

$$e^{-E_f/kT} B_i^f \rho_\nu = e^{-E_i/kT} (A_f^i + B_f^i \rho_\nu), \qquad (4.20)$$

from which Eq. (4.19) follows, provided that $A_f^i/B_f^i = \alpha \nu^3$. Pauli now observed that Einstein's derivation does not apply to the situation of the Compton effect because free electrons cannot have spontaneous emission. However, as in the emission/absorption case the equilibrium condition in the case of scattering again requires equality of the numbers per unit time of processes $i \to f$ and their inverse (time reversed), $f \to i$. Now the kinematics of the Compton effect are completely determined by energy and momentum conservation in the scattering, $U + h\nu = U_1 + h\nu_1$ and $G + \Gamma = G_1 + \Gamma_1$, respectively, where $|\Gamma| = h\nu/c$ is the value of the photon momentum [46a,b]. Scattering from initial values G and Γ within volume elements d^3G and $d^3\Gamma$ to final values G_1 and Γ_1 within d^3G_1 and $d^3\Gamma_1$, respectively, gives rise, in analogy with Eq. (4.20), to the equilibrium condition

$$F e^{-U/kT} d^3G d^3\Gamma = F_1 e^{-U_1/kT} d^3G_1 d^3\Gamma_1 \qquad (4.21)$$

where F must be chosen such that Eq. (4.21) satisfies relativistic invariance. Pauli arrived at the conclusion that in order to satisfy Eq. (4.21) one must choose

$$F = A\rho_\nu + B\rho_\nu \rho_{\nu_1} \qquad (4.22)$$

where, in analogy with Einstein's case, $A/B = \alpha\nu_1^3$, and he interpreted the second term in (4.22) as interaction between radiation bundles.

Shortly after Pauli's paper [45] Einstein and Ehrenfest published a different interpretation of Eq. (4.22) [48]. By writing $F = b\rho_\nu(a_1 + b_1\rho_{\nu_1})$ scattering may be understood as a composite process consisting of the absorption of a quantum ν followed by the emission of a quantum ν_1. Pauli has given a beautiful account of this entire subject in Ziff. 5 of his 'Quantentheorie' [18]. There he concludes: 'In order to maintain the connection between emission and absorption on the one hand and scattering on the other hand also in quantum theory it seems therefore natural in quantum theory to assume *always* scattering processes as consisting of two partial processes. ... Although in the case of free electrons there is no case of emission and absorption we will have to hold on to the decomposition of the scattering processes into two partial processes' (translated from Ref. [18], p. 28).

Notes and references

[1] Translated from: Max Born, letter of 29 November 1921 to Albert Einstein, in Ref. [2], p. 93 and comment, p. 95.

[2] Albert Einstein, Hedwig und Max Born, *Briefwechsel 1916–1955 kommentiert von Max Born* (Nymphenburger Verlagshandlung, Munich, 1969).

[3] Documents from the *Pauli Archive*: (a) Extract from military certificate, PLC Bi 50; (b) *Protokoll über das Examen rigorosum*, 25. Juli 1921 (handwritten; copy), PLC Bi 95; (c) Abgangszeugnis PLC Bi 56.

[4] A. Hermann, *Heisenberg* (Rowohlt Taschenbuch Verlag, Reinbeck bei Hamburg, 1976). The photograph is in the *Pauli Archive*.

[5] (a) J. Mehra and H. Rechenberg, *The Historical Development of Quantum Mechanics. Volume 1. The Quantum Theory of Planck, Einstein, Bohr and Sommerfeld: Its Foundation and the Rise of Its Difficulties, 1900–1925* (Springer, New York, 1982); (b) D. C. Cassidy, *Uncertainty. The Life and Science of Werner Heisenberg* (Freeman, New York, 1992); German translation by A. and G. Kleinert: *Werner Heisenberg: Leben und Werk* (Spektrum Akademischer Verlag, Heidelberg, 1995).

[6] (a) M. Born, *Vorlesungen über Atommechanik* (Julius Springer, Berlin, 1925), Erster Band; (b) M. Born, *My Life: Recollections of a Nobel Laureate* (Scribner's Sons, New York, 1978); German translation by H. Degner, with I. Krewinkel and W. Rau, *Mein Leben. Die Erinnerungen des Nobelpreisträgers* (Nymphenburger Verlagshandlung, Munich, 1975); (c) W. Moore, *Schrödinger. Life and Thought* (Cambridge University Press, Cambridge, 1989); (d) K. von Meyenn (ed., with the cooperation of A. Hermann and V. F. Weisskopf), *Wolfgang Pauli. Scientific Correspondence with Bohr, Einstein, Heisenberg, a.o., Volume II: 1930–1939* (Springer, Berlin and Heidelberg, 1985).

[7] W. Pauli jr., 'Über das Modell des Wasserstoffmolekülions', *Annalen der Physik* (4) **68**, 177 (1922). Reprinted in Ref. [8], Vol. 2, pp. 70–133.

[8] R. Kronig and V. F. Weisskopf (eds), *Collected Scientific Papers by Wolfgang Pauli. In Two Volumes* (Wiley Interscience, New York, 1964).

[9] A. Hermann, K. von Meyenn and V. F. Weisskopf (eds), *Wolfgang Pauli. Scientific Correspondence with Bohr, Einstein, Heisenberg, a.o., Vol. I: 1919–1929* (Springer, New York, 1979).

[10] (a) W. Heisenberg, 'Zur Quantentheorie der Linienstruktur und der anomalen Zeemaneffekte', *Zeitschrift für Physik* **8**, 273 (1922). Reprinted in: W. Blum, H.-P. Dürr, and H. Rechenberg (eds), *Werner Heisenberg Collected Works. Series A/Part I. Original Scientific Papers* (Springer, Berlin and Heidelberg, 1985), pp. 134–58; (b) A. Sommerfeld, *Atombau und Spektrallinien* (Vieweg, Braunschweig, Vierte umgearbeitete Auflage, 1924).

[11] (a) W. Gerlach und O. Stern, 'Der experimentale Nachweis der Richtungsquantelung im Magnetfeld', *Zeitschrift für Physik* **9**, 349 (1922); (b) A. Pais, *'Subtle is the Lord . . . '. The Science and the Life of Albert Einstein* (Oxford University Press, Oxford, 1982).

[12] M. Born und W. Pauli jr., 'Über die Quantelung gestörter mechanischer Systeme', *Zeitschrift für Physik* **10**, 137 (1922). Reprinted in Ref. [8], Vol. 2, pp. 48–69.

[13] (a) H. Poincaré, *Méthodes nouvelles de la mécanique céleste* (Gauthier-Villars, Paris, 1892–1899), Vols I–III; (b) W. Pauli, letter [1205] to Marie-Louise von Franz of 22 February 1951, in: K. von Meyenn (ed.), *Wolfgang Pauli. Scientific Correspondence with Bohr, Einstein, Heisenberg, a.o., Volume IV, Part I: 1950–1952* (Springer, Berlin and Heidelberg, 1996), p. 268; (c) W. Pauli jr., *Störungstheorie*, in: A. Berliner and K. Scheel (eds), *Physikalisches Handwörterbuch* (Julius Springer, Berlin, 1924), pp. 752–6. Reprinted in Ref. [8], Vol. 1, pp. 264–8.

[14] P. S. Epstein, 'Simultaneous action of an electric and a magnetic field on a hydrogen-like atom', *Physical Review* **22**, 202 (1923).

[15] N. Bohr, 'On the quantum theory of line spectra. Part I: On the general theory, Part II: On the hydrogen spectrum, Part III: On the spectra of elements of higher atomic number', *Videnskabernes Selskab Skrifter, 8. Raekke* **IV.1**, pp. 1 (1918), 37 (1918), 101 (1922). Reprinted in Ref. [16b], pp. 67–102, 103–66, 167–84.

[16] L. Rosenfeld (general ed.), *Niels Bohr, Collected Works*, (a) U. Hoyer (ed.), *Volume 2: Work on Atomic Physics (1912–1917)* (North-Holland, Amsterdam. New York. Oxford, 1981); (b) J. R. Nielsen (ed.), *Volume 3: The Correspondence Principle (1918–1923)* (North-Holland, Amsterdam, 1976); (c) J.R. Nielsen (ed.), *Volume 4: The Periodic System (1920–1923)* (North-Holland, Amsterdam, 1977).

[17] (a) O. Klein, 'Über die gleichzeitige Wirkung von gekreuzten homogenen elektrischen und magnetischen Feldern auf das Wasserstoffatom. I',

Zeitschrift für Physik **22**, 109 (1924); (b) W. Lenz, 'Über den Bewegungsverlauf und die Quantenzustände der gestörten Keplerbewegung', *Zeitschrift für Physik* **24**, 197 (1924).

[18] W. Pauli, 'Quantentheorie', in: H. Geiger and K. Scheel (eds), *Handbuch der Physik* (Springer, Berlin, 1926), Vol. 23, pp. 1–278. Reprinted in Ref. [8], Vol. 1, pp. 271–548.

[19] (a) W. Pauli jr., 'Über das Wasserstoffspektrum vom Standpunkt der neuen Quantenmechanik', *Zeitschrift für Physik* **36**, 336 (1926). Reprinted in Ref. [8], Vol. 2, pp. 252–79. (b) W. Pauli, 'Über die Intensitäten der im elektrischen Feld erscheinenden Kombinationslinien', *Kgl. Danske Videnskabernes Selskab, Math.-Fys. Meddelelser* **7**, 3 (1925). Reprinted in Ref. [8], pp. 233–50.

[20] M. Born, *Atomtheorie des festen Zustandes (Dynamik der Kristallgitter)*, *Encyklopädie der mathematischen Wissenschaften*, Vol. V/3, pp. 527–781 (1923).

[21] (a) H. A. Kramers, 'Über das Modell des Heliumatoms', *Zeitschrift für Physik* **13**, 312 (1923); (b) M. Born und W. Heisenberg, 'Die Elektronenbahnen im angeregten Heliumatom', *Zeitschrift für Physik* **16**, 229 (1923).

[22] (a) A. Sommerfeld, *Atombau und Spektrallinien. Wellenmechanischer Ergänzungsband* (Vieweg, Braunschweig, 3. umgearbeitete und erweiterte Auflage, 1951), Chapter 9; (b) W. Heitler, *Elementary Wave Mechanics with Application to Quantum Chemistry* (Clarendon Press, Oxford, 1956), 2nd edn.

[23] M. Dresden, *H. A. Kramers. Between Tradition and Revolution* (Springer, New York, 1987).

[24] 'Dedication' in: W. Pauli, 'Exclusion Principle, Lorentz Group and Reflection of Space-Time and Charge', in : W. Pauli (ed., with the cooperation of L. Rosenfeld and V. Weisskopf), *Niels Bohr and the Development of Physics* (Pergamon, London, 1955), pp. 30–51, here p. 30. Reprinted in Ref. [8], Vol. 2, pp. 1239–60.

[25] (a) N. Bohr, 'Foreword', in: M. Fierz and V. F. Weisskopf (eds), *Theoretical Physics in the Twentieth Century. A Memorial Volume to Wolfgang Pauli* (Wiley Interscience, New York, 1960); (b) K. von Meyenn (ed.), *Wolfgang Pauli. Scientific Correspondence with Bohr, Einstein, Heisenberg, a.o., Volume III: 1940–1949* (Springer, Berlin and Heidelberg, 1993); (c) S. Rozental (ed.), *Niels Bohr. His life and work as seen by his friends and colleagues* (North-Holland, Amsterdam, 1967, 1968).

[26] (a) L. Rosenfeld and E. Rüdinger, *The Decisive Years 1911–1918*, in Ref. [25c], pp. 38–73; (b) A. G. Debus (ed.), *World Who's Who in Science* (Marquis–Who's Who, Chicago, 1968), p. 1461; (c) L. Badash, 'Rutherford Ernest', in: C. C. Gillispie (ed.), *Dictionary of Scientific Biography* (Scribner's Sons, New York, 1975), Vol. XII, pp. 25–36.

[27] N. Bohr, (a) 'On the constitution of atoms and molecules', *Philosophical Magazine* **26**, 1 (Part I), 476 (Part II), 857 (Part III) (1913). Reprinted in Ref. [16a], pp. 159–233; (b) *On the constitution of atoms and molecules: Papers of 1913 reprinted from the Philosophical Magazine with an Introduction by L. Rosenfeld* (Benjamin, New York, 1963).

[28] (a) J. L. Heilbron and T. S. Kuhn, 'The genesis of the Bohr atom', *Historical Studies in the Physical Sciences* **1**, 211 (1969); (b) A. Pais, *Inward Bound. Of Matter and Forces in the Physical World* (Oxford University Press, Oxford, 1986), Chapter 9: 'Atomic Structure and Spectral Lines'; (c) A. Pais, *Niels Bohr's Times, In Physics, Philosophy, and Polity* (Oxford University Press, Oxford, 1991, 1993).

[29] (a) *Note*: The Wolfskehl Lectures were financed essentially by the interests of a bequest of 100 000 marks from the mathematician Paul Wolfskehl of Darmstadt to the Göttingen Academy of Sciences in 1906 for the *first* person who, during the next 100 years, would publish a *complete* proof of Fermat's last theorem (see Ref. [5a], pp. 278–80). This proof was eventually published in the *Annals of Mathematics* of 1995 but by two authors: Andrew Wiles and Richard Taylor, see e.g. D. Goldfeld, 'Beyond the Last Theorem', *The Sciences* **35**, 34 (1996); (b) R. Courant, *Fifty Years of Friendship*, in Ref. [25c], pp. 301–5.

[30] W. Heisenberg, *Der Teil und das Ganze. Gespräche im Umkreis der Atomphysik* (Piper, Munich, 1969); English translation by A. J. Pomerans, *Physics and Beyond* (Allen & Unwin, London, 1971).

[31] H.A. Kramers, 'Über den Einfluss eines elektrischen Feldes auf die Feinstruktur der Wasserstofflinien', *Zeitschrift für Physik* **3**, 199 (1920).

[32] A. Sommerfeld, 'Die Feinstruktur der Wasserstoff- und der Wasserstoffähnlichen Linien', *Sitzungsberichte der Mathematisch-physikalischen Klasse der Königlich Bayerischen Akademie der Wissenschaften zu München* (1915), p. 459; 'Zur Quantentheorie der Spektrallinien', *Annalen der Physik* **51**, 1, 125 (1916).

[33] (a) *Note*: In the fifth lecture Bohr came back to his discussions with Heisenberg, commenting on the choice of an angular momentum 1/2 for the core of a doublet atom (see page 78): 'It is difficult to justify Heisenberg's assumptions' (Ref. [16c], p. 391). (b) *Note*: This statement was brillantly confirmed a few months later by an experiment executed in Bohr's institute by Dirk Coster (1889–1950) and Georg von Hevesy (1885–1966) who called the new element *hafnium* in honour of Copenhagen (*Hafniae* in Latin), see Ref. [5a], Section III.5 and Ref. [28c], Section 10e. Hevesy received the Nobel Prize in Chemistry of 1943 'for his work on the use of isotopes as tracers in the study of chemical processes'.

[34] W. Pauli, 'Remarks on the history of the exclusion principle', *Science* **103**, 213 (1946). Reprinted in Ref. [8], Vol. 2, pp. 1073–5.

[35] (a) O. Klein, 'Wolfgang Pauli. Mâgra Minnesord', *Kosmos. Fysika Uppsatser* **37**, 9–12 (1959); German translation: Oskar Klein, *Wolfgang Pauli. Einige Worte zu seinem Gedächtnis*, in: C. P. Enz and K. von Meyenn (eds), *Wolfgang Pauli. Das Gewissen der Physik* (Vieweg, Braunschweig, 1988), pp. 49–52; see also Ref. [9], p. 370; (b) S. Richter, *Wolfgang Pauli. Die Jahre 1918–1930. Skizzen zu einer wissenschaftlichen Biographie* (Verlag Sauerländer, Aarau, 1979); (c) V. L. Telegdi, *Pauli-Anekdoten*, as under (a), pp. 115–20; (d) H. B. G. Casimir, *Haphazard Reality. Half a Century of Science* (Harper & Row, New York, 1983).

[36] Translated from: W. Pauli, letter [37] of 6 June 1923 to A. Sommerfeld, Ref. [9], p. 97.

[37] A. Landé, 'Über den anomalen Zeemaneffekt', (a) (Teil I), *Zeitschrift für Physik* **5**, 231 (1921); (b) (Teil II), *Zeitschrift für Physik* **7**, 398 (1921).

[38] E. U. Condon and G. H. Shortley, *The Theory of Atomic Spectra* (Cambridge University Press, Cambridge, 1935–63).

[39] (a) A. Rydberg, 'On triplets with constant differences in the line spectrum of copper', *Astrophysical Journal* **6**, 239 (1897); (b) A. Sommerfeld, 'Über die Deutung verwickelter Spektren (Mangan, Chrom usw.) nach der Methode der inneren Quantenzahlen', *Annalen der Physik* **70**, 32 (1923).

[40] C. P. Enz (ed.), *Pauli Lectures on Physics. Volume 1. Electrodynamics*, translated by H. R. Lewis and S. Margulies (MIT Press, Cambridge, MA, 1973).

[41] W. Pauli jr., 'Über die Gesetzmässigkeiten des anomalen Zeemaneffektes', *Zeitschrift für Physik* **16**, 155 (1923). Reprinted in Ref. [8], Vol. 2, pp. 151–60.

[42] A. Landé, 'Termstruktur und Zeemaneffekt der Multipletts', (a) *Zeitschrift für Physik* **15**, 189 (1923); (b) (*Zweite Mitteilung*), *Zeitschrift für Physik* **19**, 112 (1923).

[43] *Note:* Since $j \geq |m|$ the left-hand side of Eq. (4.17) is modified as compared with Eq. (4.13) if $m > |l - s|$. Consider now $m > l - s \geq 0$ and define

$$x_i \equiv g(j; l, s) \qquad \text{(A)}$$

with $1 \leq i \equiv l + s + 1 - j \leq n$ and $n \equiv l + s + 1 - m < 2s + 1$ where the inequalities follow, respectively, from $l + s \geq j \geq m > l - s$. The right-hand side of Eq. (4.17) may be written

$$a_n \equiv \sum_{r=0}^{n-1} \left(1 + \frac{m_s}{m}\right) \qquad \text{(B)}$$

with $0 \leq r \equiv s - m_s \leq n - 1$ where the inequalities follow, respectively, from $m_s \leq s$ and $m - m_s \leq l$. Because $n - 1 < 2s$, the upper limit of the sum in (B) is also a restriction compared with the limits (4.15), namely

$m_s > -s$. Assuming that the equality (4.17) still holds for $m > l - s$ one obtains the system of equations

$$\sum_{i=1}^{n} x_i = a_n; \ n = 1, 2, \ldots, 2s + 1 \tag{C}$$

whose solution is $x_n = a_n - a_{n-1}$ with $a_0 \equiv 0$, or

$$g(j; l, s) = a_i - a_{i-1}. \tag{D}$$

From (B) one obtains with the above definitions

$$
\begin{aligned}
a_i &= i \left(1 + \frac{s}{j} \right) - \frac{(i-1)i}{2j}; \\
a_{i-1} &= (i-1) \left(1 + \frac{s}{j+1} \right) - \frac{(i-2)(i-1)}{2(j+1)}.
\end{aligned}
\tag{E}
$$

Substitution of (E) into (D) readily yields the expression (4.11). A similar calculation holds for $l - s \leq 0$.

[44] W. Pauli jr., 'Zur Frage der Zuordnung der Komplexstrukturterme in starken und in schwachen Feldern', *Zeitschrift für Physik* **20**, 371 (1924). Reprinted in Ref. [8], Vol. 2, pp. 176–92.

[45] W. Pauli jr., 'Über das thermische Gleichgewicht zwischen Strahlung und freien Elektronen', *Zeitschrift für Physik* **18**, 227 (1923). Reprinted in Ref. [8], Vol. 2, pp. 161–75.

[46] (a) A. H. Compton, 'A Quantum Theory of the Scattering of X-Rays by Light Elements', *Physical Review* **21**, 483 (1923); (b) P. Debye, 'Zerstreuung von Röntgenstrahlen und Quantentheorie', *Physikalische Zeitschrift* **24**, 161 (1923).

[47] A. Einstein, 'Zur Quantentheorie der Strahlung', *Physikalische Zeitschrift* **18**, 121 (1917).

[48] A. Einstein und P. Ehrenfest, 'Zur Quantentheorie des Strahlungsgleichgewichts', *Zeitschrift für Physik* **19**, 301 (1923).

5
The Hamburg years

The curious history of spin

I did hit upon Kepler as trinitarian and Fludd as quaternarian—and with their polemic, I felt an inner conflict resonate with myself. I have certain features of both. . . . By the way, I wish to remark that once (in Hamburg) my path to the Exclusion Principle had to do precisely with the difficult transition from 3 to 4: namely with the necessity to attribute to the electron a fourth *degree of freedom (which soon after was explained as 'spin') instead of the* three *translations. To convince myself that, contrary to the naive 'conception' (Anschauung) the fourth quantum number also is a property of* one *and the same electron (on a par with the well-known* three *quantum numbers now designated as* n_r, l, m_l*)—that was really the main work. [I had to wrestle so much with the then accepted theories which attributed the fourth quantum number to the* atomic core *(Rumpf)]. [1]*

The anomalous Zeeman effect continued to bother Pauli in Hamburg. But then he had a revolutionary idea: what if the cause of all the trouble with the complex structure did not reside in the atomic core at all but in the valence electron itself? And what evidence was there for such a possibility? That the second question was decidable by looking for a relativistic velocity dependence of the mass of the core electrons was Pauli's clever idea which he discussed in his next paper [3a]. This paper is quite elementary, so much so that Pauli hesitated to publish it. Indeed, on 10 November 1924 he wrote to Landé: 'I would like to ask your advice about a funny (*komische*) reflection concerning the Zeeman effect.' And he closes with the question: 'Do you think that this reflection is sufficiently interesting to warrant publication?' (translated from letter [68], Ref. [5], pp. 169 and 172). Four days later he sent Landé a postcard saying: 'Since you seem to be interested by this matter I think I will publish it anyway' (translated from Ref. [5], p. 173). This paper [3a] was submitted on 2 December 1924, but in the cited letter Pauli did not mention his revolutionary thesis. It reads:

The closed electron configurations will contribute nothing to the magnetic moment and to the angular momentum of the atom. In particular with the alkalis the values of the [angular] momenta of the atom and its energy change in an external magnetic field are considered as being essentially the sole effect of the valence electron, which is also considered to be the seat of the magneto-mechanical anomaly. According to this point of view the doublet structure of the alkali spectra, as also the piercing (Durchbrechung) of the Larmor theorem, comes about by a peculiar, classically not describable kind of two-valuedness (Zweideutigkeit) of the quantum theoretical properties of the valence electron. (Translated from Ref. [3a], p. 385)

Here the translation of '*Zweideutigkeit*' as 'two-valuedness' (*Zweiwertigkeit*), and not as 'ambiguity', is that of Pauli's Nobel Lecture (Ref. [6a], p. 133). In the celebrated companion paper [3b] containing the announcement of the exclusion principle (see the following section), which was submitted on 16 January 1925, Pauli writes: 'Apparently, the occurrence of half (effective) quantum numbers and the value $g = 2$ of the splitting factor which thereby is formally implied with the s-term of the alkalis, is most closely related with the two-foldedness (*Zweifachheit*) of the term levels' (translated from Ref. [3b], p. 767). The content of the above thesis is more spontaneously expressed in the letter of 24 November 1924 to Landé containing Pauli's new ideas. At the beginning of the letter Pauli begs Landé not to mention in any of his papers any of the ideas expressed in the letter and, alluding to the instance when Landé had published some of Pauli's ideas (see letter [46] quoted on page 97), Pauli adds in a footnote: 'Once you indeed much helped me in this way but, please, not this time!' But then Pauli writes: 'With the alkalis the valence electron alone makes the complex structure as well as the anomalous Zeeman effect. There is no question of a cooperation of the inert-gas core (also with the other elements). In a mysterious unmechanical manner the valence electron succeeds in running in two states (with the same $k[= l + 1]$) with different [angular] momentum.' And later in the letter Pauli writes: 'Now we consider again in particular the alkalis, and this time the case of strong fields (Paschen–Back effect). Here we must associate with the valence electron, in addition to the three mentioned quantum numbers $[n, l, j = |l \pm 1/2|]$, the [angular] momentum quantum number m $[= m_l \pm 1/2]$' (translated from letter [71], Ref. [5], pp. 176, 177).

The notion of '*Komplexstruktur*' and '*Zweideutigkeit*' may be traced back to an important review article by Bohr in the previous year [7]. In this article Bohr, after observing that the necessity of introducing half-integer quantum numbers has rendered the use of the theory of multiply periodic systems questionable (p. 253), states in his well-known cautious style: 'We rather are led to the view that, because of the mechanically indescribable stability properties of the atoms, the coupling of the series [valence] electron to the atomic core (*Atomrest*) carries along a *force* (*Zwang*) which is not analogous to the action of external fields of force, and which implies that the atomic core, instead of having one single possibility of alignment in a constant external field, is forced upon (*aufgezwungen*) two different alignments while, on the other hand, because of the same force (*Zwang*) the outer electron disposes in the atomic compound of only $2k - 1$ possibilities of alignment, instead of the $2k$ possibilities of alignment in an external field' (translated from Ref. [7], p. 276). More than anything else this 'unmechanical *Zwang*' signalled the changes to come (see

Ref. [8]).

But let us now look at this paper [3a] in more detail. Pauli first calculates the ratio of the magnetic moment $\vec{\mu}$ to the angular momentum **L** for an orbiting electron (he writes German M and J, respectively, for these two vectors). This ratio, which according to Eq. (4.8) defines Landé's g-factor, was calculated in Eq. (3.1). However, the electron mass which appears there is now velocity dependent, $m = m_0/\sqrt{1 - v^2/c^2}$, so that

$$\frac{|\vec{\mu}|}{|\mathbf{L}|} = \gamma \frac{e}{2m_0 c} \tag{5.1}$$

where $\gamma = \overline{\sqrt{1 - v^2/c^2}}$ is the g-factor of classical relativistic mechanics and the bar means a time average which is needed for non-circular orbits. In order to obtain γ from experiment the velocity v must be expressed in terms of the total energy $E = E_{kin} + E_{pot}$, where now $E_{kin} = (m - m_0)c^2$ and, for Coulomb forces according to Eq. (3.10), $\overline{E_{pot}} = -\overline{mv^2}$; hence $\gamma = 1 + E/m_0 c^2$. Using Sommerfeld's celebrated formula for the relativistic fine structure of the term formula (4.6), generalized to arbitrary nuclear charge Z [9], which Pauli derives in Ziff. 21 of his 'Quantentheorie' [10a,b], the following expression is obtained:

$$\begin{aligned} \gamma = 1 + \frac{E_{n,k}}{m_0 c^2} &= \left\{ 1 + \frac{\alpha^2 Z^2}{(n-k+\sqrt{k^2 - \alpha^2 Z^2})^2} \right\}^{-1/2} \\ &= \quad 1 - \frac{\alpha^2 Z^2}{2n^2} - \frac{\alpha^4 Z^4}{2n^3}\left(\frac{1}{k} - \frac{3}{4n}\right) + \cdots . \end{aligned} \tag{5.2}$$

Here the last expression is a development in powers of $\alpha^2 Z^2$, and $\alpha = e^2/\hbar c \simeq 1/137$ is the famous fine structure constant of Sommerfeld [9].

For estimates of γ the second term in the development of formula (5.2) is sufficient, and it is only for large Z-values that γ becomes appreciably different from one. In addition, the core electrons are the most tightly bound and therefore have lowest n-values. Now it had been Bohr's main endeavour to explain the periodic table of the chemical elements in terms of his atomic model by adding the electrons one by one and simultaneously increasing the atomic number Z, so as to maintain the atoms electrically neutral. This problem, which Bohr called the building-up principle (*Aufbauprinzip*), had been the main theme of his Wolfskehl Lectures in 1922 (see page 86). The experimental information about this shell structure of the electrons in the atoms came mainly from X-ray spectra which are produced according to Eq. (4.1) by some outer electron falling into an inner orbit, from which another electron had previously been knocked out by absorbing an X-ray photon, or also by collision. Thus it was possible to classify the $K-, L-, M-, \ldots$ shells in terms of the principal quantum number $n = 1, 2, 3, \ldots$, respectively, each shell closing with a chemically inert element, helium, neon, argon, $\ldots$, and accommodating $2n^2$ electrons. This last number was, however, purely empirical and was gained from chemical and spectroscopic evidence.

In his paper [3a] Pauli now argued as follows. The building-up principle says that once a shell is complete it stays the same for all elements with higher

Z, and 'If it is at all possible to associate a non-vanishing moment to inner closed shells one consequently arrives at assigning it to the K-shell, as soon as one assumes the constitution of the given shell to be of the same kind for all elements; since indeed, e.g. with Li, this shell is alone present in the atomic core' (translated from Ref. [3a], p. 379). Therefore we may put $n = k = 1$, so that Eq. (5.2) yields $\gamma \simeq 1 - \alpha^2 Z^{*2}/2$ where the effective atomic number Z^* results from the shielding effect on one of the two K-electrons, i.e. $Z^* = Z - 1$. In this way Pauli estimated, for Ba ($Z = 56$), Hg ($Z = 80$), and Tl ($Z = 81$), γ-values of $0.924, 0.817$, and 0.812, respectively, while for the valence electron $\gamma \simeq 1$ since $Z^* \ll Z$. Because of the lack of a theory it is not evident how the anomalous g-factor of the core is affected by γ; calling the former κ it may be considered either as simple doubling, $\kappa = 2\gamma$, or as addition to the normal value one, $\kappa = 1 + \gamma$. When applied to the anomalous Zeeman effect this results in a modification of Eqs (4.16) and (4.11), and Pauli concludes that the Zeeman splittings following from these modified formulae, 'which contain a systematic dependence on the atomic number, are not in accord with the observations, however' (translated from Ref. [3a], p. 383).

An even more important reason against the core interpretation of the anomalous Zeeman splittings is given in letter [71] from Pauli to Landé quoted above: 'The most important argument seems to me your (and Millikan's) finding (*Befund*) [Refs [11a,b,c]] that the energy difference of the doublets in the alkalis may be represented by a relativistic formula. For, it [the finding] shows that it is just not possible to interpret this energy difference as interaction energy with the atomic core (in the different orientations). This speaks quite generally against a cooperation of the atomic core (*Atomrest*) in the complex structure and Zeeman effect of the alkalis' (translated from Ref. [5], p. 177, see also Ref. [3a], p. 384). Now the energy difference of the relativistic doublet is defined as $h\Delta\nu_{rel} = E_{n,k} - E_{n,k-1}$ where $E_{n,k}$ is the fine structure formula (5.2) whose development in powers of $\alpha^2 Z^2$ yields

$$h\Delta\nu_{rel} = R_\infty \frac{\alpha^2 Z^4}{n^3 k(k-1)} \tag{5.3}$$

where $R_\infty = \alpha^2 mc^2/2$ is the Rydberg constant (see page 64). This is Eq. (28) on p. 205 of Pauli's 'Quantentheorie' [10a] where, however, k, called k_1, is replaced by $k_2 = j + 1/2 = k$ or $k - 1$, and Z is modified by a shielding number d_{n,k_1,k_2}.

On the other hand, 'the interaction energy with the atomic core' is given by the third term in Eq. (4.10). The field H_i produced by the valence electron at the core is determined by Biot–Savart's law (see e.g. Ref. [12], p. 71) which, for a circular orbit of radius r and velocity $v = \omega r$, is $H_i = (J/c)2\pi r/r^2$, and the current is $J = e\omega/2\pi$, as on page 57 where Eq. (3.3) may be used to write $H_i = 2\mu_B k \overline{r^{-3}}$. Here $\mu_B = e\hbar/2mc$ is the Bohr magneton and the bar means, in the case of an ellipse, $k < n$, averaging over the orbit, which yields $\overline{r^{-3}} = b^{-3}$ where $b = a_B nk/Z$ is the smaller semi-axis and $a_B = \hbar/\alpha mc$ the Bohr radius (see page 64). With these values one finds $H_i = 2R_\infty em\alpha Z^3/\hbar^2 n^3 k^2$.

The difference between the core magnetic moments in the two positions of the doublet may be written $\Delta\mu_{core} = \mu_B\Delta m$ with Δm the quantum number difference. Hence the energy difference due to the magnetic interaction with the atomic core is (see Ref. [11b], p. 95)

$$h\Delta\nu_{magn} = \Delta\mu_{core}H_i = R_\infty\frac{\alpha^2 Z^3 \Delta m}{n^3 k^2}. \qquad (5.4)$$

Compared with Eq. (5.3) this expression is seen to have the wrong Z-dependence 'which at high nuclear charges is much too small' (translated from Ref. [10a], p. 215).

As in previous years Pauli spent the Christmas vacation of 1924 with his family in Vienna. On 25 December he informed Landé by postcard of his one-day visit to Tübingen on 9 January. He wished to discuss his exclusion principle and check with the data from Paschen's institute (Ref. [5], p. 196). It so happened that the young Ralph Kronig, who had just got his Ph.D. at Columbia University in New York, was also visiting Landé who let him read Pauli's letter containing the new ideas. Much later Kronig wrote up his recollections of his encounter with Pauli and his new ideas in an essay with the appropriate title 'The Turning Point' [13]. He writes: 'Pauli's letter made a great impression on me and naturally my curiosity was aroused as to the meaning of the fact that *each individual* electron of the atom was to be described in terms of quantum numbers familiar from the spectra of the alkali atoms, in particular the two angular momenta l and $s = 1/2$ encountered there. Evidently s could now no longer be attributed to a core, and it occured to me immediately that it might be considered as an intrinsic angular momentum of the electron. In the language of the models which before the advent of quantum mechanics were the only basis for discussion one had, this could only be pictured as due to a rotation of the electron about its axis. Such a picture, it is true, was subject to a number of serious difficulties. . . . Yet it was a fascinating idea and the same afternoon, still quite under the influence of the letter I had read, I succeeded in deriving with it the so-called relativistic doublet formula' (Ref. [13], pp. 19–20).

Kronig then describes the idea that led him to this result: 'Now if, in conjunction with its intrinsic angular momentum $s = 1/2$, an intrinsic magnetic moment of one Bohr magneton was ascribed to the electron, then it would not only suffer an electrostatic attraction in its motion around the (screened) nucleus, but also a magnetic coupling. Indeed, in passing from a reference system in which the nucleus is at rest to one moving with the velocity $\mathbf{v}$ of the electron, a Lorentz transformation applied to the radial electric field $\mathbf{E}$ of the (screened) nucleus gives rise to a magnetic field

$$\mathbf{H} = \frac{\mathbf{E} \times \mathbf{v}}{(c^2 - v^2)^{1/2}} \dots' \qquad (5.5)$$

Inserting in this formula, which is contained in Eqs (204) in Section 28 of Pauli's *Encyklopädie Article* [15], the electric field $\mathbf{E} = Zer/r^3$ one finds, with $v^2/c^2 \simeq 0$, $\mathbf{H} = (Ze\mathbf{L}/mc)\overline{r^{-3}}$. Using the above value for the average of r^{-3} and $\Delta\mu_{el} = 2\mu_B$ for the difference between the electron's magnetic moment $\mu_{el} = \pm 1\mu_B$ in its up and down positions, one obtains for the spin–orbit splitting

$$h\Delta\nu_{spin} = \Delta\mu_{el}H = 2R_\infty \frac{\alpha^2 Z^4}{n^3 k^2}. \qquad (5.6)$$

This result, which according to the above account was deduced by Kronig in January 1925, differs by a factor $2(k-1)/k$ from the relativistic splitting (5.3). Here the factor 2 soon acquired celebrity in an enormous confusion and controversy.

Kronig continues his recollections as follows: 'On the next day we went to fetch Pauli from the station. For some reason I had imagined him as being much older and as having a beard. He looked quite different from what I had expected, but I felt immediately the field of force emanating from his personality, an effect fascinating and disquieting at the same time. At Landé's institute a discussion was soon started, and I also had occasion to put forward my ideas. Pauli remarked: "*Das ist ja ein ganz witziger Einfall*" [this is indeed quite a witty idea], but did not believe that the suggestion had any connection with reality' (Ref. [13], p. 21). After this encounter with Pauli, Kronig went on to Göttingen, Berlin, and Copenhagen where Bohr, Kramers, and Heisenberg also talked him out of the spin idea. Thus Kronig, who later had his own objections, did not publish anything and returned to his intensity calculations for the Zeeman spectra in which he was an expert. Pauli the great sceptic, on the other hand, did not know what to do with his new ideas. On 21 May 1925 he wrote to Kronig: 'At the moment physics is again very much in disarray (*verfahren*), for me anyhow it is much too difficult and I wish I were a movies comedian (*Filmkomiker*) or something like that and had never heard anything of physics!" (translated from letter [89], Ref. [5], p. 216). At that time Charlie Chaplin films were very popular both with Pauli and with Heisenberg (see Ref. [14c], p. 196).

Born in Dresden, Germany, on 10 March 1904 Ralph de Laer Kronig received his training as a physicist at Columbia University in New York where, returning from his trip to Europe in December 1925, he became an enthusiastic proponent of the new quantum mechanics and where in 1927 he became an assistant professor. But then Pauli, who in spite of his negative reaction had an excellent opinion of Kronig, invited him to be his first assistant when he assumed the chair at ETH in Zurich on 1 April 1928. Kronig soon developed both a deep friendship with Pauli and a liking for Zurich. His impressive career, however, was made in the Netherlands. After a short stay in England as a lecturer in Cambridge and then as assistant professor at Imperial College, London, he moved to Holland, first as a lecturer at the University of Groningen and, from 1939 until his retirement in 1969, as professor, and 1959–62 as rector, of the Technical University of Delft.

For Kronig's appointment in Groningen Pauli had written a favourable letter of recommendation to Dirk Coster (letter [251] in Ref. [16a]). Before settling at Delft, however, Kronig had the frustrating experience of being passed over because he was not Dutch when Uhlenbeck was nominated at Utrecht in 1935. He complained to Pauli who conceded that Kronig was a better physicist but that he should rather consider going back to the United States. Pauli also took this opportunity to write: 'it gives me pleasure that, in spite of my former remark uttered to you in Tübingen about the spin question, you still consider me worthy to receive a letter' (translated from letter [406] of 5 April 1935, Ref. [16a], p. 384). But for his retirement Kronig returned to Switzerland where he lived with his wife in Erlenbach on lake Zurich and where he celebrated his 80th birthday. Then he returned to Holland where he died on 16 November 1995 [16b,c]. Kronig's name is related to many achievements in physics, in particular with the dispersion relations named after him and Kramers.

Unexpectedly and independently, the idea of the electron spin surfaced again nine months after Kronig's first meeting with Pauli when two young Dutchmen who were students of Ehrenfest, George E. Uhlenbeck (1900–88) and Samuel A. Goudsmit (1902–78), published on 17 October 1925 a one-page note 'Replacement of the hypothesis of the unmechanical *Zwang* by a requirement concerning the inner behaviour of each electron' [17], in which they proposed an alternative to Landé's vector model: 'It is natural now to associate to the electron with its 4 quantum numbers also 4 degrees of freedom. One then may give, e.g., the following meaning to the quantum numbers: n and k remain as before the principal and the azimuthal quantum numbers of the electron in its orbit. To R, however, one should associate a proper rotation of the electron.' And in footnote 2 it is emphasized that 'R [$= s + 1/2$] therefore has for each electron [of the alkali spectra] only the value 1'. Furthermore: 'The ratio of the magnetic moment of the electron to the mechanical one must, for the proper rotation, be double that for the orbital motion.' The note closes with the remark: 'The different orientations of R with respect to the orbital plane of the electron [$m_s = \pm 1/2$] should be able to provide an explanantion for the relativity doublet' (translated from Ref. [17]).

The last sentence shows that Uhlenbeck and Goudsmit had not done the calculation of Eq. (5.6). In his recollections Kronig writes: 'The type of coupling which was sketched above in connection with my experiences in Tübingen came to their [Uhlenbeck and Goudsmit's] notice through Heisenberg' (Ref. [13], p. 27). And on 26 March 1926 Bohr wrote to Kronig: 'Now I even suspect that it is from yourself that the understanding of the mutual coupling between spin and orbital motion has come to the notice of the physicists. In fact, when on my return from Leiden I discussed the problems in Göttingen with Heisenberg, from whom the Leiden people had learnt the cause of the coupling, he told me that he had heard it himself a year before from somebody, but he could, unfortunately, in no way remember who it had been' (Ref. [13], p. 27). Heisenberg did derive Eq. (5.6) in his letter to Pauli of 24 November 1925 (letter [108], Ref. [5], p. 65) while the result already figures on his postcard of 21 November where the following comment may be found: 'That in the denominator [of Eq. (5.6)] stands k^2 instead of $k(k-1)$ would bother nobody; but I cannot get rid of the factor two' (translated from Ref. [5], p. 262). Pauli, on the other hand, became more and more desperate. 'One half, one half, a kingdom for the factor 1/2!' he exclamed in his letter to Heisenberg of 31 January 1926 (translated from letter [118], Ref. [5], p. 284).

Bohr had been invited by Ehrenfest to participate in the celebration of the 50th anniversary of Lorentz's doctorate on 11 December 1925. In the quoted letter to Kronig, Bohr writes about this visit to Leiden: 'Einstein asked the very first moment I saw him what I believed about the spinning electron. Upon my question about the cause of the necessary mutual coupling between the spin axis and the orbital motion, he explained that this coupling was an immediate consequence of the theory of relativity. This remark acted as a complete revelation to me, and I have never since faltered in my conviction that we at last were at the end of our sorrows' (Introduction to Part II, in Ref. [18], p. 229). On his way back Bohr was met at the Hamburg railway station by Pauli and Otto Stern who both warned him against accepting the spin hypothesis. Bohr then tried to convince Heisenberg and Jordan in Göttingen and also Pauli whom he met again in Berlin (Ref. [18], p. 229). On 24 December 1925 Heisenberg wrote to Pauli: 'Me too, I have been so influenced by Bohr's optimism with respect to Goudsmit's theory that I would much like to believe in the magnetic electron' (translated from letter [112], Ref. [5], p. 271). Pauli, however, remained steadfast in his opposition against the electron spin model.

There were now two factors of 2 to be explained. Concerning the value 2 of the electron's g-factor, Uhlenbeck writes in his 'Personal Reminiscences': 'I remember that he [Goudsmit] wrote me a postcard from Amsterdam [where he was a part-time "house theoretician" in Zeeman's laboratory] ... in which he asked whether I was sure that the gyromagnetic ratio had to be $e/2mc$ classically—perhaps it was different for the rotation of an extended charged body. I showed this postcard to Ehrenfest, who then recalled a paper by Max Abraham. ... I studied this paper very hard and found there to my great satisfaction that if the electron has only surface charge the gyromagnetic ratio was $2e/2mc$ just as we had postulated!' (Ref. [19], pp. 46–7). The purpose of Abraham's paper [20a] had been to show that it is possible to understand the electron's mass as an electromagnetic self-energy. He indeed showed that a surface charge on a sphere gives rise to an electromagnetic moment of inertia half as large as that produced by a material surface mass on the same sphere (Ref. [20a], Eq. (25h), p. 171). However, as with material mass, arbitrary deviations of the gyromagnetic ratio from the classical value $e/2mc$—which is a consequence of Larmor's theorem—may be obtained when the distributions of charge and mass are not identical (see *Note* [20b]).

So the value $g = 2$ is not really a problem since it may be realized, as observed by Uhlenbeck and Goudsmit, in a model of rotating mass and charge distributions for the *free* electron. The real problem is that for an extended *mass* distribution of radius a it follows from the value $\hbar/2$ of the spin either that, for a reasonable radius a of the order of the classical electron radius, $a = \alpha\hbar/mc$, a peripheral velocity more than 200 times larger than the speed of light c results or that, for $v/c < 1$, the electron's radius a becomes larger than $1/100$ of

a_B (see *Note* [20b]). Pauli, in the *Nachtrag* (Appendix) written in 1928 to his contribution 'General Foundations of the Theory of the Atomic Structure', completed at the end of 1924, indeed in Müller-Pouillet's treatise [21] offers the following comment concerning the 'Magnetelektron': 'In a manner which considers more the kinematical circumstances one also speaks of a "rotating electron" [English "spin-electron"]. However, we consider the notion of a rotating material entity unimportant and it is also not advisable because of the larger-than-light velocities that one then has to take into account. The designation "Magnetelektron" shall direct the emphasis to the electromagnetic field of the electron' (Ref. [21], footnote 2, pp. 1794–5, quoted in Ref. [22a], p. 771).

The second factor of 2 was elucidated by a then still unknown young theoretical physicist from Cambridge who spent the year 1925/6 in Bohr's group, namely Llewellyn H. Thomas (1903–1992). Having completed his graduate work on the interaction between charged particles and matter with Ralph Howard Fowler (1889–1944) [22b], he was familiar with relativistic kinematics. Thomas's idea was [24a] that in the rest frame of the electron the atomic core not only had a velocity v but also special relativity gives rise to an additional rotation with angular velocity $\delta\vec{\omega} = \mathbf{v} \times \dot{\mathbf{v}}/2c^2$ which contributes to the equation of motion of the magnetic moment as follows:

$$\frac{d}{dt}\vec{\mu} = \frac{e}{mc}\vec{\mu} \times \mathbf{H} - \delta\vec{\omega} \times \vec{\mu}. \qquad (5.7)$$

But since $m\dot{\mathbf{v}} = e\mathbf{E}$ it follows for the magnetic field of Eq. (5.5) that

$$\frac{d}{dt}\vec{\mu} = \frac{e}{2mc}\vec{\mu} \times \mathbf{H}. \qquad (5.8)$$

The angular velocity $\delta\vec{\omega}$ is obtained by two successive Lorentz transformations (see e.g. Eqs (1a) in Section 4 of Pauli's *Encyklopädie Article* [15]), the first by $-\mathbf{v}$ and the second by $\mathbf{v} + \dot{\mathbf{v}}dt$. The result for $v \ll c$ is

$$\mathbf{r}' = \mathbf{r} - (\delta\vec{\omega} \times \mathbf{r})dt - (\dot{\mathbf{v}} + \delta\mathbf{a})tdt \qquad (5.9)$$

with the additional acceleration $\delta\mathbf{a} = (\mathbf{v} \cdot \dot{\mathbf{v}}/2c^2)\mathbf{v}$ (for arbitrary values of v see Eqs. (2.2) of Ref. [24b]). In other words, the dynamically relevant magnetic moment of the electron to be used in Eq. (5.6) is $\mu_{el} = \pm(1/2)\mu_B$.

Not surprisingly, Pauli was not satisfied with this argument. He was first informed of Thomas's calculation in a letter from Bohr of 20 February 1926 where Bohr writes: 'even your being alarmed about my frivolous enthusiasm for the "new heresy" (*Irrlehre*) didn't discourage me.' And he continues, not without irony: 'Now comes the surprise for the learned theoreticians of relativity and responsibility-carrying scientists. Indeed, a young Englishman, Thomas, who during the last half year was in Copenhagen, has found out these days that the whole question of the unlucky factor 2 probably is due exclusively to a mistake in the calculation of the relative motion of the electron and the nucleus' (translated from letter [121], Ref. [5], p. 295). Having got Thomas's manuscript [24a] from Bohr, Pauli replied on 26 February. His rejection was complete. First, he said, with justification, 'Thomas's considerations do not contain that *energy* [Eq. (6)] at all but refer only to the *angular momentum* of the electron'. And, secondly, the energy of Eq. (5.6) refers to the coordinates in which the nucleus is at rest—which is also true. Thus Pauli concludes: 'In any case I consider the publication of the present note by Thomas in "Nature" to be a blunder (*Missgriff*) and would be glad if you could block it or else see to it that essential modifications are made in the text of the note' (translated from letter [122], Ref. [5], p. 297).

But Bohr was unwavering in his optimism (letter [123] of 3 March 1926, Ref. [5]), while Pauli repeated his criticism in long letters to Bohr on 5 March and to Kramers on 8 March 1926. In the latter Pauli writes: 'Today Herr Goudsmit was here, and I was delighted to see him. What he told me about the calculations of Thomas and the manuscripts of Thomas which he brought along have utterly strengthened me in the view that with my objections . . . I am totally right' (translated from letter [125], Ref. [5], p. 304). It was in this letter also that Pauli referred to himself as 'God's whip' (see page 89) striking the Copenhagen Institute again. In visiting Pauli in Hamburg, Goudsmit was on his way back to Leiden from Copenhagen where, at the beginning of 1926, he had spent several weeks at Bohr's invitation. At that occasion Bohr had added a postscript to Uhlenbeck and Goudsmit's second note, which interprets the spectrum of hydrogen—considered as an alkali atom without a core electron— in terms of the spinning electron model [25]. In this interpretation the authors made use of the doublet formula (5.6) communicated to them by Heisenberg.

Meanwhile, Thomas's problem was taken up by Jakov (Yakov) I. Frenkel (1894–1952), a lecturer in theoretical physics at the Polytechnical Institute in Leningrad, who came to Germany on a Rockefeller grant. From 1 December 1925 to 17 April 1926 he was in Hamburg where he had close contact with Otto Stern and, particularly, with Pauli who introduced him to Thomas's paper (see Chapter 3 in Ref. [24c]). Frenkel did the only sensible thing: that is, use the analogy with the electromagnetic field to enlarge the magnetic moment vector $\vec{\mu}$ into a covariant antisymmetric tensor by combining it with an electric dipole moment and to develop the associated equation of motion [24d]. This work, which confirmed a Thomas factor of 2, was submitted for publication on 2 May 1926. In January of the following year a second paper by Thomas appeared in which essentially the same formalism was developed [24b]. But Pauli had been convinced of the correctness of Thomas's result already before. Indeed, on 12 March he wrote to Bohr: 'Now there is nothing else I can do than to *capitulate completely*!' (translated from letter [127], Ref. [5], p. 310). And to Goudsmit he wrote on 13 March: 'Today I write you, first, to inform you that, based on recent informations from Copenhagen, I now indeed have come to the conviction that I was wrong with my objections against Thomas and that his relativistic consideration may be brought into a totally correct and unobjectionable form. The question of the fine structure therefore is now truly satisfactorily clarified' (translated from postcard [129], Ref. [5], p. 313).

Looking back it is obvious that Pauli's 'two-valuedness' referred to the spin, the electron's fourth degree of freedom which became of fundamental importance in the announcement of the exclusion principle (see the following section). Indeed, some two years later Pauli showed in another fundamental paper [26] that it was possible to describe the spinning electron by *two-component wave functions* (see page 145). With this, the problem of the electron spin was completely solved, at least in the framework of non-relativistic wave mechanics, and it yielded as another manifestation of the two-valuedness the well-known Pauli matrices. But who should now be considered the discoverer of the electron's spin, Pauli, Kronig, or Uhlenbeck and Goudsmit?

Contrary to widespread belief, the last two authors did not receive the Nobel Prize and, until 1946, were never proposed for it. There was, however, a nomination of Pauli and Goudsmit, both for the electron spin and for the nuclear spin, by Léon Brillouin (1889–1979) in 1934 [27a]. Uhlenbeck, on the other hand, although the initiator of the spinning electron idea and the first author in both notes [17] and [25], was unknown at the beginning of the discussions among the experts. In the summer of 1925 Uhlenbeck, although one and a half years older than Goudsmit, in the summer of 1925 was a novice in physics research, while Goudsmit had already written a number of papers on complex spectra and the vector model and had an extensive knowledge of the spectroscopic literature.

Uhlenbeck, who had spent his childhood in Padang, Sumatra, was for the three winters of 1922 to 1925 a private tutor of science to the youngest son of the Dutch ambassador in Rome—and there became a good friend of Enrico Fermi (1901–54). Goudsmit and Uhlenbeck both got their Ph.D.s under Ehrenfest in Leiden on the same day, 7 July 1927, after which they set out for the United States where both started their academic careers at the University of Michigan in Ann Arbor. Uhlenbeck returned to Holland in 1935 as a professor of theoretical physics at the University of Utrecht, returning to Ann Arbor in 1939. After one year as president of the American Physical Society which he spent at the Institute for Advanced Study in Princeton, he joined the Rockefeller University in New York in 1961. Goudsmit, who grew up in a Jewish family in the Hague remained at Ann Arbor until after the war during which time he served as Chief Scientific Officer of the Alsos Project, the Allied unit for secret scientific information. He joined Brookhaven National Laboratory in 1946 and there chaired the physics department and was managing editor of *Physical Review* and initiator of the *Physical Review Letters* (see Refs [27b], pp. 695 and 697, and Ref. [27c], Section 13(c)).

Returning to the question of the Nobel Prize, in his recollections Goudsmit makes the following comment: 'we each [Uhlenbeck and Goudsmit] received the Max–Planck medal from the German Physical Society. The Nobel Prize was not for us—there were too many physicists who made more important contributions at that time. . . . The discovery of spin was the main factor of our being offered, in 1926, jobs at the University of Michigan as instructors. That was for me a far more significant award than a Nobel Prize' (Ref. [19b], p. 42). To this Uhlenbeck adds: 'I do know, however, from a long conversation with Pauli in the 1950's during a summer school in Les Houches [1951], that he blamed himself about the whole episode—"*Ich war so dumm wenn [als] ich jung war!*" (I was so stupid when I was young!)' (Ref. [19a], p. 44). At the meeting in Trieste where I gave the lecture of Ref. [22a] Uhlenbeck told me that Pauli had apologized for the fact that he, Uhlenbeck, and Goudsmit had not received the Nobel Prize.

This shows that the priority question raised above left behind some bitterness and hard feelings. Kronig wrote to Kramers on 6 March 1926 from New York: 'I somewhat regret that due to the adverse criticism I did not publish anything at the time, as a simple and concrete idea like the spinning electron would have had considerable advertising value in this country. . . . In future I shall trust my own judgment more and that of others less.' And again on 11 March: 'Should you find it hard to believe what I told you in my last letter, I must refer you to Landé, who surely will remember my conversation with him; Pauli, too, could hardly have forgotten it. If the memory of these gentlemen fails them, then I am at the mercy of your good opinion, as the "dear father above" cannot be employed as witness. Anyhow, I learned a little life-wisdom from this tragicomedy' (Ref. [18], p. 233). Thomas wrote to Goudsmit on 15 March 1926: 'I believe that you and Uhlenbeck had the good luck to see your spinning electron published and discussed before Pauli knew of it. More than a year ago Kronig had conceived the rotating electron and had developed his idea; Pauli was the first person to whom he showed his paper [*sic*!]. Pauli told him that it was a ridiculous assumption, and so it happened that the first person to see Kronig's paper was also the last one.' And Thomas concludes: 'All this shows that God's infallibility is not extended to the one who had nominated himself to be his vicarial on earth' (Ref. [28a], p. 73). In Germany the following verse circulated: '*Der Kronig hätt' den Spin entdeckt, hätt' Pauli ihn nicht abgeschreckt*' (Kronig would have discovered the spin if Pauli had not discouraged him).

Even among the historians writing about this episode there is disagreement. Goudsmit, referring to the *Memorial Volume* [14a], observes: 'It contains also an article written by an author who tries to "interpret" history and apparently attempts to read Pauli's thoughts from published papers and from some old and new correspondence. That such an analysis is neither objective nor useful is borne out, for example, by one impression given by that article—namely, that the concept of electron spin was, in 1925, one of the most stupid ideas in physics and that therefore all credit and praise should go to those who were then against accepting the hypothesis' (Ref. [28b], pp. 18–19). The article criticized by Goudsmit is the contribution [29a] by Bartel L. van der Waerden (1903–96), the inventor of spinor analysis [29b], a colleague of Pauli in the mathematics department of the University of Zurich, and a noted historian of science. In his acceptance speech for the Max-Planck Medal of the German Physical Society on the 4 October 1965 in Frankfurt, Goudsmit directed his criticism against

Pauli: 'In particular I would like to avoid as much as possible to divine the thoughts of others. And along these lines I admit that I do not understand Pauli who in his Nobel Prize address declares that his note of 1924 [30] on the interpretation of the hyperfine structure has inspired Goudsmit and Uhlenbeck in the establishment of the hypothesis of the electron spin. ... Incidentally, Pauli's note of 1924 cannot at all be considered as asserting the concept of spin as a fundamental property of an elementary particle. He only considered the angular momentum of a composite nucleus, due to the orbital motions of the constituting particles' (translated from Ref. [28c], p. 125).

Pauli in his Nobel Lecture indeed writes: 'I may include the historical remark that already in 1924, before the electron spin was discovered, I proposed to use the assumption of a nuclear spin to interpret the hyperfine structure of spectral lines [30]. This proposal met on the one hand strong opposition from many sides but influenced on the other hand Goudsmit and Uhlenbeck in their claim of an electron spin. It was only some years later that my attempt to interpret the hyperfine structure could be definitely confirmed experimentally by investigations in which also Zeeman himself participated and which showed the existence of a magnetooptic transformation of the hyperfine structure as I had predicted it' (Ref. [6a], pp. 137–8).

This quotation is remarkable for two reasons. First, the mention, in relation with this experiment, of Zeeman's name, but not of Goudsmit's, is Pauli's revenge for not having been quoted by Goudsmit. In fact, the latter writes in his reminiscences concerning nuclear spin: 'For a number of years, whenever I met Pauli, he would remark cryptically that he "could afford not to be quoted". It was only in the late thirties that I found out to what he referred.' And speaking of his experimental work with Ernst Back (1881–1957), which in part was indeed done in Zeeman's laboratory in Amsterdam, Goudsmit continues: 'We had completely overlooked Pauli's most significant pioneering contribution to this new field which introduced nuclear physics into spectroscopy.' As to this pioneering contribution, Goudsmit finds with good reason that 'it is regrettable that in the above-mentioned book [the *Memorial Volume*] it merely rates a footnote in the bibliography' (Ref. [28b], pp. 19 and 20). In fact, that footnote on page 306 of Ref. [14a] was my personal contribution [31a]!

The second reason why the above quotation from Pauli's Nobel Lecture is remarkable is because of the statement that his note of 1924 had 'influenced ... Goudsmit and Uhlenbeck in their claim of an electron spin'. It seems to indicate that Pauli was convinced that in summer 1925 Goudsmit was aware of his note of 1924, contrary to Goudsmit's assurances in the above reminiscences. In Goudsmit's works quoted in the authoritative treatise by Hans Kopfermann (1895–1963) [31b] Pauli's note of 1924 does not appear, while, on the other hand, Kopfermann gives credit to Pauli on page 1 of his book. The first mention of Pauli's note by Goudsmit seems to be the one on page 203 of the book he co-authored in 1930 with Linus C. Pauling (1901–94) [31c]. By spring 1931, however, Pauli's authorship of the explanation of hyperfine structures was well established. He himself made two oral contributions, the first in the form of a discussion remark after the report 'Hyperfine structure and nuclear moments' by Hermann Schüler (b. 1894) from Potsdam at the ETH *Vortragswoche* (week of lectures) in May [31d], and by his talk *Problems of Hyperfine Structure* at the Symposium held at the California Institute of Technology in Pasadena in June, which was followed by a talk of Goudsmit on *Hyperfine Structure and Nuclear Magnetism* (see Ref. [31e]).

In addition, Pauli's use in his Nobel Lecture of the term 'nuclear spin' had irritated not only Goudsmit (see above) but also historians after him. Thus Belloni writes: 'Pauli had not proposed a spinning nucleus (in the modern sense) and hence could reject a spinning electron without contradiction or special pleading. It also shows the inappropriateness of Enz's solution to the riddle' (Ref. [31f], pp. 462–3). This refers to my Trieste review [22a], in which I had stated that 'Pauli could at the same time propose the idea of the nuclear spin and reject the apparently identical idea of the electron spin' (Ref. [22a], p. 771), because of the peripheral velocities which are large compared with c for the latter (see *Note* [20b]), but smaller for the nucleus. Now Pauli's wording in his note of 1924, in which the term *Hyperfeinstruktur* is introduced, is as follows: 'We wish to assume in addition that the nucleus . . . possesses a non-vanishing resultant angular momentum. Then the nuclear architecture and the system of external electrons must . . . , as a consequence of the interaction forces reigning among them, assume different, quantum-wise determined orientations against each other' (translated from Ref. [30], p. 742). This shows that Pauli considered the resultant nuclear angular momentum to obey directional quantization.

That in the history of modern physics the idea of spin has stirred up so much controversy is a strange fact. It is in referring to this episode—apart from that other controversial subject of the same period, namely the unfortunate radiation theory of Bohr, Kramers, and Slater [32] (see page 158)—that Pauli, 30 years later, in the dedication quoted on page 84, speaks of 'a brief period of spiritual and human confusion, caused by a provisional restriction to "*Anschaulichkeit*" (visualizability).' (Ref. [33], p. 30). No wonder then that the literature concerning the history of spin is quite considerable. In addition to the accounts already quoted I should also mention Mehra and Rechenberg's *History*, Ref. [34a], Section V.5, and Richter's *Skizzen*, Ref. [34b], Sections IV and VII, as well as the recent book by Tomonaga [34c].

The exclusion principle and stability of matter

With the help of the exclusion principle, we now understand the stability of matter on two scales, first the atomic scale . . . and second the scale of ordinary human experience with matter in bulk. It remains only to discuss the question of stability on a third scale, the scale of astronomy and cosmology. [35a]

The far-reaching importance of the exclusion principle for understanding the material world emphasized here by Freeman Dyson (b. 1923) was of course not realized by Pauli when he formulated the principle in late 1924. However,

Ehrenfest in letter [152] of 24 January 1927 to 'dear dreadful Pauli' made the following visionary observation: 'For a very long time already I have had the feeling that it is your prohibition which prevents at first the atoms and thereby the crystals from shrinking (*einschrumpfen*)' (translated from Ref. [5], p. 371, see also Ref. [35b]). But what was even less evident is that, on the third scale of stability mentioned by Dyson, namely mass concentrations substantially larger than that of our sun, the exclusion principle, as well as Coulomb repulsion, become powerless against gravitational attraction which, promoted by the curving of space–time, leads to collapse into black holes.

Compared with the preceding paper [3a] (see the previous section) this Nobel Prize work 'On the connection of the closure of the electron groups in the atom with the complex structure of the spectra' [3b] is rather difficult to access, and this in spite of the almost total absence of formulae. Even the notation is unfamiliar, as the following 'dictionary' for the case of the alkali spectra ($s = |m_s| = 1/2$) shows:

$$k_1 = k = l + 1, \; k_2 = j + \frac{1}{2} = |l \pm s| + \frac{1}{2} = k_1 - 1 \text{ or } k_1$$
$$m_1 = m_j = m_l + m_s, \; m_2 = m_1 \pm \frac{1}{2} = m_l + 2m_s. \tag{5.10}$$

Here k_1 and k_2, mentioned in the last section in relation to the relativistic doublet formula (5.3), were introduced by Bohr and Coster following a paper by Wentzel (see the 'Quantentheorie' [10a], p. 204); m_1 and m_2 are the values of the electron's angular momentum in units of $\hbar$ and of the energy in a magnetic field H in units of $\mu_B H$, respectively. The 'closure of the electron groups in the atom' in the title indicates that, according to the *Aufbauprinzip* (building-up principle), these groups can accommodate exactly $2n^2$ electrons. And, as already argued in the preceding paper [3a], this presupposes permanence of the quantum numbers in the sense that once a state is occupied the associated quantum numbers no longer change when more electrons and nuclear charges are added.

Consider now the building-up from the alkalis, which have one valence electron and in a magnetic field exhibit doublets, to the alkaline earths which have two valence electrons and exhibit singlets and triplets. In the new notation a valence electron always has $2k_1 - 1$ positions in a magnetic field. On the other hand, in the magnetic interaction model the atomic remainder (core) has two positions in the case of the alkalis, even though it contains no valence electron, and it has one or three positions in the case of the alkaline earths. These numbers of core positions cannot be understood in terms of a conventional coupling between valence electron and core, a fact which induced Bohr to speak of 'unmechanical *Zwang*' on page 276 of Ref. [7] (see the last section). With Pauli's 'classically not describable two-valuedness of the quantum-theoretical properties of the valence electron', however, the situation becomes clear: in the case of the alkalis the valence electron has $2 \times (2k_1 - 1)$ positions. But in the case of the alkaline earths the 'inner' valence electron gives rise, in addition, to two positions of the core; hence there are in all $4 \times (2k_1 - 1) = (1 + 3) \times (2k_1 - 1)$ states. In other words, the combination of the permanence of the quantum numbers and the two-valuedness leads to the correct counting.

Further, since the two-valuedness is located at the valence electron and not at the core, this procedure is *additive*. It may be generalized to an atom with N states (positions) in a magnetic field, to which a further electron is added. Indeed, this results in a core with N states plus a valence electron with $2 \times (2k_1 - 1)$ states, and hence in $2N \times (k_1 - 1)$ states in all. Pauli concludes: 'Furthermore, according to the interpretation (*Auffassung*) proposed here, Bohr's *Zwang* manifests itself not in a piercing (*Durchbrechung*) of the permanence of the quantum numbers in the coupling of the valence electron to the atomic remainder but only in a strange (*eigentümliche*) two-valuedness of the quantum theoretical properties of the individual electrons in the stationary states of the atom' (translated from Ref. [3b], p. 768). This additivity may be written as a sum over all valence electrons which, according to Eqs (5.10), also holds for the energy in the magnetic field (see Eqs (1) of Ref. [3b]),

$$\overline{m}_1 = \sum m_1; \; \overline{m}_2 = \sum m_2. \tag{5.11}$$

Taking again the example of the alkaline earths one has, for s-terms ($l = m_l = 0$), $m_1 = \pm m_s = \pm 1/2$; $m_2 = \pm 2m_s = \pm 1$ for both valence electrons and hence $\overline{m}_1 = -1, 0, +1$ and $\overline{m}_2 = -2, 0, +2$, as indicated on p. 769 of Ref. [3b].

In order to arrive at the formulation of the new principle, Pauli again discusses the alkaline earths as an example. It was known that in the ground state (*Normalzustand*) of these elements the two valence electrons are in an s-state and form a singlet (antiparallel spins) 'while in those stationary states of the atom which belong to the triplet system the valence electrons are never equivalently bound, in that the largest triplet s-term possesses a principal quantum number which is larger by one as compared to the ground state' (translated from Ref. [3b], p. 772). The other example discussed by Pauli is the neon spectrum which, however, is too technical and unilluminating to be presented here (see the 'Quantentheorie' [10a], Ziff. 46). Pauli concludes this discussion in saying 'Generally we therefore may expect that with those values of the quantum numbers n and k_1 for which there are already electrons in the atom, certain multiplet terms of the spectra will be missing or coincident. And here the question arises, by which quantum theoretical rules this behaviour is governed' (translated from Ref. [3b], p. 772).

The first step in answering this question is the determination of the substructure of the groups of $2n^2$ electrons with the principal quantum number $n = 1, 2, 3, \ldots$ corresponding to the $K, L, M, \ldots$ shells. Bohr, designating the orbits as n_k where the azimuthal quantum number takes the values $k = 1, 2, \ldots, n$, was led tentatively to distribute the $2n^2$ electrons *evenly* over these n subgroups, each therefore accommodating $2n$ electrons (see the 'Quantentheorie' [10a], Ziff. 31). The result is shown in table 1 on p. 774 of Ref. [3b] (see also the more detailed table on p. 260 of Ref. [7]). This is not, however, a natural subdivision since the $2(2k - 1)$ electrons having the same symbol n_k, and which are called *equivalent* on p. 186 of the Quantentheorie [10a] and in Pauli's rule below, are then not all in the same subgroup.

Now Sommerfeld had noticed a paper by a certain Edmund C. Stoner (1889–1973) in Cambridge, in which a different distribution was proposed [36a]. Pauli, who saw this work mentioned in the preface to the fourth edition of Sommerfeld's *Atombau und Spektrallinien*, was enthusiastic about it. 'There a subdivision of Bohr's subgroups according to k_1 and k_2 [Stoner's 'azimuthal' and 'inner' quantum numbers k and j, respectively] is assumed and the very interesting point of view developed that the number of electrons in a closed $n_{k_1 k_2}$-group should coincide with the number of "positions" of the $n_{k_1 k_2}$- "orbit" that the alkalis have in the magnetic field' (translated from letter [71] to Landé on 24 November 1924, Ref. [5], p. 180). This description of Stoner's ideas given here by Pauli already contains the important rule answering the above question, namely the equality of the number of electrons in the subgroup with the number of positions available to their orbits.

The subgroups $2_{11}, 2_{21}, 2_{22}, 3_{11}, \ldots, 3_{33}$, etc., of Stoner correspond to the subshells $L_I, L_{II}, L_{III}, M_I, \ldots, M_V$, etc. (see Ref. [36a], Table I), familiar from X-ray spectra. Stoner notes: 'In the classification adopted, the remarkable feature emerges that the number of electrons in each completed level is equal to double the sum of the inner quantum numbers as assigned, there being in the K, L, M, N levels, when complete, 2, 8 (2+2+4), 18 (2+2+4+4+6), 32 . . . electrons' (Ref. [36a], p. 722). For a given value of n these are $2n - 1$ subgroups which together indeed accommodate

$$\sum_{l=0}^{n-1}[2l + 2(l + 1)] = 2n^2 \tag{5.12}$$

electrons. This ordering is exhibited in Stoner's Table II and in table 2 of Ref. [3b], a more precise form of which is given in table 17 on p. 268 of the 'Quantentheorie' [10a]. A quite explicit formulation of these findings may be found on p. 726 of Stoner's paper: 'If electrons in the atom are distributed according to the present scheme, the interesting point is suggested that all electrons bound in the atom forming constituents of completed groups are to be regarded as having the same statistical weight, namely unity (or h^3); for there is then one electron in each possible equally probable state.' This last formulation is quite close to Pauli's rule which is expressed as a prohibition:

There can never be two or more equivalent electrons in the atom, for which, in strong fields, the values of all quantum numbers n, k_1, k_2, m_1 (or, what amounts to the same, n, k_1, m_1, m_2) coincide. If one electron is present in the atom, for which these quantum numbers (in the external field) have definite values, then this state is 'occupied'. (Translated from Ref. [3b], p. 776)

To this Pauli adds the following important comments: 'We are unable to give a closer justification for this rule, but it seems to offer itself automatically in a natural way. To start with it refers, as mentioned, to the case of strong fields. However, from thermodynamical arguments (invariance of the statistical weights in adiabatic transformations of the system) . . . the number of stationary states of the atom, for given values of the numbers k_1 and k_2 for the individual electrons and of the number $\overline{m}_1 = \sum m_1$ [see Eq. (5.11)] for the entire atom, must coincide in strong and in weak fields' (translated from Ref. [3b], p. 776). It is evident from Eqs (5.10) that in the above wording of Pauli's rule the fourth (spin) quantum number implicit in the quantum numbers k_2 and m_1, or m_1 and m_2, is an essential ingredient. In this sense the two papers [3a] and [3b] are logically inseparable.

The remainder of paper [3b] is devoted to the testing of this prohibition rule against the available empirical spectroscopic material, which was also the reason for Pauli's visit to Tübingen at the beginning of January 1925, as we saw in the previous section. In the course of this test Pauli introduces another important notion, namely the reciprocity law which states that 'To each arrangement of the electrons there is a conjugate arrangement, for which the gap values (*Lückenwerte*) of m_1 and the occupied values of m_1 are interchanged' (translated from Ref. [3b], p. 778). Here the gap value of m_1 is defined as $-\overline{m}_1$. Examples of conjugate chemical elements are B, F and C, O. The symmetry in the magnetic effects of conjugate elements was likened by Stoner to Babinet's principle in optics (see Ref. [36a], p. 733). This reciprocity law may be viewed as an early form of particle–hole symmetry. Pauli concludes the paper with the following outlook: 'Based on the gained results there do not, however, seem to be any signs for the existence of a connection between the problem of the closing of the electron groups in the atom and the correspondence principle as was suspected by Bohr. It is likely that the problem of the more detailed justification of the general rule concerning the existence of equivalent electrons in the atom supposed here might be successfully addressable (*angreifbar*) only after a further deepening of the fundamental principles of quantum theory' (translated from Ref. [3b], p, 783).

Stoner's career was very much centred around magnetism. He studied in Cambridge where in 1925 he obtained his doctorate at the Cavendish Laboratory under Rutherford who had moved there from Manchester in 1919. In 1924 Stoner went to Leeds as a lecturer and there later held the chair of theoretical physics until his retirement in 1963, becoming a Fellow of the Royal Society in 1937 (see Ref. [27b], footnote 1003, p. 666). Stoner is best known for his theory of ferromagnetism of itinerant electrons in metals (see e.g. Section 32 of Ref. [36b]). In spite of the fact that in his paper [36a] Stoner presents X-ray, optical, chemical, and magnetic evidence for his scheme, Bohr had been rather sceptical towards this work. On 10 December 1924 he wrote to his former collaborator Coster: 'I have from the first understood the formal beauty and simplicity of his classification of the levels; however, in the discussions here I have merely emphasized that, from quantum-theoretical points of view, it cannot mean a final solution of the problem, since we do not yet possess any possibility of connecting the classification of levels in a rational manner with a quantum-theoretical analysis of electron orbits' (Introduction to Part I, in Ref.

[37a], p. 41). In the last reference, as well as in Pauli's reviews [10a], p. 268 and [21], p. 1788, a paper by a certain J. D. Main Smith is mentioned in which Stoner's distribution of electrons is proposed independently.

Bohr and Heisenberg in Copenhagen were the first to read Pauli's manuscript which he had sent with his letter to Bohr of 12 December 1924. In this letter Pauli assures Bohr that 'what I do here, is *not a bigger* nonsense than the hitherto existing perception of the complex structure. My nonsense is conjugate to the hitherto customary one' (translated from letter [74], Ref. [5], p. 188). Bohr's and Heisenberg's reactions were swift. Heisenberg sent a postcard on 15 December 1924 exclaiming in his typical youthful exuberance: 'Today I have read your new work, and it is certain that I am the one who *rejoices most* about it, not only because you push the swindle to an unimagined, giddy height (by introducing *individual* electrons with 4 degrees of freedom) and thereby have broken all hitherto existing records of which you have insulted me, but quite generally, I triumph that you too (*et tu, Brute!*) have returned with lowered head to the land of the formalism pedants (*Formalismusphilister*); but don't be sad, there you will be welcomed with open arms. And if yourself you think to have written something against the hitherto existing kinds of swindle then, of course, this is a misunderstanding; for, swindle × swindle does not yield something correct and, therefore, two swindles can *never* contradict each other. Therefore I congratulate!!!!!!!! Merry Christmas!!' (translated from card [76], Ref. [5], pp. 192–3).

From this and also from Bohr's response one may guess that they had a good laugh over Pauli's manuscript. For, usually when Bohr writes something like (see his letter to Coster quoted above) 'we are all enthusiastic of the many new beauties that you have brought to light', as he does in his letter to Pauli of 22 December, it means that he is rather sceptical. However, the laugh notwithstanding, Bohr and Heisenberg of course realized that there must be something important behind this nonsense, for Bohr later writes in the letter: 'I have the impression that we stand at a decisive turning point, now that the extent of the whole swindle has been characterized so exhaustively' (translated from letter [77], Ref. [5], pp. 194 and 195).

The point on which Bohr specifically expressed doubts in his letter was Pauli's concluding remark concerning the role of the correspondence principle as a possible explanation of Pauli's rule. Bohr originally had hoped that this principle might explain the closure of the electron groups, because it implies a restriction on the transition processes (see Ref. [10a], p. 187). Bohr's correspondence principle indeed states that transition processes between states characterized by quantum numbers n and n' such that $|n' - n| \ll n$ may be calculated classically. And this calculation may be done largely independently of the details of the electronic orbits, to yield frequencies, polarizations, selection rules and intensities of emission and absorption processes (see the 'Quantentheorie' [10a], Ziff. 9, in particular p. 49). But Pauli responded to Bohr: 'I personally do not believe, however, that the correspondence principle will lead to a foundation of the rule (I cannot prove this rigorously, it is only a scepticism based on my physical feeling). . . . I see the promising clarification (also that of the coupling problem in general) rather in a physical analysis of the notions of motion and of force in the sense of the quantum theory' (translated from letter [79] from Pauli to Bohr of 31 December 1924, Ref. [5], pp. 197 and 198).

Pauli's prohibition rule was perceived by most physicists as too abstract to be befriended easily. As with the Mosaic Ten Commandments the adaptation process took the road of exegesis, and not at all the road of proof. Pauli himself had hoped in vain to succeed along the second road, as he recalls in his Nobel Lecture: 'Already in my original paper I stressed the circumstance that I was unable to give a logical reason for the exclusion principle or to deduce it from more general assumptions. I had always the feeling and I still have it today, that this is a deficiency. Of course in the beginning I hoped that the new quantum mechanics, with the help of which it was possible to deduce so many half-empirical formal rules in use at that time, will also rigorously deduce the exclusion principle. Instead of it there was for electrons still an exclusion: not of particular states any longer, but of whole classes of states, namely the exclusion of all classes different from the antisymmetrical one. The impression that the shadow of some incompleteness fell here on the bright light of success of the new quantum mechanics seems to me unavoidable' (Ref. [6a], p. 136).

Pauli's belated Nobel Prize itself might be seen as a sign of perplexity among the physics community, although some early nominations for the exclusion principle exist. Apart from Brillouin, mentioned in the last section, these are the early nominators: Carl Wilhelm Oseen (1879–1944) from Stockholm in 1933 who nominated Pauli with Heisenberg and Schrödinger (Heisenberg got the Prize of 1932 and Schrödinger with Dirac the Prize of 1933 in 1933); and D. Coster from Groningen in 1934 who nominated Pauli with Born and Sommerfeld (Born got the Prize in 1954). There is also a nomination by M. Planck, M. von Laue, and D. A. Deissmann from Berlin in 1935. Among the later nominations I wish to mention those by Pauli's friend Gregor Wentzel in 1940, 1941, 1943, and 1944. Wentzel, who during the war had agreed to take over Pauli's teaching at ETH when Pauli was in the United States, included in his nomination, not only the exclusion principle, but also the nuclear spin, the neutrino, and the spin-statistics theorem. But the event that probably triggered the decision of the Nobel Committee for Physics to award the Prize to Pauli in 1945 'for the discovery of the Exclusion Principle, also called the Pauli Principle' was the telegram from Einstein received by the Nobel Committee on 19 January 1945, just in time to be effective: 'nominate wolfgang pauli for physics-prize stop his contribution to modern quantumtheory consisting in so called pauli or exclusion principle became fundamental part of modern quantumphysics being independent from the other basic axioms of that theory stop albert einstein' [27a] (see also Ref. [37b], p. 517).

By characterizing the exclusion principle as 'being independent from the other basic axioms', Einstein evidently expressed the opinion that there was no way of 'proving' it. On the other hand, Pauli's emphasis on 'equivalent' for the electrons meant 'A Loss of Identity', which is the title Pais gives to Chapter 23 referring to 'The Birth of Quantum Statistics' in Ref. [37b]. As one reads there on page 423, in June 1924 the Bengali Satyendra Nath Bose (1894–1974) wrote to Einstein, sending him a manuscript which he asked Einstein to submit to *Zeitschrift für Physik*, provided he agreed with its content. Einstein himself translated the paper into German and sent it to the *Zeitschrift* [38a] with a translator's note saying 'In my opinion Bose's derivation of the Planck formula constitutes an important advance. The method used here also yields the quantum theory of the ideal gas, as I shall discuss elsewhere in more detail' (translation of Ref. [37b], p. 423). The last sentence shows that Einstein immediately saw the general applicability of Bose's method to any kind of *identical* and *indistinguishable* particles of arbitrary mass with the property that they may populate quantum cells ω_{si} ($i = 1, 2, ...$) in phase space, each of volume h^3, in arbitrary numbers [38b]. To obtain their statistics one then counts the numbers N_s of particles contained in the energy shells $\Omega_s = \sum_i \omega_{si}$ having average energy ε_s and constant width $\delta\varepsilon$.

The development of course was not as straightforward as described here. Already the assumption of a constant density h^{-3} in phase space introduced by Otto Sackur (1880–1914) [38c], and by Planck (see Ref. [38d], p. 27), was far from evident, at least until de Broglie gave it a wave interpretation (see the next section). But once accepted, Bose's method becomes relevant here because it also yields an ideal quantum gas in the case of particles obeying Pauli's exclusion principle, that is, particles which can populate quantum cells ω_{si} at most one at a time. But before showing this, let us quickly turn to the classical ideal gas treated by Boltzmann. In this case N_s identical but *distinguishable* particles are distributed directly on the energy shells Ω_s. And a state is specified by the particle numbers N_s, for fixed total particle number and total energy,

$$N = \sum_s N_s = \text{const.}, \quad E = \sum_s N_s \varepsilon_s = \text{const.} \tag{5.13}$$

The probability of this state is then, up to a normalization factor,

$$W \propto \frac{N!}{\prod_s N_s!}. \tag{5.14}$$

On the other hand, the essence of Bose's method is the idea that indistinguishability may be expressed by counting the numbers Z_{sr} of quantum cells ω_{si} containing $r = 0, 1, ..., p$ particles. Here the maximum occupation number is $p = \infty$ for particles satisfying Bose–Einstein (BE) statistics, but $p = 1$ for particles obeying the exclusion principle. And a state is specified by the cell numbers Z_{sr}, for fixed shell numbers

$$Z_s = \sum_r Z_{sr} = \text{const.} \tag{5.15}$$

and under the conditions (5.13), where

$$N_s = \sum_r r Z_{sr} \qquad (5.16)$$

(Bose's notation in Ref. [38a] is A^s for Z_s and p_r^s for Z_{sr}). The probability of this state is

$$W \propto \frac{Z_s!}{\prod_{sr} Z_{sr}!}. \qquad (5.17)$$

Making use of the approximate formula $\delta \log N! = \log N \delta N$ the most probable state with the supplementary conditions (5.13), and in Bose's case also with condition (5.15), is found to be characterized by the average occupation numbers of the energy shells Ω_s. In Boltzmann's case the result is, with $n_s = N_s/N$,

$$\langle n_s \rangle = e^{-\beta(\varepsilon_s - \mu)} \qquad (5.18)$$

where $\beta = kT$ and μ is the chemical potential. In Bose's case one finds, with $n_s = N_s/N$,

$$\langle n_s \rangle = \frac{1}{e^{\beta(\varepsilon_s - \mu)} \mp 1} \qquad (5.19)$$

where the upper sign holds for particles satisfying BE statistics and the lower sign for particles satisfying the exclusion principle. Photons are particular BE particles in that they are massless and therefore their total number cannot be fixed. Hence the first condition (5.13) does not hold, which amounts to setting $\mu = 0$.

The confusing thing about these quantum statistics is that there are several ways to arrive at the result (5.19). Einstein himself was not entirely happy with Bose's derivation [38a], which was limited to the case of photons. Inspired by de Broglie's idea of matter waves (see the next section), Einstein devised his own method, valid for any particles obeying BE statistics, showing in the second paper that for massive BE particles there is condensation into the state of zero momentum [38b]. It was only in 1926 that Enrico Fermi (1901–54) in March and Paul A.M. Dirac (1902–84) in August submitted papers deriving the statistics for particles that obey the exclusion principle [39a,b]. They both independently used in Eq. (5.17) the expressions $Z_{s1} = N_s$ and $Z_{s0} = Z_s - N_s$ which follow from (5.15) and (5.16). This counting method is now called Fermi–Dirac (FD) statistics. Pauli, on the other hand, in his lectures on the subject apparently did not like this confusing variety and derived both statistics from the Gibbs grand-canonical potential, thus avoiding counting methods altogether. He characterized the particles obeying BE statistics, called bosons today, physically by saying that they 'have the tendency to condense into groups [*sich zusammenballen*]', and the particles obeying FD statistics, called fermions, by saying that they 'repel [*fliehen*] one another'. In both cases 'this has the consequence that the particles are no longer statistically independent' (Ref. [40], pp. 98 and 99).

The essence in all these derivations is the indistinguishability of identical particles, which has become one of the cornerstones of modern quantum theory. Concerning this important feature, Hermann Weyl, speaking of the electrons obeying Pauli's principle wrote in 1927: 'The upshot of it all is that the electrons satisfy Leibniz's *principium identitatis indiscernibilium*, or that the electron gas is a "monomial aggregate" (Fermi-Dirac statistics) . . . neither to the photon nor to the (positive and negative) electron can one ascribe individuality. As to the Leibniz–Pauli exclusion principle, it is found to hold for electrons but not for photons' (Ref. [41a], p. 247). In reading this passage in Weyl's book [41a] Pauli later observed to Fierz: 'This sounds like a *philosophical* principle and then, it seems to me, there are only two possibilities: a) as such it is wrong; b) it is correct, but *nothing* follows from it for physics.' And he continues: 'This would really be a strange principle in the philosophy of Leibniz, *which does not hold for all objects* (e.g. not for photons, as Weyl explicitly states) but only for *some*' (translated from letter [1052] of 15 October 1949, Ref. [41b], p. 697; see also Pauli's letters [1054] to Weyl and [1055] to Fierz). It indeed appears somewhat strange that Weyl singles out fermions. Now, one reading that Leibniz gave to his principle is that 'there are no two indistinguishable single objects' (Ref. [41c], column 283). If one includes the 'position' among the distinguishing characteristics of an object—what Leibniz seems to have admitted (Ref. [41c], column 285)—then the identity of two objects means their coincidence and, hence, their indistinguishability, and the principle says that they are one and the same. Replacing 'position' by the quantum-theoretical 'state' one indeed arrives at the exclusion principle. This, however, does not leave room for bosons, so that Pauli is right in finding this 'Leibniz–Weyl principle' 'strange' (see also Refs. [41d], p. 350 and [41e], pp. 338–45).

The lack of statistical independence of particles satisfying this principle of identity and indistinguishability is particularly telling when expressed in the form of Schrödinger's wavefunctions (which will be considered in the next section). This was shown by Dirac in his remarkably rich paper [39b]—of which the derivation of FD statistics mentioned above is another part—which makes use of the wave aspect of the particles by calling the cell numbers Z_s 'the number of waves in the sth set' A_s (Ref. [39b], p. 672). Dirac observed that what was called BE and FD statistics above amounts to symmetrizing and antisymmetrizing the uncorrelated products $\psi_{r_1}(1)...\psi_{r_N}(N)$ of one-particle wavefunctions $\psi_r(i)$ including spin,

$$\Psi_{\pm} = \sum_P (\pm)^{\eta_P} \psi_{r_1}(i_1)...\psi_{r_N}(i_N), \tag{5.20}$$

respectively, where $i_1...i_N = P(1...N)$ are the $N!$ permutations of the particle numbers $1...N$ and η_P is the number of transpositions that generate the permutation P. And this linear combination introduces correlation, i.e. loss of statistical independence. It is here that Dirac coined the term 'exclusion principle'. Since Ψ_- is just a determinant, Pauli's exclusion principle follows immediately from the fact that $\Psi_- = 0$ if any two states among $r_1...r_N$ are the same. The decisive feature in Eq. (5.20) is that only coefficients $(\pm)^{\eta_P}$ are allowed, and not more general numbers A_P corresponding to the symmetry classes defined by the irreducible representations of the group of permutations of N elements (see Section 14 of Ref. [43]). In the above quotation from his *Nobel Lecture* this is what Pauli calls, for electrons or more generally for FD particles, 'the exclusion of all classes different from the antisymmetrical one'.

Again Dirac was not the only one to cast the exclusion principle into determinantal form Ψ_-. Indeed, Heisenberg had arrived at the same result in May 1926 by applying Pauli's prohibition rule—he called it *Pauli Verbot*—to the problem of singlet–triplet splitting which he studied at the model of two coupled harmonic oscillators, using symmetry arguments to obtain transition probabilities [44]. Informing Pauli by postcard [132] of Ref. [5], he wrote from Copenhagen on 5 May 1926, 'that we have found a rather decisive argument that your prohibition of equivalent orbitals is related to singlet-triplet spacing' (translation of Ref. [46], p. 508). By also including the electron's spin this led Heisenberg to a qualitative explanantion of ortho and para helium (see page 83). Towards the end of this paper [44] Heisenberg considers the generalization to an arbitrary number 'n of identical partial systems', constructing in his Eq. (20) the wavefunction Ψ_- of Eq. (5.20) above. For an account of how Heisenberg introduced the decisive notion of exchange in later papers and applied it both to the helium problem and to ferromagnetism see Ref. [46].

Setting out for the New Land

Recently I saw Heisenberg on the occasion of a physicists' meeting in Braunschweig. With him I always feel very strange. When I think about his ideas then I find them dreadful, and I swear about them internally. For he is very unphilosophical, he does not pay attention to clear elaboration of the fundamental assumptions and their relation with the existing theories. However, when I talk with him he pleases me very much and I see he has all sorts of new arguments—at least in his heart. Apart from the fact that personally he is also a very nice human being—I then consider him as very important, even genial, and I believe that some time in the future he even will greatly further the Science. [47]

Pauli expressed this opinion to Bohr a few days before Heisenberg's first visit to Copenhagen, from 15 to 27 March 1924. In an intense exchange of letters over the following three years Pauli acted as the stimulating critical mind, purifying so to speak Heisenberg's 'dreadful ideas' and guiding him to his ground-breaking discoveries of quantum mechanics in June 1925 and the uncertainty relations in March 1927. While there are 61 letters and postcards from Heisenberg to Pauli from the 1920s, unfortunately, only three of Pauli's letters to Heisenberg over the same period have been conserved, a large part of them having been destroyed by fire during the war (see Ref. [5], footnote 3, p. 340).

As mentioned on page 70, the main reason why Pauli did not take a more active part in this 'Setting out for the New Land'—the title of Chapter 6 in Ref. [48]—was the enormous work he accomplished with his 'Old Testament', the 'Quantentheorie' [10a], on which the preceding sections relied so heavily. On

9 October 1925 Pauli wrote to Kronig: 'The mentioned article now converges, thank God! towards its end and I hope to be freed of this burden in 2 to 3 weeks! I also have a great desire for reasonable scientific activity!' (translated from letter [100], Ref. [5], p. 247). This work, however, appeared only in July 1926, as can be seen from the letter in which Sommerfeld thanks Pauli for sending him a copy: 'With your Handbuch article you have again accomplished an enormous work, in its reliability it almost calls to mind the famous *Enzyklopädie* article. Accept my cordial thanks for sending it and particularly for the kind dedication' (translated from letter [141] of 26 July 1926, Ref. [5], p. 337). Also Heisenberg expressed his thanks 'for your beautiful book in which I have read with much pleasure, although with a critical mind and unforgivingly. Obviously, it's an exact exposition of the physical connections which were known before the mess of last year, and reading it was a true recreation after Schrödinger's lectures here in Munich' (translated from letter [142] of 28 July 1926, Ref. [5], p. 337). This last remark shows Heisenberg's sceptical and even hostile attitude towards Schrödinger's wave mechanics. In an earlier letter he wrote to Pauli from Copenhagen on 8 June 1926: 'The more I think about the physical part of Schrödinger's theory the more I find it dreadful. Imagine the spinning electron whose charge is distributed over the entire space with the axis in a 4th and 5th dimension' (translated from letter [136], Ref. [5], p. 328).

In the letter to Bohr accompanying his manuscript on the exclusion principle quoted in the last section Pauli is quite convinced 'that not only the dynamical notion of force but also the kinematical notion of motion of classical theory will have to undergo profound modifications. ... I believe that energy and momentum values of the stationary states are something much more real than "orbits"' (translated from letter [74] of 12 December 1924, Ref. [5], pp. 188, 189). The reason is that the energy values E_n of the stationary states, Eq. (4.6), and the momenta p_k averaged over the orbits, Eq. (3.12), are directly related to the respective quantum numbers. On the other hand, the coordinates $q_k(t)$ describing the 'orbits' are complicated multiply, or conditionally, periodic functions of the time t whose justification, according to Bohr (Ref. [7], p. 253), had become questionable. But, as described on pages 65–6, this periodicity gives rise to a development of $q_k(t)$ into Fourier series in which the time dependence appears explicitly in the exponentials $\exp(2\pi i \tau_k \nu_k t)$. Here $\nu_k = dw_k/dt$ are the orbital frequencies and τ_k are integers which label the Fourier amplitudes $A(\tau_1, \tau_2, ..., \tau_f)$, f being the number of degrees of freedom.

While both the frequencies ν_k and the amplitudes A are unobservable, in June 1925 Heisenberg succeeded by a magical '*Umdeutung*' (reinterpretation) of these frequencies and amplitudes of evoking a vision of the new theory. The first glimpse of this vision is contained in the letter from Heisenberg to Kronig of 5 June 1925 reproduced in Ref. [13]. But then spring 'adorned Göttingen, the pleasant small town of the villas and gardens on the slope of Hainberg, with innumerable blooming bushes, roses and flowerbeds' (translated from Ref. [48], p. 59, as quoted on page 88), and Heisenberg suffered a severe attack of hay fever, so that he had to ask Born for a two-week recess, leaving Göttingen by night train on 7 June (Ref. [49], p. 248). That was the celebrated vacation on the island of Helgoland in the North Sea, north-west of Hamburg, where the new quantum mechanics was born, as described by Heisenberg in his recollections (Ref. [48], pp. 88–90).

On his way back Heisenberg spent a few hours in Hamburg to report his findings to Pauli (see letter [91], Ref. [5], p. 219). Back in Göttingen, where he had been a Privatdozent since July 1924 (Ref. [48], p. 87), he immediately wrote to Pauli: 'My attempts to fabricate a quantum mechanics advance only slowly; but I will not care at all how much I deviate from the theory of the conditionally periodic systems' (translated from letter [91] of 21 June 1925, Ref. [5], p. 221). And three days later he wrote again: 'My own works give me almost no pleasure to write about because all is still unclear, and I only guess about how it will become; but perhaps the basic thoughts are yet correct. *The principle is: In the calculation of any quantities such as energy, frequency, etc. only relations between quantities which are controllable in principle shall occur*' (translated from letter [93] of 24 June 1925, Ref. [5], p. 227, italics added). Since for this exploratory endeavour even the one-electron problem of the hydrogen atom was too complicated, Heisenberg discussed as an example the equation of motion of the anharmonic oscillator

$$\ddot{q} + \omega_0^2 q + \lambda q^2 = 0. \tag{5.21}$$

In letter [91] as well as in the two following postcards [94] of 29 June and [95] of 4 July the main emphasis is directed towards the quantized energy following from Eq. (5.21), for which Heisenberg found the expression $E = (n+1/2)\hbar\omega_0 + \beta(n+1/2)^2$. He asked for Pauli's comments on this result which is of particular interest because the zero-point energy $(1/2)\hbar\omega_0$ occurring in this formula was a favourite discussion topic between Pauli and Stern in Hamburg (see the following section). In the following letter [96] of 9 July Heisenberg informed Pauli that he was sending him his manuscript 'because I believe that, at least in its critical, i.e. negative, part it contains real physics' (translated from Ref. [5], p. 231). This is the celebrated paper 'On quantum theoretical reinterpretation of kinematical and mechanical relations' [50], submitted from Göttingen on 29 July 1925.

The essence of the '*Umdeutung*' proposed in this paper was a new notation whose magic was recognized by Born as describing matrices. It went like this. First, write Bohr's condition (4.1) as $\hbar\omega(n, n - \alpha) = E(n) - E(n - \alpha)$, which in the limit $n \gg \alpha$ of the correspondence principle assumes the classical form $\omega(n, n - \alpha) = \alpha\omega(n) = \alpha dE/\hbar dn$. Secondly, consider a classical periodic motion described by a Fourier series

$$x(t) = \sum_{\alpha=-\infty}^{+\infty} A_\alpha(n)e^{i\alpha\omega(n)t}. \tag{5.22c}$$

Quantum-theoretically this becomes in Heisenberg's new notation

$$x(t) = \sum_{\alpha=-\infty}^{+\infty} A(n, n-\alpha)e^{i\omega(n,n-\alpha)t}. \tag{5.22q}$$

$A(n, n-\alpha)$ now is, for $\alpha \neq 0$ and any n, the 'amplitude of the transition from the state n to the state $n-\alpha$', while $A(n, n)$ is the 'value of x in the stationary state n', for which $\omega(n, n) = 0$. Then, if $x(t)$ and $y(t)$ have Fourier coefficients $A_\alpha(n)$ and $B_\beta(n)$, respectively, the Fourier coefficients of the product $x(t)y(t)$ are classically

$$C_\gamma(n) = \sum_{\alpha=-\infty}^{+\infty} A_\alpha(n)B_{\gamma-\alpha}(n), \tag{5.23c}$$

while quantum-theoretically

$$C(n, n-\gamma) = \sum_{\alpha=-\infty}^{+\infty} A(n, n-\alpha)B(n-\alpha, n-\alpha-\gamma). \tag{5.23q}$$

Third, consider the quantum condition (3.12) for one degree of freedom, $J = \int p\,dq = m \int \dot{x}^2 dt = 2\pi n\hbar$. Since, according to the correspondence principle, an arbitrary constant may be added to J (see, however, letter [103], Ref. [5], p. 253), the differentiated classical form

$$\hbar = m \sum_{\alpha=-\infty}^{+\infty} \alpha \frac{d}{dn}(\alpha\omega(n)|A_\alpha(n)|^2) \tag{5.24c}$$

must be used which quantum-theoretically reads

$$\hbar = \frac{m}{2} \sum_{\alpha=-\infty}^{+\infty} [\omega(n, n-\alpha)|A(n, n-\alpha)|^2 - \omega(n, n+\alpha)|A(n, n+\alpha)|^2]. \tag{5.24q}$$

This quantum condition (5.24q) is now applied to the anharmonic oscillator (5.21) which, because of time-reversal symmetry, may be solved by the power series

$$q = \lambda a_0(n) + \sum_{\alpha=1}^{\infty} \lambda^{\alpha-1} a_\alpha(n) \cos(\alpha\omega_o t) \tag{5.25}$$

with real coefficients $a_\alpha(n)$ that may still have a higher-order λ-dependence. Using in Eq. (5.24q) $A(n, n-\alpha) = \lambda^{\alpha-1}a_\alpha(n)/2$, $\alpha > 0$, one obtains to lowest order in λ

$$a_1^2(n) = a^2(n, n-1) = \frac{2n\hbar}{m\omega_0}. \tag{5.26}$$

The same result also follows from Eq. (5.24c) (see Ref. [5], p. 227 and Ref. [50], p. 889). Making use of Eq. (5.26) and of $a_0(n) = -a_1^2(n)/2\omega_0^2$ in Eq. (5.25), this also yields the quantum expression for the classical energy $E(\lambda) = (m/2)(\dot{q}^2 + \omega_0^2 q^2 + (2\lambda/3)q^3)$ by selecting the non-oscillating terms.

Note that proving energy conservation means showing that these are the only terms. Thus one finds for the harmonic oscillator

$$E(0) = m\omega_0^2[|A(n, n-1)|^2 + |A(n+1, n)|^2] = \left(n + \frac{1}{2}\right)\hbar\omega_0, \quad (5.27)$$

while for the anharmonic part the expression written before follows to lowest order in λ, with $\beta \propto \lambda^2$. This is, after some notational cleaning, the essence of Heisenberg's paper [50].

Pauli's reaction to this *tour de force* of Heisenberg is well documented in the letter he wrote to Kramers in Copenhagen on 27 July 1925, namely 'that there is great hope to advance also *positively* on this path. In particular, I have greeted Heisenberg's bold attempts (*Ansätze*) . . . with jubilation. Certainly, one is still very far from being able to say something definitive, and here we are only at the very first beginnings. But what has given me so much pleasure with Heisenberg's reasonings is the *method* of his procedure and the *endeavour* out of which he has made these reasonings. Anyhow, I think with respect to my scientific views that I now have come very close to Heisenberg, and that we have concurring opinions about almost everything, as much as this is at all possible between two people of independent thinking. With pleasure I have also noticed that with Bohr in Copenhagen Heisenberg has learnt somewhat the philosophical reasoning and has yet turned away appreciably from the purely formal. I therefore wish him heart-felt success in his endeavour! Thus I now feel less lonely than about half a year ago when I found myself (spiritually and spatially) fairly alone between the Scylla of the number-mystical Munich School and the Charybdis of the reactionary Copenhagen coup (*Putsch*) propagated with zealotic excesses by yourself!' (translated from letter [97], Ref. [5], pp. 233–4). Pauli designated as the '*Kopenhagener Putsch*' the ill-conceived and incorrect paper [32] by Bohr, Kramers, and Slater alluded to on pages 119 and 158.

As already mentioned, it was Born who recognized that the quantum condition (5.24q) is nothing more than the diagonal element in the matrix form of the commutation relation [52a,b]

$$i(pq - qp) = \hbar 1 \qquad (5.28)$$

which is the cornerstone of the new quantum mechanics. Helped by his new 'private' assistant Pascual Jordan (1902–80) (see Ref. [14b], p. 198 and Ref. [51], Section 5) he presented his ideas in a paper of 31 pages, submitted on 27 September 1925. In this paper Born and Jordan developed matrix calculus and applied it to Hamiltonian dynamics based on Eq. (5.28), and treated as applications the anharmonic oscillator (5.21) and the free electromagnetic field [52a]. In an even longer follow-up paper of 59 pages written in collaboration with Heisenberg and submitted on 16 November 1925, the whole programme of quantum mechanics was laid down in all its mathematical detail. This fundamental paper has gone down in the history of physics as the 'Three-Man-Paper' (*3-Männerarbeit*) [52b].

At the beginning Pauli was not happy about this development. In the letter to Kronig quoted above he expressed the opinion that 'To start with, one should try to liberate Heisenberg's mechanics somewhat from the Göttingen effusion of erudity (*Göttinger Gelehrsamkeitsschwall*) and expose its physical core somewhat better' (translated from letter [100], Ref. [5], p. 247). Born had a more negative recollection of Pauli's reaction. Meeting Pauli in the train from Göttingen to Hanover—which is on the way to Hamburg—Born remembers: 'I joined him in his compartment, and absorbed by my new discovery [namely, Eq. (5.28)], I at once told him about the matrices and my difficulties in finding the value of those non-diagonal elements [on the right-hand side of Eq. (5.28)]. I asked him whether he would like to collaborate with me in this problem. But instead of the expected interest, I got a cold and sarcastic refusal. "Yes, I know you are fond of tedious and complicated formalism. You are only going to spoil Heisenberg's physical ideas by your futile mathematics", and so on' (Born's recollections quoted in Ref. [34b], p. 60).

Independently of Born the young Dirac at Cambridge had arrived at essentially the same results in his first publication on the subject. Dirac met Heisenberg in one of his seminars in the 'Kapitza Club' at Cambridge on 28 July 1925. These were informal weekly seminars sponsored by Piotr Leonidovich Kapitza (1894–1984) who from 1924 to 1932 was assistant director of magnetic research at the Cavendish Laboratory. In 1934 he returned to Moscow where he founded the Institute for Physical Problems. He shared the Nobel Prize of 1978 'for his basic inventions and discoveries in the area of low-temperature physics' [53a]. However, it was only in September that Dirac got Heisenberg's paper from Rutherford's son-in-law Ralph Fowler (see Ref. [5], p. 235). Dirac's paper was submitted by Fowler on 7 November 1925 [53b]. But Dirac developed a quite different formalism, analysing the general properties of non-commuting quantities that satisfy Heisenberg's multiplication rule (5.23q). Studying in particular the rules of differentiation he found that what was to be called the commutator $[x, y] \equiv xy - yx$ had the same properties as the classical *Poisson bracket*

$$\{x, y\} \equiv \sum_r \left\{ \frac{\partial x}{\partial q_r} \frac{\partial y}{\partial p_r} - \frac{\partial y}{\partial q_r} \frac{\partial x}{\partial p_r} \right\}. \tag{5.29}$$

Postulating the equality $[x, y] = i\hbar\{x, y\}$ then led him to the commutation relation (5.28).

Now the time had come to test the new formalism on the hydrogen atom. Dirac, in a second paper, had solved a two-dimensional hydrogen model [53c]. But the real thing apparently was too difficult even for Born. Heisenberg wrote in his contribution to the *Memorial Volume* for Pauli: 'At that time I myself was somewhat unhappy that I did not succeed in deriving even the simple hydrogen atom' (translated from Ref. [54], p. 43). Liberated from his work on the 'Quantentheorie' [10a] Pauli had gone into this problem at full pelt, and on 17 January 1926 he submitted for publication his paper 'On the hydrogen spectrum from the point of view of the new quantum mechanics' [55a]. Already on 3 November 1925 Heisenberg had written to Pauli: 'I probably don't have to tell you how much I rejoiced at your new theory of hydrogen and how much I admire that you have made out this theory so fast', and he closed the letter with 'hearty congratulations for your theory' (translated from Ref. [5], pp. 252 and 253).

In Section 1 of his paper [55a] Pauli gives a concise description of Heisenberg's ideas elaborated upon in the 'Three-Man-Paper' [52b]. Then he develops in Section 2 the 'special integration method of classical mechanics applicable with Coulomb forces which was used already by Lenz' (translated from Ref. [55a], p. 345). The Lenz vector **a** defined in Eq. (4.2) satisfies the equation $\mathbf{a} \cdot \mathbf{r} = r - s$ where $s = \mathbf{L}^2/Ze^2m$. Since both vectors **L** and **a** are constants of the motion this equation describes a Kepler ellipse with excentricity $\varepsilon = |\mathbf{a}|$ and with **a** pointing from the nucleus in the direction of the aphelion. In addition one verifies with the help of Eq. (3.10) that the energy $E = \overline{E_{pot}}/2 = -Ze^2/2r$ satisfies the relation $1 - \mathbf{a}^2 = -2Es/Ze^2$. This shows the degeneracy of the Kepler problem, for given energy E the eccentricity ε of the ellipse is not determined. The natural variables are of course r and φ defined by $\mathbf{a} \cdot \mathbf{r} = \varepsilon r \cos \varphi$. However, as matrix functions are defined through power series, these variables are not immediately transposable into matrices. Thus Pauli postulates in Section 3 the existence of a matrix r satisfying all algebraic relations connecting it with the matrices for the Cartesian coordinates x, y, z of **r** and p_x, p_y, p_z of **p**, for which the commutation relations $i[p_x, x] = \hbar 1$ etc. analogous to Eq. (5.28) hold.

A difficulty now arises because of the mentioned degeneracy. Pauli writes: 'As thoroughly discussed by Born, Jordan and Heisenberg [52b], with such a system the amplitudes of the various partial oscillations belonging to transitions with given energy values are not uniquely determined; also matrices which are constant in time do not in general have to be diagonal matrices, as their elements may be different from zero at such places (n, m), to which corresponds a vanishing frequency $\nu_m^n = (E_n - E_m)/h = 0$' (translated from Ref. [55a], p. 352). He concludes that, in principle, it should be possible to derive from the algebraic relations satisfied by the vector matrices **L** and **a**, 'without further specializing assumptions concerning the type of solution, the Balmer terms and associated statistical weights. This, however, we have not been able to do, and we will in the following choose a different path for the solution of these equations, by introducing (in several ways) additional conditions which render the solution . . . unique' (translated from Ref. [55a], pp. 352–3).

These additional conditions are obtained by adding various perturbing terms E_1 to the energy matrix E. Pauli considers three options for E_1: (a) a weak magnetic field along the z-axis; (b) a weak electric field along the z-axis; (c) weak crossed electric and magnetic fields in arbitrary directions, **E** and **H**, respectively. In case (a) one has $E_1 = \vec{\mu} \cdot \mathbf{H}$ where $\vec{\mu} = (e/2mc)\mathbf{L}$ is the orbital magnetic moment (3.1) of the electron. This implies that if **H** is along the z-axis both $\mathbf{L}^2$ and L_z are constant and hence must be represented by diagonal matrices. In case (b) $E_1 = \vec{\pi} \cdot \mathbf{E}$ where $\vec{\pi}$ is the time-averaged electric dipole moment which for the above Kepler ellipse is found to be $3e/2$ times the vector $\mathbf{a}s/(1 - \varepsilon^2)$ from the nucleus to the centre of the ellipse. Hence if $\mathbf{E}$ is along the z-axis, the matrix a_z must be chosen to be diagonal. The additional degeneracy of the Stark effect may be lifted by adding a parallel magnetic field, so that L_z is also diagonal. In case (c) E_1 is the sum of the expressions of (a) and (b), and one has to determine appropriate projections of combinations of $\mathbf{L}$ and $\mathbf{a}$ to be diagonal.

Case (a) is carried through in detail and yields the Balmer term $E_n = -R_\infty Z^2/n^2$ which, for $Z = 1$, is Eq. (4.6). This is the triumph which gave Heisenberg's ideas the necessary boost of credibility. But also case (c) is noteworthy since the result (84), or (15), of Ref. [55a] exactly reproduces the formula obtained by Klein and by Lenz (see pages 80–1) but without the difficulty of the old theory, namely that the exclusion of straight orbits ('*Pendelbahnen*', see page 66) through the nucleus cannot be implemented without contradiction. This problem, which is discussed in the introductory Section 2 (Ref. [55a], p. 343), had already been described in detail in Ziff. 25 of Pauli's 'Quantentheorie' (Ref. [10a], p. 164) where also Eq. (15) of Ref. [55a] appears (Eq. (135) of Ref. [10a]).

As we have seen, Pauli had realized that, in principle, it should be possible to solve the degeneracy problem of the hydrogen atom 'without further specializing assumptions concerning the type of solutions' but that this 'we have not been able to do' (see above). The reason was that he did not have at his disposal at the time the necessary mathematical means, namely group theory. It took, in fact, 10 years until the problem was addressed in this way by Vladimir Fock (1898–1974) in Leningrad and by Valentin Bargmann (1908–89) who spent several months in 1936 as Pauli's collaborator in Zurich, after having obtained his Ph.D. under Gregor Wentzel at the university there.

Fock showed that in momentum space the Schrödinger equation for negative energy, $E < 0$, is identical to the integral equation for the spherical functions in four dimensions—which explains the degeneracy in the azimuthal quantum number l [55b]. Bargmann then showed that for $E < 0$ the commutation rules satisfied by the six components of $\mathbf{L}$ and $\mathbf{a}$ are those of the generators of the four-dimensional rotation group. In the case $E > 0$, Bargmann's parameter $p_0 = \sqrt{-2mE}$ becomes imaginary and, as he shows, the group becomes the Lorentz group [55c]. Pauli discussed these papers in his *Spezialvorlesung* [55d] at ETH during the summer term of 1954. The algebraic part of these lectures gave rise to the lectures [55e] that Pauli gave at the new Theory Division of CERN in Copenhagen in September 1955. There he discussed the representations of the four-dimensional rotation group explicitly.

An important omission in Pauli's paper [55a] is the electron spin. Only in the final Section 6 is Uhlenbeck and Goudsmit's hypothesis mentioned. And Pauli writes: 'Whether this assumption suffices to explain, in conjunction with the new quantum mechanics, all experimental results might be decidable only when also the calculation of the relativistic fine structure is performed on the basis of the new mechanics. The latter for the moment is beyond reach because we have not yet succeeded in doing the required calculation of the time average $\overline{1/r^2}$' (translated from Ref. [55a], p. 361). This of course concerns the discrepancy of the Thomas factor of 2 between Eqs (5.3) and (5.6) which, as we saw earlier, Pauli accepted only in March 1926, some four months after writing this paper.

Ten days after Pauli's submission of his paper on the hydrogen atom that other Austrian, Erwin Schrödinger, submitted from Zurich a paper containing an entirely different approach to the same problem, also culminating in the Balmer formula. It was the 'First Communication' of a theory Schrödinger called 'Quantization as eigenvalue problem' [56a]. In this paper Schrödinger acknowledges that he had got the idea to describe the electron as a wave from the 'thoughtful (*geistvollen*) Thèses of Mr. Louis de Broglie' [58b]. And in his 'Second Communication' he says that 'as soon as the radii of curvature and dimensions of the [classical] orbit are no longer large with respect to a certain wavelength which attains real meaning in q-space, then one has to look for an "undulatory mechanics"' (translated from Ref. [56b], p. 497), Schrödinger refers to Einstein's second communication to the Berlin Academy (Ref. [38b], 1925, p. 3). It is likely that Schrödinger got the information about de Broglie's *Thèses* from this reference (see Ref. [59a], p. 403). Einstein, in fact, was quite impressed by de Broglie's ideas, writing to Paul Langevin (1872–1946), one of de Broglie's thesis advisors, 'He has lifted a corner of the great veil' (Ref. [59b], p. 187).

In 1913, after obtaining his *licence ès sciences*, Louis V. P. R. Prince and Duke de Broglie (1892–1987) entered the Engineering Corps of the French Army where he became associated with wireless telegraphy and familiar with electromagnetic waves. When the war broke out he was assigned to the radio telegraphy unit at the Eiffel Tower where he served until August 1919 [58c] (see also Ref. [27b], p. 582, Ref. [59b], p. 185 and Ref. [59c], VI. *Articles biographiques*). Starting in 1922 de Broglie developed a theory of light consisting of massive, though very light (less than 10^{-50} g) photons. But in a communication submitted by Jean Perrin (1870–1942) to the session of 10 September 1923 of the Paris Academy, de Broglie turned the problem around, associating a fictitious wave with the electron, such that if the electron moves with velocity $v = \beta c < c$ the wave propagates in the same direction with velocity $c/\beta > c$ and with frequency $\nu = mc^2/h$, where $m = m_0/\sqrt{1-\beta^2}$ is the relativistic mass of the electron. For a bound electron on a circular orbit of period τ de Broglie assumed Bohr's quantization condition (4.5), $mv^2\tau = nh$ with $n = 1, 2, \dots$. [60a] From this one obtains, for the de Broglie wavelength, $\lambda = c^2/(v\nu) = h/(mv) = v\tau/n = 2\pi R/n$ where R is the radius of the orbit. In other words, the electron is accompanied by a closed wave train of n wavelengths. In his *Thèses* de Broglie uses this 'resonance condition' to justify Bohr's quantization condition (see Ref. [58a], p. 44).

In a second communication, submitted to the Paris Academy in the session of 24 September 1923 by Jean Perrin, de Broglie suggested that there is a whole group of waves—which he now calls 'phase waves'—with velocities centred at c/β and such that their group velocity is the velocity v of the particle, which follows 'the ray of its phase wave, that is (in an isotropic medium) the normal to the surfaces of equal phase'. And later he says: '*The new dynamics of the free material point is to the old dynamics (including the one of Einstein) what the undulatory optics is to the geometrical optics*' (translated from Ref. [60b], p. 549, original italics). This analogy between mechanics and optics was expounded in detail by Schrödinger in his 'Second Communication' [56b]. But the truly visionary statement in de Broglie's second communication [60b] is the following: 'A flux of electrons passing through a sufficiently small aperture would show diffraction phenomena. It is in this direction that perhaps one should look for experimental confirmations of our ideas' (translated from Ref. [60b], p. 549). This hypothesis was indeed verified experimentally in 1927, by C. J. Davisson (1881–1958) and L. H. Germer (1896–1971) in the United States and by G. P. Thomson (1892–1975), the son of J. J., and A. Reid in Scotland, with an electron beam traversing a crystal [61a,b]. Davisson and Thomson shared the Nobel Prize of 1937 'for their discovery of the diffraction of electrons by crystals'.

De Broglie gave this particle–wave duality a more systematic, relativistic form in a fourth communication [60d] submitted to the Paris Academy on 7 July 1924 by his elder brother, Maurice L. C. V. Duke de Broglie (1875–1960), connecting through Planck's constant $h = 2\pi\hbar$ the particle properties, energy E and momentum $\mathbf{p}$, with the wave properties, frequency $\nu = \omega/2\pi$ and wavenumber $2\pi/\lambda = |\mathbf{k}|$, where $\mathbf{k}$ is the wavevector. This connection may be written in the general form

$$E = \hbar\omega, \ \mathbf{p} = \hbar\mathbf{k} \tag{5.30}$$

which is the fundamental relation (I) at the outset of Pauli's 'Wellenmechanik' of 1933 (Ref. [43], p. 86; see also Ref. [40], p. 89). De Broglie had already introduced the second relation (5.30)—which goes significantly beyond the first relation due to Planck—in his theory of heavy photons in 1922. It was this relation that seems to have guided him to the idea of 'phase waves' (see Ref. [27b], p. 586). Finally, de Broglie's third communication [60c], submitted to the Paris Academy by Henri A. Deslandres (1853–1948) in the session of 8 October 1923, is of particular interest because there de Broglie derives the Planck distribution in the wave picture. It was probably mainly this calculation in de Broglie's *Thèses* [58a] that stimulated Einstein's communications [38b].

Starting his note [60c] with the hypothesis of a constant density h^{-3} in phase space (see the last section), de Broglie declares that 'today' he is able to justify it. He does it by deriving the density of states n_ν for standing waves in a box of volume L^3. Using the second relation (5.30) one deduces from this hypothesis that the phase–space volume containing just one wave mode satisfies $1 = \Delta^3 p (L/h)^3 = \Delta^3 k (L/2\pi)^3$. Denoting Z_ν as the number of modes per unit volume for which $|\mathbf{k}|(c/\beta) < 2\pi\nu$ where c/β is the velocity of de Broglie's phase waves, one has $Z_\nu = \sum'_{\mathbf{k}} L^{-3} = \sum'_{\mathbf{k}} \Delta^3 k/(2\pi)^3$, where the prime stands for the indicated restriction on $|\mathbf{k}|$. Going over to an integral by letting $\Delta^3 k$ become infinitesimal, this expression is easily evaluated to yield

$Z_\nu = (4\pi/3)(\beta\nu/c)^3$. To get de Broglie's expression $n_\nu = 4\pi\beta\nu^2/c^3$ one has to make the not very convincing assumption that $dZ_\nu/d\nu = n_{\beta\nu}$. However, for light waves ($\beta = 1$) it gives the correct value. Then, noting that each wave of frequency ν may 'transport' $n = 0, 1, 2, \ldots$ atoms, de Broglie calculates the average number at temperature T, $\langle n \rangle = \sum n \exp(-nx)/\sum \exp(-nx) = (e^x - 1)^{-1}$ where $x = h\nu/kT$. For light waves this leads to Planck's spectral density $\rho_\nu = 2n_\nu\langle n \rangle h\nu$ where the factor 2 stands for the two polarizations. An excellent presentation of this problem is contained in Section 19 of Pauli's *Lectures on Statistical Mechanics* [40] (see also Ref. [61c], Chapter 6, in particular Dirac's comment in Note 61). De Broglie was awarded the Nobel Prize of 1929 'for his discovery of the wave nature of electrons'.

Exhausted by frequent moves after his marriage in 1920, from Vienna to Jena, to Stuttgart, and to Breslau, Schrödinger and his wife Anny at last settled in Zurich where he was appointed to the chair of theoretical physics at the university in the autumn of 1921. This position had been vacant after the successive short stays of Einstein, Debye, and Max T. F. von Laue (1879–1960) before the war. But prior to assuming his duties, Schrödinger spent the summer trying to cure some respiratory problems at the resort of Arosa in the eastern Swiss Alps of Grisons. Schrödinger's colleagues at the university were two experimentalists, the director of the physics department Edgar Meyer (1879–1960) and the highly competent but modest dozent Richard Bär (1892–1940) from the Zurich banking family of Julius Bär (see Ref. [62a]). Besides the State (Kanton) University of Zurich, Switzerland's largest founded in 1833, there is also ETH in Zurich. The proximity of the two institutions favoured lively contacts such as the common physics colloquium. But the more prestigious ETH was said to be a 'waiting room first class' for ambitious professors who used it as a springboard to even more glamorous positions (see Ref. [62a], p. 217). At ETH Schrödinger's colleagues were Pieter Debye, who had returned to Zurich in 1920 with his Swiss collaborator from Göttingen, Paul Scherrer (1890–1969), and Hermann Weyl in mathematics. There was also an active and pleasant social life among these physicists and their spouses (see the section 'Life in Zürich' in Ref. [59b]). But this constellation only lasted until 1927 when Schrödinger moved to Berlin and Debye to Leipzig, and in 1930 Weyl left to become the successor of David Hilbert in Göttingen.

It is in the 'First Communication' [56a], in Eq. (5), that the famous Schrödinger equation

$$\Delta\psi + \frac{2m}{\hbar^2}(E - V)\psi = 0 \tag{5.31}$$

appears for the first time, where the potential energy of the hydrogen atom, $V = -e^2/r$, and $\hbar = K$ are used, and $\Delta \equiv (\partial/\partial\mathbf{r})^2$ is the Laplacian operator. (Since it is a non-relativistic equation we write again m for m_0.) The deductive form of this paper and the impressive mastery of the mathematical problems posed by the solution of Eq. (5.31), for which Schrödinger acknowledges the help of his friend Hermann Weyl, make it obvious that he had done much preliminary work. For historians this has posed a problem due to the extremely short time span between his reading de Broglie's *Thèses* and submitting the 'First Communication' on 27 January, and the scarcity of the existing unpublished and mostly undated notes on this work (see Section III.2 of Ref. [59a] and the section 'Before the Breakthrough' in Ref. [59b], also Ref. [62a]). Debye had suggested that Schrödinger give a talk on de Broglie's work in the colloquium; most likely it took place on 23 November 1925 (see Ref. [59b], p. 192). A few weeks later Schrödinger, in another colloquium talk, declared 'My colleague Debye suggested that one should have a wave equation; well, I have found one!' (Ref. [62b], p. 24). Presumably in between there was the decisive Christmas vacation which Schrödinger again spent in Arosa [62c].

The first wave equation studied by Schrödinger had a relativistic form. It may be obtained from Eq. (5.31) as follows. The factor $2m(E - V)$ multiplying ψ in the second term of Eq. (5.31) is just the non-relativistic approximation of $\mathbf{p}^2 = [(m_0c^2 + E - V)^2 - (m_0c^2)^2]/c^2$. Thus, writing the total energy as $\hbar\omega = m_0c^2 + E$, the relativistic generalization of Eq. (5.31) reads

$$\Delta\psi + \frac{(\hbar\omega - V)^2 - (m_0c^2)^2}{(\hbar c)^2}\psi = 0 \qquad (5.32)$$

(see Ref. [59a], p. 425 and Ref. [59b], p. 198). As mentioned on p. 372 of the 'First Communication' [56a], this relativistic equation gave the Balmer formula with half-integer quantum numbers or, more precisely, in Sommerfeld's form (5.2) but with k replaced by $k + 1/2$ (see Ref. [59a], p. 430). Since this did not agree with the spectroscopic data, Schrödinger did not publish it. Remarkably, in this early unpublished version Schrödinger had already used the symbol ψ to designate the wavefunction (see Ref. [59b], p. 193); this letter has since become the very symbol of the quantum world.

Since there were now two completely distinct theories leading to the same Balmer formula for the hydrogen atom, the problem of the relation between the two urgently demanded clarification. Einstein offered the following comment: 'Now hear that! Until now we had no exact quantum theory, and now we suddenly have two. You will agree with me that the two exclude each other. Which theory is correct? Perhaps neither is.' But Gordon informed him that Pauli had shown both theories to be identical (translated from Ref. [34b], p. 64). The question was answered independently and almost simultaneously by Schrödinger in a paper submitted on 18 March 1926 entitled 'On the relation of the Heisenberg–Born–Jordan quantum mechanics to mine' [63], and by Pauli in a letter to Pascual Jordan dated 12 April 1926 and written in Copenhagen (see Ref. [5], p. 322), where he says at the outset: 'I believe that this work [56a] counts among the most important of what has been written recently' (translated from letter [131], Ref. [5], pp. 315–16). For the association of a matrix to any function Schrödinger made use in paper [63] of an arbitrary complete set of orthogonal and normalized functions in the domain of q-space defined by the Hamiltonian operator H of the problem—which in the particular case of Eq. (5.31) is $H = -(\hbar^2/2m)\Delta + V$—while using the special set of eigenfunctions of the Hamiltonian only to derive Heisenberg's condition that the matrix representing the operator H be a diagonal matrix E.

Pauli, on the other hand, only verified this second step of the equivalence between Schrödinger's work and the 'Göttingen Mechanics'. However, he also gave the derivation of the Schrödinger equation (5.31) from de Broglie's relations. While in his 'First Communication' [56a] Schrödinger had obtained it from the Hamilton–Jacobi equation, Pauli started with the equation for de Broglie's phase waves which, as we have seen, propagate with the velocity c^2/v (called V by Pauli), so that $\Delta\psi - (v/c^2)^2 \partial^2\psi/\partial t^2 = 0$. Since $c^2/v = \hbar\omega/|\mathbf{p}|$ where, as before, $\hbar\omega$ is the total energy, insertion of the above expression for $\mathbf{p}^2$ into this wave equation leads to Eq. (3) of Pauli's letter [131]. And assuming $\psi \propto \exp(-i\omega t)$ then directly yields Eq. (5.32), which is Pauli's Eq. (4), from which Eq. (5.31) follows in non-relativistic approximation. This derivation of Eq. (5.31) by Pauli is how Schrödinger himself most likely proceeded in his unpublished notes. Equations (1) and (1') of Schrödinger's 'Fourth Communication' [56d], submitted on 21 June 1926, are in fact the non-relativistic approximation of Pauli's Eqs. (3) and (4), respectively, where units are chosen such that $m_0 = c = 1$. And in Eq. (4'') of Ref. [56d] the time-dependent Schrödinger equation

$$\Delta\psi - \frac{2m}{\hbar^2}V\psi = -\frac{2mi}{\hbar}\frac{\partial\psi}{\partial t} \tag{5.33}$$

appears for the first time.

Pauli manifestly attached some importance to this letter [131] since, contrary to his habit of writing his correspondence by hand, this letter is typed, and Pauli kept a signed carbon copy in a special envelope (see Ref. [22a], p. 773). His proof of the equivalence between the matrix and the wave approaches to quantum mechanics was brought to the attention of the physics community by van der Waerden at the symposium held in Trieste from 18 to 25 September 1972 [64]. In his comments van der Waerden observes that Pauli's 'Equation (3) is essentially the Klein–Gordon equation,' (Ref. [64], p. 282; see also Note b in Ref. [5], p. 357), where 'essentially' must mean 'for a free particle' ($E_{pot} = 0$ in Pauli's notation) or 'in a scattering state of the hydrogen atom' where $E \gg e^2/r$. This is not, however, true for Pauli's Eq. (3) but for his Eq. (4), namely the above Eq. (5.32) (see Note 33 in Ref. [22a], p.794). Indeed, with $V = 0$ and $\psi \propto \exp(-i\omega t)$ Eq. (5.32) may be written as

$$\Box\psi \equiv \Delta\psi - \frac{1}{c^2}\frac{\partial^2\psi}{\partial t^2} = \left(\frac{m_0 c}{\hbar}\right)^2\psi, \tag{5.34}$$

which is the Klein–Gordon equation where $\Box$ is called the d'Alembertian operator. Later Pauli called Eq. (5.32) the equation 'with the many fathers' (letter [147] to Schrödinger of 22 November 1926, Ref. [5], p. 356), because it was independently written by many authors (see the footnote 2 on p. 216 of Pauli's 'Wellenmechanik' [43]). He also noted that it 'has an unpleasant property. Namely it is not self-adjugate, the associated eigenfunctions ψ_n are not orthogonal' (translated from letter [140] of 5 July 1926 to Wentzel, Ref. [5], p. 333).

Now, for $V = 0$, the essential difference between Pauli's Eq. (4) and his wave equation (3) lies in the form of the dispersion relation that results for a plane wave $\psi \propto \exp[i(\mathbf{k} \cdot \mathbf{r} - \omega t)]$. In the case of Eq. (3) it reads $\omega = (c^2/v)|\mathbf{k}|$ which is a curve passing through the origin, whereas in the case of Eq. (5.34) it has the form $\omega = c\sqrt{k_c^2 + \mathbf{k}^2}$ and is represented by a curve with a gap $k_c = m_0 c/\hbar$. This gap introduces a length into the problem, namely the Compton radian length $\lambda_c = k_c^{-1} = 3.86 \times 10^{-11}$ cm, which, for $\omega < ck_c$, leads to a localized wavefunction, as in the case of the Meissner effect (see e.g. Ref. [36b], p. 214). This is the reason why in a scattering state of the hydrogen atom ($E \gg e^2/r$, i.e. for a free particle) Pauli's Eqs (3) and (4) describe physically quite different situations. In the case of a bound state, however, a new length becomes dominant, which for the hydrogen atom in the ground state is the Bohr radius $a_B = \hbar^2/me^2$ that results from combining Eqs (3.10) and (4.5) for $n = 1$; and the passage between Pauli's Eqs (3) and (4) is entirely legitimate. The name of Eq. (5.34) refers to the papers [65a] of Oskar Klein and [65b] of Walter Gordon (1893–1940), in which also electromagnetic potentials ϕ_α ($\alpha = 1, ..., 4$) are included by adding $(ie/\hbar c)\phi_\alpha$ to $\partial/\partial x_\alpha$. However, as van der Waerden observes (Ref. [64], p. 282), several other people have introduced it independently (see footnote 2 on p. 216 of Pauli's 'Wellenmechanik [43]').

Towards the end of letter [131] Pauli mentions a rather unknown physicist, saying: 'Anyway, I believe that Lánczos's Ansatz does not have much value' (translated from Ref. [5], p. 319). Heisenberg had sent Pauli two papers, one by Lánczos [66a], which he wished to discuss with Pauli on his forthcoming visit to Hamburg on 20 January 1926. The day before, Heisenberg had to give a popular talk in Stade, some 40 km west of Hamburg, and on the way back to Göttingen he visited Pauli (see letters [116] and [112] in Ref. [5]). In his letter of 27 January Heisenberg thanks Pauli 'for your invitation and the cosy (*gemütliche*) evening!' (translated from letter [117], Ref. [5], p. 281). Lánczos's work [66a] is also favourably mentioned in a footnote on p. 754 of Schrödinger's paper [63]. It so happened that Cornelius (Kornel) Lánczos (1893–1974) was in the audience when van der Waerden explained Lánczos' method in his talk (Ref. [64], pp. 282–5), and the latter got an applause in the discussion (see Ref. [64], p. 285).

This applause was, in fact, justified since Lánczos's name had become known, first for the many papers on general relativity he had written in the 1920s, and then for an algorithm [66b] to calculate numerically certain physical quantities of finite systems (see e.g. *Appendix J. The Lanczos Method* in Ref. [66c]). After his studies in Szeged, Hungary, in 1921 Lánczos went to Frankfurt-on-Main where he became an assistant to Erwin Madelung and a Privatdozent and in 1932 an associate professor of theoretical physics. In 1928 Einstein invited him to Berlin for one year as a collaborator on problems of unified field theory. In 1933 Lánczos relinquished his title of professor in protest against the Nazis and went to the United States, first to Purdue University in Lafayette and then to California. The time from 1952 to his death was spent at the Institute for Advanced Studies in Dublin (see Ref. [37b], p. 491 and Ref. [66d], vol. 6, p. 212).

This narration of the 'Setting out for the New Land' would not be complete without mentioning the problem of interpretation of the new theory, a problem that has polarized the community of physicists and even of philosophers ever since. Schrödinger, in his 'First Communication', had rejoiced at 'the human approaching (*Näherrücken*) of all these things', namely 'that the atom oscillates, when it does not radiate, respectively (*jeweilig*) in the form of *one* eigen-oscillation', while 'a macroscopic system ... in general furnishes a pot-pourri of its eigen-oscillations' (translated from Ref. [56a], p. 375). But it was Born who in two decisive papers [67a,b] led the way for what soon after became known as the 'Copenhagen Interpretation' (see the next section). In the second paper, submitted on 21 July 1926, he wrote: 'The matrix form of quantum mechanics founded by Heisenberg ... starts from the idea that an exact representation of the processes in space and time is impossible at all. ... Schrödinger, on the other side, seems to attribute to the waves, which he considers, following de Broglie, as carriers of the atomic processes, a reality of the same kind as possess light waves.' And he continues: 'None of these interpretations (*Auffassungen*) seems to me satisfactory.' Then referring 'to a remark of Einstein on the relation of wave field and light quanta; he [Einstein] said essentially that the waves serve the sole purpose of showing the quanta the way (*den Weg weisen*), and in this sense he spoke of a "ghost field" (*Gespensterfeld*)' (translated from Ref. [67b], pp. 803–4), which Born preferred to call 'guiding field' (*Führungsfeld*).

Now as early as 1909 Einstein had indeed been much concerned with this problem of particle–wave duality and, particularly, of a possible 'fusion of the wave and the emission theory of light' (see the Introduction, p. 148, and Document 60, pp. 564–5, in Ref. [68]). Born then argued that ψ stands for 'probability waves' and that $|\psi|^2$ represents the probability density. It is unfortunate that Born did not receive the recognition he merited for this work (see Ref. [27c], Section 12.d). However, in his 'Wellenmechanik' Pauli does mention Born in a footnote: 'It was Born [67b] in particular who pointed to the necessity of a *statistical interpretation* of the wavefunction' (translated from Ref. [43], p. 106).

The next decisive step came from Heisenberg with his paper 'On the intuitive content of the quantum-theoretical kinematics and mechanics', submitted from Copenhagen on 23 March 1927 [69]. In the second part of the long letter [143] dated 19 October 1926, the last of the three conserved letters from Pauli to Heisenberg from the 1920s, Pauli emphasized that in scattering processes the choice of canonical variables is quite arbitrary. However, 'why then is one allowed to prescribe only the p's and, anyhow, not *as well* the p's *as* the q's, both with arbitrary precision ... ? It is always the same thing: because of re-fraction there are no arbitrarily thin rays in the wave optics of the ψ-field, and one is not allowed to associate ordinary [classical] "c-numbers" to the "p-numbers" and the "q-numbers". One may view the world with the p-eye and one may view it with the q-eye, but if one opens both eyes at the same time one becomes crazy' (translated from letter [143], Ref. [5], p. 347). Heisenberg answered from Copenhagen nine days later noting that he, Bohr, Dirac, and Hund scuffled (*rauften sich*) for Pauli's letter. Then he says: 'About your consideration on the scattering processes I am *very* enthusiastic because one now understands the physical meaning of Born's formalism really much better than before' (translated from letter [144], Ref. [5], p. 349). For more details on these questions see Ref. [61c].

The main result of Heisenberg's paper [69] is the derivation in Section 2 of his famous uncertainty relation $p_1 q_1 = h/2\pi$ where p_1 and q_1 are the uncertainties in the measurement of the quantum variables p and q, respectively. He obtained it using the symmetry between the p and q alluded to in Pauli's letter [143], and by Fourier-transforming a Gaussian wave packet in q-space with width q_1 into a Gaussian wave packet in p-space with width p_1. Heisenberg, who had informed Pauli in a letter of 14 pages dated 23 February 1927 on this work (letter [154] of Ref. [5]), was surprised by Pauli's positive response, culminating in the remark 'it dawns in quantum theory' (translated from Ref. [54], p. 46). Referring to Bohr, Pauli notes at the outset of his 'Wellenmechanik' [43] that this relation follows more generally and more directly from the fundamental property $\Delta q_i \Delta k_i \sim 1 (i = 1, ..., f)$ of the Fourier transformation between a peak of widths Δq_i in q-space and a peak of widths Δk_i in k-space, and similarly $\Delta \omega \Delta t \sim 1$, just by applying de Broglie's relations (5.30). The result is Pauli's fundamental relations (II) of Ref. [43]. Heisenberg's result is not, however, the optimal form the uncertainty relation may take. Indeed, defining the average value of any quantum variable A in a state defined by the wavefunction $\psi(q)$ in Born's sense as $\langle A \rangle = \int \psi^* A \psi dq$, where ψ^* is the complex conjugate of ψ, and the uncertainty attached to a measurement of A as $\Delta A = \sqrt{\langle (A - \langle A \rangle)^2 \rangle}$, one deduces the rigorous relation (see Pauli, 'Wellenmechanik' [43], Ziff. 3 and Ref. [42], Eq. (3.31))

$$\Delta p \Delta q \geq \frac{\hbar}{2} \qquad (5.35)$$

which, more generally, is valid for any pair of canonical variables.

Pauli had been invited by the Nobel Committee for Physics to make nominations for the Prize of 1932. In his response Pauli finds Heisenberg 'to be most suited for the Prize', and he cites the two papers on the *reinterpretation* [50] and on the *uncertainty relations* [69] as the main justification. As to the 'part of other researchers in the development of modern quantum theory and its relation to Heisenberg's achievements', Pauli notes: 'An achievement of comparable scale (*Format*) is *Schrödinger's* setting up of the wave equation named after him.' Pauli justifies his preference by the following two circumstances: 'First, Heisenberg's matrix mechanics precedes Schrödinger's work in time, second, Heisenberg's creation must be valued as the still more original one since, concerning the idea, Schrödinger could build (*anknüpfen*) in an essential way on de Broglie (already Nobel Laureate). In addition, as concerns Heisenberg's second achievement, the setting up of the uncertainty principle, it is correct that its foundation has later been improved, simplified and deepened by *Bohr*. However, Heisenberg's merit to have recognized the principle for the first time, as well as to its content as also to its significance, thereby is not lessened' (translated from the letter to the Nobel Committee for Physics, dated 29 January 1932, Zurich [27a]). In 1933 Heisenberg was awarded the Nobel Prize of 1932 'for the creation of quantum mechanics, the application of which has, *inter alia*, led to the discovery of the allotropic forms of hydrogen'. Schrödinger shared the Nobel Prize of 1933 with Dirac 'for the discovery of new productive forms of atomic theory', while Born shared the Prize only in 1954 'for his fundamental research in quantum mechanics, especially for his statistical interpretation of the wave-function'.

A very last step to arrive at a general form of the non-relativistic quantum mechanics of the electron was necessary in order to incorporate Pauli's exclusion principle which, according to Heisenberg [44] and Dirac [39b], is expressed by the antisymmetrized products (5.20) of one-electron wavefunctions, including spin, the fourth degree of freedom. It was not at all obvious how the configuration space $\mathbf{r}$ of the electron should be enlarged to incorporate spin. Charles Galton Darwin (1887–1962), the grandson of the biologist and a friend of Niels Bohr from their time in Manchester with Rutherford, described spin by associating a vector wavefunction to the electron [70a]. This, however, as observed by Pauli in his paper 'On the quantum mechanics of the magnetic electron' [26], submitted on 3 May 1927, led to difficulties with the two-valuedness, to which Darwin responded in a second paper [70b] that his two components—which define a unit vector—are equivalent to Pauli's. It turned out the following year that Darwin's approach was indeed a valid non-relativistic approximation of the Dirac equation in the presence of forces (see page 185). On the other hand, Pauli also rejected the idea to associate, in analogy with orbital angular momentum, the azimuthal angle φ as a variable canonically conjugate to the spin component in a fixed direction, in units of $\hbar$, $s_z = \pm 1/2$. The reason was that the wavefunction $\exp(is_z\varphi)$ would be two-valued. He concluded that instead one may choose s_z itself, so that the wavefunction is $\psi(\mathbf{r}, s_z)$, to which, for given energy E, belong two eigenfunctions $\psi_{\alpha E}(\mathbf{r})$ and $\psi_{\beta E}(\mathbf{r})$. Writing them as a two-component column wavefunction Ψ_E, the spin components s_x, s_y, s_z are 2×2 matrices acting on Ψ_E and satisfying the general angular momentum commutation relations

$$J_x J_y - J_y J_x = iJ_z, \text{ and cyclic; } \mathbf{J}^2 = J(J+1), \qquad (5.36)$$

where the second equation is identical to Eq. (4.14)—which in the old quantum theory had only been guessed! But spin 1/2 satisfies additional relations which are conveniently expressed in terms of the celebrated Pauli matrices

$$\sigma_x = \begin{pmatrix} 0 & 1 \\ 1 & 0 \end{pmatrix}; \quad \sigma_y = \begin{pmatrix} 0 & -i \\ i & 0 \end{pmatrix}; \quad \sigma_z = \begin{pmatrix} 1 & 0 \\ 0 & -1 \end{pmatrix} \qquad (5.37)$$

that form the vector $\vec{\sigma} = 2\mathbf{s}$. The additional relations are $\sigma_x^2 = \sigma_y^2 = \sigma_z^2 = 1$ and $\sigma_x\sigma_y = -\sigma_y\sigma_x = i\sigma_z$, and cyclic. The latter were communicated to Pauli by Jordan (see Pauli's letter [157] to Jordan of 12 March 1927, Ref. [5], p. 386).

In the following mathematical and physical discussion of paper [26] the problem of the relativistic corrections for hydrogen-like atoms in Section 4b is of particular interest. In his Eq. (19) Pauli writes the spin–orbit interaction— which gave rise to the splitting (5.6)—by putting in the Thomas factor of $1/2$ 'by hand'. This is Eq. (19) of Ref. [26] which in simplified notation reads $H_{so} = Ze^2 \mathbf{L} \cdot \vec{\sigma}/(4k_c^2 r^3)$. This expression could be fully justified only in the following year in the framework of the relativistic Dirac equation; it is identical to the penultimate term in Eq. (89) of Pauli's 'Wellenmechanik' (see Ref. [43], pp. 237-8). Furthermore, Pauli noted in this discussion that in higher approximations particular forcing terms (*Zwangskräfte*) are needed in order to maintain the electron's electric dipole moment at zero, and he concludes: 'for this reason I have so far not been able to arrive at a relativistically invariant formulation of the quantum mechanics of the magnetic electron which could be considered sufficiently natural and compelling' (translated from Ref. [26], p. 619). As we shall see, this was Dirac's privilege.

Between Jungiusstrasse and Sankt Pauli

The three days of 29, 30 November and 1 December [1955] I spent in Hamburg where I had not been for a long time. There I gave a lecture on invitation, and in one newspaper my name and the hotel in which I stayed were mentioned. This gave rise to a romantic adventure: a woman whom I did know 30 years ago but whom I had completely forgotten then announced herself. I had completely lost sight of her when, as a young girl at that time, she became addicted of morphine and I considered her lost. She phoned me on 29 November at about 5 p.m., and I saw her on 1 December for two hours before I got on the fast sleeping-car train to Zurich, to which she accompanied me. An entire human life of 30 years passed by before me during these two hours, of her cure, a marriage and a divorce, with war and national-socialism as historical background. [71]

This recollection of a more human aspect of Pauli's life expressed here in a letter to C. G. Jung from 1956 is a reassuring sign that physics, symbolized by the Jungiusstrasse, the location of the university's physics department, was not all that counted. Indeed, Hamburg also harbours the famous amusement district of Sankt Pauli—what an appropriate name for our story!—particularly along the Reeperbahn avenue. However, for Pauli the quality of the entertainment was not a negligible parameter, as is evident from the postcard he wrote on 11 June 1926 to Wentzel which closes with the observation: 'The cinemas and cabarets only partially correspond to my demands but the theatres, in particular, could bear a revalorization' (translated from Ref. [5], p. 331).

Founded at the junction of the Alster with the Elbe river as a fortified border post against the Danes in the ninth century, Hamburg soon developed into a relatively independent settlement with an important harbour, joining the Hanse trading league of northern German cities in the thirteeth century and Luther's reformation in 1529. Just in time to withstand the troubles of the 30 Years War a ring of ramparts had been erected around the city, which today forms the semi-circle of the walled streets. To the north the university, founded in 1919, has its main building, from which a green belt follows the western walled streets to the Sankt Pauli landing-stage on the Elbe. This belt is crossed by the Jungiusstrasse in its northern part of the fairground and joins the Reeperbahn further south. The inner city is centred around the city hall and the Jungfernsteg bordering the Alster basin, extending east to the main railway station.

Among Pauli's colleagues Gregor Wentzel, with his way of life, was perhaps Pauli's most kindred spirit. After spending the winter term of 1925/6 in Hamburg where he stood in for Lenz who was ill (see Ref. [5], p. 297), Wentzel was often in Paris where Pauli hoped to visit him some time (see Ref. [5], p. 322). In fact, the first of Wentzel's two papers on the Compton effect [73a] was signed from Paris, which fact incited Pauli to issue the following comment: 'The question is this, whether the indication of Paris at the bottom of your paper already suffices to justify all this psychologically and whether at proof-reading you should not change it more specifically into Paris, Moulin-Rouge or something analogous' (translated from letter [162] of 16 Mai 1927, Ref. [5], p. 393). And the previous letter from Pauli to Wentzel, dated 5 December 1926 and written for the first time in the '*Du*' form, closes like this: 'Otherwise I am fine. I have noticed that drinking wine agrees very well with me. After the second bottle of wine or champagne I usually adopt the manners of a good companion (which I never have in the sober state) and then may at times (*unter Umständen*) enormously impress the surroundings, in particular if they are female' (translated from letter [149], Ref. [5], p. 363).

That, according to Otto Stern, was a habit Pauli had acquired in Hamburg. Indeed, in a recorded interview with Res Jost (1918–90) in Zurich that took place on 2 December 1961 Stern said: 'Baade. He was the main seducer of Pauli concerning wine, alcohol, yes. But when he came to Hamburg Pauli was such that he, in fact, was a strict teetotaller (*Temperenzler*), never had drunk alcohol, in any form, you know, and [he] inveighed horribly against people who [did]' (translated from Ref. [73b]). This is corroborated by the recollection of Harry Lehmann (1924–98) who much later succeeded Lenz in Hamburg, to whom Pauli once observed: 'When I came to Hamburg I passed, under the influence of Stern, directly from mineral water to champagne' [73c]. The bachelors among the Hamburg physicists regularly met for lunch, mainly Stern and Lenz and, among the younger ones, Pauli and, in early 1926, Frenkel and Gordon when in 1927 he became the assistant of Peter Paul Koch (1879–1945), the professor of experimental physics, and Jordan when he succeeded Pauli as Lenz's assistant in 1928, and the astronomer Walter Baade (1893–1960) and, less regularly, the mathematician Emil Artin (1898–1962) (see Ref. [73d], p. 59, also Ref. [24c]). As to Walter Baade, Harry Lehmann in Ref. [73c] also mentions Pauli's good relations with the astronomers at the observatory in Bergedorf, situated on the city limits to the east, where he often went to drink wine with them at full moon (when observation was impossible!).

In the autumn of 1927 two others came to Hamburg, sent by Bohr from Copenhagen to work with Pauli, namely the American Isidor I. Rabi (1898–1988) and the Japanese Yoshio Nishina (1890–1951). In his obituary for Stern, Rabi writes: 'On arrival I was very pleasantly surprised to find Stern's molecular beam laboratory in full swing. ... Hamburg University at that time was one of the leading centers of physics in Germany, therefore in the world. What characterized physics in Hamburg was the extremely close collaboration between Stern and Pauli, experiment and theory. Some of Pauli's great theoretical contributions came from Stern's suggestions, or rather questions; for example, the theory of magnetism of free electrons in metals. ... From his earlier work on entropy and Nernst's heat theorem, to his demonstration of space quantization with the glorious Stern–Gerlach experiment, the reality of matter waves and the anomalous magnetic moment of the proton, Stern was always close to the basic problems of physics as they evolved' [73e].

Wilhelm Lenz had obtained his doctorate under Sommerfeld in Munich in 1911 and became a Privatdozent there in 1914. In 1920 he became an associate professor in Rostock and one year later a full professor in Hamburg where, together with Otto Stern, he established the Centre for Atomic Physics. Lenz was a cultivated and sensitive man with delicate health [73d]. In June 1925 Pauli wrote to Sommerfeld: 'I am very sad about Lenz's health. Now he has left for a Sanatorium in the vicinity of Heilbronn. ... If only this would help him! But I am afraid that, apart from slight variations, his state will become chronic. I am awfully sorry for him. It is much less pleasant in Hamburg since he is so difficult to come by. I always had great pleasure and stimulation from discussing physical questions with him but, unfortunately, this has been possible only in a very limited measure lately' (translated from letter [92], Ref. [5], p. 224).

As we saw on page 81, Lenz was known mainly for his solution of the problem of hydrogen in external fields. But in 1924 he also suggested to his student Ernst Ising the problem of the statistical mechanics of a chain of magnetic moments $\pm\mu$ coupled between nearest neighbours, which led to the famous Ising model [74a]. This work by Ising was based on a little-known short paper by Lenz [74b] in which Lenz showed that in a solid where the atomic moments μ in a magnetic field H are restricted to point in very few symmetry directions, essentially $\pm\mu$, the result, proportional to $\tanh(\mu H/kT) \simeq \mu H/kT$, is still a Curie law. In a letter to S. G. Brush, Ising—who was just two weeks younger than Pauli—wrote: 'I discussed the result of my paper widely with Professor Lenz and with Dr. Wolfgang Pauli, who at that time was teaching in Hamburg' [74c]. Ising became a teacher; he spent the war in isolation in a small town in Luxembourg and emigrated in 1947 to the United States (see Ref [74c]). One year after Ising Lucy Mensing was Pauli's collaborator on the problem of the polarizability, and obtained her doctorate working with Wilhelm Lenz (see pages 62–3).

Emil Artin studied in Vienna and Leipzig. In 1923 he became a Privatdozent in Hamburg, in 1925 an associate professor, and the following year a full professor. In 1937 he emigrated to the United States because of his Jewish wife but returned to Hamburg in 1958. In the winter term of 1927/8 Pauli sat in on Artin's course on the 'Representation theory of semi-simple systems', taking 44 pages of notes which he consulted again in the late 1930s for his research on elementary particles (see letter [544] in Ref. [16a], p. 625; the notes are in the *Pauli Archive*). The astronomer Walter Baade got his doctorate in Göttingen in 1919 and then, like Pauli, became a 'scientific auxiliary worker' (*Hilfsarbeiter*) in Hamburg but at the observatory where in 1920 he discovered the asteroid Hidalgo. In 1931 he became an associate professor, but the following year he went to California where he became staff member at the Mount Wilson Observatory and later also at the Mount Palomar Observatory in Pasadena. He became known for certain star populations that carry his name (see Ref. [66d]). From the Mount Wilson Observatory Baade, together with Pauli, published a note 'On the radiation pressure exerted on the particles in the tails of comets' [75]. The problem treated in this note was to calculate the ratio μ of the radiation pressure to the gravitational attraction exerted by the sun on the tail particles, for which reasonable agreement with observations was found.

Returning to Otto Stern's interview with Res Jost, Stern then said: 'but, of course, it was very nice with Pauli for, although he was thus highly learned, one could all the same really discuss physics with him. And . . . you know, he was not allowed to enter our laboratory, because of the Pauli effect. Don't you know the famous Pauli effect? Jost: I know it all right, but I didn't know that this led to such consequences. Stern: Yet, now, as I said, we always went eating together, he always fetched me. But he did not enter, instead he only knocked, and I then came to the door and said I'm coming. Oh yes, we were very superstitious at that time. Jost: Did something ever happen? Stern: Alas, many things did happen. The number of Pauli effects, the *guaranteed* (*verbürgten*) Pauli effects, is enormously large. Jost: Also in your institute? Stern: No, for he was not allowed to enter there. Now, no, but I mean, the superstition of experimental physicists is indeed amazing. So, for instance . . . I don't remember who it was . . . brought his apparatus a flower every day, in order to keep it, yes . . . in good temper . . . I had a somewhat higher method. In Frankfurt [1914–21, except for military service during the war] I had a wooden hammer lying next to my apparatus. And I always threatened the apparatus. One day the wooden hammer had disappeared, whereupon the apparatus did not function, until after three days it was found again—the wooden hammer—then the apparatus functioned again. . . . Some time when there was a particularly evil series of bad luck (*Strähne Malheur*), then I invited Pauli for a meal—in Hamburg there were wonderful restaurants, one of the nice ones: "Schühmann" or "Halali", in spite of the name it was a wonderful restaurant, or "Four Seasons" or so. And then we ate very well, and drank and, this was already after the [education of Pauli in eating and drinking] . . . , it did not last long that Pauli was educated, and then I bewailed of my sufferings, and that helped. . . . No, anyhow, as I said, Pauli of course was the greatest "asset" in Hamburg' (translated from Ref. [73b]).

Concerning the Pauli effect, Markus Fierz (b. 1912), who among all physicists has the most intimate knowledge of Pauli's thinking, explains: 'Pauli himself thoroughly believed in his effect. He has told me that he senses the mischief already before as a disagreeable tension, and when the anticipated misfortune then actually hits—another one!—he feels strangely liberated and lightened. It is quite legitimate to understand the "Pauli effect" as a synchronistic phenomenon as conceived by Jung' (translated from Ref. [76a], pp. 190-1, see also Ref. [76b], p. 17). As we shall see later, synchronicity as a principle of acausal connections was not only a subject of letters and discussions between C. G. Jung and Pauli in the late 1940s but also the subject of an essay by the former [77a]. In this sense Pauli wrote to Jung on 28 June 1949: 'Experience has indeed shown me that what you call an *"event of conjunction"*, is in general favorable for the occurence of (what by "foreigners" is designated as "radioactivity") the "synchronistic" phenomenon. And, indeed, the latter predominantly occurs then, *when the pairs of opposites keep the balance among each other as much as possible.* In the I Ging this moment is characterized by the sign "Dschen" (shaking, thunder)' (translated from letter [37], Ref. [72], p. 44). In the *I Ging*, the Chinese *Book of the Changes*, 'Dschen', the shaking, frightens and brings loss, but 'after seven days' the things lost are returned (see Ref. [77b], *51. DSCHEN*) So the Pauli effect may indeed be seen as a personalized 'Dschen'.

As mentioned in the last section the zero-point energy was a favourite discussion topic between Pauli and Stern in Hamburg. Stern himself confirmed this in a letter to me, from Berkeley, dated 21 January 1960, in the following words: 'Pauli and I continually discussed the question of the zero-point energy in Hamburg in the early 1920s. We always went together from the Institute to the Kurie house [77c] for lunch, and thereby there were two problems which we talked over time and again. Pauli was already convinced at that time of the spherical symmetry of the H-atom against which I bristled violently (you should remember that this happened before Heisenberg and electron angular momentum!). I for my part always tried to convert Pauli to the zero-point energy against which he had the gravest hesitations. My main argument was that I had calculated the vapour pressure differences of the neon isotopes 20 and 22, which [Frances William] Aston [1877–1945] had tried in vain to separate by distillation. If one calculates without zero-point energy there results such a large difference that the separation should have been quite easy. The argument seemed (and seems) to me so strong because one does not assume anything else than Planck's formula and the fact that the isotopes are distinguished only by the atomic weight. I have myself not undertaken any vapour pressure measurements because Aston's experiments seemed to me quite decisive [78b]. I did not publish this matter at the time which now appears very incorrect to me. As to the gravitation of the zero-point energy Pauli has communicated this to me essentially in the form you mention, only much later, on one of my regular visits to Zurich after the war, about 1950 or later. He mentioned, however, that it concerned a calculation which he had done much earlier. When, he did not mention or I have forgotten it. Sorry. So, this is all I presume' (translation from Ref. [78a], p. 130; German handwritten original, property of Charles P. Enz).

Shortly in 1911 after Planck had introduced the zero-point energy (ZPE) $\hbar\omega/2$ of the material oscillators (assumed to be one dimensional) in equilibrium with the photon gas [79a], Stern applied the same idea to calculate the vapour pressure of a monatomic gas in equilibrium with its solid state [79b]. As Stern says in the above quotation, a decade later he used his calculation to show that only by including the ZPE of the (three-dimensional) oscillators in the solid— the phonons—could the expression for the vapour pressure p be consistent with the observed negligible (linear) dependence on the isotopic mass M. Since $\omega \propto M^{-1/2}$, omission of the ZPE in the formula for p would have resulted in an exponential effect.

In his lectures on statistical mechanics [40] it was Pauli who showed, by introducing an unspecified ZPE $e_0(\omega)$, that for $e_0(\omega) = (3/2)\hbar\omega$ the M-dependence of $\log p$ becomes logarithmic and hence negligible. This result is given on p. 85 of Ref. [40] where, however, in the argument of the second log-term k should read kT; in a more physical notation this formula reads

$$\log p = \log\left(\frac{kT}{\lambda_{th}^3}\right) + 3\left\langle \log\frac{\hbar\omega}{kT} \right\rangle - \frac{\lambda_0}{kT} + \frac{1}{kT}\left\langle e_0(\omega) - \frac{3}{2}\hbar\omega \right\rangle. \quad (5.38)$$

Here $\lambda_{th} = \hbar\sqrt{2\pi/MkT}$ is the 'thermal wavelength', $-\lambda_0$ the minimum potential energy per atom in the solid, and $\langle...\rangle$ means the average over the phonon spectrum. Taking for this spectrum the Einstein model of a single frequency and setting $e_0(\omega) = (3/2)\hbar\omega$, Eq. (6) of Stern's paper [79b] follows. Equation (5.38) is explicitly derived in Ref. [79c] and agrees with Eq. (3) there.

Stern argued that the experimentally confirmed negligible isotope dependence of the vapour pressure shows the reality of the ZPE. For Pauli, on the other hand, it was a confirmation, for material oscillators, of the value $(1/2)\hbar\omega$ per degree of freedom which follows from quantum mechanics. Note that another and perhaps even more fundamental confirmation is the absence of a permanent magnetic moment of the electron gas (neglecting spin), as shown in Ref. [78a]. However, as is evident from the examples discussed in Ref. [79c], this fact should not be taken as a formal argument in favour of the conventional order of the operators in the Hamiltonian of a material harmonic oscillator. And this is also true for the quantized free electromagnetic field which may be written in the form of a sum of oscillator terms. Indeed, in deriving this form in his 'Wellenmechanik' Pauli observes 'that it is more consistent not to introduce here a zero-point energy of $(1/2)\hbar\nu_r$ per degree of freedom, in contrast to the material oscillator. For, on the one hand, because of the infinite number of degrees of freedom this would lead to an infinitely large energy per unit volume, on the other hand it would be unobservable in principle since it is neither emitted, absorbed or scattered, hence cannot be enclosed inside walls and, as is evident from experience, it also does not produce a gravitational field' (translated from Ref. [43], p. 250; see also Ref. [42], p. 180).

Thus, contrary to Nernst's fascinating but quite fantastic speculations (see Ref. [78a], p. 126), Pauli rejected the reality of the ZPE of the radiation field, 'in contrast to the material oscillator'. In fact, as Stern confirms in the above letter to me, Pauli had calculated the gravitational action of the ZPE of the former earlier and had found that, even with a cut-off at the classical electron radius $\alpha\lambda_c$, the radius of the universe 'would not even extend to the moon' (Ref. [79c], p. 842). This difference is already emphasized in Pauli's famous paper 'On gas degeneracy and paramagnetism' submitted on 16 December 1926 [80]. In this paper Pauli proposes, following Dirac, to treat the material gas according to Fermi statistics, although this might appear to be put into question by the complete analogy between material gases and the gas of light quanta emphasized by Einstein. But he continues: 'On the other hand, one must consider, however, that, with regard to the occurrence of a zero-point energy, material systems (e.g. a crystal lattice) in general differ from the radiation and that, precisely on this point, Fermi statistics leads to more satisfactory results than the Einstein–Bose one' (translated from Ref. [80], p. 84). The last remark concerns the question whether a ZPE may also be associated to an ideal quantum gas, which Pauli had already discussed in the first part of his important letter [143] of 19 October 1926 to Heisenberg quoted in the last section.

Pauli's paper [80] was his third encounter with magnetism (see pages 56f and 91). Indeed, in Section 2 of Ref. [80] he states: 'In the following an attempt shall be made to show how on the basis of this [Fermi] statistics the fact that many metals are diamagnetic or only very weakly paramagnetic can be brought into harmony with the existence of a magnetic moment of the electron. Hereby the conduction electrons in the metals in question shall be considered as an ideal gas' (translated from Ref. [80], p. 85). Pauli had a sudden vision of this fact in the mentioned letter [143]: 'I just now notice that the application of Dirac's theory to gases of atoms with angular momentum implies that a monatomic vapour which is normally paramagnetic at high temperature must gradually cease to be paramagnetic as soon as the degeneracy sets in at diminishing temperature' (translated from Ref. [5], p. 341).

As mentioned on page 127, Pauli derived the quantum statistics from the grand-canonical ensemble. For the purpose of calculations he defined the functions

$$G_{\mp}(A) \equiv \frac{4}{3\sqrt{\pi}} \int_0^\infty \frac{x^{3/2}\,dx}{e^{x-A} \mp 1}; \quad F_{\mp}(A) \equiv \frac{dG_{\mp}}{dA} \tag{5.39}$$

introduced on p. 101 of his lectures on statistical mechanics [40], where $AkT = \mu$ is the chemical potential appearing in Eqs (5.19). Now the functions $G_+(A)$ and $F_+(A)$ appear for the first time in the paper [80] in the notation $G_+(-A) \equiv G(\alpha)$ and $F_+(-A) \equiv F(\alpha)$, and Pauli adhered to the definitions (5.39) for almost 30 years in his teaching the subject. In the following I wish to show the close connection between this paper [80] on gas degeneracy and paramagnetism and Pauli's rather austere lecture notes [40] on statistical mechanics which, as emphasized in the preface to Ref. [40], primarily served the purpose of a compendium for Pauli's personal use.

In Eq. (35') of Ref. [80] the general formula for the susceptibility of a monatomic gas with angular momentum quantum number j is expressed in terms of the function $F(\alpha)$ as follows:

$$\chi = \frac{n\mu^2}{3kT}\left(\frac{-F'(\alpha)}{F(\alpha)}\right). \tag{5.40}$$

Here n is the atomic density and $\mu = \sqrt{j(j+1)}g\mu_B$ the effective atomic magnetic moment, g being the Landé factor (4.11). In the high-temperature or dilute gas limit where Boltzmann statistics hold and $\alpha \rightarrow +\infty$, one finds $F'(\alpha)/F(\alpha) = -1$, and Eq. (5.40) becomes the well-known Curie law $\chi_C = C/T$ with a Curie constant $C = n\mu^2/3k$. The new result of the paper [80], however, was the expression valid in the zero-temperature limit $\alpha \rightarrow -\infty$. It is the celebrated Pauli spin susceptibility which in Eq. (36') of Ref. [80] is written as

$$\chi_0 = \frac{3n\mu^2}{10e_0}; \ e_0 \equiv \frac{3h^2}{40m}\left(\frac{6n}{\pi(2j+1)}\right)^{2/3}. \tag{5.41}$$

In contrast to the Curie law this susceptibility is temperature independent. In Eq. (5.41) e_0 is the *zero-point energy* per atom, E_0/N, where E_0 appears in identical form but with $j = 0$ on p. 104 of Ref. [40].

As announced in the above quotation from Pauli's paper [80], the most important application of the result (5.41) concerns the conduction electrons in metals, for which $j = 1/2$ and $g = 2$. In this case a physically more telling expression is obtained as the difference of electrons with spin parallel and antiparallel to the magnetic field. According to the exclusion principle this difference is non-zero only for states within an interval of approximate width kT at the maximum energy E_F of the filled states. This reduces the Curie susceptibility χ_C by multiplication with a small factor of approximately kT/E_F. The rigorous expression is (see e.g. Section 7.1 of Ref. [81a], also Section 31 of Ref. [36b])

$$\chi_P = \frac{3n\mu_B^2}{2E_F} = 2\mu_B^2 N(E_F), \tag{5.42}$$

which appears on p. 106 of Ref. [40] in the form $\overline{M} = HV\chi_P$. In Eq. (5.42) $N(E)$ is the density of states per spin which may be obtained from Eq. (2) of Ref. [80], namely $N(E) = dZ(E)/(V\,dE) = (2m)^{3/2}E^{1/2}(2\pi)^{-2}\hbar^{-3}$. The definition of the Fermi energy E_F or chemical potential μ at $T = 0$ leads to the following condition which, for $j = 0$, appears on p. 104 of Ref. [40]: $2Z(E_F) = n$ where $Z(E) = \int N(E)dE$ is the number of states per spin with energy smaller than E. Hence $E_F = (3\pi^2\hbar^3 n)^{2/3}/2m$. As in his first encounter with magnetism (see pages 56f) Pauli compares his result with experimental data, here for the alkali metals, writing for this purpose $\chi_P = n^{1/3}2.209 \times 10^{-14}$. And he observes, first, that χ_P is much smaller than the Langevin(–Debye) paramagnetism, Eq. (3.4), calculated with one magneton per electron and, secondly, that the order of magnitude is comparable with the expected orbital diamagnetism. The latter, in fact, was calculated three years later by Landau [81b] with the result that $\chi_L = -\chi_P/3$ (see e.g. Section 7.2 of Ref. [81a], also Section 31 in Ref. [36b]).

For electrons one also verifies that in Eq. (5.41) $e_0 = (2/n) \int_0^{E_F} N(E)E dE$ which therefore is seen to be the minimum energy of the electron gas. It is attained at zero temperature, thus justifying the name zero-point energy. The corresponding ZPE of a Bose gas, given on p. 106 of Ref. [40], is $e_0 = E_0/N_0 = (3G_-(0)/2F_-(0))kT \to 0$. This difference between FD and BE statistics manifests itself in the number fluctuation which is the same type of quantity as the general uncertainty ΔA defined before Eq. (5.35), namely $\Delta_s \equiv \sqrt{\langle (n_s - \langle n_s \rangle)^2 \rangle}$. Here $\langle n_s \rangle$ is the average occupation number, given for BE and FD statistics by Eqs (5.19). Using $\exp[-(\beta\varepsilon_s + \alpha)n_s]$ in the averaging as weights, one finds for FD and BE statistics

$$\frac{\Delta_s}{\langle n_s \rangle} = \left(\frac{1}{\langle n_s \rangle} \mp 1 \right)^{1/2}, \tag{5.43}$$

corresponding to Eq. (24a) and the following one, respectively, of Pauli's paper [80] (see also Eqs (338a,b) in Ziff. 14 of Ref. [43] and Eqs (14.20a,b) of Ref. [42]). (In the FD case the calculation is trivial since, obviously, $\langle n_s^2 \rangle = \langle n_s \rangle$.) Hence, as Pauli observes on p. 95 of Ref. [80], with FD statistics the number fluctuation is smaller than with Boltzmann statistics and vanishes at zero temperature where $\langle n_s \rangle = 1$ for $\varepsilon_s < E_F$, while the fluctuation is larger with BE statistics.

Pauli had expressed the hope that, by assuming phase relations among de Broglie waves, this behaviour of a Fermi gas could perhaps 'yield an indication for a future physical explanantion of the fundamental assumption of Fermi statistics and thereby also of the "equivalence rule" [exclusion principle]' (translated from Ref. [80], p. 96). This, however, was not to be since, as Einstein wrote in his proposition of Pauli for the Nobel Prize (see page 125), this principle is 'independent from the other basic axioms' of quantum physics. Rather, Eqs (5.43) give rise to a distinction between the respective ZPEs: in the FD case $e_0 = (3/5)E_F$ with no fluctuation, while in the BE case there is Bose condensation into the zero-momentum state $\varepsilon_0 = 0$ (see Ref. [40], p. 107) which according to Eqs (5.19) implies $\langle n_s \rangle = \Delta_s = 0$ for $\varepsilon_s > 0$ but $e_0 = 0$ with $\langle n_0 \rangle = \Delta_0 = \infty$ for $\varepsilon_0 = 0$. In letter [143] of 19 October 1926 Pauli therefore concludes that a ZPE 'may only be forced (*aufpfropfen*) artificially upon the Einstein–Bose theory' (translated from Ref. [5], p. 340). Equations (5.43) already appear in this letter where Pauli emphasizes that, because of its quantum origin, a positive second term as in the BE case does not look reasonable and that this term, 'wave-theoretically, can only be derived under the assumption that the phases of the partial waves are disordered' (translated from Ref. [5], p. 341).

Pauli's paper on gas degeneracy and paramagnetism became the point of departure for the modern electron theory of metals. On the occasion of the meeting of the Lower Saxony section of the German Physical Society held in

Hamburg on 5–6 February 1927, at which Born also participated, Sommerfeld was invited by the Science Faculty to give a lecture (see letter [149], Ref. [5]). Pauli took the oportunity to let Sommerfeld read the proofs of his paper [80]. Afterwards Sommerfeld said 'that he was very much impressed by it and that one should make further applications to other parts of metal-theory' (letter from Pauli to Rasetti of 30 October 1956 quoted on p. 69 of Ref. [34b]). During the coming summer term Sommerfeld gave a course on the subject, in which he 'arrived at noteworthy results which in the following are to be represented' (translated from Ref. [82a], p. 1). This is the long paper in two parts, 'On the electron theory of metals' [82a], which, aided by his assistant Hans Bethe (b. 1906), he later extended into the review [82b] that appeared in the same volume of the *Handbuch* as Pauli's 'Wellenmechanik' [43]. This review was predominantly the work of Bethe who obtained his Ph.D. under Sommerfeld in 1928. Pauli, on the other hand, gave a report on the 'Quantum Theory of Magnetism: The Magnetic Electron' at the Sixth Solvay Conference which was held in Brussels from 20 to 25 October 1930 (see Ref. [38d], pp. 194–9).

There is only one more published contribution by Pauli to solid-state physics. It is a communication to the meeting of the Lower Saxony section of the German Physical Society held in Göttingen on 9 February 1925. It has the title 'On the absorption of the residual rays in crystals' [83]. As must be concluded from Pauli's correspondence, between December 1923 and February 1925 he devoted much effort to the problem of heat conduction in insulating solids (see letters [50, 52, 54, 56, 58, 74, 83, 86] in Ref. [5]). On 21 March 1924 Pauli reported to Bohr: 'My calculations on the kinetic theory of the heat conduction in the solid now advance well, and I am in a joyful mood. It just gives me pleasure that, to start with, a purely classical problem is at hand here. But at low temperatures where the quanta become active also interest-ing quantum-theoretical considerations seem to emerge' (translated from letter [58], Ref. [5], p. 151). But Pauli ran into difficulties and gave up.

Thus the problem reported in his communication [83] is not that of heat conduction but of light absorbtion by a one-dimensional chain of particles. More explicitly Pauli states that 'in addition it was assumed that only neigh-bouring particles of the chain exert forces on each other and that all particles possess equal mass m (however, adjacent particles possess opposite charges)' (translated from Ref. [83]). But the quote in parantheses is easily shown to lead to an unstable chain! It is therefore advisable not to be too explicit about the physical nature of the forces and to define the model by its equation of motion, as Pauli in fact does:

$$\ddot{u}_n = \omega_0^2(u_{n+1,n} - u_{n,n-1}) + \varepsilon\omega_0^2(u_{n+1,n}^2 - u_{n,n-1}^2); \quad u_{n+1,n} \equiv u_{n+1} - u_n.$$
$$(5.44)$$

Here u_n is the elongation of the nth particle from its equilibrium position and the perturbation parameter ε has the dimension of a reciprocal length. Pauli calculated the damping constant γ due to absorption of infrared radiation. Without giving any details he states the result to lowest order in ε as $\gamma = 4\varepsilon^2 kT/m\omega_0$.

This makes it rather difficult to understand what Pauli did. I have tried to give in *Note* [84] an independent derivation of Pauli's result which, however, is at variance with Peierls's description of Pauli's notes that Peierls had seen (see Note 80 in Ref. [22a], also Ref. [14b], p. 69). Based on my calculation I venture to disagree with Peierls's statement in his excellent review of the quantum theory of solids [85a] where on p. 154 he writes: 'The published summary of this talk [83] is probably the only incorrect formula published under Pauli's name.' This conclusion by Peierls is based on the following argument (see also Ref. [22a], p. 782 and Note 80). As shown in *Note* [84], the equation of motion (5.44) gives rise to the dispersion relation

$$\omega_k = 2\omega_0 \left| \sin \frac{ak}{2} \right| ; \quad -\frac{\pi}{a} < k \leq +\frac{\pi}{a}, \tag{5.45}$$

and one can easily check that energy and momentum conservation in a three-phonon process, $\omega_k \pm \omega_{k'} = \omega_{k\pm k'}$, has only the trivial solutions $k = 0$ or $k' = 0$ (see Ref. [85a], p. 155), which implies a vanishing energy exchange and hence an infinite heat conductivity.

This difficulty is relaxed in two and, even more so, in three dimensions where, however, the conservation laws become much more complicated, rendering the problem of heat conduction in insulators enormously complex and prone to mistakes. Peierls writes: 'It seems there is no problem in modern physics for which there are on record as many false starts, and as many theories which overlook some essential feature, as in the problem of the thermal conductivity of non-conducting crystals' (Ref. [85a], p. 153). I now think that Pauli, in his struggle with the problem of the heat conductivity, was the first to realize this difficulty and that he then turned to the related problem of light absorption in order not to waste all his acquired expertise and to cut short a lingering enterprise. The reason why I think Pauli's result for γ is correct is because the phonon created by the absorption of a photon of frequency ω_{rad} is *off-shell*, i.e. virtual, and therefore the above energy conservation does not apply, but rather $\omega_k \pm \omega_{k'} = \omega_{rad}$ while the photon wavenumber $k_{rad} \simeq 0$. It might be that Pauli's notes which Peierls saw (probably in 1929, see below) referred to heat conduction and not to light absorption (see also Ref. [85b], pp. 42, 43–4).

Understandably, Pauli was not satisfied with this unfinished business of the heat conduction 'and suggested a further study of the problem' (Ref. [85a], p. 155). He proposed the problem as Ph.D. work to Rudolf Peierls (1907–95) when the latter came to Zurich for the summer term of 1929 after having studied in Berlin, in Munich with Sommerfeld, and in Leipzig with Heisenberg. The latter had left Leipzig on 1 March for a visit to the United States, Japan, and India and was back only on 5 November. Peierls writes in his recollections: 'In the spring of 1929, when I had been in Leipzig for two semesters, Heisenberg went to the United States for a time, so I moved again. On his recommendation I joined Pauli at the E.T.H.' (Ref. [85b], p. 40). However, 'One semester's residence in Zurich was not sufficient to qualify for a Ph.D., so I submitted the thesis to Leipzig, where two semesters were sufficient. For this purpose I returned to Leipzig in July' (Ref. [85b], p. 44). So Peierls still got his degree from Leipzig under Heisenberg (see Refs [14b], pp. 68–9 and [16a], p. xi). This work was published under the title 'On the kinetic theory of heat conduction in crystals' [86a] and became the point of departure for the theory of phonon transport in insulating crystals (see e.g. Ref. [86b]). Surprisingly, in this work Peierls writes: 'In contrast [to heat conduction], one already obtains with the second approximation processes that an energy exchange between neighbouring frequencies takes place, and this fact may be used for the explanation of the absorption of the residual rays. For this purpose the limitation to the linear model is thus allowed [83]' (translated from Ref. [86a], pp. 1080–1). Thus in 1929 Peierls did not think that Pauli's paper [83] was wrong, in agreement with my *Note* [84].

Later, in 1931, Peierls recalculated the light absorption of non-metallic crystals in his habilitation paper 'On the theory of absorption spectra of solid bodies' [87a] which made him a Privatdozent of ETH (see the reports I.48 and 49 by Pauli and Scherrer and the decision I.51 by the School Council of ETH in Ref. [87b]). In this paper Eq. (12) is the Schrödinger equation with the Hamiltonian H_0 and the terms (20) correspond to the anharmonic part H_1 in Eqs (A) of *Note* [84]; and Eq. (17) describes the lattice dynamics (see e.g. Section 1 of Ref. [36b]). While most of the paper [87a] is qualitative, Section 6 contains a calculation of light absorption based on the 'golden rule' formula (36) which is of the form (C) of *Note* [84]. Pauli's note [83] is not mentioned in this paper by Peierls.

Pauli later had a somewhat ambiguous attitude towards solid-state physics. This is vividly illustrated by the following remarks in letters which he wrote to his assistant Peierls during his first visit to the United States: 'the residual resistance is a dirt effect (*Dreckeffekt*) and in the dirt one should not stir.' And: 'On semiconductors one should not work, that's an obscenity (*Schweinerei*), who knows if semiconductors exist at all?' (translated from letters [279] of 1 July 1931 from Ann Arbor and [283] of 29 September 1931 from New York, Ref. [16a], pp. 85 and 94, respectively). Pauli is also quoted to have said 'I don't like this solid-state physics . . . I initiated it though' [87c]. But Peierls presents a somewhat more differentiated opinion when he says: 'He has often expressed his scorn for solid-state physics although he had contributed himself in an essential way to its foundation. But even in this domain there were problems which interested him' (translated from Ref. [14b], p. 73). Pauli's letters to Peierls quoted so negatively above show at the same time his keen interest in such matters. Examples were magnetism again, but also superconductivity. Indeed, on 2 February 1938 Pauli wrote to Kramers: 'I myself hope to learn in Leiden, particularly from Casimir, various things about superconductivity which at present interests me again very much. I have announced in the summer term a course on "theory of the solid" which has evoked great amusement among the Zurich physicists' (translated from letter [487e], Ref. [87d], p. 818).

Of course, in Hamburg Pauli could not get away without teaching. According to the list of handwritten lecture notes preserved in the *Pauli Archive* (see Ref. [5], p. 546) his official teaching activity started in the summer term of 1925 with a course on electrodynamics. But he undoubtedly had given many lectures before, as may be concluded from the letter that the dean of the Faculty of Mathematics and Natural Sciences of Hamburg University, the mathematician Erich Hecke (1887–1947), wrote to Pauli on 7 February 1924: 'Very honoured Herr Doktor, in its meeting of 30.I.1924 the Faculty has decided to admit you to the habilitation and, in acknowledgement of your scientific achievements, to dispense you with the trial lecture and the colloquium. I therefore confer on you the *venia legendi* for theoretical physics. The Faculty would very much welcome it if sometime you would give a scientific lecture in your narrower circle. With the best congratulations, yours very devoted Hecke' (translated from Ref. [88]). As one learns from an official announcement by Dean Hecke, Pauli gave his inaugural lecture on Saturday 23 February 1924 at 12 noon in the large lecture hall of the Chemistry Institute (see Ref. [5], p. 140); the title was 'Quantum theory and periodic system of the elements'.

Thus Pauli had become a Privatdozent. But this did not prevent him from travelling during vacations to meet colleagues and have discussions. He spent the Easter vacation of April 1924 in Copenhagen (see letters [58, 59, 62]) where the Bohr–Kramers–Slater radiation theory (see pages 119 and 133) was in vogue. This was a programme of how to reconcile discrete atomic levels with the emission and absorption of continuous classical radiation through a 'virtual field'. While the latter was an idea imported by the young American, John Clarke Slater (1900–76), who originally saw it as a guiding field of photons in the sense of de Broglie or Einstein, Bohr's and Kramers' goal was to do away with photons altogether because of the serious difficulties with this concept both in the limit of static fields and in dispersive media. But this was possible only by forgoing energy conservation in individual emission and absorption processes (see Ref. [89], also Refs [18], Part I, [27b], Section V.2 and [37b], Chapter 22). Half a year later Pauli wrote to Bohr: 'At that time [in April] you succeeded by your arguments in silencing my scientific conscience which much revolted against this perception. That, however, lasted only for a short time ... and today I am totally opposed to this perception of the radiation phenomena' (translated from letter [66], Ref. [5], p. 163). In the same letter Pauli gives his own arguments against the Bohr–Kramers–Slater theory as well as those also of Einstein, with whom he had discussed the matter at the meeting of the German Natural Scientists and Physicians held in Innsbruck from 21 to 27 September 1924. In Ziff. 15 of his 'Quantentheorie' [10a] Pauli made a detailed critical examination of this theory and also mentioned Geiger and Bothe's experiment which finally verified energy conservation in individual processes [90].

After his visit to Landé in Tübingen the following January (see pages 110–11) Pauli had returned to Hamburg one week late for teaching. But in the following spring vacation he went again to Copenhagen where he met his father who had been invited to give a series of lectures there (see letter [79], Ref. [5], p. 198). And in December 1925 Pauli met Bohr again, this time in Berlin where Pauli accompanied him on his way back from Leiden (see page 113). About this meeting, where the 'new "gospel"' of the electron spin was the main subject of discussion, Pauli sent Bohr the following comment from his Christmas vacation in Vienna: 'I am very glad that I have succeeded in drinking wine and rising earlier than you, and that you have not succeeded in converting me. But don't give up the idea, perhaps you will succeed in yet another time later' (translated from letter [114], Ref. [5], p. 275). As we saw on page 115, that conversion came in March 1926. Thereafter Pauli spent his Easter vacation once more in Copenhagen. We learn from a message from Bohr to Pauli that 'Miss Schultz [Bohr's secretary] has called Mrs. Schack's pension, where Fowler resided and where Thomas and Hund now live, and has retained a room for you. It is free starting Sunday, 4 April. I am not in town during Easter but am back Tuesday or Wednesday' (translated from letter [130], Ref. [5], p. 314).

Soon after, in May 1926, came Pauli's first offer of a professorship, on which he comments to Wentzel as follows: 'From the Leipzig [associate] professorship (succession of [Georg Cecil] Jaffé [1880–1965] who moved to Giessen [see Ref. [16a], p. 23]) I heard that I figure behind Heisenberg on the list and that you are the third in the group (*im Bunde*). Now Heisenberg has declined. In case fate now should catch up with me it would not be entirely excluded that I would be offered so much in Hamburg that I, too, would be able to decline. Therefore, be on your guard!' (translated from letter [133] of 8 May 1926, Ref. [5], p. 323). At the end of June Pauli declined (see letter [139] of Ref. [5]). Previously, he had been in Zurich—probably his first visit there—to take part in the well-attended week of lectures (*Vortragswoche*) from 21 to 26 June 1926 organized by the physics professors of ETH at the time, namely Debye and Scherrer. Subsequently such weeks of lectures were held every few years at ETH, in 1929, 1931, 1933, and 1936 (see Ref. [87b], p. 18). Then, on 25 November, Pauli received the official document informing him of the decision of the High Senate of 22 November 1926 that 'for the duration of his membership with the Hamburg University the official designation *Professor* is attributed to the Privatdozent Dr Wolfgang Pauli' (translated from document PLC Bi 45, *Pauli Archive*).

In 1927 the travelling plans of Pauli included a short visit to Born and Jordan in Göttingen, followed by a vacation in southern Germany in April (letters [157, 160] of Ref. [5]) and a short visit to Copenhagen in June where, unfortunately, the discussions between Heisenberg on one side and Bohr and Klein on the other had led to serious differences (see letters [163, 164, 165] of Ref. [5]). But the really big events of 1927 happened in the autumn: the important international congress commemorating the 100th anniversary of the death of Alessandro Volta (1745–1827), which was held from 11 to 20 September in Como, the Italian city on the Swiss border where Volta was born and where he died; and the Fifth Solvay Conference that took place in Brussels from 24 to 29 October. Participation in the latter was by personal invitation, and Pauli had the honour of receiving a letter of invitation by H. A. Lorentz who acted as chairman (letter [171] of Ref. [5]).

Both of these meetings were of historic importance, the first mainly because of the large international participation—only Einstein and Ehrenfest were missing!—and the grandiose setting at the Istituto Carducci on the shore of Lake Como. The Como congress was presided over by Quirino Majorana (1871–1957), professor of experimental physics at Bologna and president of the Italian Physical Society. H. A. Lorentz, M. von Laue, and R. A. Millikan were among the vice-presidents. Scientifically the main event of the conference was Bohr's lecture 'The Quantum Postulate and the Recent Development of Atomic Theory' [91a]. It became the credo of the Copenhagen Interpretation and for the first time gave a comprehensive account of Bohr's idea of complementarity. The published final version of this lecture is a considerably extended and modified form of the original text of Bohr's lecture contained in the conference proceedings (Ref. [91b], Vol. 2, pp. 565–88). It was Pauli whom Bohr had asked to go through the successive versions with his uncompromising critical mind, as is documented by letters [172, 173, 177, 179] of Ref. [5]. On this Oskar Klein recalls: 'after the congress, during a stay at Lake Como together with Bohr, [Pauli] helped him by means of a suitable mixture of criticism and sympathetic understanding to write a much fuller manuscript—now in German. . . . But soon it became clear that this, still, was only the beginning. During the whole of the autumn of 1927 and the winter of 1928 there was an almost uninterrupted correction of proofs. . . . The last proof was finished about Easter time 1928' (Ref. [91c], p. 90).

In his lecture Bohr emphasized the irrationality of the quantum postulate and saw an individuality, determined by Planck's constant, in all atomic processes such that it is impossible to ascribe independent reality neither to these phenomena nor to the means of their observation. And as for complementarity, he saw it as the fact that, in distinction to classical theory, space–time coordination and causality on the one hand and the characterization by energy and momentum in statistical laws on the other become mutually exclusive features of the quantum description. This is borne out in Bohr's perception of the uncertainty relations as the properties of Fourier transformation, as already mentioned in the last section. The conference proceedings also contain the discussions after Bohr's talk, in which Born, Kramers, Heisenberg, Fermi, and Pauli participated. In his remarks Pauli comes back again to the lack of a derivation of his exclusion principle which he calls an aesthetic failure (*Schönheitsfehler*) [91b].

The final version of Bohr's lecture also contains an infallible sign of Pauli's intervention in the following statement which, although clad in Bohrian language, reflects Pauli's deep concern with quantum electrodynamics that started in this period and with the quantum nature of the electric charge: 'A satisfactory solution of the problems touched upon would seem to be possible only by means of a rational quantum-theoretical transcription of the general field theory, in which the ultimate quantum of electricity has found its natural position as an expression of the feature of individuality characterizing the quantum theory' (Ref. [91a], p. 589). This may be compared with Pauli's view expressed in letter [173] to Bohr on 17 October 1927: 'For, I even hope that some time later it will be possible to conceive the atomistic nature of the electric charge in a way analogous to the existence of stationary states, so that the number of elementary charges in a certain domain will appear as a quantum number' (translated from Ref. [5], p. 412). The previous year also witnessed Pauli's review of the German translation of Eddington's book on relativity (see Chapter 2) which concludes with the remark that 'it is not excluded that through the more recent development of quantum theory the . . . problem of the structure of the electron will again come to the foreground' (translated from Ref. [92], p. 1393).

The elegant Solvay conferences on physics were instituted, after consultation with Walter Nernst (1864–1941), in 1910 by the Belgian chemist and industrialist Ernest Solvay (1838–1922), the inventor of the Solvay process for the production of soda. The first 'Conseil de Physique' was convened in October 1911. In 1913, after consultation with Wilhelm Ostwald (1853–1932) and William Ramsey (1852–1916), Ernest Solvay also established conferences on chemistry (see the Introduction to Ref. [38d]). The official programmme of the Fifth Solvay Conference, 'Electrons and Photons', as well as the speakers, Sir William Lawrence Bragg (1890–1971), A. H. Compton, de Broglie, Born, Heisenberg, Schrödinger, and Bohr, promised to yield high-level discussions on the new quantum theory. For Bohr it was essentially a repetition of his Como lecture. However, the presence of Einstein turned it into high drama. Indeed, 'Einstein was particularly reluctant to renounce deterministic description in principle. He challenged Bohr and the other proponents of the new quantum mechanics with arguments suggesting the possibility of taking the interaction between the atomic objects and the measuring instruments more explicitly into account. Einstein was not convinced by answers which pointed out the futility of this prospect, and he returned to these problems again at the sixth Solvay Conference in 1930' (Ref. [38d], p. 152).

For the novices attending the Solvay events, de Broglie, Dirac, Heisenberg, and Pauli (Schrödinger had already been invited in 1924), participation meant international recognition at large. It was certainly a memorable experience, as witnessed by Heisenberg's Foreword to Ref. [38d], written in 1974: 'Einstein criticized this very limitation [of the applicability of the two pictures of waves and particles] because it seemed to undermine the ideal of an objective description of Nature, which had been considered to lie at the basis of physics. Besides it introduced a statistical element into the foundations of physics, which Einstein would not admit. Einstein therefore suggested special experimental arrangements for which, in his opinion, the uncertainty relations could be evaded. But the analysis carried out by Bohr and others during the Conference revealed errors in Einstein's arguments. In this situation, by means of intensive discussions, the Conference contributed directly to the clarification of the quantum-theoretical paradoxa' (Ref. [38d], pp. v–vi).

In his interview with Res Jost [73b] quoted above, Stern gave the following account: 'Einstein came down to breakfast [at 'the sumptuous Hotel Métropole in downtown Brussels', see Ref. [14c], p. 251] and expressed his misgivings about the new quantum theory, every time [he] had invented some beautiful experiment from which one saw that [the theory] did not work. . . . Pauli and Heisenberg, who were there, did not pay much attention. . . . Bohr, on the other hand, reflected on it with care and in the evening, at dinner, we were all together and he cleared up the matter in detail' (Ref. [37b], p. 445). The Bohr–Einstein dialogue, in fact, 'continued for over two decades. Niels Bohr presented an account of his discussions with Einstein on epistemological questions at the eighth Solvay Conference in 1948, and expanded it to be included in the seventieth birthday volume presented to Einstein in 1949.' ([93], see Ref. [38d], p. 152). More details on these events and the problems involved may be found in Chapter 9 of Ref. [61c] and in Sections 4.1 and 5.1 of Ref. [94].

As Pauli once observed to Lehmann, his Hamburg years had been a beautiful time for him [73c]. But in all this social and intellectual high-life not one word about Pauli's family had been uttered—except that until 1925 Pauli regularly spent Christmas at home. For his 26th birthday his mother Bertha had sent him a gift of the essays of (Michel Eyquem de) Montaigne (1533–92) in German translation (Deutsche Bibliothek, Berlin) with the dedication '*Meinem Wolfi zum 26. Geburtstag von Mamma*' (the book is in the *Pauli Archive*). In the autumn of the same year (1926) his sister Hertha left home at age 20 to start her career as an actress in the Breslau Theatre [66d]. So life became rather lonely at Anton-Frank-Gasse. But more seriously, another woman entered the life of his father Wolfgang. In her existential doubts his mother Bertha committed suicide. For son Wolfgang the shock, scarcely detectable by the outside world, was building up. But in continuation of the letter to Jung quoted at the beginning of this section Pauli wrote: 'Alone in the fast train to Zurich I remembered how in 1928 I travelled by the same route towards my new professorship and my big neurosis' (translated from letter [69], Ref. [72], p. 150).

Notes and references

[1] Translated from: W. Pauli, letter [1286] of 3 October 1951 to Markus Fierz, in Ref. [2], p. 375.

[2] K. von Meyenn, *Wolfgang Pauli. Scientific Correspondence with Bohr, Einstein, Heisenberg, a.o., Volume IV, Part I: 1950-1952* (Springer, Berlin and Heidelberg, 1996).

[3] W. Pauli jr., (a) 'Über den Einfluss der Geschwindigkeitsabhängigkeit der Elektronenmasse auf den Zeemaneffekt', *Zeitschrift für Physik* **31**, 373 (1925); (b) 'Über den Zusammenhang des Abschlusses der Elektronengruppen im Atom mit der Komplexstruktur der Spektren', *Zeitschrift für Physik* **31**, 765 (1925). Reprinted in Ref. [4], Vol. 2, pp. 201–13 and 214–32.

[4] R. Kronig and V. F. Weisskopf (eds), *Collected Scientific Papers by Wolfgang Pauli. In Two Volumes* (Wiley Interscience, New York, 1964).

[5] A. Hermann, K. von Meyenn, and V. F. Weisskopf (eds), *Wolfgang Pauli. Scientific Correspondence with Bohr, Einstein, Heisenberg, a.o., Volume I: 1919-1929* (Springer, New York, 1979).

[6] (a) W. Pauli, 'Exclusion Principle and Quantum Mechanics', in: *Les Prix Nobel en 1946* (Norstedt & Söner, Stockholm, 1948), pp. 131–47 (also Editions du Griffon, Neuchâtel, 1947). Reprinted in Ref. [4], Vol. 2, pp. 1080–97, and in Ref. [6b], pp. 165–81. (b) C. P. Enz and K. von Meyenn (eds), *Wolfgang Pauli. Writings on Physics and Philosophy* (Springer, Berlin and Heidelberg, 1994).

[7] N. Bohr, 'Linienspektren und Atombau', *Annalen der Physik* **71**, 228 (1923).

[8] D. Serwer, 'Unmechanischer Zwang: Pauli, Heisenberg, and the Rejection of the Mechanical Atom, 1923-1925', in: R. McCormmach and L. Pyenson (eds), *Historical Studies in the Physical Sciences* (Johns Hopkins University Press, Baltimore, Md), 1977, Vol. 8, p. 189.

[9] A. Sommerfeld, *Berichte der Akademie, München*, 1915; 'Zur Quantentheorie der Spektrallinien', *Annalen der Physik* **51**, 1 (1916).

[10] (a) W. Pauli, 'Quantentheorie', in: H. Geiger and K. Scheel (eds), *Handbuch der Physik* (Springer, Berlin, 1926), Vol. 23, pp. 1–278. Reprinted in Ref. [4], Vol. 1, pp. 271–548. (b) Note that in Eqs. (47) to (50) of Ziff. 21 in Ref. [10a] Z^4 should read Z^2.

[11] (a) I. S. Bowen and R. A. Millikan, 'The Extension of the X-Ray-Doublet Laws into the Field of Optics', and R. A. Millikan and I. S. Bowen, 'Some Conspicuous Successes of the Bohr Atom and a Serious Difficulty', *Physical Review* **24**, 209, and 223 (1924); (b) A. Landé, 'Das Wesen der relativistischen Röntgendubletts', *Zeitschrift für Physik* **24**, 88 (1924); (c) A. Landé, 'Die absoluten Intervalle der optischen Dubletts und Tripletts', *Zeitschrift für Physik* **25**, 46 (1924).

[12] C. P. Enz (ed.), *Pauli Lectures on Physics: Volume 1. Electrodynamics*, translated by H. R. Lewis and S. Margulies (MIT Press, Cambridge, MA, 1973).

[13] R. Kronig, 'The Turning Point', in Ref. [14a], pp. 5–39. German translation of parts of it in Ref. [14b], pp. 54–5.

[14] (a) M. Fierz and V. F. Weisskopf (eds), *Theoretical Physics in the Twentieth Century. A Memorial Volume to Wolfgang Pauli* (Wiley Interscience, New York, 1960); (b) C. P. Enz and K. von Meyenn (eds), *Wolfgang Pauli. Das Gewissen der Physik* (Vieweg, Braunschweig, 1988); (c) D. C. Cassidy, *Uncertainty. The Life and Science of Werner Heisenberg* (Freeman, New York, 1992); German translation by A. and G. Kleinert, *Werner Heisenberg: Leben und Werk* (Spektrum Akademischer Verlag, Heidelberg, 1995).

[15] W. Pauli, *Theory of Relativity*, translated by G. Field, *With Supplementary Notes by the Author* (Pergamon Press, London, 1958). German original and *Supplementary Notes* reprinted in Ref. [4], Vol. 1, pp. 1–263.

[16] (a) K. von Meyenn (ed., with the cooperation of A. Hermann and V. F. Weisskopf), *Wolfgang Pauli. Scientific Correspondence with Bohr, Einstein, Heisenberg, a.o.. Volume II: 1930-1939* (Springer, Berlin and Heidelberg, 1985); (b) M. Dresden, 'Ralph Kronig', obituary in: *Physics Today*, March 1997, pp. 97-8; (c) R. Jost, 'Der holländische Physiker Ralph Kronig 80 jährig', *Neue Zürcher Zeitung*, Nr. 59, 10./11. März 1984, p. 51.

[17] G. E. Uhlenbeck and S. A. Goudsmit, 'Ersetzung der Hypothese vom unmechanischen Zwang durch eine Forderung bezüglich des inneren Verhaltens jedes einzelnen Elektrons', *Naturwissenschaften* **13**, 953 (1925).

[18] K. Stolzenburg (ed.), *Volume 5. The Emergence of Quantum Mechanics (Mainly 1924-1926)*, in: E. Rüdinger (general ed.), *Niels Bohr. Collected Works* (North-Holland, Amsterdam, 1984).

[19] (a) G. H. Uhlenbeck, 'Fifty Years of Spin. Personal Reminiscences', *Physics Today*, June 1976, p. 43; (b) S. A. Goudsmit, 'Fifty Years of Spin. It might as well be spin', *Physics Today*, June 1976, p. 40.

[20] (a) M. Abraham, 'Prinzipien der Dynamik des Elektrons', *Annalen der Physik* **10**, 105 (1903). (b) *Note*: Assume arbitrary spherically symmetrical mass and charge distributions $m\rho_m(r)$ and $e\rho_e(r)$, normalized in the sphere of radius a,

$$4\pi \int_0^a \rho_i(r)r^2 dr = 1; \; i = m, e. \tag{A}$$

These distributions are assumed to rotate rigidly about the z-axis with angular velocity ω. If ϑ is the angle from the north pole, the contribution to the z-axis moment of inertia of the zonal volume element $rd^2r = 2\pi r^2 \sin\vartheta dr d\vartheta$ is $d^2\theta_z = m\rho_m(r)rd^2r(r\sin\vartheta)^2$. Hence

$$\theta_z = 2\pi m \int_0^a \rho_m(r)r^4 dr \int_0^\pi \sin^3 \vartheta d\vartheta = \frac{8\pi}{3}m \int_0^a \rho_m(r)r^4 dr \tag{B}$$

and the z-axis angular momentum is $s_z = \omega\theta_z$.

Considering the z-axis magnetic moment as due to the molecular Ampère currents produced by the rotation ω (see e.g. Ref. [12], pp. 73–4), the zonal volume element rd^2r contributes $d^2\mu_z = e\rho_e(r)rd^2r(\omega/2\pi c)\pi \times (r\sin\vartheta)^2$, so that

$$\mu_z = \frac{4\pi\omega}{3c}e \int_0^a \rho_e(r)r^4 dr. \tag{C}$$

Thus the gyromagnetic ratio becomes

$$g = \frac{\mu_z}{s_z} = \frac{e}{2mc} \int_0^a \rho_e(r) r^4 dr \left[\int_0^a \rho_m(r) r^4 dr \right]^{-1}. \qquad \text{(D)}$$

Now $\rho_e = \rho_m$ means that Larmor's theorem, which is valid for point particles, applies to the whole distribution and, therefore, $g = e/2mc$. On the other hand, taking a 'metal' sphere, so that $\rho_e(r) = \delta(r - a)/4\pi a^2$, and choosing $\rho_m(r) = 1/2\pi a^2 r$, both of which distributions satisfy the normalization (A), then Eq. (D) indeed yields $g = 2$, and there is 'piercing' of Larmor's theorem.

Setting $s_z = \hbar/2$ one obtains $\omega = \hbar/2\theta_z$ and, with $\rho_m = 1/2\pi a^2 r$, from Eq. (B),

$$\frac{a\omega}{c} = \frac{3}{16\pi} \lambda_c a \left[\int_0^a \rho_m(r) r^4 dr \right]^{-1} \qquad \text{(E)}$$

where $a\omega$ is the peripheral velocity and $\lambda_c = \hbar/mc = 3.86 \times 10^{-11}$ cm is the Compton radian length. Therefore, if a is of the order of the classical electron radius $r_e = \alpha\lambda_c$ one finds $a\omega/c \simeq 3/2\alpha \simeq 206!$ On the other hand, for $a\omega/c < 1$ one has, with the Bohr radius $a_B = \lambda_c/\alpha$, $a/a_B > 3\alpha/2 = 0.011$.

[21] W. Pauli, 'Allgemeine Grundlagen der Quantentheorie des Atombaues', in: *Müller-Pouillets Lehrbuch* (Vieweg, Braunschweig, 11th edn, 1929), Vol. 2, Part 2, pp. 1709–1842. Reprinted in Ref. [4], Vol. 1, pp. 636–769.

[22] (a) C. P. Enz, 'W. Pauli's Scientific Work', in Ref. [23], pp. 766–99; (b) P. J. Price, 'Llewellyn H. Thomas', *Physics Today*, September 1994, p. 115.

[23] J. Mehra (ed.), *The Physicist's Conception of Nature* (Reidel, Dordrecht and Boston, 1973), pp. 766–99.

[24] (a) L. H. Thomas, 'The motion of the spinning electron', *Nature* **117**, 514 (1926); (b) L. H. Thomas, 'The Kinematics of an Electron with an Axis', *Philosophical Magazine* **3**, 1 (1927); (c) V. Ya. Frenkel, *Yakov Ilich Frenkel. His work, life and letters* (Birkhäuser, Basle, 1996); (d) J. Frenkel, 'Die Elektrodynamik des rotierenden Elektrons', *Zeitschrift für Physik* **37**, 243 (1926).

[25] G. E. Uhlenbeck and S. A. Goudsmit, 'Spinning Electrons and the Structure of Spectra', *Nature* **117**, 264 (1926).

[26] W. Pauli jr., 'Zur Quantenmechanik des magnetischen Elektrons', *Zeitschrift für Physik* **43**, 601 (1927). Reprinted in Ref. [4], Vol. 2, pp. 306–28.

[27] (a) Information concerning proposals for the Physics Nobel Prize until 1946. I thank Professor Anders Bárány, Secretary of the Nobel Committee of Physics, for this information. (b) J. Mehra and H. Rechenberg, *The Historical Development of Quantum Theory. Volume 1. The Quantum Theory of Planck, Einstein, Bohr and Sommerfeld: Its Foundation*

and Rise of Its Difficulties, 1900-1925 (Springer, New York, 1982); (c) A. Pais, *Inward Bound. On Matter and Forces in the Physical World* (Oxford University Press, Oxford, 1986).

[28] (a) A. Kastler, 'Masers and lasers', *European Journal of Physics* **7**, 69 (1986); (b) S. A. Goudsmit, 'Pauli and Nuclear Spin', *Physics Today*, June 1961, p. 18; (c) S. A. Goudsmit, 'La découverte du spin de l'électron', *Journal de Physique* **28**, 123 (1967); German translation in: *Physikalische Blätter* **21**, 445 (1965).

[29] B. L. van der Waerden, (a) 'Exclusion Principle and Spin', in Ref. [14a], pp. 199–244; (b) 'Spinoranalyse', *Göttinger Nachrichten (math.-phys.)* 1929, p. 100.

[30] W. Pauli jr., 'Zur Frage der theoretischen Deutung der Satelliten einiger Spektrallinien und ihrer Beeinflussung durch magnetische Felder', *Naturwissenschaften* **12**, 741 (1924). Reprinted in Ref. [4], Vol. 2, pp. 198–200.

[31] (a) 'Bibliography Wolfgang Pauli. Prepared by Charles Enz', in Ref. [14a], pp. 304–11; (b) H. Kopfermann, *Nuclear Moments*, English translation by E. E. Schneider (Academic, New York, 1958); (c) L. Pauling and S. Goudsmit, *The Structure of Line Spectra* (McGraw-Hill, New York, 1930); (d) H. Schüler, 'Hyperfeinstrukturen und Kernmomente', *Physikalische Zeitschrift* **32**, 667 (1931); (e) L. B. Loeb, 'Proceedings of the American Physical Society', *Physical Review* **38**, 579 (1931); (f) L. Belloni, 'Pauli's 1924 note on hyperfine structure', *American Journal of Physics* **50**, 461 (1982).

[32] N. Bohr, A. H. Kramers, and J. C. Slater, 'The quantum theory of radiation', *Philosophical Magazine* **47**, 785 (1924); German translation: 'Über die Quantentheorie der Strahlung', *Zeitschrift für Physik* **24**, 69 (1924).

[33] 'Dedication' in: W. Pauli, 'Exclusion Principle, Lorentz Group and Reflection of Space–Time and Charge', in: W. Pauli (ed., with the cooperation of L. Rosenfeld and V. Weisskopf), *Niels Bohr and the Development of Physics* (Pergamon, London, 1955). Reprinted in Ref. [4], Vol. 2, pp. 1239–60.

[34] (a) J. Mehra and H. Rechenberg, *The Historical Development of Quantum Theory. Volume 3. The Formulation of Matrix Mechanics and Its Modifications, 1925-1926* (Springer, New York, 1982); (b) S. Richter, *Wolfgang Pauli. Die Jahre 1918-1930. Skizzen zu einer wissenschaftlichen Biographie* (Sauerländer, Aarau, 1979); (c) S.-i. Tomonaga, *The Story of Spin*, translated by T. Oka (Chicago University Press, Chicago, 1997), Lectures 2 and 12.

[35] (a) F. J. Dyson, 'Our Stability is but Balance', Pauli Memorial Lecture, given at ETH, Zurich, on 18 February 1974 (unpublished). *Note*: A Pauli Memorial Lecture on the same subject was given at ETH on 25 January 1993 by Elliott H. Lieb, another expert on the subject: 'The Pauli Principle and the Stability of Matter, from Atoms to Stars', see:

W. Thirring (ed.), *The Stability of Matter: From Atoms to Stars. Selecta of Elliott H. Lieb* (2nd enlarged edn, Springer, Berlin and Heidelberg, 1997), Part V; (b) P. Ehrenfest, 'Besteht ein allgemeiner Zusammenhang zwischen der wechselseitigen Undurchdringlichkeit materieller Teilchen und dem "Pauli-Verbot"?', *Naturwissenschaften* **15**, 161, 268 (1927).

[36] (a) E. C. Stoner, 'The Distribution of Electrons among Atomic Levels', *Philosophical Magazine* **48**, 719 (1924); (b) C. P. Enz, *A Course on Many-Body Theory Applied to Solid-State Physics* (World Scientific, Singapore, 1992).

[37] (a) J. R. Nielsen (ed.), *Niels Bohr. Collected Works. Volume 4. The Periodic System (1920–1923)* (North-Holland, Amsterdam, 1977); (b) A. Pais, *'Subtle is the Lord ... ' The Science and the Life of Albert Einstein* (Oxford University Press, Oxford, 1982).

[38] (a) S. N. Bose, 'Plancks Gesetz und Lichtquantenhypothese', *Zeitschrift für Physik* **26**, 178 (1924); (b) A. Einstein, 'Quantentheorie des einatomigen idealen Gases', *Sitzungsberichte der Preussischen Akademie der Wissenschaften*, Berlin, 1924, p. 261 and 1925, p. 3; (c) O. Sackur, 'Die Anwendung der kinetischen Theorie der Gase auf chemische Probleme', *Annalen der Physik* **36**, 958 (1911); (d) J. Mehra, *The Solvay Conferences on Physics. Aspects of the Development of Physics since 1911. With a Foreword by Werner Heisenberg* (Reidel, Dordrecht, 1975).

[39] (a) E. Fermi, 'Sulla quantizzazione del gas perfetto monatomico', *Rendiconti dell'Academia dei Lincei* (6) **3**, 145 (1926); 'Zur Quantelung des idealen einatomigen Gases', *Zeitschrift für Physik* **36**, 902 (1926). Reprinted in: E. Segré (ed., with the cooperation of E. Amaldi, E. Persico, and F. Rasetti), *Enrico Fermi. Collected Papers (Note e Memorie). Volume 1: Italy, 1921-1938* (Univeristy of Chicago Press and Accademia Nazionale dei Lincei, Chicago and Rome, 1962), pp. 178–85, 186–95. (b) P. A. M. Dirac, 'On the Theory of Quantum Mechanics', *Proceedings of the Royal Society* A **112**, 661 (1926).

[40] C. P. Enz (ed.), *Pauli Lectures on Physics: Volume 4. Statistical Mechanics*, translated by H. R. Lewis and S. Margulies (MIT Press, Cambridge, MA, 1973).

[41] (a) H. Weyl, *Philosophy of Mathematics and Natural Science*, translated from the German edition of 1927 by O. Helmer (Princeton Univeristy Press, Princeton, NJ, rev. and augmented English edn, 1949); (b) K. von Meyenn (ed.), *Wolfgang Pauli. Scientific Correspondence with Bohr, Einstein, Heisenberg, a.o., Volume III: 1940-1949* (Springer, Berlin and Heidelberg, 1993); (c) J. E. Heyde, 'Indiscernibilien', in: J. Ritter and K. Gründer (eds), *Historisches Wörterbuch der Philosophie* (Wissenschaftliche Buchgesellschaft Darmstadt, Schwab, Basle, 1976), Vol. 4: I–K, columns 283– 286; (d) W. Totok *et al.* (eds), *Handbuch der Geschichte der Philosophie. IV. Frühe Neuzeit 17. Jahrhundert* (Vitto-

rio Klostermann, Frankfurt/M, 1981); (e) M. Jammer, *The conceptual development of quantum mechanics* (McGraw-Hill, New York, 1966).

[42] W. Pauli, *General Principles of Quantum Mechanics*, translated from Ref. [43] and edited by P. Achuthan and K. Venkatesan (Springer, Berlin and Heidelberg, 1980).

[43] W. Pauli, 'Die allgemeinen Prinzipien der Wellenmechanik', in: H. Geiger and K. Scheel (eds), *Handbuch der Physik* (Springer, Berlin, 1933) 2nd edn, Vol. 24, Part 1, pp. 83–272.

[44] W. Heisenberg, 'Mehrkörperproblem und Resonanz in der Quantenmechanik', *Zeitschrift für Physik* **38**, 411 (1926). Reprinted in Ref. [45], Part I, pp. 456–71.

[45] W. Blum, H.-P. Dürr, and H. Rechenberg (eds), *Werner Heisenberg. Collected Works. Series A. Original Scientific Papers* (Springer, Berlin and Heidelberg, 1985).

[46] C. P. Enz, 'Applications of Quantum Mechanics (1926-1933)', in Ref. [45], pp. 507–15.

[47] Translated from: W. Pauli, letter [54] of 11 February 1924 to Niels Bohr, Ref. [5], p. 143.

[48] W. Heisenberg, *Der Teil und das Ganze. Gespräche im Umkreis der Atomphysik* (Piper, Munich, 1969); English translation by A. J. Pomerans, *Physics and Beyond* (Allen & Unwin, London, 1971).

[49] J. Mehra and H. Rechenberg, *The Historical Development of Quantum Theory. Volume 2. The Discovery of Quantum Mechanics 1925* (Springer, New York, 1982).

[50] W. Heisenberg, 'Über quantentheoretische Umdeutung kinematischer und mechanischer Beziehungen', *Zeitschrift für Physik* **33**, 879 (1925). Reprinted in Ref. [45], *Part I*, pp. 382–96.

[51] B. L. van der Waerden and H. Rechenberg, 'Quantum Mechanics (1925-1927)', in Ref. [45], pp. 329–43.

[52] (a) M. Born and P. Jordan, 'Zur Quantenmechanik', *Zeitschrift für Physik* **34**, 858 (1925); (b) M. Born, W. Heisenberg, and P. Jordan, 'Zur Quantenmechanik II', *Zeitschrift für Physik* **35**, 557 (1926). Reprinted in Ref. [45], pp. 397–455.

[53] (a) S. Lundqvist (ed.), *Nobel Lectures. Including Presentation Speeches and Laureates' Biographies. Physics, 1971-1980* (World Scientific, Singapore, 1992); (b) P. A. M. Dirac, 'The fundamental equations of quantum mechanics', *Proceedings of the Royal Society* A **109**, 642 (1925); (c) P. A. M. Dirac, 'Quantum mechanics and a preliminary investigation of the hydrogen atom', *Proceedings of the Royal Society* A**110**, 561 (1926).

[54] W. Heisenberg, 'Erinnerungen an die Zeit der Entwicklung der Quantenmechanik', in Ref. [14a], pp. 40–7.

[55] (a) W. Pauli jr., 'Über das Wasserstoffspektrum vom Standpunkt der neuen Quantenmechanik', *Zeitschrift für Physik* **36**, 336 (1926). Reprinted

in Ref. [4], Vol. 2, pp. 252–79. (b) V. Fock, 'Zur Theorie des Wasser-stoffatoms', *Zeitschrift für Physik* **98**, 145 (1935); (c) V. Bargmann, 'Zur Theorie des Wasserstoffatoms. Bemerkungen zur gleichnamigen Arbeit von V. Fock', *Zeitschrift für Physik* **99**, 576 (1936); (d) W. Pauli, 'Gruppentheorie und Quantenmechanik', Notes by C. P. Enz (1954, un-published); (e) W. Pauli, 'Continuous Groups in Quantum Mechanics', CERN Publication 56-31, 1956. Reprinted in: G. Höhler (ed., with the co-operation of S. Flügge, F. Hund, and F. Trendelenburg), *Ergebnisse der exakten Naturwissenschaften* (Springer, Berlin and Heidelberg, 1965), Vol. 37, pp. 85–104.

[56] E. Schrödinger, 'Quantisierung als Eigenwertproblem', (a) 'Erste Mit-teilung', *Annalen der Physik* **79**, 361 (1926); (b) 'Zweite Mitteilung', **79**, 489 (1926); (c) 'Dritte Mitteilung: Störungstheorie mit Anwendung auf den Starkeffekt der Balmerlinien', **80**, 437 (1926); (d) 'Vierte Mit-teilung', **81**, 109 (1926). Reprinted in Ref. [57], (a) pp. 82–97; (b) pp. 98–136; (c) pp. 166–219; (d) pp. 220–50.

[57] *Erwin Schrödinger. Gesammelte Abhandlungen. Band 3. Beiträge zur Quantentheorie* (Verlag der Österreichischen Akademie der Wissen-schaften, Vieweg, Braunschweig, 1984).

[58] L. de Broglie, (a) *Recherche sur la théorie des quanta. Thèses présentées à la Faculté des Sciences de l'Université de Paris* (Masson, Paris, 1924); (b) 'Recherche sur la théorie des quanta', *Annales de Physique* **3**, 22 (1925); (c) 'Vue d'ensemble sur mes travaux scientifiques', in Ref. [59c], pp. 457–93.

[59] (a) J. Mehra and H. Rechenberg, *The Historical Development of Quan-tum Theory. Volume 5. Erwin Schrödinger and the Rise of Wave Me-chanics. Part 2. The Creation of Wave Mechanics; Early Response and Applications, 1925-1926* (Springer, New York, 1987); (b) W. Moore, *Schrödinger. Life and Thought* (Cambridge University Press, Cambridge, 1989); (c) A. George (ed.), *Louis de Broglie. Physicien et Penseur* (Albin Michel, Paris, 1953).

[60] L. de Broglie, (a) 'Ondes et quanta', *Comptes Rendus (Paris)* **177**, 507 (1923); (b) 'Quanta de lumière, diffraction et interférences', **177**, 548 (1923); (c) 'Les quanta, la théorie cinétique des gaz et le principe de Fermat', **177**, 630 (1923); (d) 'Sur la définition générale de la correspon-dance entre onde et mouvement', **179**, 39 (1924).

[61] (a) C. J. Davisson and L. H. Germer, 'The scattering of electrons by a single crystal of nickel', *Nature* **119**, 558 (1927); 'Diffraction of electrons by a crystal of nickel', *Physical Review* **30**, 705 (1927); (b) G. P. Thomson and A. Reid, 'Diffraction of cathode rays by a thin film', *Nature* 119, 890 (1927); (c) J. Hendry, *The Creation of Quantum Mechanics and the Bohr-Pauli Dialogue* (Reidel, Dordrecht, 1984).

[62] (a) G. Rasche and H. H. Staub, 'Physik und Physiker an der Universität Zürich', *Vierteljahrsschrift der Naturforschenden Gesellschaft in Zürich*

124, 205–20 (1979); (b) F. Bloch, 'Reminiscences of Heisenberg and the early days of quantum mechanics', *Physics Today*, December 1976, pp. 23–7; (c) *Note*: A discrepancy seems to exist between the account given in Ref. [59a], p. 461, which is based on much later interviews with Anny Schrödinger and 'the lady of Arosa' theory of Ref. [59b], p. 195 ('Anny remained in Zürich', Ref. [59b], p. 194).

[63] E. Schrödinger, 'Über das Verhältnis der Heisenberg-Born-Jordanschen Quantenmechanik zu der meinen', *Annalen der Physik* **79**, 734 (1926). Reprinted in Ref. [57], pp. 143–65.

[64] B. L. van der Waerden, 'From Matrix Mechanics and Wave Mechanics to Unified Quantum Mechanics', in Ref. [23], pp. 276–93

[65] (a) O. Klein, 'Elektrodynamik und Wellenmechanik vom Standpunkt des Korrespondenzprinzips', *Zeitschrift für Physik* **41**, 407 (1927); (b) W. Gordon, 'Der Comptoneffekt nach der Schrödingerschen Theorie', *Zeitschrift für Physik* **40**, 117 (1926).

[66] (a) K. Lanczos, 'Über eine feldmässige Darstellung der neuen Quantenmechanik', *Zeitschrift für Physik* **35**, 812 (1926); (b) C. Lanczos, in: *Journal of Research of the National Bureau of Standards* **45**, 255 (1950); (c) P. Fulde, *Electron Correlations in Molecules and Solids* (Springer, Berlin and Heidelberg, 1993, 2nd edn, 1993); (d) W. Killy and R. Vierhaus (eds), *Deutsche Biographische Enzyklopädie (DBE)* (Saur, Munich, 1996).

[67] M. Born, 'Quantenmechanik der Stossvorgänge', *Zeitschrift für Physik*, (a) **37**, 863 (1926); (b) **38**, 803 (1926).

[68] J. Stachel (ed.), *The Collected Papers of Albert Einstein. Volume 2. The Swiss Years: Writings, 1900-1909* (Princeton University Press, Princeton, NJ, 1989).

[69] W. Heisenberg, 'Über den anschaulichen Inhalt der quantentheoretischen Kinematik und Mechanik', *Zeitschrift für Physik* **43**, 172 (1927). Reprinted in Ref. [45], Part I, pp. 478–504.

[70] C. G. Darwin, (a) 'The electron as a vector wave', *Nature* **119**, 282 (1927); (b) 'The Electron as a Vector Wave', *Proceedings of the Royal Society* A **116**, 227 (1927).

[71] Translated from: W. Pauli, letter [69] of 23 October 1956 to C. G. Jung, Ref. [72], p. 150.

[72] C. A. Meier (ed., with the cooperation of C. P. Enz and M. Fierz), *Wolfgang Pauli und C. G. Jung. Ein Briefwechsel, 1932-1958* (Springer, Berlin and Heidelberg, 1992).

[73] (a) G. Wentzel, 'Zur Theorie des Comptoneffekts', *Zeitschrift für Physik* **43**, 1, 779 (1927). (b) Transcript of a conversation between Otto Stern and Res Jost of 2 December 1961, tape in Phonothek, Wissenschaftshistorische Sammlung, ETH, Zurich. (c) Letter dated 5 May 1998, Hamburg, from the late Harry Lehmann to Charles P. Enz. (d) P. Jordan, '50

Jahre physikalische Forschung an der Universität Hamburg', in: P. Jordan, *Begegnungen* (Stalling Verlag, Oldenburg and Hamburg, 1971), pp. 53–64; (e) I. I. Rabi, 'Otto Stern, Co-discoverer of Space Quantization, Dies at 81', *Physics Today*, October 1969, p. 103.

[74] (a) E. Ising, 'Beitrag zur Theorie des Ferromagnetismus', *Zeitschrift für Physik* **31**, 253 (1925); (b) W. Lenz, 'Beitrag zum Verständnis der magnetischen Erscheinungen in festen Körpern', *Physikalische Zeitschrift* **21**, 613 (1920); (c) G. Brush, 'History of the Lenz-Ising Model', *Reviews of Modern Physics* **39**, 883 (1967).

[75] W. Baade and W. Pauli jr., 'Über den auf die Teilchen in den Kometenschweifen ausgeübten Strahlungsdruck', *Naturwissenschaften* **15**, 49 (1927). Reprinted in Ref. [4], Vol. 2, pp. 329–30.

[76] (a) M. Fierz, 'Naturerklärung und Psyche. Ein Kommentar zu dem Buch von C. G. Jung und W. Pauli (1979)', in: M. Fierz, *Naturwissenschaft und Geschichte* (Birkhäuser, Basle, 1988), pp. 181–91; (b) C. P. Enz, 'Wolfgang Pauli (1900-1958). A Biographical Introduction', in Ref. [6b], pp. 13–26.

[77] (a) C. G. Jung, 'Synchronizität als ein Prinzip akausaler Zusammenhänge', in: C. G. Jung and W. Pauli, *Naturerklärung und Psyche* (Rascher Verlag, Zurich, 1952), pp. 1–107; also in: C. G. Jung, *Gesammelte Werke* (Walter Verlag, Olten, 1987), Vol. 8; (b) R. Wilhelm (ed.), *I Ging. Das Buch der Wandlungen* (Diederichs Verlag, Munich, 1990); (c) *Note*: The restaurant Curio-Haus, which still exists today, is situated on the Rothenbaumchaussee that runs from the university's main building to the north. It is within 15 minutes' walking distance along a footpath from the institutes at Jungiusstrasse (information contained in Ref. [73c]).

[78] (a) C. P. Enz, 'Is the zero-point energy real?', in: C. P. Enz and J. Mehra (eds), *Physical Reality and Mathematical Description* (Reidel, Dordrecht, 1974), pp. 124–32; (b) F. W. Aston, 'The Constitution of Atmospheric Neon', *Philosophical Magazine* **39**, 449 (1920).

[79] (a) M. Planck, 'Eine neue Strahlungshypothese', *Verhandlungen der Deutschen Physikalischen Gesellschaft* **13**, 138 (1911); (b) O. Stern, 'Zur kinetischen Theorie des Dampfdrucks einatomiger fester Stoffe und über die Entropiekonstante einatomiger Gase', *Physikalische Zeitschrift* **14**, 629 (1913); (c) C. P. Enz and A. Thellung, 'Nullpunktsenergie und Anordnung nicht vertauschbarer Faktoren im Hamiltonoperator', *Helvetica Physica Acta* **33**, 839 (1960).

[80] W. Pauli jr., 'Über Gasentartung und Paramagnetismus', *Zeitschrift für Physik* **41**, 81 (1927). Reprinted in Ref. [4], Vol. 2, pp. 284–305.

[81] (a) R. E. Peierls, *Quantum Theory of Solids* (Oxford University Press, Oxford, 1955); (b) L. D. Landau, 'Diamagnetismus der Metalle', *Zeitschrift für Physik* **64**, 629 (1930).

[82] (a) A. Sommerfeld, 'Zur Elektronentheorie der Metalle auf Grund der Fermischen Statistik. I. Teil: Allgemeines, Strömungs- und Austrittsvor-

gänge.', 'II. Teil: Thermo-elektrische, galvano-magnetische und thermo-magnetische Vorgänge', *Zeitschrift für Physik* **47**, 1, 43 (1928); (b) A. Sommerfeld and H. Bethe, 'Elektronentheorie der Metalle', in: H. Geiger and K. Scheel (eds) *Handbuch der Physik* Vol. 24, Part 2 (Springer, Berlin, 1933), pp. 1–290.

[83] W. Pauli jr., 'Über die Absorption der Reststrahlen in Kristallen', *Verhandlungen der Deutschen Physikalischen Gesellschaft* **6**, 10 (1925). Reprinted in Ref. [4], Vol. 2, p. 251.

[84] *Note*: The Hamiltonian which according to the canonical equations (3.11) leads to the equation of motion (5.44) consists of the two terms

$$H_0 = \frac{1}{2} \sum_{n=1}^{N} \left(\frac{p_n^2}{m} + m\omega_0^2 u_{n+1,n}^2 \right); \quad H_1 = \varepsilon \frac{m\omega_0^2}{3} \sum_{n=1}^{N} u_{n+1,n}^3 \quad (A)$$

where N is the number of lattice sites (I use periodic boundary conditions). If a is the lattice constant, discrete wavenumbers $k = 2\pi\kappa/Na$ with integers κ are defined in the Brillouin zone $-\pi/a < k \le +\pi/a$, and Fourier transformation of Eqs (A) with

$$u_n = \frac{1}{\sqrt{Nm}} \sum_k \tilde{u}_k e^{inka}; \quad p_n = \sqrt{\frac{m}{N}} \sum_k \tilde{p}_k e^{inka}$$

yields (see e.g. Section 1 of Ref. [36b])

$$H_0 = \frac{1}{2} \sum_k \left(|\tilde{p}_k|^2 + \omega_k^2 |\tilde{u}_k|^2 \right)$$

with ω_k defined in Eq. (5.45) and

$$H_1 = \frac{i\varepsilon}{3\omega_0\sqrt{Nm}} \sum_{kk'} \omega_k \omega_{k'} \omega_{k+k'} sgn(kk'(k+k')) \tilde{u}_k \tilde{u}_{k'} \tilde{u}_{-k-k'}. \quad (B)$$

Pauli's γ is now determined by the 'golden rule' (see e.g. Ref. [36b])

$$\gamma_k = 2\pi \sum_{k'} |\langle k|H_1|k', k-k'\rangle|^2 \delta(\omega_k \pm \omega_{k'} - \omega_{rad}). \quad (C)$$

Defining the phonon states in the matrix element of Eq. (C) by the creation and annihilation operators a_k^+, a_k, and $u_k = (a_k^+ + a_{-k})/\sqrt{2\omega_k}$ one finds from Eqs (B) and (C)

$$\gamma_k = \frac{\pi\varepsilon^2\omega_k}{m\omega_0^2} \frac{1}{N} \sum_{k'} \omega_{k'}\omega_{k-k'}\delta(\omega_k \pm \omega_{k'} - \omega_{rad}).$$

With the physically reasonable approximation $\omega_{k'} \ll \omega_k \simeq \omega_{k-k'} \simeq \omega_{rad}$ and with $N^{-1}\sum_{k'} \to \int d\omega_{k'}/2\pi\omega_0$ this becomes

$$\gamma_k \simeq \frac{\varepsilon^2 \omega_{rad}^2}{2m\omega_0^3} |\omega_k - \omega_{rad}|.$$

Taking a thermal average $\langle |\omega_k - \omega_{rad}| \rangle$ for $\Delta\omega \equiv |\omega_k - \omega_{rad}| \ll kT$, which is given by a Maxwell distribution proportional to $\exp(-\Delta\omega/kT)$, and setting $\omega_{rad} = 2\omega_0$ one arrives at

$$\gamma \equiv \langle \gamma_k \rangle \simeq \frac{4\varepsilon^2 kT}{m\omega_0}$$

which is Pauli's result.

[85] (a) R. E. Peierls, 'Quantum Theory of Solids', in Ref. [14a], pp. 140–60; (b) R. Peierls, *Bird of Passage. Recollections of a Physicist* (Princeton University Press, Princeton, NJ, 1985).

[86] (a) R. Peierls, 'Zur kinetischen Theorie der Wärmeleitung in Kristallen', *Annalen der Physik* 3, 1055 (1929); (b) P. G. Klemens, 'Thermal Conductivity and Lattice Vibrational Modes', in: F. Seitz and D. Turnbull (eds), *Solid State Physics. Advances in Research and Applications* (Academic Press, New York, 1958), Vol. 7, pp. 1–98.

[87] (a) R. Peierls, 'Zur Theorie der Absorptionsspektren fester Körper', *Annalen der Physik* **13**, 905 (1932); (b) C. P. Enz, B. Glaus, and G. Oberkofler, *Wolfgang Pauli und sein Wirken an der ETH Zürich* (vdf Hochschulverlag ETH, Zurich, 1997); (c) H. B. G. Casimir, 'Pauli and the Theory of the Solid State', in Ref. [14a], pp. 137–9; (d) K. von Meyenn (ed.), *Wolfgang Pauli. Scientific Correspondence with Bohr, Einstein, Heisenberg, a.o., Volume III: 1940-1949* (Springer, Berlin and Heidelberg, 1993).

[88] Letter from Dean Hecke to Pauli of 7 February 1924, PLC Bi 44, *Pauli Archive*.

[89] M. Dresden, *H. A. Kramers. Between Tradition and Revolution* (Springer, New York, 1987), Chapter 13, Section III.

[90] H. Geiger and W. Bothe, 'Über das Wesen des Comptoneffektes; ein experimenteller Beitrag zur Theorie der Strahlung', *Zeitschrift für Physik* **32**, 639 (1925).

[91] (a) N. Bohr, 'The Quantum Postulate and the Recent Development of Atomic Theory', *Nature* **121**, 580 (1928); German version: 'Das Quantenpostulat und die neuere Entwicklung der Atomistik', *Naturwissenschaften* **16**, 245 (1928); (b) *Atti del Congresso internazionale dei Fisici, 11-20 Settembre 1927, Como – Pavia – Roma* (Zanichelli, Bologna, 1928), 2 vols; (c) O. Klein, 'Glimpses of Niels Bohr as Scientist and Thinker', in: S. Rozental (ed.), *Niels Bohr. His life and work as seen by his friends and colleagues* (North-Holland, Amsterdam, 1968), pp. 74–93.

[92] W. Pauli, 'Eddington, S. A., Relativitätstheorie in mathematischer Behandlung', *Naturwissenschaften* **14**, 273 (1926). Reprinted in Ref. [4], Vol. 2, pp. 1392–3.

[93] N. Bohr, *Albert Einstein: Philosopher–Scientist* (Library of Living Philosophers, Inc., 1949; Harper & Row, New York, 1959).

[94] M. Jammer, *The Philosophy of Quantum Mechanics. The Interpretation of Quantum Mechanics in Historical Perspective* (Wiley, New York, 1974).

6
The new ETH professor

The pioneers of quantum electrodynamics

At the beginning of this semester (autumn 1928) physics was fairly remote from me, I was very lazy ... but yet very alert and in a good mood. For my proper amusement I then made a short sketch of a utopian novel which was supposed to have the title 'Gulliver's journey to Urania' and was intended as a political satire in the style of [Jonathan] Swift [1667–1745] against present-day democracy, namely against everything that even remotely smells of elections, parliaments, votes and majorities! Caught in such dreams, suddenly in January, news from Heisenberg reached me that he is able, with the aid of a trick ... to get rid of the formal difficulties that stood against the execution of our quantum electrodynamics. [1]

But, instead of a utopian novel Pauli wrote two important papers with Heisenberg. The whole activity, however, had been stimulated by Dirac who on 2 February 1927 had submitted from Copenhagen his first paper on quantum electrodynamics (QED); he thanked Bohr for 'his interest in this work and for much friendly discussion about it' [3a]. This paper, in fact, set in motion a new development that was to dominate theoretical physics for the following two decades. Heisenberg, who in May 1926 had succeeded Kramers as a lecturer in Bohr's institute and who had collaborated with Born and Jordan on the first exploration of QED in the 'Three-Man-Paper' (see page 133), got the news first hand and, as usual, relayed it immediately to Pauli. The latter's reaction was swift since on 23 February 1927 Heisenberg had already told Pauli that 'with your programme concerning electrodynamics I very much agree' (translated from letter [154], Ref. [2], p. 376). What this 'programme' consisted of, we do not know, but it is mentioned again in Heisenberg's letter [161] of 4 April, according to which Pauli had already received criticism concerning Dirac's expression for the interaction between radiation and matter, in particular the use of time as independent variable (Ref. [2], pp. 390, 391).

Dirac's plan in his paper [3a] is quite ambitious since he proposes to show that 'the interaction of the atom and the electromagnetic waves can be made identical with the Hamiltonian for the problem of the interaction of the atom with an assembly of particles moving with the velocity of light and satisfying the Einstein–Bose statistics, by a suitable choice of the interaction energy for the particles' (Ref. [3a], p. 245). In Section 2 of his paper Dirac starts quite generally with an unperturbed system described by a Hamiltonian H_0 having eigenstates ψ_r and eigenvalues W_r, which is perturbed by an interaction V. The solution of the time-dependent Schrödinger equation (5.33) with $H = H_0 + V$ may then be solved to first order in V following Dirac's own method described earlier [3b], by developing the wavefunction into the eigenfunctions of H_0, $\psi(t) = \sum_r b_r(t)\psi_r$. This leads to the standard complex conjugate equations

$$i\hbar\dot{b}_r = W_r b_r + \sum_s v_{rs} b_s; \quad -i\hbar\dot{b}_r^* = b_r^* W_r + \sum_s b_s^* v_{sr} \quad (6.1)$$

where $v_{rs} = v_{sr}^* = \langle \psi_r | V | \psi_s \rangle$ is the matrix element of V between the states ψ_r and ψ_s. Dirac actually starts in the *interaction representation* $\psi^i(t) = \exp(iH_0t/\hbar)\psi(t) = \sum_r a_r(t)\psi_r$, which two decades later becomes of central importance (see e.g. Section 21 of Pauli's lectures on *Field Quantization* [4]). In this representation $i\hbar\dot{\psi}^i = V^i\psi^i$ where $V^i = \exp(iH_0t/\hbar)V\exp(-iH_0t/\hbar)$ and $V_{rs} = \langle \psi_r | V^i | \psi_s \rangle = \exp(i[W_r - W_s]t/\hbar)v_{rs}$.

In an elegant formal argument Dirac then shows that, with respect to the new Hamiltonian $F = \sum_{rs} b_r^*(W_r\delta_{rs} + v_{rs})b_s$, $i\hbar b_r^*$ and b_r are canonically conjugate variables analogous to p_k and q_k, respectively, in Eqs (3.11) which then become Eqs (6.1). In addition, the relation $b_r = \sqrt{N_r}\exp(-i\Theta_r/\hbar)$ and its complex conjugate define a canonical transformation to new variables N_r and Θ_r which satisfies the identity

$$N_r d\Theta_r - i\hbar b_r^* db_r = \frac{\hbar}{2i} dN_r(\Theta_r, b_r) \quad (6.2)$$

called a *contact transformation*. Therefore N_r and Θ_r indeed also satisfy canonical equations of motion with F as Hamiltonian. Since $N_r = |b_r|^2 \geq 0$, N_r may be interpreted as the probable number of systems in the state ψ_r. Now the above analogy with Eqs (3.11) immediately suggests the quantization rule analogous to Eq. (5.28) between $i\hbar b_r^*$ and b_r, or

$$b_r b_r^+ - b_r^+ b_r = 1, \quad (6.3)$$

all other commutators being zero, where b_r^+ is now the operator (matrix) Hermitian conjugate to b_r. With Eq. (6.3) Dirac arrived at the creation and annihilation operators which, for the harmonic oscillator, are already implicitly contained in Heisenberg's expression of Eq. (5.26).

For the new variables N_r and Θ_r, considered as operators acting on functions of the N_r, the canonical commutation relation $N_r\Theta_r - \Theta_r N_r = \hbar/i$ implies that $\Theta_r = -(\hbar/i)\partial/\partial N_r$. And since, quite generally, $\exp(a\partial/\partial x)f(x) = f(x + a)$, the operators $\exp(\pm i\Theta_r/\hbar) = \exp(\mp\partial/\partial N_r)$ raise/lower the number N_r by one. Hence $\exp(-i\Theta_r/\hbar)f(N_r)\exp(i\Theta_r/\hbar) = f(N_r + 1)$ and, in particular,

$$b_r = \sqrt{N_r + 1}\,e^{-i\Theta_r/\hbar} = e^{-i\Theta_r/\hbar}\sqrt{N_r}$$
$$b_r^+ = \sqrt{N_r}\,e^{i\Theta_r/\hbar} = e^{i\Theta_r/\hbar}\sqrt{N_r + 1}, \quad (6.4)$$

which are Dirac's Eqs (10), so that $b_r^+ b_r = b_r b_r^+ - 1 = N_r$ and the number operator N_r has the characteristic values $0, 1, 2, \dots$. Applied to an N-particle wavefunction $\Psi(N_{r_1}, N_{r_2}, \dots, N_{r_N})$ of N_r particles in the state ψ_r, $r = 1, 2, \dots, N$, the creation and annihilation operators thus have the property

$$
\begin{aligned}
b_r^+ \Psi(\dots, N_r, \dots) &= \sqrt{N_r + 1}\,\Psi(\dots, N_r + 1, \dots), \\
b_r \Psi(\dots, N_r, \dots) &= \sqrt{N_r}\,\Psi(\dots, N_r - 1, \dots).
\end{aligned}
\tag{6.5}
$$

In Section 3 of Ref. [3a] Dirac applies this rather simple theory to an assembly of systems satisfying BE statistics, proceeding in analogy to Bose's method to derive the average particle number $\langle n_s \rangle$ of Eq. (5.19), showing that such an assembly, perturbed by an interaction V, is described by the (quantized) Hamiltonian F defined above. Written in the variables N_r and Θ_r by using Eqs (6.4), it reads

$$
F = \sum_r W_r N_r + \sum_{rs} v_{rs} \sqrt{N_r} \sqrt{N_r + 1 - \delta_{rs}}\, e^{i(\Theta_r - \Theta_s)/\hbar}.
\tag{6.6}
$$

Finally, in Section 6 this theory is applied to an assembly of light quanta interacting with electrons which may be free or bound in an atom. In this case the states ψ_r designate the Fourier components of the electromagnetic field, specified by photon momentum and polarization. And v_{rs} are the matrix elements of the coupling V between the electromagnetic field described by the vector potential $\mathbf{A}$ and the charge e of an electron described by its momentum $\mathbf{p}$ and position $\mathbf{q}$. More explicitly, V is the interaction part of the Hamiltonian obtained with the help of the correspondence principle,

$$
\frac{\mathbf{p}^2}{2m} + V = \frac{1}{2m}\left(\mathbf{p} + \frac{e}{c}\mathbf{A}(\mathbf{q})\right)^2.
\tag{6.7}
$$

The quantization of the radiation field enabled Dirac to treat quite generally transitions between states of different numbers of photons but of equal energy as, for example, the absorption or emission of a photon. He then showed that the probability of absorption from a state $|N_r\rangle$ is proportional to N_r while that of emission is proportional to $N_r + 1$ in agreement with Einstein's terms $B\rho_\nu$ and $B\rho_\nu + A$, respectively, appearing in Eq. (4.20) above.

But Dirac's theory still had two shortcomings. First, it is not relativistically covariant, as Dirac emphasized himself. And secondly, it treats the number of charges (electrons) as fixed. The first problem was addressed by Jordan and Pauli in their common paper 'On the quantum electrodynamics of charge-free fields' [5a], the second required an extension of the idea of creation and annihilation operators to particles obeying Pauli's exclusion principle. While for bosons, i.e. for particles obeying BE statistics, these operators are determined by the commutation relations (6.3), the generalization to the case of fermions, i.e. particles obeying FD statistics, of which the only examples known in the late 1920s were electrons and protons, was not immediately evident. It was Jordan who pioneered this important step in several papers [7a,b,c].

In his first note [7a] submitted from Copenhagen on 7 July 1927, Jordan started from Dirac's representation (6.4) in which, however, the number operator N_r is restricted by the Pauli principle to have only the eigenvalues 0 and 1. This two-valuedness, which is surprisingly analogous to that of the electron spin, suggests a formulation in terms of the Pauli matrices (5.37). As was mentioned after these equations, Pauli and Jordan had corresponded on the properties of these matrices in the spring of 1927, and Pauli's paper containing them was submitted just two months before Jordan's note [7a]. Thus, $N_r = (1/2)(1 - \sigma_z)$ and, as in Dirac's case, we require that $\exp(-i\Theta_r/\hbar)N_r \exp(i\Theta_r/\hbar)$ is a matrix obtained from N_r by raising or lowering the eigenvalues by one. But since the resulting matrix again can only have eigenvalues 0 and 1, this matrix must be $1 - N_r = (1/2)(1 + \sigma_z)$. Since, in addition, $N_r^2 = N_r$ and $(1 - N_r)^2 = 1 - N_r$, we also have $\sqrt{N_r} = N_r$ and $\sqrt{1 - N_r} = 1 - N_r$, so that Eqs (6.4) become in the case of the exclusion principle

$$b_r = \sqrt{1 - N_r}\,e^{-i\Theta_r/\hbar} = e^{-i\Theta_r/\hbar}\sqrt{N_r}$$
$$b_r^+ = \sqrt{N_r}\,e^{i\Theta_r/\hbar} = e^{i\Theta_r/\hbar}\sqrt{1 - N_r}. \tag{6.8}$$

These are Jordan's Eqs (13) of Section 2 in Ref. [7a]. From Eqs (6.8) it follows that $b_r^+ b_r = 1 - b_r b_r^+ = N_r$ or that b_r and b_r^+ satisfy the anticommutation relation

$$b_r b_r^+ + b_r^+ b_r = 1 \tag{6.9}$$

instead of (6.3). Because of $N_r(1 - N_r) = (1/4)(1 - \sigma_z)(1 + \sigma_z) = 0$ one also concludes from Eqs (6.8) that $b_r^2 = 0$ and $b_r^{+2} = 0$. These relations express the action of the Pauli principle; that is, to prohibit in any state ψ_r the annihilation or creation of more than one fermion. Finally, it is also possible to determine the matrix Θ_r. Indeed, the above equations imply that $\exp(i\Theta_r/\hbar)\sigma_z = -\sigma_z \exp(i\Theta_r/\hbar)$. According to the relations given after Eqs (5.37) the only 2×2 matrices having this property are σ_x and σ_y. Hence, up to a phase φ, $\exp(i\Theta_r/\hbar) = \sigma_\alpha \equiv (\cos\alpha)\sigma_x + (\sin\alpha)\sigma_y$, and since $\sigma_\alpha^2 = 1$, $\Theta_r/\hbar = (\pi/2)\sigma_\alpha + \varphi$. Jordan, in Eq. (17) of Section 1 in Ref. [7a], chose $\alpha = \varphi = 0$.

This is the essence of Jordan's first paper [7a]. Since the states ψ_r are already wavefunctions satisfying a one-particle Schrödinger equation the quantization of the amplitudes b_r and b_r^+, unlike the case of the photons, in fact amounts to a second quantization. And this is true both for BE and for FD statistics, i.e. with commutation relations (6.3) as well as with anticommutation relations (6.9). The first case, which we need not pursue further, was treated by Jordan and Klein [7b]. More care is needed when many fermions in different states ψ_r are present, as described by an N-particle wavefunction $\Psi(N_{r_1}, N_{r_2}, \ldots, N_{r_N})$. The reason is an ambiguity of an overall sign which Jordan had left open in Ref. [7a]. Note that the same problem occurs with the wavefunction Ψ_- of Eq. (5.20), the sign of which depends on the order that is chosen for the state labels $r_1, r_2, \ldots, r_N$. In second quantization this problem was addressed by Jordan and Wigner in their paper 'On Pauli's equivalence prohibition', submitted on 26 January 1928 [7c]. The first observation is that, to be consistent with Eq. (6.9), creation and annihilation operators between different states must anticommute instead of commute,

$$b_r^+ b_s^+ + b_s^+ b_r^+ = b_r b_s + b_s b_r = 0$$
$$b_r b_s^+ + b_s^+ b_r = 0, r \neq s. \tag{6.10}$$

For $r = s$ the action of the Pauli principle, $b_r^2 = b_r^{+2} = 0$, then immediately follows.

Jordan and Wigner then proposed to associate, for a given order $r_1, r_2, \ldots, r_N$, the sign $\varepsilon_n = \prod_{n' < n}(1 - 2N_{n'})$, $n = 1, 2, \ldots, N$, with the creation and annihilation operators, so that $a_r = \varepsilon_r b_r$ and $a_r^+ = b_r^+ \varepsilon_r$. And they showed that the a_r and a_r^+ satisfy the same anticommutation relation as the b_r and b_r^+, i.e. using the symbol $\{A, B\} \equiv AB + BA$,

$$\{a_r, a_s^+\} = \delta_{rs}, \{a_r, a_s\} = \{a_r^+, a_s^+\} = 0. \tag{6.11}$$

Equations (6.8) may also be retained for the a_r and a_r^+, just by adding to Θ_r a phase $\varphi = 0$ or π. With $\exp(i\Theta_r/\hbar) = \exp(-i\Theta_r/\hbar) = \varepsilon_r \sigma_x$ one then has, in analogy with (6.5),

$$a_r^+ \Psi(\ldots, N_r, \ldots) = \varepsilon_r \sqrt{1 - N_r} \Psi(\ldots, 1 - N_r, \ldots),$$
$$a_r \Psi(\ldots, N_r, \ldots) = \varepsilon_r \sqrt{N_r} \Psi(\ldots, 1 - N_r, \ldots). \tag{6.12}$$

One may then forget the b_r and b_r^+ altogether since, for a given order $r_1, r_2, \ldots, r_N$, the anticommutation relation (6.11) takes care of the sign ε_r. In what follows fermionic operators will be designated by a_r, a_r^+ and bosonic operators by b_r, b_r^+. This formalism of second quantization is described in Ziff. 14 of Pauli's 'Wellenmechanik' (Ref. [8], pp. 127–8 and Ref. [9], pp. 198–9).

The next step was, as already mentioned, a relativistically covariant formulation of the quantized radiation field in the absence of charges. According to Jordan's recollections about Pauli, it was the latter's conviction 'that the reconciling unification of light-quantum theory and wave theory of light was to be sought in the elaboration of a new notion of field. . . . But Pauli clearly recognized that the next step now should go in the direction of finding a formulation for the "field quantization" satisfying the requirements of relativistic invariance. 'A long exchange of letters between us was necessary until the proposed solution also of this question had been found in a common paper' (translated from Ref. [10], p. 38). This paper [5a] was submitted on 7 December 1927; the correspondence leading to it, however, seems to have been lost.

Pascual Jordan was of Spanish descent by his grandfather—he used to emphasize that his family name was of Spanish origin. Studying physics and mathematics first in his home town of Hanover and then in Göttingen he became Born's assistant there in 1924 and also collaborated with James Franck (see pages 69 and 133). Jordan's familiarity with matrix calculus allowed him to assume in summer 1925 a leading role in Born's programme of the mathematical formulation of the new quantum mechanics; he became a Privatdozent in 1927. Succeeding Pauli as Lenz's assistant in Hamburg the following year, Jordan became an associate professor at Rostock in 1929 and then a full professor in 1935, a position he kept until 1944 when he moved to Berlin to succeed Max von Laue. In the late 1930s quantum biology became Jordan's main field of research. At the same time he grew quite sympathetic towards and even collaborated with the National Socialists, which gave rise to unfortunate popular writings. These had an adverse effect on Jordan's post-war career, and it was not until 1953 that he was again reinstated as a full professor, this time in Hamburg where he retired in 1971. Jordan's research was then concentrated on the five-dimensional formulation of general relativity. During the Adenauer era from 1957 to 1961 Jordan was CDU member to the German Bundestag advocating peaceful uses of atomic energy [11b]. Jordan's unsteady political orientations incited Pauli to issue the remark: 'Alas, good Jordan! He has served all regimes in utmost faithfulness' (translated from Ref. [11c], p. 116; see also Ref. [11d]).

In the first part of Jordan and Pauli's common paper [5a] the operator algebra of the quantized field ensuing from the commutation relation (6.3) is developed in relativistic form. It is here that the celebrated invariant Jordan–Pauli function $D(x)$ (called $\Delta(x)$ in Ref. [5a]) appears for the first time, which determines the commutation properties between relativistic field operators whose space–time separation is given by the four-vector x. While this part is a generalization of Heisenberg's formulation of quantum theory from 'q-numbers' to 'q-functions', the second part discusses the Schrödinger equation, according to which these field operators act, and thus opens up the entire question of interpretation. This part introduces the mathematical generalization of Schrödinger's formulation of quantum theory from wavefunctions $\psi(q)$ to wave-functionals $\Psi\{q(x)\}$, which is necessary because the Ψ themselves now depend on space–time functions $q(x)$. For Pauli functional analysis was a new field of mathematics which he studied in early 1927. In his letter to Jordan of 12 March he writes: 'My principal source is a (French) book by P[ierre] Lévy [1886–1971], *Leçons d'analyse fonctionnelle*, Paris 1922 [second edition: *Problèmes concrèts d'analyse fonctionnelle*, Paris, 1951]. We shall see alright whether I bring about the quantum electrodynamics. For the moment I am in good spirits!' (translated from letter [157], Ref. [2], p. 386). As we see, the idea of a collaboration with Jordan was not yet in Pauli's mind.

The way to obtain the time-dependent commutation relations between the fields is to decompose the latter into harmonic-oscillator modes, for which the time evolution is known. This, however, is far from being a unique procedure. In fact, the details of the decomposition used by Jordan and Pauli in Section 1 of Part I of their paper [5a] differ from those of Pauli's 'Wellenmechanik' (Section 25 of Ref. [8] and Ziff. 6, Part B of Ref. [9]) and again from the explicitly relativistic form of his *Field Quantization* (Section 15 of Ref. [4]) which will be followed here. The four-potential is expressed as the sum

$$\Phi_\mu(x) = c \sum_{\lambda \mathbf{k}} \sqrt{\frac{\hbar}{2V\omega_\mathbf{k}}} e_\mu(\lambda \mathbf{k})\{b(\lambda \mathbf{k})e^{ikx} + b^+(\lambda \mathbf{k})e^{-ikx}\}, \qquad (6.13)$$

where $\lambda = 1, \ldots, 4$ and V is the volume of periodic boundary conditions giving rise to a discrete set of wavevectors $\mathbf{k}$ that satisfy $kk = 0$ with $\omega_\mathbf{k} \equiv c|\mathbf{k}|$ and $kx \equiv \mathbf{k} \cdot \mathbf{x} - \omega_\mathbf{k}t$. Writing $\Phi_\mu = (\mathbf{A}, iA_0)$, the vector potential $\mathbf{A}$ may be decomposed into a transverse part $\mathbf{A}_{tr}$ satisfying $\nabla \cdot \mathbf{A}_{tr} = 0$ and a longitudinal part $\mathbf{A}_{lg} = -\nabla A$. The polarization four-vectors $e_\mu(\lambda \mathbf{k})$ satisfy orthogonality relations $e_\mu(\lambda \mathbf{k})e_\mu(\lambda' \mathbf{k}) = \delta_{\lambda\lambda'}$. A choice giving the above decomposition are the real four-vectors $e_\mu(\lambda \mathbf{k}) = (\mathbf{e}(\lambda \mathbf{k}), 0)$ satisfying the transversality condition $\mathbf{e}(\lambda \mathbf{k}) \cdot \mathbf{k} = 0$ for $\lambda = 1, 2$ and $e_\mu(3, \mathbf{k}) = (\hat{k}, 0)$; $e_\mu(4, \mathbf{k}) = (0, 1)$, where $\hat{k} \equiv \mathbf{k}/|\mathbf{k}|$. In the covariant formulation chosen here, quantization proceeds according to the commutation relations (6.3) which now read

$$[b(\lambda \mathbf{k}), b^+(\lambda' \mathbf{k}')] = \delta_{\lambda\lambda'}\delta_{\mathbf{k}\mathbf{k}'}. \qquad (6.14)$$

These are called 'strong' in Ref. [4], Section 15, in order to distinguish them from those satisfied by the components of $\mathbf{A}_{tr}$. In terms of the latter the radiating electric and magnetic fields are given by $\mathbf{E} = -\dot{\mathbf{A}}_{tr}/c$ and $\mathbf{H} = \nabla \times \mathbf{A}_{tr}$, respectively, for which the decomposition may be expressed in terms of the auxiliary vector

$$\mathbf{F}(\mathbf{k}, t) = \sqrt{\frac{\hbar\omega_\mathbf{k}}{2V}} \sum_{\lambda=1,2} \mathbf{e}(\lambda \mathbf{k})b(\lambda \mathbf{k})e^{-i\omega_\mathbf{k}t},$$

and its Hermitian conjugate, namely (see Refs. [8, 9])

$$\mathbf{E}(x) = i \sum_\mathbf{k} (\mathbf{F}(\mathbf{k}, t) - \mathbf{F}^+(-\mathbf{k}, t))e^{i\mathbf{k}\cdot\mathbf{x}},$$

$$\mathbf{H}(x) = i \sum_\mathbf{k} \hat{k} \times (\mathbf{F}(\mathbf{k}, t) + \mathbf{F}^+(-\mathbf{k}, t))e^{i\mathbf{k}\cdot\mathbf{x}}. \qquad (6.15)$$

These are Hermitian field operators whose $\mathbf{F}^+$- and $\mathbf{F}$-parts describe incoming and outgoing, respectively. Making use of Eqs (6.15), the total field energy in the volume V becomes a sum of energy quanta of the modes $\lambda \mathbf{k}$,

$$\frac{1}{2} \int_V d^3x(\mathbf{E}^2(x) + \mathbf{H}^2(x)) = \sum_{\lambda \mathbf{k}} \hbar\omega_\mathbf{k}\left(b^+(\lambda \mathbf{k})b(\lambda \mathbf{k}) + \frac{1}{2}\right). \qquad (6.16)$$

This expression includes the divergent sum of the zero-point energies $\hbar\omega_\mathbf{k}/2$, on which the authors' comment in the introduction of the paper [5a] is almost identical to the one in Pauli's 'Wellenmechanik' (Ref. [8], p. 180 and Ref. [9], p. 250): 'in contrast to the eigen-oscillations in the crystal lattice ... with the eigen-oscillations of the radiation, no physical reality is associated to that "zero-point energy" $h\nu/2$ per degree of freedom.' And this 'because that "zero-point radiation" cannot be absorbed nor scattered nor reflected' (translated from Ref.

[5a], p. 154). Here the authors implicitly refer to Nernst who had stated that 'only with the use of mirrors that are efficient also at very short-wavelength radiation would the zero-point radiation manifest itself' (translated from Ref. [11e], p. 90). In the 'Wellenmechanik' Pauli observes in addition that the ZPE of radiation 'produces no gravitational field' (see page 150). This shows that this problem was very much Pauli's concern.

From Eqs (6.13) and (6.14) and the closure relation $\sum_\lambda e_\mu(\lambda\mathbf{k})e_\nu(\lambda\mathbf{k}) = \delta_{\mu\nu}$ one finds the commutator

$$[\Phi_\mu(x), \Phi_\nu(x')] = -i\hbar c^2 \delta_{\mu\nu} D(x - x'), \qquad (6.17)$$

where, after taking the limit of continuous wavevectors $\mathbf{k}$ by $V^{-1}\sum_\mathbf{k} \rightarrow (2\pi)^{-3}\int d^3k$,

$$D(x) = -\int \frac{d^3k}{(2\pi)^3} e^{i\mathbf{k}\cdot\mathbf{x}} \frac{\sin\omega_\mathbf{k} t}{\omega_\mathbf{k}}, \qquad (6.18)$$

which is the invariant Jordan–Pauli function (I use the definition of Section 13 of Ref. [4]). This function is discussed in Section 2 of Part I of paper [5a] where it is emphasized that $d^3k/\omega_\mathbf{k}$ is invariant under Lorentz transformations and where the following evaluation is given, which is easily obtained from Eq. (6.18):

$$D(x) = \frac{1}{(2\pi)^2 cr} \lim_{K\to\infty} \left(\frac{\sin K(r + ct)}{r + ct} - \frac{\sin K(r - ct)}{r - ct} \right)$$
$$= \frac{\delta(r + ct) - \delta(r - ct)}{4\pi cr}, \qquad (6.19)$$

here $r \equiv |\mathbf{x}|$. The last expression shows that $D(x)$ is a singular function with values concentrated on the 'incoming' and 'outgoing' light cones, $r + ct = 0$ and $r - ct = 0$, respectively. From Eq. (6.18) one also deduces the following properties:

$$\frac{\partial D(x)}{\partial t}\bigg|_{t=0} = -\delta(x); \quad \Box D(x) = 0, \qquad (6.20)$$

where the second relation is the Klein–Gordon equation (5.34). With the help of

$$[F_i(\mathbf{k}, t), F_j^+(\mathbf{k}', t')] = (\delta_{ij} - \hat{k}_i\hat{k}_j)\frac{\hbar\omega_\mathbf{k}}{2V} e^{-i\omega_\mathbf{k}(t-t')}\delta_{\mathbf{k}\mathbf{k}'} \qquad (6.21)$$

one finally deduces from (6.15) the equations

$$[E_i(x), E_j(x')] = [H_i(x), H_j(x')] = i\hbar c^2 \left(\frac{\partial^2}{\partial x_i \partial x_j} - \delta_{ij}\frac{\partial^2}{c^2\partial t^2} \right) D(x - x'),$$

$$[E_i(x), H_j(x')] = i\hbar c^2 \frac{\partial^2}{\partial x_l c\partial t} D(x - x'), \quad (i, j, l) \text{ cyclic}, \qquad (6.22)$$

which are given in Eqs (III) of Ref. [5a]. These equations are made relativistically covariant by first introducing, following Lorentz, the covariant notation for the fields that Pauli had already described in Section 28 of his *Encyklopädie*

Article [13]. One starts with the four-vector potential (6.13) which, however, is again determined only up to a gauge, here the *Lorentz* condition

$$\frac{\partial \Phi_\alpha}{\partial x^\alpha} = 0, \tag{6.23}$$

where the Greek indices run from 1 to 4 and $x^4 = ict$, repeated indices being summed. Expressed as a skew-symmetric tensor $F_{\alpha\beta} = -F_{\beta\alpha}$ such that $F_{4k} = iE_k$ and $F_{ij} = H_l$, (ijl) cyclic, the fields are then given by (see Eqs (2.9))

$$F_{\alpha\beta} = \frac{\partial \Phi_\beta}{\partial x^\alpha} - \frac{\partial \Phi_\alpha}{\partial x^\beta} \tag{6.24}$$

and satisfy the two sets of *Maxwell equations*

$$\frac{\partial F_{\alpha\beta}}{\partial x^\beta} = 0; \quad \sum_{(\alpha\beta\gamma)\text{cycl}} \frac{\partial F_{\alpha\beta}}{\partial x^\gamma} = 0. \tag{6.25}$$

These fields $F_{\alpha\beta}$ now satisfy the modified commutation relations (III′) of Ref. [5a], for which the compatibility with the Maxwell equations (6.25) may be verified.

In Part II of their paper [5a] Jordan and Pauli study the action of q-functions like the fields $\mathbf{E}(x)$ and $\mathbf{H}(x)$ on Ψ-functionals depending on a set of functions $g_i(x)$ which commute among themselves but have with functions $f_i(x)$ commutation relations

$$[f_i(x), g_j(y)] = i\hbar d_{ij}(x - y) \tag{6.26}$$

with some singular functions d_{ij} whose values are concentrated on the light cone. Letting Eqs (6.26) act on Ψ and making use of the product rule of differentiation, one then deduces in the usual way that, by defining a functional derivative as

$$\frac{\delta \Psi\{g_j(y)\}}{\delta g_i(x)} \equiv \lim_{\substack{\delta g_i(x'-x)\to \alpha d_{ij}(x'-x); \\ \alpha \to 0}} \frac{\Psi\{g_j(y) + \delta g_i(x'-x)\} - \Psi\{g_j(y)\}}{\int \delta g_i(x'-x)dx'}, \tag{6.27}$$

the action of $f_i(x)$ on Ψ takes the familiar form $i\hbar\delta\Psi\{g_j(y)\}/\delta g_i(x)$. This is the result (31) given in the last paragraph of Ref. [5a] (but see [5b]). The authors, however, close by noting that 'the equations quoted in the last paragraph of this Part II, for which, incidentally, direct integration methods are not yet available, must be considered as preliminary to an even higher degree than the considerations on q-functions developed in Part I. On the other hand, we consider the introduction of functionals into a consistent quantum-theoretical reinterpretation of the classical field physics in general as appropriate, in spite of many unsolved problems in their particular realization' (translated from Ref. [5a], p. 173).

This last remark shows that the authors considered their work as the beginning of a more important endeavour. For Pauli it was the first step in a lifelong effort. But before reporting on his next step we must again turn to Dirac who on 2 January 1928 had submitted the paper on the celebrated Dirac equation [14] which solved all the remaining problems related to the electron spin and relativistic properties—and raised new ones as well, as will be seen. For the QED endeavour this equation opend the way towards a consistent second quantization of fermions. The sought-for equation, Dirac argued, must account in a relativistically covariant way for the two-valuedness due to the spin—he calls it the 'duplexity phenomena'—expressed by the Pauli matrices (5.37). The difficulty with the only relativistic equation known at that time, the Klein–Gordon (KG) equation (5.34), is that in order to have conservation and invariance of probability one must, following Gordon and also Klein [15a], define the probability density by $\rho = -\psi^* \partial\psi/c\partial t + \psi\partial\psi^*/c\partial t$. However, since the KG equation contains the second time derivative, initial values of both ψ and $\partial\psi/\partial t$ may be arbitrarily prescribed, which implies that ρ is not a good probability density because $\rho \geq 0$ is not ensured. This is the reasoning given in Pauli's "Wellenmechanik' (Ref. [8], Section 18 and Ref. [9], Part B, Ziff. 2) which I follow in essence.

Therefore, Dirac concluded that the sought-for equation must be linear in $\partial\psi/\partial t$. Relativistic covariance then implies that the equation must also be linear in $\nabla\psi$,

$$\frac{1}{c}\frac{\partial\psi}{\partial t} + \vec{\alpha} \cdot \nabla\psi + ik_c\beta\psi = 0 \qquad (6.28)$$

where $k_c \equiv m_0c/\hbar$. Just as Pauli did with his non-relativistic equation (see Section 5-3), Dirac made use of multi-component wavefunctions so that $\vec{\alpha} = (\alpha_1, \alpha_2, \alpha_3)$ and β are constant Hermitian matrices. But since there are four such matrices while there are only three Pauli matrices, two components are not sufficient to describe ψ. Dirac showed that the theory works if the Pauli matrices are doubled into 4×4 matrices. Multiplication of Eq. (6.28) from the left by $-\partial/c\partial t + \vec{\alpha} \cdot \nabla + ik_c\beta$ then leads to the KG equation, provided that these matrices satisfy the conditions

$$\alpha_i\alpha_j + \alpha_j\alpha_i = 2\delta_{ij}, \; \alpha_i\beta + \beta\alpha_i = 0, \; \beta^2 = 1. \qquad (6.29)$$

The Dirac equation (6.28) may also be written in the form of the Schrödinger equation (5.33) by defining the Hamiltonian as $H = c(\vec{\alpha} \cdot \mathbf{p} + \beta mc)$.

Dirac proved the relativistic covariance of Eq. (6.28) by first choosing the appropriate notation. Defining $\gamma^j = -i\beta\alpha_j, j = 1, 2, 3, \gamma^4 = \beta$ which satisfy the relations

$$\gamma^\mu\gamma^\nu + \gamma^\nu\gamma^\mu = 2\delta_{\mu\nu}, \qquad (6.30)$$

the Dirac equation and its Hermitian conjugate assume the compact form

$$\gamma^\mu\frac{\partial\psi}{\partial x^\mu} + k_c\psi = 0; \; \frac{\partial\overline{\psi}}{\partial x^\mu}\gamma^\mu - \overline{\psi}k_c = 0 \qquad (6.31)$$

where $\overline{\psi} \equiv \psi^*\gamma^4$ (I use here the notation of Pauli's *Field Quantization*, Ref. [4], Section 3). Now it is easy to show that to each Lorentz transformation there exists a unitary transformation of the γ (see Refs [8,9]). Defining the charge density by the scalar product $-e\rho = -e(\psi^*\psi)$ and the current density by $\mathbf{i} = -ec(\psi^*\vec{\alpha}\psi)$ (the charge $-e$ refers to the electron) these quantities form a four-current

$$j^\mu = \left(\frac{1}{c}\mathbf{i}, -ie\rho\right) = -ie(\overline{\psi}\gamma^\mu\psi) \qquad (6.32)$$

satisfying the continuity equation

$$\frac{\partial j^\mu}{\partial x^\mu} = 0. \qquad (6.33)$$

The coupling to the electromagnetic field is obtained as in Eq. (6.7) by adding to the derivatives in Eqs (6.31) the appropriate terms with the potentials Φ_μ,

$$\gamma^\mu\left(\frac{\partial}{\partial x^\mu} + \frac{ie}{\hbar c}\Phi_\mu\right)\psi + k_c\psi = 0 \qquad (6.34)$$

and Hermitian conjugate. Dirac also derived in paper [14] several physical consequences from this equation. He showed that for a bound electron it is the angular momentum $\mathbf{l} + (1/2)\vec{\sigma}$ which is conserved, that the Zeeman energy of an electron in a magnetic field $\mathbf{H} = \nabla \times \mathbf{A}$ is indeed $H_Z = \mu_B\mathbf{H}\cdot\vec{\sigma}$, and that the spin–orbit interaction of an electron bound in a potential $ie\Phi_4 = e^2/r$ is, with the correct Thomas factor (see pages 114 and 137), found to be $H_{so} = e^2\mathbf{l}\cdot\vec{\sigma}/(4k_c^2r^3)$. These terms are derived by Pauli in his 'Wellenmechanik', making use of a non-relativistic approximation due essentially to Darwin [15b], in which the four components of the Dirac ψ are split into two 'large' and two 'small' components of the order $(1/c)^0$ and $(1/c)^1$, respectively (see Eq. (22.4) of Ref. [8] and Eq. (89) on p. 237 of Ref. [9]).

Now all the ingredients were assembled for an attack on the full QED. As we saw, Heisenberg and Pauli corresponded about the 'programme' for this enterprise in spring 1927. However, both had more urgent business to do. Heisenberg, after his uncertainty paper, devoted his energy to the problem of ferromagnetism, the decisive paper on which was submitted on 20 May 1928 [16a]. Pauli, on the other hand, was collaborating with Jordan and later had to be installed in Zurich. Then, however, an unexpected obstacle surfaced, causing a further delay and turning Pauli's interest away from physics to utopia and to the pleasures of life in Zurich (see the next section). He expressed his discontent to Bohr in the following terms: 'Unfortunately, at present my own works do not succeed at all (also on the question of superconductivity I was unable to reach any definite results). For this I am so ashamed that I have always postponed to write you. At my place you would now add that, first, you have no time and second, that you were very tired. But let me be quite honest for once: First, I have plenty of time and, second, I am not tired at all. (By the way, at Christmas I recovered very well at Pontresina [in Engadin, the Inn valley in Eastern Grisons].) I am only stupid and lazy. I think that somebody should thrash (*durchprügeln*) me a little every day! But since, unfortunately, nobody is doing this, I must look for other means to reactivate my interest in physics:

Friday to Monday I travel to Leipzig where, as it happens, a so-called regional meeting (*Gautagung*) takes place' (translated from the Postcard [214] of 16 January 1929, Ref. [2], p. 485).

The physical problem, as we shall see, was a serious one; it had to do with the vanishing mass of the photon and stayed with QED ever since. Therefore the first of the two papers [17a,b] by Heisenberg and Pauli, 'On quantum dynamics of the wave fields', was submitted only on 19 March 1929. Heisenberg's place of submission was Leipzig where in October 1927 he had become, at the age of 25, the professor of theoretical physics. For Pauli, on the other hand, it was the first paper (apart from the H-theorem paper discussed in the next section) he submitted as 'new' professor at ETH in Zurich, and on that occasion he jettisoned the designation 'jr.' after his name. The paper was actually submitted by Pauli since Heisenberg had left for the United States on 1 March. These two papers of 61 and 23 pages, respectively, indeed constituted the 'programme' in which quantum field theory was developed during the following two decades.

The first of these two papers contains the following three parts: 'I. General Methods', 'II. Setting up the fundamental equations of the theory for electromagnetic fields and matter waves', 'III. Approximation methods for the integration of the equations and physical applications'. In analogy with the mechanics of mass points the system is defined by a 'Lagrangian density' $\mathcal{L}$ which depends on some space–time functions $Q_\alpha(\mathbf{x}, t)$ describing the fields and on their derivatives. The field equations are

$$\frac{\partial \mathcal{L}}{\partial Q_\alpha} = \nabla \cdot \frac{\partial \mathcal{L}}{\partial (\nabla Q_\alpha)} + \frac{\partial}{\partial t} \frac{\partial \mathcal{L}}{\partial \dot{Q}_\alpha} = \left(\frac{\partial \mathcal{L}}{\partial Q_{\alpha,\beta}} \right)_{,\beta}, \tag{6.35}$$

where $f_{,\alpha} \equiv \partial f / \partial x^\alpha$. Eqs (6.35) follow from the variational principle

$$\delta \int \mathcal{L}(Q_\alpha, \nabla Q_\alpha, \dot{Q}_\alpha) d^3 x dt = 0. \tag{6.36}$$

Again in analogy with the mechanics of mass points a Hamiltonian formulation is obtained by defining momenta P_α and a Hamiltonian density $\mathcal{H}$ by

$$P_\alpha = \frac{\partial \mathcal{L}}{\partial \dot{Q}_\alpha}; \quad \mathcal{H}(P_\alpha, Q_\alpha, \nabla Q_\alpha) = P_\alpha \dot{Q}_\alpha - \mathcal{L}. \tag{6.37}$$

The dynamics is then described by the canonical field equations (Eqs (10) of Ref. [17a]) analogous to Eqs (3.11),

$$\dot{Q}_\alpha = \frac{\partial \mathcal{H}}{\partial P_\alpha}; \quad \dot{P}_\alpha = -\frac{\partial \mathcal{H}}{\partial Q_\alpha} + \nabla \cdot \frac{\partial \mathcal{H}}{\partial (\nabla Q_\alpha)}. \tag{6.38}$$

The generalization of the commutation relations (5.28) are the canonical commutation relations (CCR)

$$[P_\alpha(\mathbf{x}, t), Q_\beta(\mathbf{x}', t)] = \frac{\hbar}{i} \delta_{\alpha\beta} \delta(\mathbf{x} - \mathbf{x}');$$

$$[Q_\alpha(\mathbf{x}, t), Q_\beta(\mathbf{x}', t)] = [P_\alpha(\mathbf{x}, t), P_\beta(\mathbf{x}', t)] = 0 \tag{6.39}$$

which are defined only at equal times. From the point of view of relativistic covariance this is of course a major handicap. In the Jordan–Pauli paper the problem of the transition from the equal-time commutation relations (6.14) to the relativistic form (6.22) was achieved by the development (6.15), (6.16) into eigenmodes which satisfy relativistic equations of motion. Here, however, the authors wish to keep the field equations (6.38) unspecified and require only that the Lagrangian density $\mathcal{L}$ be invariant. But the price to pay for this generality is high. The argument goes like this: because of the group property of the Lorentz transformations it suffices to consider infinitesimal transformations, for which the field equations (6.38) may be used. But one has to consider the transformation of both the coordinates $x^{\mu} = (\mathbf{x}, ict)$ and the fields P_{α}, Q_{α}. This proof is carried out on p. 19 to 23 of Ref. [17a], and is so heavy that Pauli himself used to say 'I warn the curious!' (see Ref [18], p. 51).

This proof was a major achievement, so the disappointment of the authors was all the more painful when they realized that the first application of their general theory, namely to QED, failed. But then, in January 1929, Heisenberg was able, with the aid of a trick, to get rid of the formal difficulties (see the beginning of this section). During the weekend of 19–20 January the regional physicists' meeting took place in Leipzig. Pauli took the opportunity to visit Heisenberg from Friday to Monday. 'Heisenberg has an idea of how our ap-proach to the many-body problem *perhaps* might still be carried out' he wrote to Bohr in the postcard of 16 January quoted above. And Heisenberg wrote to Jordan on 22 January: 'the trick consists in that one doesn't start with the correct Lagrange function à la Maxwell–de Broglie but with a modified one, and that one returns to the limit of the unmodified one only in the result' (translated from footnote *d*, Ref. [2], p. 486). To understand the problem and its resolution let us turn to Part II of the paper [17a]. It begins with the Maxwell equations (6.25) which follow from the variational principle (6.36) with the *field* Lagrangian density

$$\mathcal{L}_f(\nabla \Phi_{\alpha}, \dot{\Phi}_{\alpha}) = -\frac{1}{4} F_{\alpha\beta} F_{\alpha\beta} = \frac{1}{2}(\mathbf{E}^2 - \mathbf{H}^2). \qquad (6.40)$$

According to Eqs (6.37), the momenta conjugate to Φ_{α} and the *field* Hamilto-nian density are found to be

$$P_k = \frac{i}{c} F_{4k} = -\frac{1}{c} E_k; \quad P_4 = 0; \quad \mathcal{H}_f = \frac{1}{2}(\mathbf{E}^2 + \mathbf{H}^2) + \mathbf{E} \cdot \nabla A_0 \qquad (6.41)$$

(the authors use the notation $P_{\alpha 4} = ic P_{\alpha}$ defined in Eq. (30) of Ref. [17a]). Therefore, the P_{α} are not independent functions that can be arbitrarily given on a 'world cut' (*Weltschnitt*) $t = 0$ but arbitrary $\mathbf{x}$, not even the P_k since, according to the first set of Maxwell equations (6.25), $\nabla \cdot \mathbf{E} = 0$. This last relation also shows that in the absence of charges the total field energy obtained with the Hamiltonian density (6.41), after partial integration of the third term, is given by Eq. (6.16). But, in addition, the conditions $P_4 = 0$ and $\nabla \cdot \mathbf{E} = 0$ are incompatible with the CCR (6.39) since, for $t = t'$, $[P_4(x), \Phi_4(x')] = 0$ and $[\partial P_k(x)/\partial x^k, \vec{\Phi}(x')] = 0$. Thus the general scheme of canonical quantization developed in Part I fails for electrodynamics.

In order that $P_4 = \partial \mathcal{L}_f / \partial \dot{\Phi}_4 \neq 0$ one must include $\dot{\Phi}_4$ in $\mathcal{L}_f$, and it must be done in a Lorentz-invariant way. This suggests adding a term $(\varepsilon/2)(\partial \Phi_\mu / \partial x^\mu)^2$ to the Lagrangian density (6.40) so that $P_4 = (\varepsilon/ic)\partial \Phi_\mu / \partial x^\mu$, and it is postulated (see Ref. [17a], p. 31) that at the end of a calculation the limit $\varepsilon \to 0$ is to be taken. This is Heisenberg's trick, implemented in Eq. (58) of Ref. [17a]. After publishing their first paper [17a] but before submitting the second [17b], Heisenberg and Pauli learned of an independent approach to this problem contained in a paper by Fermi [20a] (see also Refs. [20b,c]) in which, however, $\varepsilon = -1$ (I follow here the presentation given in Section 16 of Wentzel's masterly monograph [21b]). If one calculates the action integral $\int \mathcal{L}d^3xdt$ with this Lagrangian density (Eq. (16.5) of Ref. [21b]) by doing two partial integrations, and assumes the boundary terms vanish, the result may be written as the action due to the *Fermi* density $\mathcal{L}_F = -\sum_{\mu\nu}(\partial\Phi_\mu/\partial x^\nu)^2/2$ (note the wrong sign in the footnote on p. 171 of Ref. [17b]).

Fermi's original argument, developed for non-relativistic electrons in Ref. [20a] before Heisenberg and Pauli's paper [17a] and extended to Dirac electrons in Ref. [20b], was that the radiation field used by Dirac in Ref. [3a] and by Jordan and Pauli in Ref. [5a] was inapt to describe the effects in the neighbourhood of the charges because it neglects both the scalar potential A_0 and the longitudinal part $\mathbf{A}_{lg}$ of the vector potential, for which $\nabla \cdot \mathbf{A}_{lg} \neq 0$ but $\nabla \times \mathbf{A}_{lg} = 0$. His idea in Ref. [20a] was to satisfy the Lorentz condition (6.23) by the initial conditions $P_4 = 0$ and $\dot{P}_4 = 0$ on a 'world cut' Ψ at $t = 0$,

$$\frac{\partial \Phi_\mu}{\partial x^\mu}\Psi = 0, \tag{6.42}$$

and use the fact that the field equations (6.35) following from the variational principle (6.36) with the density $\propto \Phi_{\mu,\nu}^2$ are $\Box\Phi_\mu = 0$. Since this implies that $\Box P_4 = 0$ also, all time derivatives are seen successively to vanish on the cut $t = 0$, and hence $P_4(x) = 0$ for all x. To see that the above difficulty is related to the vanishing photon mass I present in *Note* [22] a theory of massive photons proposed by Alexandre Proca (1897–1955) in 1936 [23a] which became important for the description of (charged) vector mesons (see Section 12 of Ref. [21b], also Ref. [23b], p. 433). A more sophisticated theory for massless photons is developed in the second paper [17b] by Heisenberg and Pauli. But before describing it we must introduce the matter waves into the theory—the authors of paper [17a] actually do this before introducing Heisenberg's trick.

Choosing the *matter* Lagrangian density as

$$\mathcal{L}_m = -\hbar c \overline{\psi}\left\{\gamma^\mu\left(\frac{\partial}{\partial x^\mu} + \frac{ie}{\hbar c}\Phi_\mu\right) + k_c\right\}\psi \tag{6.43}$$

the Dirac equation (6.34) and its Hermitian conjugate follow by variation of $\overline{\psi}$ and of ψ, respectively, while variation of $\partial\Phi_\mu/\partial x^\nu$ yields the four-current (6.32). Hence, the field equations (6.35) resulting from the total density

$\mathcal{L} = \mathcal{L}_m + \mathcal{L}_f$, including the term $(\varepsilon/2)(\partial \Phi_\mu / \partial x^\mu)^2$, are the modified Maxwell equations in the presence of charges,

$$\frac{\partial F_{\mu\nu}}{\partial x^\nu} + ic \frac{\partial P_4}{\partial x^\mu} = j^\mu, \tag{6.44}$$

given in Eq. (59) of Ref. [17a] (Heisenberg and Pauli use the notation $\psi^+ = i\overline{\psi}$, see Eq. (56) of Ref. [17a]). Concerning the CCR for the ψ, they may be formulated with commutators (bosons) or with anticommutators (fermions), designated here as $[A, B]_\mp \equiv AB \mp BA$, i.e. for $t = t'$,

$$\begin{aligned} [\psi_\alpha(x), \psi_\beta^*(x')]_\mp &= \delta_{\alpha\beta}\delta(x - x'), \\ [\psi_\alpha(x), \psi_\beta(x')]_\mp &= [\psi_\alpha^*(x), \psi_\beta^*(x')]_\mp = 0. \end{aligned} \tag{6.45}$$

The importance of the third part of paper [17a] is the demonstration that the physical effects resulting from the interaction contained in the Lagrangian density (6.43) are indeed calculable. First it is shown that the scalar potential A_0 in the absence of any radiation leads to the well-known electrical forces between the electrons. This is done by developing both the potentials Φ_μ and the second-quantized Dirac ψ into eigenmodes of a finite volume, $v_\mu^{r\lambda}$ and u^s, respectively. Now it may be verified by writing $\mathbf{A}_\lambda = \sum_k \mathbf{a}_{\lambda k} \exp(i\mathbf{k} \cdot \mathbf{x} - \omega t)$ with $\mathbf{a}_{\lambda k} \perp \mathbf{k}$ for $\lambda = 1, 2$ and $\mathbf{a}_{3k} \parallel \mathbf{k}$ and similarly for A_0, that the free field equations (6.44) with $j^\mu = 0$ have frequencies $\omega = ck$ for all the modes. However, for the longitudinal and scalar potentials, A_3 and A_0, respectively, double solutions $(k^2 - \omega^2/c^2)^2 a_\lambda = 0$, $\lambda = 0, 3$, are found. To lift this degeneracy, more additional terms depending on a parameter δ are introduced in the field Hamiltonian (83) of Ref. [17a], leading by a complicated procedure to Eq. (95), which consists of a sum of oscillator terms in which, however, the $\lambda = 0$ mode has an unphysical negative frequency!

Following Dirac, the Schrödinger equation is now written in the representation by particle numbers, N_s for the electrons and $M_{r\lambda}$ for the photons. Taking as unperturbed state $\varphi_0(\ldots N_s \ldots; \ldots M_{r\lambda} \ldots)$ such that $\varphi_0 = 1$ if all $N_s = N_s^0$ and all $M_{r\lambda} = 0$ and $\varphi_0 = 0$ otherwise, the authors then calculate the first-order correction $\varphi_1(N_1^0, \ldots, N_s^0 + 1, \ldots, N_t^0 - 1, \ldots; 0, \ldots, M_{r\lambda} = 1, 0 \ldots)$ due to the interaction with the potentials Φ_μ. They find $\varphi_1 \propto d_{st}^{r\lambda} - ic_{st}^{r\lambda}$ where $d_{st}^{r\lambda} = \int (u^{*s} u^t) v_0^{r\lambda} d^3 x$ and $c_{st}^{r\lambda} = \int (u^{*s} \vec{\alpha} u^t) \cdot \mathbf{v}^{r\lambda} d^3 x$ are the *charge* and *current* matrix elements coupling to A_0 and $\mathbf{A}$, respectively. The energy perturbation to second order follows from this, neglecting the magnetic c-terms and taking the limit $\delta \to 0$,

$$E^{(2)} = \frac{e^2}{2} \sum_{st} (1 \pm N_s^0) N_t^0 A_{st,ts} + \sum_{st} N_s^0 N_t^0 A_{ss,tt} \tag{6.46}$$

where the upper/lower sign is valid for BE and FD statistics, respectively, and

$$A_{st,s't'} \equiv \int \int d^3x d^3x' \frac{(u^{*s}(\mathbf{x})u^t(\mathbf{x}'))(u^{*s'}(\mathbf{x}')u^{t'}(\mathbf{x}'))}{|\mathbf{x} - \mathbf{x}'|}. \tag{6.47}$$

This shows that $A_{ss,tt}$ is the usual Coulomb integral while $A_{st,ts}$ is the exchange integral which Heisenberg had found to explain both ortho- and para-helium (see page 82) and ferromagnetism (see Ref. [16a]). Owing to the closure relation $\sum_t u^t(\mathbf{x}) \otimes u^{*t}(\mathbf{x}') = \delta(\mathbf{x} - \mathbf{x}')$ and to the normalization $\int (u^{*s}u^s) d^3x = 1$ the first sum of Eq. (6.46) contains a divergent term $\sum_t A_{st,ts} = 1/r$, $r = |\mathbf{x} - \mathbf{x}| = 0$, representing the self-energy of the electron due to charge coupling. This, even more than the ZPE, has haunted researchers in QED. But, in addition, the result (6.46) shows complete symmetry between the two statistics which led the authors to conclude that 'a satisfactory explanation of the preference by Nature for the second [fermionic] possibility therefore cannot be given' (translated from Ref. [17a], pp. 29–30). This problem occupied Pauli for the next decade; it is known by the name of Pauli's spin-statistics theorem (see Ref. [24b]).

Finally, in Section 9 of paper [17a] 'On the light emission expected according to the theory at the passage of an electron through potential wells' Heisenberg and Pauli also calculate what today is called *Bremsstrahlung*. Although it is a problem to which Pauli returned later, it is of particular interest here because of the following concluding remark: 'If one applies this result to the Gamov–Gurney–Condon theory of the radioactive decay of the nuclei [24c] one will conclude that primary β-ray spectra never can be sharp because all radiation effects of the electrons in the nucleus are of the order of magnitude 1. Of course, according to the theory proposed here the associated continuous γ-ray spectra should also always occur since in this theory the validity of the energy theorem is always maintained. This theory thus does not give any hint about the difficulties which are connected with the apparent non-existence of those γ-ray spectra' (translated from Ref. [17a], p. 61). Less than two years later these difficulties led to Pauli's proposal of the neutrino. A concise description of the genesis and the essentials of this first paper is contained in Pauli's letter [216] to his friend Oskar Klein of 18 February 1929 in Ref. [2].

The second paper [17b] by Heisenberg and Pauli was submitted on 7 September 1929, again by Pauli for Heisenberg was in Japan to where he had sailed from the United States together with Dirac, and in mid-September he continued his world tour in the direction of Calcutta, returning to Leipzig on 5 November. This information is contained in letter [232] from Heisenberg to Pauli, written in the United States on 20 July 1929. There is only one more letter, [234], to Pauli from this journey, written in Chicago on 1 August, in which Heisenberg looks forward to his train ride to the west coast (see Ref. [2]). The paper, therefore, was mainly the work of Pauli, although, according to letter [232], Heisenberg had done some calculations on the 'formulation of the theory without additional terms' which Pauli incorporated as Section 6 in Ref. [17b]. These additional terms were both the ε-terms of Heisenberg's trick and the δ-terms of Eq. (83) of Ref. [17a] mentioned above. It is easy to see that, in the gauge $A_0 = 0$ (see below), the longitudinal part A_3 is a zero-frequency mode, $\dot{A}_3 = 0$, which, when integrated, gives rise to the linear time dependence of Eq. (47) of Ref. [17b]. Again the second-order energy perturbation $E^{(2)}$ may be calculated as in the first paper [17a]. For the electric d-terms Eq. (6.46) is then recovered (Eq. (56) of Ref. [17b] where, however, only the FD case is given), while the magnetic c-terms were calculated by the American Gregory Breit (1899–1981) [25a], who towards the end of 1928 was visiting Pauli in Zurich (see postcard [212] from Pauli to Ehrenfest of 24 December 1928, Ref. [2], p. 478; see also the next section).

The most important part of this second paper, however, is the formulation, in the first five paragraphs, of the invariance properties of the Hamiltonian $H = \int \mathcal{H} d^3 x$. This work not only carries Pauli's mark but also shows the influence of Hermann Weyl who until 1930 was Pauli's colleague at ETH (see page 139) and whose recent book *Theory of Groups and Quantum Mechanics* [25b] is quoted in the paper. The simplest example, treated in Section 1, is the invariance of H under translations $\mathbf{x} \to \mathbf{x} - \delta\mathbf{x}$ which imply $Q_\alpha \to Q_\alpha - \nabla Q_\alpha \cdot \delta\mathbf{x}$. Recalling the definition (6.27) of a functional derivative, the effect on the Hamiltonian may be expressed as $H \to H + \delta H$ with $\delta H = -\delta\mathbf{x} \cdot \int d^3 x (\delta H/\delta Q_\alpha) \nabla Q_\alpha$. Making use of the CCR (6.39) one may write $\delta H/\delta Q_\alpha = (i/\hbar)[P_\alpha, H]$ and hence $\delta H = (i/\hbar)\delta\mathbf{x} \cdot [\mathbf{G}, H]$ where

$$\mathbf{G} = -\int d^3 x P_\alpha \nabla Q_\alpha \qquad (6.48)$$

is the momentum integral which forms a four-vector with H. Therefore, translation invariance of H implies that $i\hbar\dot{\mathbf{G}} = [\mathbf{G}, H] = 0$ which is momentum conservation $\mathbf{G} = $ const. This is a more elegant way to derive Eq. (24) of the first paper [17a].

An even more important case, investigated in Section 3 of paper [17b], is the invariance under the gauge transformations (2.12), which Weyl discussed in his book [25b] (see pp. 100 and 214 of the English translation). Heisenberg and Pauli write them (Eqs (13) of Ref. [17b]) as

$$\Phi_\mu \to \Phi_\mu + \frac{\partial\chi}{\partial x^\mu}; \; \psi \to e^{-i(e/\hbar c)\chi}\psi; \; \psi^* \to \psi^* e^{i(e/\hbar c)\chi} \qquad (6.49)$$

(see also Ref. [8], pp. 30–1, 164 and Ref. [9], pp. 110–11, 232–3, also Ref. [23b], Section 15(e)). Here the global transformations with arbitrary constant χ, which leave the potentials untouched, are already of interest since invariance ensures conservation of charge. This special case is discussed in Section 2 of paper [17a]. Concerning the general local transformations, Pauli later called the second part of Eqs (6.49), namely the multiplication of ψ by a phase factor, 'gauge transformations of the first kind', and the first part, namely the addition to Φ_μ of a gradient, 'gauge transformations of the second kind' (see the footnote on p. 718 of Ref. [24b]).

The local transformations allow $A_0 = 0$ to be fixed, which, however, restricts the gauge transformations by the condition $\dot\chi = 0$. In H an infinitesimal transformation $\delta\chi$ gives rise, after partial integration of the first term, to

$$\delta H = -\int d^3 x \Big\{ \nabla \cdot \frac{\delta H}{\delta \mathbf{A}} + \frac{ie}{\hbar c}\Big(\psi\frac{\delta H}{\delta\psi} - \bar\psi\frac{\delta H}{\delta\bar\psi}\Big) \Big\} \delta\chi.$$

According to the canonical field equations (6.38), $\dot{\mathbf{P}} = -\delta H/\delta \mathbf{A} = (i/\hbar)[H, \mathbf{P}]$ and $\dot{\pi} = -\delta H/\delta \psi = (i/\hbar)[H, \pi]$ (note that $\overline{\dot{\psi}} = \partial \mathcal{L}_m/\partial \overline{\dot{\psi}} = 0$!), this variation may be expressed by the gauge operator

$$\delta\Gamma = \int d^3x \delta\chi \left\{ \nabla \cdot \mathbf{P} + \frac{ie}{\hbar c}\pi\psi \right\} = -\frac{1}{c}\int d^3x \delta\chi \{\nabla\cdot\mathbf{E} + e(\psi^*\psi)\} \quad (6.50)$$

as $\delta H = [\delta\Gamma, H] = i\hbar\delta\dot{\Gamma}$. *Gauge invariance* then implies that $\delta\dot{\Gamma} = 0$, or that $\nabla \cdot \mathbf{E} + e(\psi^*\psi) = C(\mathbf{x})$ is an arbitrary function.

This argument explicitly assumes that there are no 'additional terms', i.e. $\varepsilon = \delta = 0$. The CCR (6.39) and (6.45) imply for any field $f = \mathbf{A}, \psi, \psi^*$ the infinitesimal transformation $f \to f + (i/\hbar c)[\delta\Gamma, f]$ which generalizes to finite gauge transformations as follows: $f \to \exp(i\Gamma/\hbar) f \exp(-i\Gamma/\hbar)$, and the gauge-invariant expressions $G = F_{\mu\nu}, \psi_\rho^*\psi_\sigma, \psi_\rho^*(\hbar c\partial\psi_\sigma/\partial x^\mu + ie\psi_\sigma\Phi_\mu)$, $(\hbar c\partial\psi_\rho^*/\partial x^\mu - ie\Phi_\mu\psi_\rho^*)\psi_\sigma$ satisfy $[\Gamma, G] = 0$ and $[C(\mathbf{x}), G] = 0$, for all $\mathbf{x}$. From this result the authors (Pauli) draw(s) the conclusion: 'Since directly measurable physical quantities always are gauge invariant, one may give the constant C a numerical value. If, in particular, $C = 0 \ldots$ then the fourth component of the Maxwell equations also holds, although not in general as a q-number relation but indeed for all gauge-invariant relations. $C = 0$ means that the operator (6.50) acting on a Schrödinger functional $F(\psi_\rho, \Phi_i)$ of any stationary states of the system yields zero, i.e. $C = 0$ selects those solutions for which the Schrödinger functionals are likewise invariant under (6.49) [reference is made to the infinitesimal transformations]' (translated from Ref. [17b], p. 174).

The next problem analysed in this second paper is the invariance under Lorentz transformations. This is the content of Section 4 where in a footnote on p. 174 the help of the mathematician John von Neumann (1903–57) 'for essential parts of this paragraph' is acknowledged. As a consequence of this analysis, the invariance of the CCR under Lorentz transformations again follows, which gives rise to the comment 'The invariance proof carried out here is probably somewhat simpler than the one indicated [*angegebene*] in [the first paper] I' (translated from Ref. [17b], p. 177). This general invariance proof now also allows relaxation of the special gauge $A_0 = 0$ used in Eqs (6.15) and (6.50). Since this condition was imposed in a particular coordinate system, one has to show that a Lorentz transformation induces a scalar potential A_0 in a gauge-covariant way—which is done in Section 5. Finally, in Section 7 the representation by Schrödinger functionals in configuration space and the ensuing interpretation problems are investigated. Much of this work was done by the young J. Robert Oppenheimer (1904–67) who was one of Pauli's collaborators in Zurich during the first half of the year 1929 (see letter [215] from Pauli to Ehrenfest of 15 February 1929 in Ref. [2]; see pages 205–6). Oppenheimer submitted his results to the American journal *The Physical Review* on 12 November 1929 from Berkeley, California, where that year he had become an assistant professor. In this paper [25c] Oppenheimer showed 'that it is impossible on the present theory to eliminate the interaction of a charge with its own field, and that the theory leads to false predictions when it is applied to compute the energy levels and frequency of the absorption and emission lines of an atom' (abstract to Ref. [25c]). He also thanks both Heisenberg and Pauli, 'for their valuable criticism and advice'.

With this monumental work of 84 pages of the two papers combined, Heisenberg and Pauli set the stage for two decades for numerous applications. No other common paper by the two friends has appeared; their only other collaboration happened almost three decades later, in the last year of Pauli's life, and did not lead to a published paper. For the content of the present section see also Chapter 15, Sections (c) to (e), of Ref. [23b] and Sections 1.2 to 1.5 of Ref. [25d].

Life in down town Zurich—physics at Gloriastrasse

In April 1928 I arrived in Zurich as a new professor, quite touristically dressed with a rucksack on my back which did not agree well with the elegant form of my being fetched at the railway station that completely surprised me. [26a]

Pauli's debut in Zurich, recalled here in his allocution for Hermann Weyl's 70th birthday in 1955, quoted on page 51, is beautifully complemented by Pauli's first assistant Ralph Kronig: 'Pauli began his new task at the end of April 1928, arriving a day before his birthday, and Scherrer cheered up his room on the occasion by having put some flowers on his writing desk. It was the best part of the year to be in Zurich, and as the days got warmer Scherrer, Pauli and myself often went swimming at the Strandbad, eating our lunch in bathing suits. One of my tasks, not agreed upon beforehand, was to watch out that Pauli should limit his consumption of ice cream at Sprüngli's Konditorei at the Paradeplatz where we often went in the afternoon. ... Such expeditions were often interspersed with discussions on subjects from physics, in which Pauli was unwilling to tolerate sloppy thinking, but was always ready to give honours where honours were due and to admit a mistake on his part if one could counter him with good arguments. Also he was quite willing to let you compensate the advantages he had when swimming in the Zürichsee by coming along on a Sunday walking tour where he had a handicap against persons of a lighter built.' Kronig, who considered his time spent in Zurich 'not only as one of the most instructive, but also as one of the most exhilarating periods' of his life, wrote these lines on the eve of Pauli's death, on 14 December 1958, unaware of the coincidence (see Ref. [26b], p. 37).

The year 1927 was critical for the physics programme offered to students in Zurich. Debye, the director of the physics department at ETH, had a tempting offer from Leipzig; Schrödinger, the professor of theoretical physics at Zurich University, was seduced by the prestigious call to succeed Max Planck in Berlin. So Debye's colleague and former collaborator Paul Scherrer (1890–1969) took the initiative to write a letter to Professor Arthur Rohn (1878-1956), the president of the governing body of ETH, the Swiss School Council.

In this letter, dated 27 July 1927, Scherrer examines the possibility 'to meet the great difficulties that result from the simultaneous calls (*Wegberufung*) to Herrn Professor Debye and Herrn Professor Schrödinger on physics in Zurich by creating a double professorship for theoretical physics. It is likely that the collaboration of ETH and Zurich University would make it possible to preserve for Zurich one of the two gentlemen in spite of the high offers abroad' (translated from document I.1, Ref. [27], p. 21).

This was not to be. In spite of considerably improved conditions offered Debye by President Rohn in agreement with the School Council, ETH could not compete with the offer of the Saxonian Ministry in Dresden. So, after an unsuccessful advertisement for Debye's succession in the Swiss government journal and in *Neue Zürcher Zeitung* the School Council decided to proceed with a call. In the meantime Hermann Weyl offered to help the physicists by replacing his course on a mathematical topic planned for the winter term of 1927/8 with a three-hour course on 'Quantum Mechanics and Group Theory'. In the archive of the School Council there is a note containing the names of 44 physicists, among whom Pauli figures with the cryptic remark 'with external disadvantages' (see Ref. [27], pp. 11–12). The star candidate was Heisenberg who on 13 September 1927 wrote from the Volta congress in Como (see pages 160–70): 'Highly revered Mister President! Since I now approximately know the programme of the following days I would like to ask you if on Saturday 17.9. I may visit the institute at the Polytechnic and speak with you and Herrn Professor Scherrer about the details. . . . In sincere devotion Werner Heisenberg' (translated from document I.2, Ref. [27], p. 22). On 3 November Heisenberg wrote again: 'Highly revered Mister President! Concerning the professorship of theoretical physics at the Institute of Technology in Zurich I was unfortunately unable to give you any news for so long because my negotiations with Dresden were very much delayed. However, after rather long reflection I now have decided anyhow to accept the professorship in Leipzig because I prefer to stay in Germany and since the collaboration with Debye tempts me very much . . . ' (translated from document I.4, Ref. [27], p. 23).

On 11 November 1927 Pauli cabled President Rohn: 'Am available to speak Monday from 12 to 2 and from 5 to 7 at Physics Department Hamburg Jungiusstrasse 9. Pauli' (translated from document I.5, Ref. [27], p. 23). In the session of 17 December 1927 President Rohn reported to the School Council about his negotiations: 'The information gathered in France and Germany showed that of the most gifted young modern physicists Heisenberg (Copenhagen), Pauli (Hamburg), de Broglie (Paris), Wentzel (Leipzig), Hund (Rostock), Brillouin (Paris) and [Fritz] Zwicky [1898–1974] (California) merit to be named. . . . The negotiations conducted at first with Heisenberg were crossed by Debye who attempted to win Heisenberg for Leipzig. The ETH could not compete with this offer. Thereafter the president turned to Pauli. . . . Pauli is considered to be a very competent physicist who belongs to the school Broglie–Heisenberg–Schrödinger (quantum and atomic physics). Since, however, he is credited with certain unfavourable features in lecturing the president set out for Hamburg in order to hear him. The scientific impression was good; Pauli doubtlessly is a researcher of the best kind. In lecturing, however, he has a certain awkwardness of which he is aware and which he strives to get rid of, which should still be possible at his age' (translated from document I.9, Ref. [27], p. 27). After discussion the School Council decided to propose to the Swiss Federal Council via the Department of the Interior the nomination of Wolfgang Pauli as full professor of theoretical physics at ETH, starting 1 April 1928. The basic yearly salary was 15 000 francs plus fees and age supplements. As usual this nomination was for 10 years. Pauli was informed of his nomination by telegram from President Rohn on 19 December 1927 (Document I.10, Ref. [27]). The official document of Pauli's nomination by the Federal Council dated 10 January 1928 is in the *Pauli Archive* (file PLC Bi 41).

As to the 'awkwardness' in Pauli's lecturing mentioned by President Rohn, it was well known among the students that Pauli was not a brilliant lecturer (see e.g. Ref. [28a]). However, Markus Fierz has given the following pertinent characterization: 'He [Pauli] belonged to these types of teachers who, as W. Ostwald [28b] has observed, have the defect to think about their subject during lecturing. With those the word is directed not to the outside at the listener but the latter rather participates in a kind of soliloquy which, because not really addressed to him, will often scarcely be intelligible. But he who in Pauli's lectures honestly strove to follow the line of thought of the master, therefore learnt to know not primarily the results of physical research, he learnt above all to think critically about a theory' (translated from Ref. [28c], p. 188).

Pauli did not accept the offer of President Rohn immediately. He contacted Richard Kuhn, his schoolmate from the Döbling Gymnasium (see page 16) who from 1926 to 1929 was professor of general chemistry at ETH (see Refs [27], p. 12 and [28d]). And on 26 November 1927 he wrote to Scherrer about his main concern, namely to obtain a decently paid assistant position: 'As President Rohn said, the salary of the assistant is roughly 300 francs, and this really is a bit small. Now, would it be possible *to raise the revenue of the assistant* by a paid teaching commission (*Lehrauftrag*)? (I consider first Dr Kronig who at present is in Copenhagen but do not know if I can get him). . . . I thus hope fairly definitively to work in the summer term in your institute in Zurich and am already looking forward very much to being with you. Till then again cordial thanks and kind regards from your W. Pauli.' (Translated from document I.7, Ref. [27], p. 24).

On 22 November Pauli had written to Kronig: 'I recently have received a call to the Zurich Polytechnic for the summer term. Thereby I have set as a condition the grant of an assistant position. . . . Now I would like to ask you, for the moment quite tentatively, if in principle you would agree to accept this position. Annoying duties would hardly be attached to it; your task would be: 1. Every time I say something to contradict me with detailed arguments. 2. To animate somewhat the scientific activity with modern ideas (particularly as concerns possible Ph.D. students and the colloquium). . . . We in fact always got along very well in Copenhagen, and I believe that the Zurich arrangement would be advantageous for both of us. Besides, the general milieu in Zurich is *very nice*. Indeed, Weyl is mathematician at the Polytechnic and has great interest in quantum mechanics; in addition, Scherrer (experimental physicist and now also director of the physical institute) cares a lot for modern physics.— At the University of Zurich also a theoretical professorship is open. . . . I have a faint hope that one will call Wentzel. If Wentzel, you and me could be together in Zurich this (+ Scherrer + Weyl) would really be a nice enterprise' (translated from letter [175], Ref. [2], pp. 415–16).

This is a fairly accurate vision of Pauli's near future. However, it almost failed to materialize because of Pauli's insistence on more favourable conditions for his assistant. On 31 January 1928 President Rohn wrote to Pauli: 'When I visited you in Hamburg I asked you to get yourself accurately informed in Zurich about the conditions of our institution and of our physics institute. Now today you must make clear whether you want to trust us or other influences— which I suspect and disregard. In this respect no doubt must subsist. In spite of your preliminary acceptance of the nomination under the set conditions and in spite of the nomination effected by the Swiss Federal Council I prefer to free you today from the final acceptance of your nomination if you do not convey to us the full confidence which is necessary for the unobjectionable execution of your task and if you are unable to approach the latter with joy' (translated from document I.16, Ref. [27], p. 31). Pauli rushed to confirm his acceptance and presented President Rohn with his excuses for the delay (document I.17, Ref. [27]). Thus Pauli's destiny was sealed for the next decade and a half. But as we shall see later, complications in the otherwise quite sympathetic relationship between President Rohn and Pauli were unavoidable, unfortunately, in spite of considerable goodwill on both sides. Pauli was now Debye's successor; his office 4c on the second floor of the imposing physics building on the bend of Gloriastrasse was close to the lecture room 6c and to Scherrer's and the secretary's offices (Ref. [27], p. 13). Privately he settled in the vicinity at Schmelzbergstrasse 34 (Ref. [27], p. 15), a few minutes' walk from the institute.

As Pauli explains in his proposal to name Kronig his assistant (document I.18, Ref. [27]) the latter at the time was assistant to Kramers in Utrecht. But since Kramers planned to go to the United States in April, Kronig could at least be with Pauli during the summer term. Based on Pauli's proposal, on 13 February 1928 President Rohn decreed the nomination of Ralph de Laer Kronig starting 1 May until the end of the summer term 1928, i.e. the end of September, with a monthly salary of 500 francs (document I.19, Ref. [27]). It seems that both, Pauli and Kronig immediately felt at ease in Zurich. Not only did they go swimming in the lake and eat ice cream at Sprüngli's but, under the expert guidance of Scherrer, they also started to explore the Zurich night life. On 4 June the three of them wrote the following postcard to 'PQ-QP Jordan': 'Lieber Herr Jordan! We are about to study the Zurich night life and try to improve it following a new method of Pauli: by comparison. Many greetings Kronig. This method, however, may also be used to make things worse!—Greetings Pauli. I

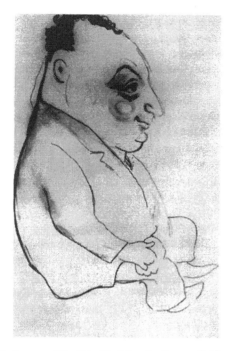

Fig. 6.1 Caricature of Pauli by Zurich artist Gregor Rabinovitch, 1930

have also heard so many bad things about you that I would like to make your acquaintance. Scherrer' (translated from postcard [199], Ref. [2]).

Around Bellevue Square, situated at the outflow of the Zürichsee into the Limmat river where the traffic along Utoquai and Limmatquai on the east side of the lake and river crosses the traffic down Rämistrasse from Zurich University and ETH and over Quaibrücke to the west side, are located most of the places frequented by Pauli and friends. To the left on Rämistrasse is the distinguished Kronenhalle where Pauli and his collaborators used to have white Aigle wine and Viennese sausages with horse-radish after concerts (see Ref. [27], p. 195); across Rämistrasse from Kronenhalle is Odeon, the artists café where in 1938 Pauli's sister Hertha paused on her escape to Paris (see page 19). And, as noted above by Stern, across Limmatquai from Odeon is the café-dancing Terrace, while the Bauschänzli beer garden is situated on an island across the Limmat, downstream from Quaibrücke. But of course, the night life of Scherrer and company comprised other locations like the Schifflände-Bar down the Limmatquai from Odeon or Mary's Old-Timer Bar [29a] across the river towards the main shopping district of the elegant Bahnhofstrasse where the chic café Sprüngli is situated. See also the section 'Life in Zurich' in Ref. [29b].

Twice a week Scherrer was unavailable for the nightly outings, namely the eve of his '*Grosse Vorlesung*', the two hour demonstration lecture, when he set up with utmost care, aided by his assistants, the experiments to be shown the next morning; Pauli often paid him a visit late at night (see Ref. [27], p. 20). These lectures of Scherrer, held in the large special auditorium of the Physics Institute at Gloriastrasse, were famous for their non-stop stunning effects, their unfailing precision, the virtuosity in the use of the blackboard, and the ease of Scherrer's delivery. Even people from downtown Zurich— especially women—sometimes attended the 'Scherrer circus', as these lectures were called, and the students had to run to get good seats in the crowded auditorium. Scherrer used to say '*das ist doch ganz einfach!*' (but this is quite simple!), but the students often learnt otherwise when they tried to work out the problems prepared by Scherrer's assistants to accompany the lectures. To one of Scherrer's enthusiastically presented simple explanations Pauli once answered '*Ja*, simple it is alright but also wrong' (see V. F. Weisskopf, in Ref. [12], p. 88).

Paul Scherrer was born in the north-eastern Swiss city of St Gallen (Saint Gall) on 3 February 1890. There he attended the commercial school but soon discovered that his interests were in the natural sciences. After taking courses in a private school in Zurich he passed the entrance exam to ETH in 1908 where he first studied botany for two semesters but then changed to mathematics and physics. Spending one semester in Königsberg (today Kaliningrad, Russia) he went to Göttingen in the autumn of 1912 where Peter Debye, Woldemar Voigt (1850–1919), and David Hilbert were his illustrious masters and his best friends were Walter Baade (see page 147) and the Swiss Humm (see Ref. [30a], p. 22), but where also, during the hardships of World War I, he suffered a serious attack of pneumonia. Rudolf Jakob Humm (1895–1977) collaborated with Hilbert for two years on Einstein's new theory of gravitation, first as a student and then as Hilbert's assistant (Ref. [30d], pp. 89 and 92), but after the war Humm gave up the sciences and became a well-known author in Zurich. In Göttingen Scherrer worked with Debye on the Faraday effect of the hydrogen molecule and in February 1916 earned his Ph.D. with this work. The same year he developed, together with his more theoretically minded teacher, the celebrated 'Debye–Scherrer powder method' of x-ray analysis applied to microcrystalline samples which made Scherrer's name known world-wide. Two years later Scherrer became a Privardozent, and when in 1920 Debye accepted the call to ETH, one of his conditions was that his collaborator Scherrer be nominated simultaneously. This was the beginning of the 40 years of Scherrer's extraordinary career at ETH, during which he not only created a physics department with a world reputation but also, together with Pauli, laid the foundations of modern physics in Switzerland [30a,b,c].

In 1922 Scherrer married Ina Sonderegger [30e], with whom he had two daughters. Ina's father was the owner or the director of the *Quellenhof*, a luxury hotel in the spa of Bad Ragaz in the Saint Gallese Rhine valley [30f]. He had opposed this marriage to a 'poor' professor (see Ref. [29b], p. 149). Scherrer was a very jovial and charming person who was a master of presentation. Respected and even feared as boss, he was often hesitant in making decisions.

Remarks like *'Wa meinet si?'*, *'Wa wömmer säge? Mer mönd doch öppis säge!'*, Saint-Gallese for 'What do you think?', 'What should we say? But we should say something!' were so well known in Scherrer's institute that even Pauli in a sketch presented at the Christmas party in 1948 made a parody of it (recollections of ETH Professor Heini Gränicher (b. 1924) in Ref. [27], p. 190). Independently of Debye, Scherrer continued in Zurich his experimental research on the structure of matter which led him to solid-state physics. This subject, particularly ferroelectricity, was further developed by Georg Busch (1908–2000), Scherrer's first pupil to obtain a chair at ETH, who introduced metal physics and, even more prominently, the physics of semiconductors in Switzerland, creating important links with industry. From 1934 Scherrer, on the other hand, moved on to nuclear physics in which, before World War II, ETH became a leading centre [30a,b,c].

Scherrer earned honorary degrees from many European universities [30e]. And his excellent contacts in the United States allowed him shortly after the war to get classified information about the American nuclear enterprise, so that on 28 November 1945 he was able to write a much noted article on nuclear energy in the technical supplement of *Neue Zürcher Zeitung*. In 1946 the Swiss Federal Council created the Committee for the Study of Atomic Energy, of which Scherrer was chairman. In 1958 this committee was replaced by the Committee for Atomic Sciences, presided over again by Scherrer and attached to the Swiss National Fund for the Promotion of Scientific Research which had been created in 1952 (Ref. [27], p. 286). Being not only a Swiss patriot but also a good European, Scherrer also played a part in the creation in 1952 of the Conseil Européen pour la Recherche Nucléaire, better known as CERN, on the periphery of Geneva. At the Atoms for Peace conference held in 1955 on the Geneva site of the United Nations the United States exhibited a 'Swimming-pool' reactor which, mainly thanks to the insistence of Scherrer, was acquired for Switzerland by the newly founded Swiss company Reaktor AG and, under the name Saphir, became the centre of the Swiss Institute for Reactor Research in Würenlingen, 30 km north-west of Zurich (Refs [30b] and [27], p. 192). An impressive number of Scherrer's pupils obtained chairs of experimental physics in Switzerland, in Europe, and in the United States and many went into industry. Still in good health after his statutory retirement from ETH in 1960 Scherrer continued his lecturing at the University of Basle. He died on 25 September 1969 after a horse-riding accident [30b,c].

In his first letter to Bohr from Zurich, dated 16 June 1928, Pauli excused himself for answering Bohr's last letter so late. Bohr had written that after Pauli's last visit to Copenhagen in March he felt 'how close to each other we have come', and he announced that he will 'soon write more thoroughly about the philosophy' (translated from letter [197] of 14 May 1928, Ref. [2], p. 456). By 'philosophy' Bohr meant elaborations on his conception of complementarity which he thought was able to reconcile contradicting views in biology, philosophy, and human culture by showing that the contradiction is actually an exclusiveness between two opposite views of equal value, as in the dual notions of 'particle and wave' or 'instinct and reason' or 'subject and object' (see e.g. Ref. [31]). In particular, Bohr had in mind 'the one-sidedness of the direction of time'. Concerning this problem Pauli wrote in his letter that recently he has had some ideas 'about the question, under what assumptions and in what generality an "H-theorem" of the entropy increase can be deduced in quantum mechanics.'I had postponed the answer to your letter again and again because I first wanted to arrive at some clarity in this. Now I believe I am able to give the following answer to the H-theorem question' (translated from letter [201] of 16 June 1928, Ref. [2], p. 462). This answer consisted of two points: first, that the entropy increase can be shown for general, unspecified systems and, second, that an assumption of elementary randomness analogous to the classical *Stosszahlansatz* (collision frequency expression) is still unavoidable in the sense that some phases must be considered to be random.

Pauli published his results in the paper 'On the H-theorem of the entropy increase from the point of view of the new quantum mechanics', which appeared in the volume dedicated to Sommerfeld on his 60th birthday in December 1928 by his pupils [32]. It is the last of Pauli's papers carrying the designation 'jr.' (junior). As explained in Sections 1 to 4 of Pauli's *Lectures on Statistical Mechanics* [33a], the H-theorem, which Boltzmann had derived from his collision frequency expression for classical particles [33b] (see also Sections 5.1 and 8.2, and Appendices 5.2 and 8.1, of Ref. [33c]), states that Boltzmann's H-function satisfies the inequality $dH/dt \leq 0$, where the equality sign holds if and only if the particle distribution function $f(\mathbf{x}, \mathbf{v}, t)$ has the form of a stationary local Maxwell distribution. This situation is called local equilibrium. But the remarkable thing about this theorem is that it is a statement concerning non-equilibrium situations. And since the H-function is related to the entropy S by $H = -kS + \text{const.}$ where k is Boltzmann's constant (see Section 6 of Ref. [33a]), the H-theorem is a kinematical proof of the second law of thermodynamics, thus establishing 'the one-sidedness of the direction of time' mentioned above.

Pauli's paper consists of two parts, in the first ('Kapitel I') he derives in Section 1 the H-theorem for BE and FD statistics by dividing the phase space of the particles (molecules) into elementary regions μ containing g_μ quantum states (note that in deriving Eq. (5.19) $g_\mu = 1$ had been chosen). If n_μ is the number of particles in the elementary region μ then $f_\mu = n_\mu/g_\mu$ is the average number per state in μ. Then the collision frequency for the scattering of a particle in κ with one in λ into a particle in μ and one in ν is

$$Z_{\kappa\lambda,\mu\nu} = A_{\kappa\lambda,\mu\nu} g_\kappa g_\lambda g_\mu g_\nu f_\kappa f_\lambda (1 + \alpha f_\mu)(1 + \alpha f_\nu), \qquad (6.51)$$

where $\alpha = +1$ and -1 for BE- and FD-statistics, respectively, and the rate of change of n_κ is

$$\frac{dn_\kappa}{dt} = \sum_{\lambda(\mu,\nu)} (Z_{\mu\nu,\kappa\lambda} - Z_{\kappa\lambda,\mu\nu}), \qquad (6.52)$$

where (μ, ν) means summation over pairs μ, ν. Equations (6.51) and (6.52) imply that $\sum_\kappa dn_\kappa/dt = 0$, i.e. conservation of the particle number.

Defining now the entropy in the usual way as (see e.g. Eq. (9), Section 38 in Volume V of Sommerfeld's famous lectures [33d])

$$\frac{1}{k}S = \sum_\kappa g_\kappa \{\alpha(1 + \alpha f_\kappa) \log(1 + \alpha f_\kappa) - f_\kappa \log f_\kappa\}, \qquad (6.53)$$

which is the same as Pauli's Eq. (3) in Ref. [32] (note that $\alpha = 0$ gives the classical statistics). Assuming microscopic reversibility $A_{\mu\nu,\kappa\lambda} = A_{\kappa\lambda,\mu\nu}$ it is then straightforward to show that $dS/dt \geq 0$, and that the equality sign only holds if, for $A_{\kappa\lambda,\mu\nu} \neq 0$,

$$\left(\frac{1}{f_\mu} + \alpha\right)\left(\frac{1}{f_\nu} + \alpha\right) = \left(\frac{1}{f_\kappa} + \alpha\right)\left(\frac{1}{f_\lambda} + \alpha\right). \qquad (6.54)$$

But according to (6.51) this means that detailed balance $Z_{\mu\nu,\kappa\lambda} = Z_{\kappa\lambda,\mu\nu}$ holds or, according to (6.52), that $n_\kappa = \text{const}$. Conservation of energy $E_\mu + E_\nu = E_\kappa + E_\lambda$ then implies, following Boltzmann's classical argument (see Section 4 of Ref. [33a]) that, with a Lagrange multiplier β, $\log(1/f_\kappa + \alpha) = \beta E_\kappa + \text{const.}$, which corresponds to Pauli's relations (8) in Ref. [32]. Determining β with the help of the thermodynamic relation $T dS = dE$, this leads, for f_κ, to the distribution function (5.19).

In Section 2 of paper [32] Pauli considers the phase space of the global system. For given elementary regions μ such that the g_μ are fixed (e.g. $g_\mu = 1$), a distribution $n = (n_1, n_2, \ldots)$ of particles determines an elementary region of the global system containing G_n stationary states. Since there is just one global system and since $G_n \gg 1$ (excluding Bose condensation in the BE case), we may use Eq. (6.53) by substituting n (fixed) for κ, G_n for g_κ, and 1 for n_κ, i.e. $1/G_n$ for f_κ, so that $S/k = \log G_n$. However, because of scatterings this situation will not stay this way; there is a probability $W_n(t)$ for all possible distributions n. Hence W_n takes the place of n_κ and $\sum_n W_n = 1$. Equation (6.53) then becomes

$$\frac{1}{k}S = \sum_n W_n(\log G_n - \log W_n) \qquad (6.55)$$

which is Pauli's Eq. (17) in Ref. [32]. Similar to Eq. (6.51), the frequency of transition from a distribution n to m is proportional to the probability W_n of the initial state and to the weight G_m of the final state,

$$Z_{nm} = A_{nm} G_m W_n \qquad (6.56)$$

and, in analogy to (6.52),

$$\frac{dW_n}{dt} = \sum_m (Z_{mn} - Z_{nm}). \qquad (6.57)$$

Making use of the microscopic reversibility $A_{nm} = A_{mn}$, Pauli deduces from the last three relations that $dS/dt \geq 0$. And he concludes this first part by noting that Eqs (6.52) and (6.57), which are of the 'type of radioactive decay laws, in their generalization for multiple processes ... already guarantee a degree of randomness of the microscopic happenings as is necessary for the statistical interpretation of the second law of thermodynamics' (translated from Ref. [32], p. 39).

The second part ('Kapitel II') of paper [32] is noteworthy because it contains in Section 3 a sample of Pauli's version of the Copenhagen Interpretation: 'For the statistical interpretation of the more recent wave- or quantum mechanics, on which the following discussion is based ... two types of circumstances are characteristic. First, the paradoxes which arose in the previous form of the quantum theory from the validity of the conservation laws of energy and momentum, together with the existence of the superposition principle of the waves (interference), are removed by the reciprocal uncertainty relations between space–time localizability of the individuals and the definability of energy and momentum. For this the universality of the dualism between "waves" and "corpuscules" ... is essential. ... Second, the discontinuities characterized by Planck's quantum of action are taken care of in a fundamentally different manner than in the previous quantum theory. In the place of the elementary processes, whose space-time description the older theories had to renounce in principle, comes in the more recent theory the view (*Auffassung*) that the observations alone are the cause for the need of a discontinuous element in the description of Nature, namely by making a selection out of a manifold of possibilities which are calculable from exactly to be given laws of nature'. Pauli emphasizes that the 'necessary separation between observed object and measuring device is of a much more trenchant sort in quantum theory than it was the case in the classical physics' (translated from Ref. [32], p. 39).

Formally, this second part of paper [32] is based on the 'method of the variation of the constant' introduced by Dirac in Ref. [3a]. Pauli's Eq. (28) is the second Eq. (6.1) of the previous section, written in interaction representation which, after integration between 0 and t, becomes, in Pauli's notation,

$$a_r(t) = \sum_s V_{rs} \frac{e^{i(E_s - E_r)t/\hbar} - 1}{E_s - E_r} a_s(0). \tag{6.58}$$

In Section 4 this equation is applied to the case where the global system consists of an atom with and without radiation or a detached electron. One of the elementary regions is then the single state n of an atom without radiation or electron, the other (index m) consists of a group of states α describing an atom in the presence of a photon or an electron, and energy conservation $E_n = E_m$ is known only with an uncertainty ΔE_m. The initial conditions are $a_n(0) = 1$, $a_\alpha(0) = 0$ and $a_n(0) = 0$, $\sum_\alpha |a_\alpha(0)|^2 = 1$ for emission and absorption of a photon or an electron, respectively.

In the case of emission, $r = \alpha$ and $s = n$ in Eq. (6.58), and $W_m(t) = \sum_\alpha |a_\alpha(t)|^2$. Here introduction of the density of states σ_m allows the sum over α to be replaced by an integral over the energy E_α from E_n to $E_n +$

ΔE_m. For large enough times t such that $\Delta E_m t \gg h$ this integral becomes $\int_0^\infty (\sin x/x)^2 dx = \pi/2$, where $x = (E_\alpha - E_n)t/2\hbar$. Hence

$$W_m(t) = \frac{\pi}{\hbar}|V_{mn}|^2 \sigma_m t \qquad (6.59)$$

(see, however, [34]). Since $W_n(0) = 1, W_m(0) = 0$, Eqs (6.56), (6.57) yield in this case $\dot{W}_m \equiv dW_m/dt = A_{nm}G_m = \pi|V_{mn}|^2\sigma_m/\hbar$, which is Eq. (33) of Ref. [32]. It is interesting to note that Dirac in Ref. [3a] used the probability per unit time $\dot{W}_m$ to derive in his Eq. (26) an expression for the scattering cross-section.

In the case of absorption Eq. (6.58) yields, with $r = n, s = \alpha, \beta$,

$$W_n(t) \equiv |a_n(t)|^2 = \sum_{\alpha\beta} V_{n\alpha}V_{n\beta}^* \frac{e^{i(E_\alpha - E_n)t/\hbar} - 1}{E_\alpha - E_n} \times$$
$$\frac{e^{-i(E_\beta - E_n)t/\hbar} - 1}{E_\beta - E_n} a_\alpha(0)a_\beta^*(0) \qquad (6.60)$$

Here the sums cannot be evaluated without a further assumption. Now $a_\alpha(0) = |a_m(0)| \exp(i\varphi_\alpha)$ where, according to the 'hypothesis of elementary disorder' the *phases* φ_α are random. Therefore, averaging over the φ_α yields the random phase relation

$$\langle a_\alpha(0)a_\beta^*(0)\rangle = \frac{1}{G_m}\delta_{\alpha\beta} \qquad (6.61)$$

which, because of the Hermiticity $V_{nm} = V_{mn}^*$, leads, for $W_n(t)G_m$, to the expression (6.59), or to $\dot{W}_n G_m = \dot{W}_m = A_{nm}G_m$. Since, on the other hand, $G_n = 1$ and, in the present case, $W_n(0) = 0, W_m(0) = 1$, Eqs (6.56), (6.57) yield $\dot{W}_n = A_{mn}$, so that $A_{mn} = A_{nm}$. In other words, for microscopic reversibility and, hence, for the H-theorem to hold, the random phase assumption is necessary. In his letter of 14 July 1928 Pauli wrote to Bohr: 'As the most important result of Chapter II [of this paper] I consider that also in wave mechanics a special assumption of "disorder" is still required for the justification of the H-theorem. This was not at all obvious to me, I originally had thought the opposite' (translated from letter [203], Ref. [2], p. 465).

In the same letter Pauli tells Bohr about his Zurich colleagues: 'Herr Scherrer ... really is a quite excellent physicist and person, and he and I have in a short time already become good friends. In addition I have learned a lot of group theory from Weyl, so that I now can really understand the papers of Wigner, as well as those of Heitler and London which are also important for Heisenberg's new theory of the ferromagnetism; although Weyl's philosophy and style of life is not really my taste' (compare page 38). And looking forward to the long summer vacation he continues: 'Now we have ended the term here and we may set out on vacations. I will stay on for some 12 days, then I meet with my sister Hertha, about mid-August I am supposed to meet Koch, Gordon and others on the island of Neuwerk (in the estuary of the Elbe); on 16 September there is the meeting of the [German] natural scientists [and physicians] in Hamburg. If it were agreable with you, I might perhaps come to Copenhagen at the beginning of September and live again in the room in the Institute so kindly offered by you. When do you come back to town from the country-side? (You know I don't like the country-side and even less your house in Tisvilde which is arranged so dreadfully primitively!)' (translated from letter [203], Ref. [2], p. 465).

The traditional summer vacation of the Hamburg physicists on the island of Neuwerk is also fondly remembered by Jordan: 'Among the most vivid remembrances of my then short stay in Hamburg are two summer vacations, in which I participated in the common journey of several Hamburg physicists to the North Sea. For Stern and Lenz this enterprise was somewhat outside of their inclinations; but Pauli had come from Zurich to participate in it' (translated from Ref. [35a], p. 61). In two postcards ([206], Ref. [2]) sent to Bohr on 10 August 'between the aestival journey to the mountains and that to the sea' Pauli refers to his forthcoming stay on Neuwerk (at Hotel *Meereswoge*) and the hoped-for visit to Copenhagen (Ref. [2], p. 571), from where he later informs Ehrenfest on 16 September of his planned visit to Holland at the end of September (letter [208], Ref. [2], p. 473). On 22 September Ehrenfest urges Pauli to come and invites him to live in his house, but proposes to eat out at least for dinner because food at home is rather bad (letter [209], Ref. [2]). It is on this occasion that Pauli put his signature, dated 25 September 1928, on the famous 'sacred wall' of the institute in Leiden (see Ref. [35b], p. 82).

For the winter term Kronig went back to Utrecht to work with Kramers again, so Pauli had to look for a new assistant starting 1 October 1928. This time he considered a local candidate, Dr Felix Bloch, living at Seehofstrasse 3 in Zurich, a small side street of Utoquai, behind the Opera house (document I.20, Ref. [27]). But, as Pauli wrote to President Rohn on 22 October, Bloch declared that 'if his pay would correspond only to a b)-position, he would be obliged to procure secondary earnings elsewhere, e.g. by giving private lessons, which would imply that only part of his time would remain for the fulfilment of his function as assistant. Now the task of my assistants, probably in contrast to other assistant positions of ETH, is a double one since it consists not only in aid with my teaching activity (e.g. with the problem series related with my courses) but has also a purely scientific side (besides independent works also giving talks in the colloquium, aid with the verification of papers by others, etc.)' (translated from document I.21, Ref. [27]). A week later the President decreed the nomination of Dr Felix Bloch for the winter term of 1928/9 with pay of 1800 francs, i.e., 300 francs per month, corresponding to a 'b)-position'. So Pauli's petition had failed—mainly because Bloch was an ETH student only 23 years old and a Swiss citizen with his family living in town (document I.22, Ref. [27]). This again shows the importance that Pauli attached to his assistant's position, but also the typical problems he faced. However, for the summer term of 1929 Bloch got an 'a)-position' with six monthly salaries of 500 francs (document I.27, Ref. [27]).

Felix Bloch (1905–1983), who with Edward M. Purcell (b. 1912) won the

Nobel Prize in 1952 'for the development of new methods for nuclear magnetic precision measurements and the discoveries in connection therewith', entered ETH as an enginering student in 1924. But 'after a year and a good deal of soul searching I decided, against all good sense, to switch over to the "entirely useless" field of physics' (Ref. [35c], p. 23). As a physics student at ETH Bloch witnessed Schrödinger's discovery of the wave equation through the latter's colloquium talk (see page 140). But in October 1927 he followed Debye to Leipzig and became Heisenberg's first student. The following year he wrote his celebrated thesis 'On the quantum mechanics of the electrons in crystal lattices' [35d] under Heisenberg. After one year with Pauli, Bloch went to Holland as a Lorentz Fellow and in the autumn of 1930 returned to Leipzig to be Heisenberg's assistant for one year and for a second year in 1932/3, 'until Hitler succeeded in forming a new Germany in his own frightful image'. In spite of this sad end Bloch's 'memories of Heisenberg belong to the happier time before those events' (Ref. [35c], p. 27). He went to Paris where he lectured at the Institut Henri Poincaré and then visited Fermi in Rome on a Rockefeller Fellowship, after which he definitively left Europe and settled at Stanford University in California.

In the winter term of 1928/9 Pauli asked for a supplement to pay a scientific collaborator, for whom the School Council granted him a contribution of 1200 francs (document I.23, Ref. [27]). The first such collaborator was Gregory Breit whom Ehrenfest had sent to Pauli in November. It is likely that Breit came with American money and that Pauli used the supplement to pay Bloch an 'a)-position'. When Breit returned to Leiden in mid-January, it was Oppenheimer's turn to come to Zurich (see Ref. [2], pp. 476, 478 and letters [212], [215]). This was the time when many bright American students came to Europe to learn the new physics at the centres of Bohr in Copenhagen, Born in Göttingen, Ehrenfest in Leiden, Fowler and Rutherford in Cambridge, and Sommerfeld in Munich. Gregory Breit belonged to a Russian family that had emigrated to the United States in 1915. After his studies at Johns Hopkins University he came to Leiden for the first time during 1921/2 on a National Research Fellowship. Back in the United States he became known for his many papers on quantum theory which allowed him to return to Europe a second time and this time also to visit Pauli in Zurich (see Ref. [2], p. 478).

J. Robert Oppenheimer was born in New York on 22 April 1904, the son of Julius Oppenheimer, a successful businessman who had emigrated from Germany. He came to Europe after obtaining his Bachelor's degree *summa cum laude* from Harvard in 1925. After a short stay at Cambridge University and several months at Leiden he accepted an invitation from Max Born who was just back from a tour of the United States. By the end of 1926 Oppenheimer was able to present his Ph.D. work 'On the quantum theory of continuous spectra' [36a]. Out of this collaboration with Born sprang the common paper 'On the quantum theory of molecules' which contains the well-known Born–Oppenheimer approximation that takes advantage of the fact that the mass ratio between the electron and a nucleus is always smaller than $1/2000$ [36b].

With his superior intelligence and his mental instability Oppenheimer was a difficult pupil, as Born, as well as Ehrenfest and Pauli after him, experienced (see Ref. [2], pp. 475–6). After spending one year on a National Research Fellowship in the United States at Harvard, Berkeley, and the California Institute of Technology (Caltech) at Pasadena he returned to Leiden in September 1928. On 26 November 1928 Ehrenfest lamented to Pauli that he does not understand Oppenheimer's spirited (*witzigen*) ideas but 'that for the full development of his (great) scientific gifts, Oppenheimer should, still *in time*, be somewhat (*a bisserl*) (!) *affectionately* thrashed (*zurechtgeprügelt*). But in contrast to many other youths he *merits* this favour for, as a personality he is a rare dear and fine chap! . . . But in physics (*rebus physicis*) there is only *one* God's whip (*Geissel Gottes*, see page 89)' (translated from letter [211], Ref. [2], p. 477). Therefore, Ehrenfest says, he wished that Pauli would take care of Oppenheimer for some time.

Pauli answered on 15 February 1929 that, although he is certainly not on Ehrenfest's list of physicists that love Oppenheimer—Dirac is in first place—he believes 'that Oppenheimer feels quite well in Zurich, that with us here he is able to work well and that scientifically anyhow much good may still be gotten out of him. . . . he has much good will and is not stubborn. Since he is interested in questions of conductivity he discusses a lot with Herrn Bloch (my present assistant). The latter has much less ideas than Oppenheimer but is enormously clear and thorough.' However, Pauli continues: 'Unfortunately, Oppenheimer has a very bad quality: he approaches me with a fairly absolute authority cult (*Autoritätsglaube*) and considers all I say . . . as last and definitive truth. I know alright how with him the need for foreign authorities comes about. Let them solve his problems and answer his questions so that he need not do it himself. (This connection of course is not *consciously* clear to him but is only latently with him in the unconscious.) But how I have to wean him from it I do not know. In vain I point to well-documented cases where I have claimed some incorrect thing and obviously was wrong!—At first I help myself so that, before he has not really thoroughly occupied himself with one of the questions he has asked, instead of superficially guessing anything, I avoid if possible at all to utter any views or opinions about this question. One thing, however, I hope to have achieved soon: that O[ppenheimer], at least in relation *to me*, adopts *my* manners! This is absolutely necessary if I should not consider myself a rascal. For, my view about formal politeness, as of the great heresy which has to be unconsiderately (*rücksichtslos*) eradicated from the relations between humans, with me is an unshakable dogma!' And he signs with 'God's whip' (translated from letter [215], Ref. [2], pp. 486–7).

Oppenheimer stayed in Zurich until the beginning of summer 'and mentally has come into bloom' (translated from letter [226] from Pauli to Kronig, Ref. [2]). In the autumn of 1929 he went to the University of California at Berkeley as assistant professor. There he became an associate professor in 1931 and was a full professor from 1936 to 1947. Between 1929 and 1942 he held the same positions also at Caltech and thus 'created the greatest school of theoretical physics that the United States has ever known' (Ref. [23b], p. 369). From 1943 to 1945 Oppenheimer was director of the Manhattan Project to develop the atomic bomb in Los Alamos, New Mexico, and in 1946–52 he was chairman of the General Advisory Committee of the US Atomic Energy Commission. In 1947–67 he was director of the Institute for Advanced Study in Princeton, New Jersey. Oppenheimer earned a honorary degree from Harvard, the US Medal for Merit and the Enrico Fermi Award for 1963. The last, which is a prize of high prestige awarded by the Atomic Energy Commission, was a gesture of reconciliation after the dramatic censuring of Oppenheimer for his opposition to the hydrogen bomb programme. Oppenheimer died in Princeton on 18 February 1967 of throat cancer (see Refs [23b], Section 16 (b), and [36c,d]).

As for Bloch's research work, Pauli told his friend Klein that Bloch has 'made progress in the theory of superconductivity; I do *not* want to assert yet that he has succeeded in finding an explanation of superconductivity but his results so far give rise to justified hopes' (translated from letter [218], Ref. [2], p. 495). But in June Pauli wrote to Kronig: 'Bloch does superconductivity and changes his theory daily (thank God, before publication!)' (translated from letter [226], Ref. [2]). Instead, on 15 December 1928 Bloch submitted a paper 'On the susceptibility and resistance change of the metals in the magnetic field' [37a], in which he thanked Pauli 'cordially for many suggestions and clarifying remarks during the conception of this work' (translated from Ref. [37a], p. 227). This shows that Pauli's attitude towards solid-state physics was much more constructive than the quotations from his letters to Peierls on page 157 would have us believe. Bloch submitted a second paper from Zurich on 21 June 1929, 'Remarks on the electron theory of ferromagnetism and of the electronic conductivity' [37b], which is noteworthy for containing the spin Hamiltonian formulation of ferromagnetism first introduced by van Vleck (see e.g. Ref. [37c]).

On 16 May 1929 Pauli wrote to Sommerfeld: 'I now have a fairly big enterprise in Zurich. Herr Bloch at present is busy with elaborating a theory of superconductivity. The matter is not finished but seems to work. Herr Peierls does theory of heat conduction in solid bodies.' And as a postscript he added: 'Wentzel sends greetings, he will write you soon. Last night I drunk brandy with him till midnight' (translated from letter [225], Ref. [2], p. 503). In the autumn of 1928 Gregor Wentzel had become Schrödinger's successor at Zurich University, thus fulfilling Pauli's vision of the future of theoretical physics in Zurich. Together, Pauli and Wentzel instituted the common Monday afternoon theoretical seminar traditionally held at the University, thus establishing the world-wide reputation of Zurich in this field. This tradition still exists today.

Gregor Wentzel was born on 17 February 1898 in Düsseldorf. His studies of physics at the Universities of Freiburg, Greifswald, and Munich from 1916 to 1921 were interrupted at the end of World War I by a year of military service. After his doctorate on X-ray spectra under Sommerfeld in 1921 he acted as an assistant when Pauli and Heisenberg were students in Munich (see page 52). In

1922 Wentzel became a Privatdozent there and in 1926 an associate professor in Leipzig. During World War II when Pauli was detained in the United States, Wentzel stood in for his friend in teaching at ETH. After Pauli's return to ETH in 1946 Wentzel, much to Pauli's regret, left Zurich to become a professor at the University of Chicago in 1948, from where he retired in 1968. He spent his last years in Ascona on Lago Maggiore (bigger lake) in southern Switzerland where he died on 12 August 1978 (see Ref. [38a], footnote 609, p. 356).

On Pauli's 'enterprise' Peierls wrote in his recollections: 'In retrospect, the period from April to July appears rather short for all that seems to have happened.... There was plenty of time for concerts and cinema, and for sailing. It was then easy to rent a sailboat for a few hours, and I liked to take friends out on the lake. I even persuaded Pauli to come sailing—I cherish a photo showing him, Robert Oppenheimer, and I. I. Rabi on the boat' (Ref. [38b], pp. 44 and 178–9). Further on Peierls writes: 'Before I left Zurich [in July 1929], Pauli had offered me the job of *Assistent* for the following year, and I accepted with pleasure.' So Peierls was back in the autunn, now with a monthly salary of 400 francs (Ref. [27], document I.30), and getting 'to know Pauli and his famous biting remarks' (Ref. [38b], p. 46), of which one of the nicest I know is Pauli's characterization of Peierls himself: 'He talks so fast that by the time you understand what he is saying, he is already asserting the opposite' (Ref. [38b], p. 47).

Peierls continues: 'One of the first visitors in the autumn of 1929 was Lev [Davidovich] Landau [1908-68], and we became friends almost immediately. ... He had come with a Soviet government fellowship, and this caused some trouble, because there were then no diplomatic relations between Switzerland and the Soviet Union. So he was given a permit to stay for only a very short time.... He came back the following year with a Rockefeller Fellowship, and had no trouble with his stay' (Ref. [38b], p. 49, also Ref. [39a], pp. xiii and 10). During the summer term of 1930 'There were new faces in Zürich. Léon Rosenfeld [1904–74] arrived from Belgium, a stubby young man with a serious round face. He was given to philosophical contemplation. ... Another visitor was George Gamow [1904–68]—like Landau, a Russian from Leningrad. He was already famous for his explanation of alpha decay ... and perhaps he was even more famous for his funny drawings and his jokes' (Ref. [38b], p. 60, also Ref. [39a], footnote 8, p. 18).

Rudolf Peierls grew up in Berlin as the youngest of three children of a Jewish family. At 18 he enrolled in physics at the university of his hometown. There he took lectures by Planck and by Nernst, of whom he observed: 'Planck's lectures were, I think, the worst I have ever attended' and: 'Nernst was a great physicist of rather small stature and an even smaller sense of humility' (Ref. [38b], p. 19). After two semesters Peierls went to Munich to study with Sommerfeld, where for him 'the most important of the other students was undoubtedly Hans Bethe. ... He was tall and heavily built, and spoke very slowly with a deep and sonorous voice. ... He did speak on most subjects with great authority' (Ref. [38b], pp. 23–4). When in spring 1928 Sommerfeld went to the United States for a year he suggested Peierls to go to Leipzig.

After three years as Pauli's assistant Peierls left Zurich, with a Russian wife, Eugenia Nikolaevna Kannegiser, a classmate of Landau's in Leningrad he had met at the Odessa Conference (see the following section) and married in March 1931, and with the title of Privatdozent of ETH (Ref. [27], documents I. 48, 49, 51). He went to Rome and Cambridge on a Rockefeller Fellowship and, after two years in Manchester with Hans Bethe, became professor of theoretical physics in Birmingham in 1937, and in Oxford in 1963. After his retirement in 1974 Peierls held invited professorships at various universities in Europe, the United States and Australia. In spring 1940 Peierls and Otto Robert Frisch (1904–79) had written their fundamental memorandum on the possibility of a nuclear chain reaction, and from 1943 to 1945 both were engaged in the Manhattan Project to develop the atom bomb in Los Alamos. Peierls received many honours, the Lorentz Medal in 1962, the Planck Medal in 1963, and a knighthood in 1968 (see Ref. [38c]).

A loose marriage and a new particle: the neutrino

Dear Pauli!
  Thanks for your card. I do appreciate as evidence of your confidence that you told me candidly about your conjugal difficulties. How strangely things shift! Your wife longed for America, you didn't. Yet now you have no spouse but long for America. Frankly, I'll prefer to travel with you alone rather then with the two of you. [38d]

These lines of June 1930 from Pauli's teacher Sommerfeld who planned to attend the summer school of 1931 in Ann Arbor, Michigan, do not bode well for Pauli's marriage on 23 December 1929 in Berlin to Kate Deppner. Luise Margarete Käthe (Kate) Deppner was born in Leipzig on 29 August 1906 (see Ref. [39a], pp. 5 and 726). Pauli seems to have met Kate in the house of his friend Adolf Guggenbühl (1896–1971) in Zurich [29a]. When he met with her in the late 1920s on his frequent visits to Berlin, Kate was a performer trained at the Max–Reinhardt School for Film and Theatre (Ref. [38a], footnote 636, p. 378). This is the same Max Reinhardt who had fetched Pauli's sister Hertha to Berlin at about that time (see pages 17–18). Since Hertha and Kate's stays in Berlin probably overlapped, it is not excluded that they met. Anyway, in Zurich Kate performed with the dance school of Trudy Schoop, [29a] who was one of Paul Scherrer's girl friends [30f]. Trudi Schoop (b. 1903) was particularly successful in the art of pantomime, and toured Europe and the United States. Back in Zurich during World War II she joined the well-known Cabaret Cornichon but afterwards went back to California [39b]. Pauli, of course, knew Trudi Schoop through Kate. He visited Trudi Schoop again in the winter of 1939/40 when she was back in Zurich: 'she was very nice. But, of course, I couldn't refrain from making fun of her somewhat about certain events in the Physics Department at Stanford [California] which, naturally, had penetrated up to me' (translated from Ref. [39c], letter [614] from Pauli to Hans Staub, Princeton, 17 December 1940). Much later Pauli told Jean Weiglé, a friend of Max Delbrück, in a letter written in Zurich on 16 June 1958 that Trudi Schoop 'was such a help to me in the time of my great crisis'.

However, even before the marriage Kate fell in love with a Polish Jew in Berlin, the chemist Paul Goldfinger, whom she married after she had left Pauli [29a]. The couple later moved to Brussels where Goldfinger became professor of chemistry at Université Libre. In a conversation with Jagdish Mehra on 12 March 1974 Kate Goldfinger recalled: 'When Pauli and I got married in Zurich he always told me that he was someone really and profoundly important in the world of physics. He used to get letters from many physicists, especially Heisenberg. He used to walk around like a caged lion in our apartment, formulating his answers in the most biting and witty manner possible. This gave him great satisfaction' (Ref. [38a], footnote 33, p. xxxvii).

On 1 October 1929 Pauli had to leave his apartment at Schmelzbergstrasse 34 (see letter [235] to Weyl, Ref. [2], p. 520). Pauli and Kate moved into an apartment at Hadlaubstrasse 41 [26c] which is somewhat further up the slope of Zürichberg, in a direction north of the institute. Previously, Pauli had lived for a short time at Germaniastrasse 6, which is still further up Zürichberg (see Ref. [27], p. 15). Even before the marriage the prospect of Pauli being a husband gave rise to many rumours which Pauli had to correct. To Weyl he had to explain that he rejected the term 'bride' as being too bourgeois. And he continued: 'What the *Ordinarien* (full professors) now do in Zurich is still not the mere repetition of what once made the Göttingen Privatdozenten proverbial (you surely know the anecdote of Lindenmann and the girl student)' (translated from letter [235], Ref. [2], p. 520). This is an allusion to the liberal socializings of a circle of Zurich professors (see Ref. [29b], pp. 175–6). To Ehrenfest Pauli observed: 'That I marry an American I first heard from your card, up to now it was unknown to me!—Since when are you active in fortune telling and card laying?' (translated from letter [237], Ref. [2], p. 524). Bohr congratulated him on 26 December: 'to your marriage, of which we have learned in different ways just these days. We much hope that we shall soon see you both at one of your visits up here which are always expected with great joy' (translated from letter [241], Ref. [2], p. 529).

To his friend Oskar Klein Pauli wrote on 10 February 1930: 'In case my wife should at one time run away, you (as also all my other friends) will receive a printed notice.' And again on 10 March: 'My wife presumably doesn't join me; even if I am married it is at least in a loose way!' (translated from letters [242] and [243a], Ref. [39a], pp. 4 and 7). Hermann Weyl, on 28 May 1930, wrote to his colleague and friend Erich Hecke in Hamburg: 'With respect to marriage, Pauli is not well; for the moment his wife lives again together with her former friend in Berlin but surely will soon again come to be with him a little. I am not close enough to him to be able to judge how threatening the whole is. Anyhow—in spite of his theories he is, like other mortals, at times vehemently plagued by jealousy' (translated from Ref. [39d], p. XXI).

This unfortunate relationship was bound to break up; it lasted less than a year. On 26 November 1930 the divorce was announced in Vienna (Ref. [39a], p. 38). For Pauli it was a crushing defeat; drinking became a problem and smoking developed into a habit—later it was exclusively the pipe, while he used to say of cigarettes 'I don't smoke paper'. What Pauli resented most was not that Kate had fallen in love with another man without telling him and that she eventually married that man, but that she had exchanged him, Pauli, for a mediocre chemist. 'If it had been a bullfighter—with someone like that I could not have competed—but such an average chemist!' [29a]. However, as expressed by the introductory quotation on page 146, the root of Pauli's crisis was deeper and more remote. An indication that his mother's suicide on 15 November 1927 had a profound effect on him is given by his apparently unmotivated decision to leave the Catholic Church (document dated Vienna, 6 May 1929, PLC Bi 43, *Pauli Archive*), thus imitating his parents who, very likely with entirely different motives, had left the Catholic Church back in 1911 (see page 10). But Pauli's conjugal failure, like his mother's suicide, had no noticeable influence on his scientific performance.

After a fortnight's stay in Vienna in the second half of March 1929, where he resided at Weimarerstrasse, a street parallel to Gymnasiumstrasse and crossing the Anton-Frank-Gasse of the parental home (see page 5), Pauli spent a few days in Berlin. There he hoped 'to talk to Miss Meitner for the purpose of gathering empirical material against the Copenhagen theoretical nuisance (*Unfug*)' (translated from letter [218] to Klein of 16 March 1929, Ref. [2], p. 494). This *Unfug*, in Pauli's eyes, was Bohr's insistence that conservation of energy holds only statistically in β-decay. Already in his letter of 18 February 1929 he had lamented to his friend Oskar Klein about his, Pauli's, 'general helplessness in the discussion of experiments' with Bohr who with his idea of a violation of energy conservation in β-decay is, according to Pauli, on an 'entirely wrong track'. And he refers to his first paper with Heisenberg [17a] where it is shown (see page 190) 'that then, even with a sharp [energy] state of the nucleus the emerging electrons, because of additional γ-ray emission, *must have a continuous velocity spectrum!*' (translated from letter [216], Ref. [2], p. 490). This problem was also vividly discussed at the Copenhagen meeting that took place from 8 to 15 April (see letter [218]). In the published version [40] of the memorable lecture held at the Zurich Society of Natural Scientists in the evening of 21 January 1957, the day Pauli had received the first news of parity violation, he writes: 'The continuous energy spectrum of beta rays, discovered by J. Chadwick in 1914 [42a], immediately raised difficult problems of theoretical interpretation. Was it to be ascribed directly to the primary electrons emitted by the radioactive nucleus, or to secondary processes? The first view, which turned out to be the correct one, was advocated by C. D. Ellis [42b], the second by L. Meitner [42c]. The latter appealed to the fact, known from alpha and gamma-rays, that nuclei possess discrete energy levels' (Ref. [40], p. 194).

Lise Meitner had begun her investigations of β-decay with Otto Hahn (1879–1968) in Berlin in October 1907. She had come from Vienna where she grew up in an intellectual Jewish family of nine children and studied theoretical physics with Boltzmann whom she revered. After her doctorate under Franz Exner in 1906 she worked on radioactivity. But in the aftermath of Boltzmann's suicide in September 1906 Meitner decided to go to Berlin to study under Max Planck, and she also started her collaboration with Hahn who in 1906 had come from Montreal, Canada, where he had worked with Rutherford (see page 85). Hahn and Meitner's working hypothesis was that, in analogy with α-decay, pure radioactive substances emit unique lines of β-rays (see Ref. [23b], Section 8 (d)). In 1913 James Chadwick had come on a studentship from Rutherford's institute in Manchester to work with Geiger at the Physikalisch Technische Reichsanstalt in Berlin. Using counters instead of photographic plates Chadwick found that the β-spectrum of RaB and RaC (Pb^{214} and Bi^{214}, respectively, see e.g. the reference given in *Note* [43]), was continuous, with some lines due to internal conversion of γ-rays into extra-nuclear electrons. During his experiments war broke out and Chadwick, together with Charles Drummond Ellis (1895–1980), a British Army officer, were interned at Ruhleben near Berlin until the end of the war, where he could continue his research activity at a reduced pace (see Ref. [23b], Section 8(h)).

Under Chadwick's influence at Ruhleben, Ellis gave up his plans for a military career and in 1919 became Rutherford's student in Cambridge. In 1925 Ellis and William Alfred Wooster (b. 1903) started their ambitious calorimetric experiment of measuring the total energy released in a β-decay, which took them two years [42d]. They argued that primary electrons, which are *directly* emitted by the nuclei, should yield the average over the continuous energy spectrum, while secondary electrons acquiring the continuous spectrum only in secondary processes *outside* the nuclei, should yield the maximum energy. Ellis and Wooster chose RaE [43] for their experiment because the absence of a line spectrum avoids the possibility of mistaking secondary electrons for electrons from internal conversion (see Ref. [23b], Section 14(c)). The result, which was confirmed with improved precision by Meitner and Orthmann, whose paper was submitted on 18 December 1929 [42e], clearly was in favour of a *primary* spectrum.

On 25 April 1929 Pauli had reported to Bohr on a physicists' meeting in Zurich [44] where Lise Meitner also participated: 'I again have spoken with her about the nuclear β-rays and their continuous velocity spectrum, and it turned out that in Copenhagen we have overlooked an essential point concerning the heat experiment of Ellis and Wooster: With these experiments the calorimeter was constructed such that it is pervious to γ-rays; hence with this experiment, only the heat developed by the β-rays is measured, that due to possibly existent γ-rays would not be included (so says Frl. Meitner).' And Pauli continued: 'A great difficulty for a theory which holds fast to the energy principle, however, are the experiments on ThC [45] which seem to show the absence of a γ-radiation. I personally continue to believe that there *still* are [is] some. The heat experiment which before we dreaded so much therefore does *not at all* speak against it!' (translated from postcard [219] to Bohr of 25 April 1929,

Ref. [2], pp. 495–6). After the Zurich week of lectures (*Vortragswoche*) which took place at ETH from 1 to 4 July 1929 (see letters [221–6, 229–31] in Ref. [2]) Pauli again wrote to Bohr: 'In Zurich Frl. Meitner has presented us a nice lecture on the experimental side of the question and she has *almost* convinced me that one can *not* explain the continuous β-spectrum by secondary processes (γ-ray emission etc.). Hence we really do *not* know what happens here! Neither do you, you are able only to indicate reasons for *why* we understand nothing' (translated from letter [231] of 17 July 1929, Ref. [2], p. 513).

In their paper [42e] Meitner and Orthmann indeed reported experiments with counters ('*Geiger–Müllersches Zählrohr*') on the γ-radiation of RaE. They concluded the paper as follows: 'Herewith it seems to be quite certain that Radium E does not possess any continuous γ-radiation that could form an energy compensation for the inhomogeneity of the primary β-rays' (translated from Ref. [42e], p. 154). In his Zurich lecture of 1957 Pauli says: 'According to these experimental results only two theoretical possibilities remained for the *interpretation of the continuous beta spectrum.*

1) Energy conservation holds only statistically in those interactions which are responsible for beta-radioactivity.

2) The energy law is strictly valid for each primary individual process, but in this process another very penetrating radiation, consisting of *new neutral particles*, is emitted along with the electrons.

The first possibility was advocated by Bohr the second by myself' (Ref. [40], p. 196).

Doubts about energy conservation of course were not new, as Pais in Section 14(d) of Ref. [23b] exhaustively shows. But in his letters to Copenhagen in Spring 1929 Pauli inveighed against Bohr's 'mishandling of the energy theorem' (letter [217], Ref. [2], p. 493). Bohr, on the other hand, had good reason not to trust quantum theory when applied to nuclei since before the discovery of the neutron by Chadwick in January 1932, the view that nuclei were formed by protons and electrons gave rise to the following serious difficulty, known as the 'wrong' statistics.

Calling the numbers of protons and electrons in the nucleus N_p and N_e, respectively, its charge number, i.e. the atomic number, $Z = N_p - N_e$, while its mass number $A = N_p$. Since both electrons and protons, by then were recognized as being fermions, the total number of fermions in the nucleus is $N_f = N_p + N_e = 2A - Z$, so 'that the parity of the charge number should determine the symmetry character of the nucleus'. The statistics of the nucleus now was determined by the following 'alternation law' (*Wechselsatz*): 'Composite particles are fermions or bosons according as the number of fermions they contain is odd or even' (Ref. [40], p. 197, also footnote 13, p. 198). This law, however, sometimes led to the 'wrong' statistics. 'The first counter example was the "nitrogen anomaly" as we called it at the time; for it was shown by R. Kronig [46a] and by W. Heitler and G. Herzberg [46b] from band spectra that nitrogen with charge number 7 and mass number 14 has spin 1 and Bose statistics. Other analogous cases followed later, such as Li 6 (charge 3, mass

6) and the deuteron (charge 1, mass 2), both likewise having spin 1 and Bose statistics. The final result was that it is the parity of the mass number, not that of the charge number, that determines the symmetry character of the nuclei' (Ref. [40], p. 197).

Pauli's diploma student Paul Güttinger (b. 1908) also studied the case of Li^7, not on the basis of band spectra but of the hyperfine splitting predicted by Pauli in 1924 (see pages 118–9). In his second paper published with Pauli [46c], the Li^6 is also discussed at the end. In 1932 Güttinger became Pauli's second Ph.D. student with a thesis on scattering processes of higher order [46d] (see Pauli's report I.54 in Ref. [27] and p. 435); the first, in fact was Rudolf Peierls (see page 156). Pauli had a good relationship with Güttinger [46e]. Already in his diploma paper of 1931 the latter, on Pauli's suggestion, had treated the problem of atoms in alternating magnetic fields (see Ref. [27], p. 434). More specifically, he had calculated the effect on a fixed magnetic moment perpendicular to an axis, about which a magnetic field is rotating [46f]. This problem probably had been mentioned to Pauli by Rabi in relation with the latter's molecular beam experiments. Indeed, Rabi later quoted this paper by Güttinger [46g]. But in spite of a quite positive appreciation of his thesis by Pauli, Güttinger did not pursue a scientific career but went to work with the Swiss life insurance company Patria in Basle (see document I.65 in Ref. [27], also p. 449). On 6 February 1940 Güttinger sent Pauli greetings from Basle with a reprint, saying that 'the love for physics has by no means left me' (translated from Ref [39c], letter [591]).

Bohr's idea of only statistical energy conservation in β-decay was an expression of his expectation that the step from the physics of the atoms to that of the nuclei may involve the sacrifice of some fundamental principle, namely energy conservation, in analogy to the step from classical to atomic physics, in which the principle of space–time causality had to be given up. Bohr expressed this idea in his Faraday Lecture [47a] on 29 August 1931 in London (see Ref. [39a], footnote a, p. 156) as follows: 'At the present stage of atomic theory, however, we may say that we have no argument, either empirical or theoretical, for upholding the energy principle in the case of β-ray disintegrations, and are even led to complications and difficulties in trying to do so. . . . Still, just as the account of those aspects of atomic constitution essential for the explanation of the ordinary physical and chemical properties of matter implies a renunciation of the classical idea of causality, the features of atomic stability, still deeper-lying, responsible for the existence and the properties of atomic nuclei, may force us to renounce the very idea of energy balance' (quoted from Ref. [40], p. 199). The similarity of this attitude with the one Bohr had in the unlucky Bohr–Kramers–Slater paper (see page 158) is too obvious to be ignored. There it was the discrepancy between the classical, continuous radiation and the quantum, discrete atomic energy levels that was Bohr's motivation for giving up strict energy conservation. In both cases, however, the remedy adopted—at least unconsciously—was the same [47b].

The first written, though indirect, echo of Pauli's hypothesis 2) above seems to be contained in Heisenberg's letter to Pauli of 1 December 1930 where we read: 'As to your neutrons I would like to observe: There should also be nuclei with even electron number but "wrong" statistics; such ones have as yet never been observed. But perhaps the experimental material is insufficient' (translated from letter [258], Ref. [39a], p. 37; see also [48a]). Three days later, on 4 December 1930, Pauli takes the unusual step of writing an 'Open Letter to the Group of Radioactives at the Meeting of the Regional Society in Tübingen', 'at which Geiger and L. Meitner were present' (Ref. [40], pp. 197–8). It reads:

> *Dear Radioactive Ladies and Gentlemen,*
>
> *As the bearer of these lines, for whom I pray the favour of a hearing will explain in more detail, I have, in connection with the 'wrong' statistics of the N and Li6-nuclei as well as the continuous β-spectrum, hit upon a desperate remedy for rescuing the 'alternation law' . . . of statistics and the energy law. This is the possibility that there might exist in the nuclei electrically neutral particles, which I shall call neutrons, which have spin 1/2, obey the exclusion principle and moreover differ from light quanta in not travelling with the velocity of light. The mass of the neutrons would have to be of the same order as the electronic mass and in any case not greater than 0.01 proton masses.— The continuous β-spectrum would then be understandable on the assumption that in β-decay, along with the electron a neutron is emitted as well, in such a way that the sum of the energies of neutron and electron is constant.*
>
> *There is now the further question of what forces act on the neutrons. The most likely model for the neutron seems to me, on wave-mechanical grounds (the bearer of these lines knows more about this) to be that the stationary neutron is a magnetic dipole with a certain moment μ. The experiments of course require that the ionising action of such a neutron cannot be greater than that of a γ-ray, and then μ could very likely not be greater than $e \times 10^{-13}$ cm.*
>
> *I do not in the meantime trust myself to publish anything about this idea, and in the first place turn confidently to you, dear radioactive folk, with the question—how would things stand with regard to the experimental detection of such a neutron if it possessed an equal or perhaps ten times greater penetrating power than a γ-ray?*
>
> *I admit that my remedy may perhaps appear unlikely from the start, since one probably would long ago have seen the neutrons if they existed. But 'nothing venture, nothing win', and the gravity of the situation with regard to the continuous β-spectrum is illuminated by a pronouncement of my respected predecessor in office, Herr Debye, who recently said to me in Brussels 'Oh, it's best not to think about it at all—like the new taxes'. One ought therefore to discuss seriously every avenue of rescue.—So, dear radioactive folk, put it to the test and judge.—Unfortunately I cannot appear personally in Tübingen, since on account of a dance which takes place in Zürich on the night of 6 to 7 December I cannot get away from here.*
>
> *With many greetings to you, likewise also to Herr Back.*
>
> *Your most humble servant,*   *W. Pauli*

The dance in Zurich that Pauli so urgently had to attend, most probably, was the Ball of the Studenti Italiani, the Italian students of Zurich, which took place on the night of Saturday, 6/7 December, at *Baur au Lac*, the most distinguished hotel in downtown Zurich, facing the lake and bordering the outflow of Schanzengraben which, together with the Limmat, makes the Bahnhofstrasse district an island surrounding the elevation of Lindenhof (see the loving description of this part of Zurich by the Zurich writer Kurt Guggenheim (1896–1983) in Ref. [48b]). This dance was advertised by notices of increasing size in the *Neue Zürcher Zeitung* (Morgenausgabe, Blatt 2) of 4, 5, and 6 December 1930, leaving no doubt that it was one of the main social events of the winter 1930/31. Pauli's urgency sheds light on his desire to get over his divorce only eight days earlier as quickly as possible.

In footnote 12 of Ref. [40] Pauli expresses his great indebtedness 'to L. Meitner for letting me have a copy of this letter, which she had kept'. While 'the bearer of these lines' is unknown (see Ref. [39a], footnote *b*, p. 40), it is obvious that the main addressee was Lise Meitner who unquestionably was the foremost expert at the Tübingen meeting. But a more hidden motive for this was the very personal sympathy between Pauli and Meitner which is subtly expressed in Meitner's letter from Stockholm of 22 June 1959 to Pauli's widow Franca already quoted on page 25. There Lise Meitner writes: 'I had been introduced to your husband already before 1921 [probably at, or even before, the Nauheim meeting of September 1920, see page 37] and, in spite of the large difference in age and my purely experimental attitude, we always also had a strong human contact, besides the occasionally parallel-running physical interests. The true basis of our relationships never appeared quite clearly to me but they were always dear and precious to me and remain so in the recollection' (letter in PLC, *Pauli Archive*). This letter reflects Meitner's warm humanity which, in spite of the grave difficulties and deceptions she met in her impressive career, she never lost. These difficulties sprang from her triple handicap of being a woman, a Jew (though she had converted to Protestantism in 1908), and from the interdisciplinary nature of her collaboration with Hahn.

Hahn directed a section of the newly founded Kaiser-Wilhelm Institute for Chemistry in Berlin where he welcomed Meitner as a guest in 1912, the year she also became an assistant to Planck for three years. Hahn and Meitner's first interdisciplinary discovery was the element protactinium (Pa, $Z = 91$) in 1917. But Meitner also worked in pure physics, verifying with Geiger–Müller counters the Klein–Nishina formula describing the Compton effect (see Ref. [39a], pp. 17–18). The discrepancy revealed by these measurements gave rise to a letter of Pauli, who on 1 August 1930 wrote Meitner: 'The surplus of the scattered radiation over that calculated from the Klein–Nishina formula therefore most likely is a genuine nuclear effect' (translated from letter [248], Ref. [39a], p. 20). And he invokes Gamow's argument, that heavier nuclei may contain increasing numbers of α-particles, to argue that 'free' electrons, not built into the latter, might contribute to the scattering.

In 1922 Meitner became a Privatdozent for physics and in 1926 was given the title of professor. Like Pauli she lost her Austrian citizenship in the spring of 1938 when her beloved country was annexed by the Nazis. At that very moment the experiments of Hahn, Meitner, and Hahn's pupil Fritz Strassmann (1902–80) on neutron irradiation of uranium (U, $Z = 92$) led to crucial results. But in summer Meitner's stay in Berlin became too dangerous. On the morning of Wednesday, 13 July 1938, Paul Rosbaud (1896–1963), the scientific adviser of the publishing house of Julius Springer, drove her to the train station where Dirk Coster (see page 88) waited for her in the train to Groningen, from where, thanks to Bohr's intervention, she went via Copenhagen to the Nobel Institute in Stockholm (see Ref. [49a], Box 1: 'Meitner's Flight from Nazi Germany', an epitome from Ref. [49b]). But Meitner's collaboration with Hahn continued by correspondence. In a crucial meeting in Copenhagen in November 1938 Meitner urged Hahn to verify the latest findings. A few weeks later Hahn and Strassmann identified an isotope of barium (Ba, $Z = 56$). It was Paul Rosbaud who pushed for immediate publication in *Naturwissenschaften* because he foresaw the catastrophic development in Germany and wanted this important news to be known outside of the Third Reich (see the Preface in Ref. [41], p. 2). The paper was published on 6 January 1939 [49c]. Lise Meitner was not a co-author; by now she was a non-person in Nazi Germany. But the explanation of this result was given by Meitner and her nephew Otto Frisch in Stockholm as a break-up—which they called 'fission'—of the uranium nucleus considered as a liquid drop. And they also estimated the release of the enormous energy of 200 MeV. They submitted these results to *Nature* on 16 January 1939 [49d].

Otto Hahn was awarded the Nobel Prize in Chemistry of 1944 in 1945 'for his discovery of the fission of heavy nuclei'. The sad fact is that neither Meitner nor Frisch, nor Strassmann, were ever honoured with a Nobel Prize (see Ref. [49a]). But Lise Meitner received many other important honours, in post-war Germany the *Ordre pour le mérite* and, together with Hahn, the Max-Planck Medal in 1949; and in the United States she shared with Hahn and Strassmann the Enrico Fermi Award of the United States Atomic Energy Commission. In 1960, at age 82, Lise Meitner moved to Cambridge, England, to be near her nephew Otto Frisch and his family. There she died eight years later (see also Ref. [49e]).

The response to Pauli's 'Open Letter' soon came from Geiger who reported that he had discussed Pauli's 'question with the others in Tübingen as well, especially with L. Meitner.' Although, unfortunately, Geiger's letter is lost, Pauli remembered 'that his [Geiger's] reply was positive and encouraging: from the experimental point of view· my new particles were quite possible' (Ref. [40], p. 199). Pauli's proposal was indeed quite convincing since, on the one hand he said 'that there might exist in the nuclei electrically neutral particles which I shall call neutrons' and, on the other, 'that in β-decay, along with the electron a neutron is emitted as well' (see Ref. [40], p. 198). The first statement means that, in analogy with the electrons in the atom, Pauli-neutrons—the neutrinos—exist in the nucleus alongside protons and electrons and are held together by some force. This means that electrons and Pauli-neutrons are not *created* in the β-decay, which is quite different from quantum field theory where particles are created and annihilated. The second statement means that in the nucleus electrons and Pauli-neutrons must occur in pairs. Since Pauli assumed his neutrons to be fermions, the total number of fermions in the nucleus is thus $N_f = N_p + 2N_e = 3A - 2Z$. Therefore the parity of the mass number now determines the symmetry character of the nuclei, in agreement with the results for N^{14}, Li^6 and H^2 mentioned above (see, however, *Note* [50]). Together with the explanation of the continuous spectrum of β-decay Pauli's proposal therefore was quite impressive. It was not a complete success, however, because the charge number was still $Z = N_p - N_e$ instead of the relation $Z = N_p$ which holds in a model of nuclei consisting of protons and Chadwick-neutrons and of a field-theoretic β-decay mechanism.

As to the forces holding his neutrons in the nucleus, Pauli in his Tübingen letter saw as 'most likely model ... that the stationary neutron is a magnetic dipole with a certain moment μ'. This force is provided by the inhomogeneous magnetic field H in the direction of μ, i.e. $\mathbf{F} = -\mu \nabla H$. Since according to Pauli the mass of his neutron should be 'not greater than 0.01 proton masses', i.e. $m_\nu \leq 0.01 m_p$, the magneton (see page 57) corresponding to it would be $\mu_\nu = e\hbar/2m_\nu c \geq e \times 10^{-12}$cm. But since 'the ionizing action of such a neutron cannot be greater than that of a γ-ray ... μ could very likely not be greater than $e \times 10^{-13}$cm', or $\mu \leq 0.1\mu_\nu$ (see letter quoted above). In a letter to Oskar Klein of 12 December 1930, in which all the arguments of the Tübingen letter reappear, Pauli gave an explicit form of the equation he thought his neutron should satisfy in an external electromagnetic field $F_{\mu\nu}$ given by Eq. (6.24), namely the Dirac equation (6.34), in which instead of the vanishing charge coupling there is a magnetic-moment coupling:

$$\gamma^\mu \frac{\partial \psi}{\partial x^\mu} - \frac{\mu}{\hbar}\gamma^\mu\gamma^\nu F_{\mu\nu}\psi + k_c\psi = 0 \qquad (6.62)$$

(see Ref. [39a], letter [261], p. 45). Such a Lorentz-invariant magnetic-moment term may always be added to the Dirac equation (6.34) and, quite generally, describes an anomalous magnetic moment. Pauli introduced it in print only 10 years later in a review of relativistic field theories [51a] where it appears as Eq. (91). This review, like Ref. [24b], was originally prepared for the Solvay Congress of 1939 which, however, because of the outbreak of World War II, had been cancelled. The additional magnetic-moment term in this equation is known today as the *Pauli term*.

In his lecture of January 1957 Pauli comments on the above comparison of the ionizing action of his particle with that of γ-rays as follows: 'Actually the penetrating power of the particles nowadays called neutrinos is about 100 light-years Pb instead of 10 cm, the factor as compared with gamma-rays 10^{16} or 10^{17} instead of 10, while the rest mass and the magnetic moment are theoretically 0, and the experimental upper limits are 0.002 electron masses and 10^{-9} Bohr magnetons' [51b] (Ref. [40], pp. 198–9). An early argument in favour of a vanishing neutrino mass m_ν was forwarded by Francis Perrin (1901–92) in the session of the Paris Academy on 18 December 1933. He concluded from the kinematics of the emission of an electron and a neutrino that equating the resulting electron energy to the observed one, interpreted as an *average* value, is only possible with $m_\nu = 0$. Perrin also argued that with a vanishing mass the neutrino would not have to be pre-existent in the nucleus [52a]. However, as will be discussed later, even this last statement was not the final word because the mass of the neutrino has remained a controversial subject to this day.

More fundamental is the comment that follows in Ref. [40]: 'I had soon given up again the idea that the neutral particles emitted in beta-decay are at the same time constituents of the nuclei, on account of the empirical values of the nuclear masses.' This important realization freed the way for a field-theoretic description of β-decay as creation and annihilation processes which was soon taken up by Fermi, as I shall describe in a moment. But, on the other hand, it also withdrew the basis for the explanantion of the 'wrong' statistics discussed above. Therefore Pauli's doubts lingered on: 'In a lecture in Pasadena in June 1931 on the occasion of a meeting of the American Physical Society, I gave the first public report on my idea of new highly penetrating neutral particles in beta decay. I no longer regarded them as building stones of the nucleus, and for this reason I no longer called them neutrons on this occasion, but used no special name for them. However, the matter still seemed to me very uncertain, and I did not have my lecture printed' (Ref. [40], p. 199).

This reminds us that Pauli was also an accomplished traveller, although compared with Heisenberg's world tour of 1929 he started relatively modestly with a trip to Odessa in August 1930. On that occasion invitations were extended to many Western physicists by their Soviet colleagues for the 7th Physicists' Congress since the Revolution which at the same time was the first All-Union Congress, in which Sommerfeld and Pauli's assistant Peierls also took part. With these invitations to their congresses, as well as by sending promising young researchers to the West, Soviet scientists strove to overcome their isolation caused by the Revolution (see Ref. [39a], pp. 20–3). This congress, which lasted from 19 to 24 August 1930, is described on pp. 57–8 of Ref. [52b]. Peierls wrote in his recollections: 'The main attraction and adventure of the summer of 1930 turned out to be a conference in Odessa.... This was a national conference, with only a small number of foreign visitors, so it was an honour to be invited. I probably owed the invitation to Ya. I. Frenkel, the Leningrad theoretician who had seen some of my papers and was interested in my work, and perhaps to a recommendation by Pauli, who was also invited. The trip promised to be an adventure, because one knew very little about life in the Soviet Union' (Ref. [38b], p. 61).

Peierls continues: 'I set out by train for Odessa via Poland (of course by third class and of course overnight) and arranged to meet Pauli at the border. We had about an hour's wait and decided to have dinner. The manager of the station restaurant urged us to have a big meal, because, he assured us, this was going to be our last decent meal for some time. Then our train pulled slowly across the border, with one carriage and three passengers: Pauli and I and a Polish physicist who was bound for the same conference. The train was supposed to reach Odessa in the morning, but it did not arrive until the afternoon. There was no dining car and we became very hungry, thus partly confirming the prediction of the Polish restaurateur.' They arrived on 18 August, and Pauli and Tamm gave talks on the 20th (Ref. [52c], pp. 147–8). Peierls found the conference 'lively and interesting', and he met Frenkel and Tamm, 'one of the most charming personalities in physics. He had an agile mind, and an equally agile body, and the first impression he gave was of never standing still.' However: 'A good deal of our time was spent at Luzanovka, the beach in Odessa' (Ref. [38b], pp. 62, 63), where Peierls took a picture of Pauli with Frenkel and Tamm in bathing suits (see Ref. [38b], pp. 178–9). This congress, in fact 'was a grandiose one: 800 participants, 200 talks, plenary sessions in the City Council building. The opening session was broadcast on radio, and the participants got to ride the trams free. And of course, there were Odessa's splendid Black Sea beaches' [52d].

Peierls continues: 'After the conference, all participants were taken by boat across the Black Sea', and he 'was sharing a cabin with Pauli' (Ref. [38b], p. 63). From the boat Pauli, Sommerfeld and an unidentified Russian sent a postcard to Ehrenfest: 'Greetings from the 'Grusia' [Gruzija] to the friend of Russian physics [in Cyrillic]. The Pauli effects so far have been quite harmless. Many greetings from God's whip. Sommerfeld' (translated from postcard [249], dated 'Grusia', 25 August 1930, Ref. [39a], p. 23). The boat called at Yalta on the Crimean peninsula and then sailed on to Batumi in Georgia, from where Pauli travelled through Turkey to Istanbul and on to Vienna, returning to Zurich only on 26 September (see Ref. [39a], postcard [250] to Jaffé and letter [251] to Coster, see page 111, written in Vienna on 18 and 19 September, respectively). To be more precise, from Batumi Pauli went to Tbilisi, the capital of Georgia, situated in the foothills of the Caucasus. He was accompanied by the experimental physicist Fritz Houtermans from Berlin and his girl friend from their student days in Göttingen, Charlotte Riefenstahl, herself a physicist. Fritz and Charlotte decided to get married in Tbilisi and asked Pauli to be the best man [52d]. Thus started a lifelong friendship between Pauli and Houtermans to which we will have occasion to come back repeatedly.

From 20 to 25 October 1930 the Sixth Solvay Conference took place in Brussels, the second Pauli attended. Its theme was Magnetism, and Pauli was among the invited speakers. In his lecture, 'The quantum theories of magnetism. The magnetic electron' [53a], Pauli gave an extended review of all the known aspects of magnetism. The first part, 'Theory of the magnetic properties of the solid bodies', again shows Pauli's total dominance of, but also his genuine motivation for, this complicated subject. Of particular interest here are the theories of orbital diamagnetism (Section 3) and of ferromagnetism (Sections 4 and 5).

Pauli started Section 3 with the observation that classical statistical mechanics gives zero orbital magnetism, the argument being identical to that described on page 59 where Pauli showed that 'elementary Ampère currents' cannot yield any paramagnetism, and where he interpreted the result as a cancellation due to reflections of the electrons on the walls of the container. These walls are also the cause of the difficulties in Landau's calculation which Pauli carefully discussed. As indicated in the references, Pauli's presentation was based on oral communication by Landau since the latter had sent his paper [53b] for publication from Cambridge, just before coming to Zurich, and Pauli could see it in print only after returning from the Solvay conference. Landau's result for the diamagnetic susceptibility, as already mentioned page 154, is one-third of that of the spin paramagnetism. But in addition Pauli showed that this not only holds for Fermi statistics but also for Boltzmann statistics.

In Section 4 Pauli gave a critical perturbative derivation of the interaction energy between the electrons which is a generalization of the Heitler–London procedure to a large number of electrons. In Section 5a he presented in the form of 'Weiss's molecular field, Heisenberg's application to ferromagnetism. Section 5b, in which a closer examination of the vicinity of saturation is made, probably contains the first mention of the fact that the ferromagnetic (as any other) *phase transition* depends on dimensionality. Indeed, Pauli states 'that the three-dimensionality of the lattice is necessary for the appearance of ferro-magnetism'; and he makes the connection with the Ising model (see page 148): 'There is in fact a very close relationship between the problem of Ising and the one we just have treated.' And in spite of the fact that in the latter the spin components do not commute 'it is quite likely that an extension of the theory of Ising to the case of a lattice of three dimensions would yield ferromagnetism even from the classical point of view.' (translated from Ref. [53a], pp. 209, 210). This prediction of Pauli was confirmed by Lars Onsager (1903–1976) who in 1944 showed that the Ising model already has a phase transition at finite temperature in two dimensions [53c].

In the second part, 'Relativistic quantum mechanics of the electron', Pauli examined in Section 1 various experimental set-ups to test Bohr's conjecture 'that it is impossible to establish the spin moments of free electrons and protons by experiments based on the usual concept of the particle motions (notion of trajectory in classical mechanics or rays in geometrical optics)' (translated from Ref. [53a], p. 217). In Section 3 he discussed the possibilities of obtaining polarized electron beams, and Section 4, finally, is devoted to the discussion of negative-energy states of the Dirac equation which, as we shall see later, intrigued Pauli enormously.

Spring 1931 was marked by short visits to Copenhagen and to Berlin, perhaps to meet his divorced wife (see Ref. [39a], p. 62 and letters [270] and [274]) and by the *Vortragswoche* (week of lectures) on nuclear physics held at ETH from 20 to 24 May. On this occasion Hermann Schüler from Potsdam, aided by Güttinger, gave a report on 'Hyperfine structures and nuclear moments' [54a], in which he described his experiments and their interpretation based on the ideas proposed by Pauli in 1924 (see page 118). This presentation was followed by discussion remarks from Pauli, in which he repeated the basic assumptions, namely the existence of a nuclear spin and the magnetic nature of its interaction, and discussed the cases that presented difficulties (Ref. [54a], pp. 669–70).

In summer it was Pauli's turn to discover America. On 15 April 1931 he asked the president of ETH for leave of absence starting on 24 May, to participate at a physics meeting in Pasadena, California, and subsequently to follow an invitation by the University of Michigan to lecture at the summer school in Ann Arbor (Ref. [27], document I.40). This 'belated request' was granted (Ref. [27], document I.43). Through Sommerfeld Pauli had become acquainted with an American couple, Walter Francis Colby (1880–1970) and his wife Martha, from Ann Arbor where Colby had been a physics professor at the University of Michigan since 1915. He had been in Vienna from 1902 to 1904 and in Munich from 1912 to 1913 [54b]. He and Harrison McAllister Randall (1870–1969) had since 1923, organized yearly summer schools in theoretical physics with non-resident lecturers. Randall had been a physics professor at the University of Michigan since 1906 and from 1917 to his retirement in 1940 was director of the laboratory there. He had worked in Tübingen with Friedrich Paschen in 1909–10 [54b]. The arrival of Goudsmit and Uhlenbeck in 1927 (see page 116) further stimulated the contacts with European guests (see Ref. [39a], p. 13 and footnote *c*, p. 19, also Ref. [23b], p. 366).

On 10 June 1930 Pauli sent a postcard to his acquaintance of many years, the philosopher Moritz Schlick (1882–1936) (see [55]): 'The world really is small: I am here on the lake of Como with acquaintances of yours who have seduced me to travel to Ann Arbor next summer (1931). Are you in California at that time? Maybe I'll see you there!', to which there is a postscript 'Cordial greetings from the seducer. W. Colby and Martha Colby' (translated from Ref. [39a], postcard [246]). In the letter quoted at the outset of this section Sommerfeld continues: 'Colby told me that you have asked him in earnest whether your company really was agreeable to me. How can you doubt this? I only have to remind you that in the previous summer I had mentioned you to Laporte in the very first place. I am convinced that we will have a very nice time together. I also promise not to want to be too "severe" with you' (translated from Ref. [39a], letter [247a]). This reflects the still somewhat submissive attitude of Pauli towards his teacher described on page 55.

The rumour of Pauli's visit to Ann Arbor reached R. A. Millikan at Caltech in Pasadena, the author of the precision measurement of the elementary electric charge (see page 16) and of the spectroscopic work mentioned on page 109. He wrote Pauli on 7 February 1931: 'The first national summer meeting of the American Association for the Advancement of Science is to be held in Los Angeles and Pasadena from June 15th to June 22. I am informed that you and Dr Sommerfeld are to be at Michigan this summer. The officers of the Association and the local committees very much hope that this means that you can be present during this meeting before your Ann Arbor commitment, and take part in a symposium on the present status of the problem of the structure of the nucleus' (Ref. [39a], letter [266]). This meeting was held in conjunction with the American Physical Society. Pauli answered by postcard on 11 March and on 4 May cabled Millikan the title of his talk: 'Problems of Hyperfine

Structure' (Ref. [39a], postcard [267] and telegram [276]). Since Pauli wanted 'absolutely' to talk to Goudsmit about this subject (Ref. [39a], postcard [272]), the two arranged to meet in Chicago and travel together to Pasadena on June 9 or 10 by rail (Ref. [39a], telegrams [277] and [278], also pp. 81–2).

Pauli arrived in New York on 6 June by boat and was met there by Gregory Breit who now was at New York University (see Ref. [39a], p. 81). Pauli's announced talk in Pasadena, followed by a similar one by Goudsmit on 'Hyperfine Structure and Nuclear Magnetism' were given at the symposium on 'The Present Status of the Problem of Nuclear Structure' [56a]. In his lecture Pauli also mentioned his neutral particle publicly for the first time, as stated before and as he recalled in the last letter he wrote to his friend Max Delbrück (1906–81) on 6 October 1958. In this letter Pauli continued: 'Unfortunately, I may say, I had to live in the home of Millikan for some while ... he wanted implicitly to convince the visitor that he would every morning transform the blood of Christ into orange juice (and on Sundays, in the afternoon, too). I was glad, when I moved out into the Athenaeum [the Faculty Club of Caltech] where Epstein ... often fetched me for some excursion. In his home the atmosphere was very different, namely with some Freudian touch. Complementary to the Millikan-house, but causally related to it, by the prohibition, were my experiences in the speakeasies of Chicago at the same trip. But this I have to omit' (quoted in Ref. [39a], p. 60). Pasadena is situated on the periphery of the Los Angeles metropolitan area some 13 km north-east of the centre, half-way towards the famous Mount Wilson astronomical observatory in the San Gabriel Mountains.

Prohibition in the United States was a nationwide effort to stop people by law from drinking alcoholic beverages. Promoted since the 1870s, mainly by hard-working, non-drinking, church-going white Protestants who considered the large cities as citadels of sin, prohibition was enforced in 1920 by the 18th Amendment to the United States Constitution which forbade the manufacture or sale of any drink with more than 0.5% alcohol and lasted until 1933 when it was repealed by the 21st Amendment. But, of course, illegal drinking could not be completely prohibited. In bars where such activity took place, customers and vendors usually talked in hushed tones, i.e. easily, about the orders—these were the 'speakeasies'. The Great Depression also occurred in this period, caused by overproduction, falling prices and wages, but mounting stock prices which triggered the crash of Wall Street, the New York Stock Exchange, on 24 October 1929 and led to the world-wide economic crisis which in the United States hit bottom in early 1932 and in general lasted until 1933.

Towards the end of June Pauli, accompanied by Oppenheimer and probably by other physicists too (Ref. [39a], p. 83), set out for Ann Arbor, situated 50 km west of the Canadian border at Detroit, exchanging the temperate climate of the west coast for the summer heat of the east. In the letter to his assistant Peierls quoted on page 157 Pauli wrote: 'In spite of the opportunity for swimming I here suffer much from the great heat. But under the "dryness" I don't suffer at all because Laporte and Uhlenbeck are superbly stored up in alcohol (one notices the vicinity of the Canadian border). Physics (and physicists) there is (are) very much (many) here but I find them too formal.—In Chicago, Detroit and New York I surely will experience still more things amusing. So far I had little time. Would have liked to stay longer in California. Greetings to you, Scherrer and the entire institute' (translated from letter [279] of 1 July 1931, Ref. [39a], p. 87). Ann Arbor presented other contrasts as well. In retrospect Pauli wrote from New York to his friend Wentzel in Zurich: 'The small-bourgeois, philistine side of America I have also been acquainted with. At the dinner of Dr Kraus, dean of the summer school of Ann Arbor, there was a prayer (*Tischgebet*) instead of coffee and cigars—not to speak at all of alcohol. (The cigars Sommerfeld missed very much.)' (translated from letter [283a] of 7 September 1931, Ref. [39c], p. 752). Edward Henry Kraus (1875–1973) had been professor of mineralogy at the University of Michigan and director of the laboratory of mineralogy there since 1908. He had obtained his Ph.D. in mineralogy in Munich in 1901 [54b].

However, the effect of alcohol on Pauli was not entirely amusing. In a second letter to Peierls, which is erroneously given the same date as the first (see footnote a, Ref. [39a], p. 90), Pauli writes: 'Unfortunately (*dummerweise*) I recently fell (in a slightly tipsy state) so unfavourably over a stair, that I have broken my shoulder and now must lie in bed until my bones are whole again—very tedious' (translated from Ref. [39a], letter [280], p. 89). This happened in the house of his host Otto Laporte (1903–71) [54b], a friend from Pauli's student days in Munich (see page 54). It was now Pauli's turn to be the target of jokes—although the true reason for Pauli's handicap of course had to be concealed from the outside world; Sommerfeld called it an inverse Pauli effect (Ref. [39a], p. 84). Pauli lectured with his injured arm high up and Uhlenbeck wrote his formulae on the blackboard for him, which had the effect that Pauli now faced the audience and gave brillant lectures (Ref. [39a], footnote k, p. 88). According to Oppenheimer, 'Professor Pauli presented the considerations which led him to the introduction and definition of the magnetic neutron at a seminar on theoretical physics in Ann Arbor in the summer of 1931' (footnote 1 in Ref. [56b]). In this seminar Pauli also presented the modified Dirac equation (6.62) which was later 'rediscovered' and published by Elsasser [56c], as Pauli recalls in a letter to Sommerfeld on 12 February 1935 (Ref. [39a], letter [404]). When the summer school ended on 18 August, Pauli wrote to Goudsmit: 'Now the summer session has come to an end and I still have a month of vacations in America. My arm is much better, and I can already move it well. Hopefully, I will still have an amusing time. First I go to Chicago, later to New York. Where I shall be in-between I don't know yet. Also I am not quite sure yet whether I will come to Rome or not, but it is probable' (translated from Ref. [39a], postcard [282]).

After a colloquium organized in Princeton on 1 October, with Pauli and Dirac as main speakers, on Dirac's idea of 'magnetic-[mono]pole neutrons' [57], which are distinct from Pauli's 'magnetic dipoles' of Eq. (6.62) (see letter [283] to Peierls, Ref. [39a], p. 94), Pauli's American adventure came to an end. In the last letter just mentioned to his assistant Peierls he writes from New

Fig. 6.2 Pauli's broken shoulder, Ann Arbor, Michigan, 1931 (CERN)

York of 29 September: 'America pleases me well. On 2 October I travel from here directly to Naples where I arrive on 11th. My arm is long since whole again' (translated from Ref. [39a], p. 94). From Naples Pauli went to Rome where an international congress on nuclear physics (see [58a]) was already in full swing, entering the lecture room 'the very moment that [Goudsmit] mentioned his name'. The latter, in fact, in his talk on 'Present difficulties in the theory of hyperfine structure' was giving the first public description in Europe of Pauli's neutron (see Ref. [58b], p. 25, also Ref. [39a], p. 93 and Ref. [23b], p. 317). In Rome Pauli also met Fermi 'who at once showed lively interest in my idea, and a very positive attitude to my new neutral particles' (Ref. [40], p. 199). He also met Delbrück, to whom he recalled this congress in his already quoted letter of 6 October 1958: 'So I don't need to describe the Congress nor Bohr with his spiritual somersaults of an only statistical validity of the energy law for the β-decay—and this long after the establishment of wave mechanics. And on the other side Fermi. But this you will find again in my lecture [40]. You will not find, however, my remembrances on the dancing party in Rome with you, Gamow, Ellis, [Patrick Maynard Stuart] Blackett [1897–1974], etc and the ladies. This may go under in peace when we die. But for me personally the history of the neutrino is unseparably connected with your—very *unsuccessful*—flirtation with Eve Curie [b. 1904] at that party' (quoted in Ref. [39a], p. 92).

In his Zurich lecture of 1957 Pauli writes: 'The following year, 1932, Chadwick discovered the long-sought neutron of charge number 0 and mass number 1 [see Rutherford's prediction on page 85], by bombarding light nuclei with alpha particles. Fermi, in seminars in Rome, thereupon named my new particle, which is emitted in beta decay, *neutrino*, to distinguish it from the heavy neutron ... this Italian name was soon in general use' (Ref. [40], p. 200). Seven months after Chadwick's discovery (see Ref. [23b], Section 17(a)) Carl David Anderson (b. 1905), a student of Millikan's at Caltech, discovered in cosmic radiation—so named by Millikan in 1925—the anti-electron with positive charge, called the *positron* (see Ref. [23b], Section 15(f)). For this discovery Anderson was awarded the Nobel Prize of 1936, together with Victor Franz Hess (1883–1964), the discoverer of cosmic radiation. In his lecture Pauli sums up: 'Thus a general clarification followed at the Solvay Conference on atomic nuclei in Brussels in October 1933, where among other speakers ot and Chadwick reported on their experimental discoveries of positron decay and of the neutron, and Heisenberg reported on nuclear structure. Fermi and Bohr too were again present. It was now evident that on the basis of this idea of nuclear structure the neutrinos, as they were now called, must be fermions, in order that the statistics may be conserved in beta decay' (Ref. [40], p. 200).

Gradually the neutrino became a general discussion topic. At the conference commemorating the 10th anniversary of Bohr's institute, in Easter 1932, the well-known *Faust Parody* was put on by the young collaborators, in which the neutrino was represented as 'Gretchen', while Ehrenfest was 'Faust' and, of course, Pauli was 'Mephisto' (see Ref. [39a], pp. 36 and 109). At the Seventh Solvay Conference Pauli reported on his ideas about the neutrino in the discussion following Heisenberg's invited talk on nuclear structure. This report appeared in French in the conference proceedings (Ref. [59], pp. 324–5, see also Ref. [12], pp. 437–8 and Ref. [40], p. 201) and is the first printed document by Pauli on the neutrino. Apart from the description already given, this report contains some noteworthy arguments concerning energy conservation, which he had already advanced in letter [261] to Oskar Klein of 12 December 1930 containing Eq. (6.62): 'To start with, the electric charge is conserved in the process [of β-decay], and I do not see why the conservation of the charge should be more fundamental than that of the energy and of the momentum. ... If the conservation laws were not valid one would indeed have to conclude from these [energetic] relations that a β disintegration is always accompanied by a loss of energy, and never by a gain; this conclusion implies an irreversibility of the processes with respect to time which seems to me scarcely acceptable' (translated from Ref. [40], p. 201).

The formal analogy between energy-momentum conservation and charge conservation emphasized here was discussed in greater depth by Pauli during his second visit to the Soviet Union in 1937 (see also page 41). The occasion was an invitation to the second All-Union Congress held from 20 to 26 September in Moscow. While Pauli only contributed discussion remarks (see Ref. [52b], p. 58, where the month is erroneously given as December), he gave a most beautiful lecture on *The Conservation Laws in Relativity Theory and in Nuclear Physics* at the *Zelinskij University* in Moscow on 27 September 1937 [60]. A write-up of this lecture in Russian had been prepared by Igor Evgenievich Tamm (1895–1971), of which Ref. [60] is a German translation (see the footnote on p. 439 of Ref. [12], where the month is erroneously given as October).

In the first part of Ref. [60] Pauli gives a very elegant proof of the theorem first suggested by Christian Huygens (1629–95) that 'momentum conservation follows from the requirement that the energy theorem be valid in an arbitrary uniformly moving reference frame' (translated from Ref. [60], p. 441). In the second part, which concerns us here, Pauli compares the field equations expressing the fields by the sources, which in electrodynamics are given by Eqs (6.44) with $P_4 \equiv 0$ and in general relativity by Eqs (2.3). Integrating the fourth of these equations over a volume V one obtains, respectively, the total charge $e = \int_V \rho dV$, where ρ is the charge density defined in Eqs (6.32), and the total energy $E = \int_V T_4{}^4 dV$ where, for simplicity, we have put the 'energy of the gravitational field' $t_4{}^4 \simeq 0$ and $\sqrt{-g} \simeq 1$ in Eqs (2.6). Taking these relations as definitions of the charge and energy content of the volume V, we may now perform partial integrations, so that e and E are expressed by integrals over the surface S enclosing V. Making use again of the field equations, it is then possible to express the time variations de/dt and dE/dt as integrals over certain fluxes through S. In particular, $de/dt = - \oint_S i_n dS$, where i_n is the component normal to S of the current density i defined in Eqs (6.32). A similar but more complicated relation may be derived for dE/dt.

The point that Pauli now makes is the following: 'One may always choose a surface S so far away that the current density vanishes on it, then the conservation of the charge in the volume V confined by it *is proven quite independently from whether the Maxwell equations are applicable to the processes* occurring *in the interior* of this volume' (translated from Ref. [60], p. 445). Similarly Pauli states in the second case: 'Extremely essential for this proof is the sufficient assumption that the equations of relativity theory hold *at the surface S* of this volume; if this is the case, then the conservation of the total energy and of the momentum in the volume V is proven quite independently of the fact whether the equations of the theory are applicable in the interior of V.' And Pauli adds: 'This conclusion is due to Einstein and represents a great progress.' He then observes that in a five-dimensional formulation, 'The analogy between the theory of electricity and the theory of gravitation may be continued even further' (translated from Ref. [60], p. 446).

This beautiful argument shows most eloquently the fundamental difference in attitude between Pauli and Bohr: Pauli approached the new realm of the nucleus by adhering to the already established *laws* of energy conservation and of quantized angular momentum, making the necessary adjustments in the *objects*, namely by proposing as working hypotheses the neutrino and the nuclear spin. Pauli wrote to Heisenberg: 'With nuclear processes, even almost more important than the conservation laws of energy and momentum for me are the conservation laws of all the discretely quantized quantities', namely the total angular momentum and the symmetry character (FD or BE statistics). 'For the time being (*zunächst*) I want to hold fast unconditionally to these assumptions and to pursue them in their consequences, before I modify them.' (translated from letter [314] of 14 July 1933, Ref. [39a], p. 184). For Bohr, on the other hand, the new *object* of the nucleus was full of mystery and did not warrant the validity of the *laws* established outside of it; and 'no one knew what surprises might still be in store for us in this field' (Ref. [40], p. 202).

Thus ends the first act of the neutrino saga. As we shall discover in the sequel to this tale, Bohr's last quoted statement did harbour some truth. He admitted for the first time the possibility of the 'real existence of the neutrino' in his letter to Pauli of 15 March 1934 (translated from letter [366], Ref. [39a], p. 310). For now let us take a glimpse into the future by looking at Fermi's field-theoretic description of β-decay.

In the introduction to his paper [61c] Fermi argues against 'nuclear electrons' by saying that 'the present relativistic theories of the light particles (electrons or neutrinos) are not able to explain in an unobjectionable manner how such particles could be bound into orbits of nuclear dimensions'. Therefore, he says, it 'seems to be more expedient to assume with Heisenberg [62] that a nucleus consists only of heavy particles, protons and neutrons. In order to still understand the possibility of the β-emission we shall try to build up a theory of the emission of light particles from a nucleus in analogy with the theory of the emission of a light quantum from an excited atom in the ordinary radiation process' (translated from Ref. [61c], p. 161). Pauli was informed very early of Fermi's ideas by Felix Bloch, as he reports to Heisenberg on 7 January 1934 (Ref. [39a], letter [338]).

These ideas are implemented by the assumption 'that to each transition from neutron to proton is associated the creation of an electron and a neutrino' (translated from Ref. [61c], p. 162). This transition is described by a matrix $Q = (\sigma_x + \sigma_y)/2$ where σ_x and σ_y are spin matrices (5.37), introduced by Heisenberg in his first paper on nuclear structure [62] and which Wigner in 1937 called *isotopic spin* (see Ref. [23b], Section 17(f)). But rather than continue along this historic path, it is more useful to write down the general form of the β-interaction, as it will be used in later chapters. There are 16 independent 4×4 matrices which may be formed from 1 and the products of the Dirac matrices γ^μ, and which will be called $O_i, i = 1, \ldots, 16$. Then the process $n + \nu \rightarrow p + e$ is, in obvious notation, described by the *interaction Hamiltonian*

$$H' = \int d^3x \left\{ \sum_{i=1}^{16} g_i (\overline{\psi}_n O_i \psi_p)(\overline{\psi}_\nu O_i \psi_e) + \text{Hermitian conj.} \right\}. \quad (6.63)$$

Here the ψ and $\bar{\psi}$ are Dirac spinors satisfying Eqs (6.31) with $k_c = mc/\hbar$, and $m = m_n, m_p, m_\nu, m_e$ are the masses of the particles involved. The 16 terms in Eq. (6.63) have definite *tensor character*. Thus $i = 1$ is a *scalar* (S), $i = 2, \ldots, 5$ a *vector* (V), $i = 6, \ldots, 11$ a *skew tensor* (T), $i = 12, \ldots, 15$ an *axial vector* (A), and $i = 16$ a *pseudoscalar* (P).

The transition probability for the process of neutron decay $n \to pe\bar{\nu}$ into a proton, an electron and an (anti-)neutrino in the nucleus may be obtained in analogy to Eqs (6.60), (6.61). Formally we write the equation for the amplitude of the inverse (absorption) process $pe\bar{\nu} \to n$ in the form (6.1) as

$$i\hbar\dot{a}(t) = \sum_{\mathbf{pk}} \langle n|H'|m, \mathbf{p}, \mathbf{k}\rangle a_{\mathbf{pk}}(0)e^{iwt}. \tag{6.64}$$

Here $\mathbf{p}$ and $\mathbf{k}$ are the wavevectors of the electron and (anti-) neutrino, respectively, and $\hbar\omega = E_{m\mathbf{pk}} - E_n$, $m = m_e$. The random phase assumption (6.61) now reads

$$\langle a_{\mathbf{pk}}(0)a_{\mathbf{p'k'}}\rangle = \frac{1}{G}\delta_{\mathbf{pp'}}\delta_{\mathbf{kk'}}, \tag{6.65}$$

and the transition probability (6.60) is found to be

$$W(t) \equiv \langle |a(t)|^2\rangle = \frac{1}{G}\sum_{\mathbf{pk}}|\langle n|H'|m, \mathbf{p}, \mathbf{k}\rangle|^2 \frac{|ei\omega^{i\omega t} - 1|^2}{\hbar^2\omega^2}. \tag{6.66}$$

On defining the weight G by the transition from discrete to continuous wavevectors, the time derivative of Eq. (6.66) becomes, for large time t, and using the fact that the integral over the whole line $\int (\sin x/x)^2 dx = \pi$,

$$\dot{W} = \frac{2\pi}{\hbar^2}\int d^3p \int d^3k |\langle n|H'|m, \mathbf{p}, \mathbf{k}\rangle|^2\delta(\omega). \tag{6.67}$$

The energies of the light particles are $E = \hbar c\sqrt{k_c^2 + p^2}$ and $\varepsilon = \hbar c\sqrt{\kappa_c^2 + k^2}$ where $k_c = mc/\hbar$ and $\kappa_c = m_\nu c/\hbar$. Introducing the upper limit E_0 of the electron energy E one finds $\hbar\omega = E + \varepsilon - E_0 - m_\nu c^2$. Taking first the average over the direction of $\mathbf{k}$ and designating it by an overbar, one finally arrives at

$$\dot{W} = \frac{32\pi^3}{\hbar^4 c^3}\int \rho(E)\overline{|\langle n|H'|m, \mathbf{p}, \mathbf{k}\rangle|^2}dE. \tag{6.68}$$

Here

$$\rho(E) \equiv \frac{pEk\varepsilon}{\hbar c}\Big|_{\varepsilon = E_0 - E - m_\nu c^2}$$
$$= \frac{pE}{\hbar^2 c^2}\sqrt{(E_0 - E)(E_0 - E + 2m_\nu c^2)}(E_0 - E + m_\nu c^2) \tag{6.69}$$

is the *shape function* of the β-spectrum given in Eq. (5) of Ref. [40] and in Eq. (36) of Ref. [61c]. This equation shows that $\rho(E)$ vanishes at E_0 with a vertical and a horizontal tangent for $m_\nu > 0$ and $m_\nu = 0$, respectively. From the empirical data Fermi concludes 'that the rest mass of the neutrino is either zero or anyway very small with respect to the electron mass' (translated from Ref. [61c], p. 171).

We still have to calculate the matrix element of H'. For this purpose we need the eigenfunctions of the electron and the neutrino, $v_{sl}(\mathbf{p})$ and $u_{\sigma\lambda}(\mathbf{k})$, respectively, where $l = \pm$ and $\lambda = \pm$ designate the positive/negative energy states. In this notation $\psi_s(x) = \sum_{\mathbf{p}} v_{s+}(\mathbf{p}) \exp(ipx)$ and $\varphi_\sigma(x) = \sum_{\mathbf{k}} u_{\sigma+}(\mathbf{k}) \exp(ikx)$. Since $\varphi(x)$ satisfies the Dirac equation (6.31) with mass m_ν, $u_{\sigma\lambda}$ satisfies the equation

$$(i\vec{\gamma} \cdot \mathbf{k} - \lambda\gamma^4 k_0 + \kappa_c)u_{\sigma\lambda}(\mathbf{k}) = 0 \qquad (6.70)$$

where $k_0 = \varepsilon/\hbar c$, and similarly for v_{sl}. Since $\overline{\psi}_s(x)$ and $\varphi_\sigma(x)$ vary little over the extension of the nucleus, taking the latter at the position $\mathbf{x} = 0$ and choosing $O_i = \gamma^4$, the matrix element of H' now reads $\langle n|H'|m, \mathbf{p}, \mathbf{k}\rangle = gQ_{mn}M_{s\sigma}^*$ where Q_{mn} is the nuclear matrix element, and we have to calculate

$$\sum_\sigma |M_{s\sigma}|^2 = (v_{s+}^*(\mathbf{p})(\sum_\sigma u_{\sigma+}(\mathbf{k}) \otimes u_{\sigma+}^*(\mathbf{k}))v_{s+}(\mathbf{p})). \qquad (6.71)$$

Here the matrix in the inner parentheses may be calculated with the aid of both the closure relation

$$\sum_{\sigma\lambda} (u_{\sigma\lambda}(\mathbf{k}) \otimes u_{\sigma\lambda}^*(\mathbf{k})) = \mathbf{1} \qquad (6.72)$$

and the projection operator

$$P_\lambda(\mathbf{k}) \equiv \frac{\lambda}{2k_0}(i\gamma^4\vec{\gamma} \cdot \mathbf{k} + \gamma^4\kappa_c + \lambda k_0) \qquad (6.73)$$

which has the properties $P_\lambda^2 = P_\lambda$ and

$$P_\lambda(\mathbf{k})u_{\sigma\lambda'}(\mathbf{k}) = u_{\sigma\lambda}(\mathbf{k})\delta_{\lambda\lambda'}. \qquad (6.74)$$

Applying (6.73) to Eq. (6.72) one finds with (6.74)

$$\sum_\sigma (u_{\sigma+}(\mathbf{k}) \otimes u_{\sigma+}^*(\mathbf{k})) = \frac{1}{2k_0}(i\gamma^4\vec{\gamma} \cdot \mathbf{k} + \gamma^4\kappa_c + k_0). \qquad (6.75)$$

Finally, averaging over the direction of $\mathbf{k}$ yields

$$\overline{\sum_\sigma |M_{s\sigma}|^2} = \frac{1}{2}\left(v_{s+}^*(\mathbf{p})\left(\gamma^4\frac{\kappa_c}{k_0} + 1\right)v_{s+}(\mathbf{p})\right). \qquad (6.76)$$

Inserting this result into Eq. (6.68) the expression for $\dot{W}$ essentially agrees with Eqs (32), (36) of Ref. [61c]. In his paper Fermi concludes from various β-decay data that the coupling constant is very small, namely of magnitude $g = 4 \times 10^{-50} \mathrm{cm}^3\mathrm{erg}$. He also introduces the important notion of *forbidden transitions*, defined by the condition $Q_{mn} = 0$, and discusses the situations in which such processes may contribute.

The neutrino also had an unexpected application, in that Louis de Broglie, who in 1932 had developed a model of the photon as a composite of two light Dirac particles [63a], proposed two years later a *neutrino theory of light* [63b]. Pauli at first found this an interesting idea. In a letter to Heisenberg he comments on 19 January 1934: 'The main problem seems to me to be a reasonable formulation of the *interaction terms of neutrinos and electrons* in the Hamiltonian. One cannot understand a priori that the special pairs of neutrinos that hold together building photons should form so much easier than any 2 neutrinos of different direction of the momentum and different energy' (translated from letter [342], Ref. [39a], p. 254). Instigated by Pauli many of his friends started to address this problem.[64a-d]. But none of these works completely satisfied him. He observed to Kronig that the coupling constant *g* should be considered in relation to that of photons, namely the fine structure constant $\alpha = e^2/\hbar c$ (letter [406] to Kronig, Ref. [39a], p. 385), and to Peierls he wrote on 13 June 1935: 'To sum up I then wish to express my profound *grief* about the present state of the so-called "neutrino theory of light", but would like to observe that your note did *not* make me more sad!' (translated from letter [411a], Ref. [39a], p. 401).

Notes and references

[1] Translated from: W. Pauli, letter [216] of 18 February 1929 to Oskar Klein, in Ref. [2].

[2] A. Hermann, V. F. Weisskopf, and K. von Meyenn (eds), *Wolfgang Pauli. Scientific Correspondence with Bohr, Einstein, Heisenberg, a.o., Volume I: 1919-1929* (Springer, New York, 1979).

[3] P. A. M. Dirac, (a) 'The Quantum Theory of the Emission and Absorption of Radiation', *Proceedings of the Royal Society* A **114**, 243 (1927); (b) 'On the Theory of Quantum Mechanics', *Proceedings of the Royal Society* A **112**, 661 (1926).

[4] C. P. Enz (ed.), *Pauli Lectures on Physics: Volume 6. Selected Topics in Field Quantization*, translated by S. Margulies and H. R. Lewis (MIT Press, Cambridge, MA, 1973).

[5] (a) P. Jordan and W. Pauli jr., 'Zur Quantenelektrodynamik ladungsfreier Felder', *Zeitschrift für Physik* **47**, 151 (1928). Reprinted in Ref. [6], Vol. 2, pp. 331–53. (b) Note on p. 171 of this paper the confusion between *f* and *g* in and after Eq. (31); note also on p. 172 the missing δ-sign in front of *F* on the left-hand side of Eq. (32).

[6] R. Kronig and V. F. Weisskopf (eds), *Collected Scientific Papers by Wolfgang Pauli. In Two Volumes* (Wiley Interscience, New York, 1964).

[7] (a) P. Jordan, 'Zur Quantenmechanik der Gasentartung', *Zeitschrift für Physik* **44**, 473 (1927); (b) P. Jordan and O. Klein, '*Zum Mehrkörperproblem*

der Quantentheorie', *Zeitschrift für Physik* **45**, 751 (1927); P. Jordan, 'Über Wellen und Korpuskeln in der Quantentheorie', *Zeitschrift für Physik* **45**, 766 (1927); (c) P. Jordan and E. Wigner, 'Über das Paulische Äquivalenzverbot', *Zeitschrift für Physik* **47**, 631 (1928).

[8] W. Pauli, 'General Principles of Quantum Mechanics', translated from Ref. [9] and edited by P. Achuthan and K. Venkatesan (Springer, Berlin and Heidelberg, 1980).

[9] W. Pauli, 'Die allgemeinen Prinzipien der Wellenmechanik', in: H. Geiger and K. Scheel (eds), *Handbuch der Physik* (Springer, Berlin, 2nd ed, 1933), Vol. 24, Part 1, pp. 83–272.

[10] P. Jordan, *Wolfgang Pauli*, in Ref. [11a], pp. 29–42.

[11] (a) P. Jordan, 'Begegnungen' (Stalling Verlag, Oldenburg and Hamburg, 1971); (b) K. von Meyenn, '*Jordan, Ernst Pascual*', in: F. L. Holmes (ed.), *Dictionary of Scientific Biography* (Scribner's Sons, New York, 1990), Vol. 17, Supplement II, pp. 448–454; (c) V. Telegdi, 'Pauli-Anekdoten', in Ref. [12], pp. 115–120; (d) E. L. Schucking, 'Jordan, Pauli, Politics, Brecht, and a Variable Gravitational Constant', *Physics Today*, October 1999, p. 26; (e) W. Nernst, 'Über einen Versuch, von quantentheoretischen Betrachtungen zur Annahme stetiger Energieänderungen zurückzukehren', *Verhandlungen der Deutschen Physikalischen Gesellschaft* **18**, 83 (1916).

[12] C. P. Enz and K. von Meyenn (eds), 'Wolfgang Pauli. Das Gewissen der Physik' (Vieweg, Braunschweig, 1988).

[13] W. Pauli, 'Theory of Relativity', translated by G. Field, *With Supplementary Notes by the Author* (Pergamon Press, London, 1958). German original and *Supplementary Notes* reprinted in Ref. [6], Vol. 1, pp. 1–263.

[14] P. A. M. Dirac, 'The Quantum Theory of the Electron', *Proceedings of the Royal Society* A **117**, 610 (1928).

[15] (a) W. Gordon, 'Der Comptoneffekt nach der Schrödingerschen Theorie', *Zeitschrit für Physik* **40**, 117 (1926); O. Klein, 'Elektrodynamik und Wellenmechanik vom Standpunkt des Korrespondenzprinzips', *Zeitschrift für Physik* **41**, 407 (1927); (b) C. G. Darwin, 'The Wave Equation of the Electron', *Proceedings of the Royal Society* A **118**, 654 (1928).

[16] (a) W. Heisenberg, 'Zur Theorie des Ferromagnetismus', *Zeitschrift für Physik* **49**, 619 (1928). Reprinted in Ref. [16b], Part I, pp. 580–96; (b) W. Blum, H.-P. Dürr and H. Rechenberg (eds), *Werner Heisenberg, Collected Works. Series A. Original Scientific Papers* (Springer, Berlin and Heidelberg, 1985).

[17] W. Heisenberg and W. Pauli, (a) 'Zur Quantendynamik der Wellenfelder', *Zeitschrift für Physik* **56**, 1 (1929); (b) 'Zur Quantentheorie der Wellenfelder. II', *Zeitschrift für Physik* **59**, 168 (1930). Reprinted in Ref. [6], Vol. 2, (a) pp. 354–414, (b) pp. 415–37.

[18] G. Wentzel, 'Quantum Theory of Fields (until 1947)', in Ref. [19], pp. 48–77.

[19] M. Fierz and V. F. Weisskopf (eds), *Theoretical Physics in the Twentieth Century. A Memorial Volume to Wolfgang Pauli* (Wiley Interscience, New York, 1960).

[20] E. Fermi, 'Sopra l'elettrodinamica quantistica', *Rendiconti dell'Accademia dei Lincei*, (a) **9**, 881 (1929); (b) **12**, 431 (1930); (c) 'Quantum Theory of Radiation', *Reviews of Modern Physics* **4**, 87 (1932). Reprinted in Ref. [21a], (a) pp. 305–10, (b) pp. 386–90, (c) pp. 401–45.

[21] (a) E. Segré (ed., with the cooperation of E. Amaldi, E. Persico and F. Rasetti), *Enrico Fermi. Collected Papers (Note e Memorie). Volume I: Italy, 1921–1938* (University of Chicago Press and Accademia Nazionale dei Lincei, Chicago and Rome, 1962); (b) G. Wentzel, *Einführung in die Theorie der Wellenfelder* (Deuticke, Wien, 1943; English translation, Wiley Interscience, New York, 1949).

[22] *Note*: Choose the Proca-type Lagrangian density

$$L_\kappa = -\frac{1}{4} \sum_{\mu\nu} \left(\frac{\partial \Phi_\nu}{\partial x^\mu} - \frac{\partial \Phi_\mu}{\partial x^\nu} \right)^2 - \frac{1}{2} \kappa^2 \sum_\mu \Phi_\mu^2 \qquad (A)$$

such that $\hbar\kappa/c$ is the mass of a (real) field Φ_μ. The associated field equations (6.35) are then, in the notation of Eqs. (6.35),

$$-\kappa^2 \Phi_\mu = (\Phi_{\nu,\mu} - \Phi_{\mu,\nu})_{,\nu} = \Phi_{\nu,\nu,\mu} - \Box \Phi_\mu. \qquad (B)$$

Applying $\partial/\partial x^\mu$ to (B) one finds zero because the bracket is antisymmetric. This results in the *Lorentz condition* (6.23). Using it in (B) yields the *Klein-Gordon equation* (5.34),

$$\Box \Phi_\mu = \kappa^2 \Phi_\mu \qquad (C)$$

which in the limit $\kappa \to 0$ of zero mass becomes the first set of the Maxwell equations (6.25).

[23] (a) A. Proca, 'Sur la théorie ondulatoire des électrons positifs et négatifs', *Journal de Physique et le Radium* **7**, 347 (1936). Reprinted in Ref. [24a]. (b) A. Pais, *Inward Bound. Of Matter and Forces in the Physical World* (Oxford University Press, Oxford, 1986).

[24] (a) G. A. Proca (ed.), *Alexandre Proca (1897–1955). Oeuvre scientifique publié* (G. A. Proca, Paris, 1988); (b) W. Pauli, 'The Connection Between Spin and Statistics', *Physical Review* **58**, 716 (1940). Reprinted in Ref. [6], Vol. 2, pp. 911–17. (c) G. Gamov, 'Zur Quantentheorie des Atomkerns', *Zeitschrift für Physik* **51**, 204 (1928); R. W. Gurney and E. V. Condon, 'Wave Mechanics and Radioactive Disintegration', *Nature* **122**, 439 (1928).

[25] (a) G. Breit, '*The effect of retardation on the interaction of two electrons*', *Physical Review* **34**, 553 (1929); (b) H. Weyl, *Gruppentheorie und Quantenmechanik* (Hirzel, Leipzig, 1928); English translation by H. P. Robertson, *The Theory of Groups and Quantum Mechanics* (Dover,

New York, 1950); (c) J. R. Oppenheimer, 'Note on the theory of the interaction of field and matter', *Physical Review* **35**, 461 (1930); (d) S. S. Schweber, *QED and the Men Who Made It: Dyson, Feynman, Schwinger, and Tomonaga* (Princeton University Press, Princeton, NJ, 1994).

[26] (a) Translated from: W. Pauli, Allocution on the occasion of the 70th birthday of Hermann Weyl on 9 November 1955, item [2183] in Ref. [26d], quoted in Ref. [2], p. 443; (b) R. Kronig, 'The Turning Point', in Ref. [19], pp. 5–39; (c) Notes of my conversation with Franca Pauli on 8 April 1984; (d) K. von Meyenn (ed.), *Wolfgang Pauli. Scientific Correspondence with Bohr, Einstein, Heisenberg, a.o., Volume IV, Part III: 1955–1956* (Springer, Berlin and Heidelberg, 2001).

[27] C. P. Enz, B. Glaus, and G. Oberkofler (eds), *Wolfgang Pauli und sein Wirken an der ETH Zürich* (vdf Hochschulverlag ETH, Zurich, 1997).

[28] (a) A. Thellung, 'Pauli als Lehrer', in Ref. [12], pp. 95–104. (b) *Note:* Wilhelm Ostwald (1853–1932), winner of the Chemistry Nobel Prize in 1909, wrote a book *Grosse Männer* (3rd and 4th edns, Akademischer Verlag, Leipzig, 1910), in which he distinguished between the 'classical' and the 'romantic' type of scientists. I thank Professor Markus Fierz for this information. (c) M. Fierz, 'Wolfgang Pauli (1900–1958), in: *Neue Österreichische Biographie ab 1815. Grosse Österreicher* (Amathea-Verlag, Vienna, 1963), Vol. 15, pp. 187–9; (d) *Eidgenössische Technische Hochschule. Ecole Polytechnique Fédérale, 1855–1955* (Buchverlag Neue Zürcher Zeitung, Zurich, 1955).

[29] (a) Notes of my conversation with Franca Pauli on 21 March 1971; (b) W. Moore, *Schrödinger. Life and Thought* (Cambridge University Press, Cambridge, 1989, paperback 1992), Chapter 5. *Zurich.*

[30] (a) O. Huber, 'Leben und Wirken von Paul Scherrer', in: K. Alder (ed.), *Paul Scherrer (1890–1969). Vorträge und Reden gehalten anlässlich der Gedenkveranstaltung zum 100. Geburtstag am 3. Februar 1990* (Paul Scherrer Institut, PSI, Villigen, 1990), pp. 15–30; (b) P. Huber, 'Prof. Dr. Paul Scherrer (1890–1969)', *Helvetica Physica Acta* **43**, 5 (1970); (c) V. L. Telegdi, 'Paul Scherrer—1890–1969. Zum 100. Geburtstag des bedeutenden Schweizer Physikers', *Neue Zürcher Zeitung. Forschung und Technik*, 31 January 1990, p. 77; (d) R. J. Humm, *Ich bin ein Humm* (GS-Verlag, Zurich, 1982); (e) *Who's Who in Switzerland 1968/69* (Nagel, Geneva, 1969), p. 574; (f) I thank Professor Markus Fierz for this information.

[31] L. Rosenfeld, 'Niels Bohr in the Thirties. Consolidation and extension of the conception of complementarity', in: S. Rozental (ed.), *Niels Bohr. His life and work as seen by his friends and colleagues* (North-Holland, Amsterdam, 1968), pp. 114–36.

[32] W. Pauli jr., 'Über das H-Theorem vom Anwachsen der Entropie vom Standpunkt der neuen Quantenmechanik', in: P. Debye (ed.), *Probleme*

der modernen Physik, Arnold Sommerfeld zum 60. Geburtstage, gewidmet von seinen Schülern (Hirzel, Leipzig, 1929), pp. 30–45. Reprinted in Ref. [6], Vol. 1, pp. 549–64.

[33] (a) C. P. Enz (ed.), *Pauli Lectures on Physics: Volume 4. Statistical Mechanics*, translated by S. Margulies and H. R. Lewis (MIT Press, Cambridge, MA, 1973); (b) L. Boltzmann, 'Weitere Studien über das Wärmegleichgewicht unter Gasmolekülen', *Sitzungsberichte der Österreichischen Akademie der Wissenschaften* **66**, 275–370 (1872). This paper, in which a single species of molecules is considered and the H-function is called E, is reprinted as paper 22 in volume I of: F. Hasenöhrl (ed.), *Wissenschaftliche Abhandlungen von Ludwig Boltzmann* (Reprint, Chelsea, New York, 1968), 3 vols. See also paper 42 of 1877 (volume II) where the H-function is still called E, and the papers 112, 119, 120 of 1895/97 (volume III). (c) C. Cercignani, *Ludwig Boltzmann. The Man Who Trusted Atoms* (Oxford University Press, Oxford, 1998); (d) F. Bopp and J. Meixner (eds), *Arnold Sommerfeld. Thermodynamik und Statistik* (Dieterich, Wiesbaden, 1952); English translation by J. Kestin: *Thermodynamics and Statistical Mechanics* (Academic Press, New York, 1956).

[34] *Note*: Equation (6.59) disagrees by a factor of 2 with Pauli's result.

[35] (a) P. Jordan, '50 Jahre physikalische Forschung an der Universität Hamburg', in Ref. [11a], pp. 53–64; (b) H. B. G. Casimir, *Haphazard Reality. Half a Century of Science* (Harper & Row, New York, 1983); (c) F. Bloch, 'Reminiscences of Heisenberg and the early days of quantum mechanics', *Physics Today*, December 1976, pp. 23–7; (d) F. Bloch, 'Über die Quantenmechanik der Elektronen in Kristallgittern', *Zeitschrift für Physik* **52**, 555 (1928).

[36] (a) J. R. Oppenheimer, 'Zur Quantentheorie kontinuierlicher Spektren', *Zeitschrift für Physik* **41**, 268 (1927); (b) M. Born and J. R. Oppenheimer, 'Zur Quantentheorie der Molekeln', *Annalen der Physik* **84**, 457 (1927); (c) R. Peierls, 'Oppenheimer, J. Robert', in: C. C. Gillispie (ed.), *Dictionary of Scientific Biography* (Scribner's Sons, New York, 1974), Vol. X, pp. 213–18; (d) J. C. Poggendorff *Biographisch-Literarisches Handwörterbuch der exakten Naturwissenschaften* (Akademie-Verlag, Berlin, 1980), Vol. VIIb. Part 6: N–Q.

[37] (a) F. Bloch, 'Zur Suszeptibilität und Widerstandsänderung der Metalle im Magnetfeld', *Zeitschrift für Physik* **53**, 216 (1929); (b) F. Bloch, 'Bemerkungen zur Elektronentheorie des Ferromagnetismus und der elektrischen Leitfähigkeit', *Zeitschrift für Physik* **57**, 545 (1929); (c) C. P. Enz, 'Applications of Quantum Mechanics (1926–1933)', in Ref. [16b], Part I, pp. 507–15.

[38] (a) J. Mehra and H. Rechenberg, *The Historical Development of Quantum Theory. Volume 1. The Quantum Theory of Planck, Einstein, Bohr and Sommerfeld: Its Foundation and the Rise of Its Difficulties, 1900–1925*

(Springer, New York, 1982); (b) R. Peierls, *Bird of Passage. Recollections of a Physicist* (Princeton University Press, Princeton, NJ, 1985); (c) R. H. Dalitz, 'A Biographical Sketch of the Life and Work of Professor Sir Rudolf Peierls, F.R.S.', in : R. H. Dalitz and R. B. Stinchcombe (eds), *A Breadth of Physics* (World Scientific, Singapore, 1988), pp. 1–42; (d) Translated from: A. Sommerfeld, letter [247a] to Pauli of 24 June 1930, in Ref. [39a], p. 19.

[39] (a) K. von Meyenn (ed., with the cooperation of A. Hermann and V. F. Weisskopf), *Wolfgang Pauli. Scientific Correspondence with Bohr, Einstein, Heisenberg a.o., Volume II: 1930–1939* (Springer, Berlin and Heidelberg, 1985); (b) Elsie Attenhofer (ed.), *Cornichon. Erinnerungen an ein Cabaret* (Benteli Verlag, Berne, 1975), p. 180; (c) K. von Meyenn (ed.), *Wolfgang Pauli. Scientific Correspondence with Bohr, Einstein, Heisenberg a.o., Volume III: 1940–1949* (Springer, Berlin and Heidelberg, 1993); (d) K. von Meyenn (ed.), *Wolfgang Pauli. Scientific Correspondence with Bohr, Einstein, Heisenberg a.o., Volume IV/Part II: 1953–54* (Springer, Berlin and Heidelberg, 1999).

[40] W. Pauli, 'On the Earlier and More Recent History of the Neutrino', in Ref. [41], pp. 193–217.

[41] C. P. Enz and K. von Meyenn (eds), *Wolfgang Pauli. Writings on Physics and Philosophy* (Springer, Berlin and Heidelberg, 1994).

[42] (a) J. Chadwick, 'Intensitätsverteilung im magnetischen Spektrum der β-Strahlen von Radium B+C', *Verhandlungen der Deutschen Physikalischen Gesellschaft* **16**, 383 (1914); (b) C.D. Ellis, 'β-Ray Spectra and their Meaning', *Proceedings of the Royal Society* A **101**, 1 (1922); (c) L. Meitner, 'Über die Entstehung der β-Strahl-Spektren radioaktiver Substanzen', *Zeitschrift für Physik* **9**, 131, (1922); 'Über den Zusammenhang zwischen β- und γ-Strahlen', *Zeitschrift für Physik* **9**, 145, (1922); 'Über die β-Strahl-Spektra und ihren Zusammenhang mit der γ-Strahlung', *Zeitschrift für Physik* **11**, 35 (1922); (d) C. D. Ellis and B. A. Wooster, 'The Average Energy of Disintegration of Radium E', *Proceedings of the Royal Society* A **117**, 109 (1927); (e) L. Meitner and W. Orthmann, 'Über eine absolute Bestimmung der Energie der primären β-Strahlen von Radium E', *Zeitschrift für Physik* **60**, 143 (1930).

[43] *Note:* RaE (not to be confused with radium emanation, today called radon, Rn, with atomic number $Z = 86$) is the bismuth ($Z = 83$) isotope Bi^{210} and decays by β^- - and α-emission into stable lead ($Z = 82$), Pb^{206}. See e.g. C.D. Hodgman *et al.* (eds) *Handbook of Chemistry and Physics* (The Chemical Rubber Co., Cleveland, OH, 1962), Sections 'The Elements' and 'Table of Isotopes'.

[44] *Note:* This meeting, most probably, was the series of lectures held 'In the framework of the Zurich Physical Society and the Zurich Physical Colloquium' from 23 to 25 April 1929, of which abstracts of the following speakers were published in volume II (1929) of the new journal

Helvetica Physica Acta, pp. 271–91: J. Thibaud (Paris), P. Auger (Paris), P. P. Ewald, W. Gerlach (Tübingen), R. Forrer (Strasburg), P. Weiss (Strasburg), J. Franck (Göttingen), C. Manneback (Louvain), O. Stern (Hamburg), L. Brillouin (Paris), J. Errera (Brussels). No abstracts had been received from E. Bauer and F. Perrin. Lise Meitner does not figure among the speakers, but Pauli in his postcard [219] to Bohr does not say so either. The meeting mentioned here is of course not the *Vortragswoche* from 1 to 4 July of footnote *a*, p. 496 of Ref. [2]!

[45] *Note*: ThC which is Bi^{212} decays by β^-- and α-emission into the stable isotope Pb^{208}, but not without γ-radiation, see the reference given in [43].

[46] (a) R. Kronig, 'Der Drehimpuls des Stickstoffkerns', *Naturwissenschaften* **16**, 335 (1928); (b) W. Heitler and G. Herzberg, 'Gehorchen die Stickstoff-kerne der Boseschen Statistik?', *Naturwissenschaften* **17**, 673 (1929); (c) P. Güttinger and W. Pauli, 'Zur Hyperfeinstruktur von Li^+', *Zeitschrift für Physik* **67**, 743 (1931). Reprinted in Ref. [6], Vol. 2, pp. 438–60; (d) P. Güttinger, 'Über Streuprozesse höherer Ordnung', *Helvetica Physica Acta* **5**, 238 (1932); (e) Notes on my phone conversation with the late Konrad Bleuler of 28 July 1984; (f) P. Güttinger, 'Das Verhalten von Atomen im magnetischen Drehfeld', *Zeitschrift für Physik* **73**, 169 (1931); (g) I. I. Rabi, 'Quantization in a Gyrating Magnetic Field', *Physical Review* **51**, 652 (1937).

[47] (a) N. Bohr, *Faraday Lecture*, 'Chemistry and the Quantum Theory of Atomic Constitution', *Journal of the Chemical Society* Part **I**, 349 (1932). (b) *Note*: In a footnote on p. 312 of Ref. [23b] Pais rejects any such relation as unfounded!

[48] (a) *Note*: Heisenberg's remark '*Du wirst mit den β-Strahlen schon recht haben*' (surely you may be right with the β-rays), contained in the post-card [243] of 18 February 1930 to Pauli, is interpreted on p. 5 of Ref. [39a] as a possible indication that Pauli's neutron hypothesis was then already known to Heisenberg. It seems to me that, if so, Heisenberg's reaction was untypically lame! (b) K. Guggenheim, *Tagebuch am Schanzengraben* (Artemis Verlag, Zurich, 1963).

[49] (a) E. Crawford, R. L. Sime, and M. Walker, 'A Nobel Tale of Post-war Injustice', *Physics Today*, September 1997, pp. 26–32; (b) R. L. Sime, *Lise Meitner: A Life in Physics* (University of California Press, Berkeley, 1996); (c) O. Hahn and F. Strassmann, 'Über den Nachweis und das Verhalten der bei der Bestrahlung des Urans mittels Neutronen entstehenden Alkalimetalle', *Naturwissenschaften* **27**, 11 (1939); (d) L. Meitner and O. Frisch, 'Disintegration of Uranium by Neutrons: A New type of Nuclear Reaction', *Nature* **143**, 239 (1939); (e) B. Karlik, 'Lise Meitner (1878–1968)', in: *Neue Österreichische Biographie ab 1815. Grosse Österreicher* (Amathea Verlag, Vienna and Munich, 1979), Vol. XX, pp. 106–11.

[50] *Note*: The situation is actually more complicated since only 'free' electrons should be considered to occur paired with Pauli neutrons in the nucleus. According to letter [248] from Pauli to Meitner quoted earlier, Pauli understands by 'free' electrons 'those that are not built into α-particles' (translated from Ref. [39a], p. 20). Now the α-particle is the nucleus He^4 with $Z = 2$, which is stable and hence contains no Pauli neutrons. If, therefore, a nucleus (A, Z) contains a α-particles, then its content of fermions is $N_f = 6a + 3A' - 2Z'$ where $A' = A - 4a$ and $Z' = Z - 2a$. Hence $N_f = 3A - 2Z - 2a$, and the symmetry is again determined by the parity of the mass number.

[51] (a) W. Pauli, 'Relativistic Field Theories of Elementary Particles', *Reviews of Modern Physics* **13**, 203 (1941). Reprinted in Ref. [6], Vol. 2, pp. 923–52; (b) C. L. Cowan, Jr. and F. Reines, 'Neutrino Magnetic Moment Upper Limit', *Physical Review* **107**, 528 (1957).

[52] (a) F. Perrin, 'Possibilité d'émission de particules neutres de masse intrinsèque nulle dans les radioactivités β', *Comptes Rendus* (Paris) **197**, 114 (1933); (b) V. J. Frenkel, 'Pauli in der UdSSR. Zur Frühgeschichte des Neutrinos', in Ref. [12], pp. 56–67; (c) V. Ya. Frenkel, *Yakov Ilich Frenkel. His work, life and letters* (Birkhäuser, Basle, 1996); (d) I. B. Khriplovich, 'The eventful life of Fritz Houtermans', *Physics Today*, July 1992, p. 29.

[53] (a) W. Pauli, 'Les théories quantiques du magnétisme. L'électron magnétique', in: *Le Magnétisme. Rapports et Discussions du Sixième Conseil de Physique Solvay tenu à Bruxelles du 20 au 25 octobre 1930* (Gautier-Villars, Paris, 1932). Reprinted in Ref. [6], Vol. 2, pp. 502–65; Discussion pp. 566–607. (b) L. Landau, 'Diamagnetismus der Metalle', *Zeitschrift für Physik* **64**, 629 (1930); (c) L. Onsager, 'Crystal Statistics. I. A Two-Dimensional Model with an Order-Disorder Transition', *Physical Review* **65**, 117 (1944).

[54] (a) H. Schüler, 'Hyperfeinstrukturen und Kernmomente', *Physikalische Zeitschrift* **32**, 667 (1931); (b) letter dated Ann Arbor, 22 February 1999 from Professor Robert Lewis to Charles P. Enz. I thank Professor Lewis for this information.

[55] *Note*: Back in 1922 Pauli had made the following remark to Schlick concerning a book on relativistic positivism by the Machian author Josef Petzoldt: 'I thereby have again reflected very carefully upon your objections against positivism and am *no more* able to recognize them as pertinent. I now consider positivism a perfectly unobjectionable and non-contradictory world view. But of course, it is not the only one' (translated from postcard [23a], dated 21 August 1922, Hamburg, Ref. [39a], p. 692; see also the footnotes to this postcard). In 1922 Schlick became the successor of Mach (and of Boltzmann) in inductive philosophy at the University of Vienna and was the main proponent thereof the Viennese Circle (*Wiener Kreis*) to promote Machian epistemology

(see the comments in Ref. [39a], pp. 690–2). This correspondence had a sequel in 1931: see letter [265] in Ref. [39a].

[56] (a) L. B. Loeb, 'Proceedings of the American Physical Society', *Physical Review* **38**, 579 (1931); (b) J. F. Carlson and J. R. Oppenheimer, 'The Impact of Fast Electrons and Magnetic Neutrons', *Physical Review* **41**, 763 (1932); (c) W. Elsasser, 'Equations du mouvement d'un neutron', *Comptes Rendus (Paris)* **198**, 441 (1934).

[57] P. A. M. Dirac, 'Quantised Singularities in the Electromagnetic Field', *Proceedings of the Royal Society* A **133**, 60 (1931).

[58] (a) *Atti, Convegno di Fisica Nucleare, Ottobre 1931* (Reale Accademia d'Italia, Rome, 1932); (b) L. M. Brown, 'The idea of the neutrino', *Physics Today*, September 1978, pp. 23–8.

[59] *Structures et Propriétés des Noyaux Atomiques. Rapports et Discussions du Septième Conseil de Physique Solvay tenu à Bruxelles du 22 au 29 octobre 1933* (Gauthier-Villars, Paris, 1934).

[60] W. Pauli, 'Erhaltungssätze in der Relativitätstheorie und in der Kernphysik', in Ref. [12], pp. 439–53.

[61] E. Fermi, (a) 'Tentativo di una teoria dell'emissione dei raggi "beta"', *Ricerca Scientifica* **2**, fasc. 12 (1933); (b) 'Tentativo di una teoria dei raggi β', *Nuovo Cimento* **11**, 1 (1934); (c) 'Versuch einer Theorie der β-Strahlen I', *Zeitschrift für Physik* **88**, 161 (1934). Reprinted in Ref. [21a], (b): pp. 559–74, (c): pp. 575–90.

[62] W. Heisenberg, 'Über den Bau der Atomkerne I', *Zeitschrift für Physik* **77**, 1 (1932). Reprinted in Ref. [16b], Part II, pp. 197–207.

[63] L. de Broglie, (a) 'Sur une analogie entre l'électron de Dirac et l'onde électromagnétique', *Comptes Rendus (Paris)* **195**, 536 (1932); (b) 'Sur la nature du photon', *Comptes Rendus (Paris)*, **198**, 135 (1934).

[64] (a) P. Jordan, 'Zur Neutrinotheorie des Lichtes', *Zeitschrift für Physik* **93**, 464 (1935); (b) R. de L. Kronig, 'Zur Neutrinotheorie des Lichtes. I, II, III', *Physica* **2**, 491, 854, 968 (1935); (c) H. Bethe and R. Peierls, 'The "Neutrino"', *Nature* **133**, 532 (1934); (d) G. Wentzel, 'Zur Frage der Äquivalenz von Lichtquanten und Korpuskularpaaren', *Zeitschrift für Physik* **92**, 337 (1934).

7

The personal crisis

Seeking the aid of Doctor Jung

With women and me things don't work out at all, and probably it also will never succeed again. This, I am afraid, I have to live with, but it is not always easy. I am somewhat afraid that in getting older I will feel increasingly lonely. The eternal soliloquy is so tiresome. [1]

Pauli made this confession in the letter already quoted on page 224 to his friend Gregor Wentzel from New York, shortly before returning to Europe in September 1931 and meeting his personal crisis. In retrospect he gives the following description of this period to his first assistant Kronig who in the meantime had become a trusted friend: 'With your criticism of my life style you are quite right. I had great fear of everything concerning the feeling and therefore have expelled (*verdrängt*) it. This finally caused an accumulation in the unconscious of all claims concerning the feeling and a revolt of the former against an attitude of the consciousness that had become too one-sided, which manifested itself in discord, loss of value and other neurotic phenomena. After having reached a fairly low point in the winter of 1931/2, it then slowly went upward again. Thereby I also became acquainted with psychic matters that I did not know before and which I want to summarize under the name *proper activity of the soul*. That here there are things which [are] spontaneous products of growth and may be called symbols, an object-psychic thing that cannot and should not be explained from material causes, is beyond doubt for me.' This letter is signed 'Your old and new W. Pauli' (translated from letter [380] of 3 August 1934, Ref. [2b], p. 340).

Pauli's father had suggested that his son Wolfgang see the psychiatrist Carl Gustav Jung (1875–1961) in Zurich [3]. So Pauli attended Jung's conferences and read Jung's works, in particular his *Contemporary Problems of the Soul* [4a] and the important *Transmutations and Symbols of the Libido* [4b] by which Jung severed the bound to Freud's psychoanalysis. Pauli's personal copies of these books, which are kept in the Pauli Room at CERN, bear multiple marks by his hand. In January 1932 Pauli had an appointment with Jung who suggested that Pauli contact with one of his, Jung's, pupils, Erna Rosenbaum, 'a young Austrian, pretty, fullish, always laughing' (translated from Ref. [3]). Pauli contacted her on 3 February 1932: 'Very honoured Frau Doktor, I do not know who you are: whether old or young, physician or amateur-psychoanalyst, totally unknown or very famous—or anything between these extremes. I only know that Herr Jung, after having given a conference, quickly squeezed into my hand a scrap of paper with your address and told me, without that yet I had the opportunity to ask him anything else, I should write to you. This came about, because a week earlier I had consulted Herrn Jung because of certain neurotic phenomena with me which, besides, also are connected with the fact that it is easier for me to achieve academic successes than successes with women. Since with Herrn Jung rather the contrary is the case, he appeared to me to be quite the appropriate man to treat me medically. Not little was my surprise when Herr Jung explained to me that this was not so and that I should absolutely have myself treated by a woman. Indeed, I am very touchy towards women and slightly distrustful and thus have some hesitations against them. Anyhow, I want nothing to be left untried and therefore I now write you this letter' (translated from Ref. [5], no. 94, see also Ref. [6a], p. XXIII). Erna Rosenbaum had just come to work with Jung for nine months, after having completed her medical education in Munich from 1925 to 1928 and in Berlin from 1928 to 1931 (Ref. [6b], letter 3639 to Jung). Some time after Pauli's treatment she went to England (see Ref. [6b]). On 30 April 1939 she wrote to Jung from London telling him of her grief ('*Herzweh*'): 'The Jewish fate has just bleeded me (*ausgeblutet*)' (translated from Ref. [6b], letter 8183).

In his second letter, on 19 March 1932, Pauli addresses his analyst as 'Dear Ms. Rosenbaum' saying: 'Today I write to you not so much because I want to know something definite from you but primarily because the written formulation and communication of various things procures me relief. Also I send you at the same time notes of my dreams, some of which I have already told you.' And referring to 'two strange dreams of the 17th' he speaks of a woman with the name Dora: 'I had approached Dora before, when my marriage had just turned awry, the wound of the broken relation to my wife therefore still was fresh' (translated from Ref. [5], letter 95, see also Refs [6a], pp. xxiii–xxiv and [6c]). Pauli later met Dora quite accidentally on a tour to Arosa in the Grisons he had undertaken with Max Delbrück who was on a visit to Zurich in March 1932. On 21 March Pauli sent his analyst a continuation of his last report with the comment: 'Hopefully you do not consider it an importunity that I write so much. But you need only answer if you consider it necessary. Somehow it now runs forth correctly by itself—so it seems to me—and I do not need too much enlightenment at present. Many greetings from your grateful W. Pauli' (translated from Ref. [5], letter 96). He sees her regularly until July when Rosenbaum goes to Berlin, to where Pauli writes on 9 July: 'Since yesterday I am in a depressive unrest. . . . I must accept it calmly. It forces me, not to run away. I believe that if I now just keep quiet something will come out of me. *To accord the wishes of the unconscious with those of the consciousness is the remedy*' (translated from Ref. [5], letter 99). And he tells her that his sister Hertha is still in Berlin (see page 18).

But his analyst's absence makes Pauli feel lonely; on 13 July he remarks in a postscript: 'The thought that soon I should decide where I shall travel for my vacations oppresses me much, as every year so also now' (translated from Ref [5], letter 100). On 25 July he writes: 'Today I passed the driver's test and the book [the *Handbook Article*, see the next section] will certainly be finished this week. Then I drive to the vacations, probably southward. . . . In psychology and the observation of my dreams I have not "invested" very much "libido" (to express it in your language). . . . I have finally thrown out my housekeeper with a big crash. . . . By September I will already get another one.' And he adds: 'You probably have much to do in Berlin? My sister has written me a very nice letter. As to the human aspect she seems to be much farther in the mean time; I am sure that soon we will understand each other well again' (translated from Ref. [5], letter 101).

Then, on 7 August, Pauli reports from Portofino-Mare on the Mediterranian, east of Genoa, that his 'book' is finished, that the weather is fine, and that he 'swims and rows assiduously'. But he continues: 'Yet I was very depressed yesterday. I fear that I am becoming unsociable (*menschenscheu*). So far on the journey I have not yet spoken a word with anybody except with waiters and hotel owners. The danger then is that I fall into a sort of brooding which is quite unproductive. How long this state will continue I do not know. Also I find my proper hand very indistinct and uncertain. In spite of the fact that externally I am quite sunburnt and look great, I do not feel very well' (translated from Ref. [5], letter 102). On 26 August Pauli wrote to his analyst—who still is in Berlin—from Venice where he 'had some pleasant days' and again sent her 'many dreams', saying 'I do not envy you for having to read all this!' In the evening he planned to drive 'quickly to Vienna for 4 or 5 days; indeed, my sister is there and I would much like to see her again. I believe that this time it will work out fine. On 1st or 2nd September I am again in Zurich' (translated from Ref. [5], letter 104). In his penultimate letter to Erna Rosenbaum, on 11 September 1932 Pauli writes: 'It seems to me that a quite definite development manifests itself therein [in the dreams]' (translated from Ref. [5], letter 105). He hopes that towards October she will come back from Berlin. And, indeed, his last letter, dated 4 October 1932, is addressed to 'Frl. [Fräulein, Miss] Dr. Rosenbaum, Hönggerstrasse 127, Zürich' (Ref. [6b], letter 1855). Thus ends a crucial relationship for Pauli which had opened up to him the fascinating world of his own dreams, and a fateful association begins: namely, with Carl Gustav Jung.

Jung has given details of this association in several lectures. The main written account is contained in the chapter 'Individual Dream Symbolism in Relation to Alchemy' (Ref. [7a], pp. 42, 45, and 98). Jung did this while keeping Pauli's name secret, as the latter had demanded (see Pauli's letter [13] to Jung from Princeton of 20 October 1935, Ref. [8], p. 15). A second written version of this association may also be found in Jung's Terry Lecture held at Yale University in 1937 under the title 'Psychology and Religion' (Ref. [9a], p. 41). Reference [7a] is based on a lecture with the same (German) title given by Jung in the summer of 1935 at the yearly Eranos Meetings in Ascona in southern Switzerland (Ref. [9c], pp. 13–133). These aestival gatherings at the Casa Eranos on the shore of Lake Maggiore had been promoted by Olga Fröbe-Kapteyn (b. 1881) in 1933, at first with themes from the humanities but subsequently enlarged to include the natural sciences. From the beginning Jung was an active participant until 1951 [10a,b,c]. Pauli's long dreams 'The Church' and 'The House of Consecration', as well as the great vision 'The World Clock' (numbers 17, 54, and 59, respectively) are also described in the Terry Lecture (Ref. [9a], pp. 42, 52, and 83, respectively). A more spontaneous, oral, description of the association between Pauli and Jung is contained in the Tavistock Lectures [11a] that Jung gave in English in London between 30 September and 4 October 1935, just after the Eranos meeting. And another oral account may be found, also in English, in the discussion meeting with the title 'The Symbolic Life' held in London on 5 April 1939 (Ref. [11b], Section 673; there in footnote 9 Pauli's name is mentioned in the English original, but not in the German translation).

In the Discussion after the fifth Tavistock Lecture Jung said: 'I had a case, a university man, a very one-sided intellectual. His unconscious had become troubled and activated; so it projected itself into other men who appeared to be his enemies, and he felt terribly lonely. . . . Then he began to drink . . . but he got exceedingly irritable . . . and once he was thrown out of a restaurant and got beaten up. . . . He came to me to ask my advice. . . . In that interview, I got a very definite impression of him: I saw that he was chock-full of archaic material, and I said to myself: "Now I am going to make an interesting experiment to get that material absolutely pure, without any influence from myself . . . ." So I sent him to a woman doctor who was then just a beginner and who did not know much about archetypal material' (Ref. [11a], Section 402). And Jung continued: 'She told him to watch his dreams, and he wrote them all down carefully. . . . I now have a series of about thirteen hundred dreams of his.' Pauli began to draw pictures which he saw in his dreams and 'even invented active imagination for himself' (Ref. [11a], Section 403). Then Jung said: 'He was five months with that doctor, and then for three months he was doing the work all by himself, continuing the observation of his unconscious with minute accuracy. . . . In the end, for about two months [two years! See note [16a]], he had a number of interviews with me. But I did not have to explain much of the symbolism to him' (Ref. [11a], Section 404). So, Jung said, Pauli 'became a perfectly normal and reasonable person' (Ref. [11a], Section 405).

C. G. Jung spent his youth on the northern border of Switzerland. He was born in Kesswil on Lake Constance, the child of a Protestant pastor who, half a year later, moved west to the parish of Laufen on the celebrated falls of the Rhine and in 1886 still further west to Basle. Much like Pauli (see page 12), Jung was an only child until his ninth year when his sister Gertrud was born (see the family tree on p. 401 of Ref. [12]). In Basle Jung attended the classical gymnasium and in 1895 entered the university. Undecided between the humanities and the natural sciences he finally chose medicine. Obtaining his medical certificate at the end of 1900 he became an assistant at the psychiatric clinic *Burghölzli* of the University of Zurich where in 1905 he became a Privatdozent in psychiatry and head-physician. While he had to give up his medical job in 1909 because his private practice had become too important, he still kept his teaching position until 1913, lecturing, among other things, on psychopathology and on Freud's psychoanalysis (see Ref. [12], p.124). Jung's strongly emotional relationship with his patient Sabina Spielrein also falls within this period (see Ref. [10a], p. 48 and Ref. [13c]). It was only in 1933 that Jung again took up his academic pursuits, this time at ETH where in 1935 he was named associate professor (Titularprofessor). But in 1942 he again withdrew from teaching because of illness. For the same reason his nomination in 1944 to the chair of medical psychology at his home university of Basle did not materialize (see footnote 13, p. 197 of Ref. [12], also Ref. [13a]).

In 1903 Jung married Emma Rauschenbach (1882–1955) who later became a faithful collaborator. The couple had four daughters and one son; in 1908 they moved into a new house on the shore of Lake Zurich at Küsnacht. Above the entrance to this house Jung had placed the enigmatic Latin inscription *VOCATVS ATQVE NON VOCATVS DEVS ADERIT* (Called and not called, God will be there). In 1923, after his mother had died, Jung started construction of his tower at Bollingen on the shore of the idyllic upper Lake Zurich. This was a primitive construction without running water or electricity, to which he added enlargements every four years until after the death of his wife in 1955. As Jung says in his recollections (Ref. [12], Section 'Der Turm'), this tower was from the beginning a place of maturation where spiritual and manual work alternated in a meaningful rhythm and where he dwelled in solitude for extended periods.

The year 1907 was a fateful time for Jung because of his personal encounter with Sigmund Freud which became a close but complex relationship in which Freud considered Jung as his successor. Jung's path, however, led him in a different direction. Jung often discussed the problems concerning Freud with his colleague and 'fatherly friend' (Ref. [12], p. 378) Théodore Flournoy (1854–1920), professor of experimental psychology at the Science Faculty of Geneva University, whose successor was Jean Piaget [13b]. Thus Flournoy became the recipient of documents concerning Jung, Freud, and Sabina Spielrein (1885–1936 or 1937). These documents were discovered in 1977 in the basement of Palais Wilson on the lakeside at Geneva where Flournoy had had his office [13c]. They show the relationship between Jung and Spielrein between 1909 and 1912, in which also Freud became involved when the relation between Jung and Freud was deteriorating [13c].

At the congress of Munich in 1912 the rupture between Jung and Freud occurred, and by 1914 all connections with Freud's psychoanalysis were severed (see the section '*Sigmund Freud*' in Ref. [12]). As Jung explains in his recollections (Ref. [12], Section 'Zur Entstehung des Werkes') from 1913 to 1917 he was occupied with the images of his unconscious coming to him through his dreams and visions. During this period his independent path was shaped and all the subsequent works are an outflow of these initial images. At the beginning of the 1930s Jung started to realize that his analytical psychology bore remarkable parallels with alchemy. Thus he started a systematic study of this esoteric medieval wisdom—whose latest bloom is Goethe's *Faust*—developing a 'dictionary' for the strange terminology.

In getting to understand the alchemical symbolism Jung realized that the relation of the ego to the contents of the unconscious induces a transmutation (*Wandlung*) comparable to that of the alchemical opus, and this led him to his central notion of the process of individuation (see Ref. [12], p. 213). In 1944 his *Psychology and Alchemy* [7b] emerged from these efforts, in the first part of which the symbolism of dreams and visions is compared with that of alchemy [7a]. The confrontation with religious questions, on the other hand, led to his 'Psychology and Religion' [9a] in 1940, while the more specific problem of the Christ gave rise in 1951 to *Aion* [14a]. This inner development of Jung finally culminated in 1955–6—at age 80—in the two volumes of his *Mysterium Coniunctionis* [14b]. In the same year (1955), on the occasion of its 100th anniversary, ETH bestowed an honorary doctorate on Jung. Jung had a devoted circle of collaborators, mainly women who had started with an analysis: Toni Anna Wolff (1888–1953), Jolande Jacobi (1890–1983), Aniela Jaffé (1903–91), and Marie-Louise von Franz (1915–98).

Describing in the introduction to Ref. [7a] the *method* of interpretation of Pauli's dreams, Jung emphasizes that a general theory of dreams exists no more than a general theory of the consciousness and that, therefore, any interpretation may be gained only from the *context*. But there is no context attached to Pauli's notes of his dreams. The point, however, is that these dreams form connected series, and it is this characteristic that supplies a context. In addition, after a description of some initial dreams in Section 2, Jung emphasizes in Section 3 that the bulk of the dreams there is viewed under the aspect of a particular archetype, namely that of the mandala symbol. So this section begins with the definition of a mandala as a magic circle with two (perpendicular) symmetry axes, this symmetry often being expressed by a circumscribed or an inscribed square or by a cross. The original lamaist mandala, however, is an inner image reached through active imagination.

Of particular interest here is the 10th dream because it gives a telling description of Pauli's personality viewed through Jung's interpretation: 'The dreamer is in the Peterhofstatt in Zurich with the doctor, the man with the pointed beard, and the "doll woman". The last is an unknown woman who neither speaks nor is spoken to. Question: To which of the three does the woman belong?' (Ref. [7a], English translation, pp. 101–2).

St Peter's, the oldest of Zurich's churches, is located near the former Roman fortress of Lindenhof. It was first mentioned in documents in 857, when King Ludwig the German donated its revenues to the Fraumünster Abbey. The tower of St Peter in Zurich has on each of its four façades an enormous dial-plate which is divided into four and hence may represent a mandala. This fourfold symmetry is again reflected in the four characters assembled within the fourfold enclosure of St Peter's Square called 'Peterhofstatt'. Now Jung says: 'In the dream the dreamer represents his own ego, the man with the pointed beard the "employed" intellect (Mephisto), and the "doll woman" the anima. Since the doll is a childish object it is an excellent image for the non-ego nature of the anima, who is further characterized as an object by "not being spoken to". This negative element . . . indicates an inadequate relationship between the conscious mind and the unconscious, as also does the question of whom the unknown woman belongs to. The "doctor", too, belongs to the non-ego; he probably contains a faint allusion to myself, although at that time I had no connections with the dreamer. . . . The man with the pointed beard, on the other hand, belongs to the ego. This whole situation is reminiscent of the relations depicted in the diagram of functions (fig. 49)' (Ref. [7a], English translation, p. 102).

Here Jung refers to his distinction of the *four functions* which are grouped in the two pairs of opposites: thought–feeling, the 'rational' pair, and intuition–sensation, the 'irrational' pair. By arranging these functions on a crossed circle (mandala) the specificity of a person at a given period of their life, i.e. their psychological type [15], lies in the *orientation* of this circle: One of the four functions is bright and dominant in the consciousness and represents the ego, while the opposite is dark and inferior and dwells in the unconscious where it represents the non-ego. The remaining two auxiliary functions have bright and dark parts. In Pauli's younger life 'thought' was dominant and 'feeling' was his problem. This is the situation represented in fig. 49 on p. 102 of the English translation of Ref. [7a]. Jung comments on this figure as follows: 'Hence it would not be impossible for the four persons in the dream to represent the four functions as components of the total personality (i.e. if we include the unconscious). But this totality is ego plus non-ego' (Ref. [7a], English translation, pp. 102–3).

Jung's discussion in Ref. [7a] of Pauli's dreams culminates in the 'Great Vision' of the *World Clock* mentioned above:

There is a vertical and a horizontal circle, having a common centre. This is the World Clock. It is supported by the black bird.

The vertical circle is a blue disc with a white border divided into $4 \times 8 = 32$ partitions. A pointer rotates upon it.

The horizontal circle consists of four colours. On it stand four little men with pendulums, and round about it is laid the ring that was once dark and is now golden (formerly carried by the children).

The 'clock' has three rhythms or pulses:
1. The small pulse: the pointer on the blue vertical disc advances by 1/32.
2. The middle pulse: one complete revolution of the pointer. At the same time the horizontal circle advances by 1/32.
3. The great pulse: 32 middle pulses are equal to one revolution of the golden ring.

This remarkable vision made a deep and lasting impression on the dreamer, an impression of 'the most sublime harmony', as he himself puts it.

And Jung continues: 'It is a three-dimensional mandala—a mandala in bodily form signifying realization. (Unfortunately medical discretion prevents my giving the biographical details. It must suffice to say that this realization did actually take place.) Whatever a man does in reality he himself becomes' (Ref. [7a], English translation, p. 194).

This reminds us that the real purpose of Jung's analysis of Pauli's dreams was therapeutic. And indeed, Jung concludes his description in Ref. [7a] as follows: 'The vision of the "world clock" is neither the last nor the highest point in the development of the symbols of the objective psyche. But it brings to an end the first third of the material, consisting in all of some four hundred dreams and visions. This series is noteworthy because it gives an unusually complete description of a psychic fact that I had observed long before in many individual cases. . . . We have to thank not only the completeness of the objective material but the care and the discernment of the dreamer for having placed us in a position to follow, step by step, the synthetic work of the unconscious. The troubled course of this synthesis would doubtless have been depicted in even greater completeness had I taken account of the 340 dreams interspersed among the 59 examined here. Unfortunately this was impossible, because the dreams touch to some extent on the intimacies of personal life and must therefore remain unpublished. So I had to confine myself to the impersonal material' (Ref. [7a], English translation, pp. 206–7).

Jung took over from Rosenbaum sometime in October 1932. There are 10 letters between Pauli and Jung from the duration of Pauli's analysis with Jung: the first, by Jung, carries the date 4 November 1932, the last, by Jung again, is from 29 October 1934. Chronologically these are the letters [1] to [5], [29], [6], [30], [7], [8] of Ref. [8] (concerning this order see *Note* [16b]). The meetings usually were fixed at 12 noon on Mondays. Letters [1], [2], [3], and [6] only concern modifications or confirmations by Jung of this schedule. But letter [7] from Pauli is quite personal. He wrote to Jung on 26 October 1934: 'As concerns my personal destiny, there still are some unsolved problems left over. But I have a certain desire to get away from dream interpretation and dream analysis and want to see now what life brings me from outside. A development of my feeling function of course is very important, but it seems to me indeed that in the course of time it could gradually proceed through life, and should not occur solely as a result of dream analyses. After mature reflection I have indeed now come to the decision, not to continue my visits with you for the time being, unless anything particular should happen to me' (translated from Ref. [8], pp. 10–11).

Thus we may date Pauli's analysis with Jung from October 1932 to October 1934 (see *Note* [16a]). That the analysis was successful may be concluded from Jung's letter [5] of 28 April 1934 which reads: 'Very honoured Herr Professor, on your wedding I wish to perfectly congratulate you. I find it very pleasing that you have drawn this conclusion. With the best greetings, yours as always devoted C. G. Jung' (translated from Ref. [8], p. 9). On 4 April 1934 Pauli had married Franca Bertram in London. But, as I shall explain in more detail in the next chapter, Franca, far from being thankful to Jung for this turn of her destiny, distrusted psychiatrists in general and Jung in particular, so that it is not impossible that it was she who had persuaded Pauli to discontinue the visits to Jung.

But Franca notwithstanding, the contacts between Pauli and Jung continued outside of any analysis, as is witnessed by the truly fascinating collection of letters between the two in Ref. [8]. This exchange, in fact, had already begun with Pauli's letters [29], [30], and [7] and Jung's reply [8], after which there is a break of eight months. Asking Jung in letter [29] to take up the 'Monday hours' again after the break for the spring vacation, Pauli thanks him for his essay 'Soul and Death' [17] where, in particular, the remarks on telepathy caught Pauli's attention. This leads Pauli to observations concerning 'the general attitude of the modern physicist towards the life phenomenon', namely that experiments with living beings are restricted by the condition to maintain life. But this renders the testing of the laws of physics impossible, 'in such a measure, that there is no room left for the reign and action of a new kind of law of nature with the life phenomena' (translated from Ref. [8], p. 30). Here Pauli makes use of an argument that Bohr had discussed in his influential paper 'Light and Life' [18a,b], of which Pauli had sent a copy to Jung who acknowledged receipt in letter [4] on 2 November 1933. Indeed, Bohr had written: 'I think we all agree with Newton that the real basis of science is the conviction that Nature under the same conditions will always exhibit the same regularities. Therefore, if we were able to push the analysis of the mechanism of living organisms as far as that of atomic phenomena, we should scarcely expect to find any features differing from the properties of inorganic matter' (Ref. [18b], p. 458).

The remaining content of letter [29], as well as the main part of letter [30], concern Pauli's personal problems. Pauli tells Jung that in his dreams as well as in waking fantasies increasingly abstract pictures occur, often in the form of alternating bright and dark stripes but also of acoustic rhythms, 'and it will become a vital necessity for me to understand something more about the objective (communicable) meaning of these symbols than it is the case with me at present. I have certain reasons to assume that only then may I succeed in completely "subjugating" my anima complex (according to your psychology: to transform into a function)' (translated from Ref. [8], p. 30). A strange fact is that a realization of 'alternating bright and dark stripes' is the abdomen of a wasp, and Pauli suffered from a fear of wasps which he was only gradually overcoming by dissociating his anxiety from these insects. He had realized that this anxiety was caused by the danger he saw in the emergence of uncontrollable ecstatic states of the unconscious. In letter [30] Pauli returns to the wasp problem and the alternating stripes. The latter, in fact, grew into a recurring theme in his fantasies, as may be seen in Pauli's much later essay 'Hintergrundsphysik' (Background Physics) of June 1948 (Ref. [8], Appendix 3).

In the last letter [8] Jung discusses a paper by Pascual Jordan [18c] that Pauli had sent him with the request for Jung's opinion. The problem at hand is that of observation. Jung concludes: 'The last consequence of Jordan's view would lead to the assumption of an absolute unconscious space in which infinitely many observers observe the same object. The psychological version would read: in the unconscious there is only one observer who observes infinitely many objects' (translated from Ref. [8], p. 12; this letter [8] is reproduced in Ref. [19], pp. 226–8).

His analysis came to an end with what Pauli called his first 'half-life' (*Lebenshälfte*). Observing in relation to the alternating stripes that the danger for the second half-life was that he would fall from one extreme into the other, he tells Jung in letter [30]: 'In the first half-life I was towards other people a cynic, a cold devil and a fanatical atheist and intellectual "enlightener" (*Aufklärer*).—The opposite thereto would be, on the one hand, an inclination towards the criminal, towards the rowdy (which could have degenerated as far as the murderer), on the other hand, an unworldly, entirely unintellectual hermit with ecstatic states and visions' (translated from Ref. [8], p. 31). The idea of the hermit most probably was an association with the Swiss hermit Niklaus von Flüe (alias Bruder Klaus whose chapel in Melchtal, 25 km south of Lucerne, Pauli reports in the same letter [30] to have visited, probably after having read Jung's essay of Ref. [20].

In his second half-life Pauli developed the habit of recounting his dreams to Franca at breakfast and then writing them down. This occupation, according to her, became increasingly important towards the end of his life. But Franca found the dreams 'embellished' (*frisiert*) and not worth keeping, and she destroyed them [3]. But the dreams sent to Jung, of course, survived (see Ref. [8]).

The *Handbuch* Article

> When . . . the use of one *classical notion stands in an excluding relation to that of* another *one, we call these two notions . . . following Bohr, complementary. In analogy with the term 'relativity theory', one therefore might call the modern quantum theory also 'complementarity theory'.* [21]

Much like Pauli's *Encyklopädie Article* (see page 25), his *Handbuch Article* on 'The General Principles of Wave Mechanics' [22a] has become a classic. But in spite of the great authority of this work the terminology that Pauli proposes above has never been adopted by the physics community. Even the notion *complementarity* itself is not appreciated by all physicists, and is often identified depreciatively with the Copenhagen Interpretation. Pauli himself said in a radio talk in New York, on the occasion of the bicentennial of Columbia University, in October 1954: 'These [epistemological] consequences are not quite easy to understand; and the point of view called 'complementarity', which was developed by Bohr and others for this purpose, though shared by the majority of physicists, did not remain without opposition' (Ref. [24a], p.

32; see also Ref. [24b]). For this reason, Pauli's introduction of the subject, which is based on experimental facts and shuns formal arguments, may not be to everybody's taste. As already mentioned on page 69, Pauli used to call this work his 'New Testament'. And he considered it 'not quite as good as my first *Handbook* article' on the old quantum theory, his 'Old Testament' (see Ref. [22d], p. iii, and Ref. [25b], p. 423), 'but in any case better than any other presentation of quantum mechanics' (Ref. [25c], p. 51 as well as Ref. [25d], p. 78, also Ref. [2b], p. 110). Both works, as also the *Encyklopädie Article*, are characterized by the same dense structure, in which the sections follow in an uninterrupted logical succession and the existing literature related to each problem is given fair attention.

It is astonishing that, as we saw in the last section in his letters to Erna Rosenbaum, Pauli could write this demanding work at the height of his crisis. Nothing in the relation to his assistant Peierls betrays this, as the following letter of 24 March 1932 shows. This letter, in fact, reflects the typical, quite personal relationship Pauli had with his assistant, a relationship which, however, usually did not touch upon private thoughts. More specifically this letter informs us about progress on the *Handbuch Article*: 'Dear Peierls! Too bad that you are sick with this fine weather. For the insertion of the formulae many thanks. Meanwhile I have written another section; by Saturday I hope to be through also with the following one (the rotation group [Ref. [22a], Ziff. 13]).Wednesday Miss Schaufelberger [Pauli's and Scherrer's secretary, see Ref. [26b], p. 17] will come again, so I hope that in the course of the week after Easter the entire non-relativistic part of the article will be finished. Towards 3 April I want also to leave.' And Pauli closes: 'Saturday I would like to visit you, somewhere between 2 and 3 o'clock in the afternoon, in order to talk with you about the question of the energy fluctuations. Best wishes for your recovery and many greetings from your W. Pauli' (translated from Ref. [2b], letter [289]).

This letter shows that by the end of March 1932 the 'Non-Relativistic Theory' (Ziff. 1–16 of Part A in Ref. [22a]) was complete, while from his letter of 7 August to Erna Rosenbaum we learn (see the last section) that the remaining 'Relativistic Theory' (Ziff. 1–8 of Part B in Ref. [22a]) was finished by early August. The question of the energy fluctuations that Pauli wished to discuss with his assistant is examined in Ziff. 14, Part A of Ref. [22a]. There, in footnote 3 on p. 197 (Ref. [22d], footnote 1, p. 126), Pauli thanks Peierls for having done the calculation for the particle-number fluctuation given by Eqs (338 a,b), which are derived, for FD and BE statistics, in Eqs (5.43), observing that the corresponding expression for the energy demands some caution because the energy of a partial volume v, which is needed for the analogous calculation, is obtained as an integral over v of the energy density $(1/2m) \sum_r \mathbf{p}_r \delta(\mathbf{x} - \mathbf{x}_r)\mathbf{p}_r$ where $\mathbf{p}_r$ and $\mathbf{x}_r$ are the momentum and position of particle r, respectively. As this expression is singular, however, one has to replace the δ-function by a smooth function of finite width which is small compared with the diameter of v.

Since much of the *Handbuch Article* has already been reviewed in previous sections I will pick out here those passages that captured my attention when trying to understand this work as a student and later when teaching the subject myself, for the first time in early 1959, on replacing Pauli at ETH after his untimely death (see Document IV.101 in Ref. [26b]) and later in Neuchâtel and in Geneva. My personal copy which is the American reprint of 1947 mentioned in Ref. [22a] bears the marks of all these years. But my personal relation to this work was even strengthened by the fact that Pauli had asked me to proof-read the revised edition [22b] of 1958, and in 1977 Pauli's widow Franca begged me to check the English translation of Ref. [22d] which I did in great detail (see pp. v and x there). In what follows, page numbers refer to Ref. [22a] and those in parentheses to Ref. [22d].

Clearly, Pauli's expression 'complementarity theory' is a reference to Bohr who takes centre stage at the outset (Ziff. 1), but later on in the article it is a younger generation that dominates the scene, even in the delicate and controversial question of interpretation. The foundation and interpretation of the non-relativistic theory are discussed in Ziff. 2–9 of Part A, and the Copenhagen Interpretation is evident. But Pauli probably offers the most consistent and the clearest exposition of this point of view—which indeed is 'better than any other presentation of quantum mechanics'. The first question discussed in Ziff. 2 concerns the meaning in quantum mechanics of the position and momentum of a particle of velocity v and mass $m = m_0/\sqrt{1 - (v/c)^2}$. The position x may be determined by light scattering, i.e., by a Compton effect. Energy and momentum conservation then yield as the limit of accuracy of a position measurement, Δx, the wavelength of the scattered photon, $\lambda = h/mc$ which, for a particle at rest, is the Compton wavelength $\lambda_c = h/m_0 c$.

Now Pauli observes that this value of Δx 'to start with is of relevance at most for nuclei and electrons since already for atoms as a whole their dimension in general is larger than their h/mc' (translated from Ref. [22a], p. 91, see Ref. [22d], p. 10). Indeed, since the mass M of an atom is at least equal to the proton mass m_p, the 'radius' r of an atom satisfies the inequalities $r \geq a_B \gg h/m_p c \geq h/Mc$, where a_B is the Bohr radius and the inequality in the middle follows from the relation $a_B m_p c/h = (m_p/m_0)(1/2\pi\alpha) = 4.00 \times 10^4$ where α is the fine structure constant. This inequality means that it is possible to manipulate electromagnetically 'inside' atoms, e.g. by ionization or by chemical reaction, while for nuclei and electrons this is not possible. Indeed, $h/m_p c = 1.32 \times 10^{-13}$ cm essentially equals the 'radius' of the proton, while the electron may be considered to be pointlike.

But Pauli continues: 'Furthermore, whether for the first-named particles [nuclei and electrons] this limit [h/mc] has an intrinsic meaning [27a] or whether it might be circumvented by indirect methods, cannot be decided a priori on the grounds of elementary arguments and depends entirely on the question, which premises a relativistic quantum mechanics may successfully be built on' (translated from Ref. [22a], pp. 91–2, see Ref. [22d], p.10). This reservation, of course, acquired real meaning when accelerators made particle reactions possible. Another way of looking at the limit of localizability, Δx, of a particle at rest is to consider its wave nature. Using the first relation (5.30) in the form $mc^2 = hc/\lambda$, the wavelength is again the Compton wavelength λ_c. Therefore, 'subdivision' of the particle becomes qualitatively impossible when its radius $r \leq \Delta x \sim \lambda_c$. In this view protons and electrons are 'indivisibles' in the sense of the atoms of the ancient Greeks [27b], which, according to Pauli, were 'a way out of the difficulties of the relation between unity and multiplicity' (Ref. [25a], p. 140).

More important for the foundation of quantum mechanics is what Pauli calls on p. 92 (p. 10) the basic assumption (*Grundannahme*) that at any moment of time t there exists, for a free particle with position vector $\mathbf{x}$, a probability $W(\mathbf{x}, t)d^3x$ to find the particle at time t at $\mathbf{x}$ within a margin of d^3x. The basic problem then is to determine $W(\mathbf{x}, t)$. Rather than following an easy, axiomatic route, Pauli first introduces the wavefunction ψ which, for a free particle, is determined by the Schrödinger equation (5.33) with $V = 0$. Arguing on p. 97 (p. 16) that, because $\partial\psi/\partial t$ is to be considered as a new function, ψ must be complex, he finally arrives at the conclusion that $W = |\psi|^2$, provided that ψ is normalized according to $\int_V |\psi|^2 d^3x = 1$.

This whole argument is extended on p. 112 (p. 32) to several interacting particles. In this general case the question arises, under which conditions the simultaneity of the different particle positions $\mathbf{x}_r$ implicit in the form $W(\mathbf{x}_1, ..., \mathbf{x}_N, t)$ is realized. Pauli argues that for this the associated measuring times t_r must satisfy the condition $|t_a - t_b| < |\mathbf{x}_a - \mathbf{x}_b|/c$, so that simultaneity is only guaranteed in non-relativistic quantum mechanics where the velocity of light c may be considered infinite. The probability $W = |\psi|^2$ thereby introduced is what Pauli used to call a primary or a priori probability (see e.g. Ref. [25a], pp. 46, 121, and 152). It is obvious from the above argument that in non-relativistic quantum mechanics time plays a privileged role. This is also reflected in the fact mentioned in footnote 1 on p. 140 (p. 63), that it is in general not possible to construct a Hermitian time operator conjugate to the Hamiltonian H, because the commutation relation $[H, t] = \hbar/i$ would imply that H should have a continuous spectrum between $-\infty$ and $+\infty$.

While in more axiomatic formulations of the theory, a physical quantity called an observable of the system is associated to every Hermitian operator, Pauli takes 'the opposite view' on p. 133 (p. 56) (see the footnote), declaring: 'What the physical quantities of a system are and how they are to be measured, can be decided only by a further development of the theory based on experience' (Ref. [22d], p, 56). This is also the position he takes in his review of Dirac's book *The Principles of Quantum Mechanics* [27c] which he praises for its 'great elegance and generality'. But he finds that Dirac's symbolic method 'also has serious disadvantages', namely: 'The reader in principle does not learn, how the devices serving to measuring an observable look like in reality and whether for an observable chosen arbitrarily general . . . a device allowing to measure it exists at all. Also no mention is made of the fact that certain measurements, e.g. the determination, which stationary state an atom is in, require a finite least duration' In addition, Pauli criticizes the fact that no mention is made of the fundamental difficulty that radiation theory 'in its application to free electrons leads to an infinitely large self-energy of the electron' (translated from Ref. [27c]).

Since the measurement of a physical quantity in general perturbs the system in an uncontrollable way, the repetition of a measurement does not in general yield the same value. On p. 152 (p. 75) Pauli introduces the distinction between measurements of the first kind which, when repeated, yield the same value and measurements of the second kind where this is not the case. A measurement of the second kind, therefore, results in a gain of information which implies a '*reduction of the wave packet*' ψ [27d], after the interaction with the measuring device is cut (see pp. 148, 161 (pp. 71, 85)), while for a measurement of the first kind this does not happen. The simplest example of a measurement of the first kind is the determination of the energy of a closed system. Indeed, as is discussed on p. 149 (p. 72), the probability $|\psi_n|^2$ to find an energy value E_n is independent of time; this is also the most general expression of energy conservation. On the other hand, the measurement of the particle momentum depends on the duration of the measurement. Only sufficiently long measuring times guarantee reproducibility of the result (see pp. 94, 152 (pp. 12, 75)).

The two topics of Part B of Ref. [22a] are the Dirac equation and quantum electrodynamics (QED) (see pages 175f). The weakness of the former were the states of negative energy which Pauli discusses in Ziff. 5. Dirac had tried to remedy this defect of his theory by stipulating that most negative-energy states be occupied, and hence unavailable because of the exclusion principle, the remaining 'holes' being identified as antiparticles. However, the difficulty apparently remained, and it was only clarified in principle, if not in practice, when Pauli, in his paper of 1934 with assistant Weisskopf, [28a] showed that the quantization of charged spin-0 particles also leads to antiparticles but without negative energy states, thus freeing the way for a description in terms of the symmetry of charge conjugation. I therefore postpone the discussion of negative-energy states until the presentation of Ref. [28a].

In spring 1932 the Dirac equation was very much on Pauli's mind. One aspect he analysed was the limit of short wavelengths, which is the analogue of the geometrical optics of light rays and describes the limit of classical orbits. In this limit he could confirm Bohr's claim that the effect of the electron spin is of the same order of magnitude as that of diffraction, and that therefore spin is essentially non-classical. This paper [28b], in which he also discussed the states of negative energy, was the first Pauli published in the new Swiss journal *Helvetica Physica Acta*, after he reported its content in an oral communication at the spring meeting of the Swiss Physical Society held in Vevey on Lake Geneva on 7 May [28c].

At this meeting Pauli was accompanied by Peierls and his wife Genia Nikolaevna, (see page 209). In his recollections Peierls describes this event as follows: 'The town of Vevey was honoured by the presence of many learned people, to whom it gave the freedom of the municipal wine cellars. There were rows of enormous casks containing the local wine, labelled by type and vintage, and we took full advantage of the offering, fortified by French-style ham rolls. When we finally emerged, we were rather cheerful. Somehow we ended up at an elegant hotel, where a band was playing. The drummer seemed to be absent, so Pauli and Genia together took his place and had a marvellous time doing it' (Ref. [29a], pp. 77–8; see also Ref. [29b], p. 73).

Another feature of the Dirac equation discussed by Pauli in the *Handbuch Article* deserves special mention, namely the two-component equation proposed by Weyl in 1929 [29c]. Weyl's idea, in fact, was that by the reduction of the Dirac ψ from four to two components the negative-energy states could be avoided. Pauli introduces this equation on p. 226 (p. 157) in the particular representation of van der Waerden's spinor calculus (see page 117), in which $\gamma^5 = \gamma^1\gamma^2\gamma^3\gamma^4$ is a diagonal matrix with elements $-1, -1, 1, 1$. This explicit representation is not necessary, however. Indeed, following the notation of Eqs (11), (12) of Ref. [30a] one may define Dirac wavefunctions $\psi^R = (1-\gamma^5)\psi/2$ and $\psi^L = (1+\gamma^5)\psi/2$. (In the mentioned representation where γ^5 is diagonal only the upper two components of ψ^R and only the lower two of ψ^L are non-zero.) Combining the Dirac equation (6.31) with the one obtained by multiplying (6.31) from the left with γ^5 one finds the two equations

$$\gamma^\mu \frac{\partial}{\partial x^\mu}\psi^R + k_c\psi^L = 0; \quad \gamma^\mu \frac{\partial}{\partial x^\mu}\psi^L + k_c\psi^R = 0. \tag{7.1}$$

Obviously, Eqs (7.1) are covariant only for zero mass, $k_c = mc/\hbar = 0$.

Weyl's hope had been that the mass m could somehow be generated by gravitational effects. But there was another problem with these two-component equations, namely that the symmetry between left and right is violated. This is seen as follows. From Eqs (6.30) it follows that the Dirac equation (6.34) is covariant under the parity transformation $\mathbf{x} \to -\mathbf{x}$ and $\psi \to \gamma^4\psi$ which transforms a right-handed coordinate system into a left-handed one and vice versa. Now for zero mass ψ^R and ψ^L describe a 'right-handed' and a 'left-handed' particle, in the sense that its spin and momentum are parallel and antiparallel, respectively. But in Eqs (7.1), even with $k_c = 0$, ψ^R and ψ^L are interchanged in a parity transformation and, hence, these equations are not covariant. Weyl considered this parity violation to be a small price to pay for a two-component theory. Not so Pauli. He wrote: 'However, as indeed follows from their derivation, these wave equations are not invariant with respect to reflexions (exchange of left and right), *and therefore they are not applicable to the physical reality*' (translated from Ref. [22a], p. 227, italics added).

As we have already seen repeatedly, Pauli was an unconditional believer in conservation laws. This time, however, he was wrong: experiments on β- and μ-decay in early 1957 showed that parity was violated in these processes. This was a great shock to Pauli, and he decided to discard the above statement in italics in the 1958 edition [22b] (it is reproduced only in the editor's comment 54 of Ref. [22c]). But this part of Pauli's biography belongs to a later chapter. Here it suffices to note that after the discovery of parity violation in processes involving a neutrino the two-component equations (7.1) with $k_c = 0$ became the natural description of zero mass, maximally parity-violating neutrinos (see Section 5 of Ref. [30a]).

The last topic of the *Handbuch Article* on which I wish to comment is the question of the atomicity of the electric charge that, as will become evident, was central to Pauli's thinking. In the original version [22a] this question appears in three places which were all discarded by Pauli in the revised edition [22b]. The first passage reads: 'Far more satisfactory [than the two distinct basic assumptions of the wave equations for matter and for radiation] would be the understanding of these two kinds of laws from a logically unified point of view which, however, so far has not been discovered. This question might be connected with the yet unresolved problem of the electric elementary quanta' (translated from Ref. [22a], p. 115). And at the outset of Part B of Ref. [22a] (see also Ref. [22c], Appendix I) Pauli writes: 'Contrary to the non-relativistic quantum mechanics, which may be considered to be logically closed, in the relativistic domain we still face unsolved problems of principle which culminate in the question of the atomicity of the electric charge, of the mass ratio between electron and proton and of nuclear structure' (translated from Ref. [22a], p. 214). Finally, Ref. [22a] closes with a reiteration of the observation of p. 115 just quoted, followed by the statement (see also Ref. [22d], p. 204 and Ref. [22c], Appendix III): 'To this also corresponds its (failed) attempt to dissolve the whole proper field of a moving electron into photons, instead of considering it as forming, together with the electron, an indivisible whole which is inseparably tied to a definite numerical value of the dimensionless number $e^2/\hbar c$. Here a future complete, relativistic theory will have to bring a deeply reaching unification of the foundations' (translated from Ref. [22a], p. 272).

Concerning the three statements mentioned, the following question arises. Why did Pauli discard them in 1958, when the atomicity of the electric charge and, in particular its numerical value expressed by Sommerfeld's fine structure constant

$$\alpha = \frac{e^2}{\hbar c} \sim \frac{1}{137},\qquad (7.2)$$

apparently meant so much to him? I can see no explanation, except perhaps that the style of physics publishing had become more austere. One thing is clear: Pauli had not changed his mind. This is evidenced by the following passage from Pauli's *Supplementary Notes* to his *Encyklopädie Article*, which also date from 1958: 'The reader of the original text of § 67 will see that I was already at that time very doubtful regarding the possibility of explaining the atomism of matter, and particularly of electric charge, with the help of classical concepts of continuous fields alone. In this connexion it should be remembered that the atomicity of electric charge had already found its expression in the specific numerical value of the fine structure constant, a theoretical understanding of which is still missing today' (Ref. [30b], p. 225). In Section 67 of the original text of 1921 Pauli had written: 'the existence of atomicity, in itself so simple and basic, should also be interpreted in a simple and elementary manner by theory and should not, so to speak, appear as a trick in analysis' (Ref. [30b], p.

205; see also the quotation from Ref. [30b], p. 206, on page 38). There had been many tricks to explain the numerical value (7.2), the most interesting, perhaps, being the one advanced by Armand Wyler who had a doctorate in mathematics from ETH. He identified α with the volume of the group $O(5, 2)$ [30d].

But the very first time this idea appears, though still in rudimentary form, is the passage from Pauli's third paper of 1919 quoted on page 36, where he writes that 'there are no test bodies smaller than the electron itself'. In the years 1934–5, when Dirac's hole theory and the problem of the self-energy of the electron gave rise to an intense exchange of letters between Pauli and Heisenberg, Pauli again raised the question 'of the fixation of e^2/hc and of the atomicity of the electric charge' (translated from letter [352] of 7 February 1934, Ref. [2b], p. 278). Gradually, Pauli awakened Heisenberg's interest in this question, who, in turn, got involved in tricky calculations of $e^2/\hbar c$ (see letters [402], [407]–[410] from Heisenberg to Pauli).

In his first philosophical lecture on modern physics that Pauli gave in November 1934 to the Zurich Philosophical Society he said that only a formulation of quantum theory would be satisfactory which expresses the relation between the value of α and charge conservation in the same complementary way as that between the space–time description and energy–momentum conservation (Ref. [30c], p. 104). Even Pauli's Nobel Lecture of 13 December 1946 closes with the declaration: 'From the point of view of logic, my report on "Exclusion principle and quantum mechanics" has no conclusion. I believe that it will only be possible to write the conclusion if a theory will be established which will determine the value of the fine structure constant and will thus explain the atomistic structure of electricity, which is such an essential quality of all atomic sources of electric fields actually occurring in nature' (Ref. [31], p. 146, also Ref. [30c], p. 181; see also Ref. [28a], p. 713 and Ref [30c], p. 135).

In the autumn of 1932 Pauli returned to some of the fundamental questions he had discussed in the *Handbuch Article*. The motivation was a somewhat provocative paper by Ehrenfest with the title 'Some exploratory questions regarding quantum mechanics' [32a]. Ehrenfest had introduced his questions by observing that they may well be put aside as 'senseless': 'Good taste even requires that. However, then, someone somehow has to accept the odium to pose them anyway.' And he has 'the firm conviction that there still are some researchers who understand the art of answering "senseless" questions meaningfully and in a clear and simple manner' (translated from Ref. [32a], p. 555). To Pauli Ehrenfest's paper was 'a source of sheer pleasure', as he wrote to him on 28 October 1932. And he explains that the questions raised, 'in fact, belong *to those upon which I have hit myself when I wrote up my new Handbuch article on the principles of wave mechanics and which to me are of burning interest. In my opinion your having exposed them publicly is only to be welcomed*' (translated from letter [294], Ref. [2b], p. 125). He then answers Ehrenfest's questions in a detailed dicussion which covers 10 pages of letter [294]. Ehrenfest received this letter with 'enormous joy' and admitted in his response that 'for fear of Bohr and you . . . I have struggled (*herumgewürgt*) more than a year with the decision to print the few lines until, at last, despair drove me to it' (translated from letter [295], Ref. [2b], p. 135). And he encourages Pauli to publish his remarks.

Pauli published these arguments in a paper bearing the same title as Ehrenfest's [32b]. There he writes in the introduction: 'Since on the occasion of writing a review paper I partially hit upon quite similar questions [as Ehrenfest], I may be permitted to publish here some remarks thereupon' (translated from Ref. [32b], p. 573). He then analyses the following three questions: 1. On the role of the imaginary unit and the notion of the spatial density of a particle in wave mechanics; 2. The questions of the analogy between photons and electrons and of its limits and 3. On the question whether quantum mechanics may be expressed as a field theory (*Nahwirkungstheorie*). Section 1 clearly is an elaboration on the construction of the relation $W = |\psi|^2$ mentioned earlier. The interest of Section 2 lies in Pauli's distinction between the second-quantized 'big fields' Ψ, $\mathbf{E}$, $\mathbf{H}$ and the one-particle 'small fields' ψ, e, h, the latter being not directly measurable in principle. Also interesting are the following limits of the analogy, namely that for a photon there is no four-current like Eqs (6.32) and that a photon cannot have negative energy. As for Section 3, by expressing for example the Coulomb energy in the many-particle Hamiltonian as $\int \mathbf{E}^2 d^3 x / 2 = \sum_{ss'} e_s e_{s'} / 2 r_{ss'}$ one is forced to include the divergent self-energy with $r_{ss} = 0$ (compare Eqs (6.46), (6.47)). This again leads, as mentioned earlier, to 'the strange distinction of space before time' (translated from Ref. [32b], p. 586).

In the year preceding this response from Pauli to Ehrenfest's questions it had been the latter who had responded to Pauli's work. And he had done it in grand style by proposing Pauli for the Lorentz Medal. Indeed, on 31 October 1931 the Royal Netherlands Academy of Arts and Sciences in Amsterdam bestowed its Lorentz Medal upon Pauli. It was the second time this gold medal had been awarded, the first recipient being Max Planck in 1927 (see Ref. [25c], p. 77). It was Pauli's first academic honour, and it came more than two years before Heisenberg received the Planck Medal on 3 November 1933 and on 10 December the Nobel Prize of the year 1932 (see page 145). Ehrenfest had been Pauli's outspoken friend since the two had met at the memorable 'Bohr Festivals' in Göttingen back in June 1922, which was also the beginning of their 'war of wit' (see page 89). In particular, Ehrenfest was an ardent admirer of Pauli's exclusion principle. In letter [271] to Pauli on 25 March 1931 (part of which is lost) Ehrenfest mentions his plan to propose Pauli for the Lorentz Medal in the following words: 'I believe that with the adjudication of the medal this work [the exclusion principle] must be emphasized above all' (translated from Ref. [2b], p. 71). In this letter, which was meant as a first draft of his laudation, Ehrenfest writes in his typical pedagogical style: 'Why is a crystal so thick as it is? Because the atoms are thick.—Why are they thick?—Because not all the electrons fall onto inner rings.—Why don't they do it?—Because of the electrostatic repulsion of the electrons against each other?—No! For then they would still be able to gather around highly charged nuclei still much more densely.—No, they don't do it for fear of Pauli!!!! And so we may say: Pauli himself is so thick, because the Pauli prohibition holds. The wonderful, incomprehensible' (translated from Ref. [2b], pp. 70-1).

Although himself not given to formal etiquette, Ehrenfest begged Pauli to wear formal dress at the award ceremony. Pauli answered five days before the event: 'Dear Ehrenfest! Just now I ordered a new black suit at my tailor's. But I shall only put it on at Amsterdam if you promise me to thank me in public in your official allocution at the Academy, for not having saved myself the trouble of going to the tailor. . . . Please say in addition that I have still got a sense of humour whereas many physicists lose it when they grow older' (translated from postcard [284], Ref. [2b], p. 96; see also Ref. [25c], pp. 85–6). Evidently, even during his crisis Pauli had not lost his sarcastic sense of humour. So, how did Ehrenfest manage to satisfy Pauli's wishes? Opening his laudation by paying due respect to H. A. Lorentz and then announcing 'With today's—the second—award of the Lorentz Medal our Academy wishes to honour the discoverer of the profound "exclusion principle"', Ehrenfest went into the details of Pauli's work, then saying: 'You will have to concede, Herr Pauli: By a *partial* suppression of your prohibition you could deliver us from a good many troubles of every-day life, for instance the traffic problem in our streets' (translated from Ref. [32c], see also Ref. [25c], p. 86). At this point, according to Ehrenfest's assistant Casimir who was in the audience, Ehrenfest improvised something like this: 'and you might also considerably reduce the expenditure for a beautiful, new, formal black suit' (Ref. [25c], p. 86).

Casimir was to succeed Peierls as Pauli's assistant in the autumn of 1932 (see Ref. [26b], Documents I.57, 58). Originally, however, Max Delbrück (1906–81) from Berlin, who had visited Pauli during the winter of 1931/2 on a Rockefeller Fellowship, was Pauli's choice as Peierls' successor. Seeing himself suddenly in competition with Lise Meitner who had offered Delbrück an assistant position in Berlin (see the comment on p. 113 of Ref. [2b]), Pauli wrote to Meitner: 'To a certain extent your letter embarrasses me.' But he assured her that 'at present a binding agreement on that between me and Delbrück does not exist', and he encourages her to offer Delbrück the position, in which he would be more independent, 'let him then decide himself' (translated from Ref. [2b], letter [291] of 29 March 1932). Delbrück decided on Meitner.

Thirty years later the reason for this decision was recalled by Bohr. Referring to a congress on radiation therapy that had taken place in Copenhagen in August 1932, at which Bohr gave the lecture on 'Light and Life' [18a,b] mentioned in the last section and where Delbrück had also participated, Bohr wrote: 'Delbrück who at that time was working as a physicist with us at Copenhagen has followed with great attention such considerations which, as he said, have awoken his interest in biology and have stimulated him to his successful investigations in genetics' (translated from Ref. [33a], p. 725). Delbrück made good use of his independence, as he recalls in his Nobel Lecture in 1969: 'During the years 1932–37 while I was assistant to Professor Lise Meitner in

Berlin, a small group of theoretical physicists held informal private meetings, at first devoted to theoretical physics but soon turning to biology. Our principal teacher in the latter area was the geneticist Timofeeff-Ressovsky, who, together with the physicist K. G. Zimmer, at that time was doing by far the best work in the area of quantitative mutation research' [33b].

Thus Casimir became Pauli's assistant. But while Peierls had occupied this position for three years (see page 209), which Pauli found to be long enough, Casimir stayed only for one year. He recalls: 'I should have liked to stay one more year at Zurich and Pauli was quite willing to keep me there. . . . But around Eastertime there came an urgent message from Ehrenfest that he insisted on my coming back to Leiden in September 1933. Both Pauli and I felt that we had to comply with that wish. Only later did it become evident to me that already at that time Ehrenfest had almost decided to put an end to his life. And he also knew that it would take some time before his successor would come to Leiden. Therefore, he wanted to be sure that there was at least an assistant who could look after the reading room, keep the colloquium running, and so on. . . . In retrospect it is clearly a farewell, and yet he must have hesitated to the very last. We still saw him at Copenhagen in September. Then, on 25 September 1933 he carried out his fateful decision' (Ref. [25c], p. 147).

Expressing the consternation of the physicists' community, Pauli wrote in a moving obituary: 'On 25th September 1933, under tragic circumstances and to the dismay of his family and numerous friends and acquaintances, Paul Ehrenfest carried out his calamitous resolve to cast off the burden of a life grown too heavy for him to bear. It is now our task to hold in remembrance his scientific work and the picture of his personality unencumbered by those cares and feelings of inferiority which increasingly clouded his spirit in his later years. The picture is of a man of scintillating intellect and wit, intervening in discussions with mordant criticism, and at the same time with profound insight into the foundations of the scientific attitude, drawing attention to some essential point hitherto unnoticed or insufficiently regarded' (Ref. [34a], p. 80). In 1904 Ehrenfest had obtained his doctor's degree under Boltzmann who had himself taken his own life on 5 September 1905 at Duino. In his recollections Casimir raises the question: 'Did Ehrenfest somehow feel guilty?' and suggests that 'The remembrance of Boltzmann's suicide may have haunted him all his life' (Ref. [25c], p. 148; see also Ref. [34b], pp. 48 and 80).

Paul Ehrenfest was born in Vienna on 18 January 1880, the youngest of five sons of a Jewish family. In 1899 he began to study chemistry at the Institute of Technology in Vienna, but soon turned to physics and to Boltzmann at the university where he also met Lise Meitner (see pages 211–12). During the two years that Boltzmann stayed in Leipzig, Ehrenfest went to Göttingen where he studied with the mathematicians Felix Klein, David Hilbert, and Ernst Zermelo and with the physicists Johannes Stark, Max Abraham, Walther Nernst, and Leopold Schwarzschild, but also with the philosopher Edmund Husserl. His friends among the students were the young Swiss Walther Ritz (1878–1909) and the Russian Tatjana Alexeyevna Afanassjewa (1876–1964) whom he married in 1904 in Vienna. Since Ehrenfest was unable to find a position in Vienna, the young couple, after a second visit to Göttingen, moved in 1907 to Saint Petersburg where they lived on private means. Early in 1912 Ehrenfest travelled to Central Europe in search of a job. The high point of this journey was his visit to Einstein in Prague, of which the latter wrote: 'Within a few hours we were true friends—as though our dreams and aspirations were meant for each other' (Ref. [34b], p. 177). Unfortunately, owing to Ehrenfest's lack of a religious confession he could not become Einstein's successor in Prague and had to go back to Saint Petersburg empty-handed.

In the same year, however, Lorentz resigned from his chair at Leiden and recommended Ehrenfest as his successor. This choice was based mainly on the well-known review Ehrenfest had written with his wife on the foundations of statistical mechanics which, nine years before Pauli's review on relativity, was published in the *Encyklopädie der mathematischen Wissenschaften* [35]. But perhaps the most important of Ehrenfest's assets was his reputation as a brilliant lecturer. Even Sommerfeld, in a letter of recommendation written to Lorentz in 1912, said: 'He lectures *like a master*. I have hardly ever heard a man speak with such fascination and brilliance' (Ref. [34b], p. 185). Casimir comments on this choice: 'It is characteristic of Lorentz's width of vision that he should esteem and encourage a young man so entirely different from himself and so utterly at variance with then prevailing ideas about what a professor should be like' (Ref. [25c], p. 65).

So the Ehrenfests settled in Leiden with their two daughters, and in July 1914 moved into their own new house where also two sons were born. There was a happy and active family life which Einstein enjoyed enormously on his two visits in 1916 and 1919 (see Ref. [34b], pp. 304 and 314). There were also frequent invitations for meals which, however, 'were never elaborate, often irregular, and furthermore, the Ehrenfests were vegetarians. Alcoholic drinks were never served. . . . The Ehrenfests did not smoke and did not permit smoking in their home, although guests who stayed with them were allowed to smoke upstairs' (Ref. [34b], p. 205; remember Ehrenfest's warning to Pauli on page 204). In view of this happy life Ehrenfest's death is all the more incomprehensible. His lack of self-confidence, the surge of anti-Semitism in Germany, some tensions with his wife, to whom he was deeply attached, and the mongolism of his youngest son born in 1918—all this may have contributed to his decision to shoot first this son and then himself (Ref. [25c], pp. 148–149).

Excursion into unified field theory

Einstein's new field theory is dead, long live Einstein's new field theory! [36]

Imitating the French ceremonial announcement of the deceased and the new king, '*Le roy est mort! Vive le roy!*', Pauli, in a review of a review article by C. Lanczos (see page 142) on Einstein's new field theory [37], greets the latter's untiring attempts in his provocative manner. Introducing this review

Pauli writes: 'It is indeed a bold action of the editors to include in the *results* of the exact natural sciences [the title of the series] a paper on a new field theory by Einstein. As a matter of fact, his never-failing inventiveness as well as his tenacious energy in the pursuit of a certain goal presents us in recent years, on the average, with one theory per annum.' (translated from Ref. [37]; see also Ref. [38], p. 347).

What became known as unified field theory (see Chapter 17 of Ref. [38]) had been initiated in 1918 by Weyl's 'extremely profound theory' (Ref. [30b], p. 192) discussed on page 37, in which gravitation and electromagnetism were unified in a generalized space–time geometry. Inspired by this work, another mathematician, Theodor Kaluza (1885–1954), published in Königsberg, in 1921, a paper in which this unification was achieved by enlarging four-dimensional space–time to five dimensions [39a]. Defining a five-dimensional metric tensor $\gamma_{\alpha\beta}$ by the invariant line element ds as $ds^2 = \gamma_{\alpha\beta}dx^\alpha dx^\beta$ where α, β are summed from 1 to 5, this theory is characterized by what Kaluza calls the cylinder condition

$$\frac{\partial \gamma_{\alpha\beta}}{\partial x^5} = 0 \qquad (7.3)$$

and by the fact that γ_{55} is invariant (see below). Kaluza set $\gamma_{55} = 1$, so that the fifth dimension is space-like, and took as field equations the generalization of Einstein's Eqs (2.3) to five dimensions. Kaluza showed, although only for weak fields and non-relativistic velocities, that the trajectory of a charged test particle is a geodesic in the combined gravitational and electromagnetic fields (see Ref. [38], pp. 330–1).

It was Oskar Klein who analysed the general case in 1926 [39b]. Making use of the tensor property under coordinate transformations $x \to x'$,

$$\gamma_{\mu\nu} = \gamma'_{\alpha\beta} \frac{\partial x'^\alpha}{\partial x^\mu} \frac{\partial x'^\beta}{\partial x^\nu}, \qquad (7.4)$$

one shows that the cylinder condition (7.3) implies $\partial x'^i/\partial x^5 = 0, i = 1, ..., 4$ and $\partial^2 x'^5/\partial x^5 \partial x^\mu = 0$, or that, up to a constant multilpying x^5, Klein's group of transformations of Ref. [39b]

$$\begin{aligned} x'^i &= f^i(x^1...x^4); \; i = 1, ..., 4, \\ x'^5 &= x^5 + f(x^1...x^4) \end{aligned} \qquad (7.5)$$

is obtained (here and below I follow my own notes of Pauli's course of 1953, Ref. [39c], pp. 51–5). It is easy to see that holding x^5 fixed (space–time transformations), $\gamma_{55} = \gamma'_{55}$ while γ_{i5} and γ_{ik} transform like a four-vector and a four-tensor, respectively. On the other hand, holding the x^i fixed (gauge transformations) one finds $\gamma_{55} = \gamma'_{55}$ and $\partial f/\partial x^i = (\gamma_{i5} - \gamma'_{i5})/\gamma_{55}$ and, on inserting these two relations, also $g_{ik} = g'_{ik}$ where (in the notation of Ref. [39c])

$$g_{ik} \equiv \gamma_{ik} - \frac{\gamma_{i5}\gamma_{k5}}{\gamma_{55}}. \qquad (7.6)$$

g_{ik} may then be identified with the metric tensor describing gravitation in space–time. Since γ_{55} is seen to be invariant under the whole group (7.5) one may put $\gamma_{55} = J = \text{const.}$ Then

$$\Phi_i \equiv \frac{\gamma_{i5}}{\gamma_{55}} \tag{7.7}$$

may be identified as the vector potential describing electromagnetism and

$$ds^2 = \gamma_{\alpha\beta}dx^\alpha dx^\beta = g_{ik}dx^i dx^k + J(\Phi_r dx^r + dx^5)^2. \tag{7.8}$$

This form of the line element and its possible interpretation is also given in *Note 23* of Ref. [30b], Eq. (25).

Klein derived his field equations from a variational principle with the action (2.10) in five dimensions, where g is now the determinant γ and W the curvature-invariant P, formed with the $\gamma_{\alpha\beta}$. This leads to homogeneous Einstein equations in five dimensions, $\mathbf{G}_{\mu\nu} = 0$. He then shows that the contravariant components $\mathbf{G}^{\mu\nu}$ separate into

$$\mathbf{G}^{ik} = G^{ik} + \frac{J}{2}S^{ik} \tag{7.9}$$

and

$$\mathbf{G}^{i5} = \frac{\partial(\sqrt{-g}F^{ik})}{\partial x^k}, \tag{7.10}$$

where S_{ik} and F_{ik} are the covariant components of the electromagnetic energy-momentum tensor and of the field (6.24), respectively. Hence $\mathbf{G}_{ik} = 0$ are the inhomogeneous Einstein equations (2.3) with $\kappa T_{ik} = (J/2)S_{ik}$, which leads to the important identification

$$J = 2\kappa, \tag{7.11}$$

where κ is Einstein's gravitational constant. On the other hand, $\mathbf{G}_{i5} = 0$ are the Maxwell equations (6.25) in the presence of gravitational fields (see Eq. (397) and Eq. (208a) without sources in Ref. [30b]). In the notation of Ref. [39c] used here Klein's constants are $\alpha = J$ and $\beta = 1$.

Einstein had joined the unification trend in 1923 with four short notes on an elaboration of Weyl's theory by Eddington, and with his own construct in 1925. In 1927 he published two notes on Kaluza–Klein theory, but the following year he returned to four dimensions, inventing distant parallelism. In 1929 he proposed a set of field equations in the framework of the latter theory but his attempt to derive them from a variational principle failed. For the year 1930 Einstein had invited a new collaborator in the person of Walther Mayer (1887–1948) from Graz, Austria. This collaboration began immediately and was quite fruitful; it even continued for a while when Einstein, after his resignation, on 28 March 1933 from the Prussian Academy in Berlin, settled in October 1933 at the Institute for Advanced Study in Princeton, New Jersey. For more on Mayer's biography see Ref. [38], pp. 492–4. At first Mayer worked with Einstein on distant parallelism. But in a report to the journal *Science* Einstein wrote in

October 1931: 'After we both had worked more than a year on the further development of the last theory, we reached the conclusion that we were striving in the wrong direction and that the theory of Kaluza, while not acceptable, was nevertheless nearer the truth than the other theoretical approaches' [40a]. And he wrote to Pauli on 22 January 1932: 'So you were right after all, you rascal (*Spitzbube*)' (translated from letter [288], Ref. [2b], p. 109; see also Ref. [38], p. 347). These efforts of Einstein, described in more detail in Section 17e of Ref. [38], continued to the end of his life. Pais wrote: 'His work on unification was probably all in vain, but he had to pursue what seemed centrally important to him, and he was never afraid to do so. That was his destiny' (Ref. [38], p. 329).

The reason Einstein considered the Kaluza–Klein theory 'not acceptable' is given in the sequel of the report to *Science* just quoted: 'It is anomalous to replace the four-dimensional continuum by a five-dimensional one and then subsequently to tie up artificially one of these five dimensions in order to account for the fact that it does not manifest itself.' However, the report continues: 'We [Einstein and Mayer] have succeeded in formulating a theory which formally approximates Kaluza's theory without being exposed to the objection just stated.' And he explains: 'Our theory arises quite readily from consideration of five-vectors (five components) in the four-dimensional continuum' [40a]. This is the paper 'Unified theory of gravitation and electricity' by Einstein and Mayer[40b] which even aroused Pauli's interest. Agreeing fully with Einstein's criticism of the Kaluza–Klein theory voiced above (see the introduction in Ref. [41a], p. 305), Pauli now decided to jump on the unification bandwagon. His motivation was to investigate a possible conection with the Dirac equation. Klein, in the second part of Ref. [39b], had discussed a five-dimensional form of the Schrödinger equation but this discussion is unconvincing. Pauli found that Klein had mixed up the phenomenological and the quantum-mechanical points of view. He writes in a remark added as a footnote to letter [168] to Bohr on 6 August 1927: 'In his five-dimensional papers Oskar Klein might be considered as a grand master of confusion between these two points of view. Indeed, Kaluza's 5-dimensional formulation of the classical theories in reality stands entirely on the *first*, phenomenological point of view' (translated from Ref. [42], p. 403). And on 14 March 1928 Pauli wrote to Bohr that he has continued to think about 'how the new equations of Dirac are related to general relativity. Thereby connections with Klein's five-dimensional formulation have also appeared' (translated from postcard [191], Ref. [42]).

Half a year after his first paper Klein published a note in which quantization is introduced by the hypothesis that 'the five-dimensional space is assumed to be closed in the direction x^0 [$= \sqrt{J}x^5$ in the present notation] with a period l' [43]. In other words, the cylinder condition (7.3) is replaced by the requirement that the $\gamma_{\alpha\beta}$ and also f in Eqs (7.5) are periodic functions of x^5. Like Kaluza, Klein considers a test particle of mass m and charge e whose dynamics in five dimensions is determined by the Lagrangian

$$L = \frac{m}{2}\gamma_{\alpha\beta}u^{\alpha}u^{\beta}, \tag{7.12}$$

where $u^{\alpha} = dx^{\alpha}/d\tau$ is the five-velocity and $d\tau = ds/ic$ the element of proper time. The canonical five-momentum is then $p_{\mu} \equiv \partial L/\partial u^{\mu} = m\gamma_{\mu\nu}u^{\nu}$, and one finds, making use of Eq. (7.8), $p_5 = mJ(\Phi_r u^r + u^5)$ and $p_i = mg_{ik}u^k + p_5\Phi_i$. Now this last expression is of the form of the parentheses in Eq. (6.7), and hence

$$p_5 = \frac{e}{c}, \tag{7.13}$$

which is Eq. (5) of Ref. [43]. Klein now introduces quantization by the rule (3.12),

$$\oint p_5 dx^5 = \frac{l}{\sqrt{J}}p_5 = 2\pi\hbar N, \tag{7.14}$$

where l is the period mentioned above and N is an integer. With $N = 1$ a combination of Eqs (7.11), (7.13), and (7.14) now determines the universal *Klein* length given in Eq. (8) of Ref. [43],

$$l_{Kl} \equiv 2\pi\frac{\hbar c}{e}\sqrt{2\kappa} = 4\pi\sqrt{\frac{2}{\alpha}}l_{Pl} = 0.843 \times 10^{-30} \text{ cm}, \tag{7.15}$$

where α is given by Eq. (7.2). For comparison I have also introduced here the universal *Planck* length

$$l_{Pl} \equiv \frac{1}{2}\sqrt{\hbar c\kappa} = 4.05 \times 10^{-33} \text{ cm} \tag{7.16}$$

(see *Note* [44]).

Klein takes Eq. (7.15) as an explanation of the atomicity for the electric charge and as 'the origin of Planck's quantum'. And he suggests that the smallness of l_{Kl} 'may explain the non-appearance of the fifth dimension in ordinary experiments' [43]. Pauli observes that by introducing the periodicity in x^5, $\gamma_{55} = J$ becomes a new field quantity, and the resulting theory is more general than the simple unification of gravitation and electromagnetism. Also, he says, quantization of a tensor field of rank 2 introduces particles of spin 2 which, because of the loss of the gauge group, do not have only two orientations like the graviton. But such particles are not observed in nature (Ref. [39c], p. 55).

Oskar Klein(1894–1977), was five years older than Pauli and one of his closest friends; they had met at the 'Bohr Festivals' in Göttingen in June 1922 (see page 86). The third child of Sweden's First Rabbi, Klein grew up in Stockholm where in 1921 he obtained his doctorate after having already published several papers. Starting in 1918 he became a frequent visitor to and then one of the closest collaborators of Bohr in Copenhagen where in 1927 he succeeded Heisenberg as a lecturer. In 1931 Klein became professor in Stockholm. On this event Pauli wrote to him on 12 December 1930: 'Now you have reached the goal of our social class, are superbonze in Stockholm and may henceforth conduct a life (hopefully) free from material worries in the civic middle class with a home that inspires luck and satisfaction . . . [and] you are left with praying daily to the God of the middle class that he might always protect your bank account.' But then Pauli expresses the hope that Klein will develop a fruitful teaching activity. 'Your great pedagogical capabilities have always been one of your strongest sides and for them there will be a wide field of application in Sweden' (translated from letter [261], Ref. [2b], p. 43). Klein had married Gerda Koch in 1923 with whom he had six children.

Klein's name is closely linked with many important results in theoretical physics, and we have met it several times already in this account, first in relation to his elementary way of explaining the action of crossed electric and magnetic fields on hydrogen atoms (page 80) and then in writing the Klein–Gordon equation (page 141). In fact, somewhat isolated from the European development, Klein had anticipated particle–wave duality during his stay with H. M. Randall in Ann Arbor, Michigan (see page 222) from 1923 to 1925. We met Klein's name again in connection with the Jordan–Klein second quantization of bosons (page 178) and will do so further on with Klein's paradox in relation to the negative-energy states of the Dirac equation, as the Klein–Nishina formula, the relativistic expression describing Compton scattering, and in relation to the first formulation of a gauge field theory. He was a member of the Nobel Committee for Physics and in 1959 was awarded the Planck Medal. For more details on Klein's biography see Refs [46a,b].

Pauli's work on unified field theory began with an examination of Einstein and Mayer's paper [40b] in view of a generally covariant formulation of the Dirac equations. During winter 1931/2 he put his new collaborator from Paris, Jacques Solomon (1908–1942 or 1943), on the problem. Solomon, who was a committed communist and was married to Paul Langevin's charming daughter Hélène, visited Copenhagen and Zurich on a Rockefeller Fellowship. The couple were in Copenhagen in early 1931, together with Casimir and Landau (see Ref. [25c], p. 108). There he worked with Pauli's former collaborator Léon Rosenfeld (see page 208) on the quantum theory of black-body radiation, which in 1931 also became the subject of Solomon's Paris thesis. Solomon and Rosenfeld became friends. In March 1942 Jacques and Hélène Solomon, who had collaborated with the French Resistance, were seized by the Gestapo. Jacques was executed on 23 May, Hélène was deported to Auschwitz (Ref. [2a], footnote 4, p. 171). She survived and returned to Paris after the war where Pauli saw her in Langevin's house in April 1946 (Ref. [2a], letter [809]).

When Casimir arrived in Zurich in September 1932, the first task Pauli assigned his new assistant was to check the formulae in the second proof of a paper Solomon had sent Pauli from Paris, proposing that the work should be published as a joint paper by Pauli and Solomon. Pauli found this proposal justified since he had not only suggested the problem but also guided Solomon in his calculations; but he wanted to be sure that there were no major mistakes. Casimir, however, found that most of the essential formulae were wrong. So Pauli cabled to Paris that the paper had to be rewritten. But the editor of *Journal de physique*, Jean Langevin, who happened to be Solomon's brother-in-law, told Pauli with apologies that the second proofs had not come back in time (Ref. [25c], p. 140; see also Ref. [25d], p. 76). So the paper 'The unified theory of Einstein and Mayer and the Dirac equations' [47a] appeared uncorrected. (A footnote on p. 470 of Ref. [23], Vol. 2, says that in his copy of the paper Pauli had crossed out this page and the following one.) Therefore Pauli decided to write a second paper with the same title [47b], and Solomon had the benefit of two publications with the master. This paper was submitted on 30 November 1932.

Einstein's new idea, described in Section 2 of Ref. [47a], had been to associate to each point x^k ($k = 1...4$) of space–time V_4 a five-dimensional vector space V_5. Contravariant vectors (see page 30) a^μ ($\mu = 1...5$) of V_5 transform according to $a'^\mu = M^\mu_\nu(x)a^\nu$. Between them and their affine parts a^k in V_4 the following relations are induced by the basic tensor γ^k_μ:

$$a^k = \gamma^k_\nu a^\nu \tag{7.17}$$

and

$$\bar{a}_\nu = a_k \gamma^k_\nu \tag{7.18}$$

where covariant vectors (see page 30) are obtained in the usual way with the help of the respective metric tensor, $a_\mu = g_{\mu\nu}a^\nu$ and $a_i = g_{ik}a^k$. The bar in Eq. (7.18) means that $\bar{a}_\mu$ is orthogonal to the characteristic direction X^μ, i.e. $\bar{a}_\mu X^\mu = 0$, where X^μ is defined as the solution of the equations

$$\gamma^k_\mu X^\mu = 0; \quad g_{\mu\nu} X^\mu X^\nu = \varepsilon \tag{7.19}$$

where $\varepsilon = \pm 1$. Here I follow the notation of Ref. [41a], except that Pauli writes $\gamma_\mu{}^k$ instead of γ^k_μ. In Refs [40b] and [47a] X^μ is called A^μ and $\varepsilon = +1$. (Note that for X^μ to be unique the matrix γ^k_μ must have rank 4.) Defining the tensor γ^μ_k by $\gamma^k_\mu \gamma^\mu_l = \delta^k_l$ one also has

$$\bar{a}^\mu = \gamma^\mu_k a^k; \quad X_\mu \gamma^\mu_k = 0, \tag{7.20}$$

so that $a^\mu = \bar{a}^\mu + \varepsilon a X^\mu$ where $a \equiv X_\nu a^\nu$.

The next step is to establish the connection between the X^μ and the co-ordinates x^i. At this point the analogy between the Einstein–Mayer theory [40b, 47a,b] and the projective formulation [41a,b] of Pauli becomes less obvious because the latter uses the X^μ as homogeneous coordinates such that the $x^k = h^{(0)k}(X)$ are homogeneous functions of degree 0 (an obvious example is $x^k = h^{(0)k} = X^k/X^5$). In this work Pauli essentially follows a paper by Schouten and van Dantzig [48a]. Defining now the basic tensor by

$$\gamma^k_\mu = \frac{\partial x^k}{\partial X^\mu} \tag{7.21}$$

the relation $\gamma^k_\nu X^\nu = 0$ expresses this homogeneity. For, a homogeneous function of degree n, $h^{(n)}$, is defined by the relation $h^{(n)}(\lambda X) = \lambda^n h^{(n)}(X)$ which implies

$$X^\mu \frac{\partial h^{(n)}(X)}{\partial X^\mu} = nh^{(n)}(X). \tag{7.22}$$

Kaluza–Klein theory may now be viewed as a special case of this formalism of homogeneous coordinates (see Ref. [39c], p. 56 and *Note 23* in Ref. [30b], pp. 229–30). Setting $X^\mu \partial/\partial X^\mu = \partial/\partial x^5$ one has $X^\mu = \partial X^\mu/\partial x^5$ or $X^\mu = \exp(x^5) f^\mu(x)$. Solving for $x^\mu(X)$, e.g. by choosing $f^k(x) = x^k$, $f^5(x) = 1$, the cylinder condition (7.3) means that $\gamma_{\alpha\beta}$ is a homogeneous function of degree 0 in the X^μ satisfying (7.22). Similarly one verifies that the transformations (7.5) become homogeneous functions of degree 1, $X'^\mu = h^{(1)\mu}(X) = X^\lambda \partial X'^\mu/\partial X^\lambda$. In this way one avoids 'to tie up artificially one of these five dimensions', as Einstein objected. Since the continuation of the theory is much less satisfactory in Refs [47a,b], I will now switch to Refs [41a,b]. This projective formulation is quite general and elegant but Pauli's papers are very heavy; they cover 68 pages. Together with the 20 pages of the Pauli–Solomon papers [47a,b] they constitute Pauli's contribution to unification.

In Section 4 of Ref. [41a], the following fundamental skew-symmetric tensor $X_{\mu\nu}$ and its affine part X_{ik} are introduced:

$$X_{\mu\nu} \equiv \frac{\partial X_\nu}{\partial X^\mu} - \frac{\partial X_\mu}{\partial X_\nu}; \quad X_{ik} \equiv X_{\mu\nu} \gamma^\mu_i \gamma^\nu_k. \tag{7.23}$$

Proving first the orthogonality $X_{\mu\nu} X^\nu = 0$ one then concludes similarly as in Eqs (7.17), (7.18), (7.20) that also $X_{\mu\nu} = X_{ik} \gamma^i_\mu \gamma^k_\nu$ holds. Therefore, the metric is entirely determined, for given γ^i_μ, by the tensors g_{ik} and X_{ik}. The latter then is shown to satisfy

$$\sum_{(ikl)\text{cycl}} \frac{\partial X_{ik}}{\partial x^l} = 0 \tag{7.24}$$

which has the form of the second set of Maxwell equations (6.25). Hence, g_{ik} and X_{ik} describe the gravitational and the electromagnetic fields, respectively, and one may identify (see Eq. (44) of Ref. [41a])

$$X_{ik} = r f_{ik} = r\sqrt{\kappa} F_{ik} \tag{7.25}$$

where F_{ik} are the conventional electromagnetic field components and r is a dimensionless number (see [48b]).

The next problem, analysed in Section 5 'Parallel displacement of vectors. Geodesic lines' of Ref. [41a], is to derive the classical equation of motion of a particle of charge e and mass m without postulating a Lagrangian (7.12). Parallel displacement (see page 30) in the projective space V_5 may be defined by $\delta a^\mu = a^\mu_{;\nu} dX^\nu = da^\mu + \Gamma^\mu_{\nu\lambda} a^\lambda dX^\nu$. In the affine space V_4 the analogous definition is $\tilde{\delta} a^i = a^i_{;k} dx^k = da^i + \tilde{\Gamma}^i_{kl} a^l dx^k$. The following expression may then be derived (see Eqs (50) and (35a) of Ref. [41a]):

$$\delta a^\mu = \gamma^\mu_i \left(\tilde{\delta} a^i - \frac{\varepsilon a}{2} X^i_k dx^k \right) + \varepsilon X^\mu da, \tag{7.26}$$

where $a = X_\nu a^\nu$. Now the geodesic direction is defined such that $\delta a^\mu = 0$ for a displacement $dx^k = a^k d\tau = \gamma^k_\nu a^\nu d\tau$. Hence $\tilde{\delta} a^i d\tau = \tilde{\delta} dx^i =$

$d^2x^i + \tilde{\Gamma}^i_{kl}dx^l dx^k$, and the vanishing of the expression (7.26) means that both orthogonal parts, namely the term in parentheses and da vanish. Therefore, $a = \text{const.}$ and, making use of Eq. (7.25) and of the skew symmetry of F_{ik},

$$\frac{d^2x^i}{d\tau^2} + \tilde{\Gamma}^i_{kl}\frac{dx^k}{d\tau}\frac{dx^l}{d\tau} = \frac{\varepsilon a}{2}r\sqrt{\kappa}F^i_{k}\frac{dx^k}{d\tau} \tag{7.27}$$

which is Eq. (53a) of Ref. [41a]. Comparing this result with $d^2\mathbf{x}/dt^2 = (e/m)\mathbf{E}$ one finds $\varepsilon ar = 2e/m\sqrt{\kappa}$. Note that Planck's constant does not occur here. For, true to his criticism of calling Klein 'a grand master of confusion', Pauli strictly limits this first paper [41a] to the classical theory.

Since the remaining two paragraphs of this first paper are devoted to the form of the laws of nature in very general terms, we may pass directly to the second paper [41b] devoted entirely to the Dirac equations for matter waves. The astonishing fact, suggesting that a five-dimensional description indeed might be a natural one, is that there are *five* 4×4 matrices $\gamma_1...\gamma_4$ and $\gamma_5 = \gamma_1\gamma_2\gamma_3\gamma_4$ which satisfy the anticommutation relations (6.30). Pauli himself called it 'a natural point of departure for the theory' (translated from Ref. [2b], p. 190) in his letter to Einstein of 16 July 1933, the day after the submission of his two papers. Pauli generalizes this fact to a set of five 4×4 matrices $\alpha_\mu(X)$ acting on a four-component projective spinor Ψ and satisfying

$$\alpha_\mu\alpha_\nu + \alpha_\nu\alpha_\mu = 2g_{\mu\nu}. \tag{7.28}$$

For the mathematics Pauli could follow Schrödinger's work [49a]. In Section 2 he proves the existence of transformation matrices S such that

$$\alpha'_\mu = S^{-1}\alpha_\mu S; \ \Psi' = S^{-1}\Psi; \ \text{Det}|S| = 1, \tag{7.29}$$

and in Section 3 he connects them to coordinate transformations

$$X'^\mu = a^\mu_{\nu}X^\nu; \ \alpha'^\mu = a^\mu_{\nu}\alpha^\nu, \tag{7.30}$$

where $\alpha^\mu = g^{\mu\nu}\alpha_\nu$. This connection associates a four-column representation S to the rotation group (a^μ_{ν}) in the five-dimensional space of coordinates X^μ. As indicated by Eqs. (7.1), keeping α^5 fixed, which means keeping X^5 fixed, projection by $(1 \pm \alpha^5)/2$ leads to the two-component spinors of van der Waerden.

Of particular importance for physics is the existence of a non-singular matrix A which associates to each α_μ its Hermitian conjugate,

$$\alpha^+_\mu = \eta A\alpha_\mu A^{-1}, \tag{7.31}$$

where η is the sign of the determinant $g = \text{Det}|g_{\mu\nu}|$. Defining in analogy to $\bar{\psi}$ in Eq. (6.31) $\bar{\Psi} = \Psi^* A$, one obtains a real scalar $\bar{\Psi}\Psi$ and a real five-vector $\bar{\Psi}\alpha^\mu\Psi$.

Covariant differentiation of projective spinors is constructed in Section 4 of Ref. [41b] by restricting Eqs (7.29), (7.30) to infinitesimal transformations of the form $S = 1 + \varepsilon^\rho \Lambda_\rho$ with infinitesimal ε^ρ. This involves new matrices Λ_μ which are defined by the conditions

$$\alpha_{\mu;\rho} \equiv \frac{\partial a_\mu}{\partial X^\rho} - \Gamma^\sigma_{\mu\rho}\alpha_\sigma + \Lambda_\rho\alpha_\mu - \alpha_\mu\Lambda_\rho = 0 , \tag{7.32}$$

and the result is

$$\Psi_{;\rho} \equiv \frac{\partial \Psi}{\partial X^\rho} + \Lambda_\rho\Psi . \tag{7.33}$$

For the generalized Dirac equations, of course, one needs the analogous relations in the affine space V_4 derived in Section 5; they are, recalling Eqs (7.18), (7.20) to define $\alpha_k = \alpha_\mu\gamma^\mu_k$,

$$\alpha_{k;l} \equiv \frac{\partial \alpha_k}{\partial x^l} - \tilde{\Gamma}^m_{kl}\alpha_m + \tilde{\Lambda}_l\alpha_k - \alpha_k\tilde{\Lambda}_l = 0. \tag{7.34}$$

(Note that X^l in Eq. (44) of Ref. [41b] should read x^l.)

The connection between Ψ and the affine spinor ψ of Dirac theory is set as follows:

$$\Psi = \psi F^l, \tag{7.35}$$

where ψ and F are homogeneous functions of degree 0 and 1, respectively, and F is real, so that l is the degree of homogeneity of Ψ. Assuming the form of the wave equation satisfied by Ψ to be

$$\alpha^\mu(\Psi_{;\mu} + kX_\mu\Psi) = 0 \tag{7.36}$$

one deduces that, for $\varepsilon = +1$ and in order to have continuity equations of the form (6.33),

$$\frac{\partial}{\partial X^\mu}(\sqrt{g}\overline{\Psi}\alpha^\mu\Psi) = 0; \quad \frac{\partial}{\partial x^k}(\sqrt{\tilde{g}}\overline{\psi}\alpha^k\psi) = 0, \tag{7.37}$$

l in (7.35) and k in (7.36) must both be imaginary. In order to have agreement with the Dirac equation (6.34) one has to set (see Eq. (56) of Ref. [41b])

$$\frac{l}{2\pi} = \frac{i\sqrt{2}}{rl_{Kl}}; \quad k = -ik_c - l, \tag{7.38}$$

where l_{Kl} is the universal Klein length of Eq. (7.15), $k_c = mc/\hbar$ (see [48b]), and r is defined in Eq. (7.25), in particular, $r = \sqrt{2}$.

Under these conditions one arrives at the following generally covariant form of the Dirac equation (6.34):

$$i\alpha_0\alpha^k\left(\frac{\partial\psi}{\partial x^k} + \tilde{\Lambda}_k\psi - \frac{ie}{\hbar c}\Phi_k\psi\right) + k_c\psi + i\frac{r}{8}\sqrt{\kappa}F_{kl}\alpha^k\alpha^l\psi = 0 \tag{7.39}$$

where $\alpha_0 = \alpha_\mu X^\mu$ and $\alpha_0^2 = 1$. This is Eq. (58) of Ref. [41b] multiplied from the left by $i\alpha_0$. Although not mentioned by Pauli, it is remarkable that the third term in Eq. (7.39) is just the Pauli term introduced in Eq. (6.62) with an anomalous magnetic moment $\mu = i(r/8)\sqrt{\kappa}$ due to gravity. This Pauli term also appears in the bracket of expression (36) for the four-current density v^m in Pauli's second paper [47b] with Solomon. There Pauli observes in the footnote that such a supplementary term had been discussed by him as a hypothesis to describe the behaviour of a 'neutron' and he quotes Carlson and Oppenheimer (Ref. [56b] of Chapter 6) where Pauli's neutron appears in print for the first time. The remaining Section 6 of Ref. [41b] is devoted to the construction of the variational principle and the energy–momentum tensor yielding the above field equation. It is also worth mentioning that in Chapter 10 of his lectures [39c] of 1953 Pauli derived the generally covariant Dirac equation (7.39), but without the third term, directly in the framework of four-dimensional space–time, following the work of Schrödinger [49a] (where $\tilde{\Lambda}_k$ is called Γ_k) and Bargmann [49b] in Berlin.

In 1933 Valentin Bargmann (b. 1908) was Pauli's collaborator in Zurich where, following Pauli's suggestion (see letter [315] from Pauli to Einstein of 16 July 1933, Ref. [2b], p. 189), he formulated the Dirac equation for the semi-vectors introduced by Einstein and Mayer[49c]. In 1936 he obtained his doctorate under Wentzel at the university. The following year he emigrated to the United States where he collaborated with Einstein at the Institute for Advanced Study and eventually became professor of mathemathics at Princeton University (see Ref. [38], p. 496). Pauli submitted the two papers [41a,b] at the end of the summer term on 15 July 1933. But the five-dimensional unification continued to bother him, as shown in his letters to Oskar Klein of summer 1935 (letters [417], [418], and [421] of Ref. [2b]). In letter [421] he writes: 'However, I always have a vague hope, whether the purely classical five-dimensional theory might not give a hint for the quantum theory of $e^2/\hbar c$—provided that, from somewhere, a good idea will come' (translated from Ref. [2b], p. 430).

While checking Solomon's paper had no further consequences for Casimir's career, another suggestion that Pauli had made at the beginning of Casimir's assistantship contributed to the latter's fame. It was 'The algebraic proof of the complete reducibility of the representations of semi-simple Lie groups' [49d] that Weyl had not been able to give. Casimir succeeded in supplying the proof for the rotation group, but there he got stuck. So he wrote to van der Waerden (see page 117) who completed the proof, which was published under both their names [49d]. Thus Casimir's name became associated with certain quadratic forms commuting with all the infinitesimal elements of the group (see the comment in Ref. [2b], p. 398). These Casimir operators and their commutation properties were derived by Pauli quite generally in the algebraic part of his *Spezialvorlesung* [49e] at ETH during the summer term of 1954 and again in his lectures [49f] at the new Theory Division of CERN in Copenhagen in September 1955.

During the term Pauli and his assistant Casimir had discussed a quite different problem, namely the modification of the Klein–Nishina formula due to the effect of the state of the scattering electron, this formula supposing the electron to be free and at rest. Both Pauli and Casimir gave a communication of their respective results at the spring meeting of the Swiss Physical Society, held in Lucerne on 6 May 1933 [50a,b,c]. In his paper Pauli thanks Heisenberg for the suggestion to test the Klein–Nishina formula in a special frequency limit. This test was negative so that, in his letter to Heisenberg of 2 June, Pauli claims the bottle of Mosel wine they had bet (Ref. [2b], letter [311]). According to Casimir's recollections the trip to Lucerne was a memorable event mainly because Pauli was driving there in his own car and had invited Casimir together with Felix Bloch, Walter Elsasser, and David Inglis to come along (see Ref. [25c], p. 144).

As we saw at the beginning of this chapter, Pauli had obtained his Swiss driver's licence on 25 July 1932. Casimir recalls: 'Pauli had driven us—that is, David Inglis, Elsasser, Felix Bloch and myself—from Zurich to Lucerne in the morning, and apart from Pauli's slightly disconcerting habit of saying from time to time, "*Ich fahre ziemlich gut*" (I'm driving rather well)—a statement he underlined by turning around to his passengers and by releasing his hold on the wheel—nothing untoward happened. In the evening Pauli was drinking fruit juice with a wry face. Suddenly he changed his mind and ordered a whisky and soda. That was still all right, but when he had ordered a second one and showed no signs of wanting to stop there, we became really worried. So we held a council of war and decided to offer him more drinks and then Inglis could drive us home. The first part of the operation went according to plan but, when we suggested that Inglis should drive, Pauli bluntly refused. He was going to drive and we could either come along or stay in Lucerne. By then the last train to Zürich had left, and anyway we did not want to let Pauli go all by himself. So we went, with Inglis sitting beside Pauli, ready to grasp the wheel in an emergency. We even had one more passenger, Bhabha, who had missed the last train and was sitting on the floor of the car. Pauli sounded his horn several times, hit one curb, swerved to the other side of the street, where he hit the other curb, then managed to find his bearings and got going. It was a memorable trip' (Ref. [25c], pp. 144–5).

But Casimir's narration continues: 'Pauli would still say from time to time, "*Ich fahre ziemlich gut*", but when the car went screeching around curves Inglis would say sternly, "*Dass heisst nich gut fahren*" (That is not called good driving), which had a somewhat sobering effect. Once a rising moon came just over the crest of a hill and Pauli started swearing at the driver who did not dim his headlights. Once Pauli said, "Here I know a shortcut", and suddenly turned into an unpaved track. It came to an end at the wagon shed of a farm, and after some angry comments about people that put wagon sheds across his shortcut, Pauli turned round and went back to the main road. But that was all; we came safely home' (Ref. [25c], p. 145; see also Ref. [25d], pp. 78–9).

David Rittenhouse Inglis (1905–95) had come from Ohio State University to spend the year 1931/2 in Leipzig and at ETH. From there he successively went to the Universities of Pittsburgh, Princeton, and Johns Hopkins. During the war he was a collaborator on the Manhattan Project at Los Alamos from 1943 to 1945. In 1949 Inglis joined the Argonne National Laboratory and in 1969 the University of Massachusetts. He did pioneering work on nuclear structure and after the war played a leading role in nuclear disarmament [51a].

Walter M. Elsasser (1904–91) spent the summer term of 1933 with Pauli in Zurich after having been forced by Nazi law to leave Germany. On 22 May Pauli wrote to Peierls, who was himself in difficulty: 'At the moment Elsasser is here, I am endeavouring to find a position for him in Paris. Today also Fröhlich surfaced, a victim of the Third Reich and of metal physics' (translated from letter [310], Ref. [2b], p. 163). Elsasser spent three years in the company of the French physicists at the Laboratory of Frédéric Joliot-Curie in Paris. There he coined the notion of magic numbers governing the stability of nuclei. In 1936 Elsasser emigrated to the United States where he wrote his well-known memoirs [51b] (see Ref. [51c]).

Herbert Fröhlich (b. 1905) emigrated in 1933 to England where in 1948 he became professor of theoretical physics and head of department at the University of Liverpool. He was one of the leading theoreticians of solid-state physics, introducing field-theoretic methods and writing two well-known books. In 1950 Fröhlich had initiated the idea of electron–phonon interactions forming superconducting electron pairs and thereby predicted the isotope effect. He was a Fellow of the Royal Society and earned several honorary degrees; in 1972 he was awarded the Max Planck Medal [51d].

Hendrik Brugt Gerhard Casimir (1909–2000) grew up, together with two sisters, in The Hague where his father was a devoted educator at a private secondary school. At this school Casimir passed the final examination in 1926, including Greek and Latin. Hesitating between linguistics, economics, and law, guided by his father and influenced by Ehrenfest who was on friendly terms with the former, he finally chose, in the autumn of 1926, to study theoretical physics with the latter at Leiden. He was taken care of by Goudsmit and Uhlenbeck who were preparing their doctorates (see page 116). After a stay of over two years at Bohr's institute in Copenhagen, Casimir became Ehrenfest's assistant, and in November 1931 received his Ph.D. degree. Pauli had invited him to be his assistant during the academic year 1932/3 (see Ref [26b], Documents I.57, 58, 62, 63). So, after Ehrenfest and Bohr, Pauli became the third of Casimir's teachers; Ref. [25c] is dedicated to the three of them. At the end of this academic year in August 1933 Casimir married Josina Jonker, returned to Leiden, and replaced Ehrenfest (see the last section).

In 1936 Casimir became senior assistant to Wander Johannes de Haas (1878–1960) who with Wilhelmus Hendrikus Keesom (1876-1956) succeeded Heike Kamerlingh Onnes (1853–1926) at the cryogenic laboratory of the University of Leiden. The latter had been the first to liquefy helium, and until 1923 his laboratory was the only place in the world where liquid helium was available for experimentation. Casimir used it in his experimental work but he also acted as a general theoretical counsellor and an executive of the laboratory. During the academic year 1938/9 he replaced Uhlenbeck at Utrecht and in spring 1939 gave a series of lectures at the Philips Laboratories in Eindhoven. At the same time he became an associate professor at Leiden. When the German occupation rendered things difficult there, Casimir accepted a position at Philips in April 1942 and moved to Eindhoven with his family of three children (another daughter was born in 1943) but keeping his professorship at Leiden. He stayed with Philips after the war, becoming 'co-director of the research laboratory in 1946 and member of the board of management, the highest executive body of the company in 1956' (Ref. [25c], p. 247). He stayed with Philips until his retirement in 1972.

Gradually, this career of course meant 'Goodbye to Physics'. Pauli was amused and puzzled but also disappointed 'that someone who was in his opinion *'nicht ganz dumm'* (not entirely stupid) should of his own free will decide to sink down to what he considered the lower realms of human achievement and activity' (Ref. [25c], p. 244). Casimir recalls: 'Whenever I met him in later years he always addressed me as "Herr Direktor". He even told other people "If you go to Holland and should happen to meet Casimir, give him my regards and call him Herr Direktor. *Das ärgert ihn nämlich* (for that vexes him)"' (Ref. [25c], p. 244). Making fun of Casimir's engineering activities Pauli once asked him: 'I hope you still write i and not [as the engineers] j for the imaginary unit" (translated from Ref. [26a], p. 119). But in spite of his managerial career, Casimir's research in physics has left important marks: in particular, his discovery that the retarded van der Waals forces give rise to an attraction of two parallel metal plates (the Casimir effect); the two-fluid model of superconductivity he had developed with Cornelis Jacobus Gorter (1907–80), the successor of de Haas and Keesom; and his derivation of general Onsager relations.

Notes and references

[1] Translated from: W. Pauli, letter [283a] of 7 September 1931 to Gregor Wentzel, in Ref. [2a], p. 753.

[2] (a) K. von Meyenn (ed.), *Wolfgang Pauli. Scientific Correspondence with Bohr, Einstein, Heisenberg, a.o., Volume III: 1940–1949* (Springer, Berlin and Heidelberg, 1993); (b) K. von Meyenn (ed., with the cooperation of A. Hermann and V. F. Weisskopf), *Volume II: 1930–1939* (Springer, Berlin and Heidelberg, 1985).

[3] Notes of my conversations with Franca Pauli on 10/11 October 1983.

[4] C. G. Jung, (a) *Seelenprobleme der Gegenwart* (Rascher, Zurich, 1931); (b) *Wandlungen und Symbole der Libido* (Deuticke, Leipzig, 1925).

[5] W. Pauli, *Letters to Erna Rosenbaum* (1932), Wissenschaftshistorische Sammlungen (WHS), ETH-Library, Hs. 176.

[6] (a) K. von Meyenn (ed.), *Wolfgang Pauli. Scientific Correspondence with Bohr, Einstein, Heisenberg, a.o., Volume IV/Part II: 1953–1954* (Springer, Berlin and Heidelberg, 1999); (b) *Correspondence Erna Rosenbaum*, Wissenschaftshistorische Sammlungen (WHS), ETH-Library, Hs. 1056. (c) *Note*: On 4 January 1944 Pauli wrote to his intimate friend

Max Delbrück from Princeton: 'On the 28th December a very old letter reached me from Dorette Wagner (now Frau Gerber). She has now got a second child, a son, and seems to be very happy.—More secrets by word of mouth' (translated from Ref. [2a], letter [687]). The association with Dora seems compelling.

[7] (a) C. G. Jung, 'Traumsymbole des Individuationsprozesses', in Ref. [7b], pp. 57–260; English translation, *Individual Dream Symbols in Relation to Alchemy*, in Ref. [7b], pp. 39–214; (b) C. G. Jung, *Psychologie und Alchemie, Gesammelte Werke, 12* (Walter-Verlag, Olten, 4. Auflage 1984). English translation by R. F. C. Hull, *Psychology and Alchemy* (Routledge & Kegan Paul, London, 1953). Reprinted in: *Collected Works, 12* (Princeton University Press, Princeton, NJ, 1968), 2nd edn.

[8] C. A. Meier (ed., with the cooperation of C. P. Enz and M. Fierz), *Wolfgang Pauli und C. G. Jung. Ein Briefwechsel, 1932-1958* (Springer, Berlin and Heidelberg, 1992).

[9] (a) C. G. Jung, 'Psychologie und Religion', in Ref. [9b], pp. 17–125; English translation, 'Psychology and Religion', in Ref. [9b], pp. 3–105; (b) C. G. Jung, *Zur Psychologie westlicher und östlicher Religion, Gesammelte Werke, 11* (Walter-Verlag, Olten, 5. vollständig revidierte Auflage 1988); English translation, *Psychology and Religion: West and East, Collected Works, 11* (Princeton University Press, Princeton, NJ, 1969), 2nd edn; (c) O. Fröbe-Kapteyn (ed.), *Eranos Jahrbuch 1935. Westöstliche Seelenführung* (Rhein-Verlag, Zurich, 1936).

[10] (a) K. Holländer, 'Chronik von Leben und Werk', in Ref. [10d], p. 24; (b) H. H. Holz, 'Die Eranos Jünger', in Ref. [10d], p. 92; (c) M. Fordham, 'Jung, Carl Gustav', in: C. C. Gillispie (ed.), *Dictionary of Scientific Biography* (Scribner's Sons, New York, 1973), Vol. VII, pp. 189–93; (d) M. Meier (ed.), *Carl Gustav Jung. Person, Psyche und Paradox*, in the monthly *Du* (Tagesanzeiger, Zurich, August 1995).

[11] (a) C. G. Jung, 'The Tavistock Lectures', in Ref. [11c], pp. 3–177; (b) C. G. Jung, 'The Symbolic Life', in Ref. [11c], pp. 265–90; (c) C. G. Jung, *The Symbolic Life. Miscellaneous Writings* (Routledge & Kegan Paul, London and Henley, 1977).

[12] A. Jaffé (ed.), *Erinnerungen, Träume, Gedanken von C. G. Jung* (Walter-Verlag, Olten, 4. Auflage 1986).

[13] (a) H. Egner, 'Ungehobene Schätze. Die Vorlesungen Carl Gustav Jungs an der ETH Zürich', in Ref. [10d], p. 68; (b) C. Borgeaud (ed.), *Histoire de l'Université de Genève. L'Académie et l'Université au XIXe siècle. 1814-1900* (Georg, Geneva, 1934); (c) M. Guibal and J. Nobécourt, *Sabina Spielrein. Entre Freud et Jung* (Aubier Montaigne, Paris, 1981).

[14] (a) C. G. Jung, *Aion. Beiträge zur Symbolik des Selbst, Gesammelte Werke. 9,II,* (Walter-Verlag, Olten, 8. Auflage 1992); English translation, *Aion: Researches into the Phenomenology of the Self, Collected Works, 9,II* (Princeton University Press, Princeton, NJ, 1968), 2nd edn;

(b) C. G. Jung unter Mitarbeit von Marie-Louise von Franz, *Mysterium Coniunctionis. Untersuchungen über die Trennung und Zusammensetzung der seelischen Gegensätze in der Alchemie, Gesammelte Werke, 14* (Walter-Verlag, Olten, 5. Auflage 1990); English translation, *Mysterium Coniunctionis. An Inquiry into the Separation and Synthesis of Psychic Opposites in Alchemy, Collected Works, 14* (Princeton University Press, Princeton, NJ, 1970), 2nd edn.

[15] C. G. Jung, *Psychologische Typen, Gesammelte Werke, 6* (Walter-Verlag, Olten, 14. Auflage 1981); English translation, *Psychological Types, Collected Works, 6* (Princeton University Press, Princeton, NJ, 1971).

[16] (a) In the letter [69] to Jung, Ref. [8], p. 134, Pauli clearly says that his analysis with Jung ended in 1934. (b) *Note*: The originals of letters [29] and [30] only indicate day and month, but not year and place. The guess by the editor, namely '1939' and 'Zollikon-Zürich' should be replaced by '1934' and 'Zürich', respectively. This important rectification follows, for letter [29], from the fact that 7 May 1934 was indeed a Monday while 7 May 1939 was not. In addition the reference of footnote *a* was published in 1934 (see Ref. [17]). As to letter [30], reference is made in the first line to letter [6] which dates from just two days earlier.

[17] C. G. Jung, 'Seele und Tod', *Europäische Revue* X/4, 229 (1934).

[18] (a) N. Bohr, 'Licht und Leben', *Naturwissenschaften* **21**, 245 (1933); (b) N. Bohr, 'Light and Life', *Nature* **131**, 421 (1933); (c) P. Jordan, 'Über den positivistischen Begriff der Wirklichkeit', *Naturwissenschaften* **22**, 485 (1934).

[19] A. Jaffé (ed., with the cooperation of G. Adler), *C. G. Jung. Briefe. Erster Band 1906-1945* (Walter-Verlag, Olten, 1972).

[20] C. G. Jung, 'Bruder Klaus', *Neue Schweizer Rundschau* I/4, 223 (1933). Reprinted in Ref. [9b], pp. 328–34.

[21] Translated from Ref. [22a], p. 89 (see also Ref. [22d], p. 7).

[22] W. Pauli, 'Die allgemeinen Prinzipien der Wellenmechanik', (a) in: H. Geiger and K. Scheel (eds), *Handbuch der Physik. Vol. 24, Part. 1* (Springer, Berlin, 1933), 2nd edn, pp. 83–272. Reprinted by Authority of the Alien Property Custodian (Edwards Brothers, Ann Arbor, Michigan, 1947). (b) Revised edition without Ziff. 6–8 from Part B, in: S. Flügge (ed.), *Handbuch der Physik. Vol. 5, Part 1* (Springer, Berlin, 1958), pp. 1–168. Reprinted in Ref. [23], Vol. 1, pp. 771–938. (c) Ref. [22b] with Ziff. 1, part of Ziff. 5 and Ziff. 6–8 from Part B of Ref. [22a] as Appendices I–III, and equation numbers as in Ref. [22d], edited and commented by N. Straumann (Springer, Berlin and Heidelberg, 1990); (d) W. Pauli, *General Principles of Quantum Mechanics*, translated from Ref. [22b] and from Ziff. 6–8 of Part B, Ref. [22a], the latter added as Chapter X, and edited by P. Achuthan and K. Venkatesan, Preface *Pauli and the Development of Quantum Theory* by C. P. Enz (Springer, Berlin and Heidelberg, 1980).

[23] R. Kronig and V. F. Weisskopf (eds), *Collected Scientific Papers by Wolfgang Pauli. In Two Volumes* (Wiley Interscience, New York, 1964).

[24] (a) W. Pauli, 'Matter', in Ref. [25a], pp. 27-34; (b) *Note*: The Copenhagen Interpretation, whose stumbling block in general is the particle–wave duality, has more recently been defended by an expert; see A. Zeilinger's letter in *Physics Today*, February 1999, pp. 13–15.

[25] (a) C. P. Enz and K. von Meyenn (eds), *Wolfgang Pauli. Writings on Physics and Philosophy* (Springer, Berlin and Heidelberg, 1994); (b) M. Fierz, 'Pauli, Wolfgang', in: C. C. Gillispie (ed.), *Dictionary of Scientific Biography* (Scribner's Sons, New York, 1974), Vol. X, pp. 422–5; (c) H. B. G. Casimir, *Haphazard Reality. Half a Century of Science* (Harper & Row, New York, 1983); (d) H. B. G. Casimir, 'Erinnerungen aus den Jahren 1932-1933', in Ref. [26a], pp. 75–9.

[26] (a) C. P. Enz and K. von Meyenn (eds), *Wolfgang Pauli. Das Gewissen der Physik* (Vieweg, Braunschweig, 1988); (b) C. P. Enz, B. Glaus, and G. Oberkofler (eds), *Wolfgang Pauli und sein Wirken an der ETH Zürich* (vdf Hochschulverlag ETH, Zurich, 1997).

[27] (a) L. Landau and R. Peierls, 'Erweiterung des Unbestimmtheitsprinzips für die relativistische Quantentheorie', *Zeitschrift für Physik* **69**, 56 (1931); (b) C. P. Enz, 'Le rôle de l'espace et le problème de localisation en physique moderne, vus en particulier par Wolfgang Pauli', *Archives des Sciences (Geneva)* **39**, 185 (1986). Reprinted in *Epistemologia* X, 187 (1987). (c) W. Pauli, 'P. A. M. Dirac, The Principles of Quantum Mechanics', *Naturwissenschaften* **19**, 188 (1931). Reprinted in Ref. [23], Vol. 2, pp. 1397–8. (d) J. von Neumann, *Mathematische Grundlagen der Quantenmechanik* (Springer, Berlin, 1932). English translation by R. T. Beyer, *Mathematical Foundations of Quantum Mechanics* (Princeton University Press, Princeton, NJ, 1955).

[28] (a) W. Pauli and V. Weisskopf, 'Über die Quantisierung der skalaren relativistischen Wellengleichung', *Helvetica Physica Acta* **7**, 709 (1934). Reprinted in Ref. [23], Vol. 2, pp. 701–23. (b) W. Pauli, 'Diracs Wellengleichung des Elektrons und geometrische Optik', *Helvetica Physica Acta* **5**, 179 (1932). Reprinted in Ref. [23], Vol. 2, pp. 481–501. (c) 'Tagung der Schweizerischen physikalischen Gesellschaft', *Helvetica Physica Acta* **5**, 211, 233 (1932).

[29] (a) R. Peierls, *Bird of Passage. Recollections of a Physicist* (Princeton University Press, Princeton, NJ, 1985); (b) R. Peierls, 'Was ich von Pauli lernte', in Ref. [26a], pp. 68–74; (c) H. Weyl, 'Elektron und Gravitation. I', *Zeitschrift für Physik* **56**, 330 (1929).

[30] (a) W. Pauli, 'On the Earlier and More Recent History of the Neutrino', in Ref. [25a], pp. 193–217; (b) W. Pauli, *Theory of Relativity*, translated by G. Field. *With Supplementary Notes by the Author* (Pergamon Press, London, 1958). The latter are reprinted in Ref. [23], Vol. 1, pp. 238–63.

(c) W. Pauli, 'Space, Time and Causality in Modern Physics', in Ref. [25a], pp. 95–105; (d) *Physics Today*, August 1971, p. 17.

[31] W. Pauli, 'Exclusion Principle and Quantum Mechanics', in *Les Prix Nobel en 1946* (Norstedt & Söner, Stockholm, 1948), pp. 131–47 (also Editions du Griffon, Neuchâtel, 1947). Reprinted in Ref. [23], Vol. 2, pp. 1080–96.

[32] (a) P. Ehrenfest, 'Einige die Quantenmechanik betreffenden Erkundigungsfragen', *Zeitschrift für Physik* **78**, 555 (1932); (b) W. Pauli, 'Einige die Quantenmechanik betreffenden Erkundigungsfragen', *Zeitschrift für Physik* **80**, 573 (1933). Reprinted in Ref. [23], Vol. 2, pp. 608–21. (c) P. Ehrenfest, 'Address on Award of Lorentz Medal to Professor W. Pauli', *Versl. Akad. Amsterdam* **40**, pp. 121-6 (1931). Reprinted in: M. Klein (ed.), *P. Ehrenfest. Collected Scientific Papers* (North-Holland, Amsterdam, 1959), pp. 617–22.

[33] (a) N. Bohr, 'Licht und Leben—noch einmal', *Naturwissenschaften* **50**, 725 (1963); (b) M. Delbrück, 'A Physicist's Renewed Look at Biology: Twenty Years Later', *Science* **168**, 1312 (1970).

[34] (a) W. Pauli, '*Paul Ehrenfest* †', in Ref. [25a], pp. 79–84; (b) M. J. Klein, *Paul Ehrenfest. Volume 1. The Making of a Theoretical Physicist* (North-Holland, Amsterdam, 1970). This volume covers the time until about 1920. Volume 2 never appeared. (c) E. Broda, 'Paul Ehrenfest (1880-1933)', in: *Neue Östereichische Biographie ab 1815. Grosse Östereicher* (Amalthea-Verlag, Vienna and Munich, 1982), Vol. XXI, pp. 110–19.

[35] P. und T. Ehrenfest, 'Begriffliche Grundlagen der statistischen Auffassung in der Mechanik', *Encyklopädie der mathematischen Wissenschaften* (Teubner, Leipzig, 1912), Vol. IV/32, pp. 1–90.

[36] Translated from Ref. [37].

[37] W. Pauli, 'Ergebnisse der exakten Naturwissenschaften. Herausgegeben von der Schriftleitung der Naturwissenschaften. Zehnter Band', *Naturwissenschaften* **20**, 186 (1932). Reprinted in Ref. [23], Vol. 2, pp. 1399–400.

[38] A. Pais, '*Subtle is the Lord ...*' *The Science and the Life of Albert Einstein* (Oxford University Press, New York, 1982).

[39] (a) T. Kaluza, 'Zum Unitätsproblem der Physik', *Sitzungsberichte der Preussischen Akademie der Wissenschaften 1921*, p. 966; (b) O. Klein, 'Quantentheorie und fünfdimensionale Relativitätstheorie', *Zeitschrift für Physik* **37**, 895 (1926); (c) 'Probleme der Allgemeinen Relativitätstheorie, Vorlesung gehalten an der Eidg. Techn. Hochschule, Zürich, 1953, von Prof. W. Pauli. Ausgearbeitet von Ch. Enz' (handwritten, 111 pp.), *Pauli Archive*, CERN.

[40] A. Einstein, (a) 'Gravitational and Electromagnetic Fields (Report)', *Science* **74**, 438 (1931); (b) A. Einstein and W. Mayer, 'Einheitliche Theorie von Gravitation und Elektrizität', *Sitzungsberichte der Preussischen Akademie der Wissenschaften. Phys. Math. Klasse* **35**, 541 (1931).

[41] W. Pauli, 'Über die Formulierung der Naturgesetze mit fünf homogenen Koordinaten', *Annalen der Physik* **18** (1933), (a) 'Teil I: Klassische Theorie', p. 305; (b) 'Teil II: Die Diracschen Gleichungen für die Materiewellen', p. 337. Reprinted in Ref. [23], Vol. 2, (a) pp. 630–61, (b) pp. 662–97.

[42] A. Hermann, K. von Meyenn and V. F. Weisskopf (eds), *Wolfgang Pauli. Scientific Correspondence with Bohr, Einstein, Heisenberg, a.o., Volume 1: 1919-1929* (Springer, New York, 1979).

[43] O. Klein, 'The Atomicity of Electricity as a Quantum Theory Law', *Nature* **118**, 516 (1926).

[44] *Note*: Planck, at the end of his 'Fifth communication On irreversible radiation processes' which appeared in the *Reports of the Prussian Academy* on 1 June 1899—the year before the communication on 19 October 1900 of his first guess of the celebrated spectral distribution formula (Ref. [45a], p. 687)—had introduced a system of natural units of measure (Ref. [45a], p. 600), of length:

$$l_{Pl} \equiv \sqrt{\frac{hG}{c^3}} = 4.051 \times 10^{-33} \text{ cm};$$

of mass:

$$m_{Pl} \equiv \sqrt{\frac{hc}{G}} = 5.456 \times 10^{-5} \text{ g};$$

of time:

$$t_{Pl} \equiv \sqrt{\frac{hG}{c^5}} = 1.351 \times 10^{-43} \text{ s};$$

and of temperature:

$$T_{Pl} \equiv \frac{h}{k}\sqrt{\frac{c^5}{hG}} = 3.551 \times 10^{32} \text{ K}.$$

Here $c = 2.998 \times 10^{10}$ cm/s is the velocity of light, $h = 2\pi\hbar = 6.626 \times 10^{-27}$ g cm^2/s Planck's constant, $G = \kappa c^4/8\pi = 6.673 \times 10^{-8}$ cm^3/g s^2 Newton's gravitational constant (see Ref. [30b], Eq. (409), p. 163), and $k = 1.381 \times 10^{-16}$ g cm^2/s^2 K Boltzmann's constant.

Now the Compton wavelength $\lambda_c(m) = h/mc$ has been recognized on pages 251–2 as the limit of localizability of a body of mass m. On the other hand, the Schwarzschild radius $r_S(m) = 2Gm/c^2$, where according to Eqs (417), (418) of Ref. [30b] $g_{44} = 0$, may be considered as the limit of gravitational stability since at $r = r_S/2$ the gravitational energy Gm^2/r equals mc^2. It follows that the Planck mass m_{Pl} may be understood as the coincidence $\lambda_c(m) = r_S(m)/2$. At this mass $\lambda_c(m_{Pl}) = l_{Pl}$ and $\lambda_c(m_{Pl})/c = t_{Pl}$. Therefore, the Planck length and the Planck time may be understood as the limit where space–time itself ceases to be well defined (see Ref. [27b] and Ref. [45b], pp. 13 and 1200–1).

[45] (a) *Max Planck. Physikalische Abhandlungen und Vorträge* (Vieweg, Braunschweig, 1958), Vol. 1; (b) C. W. Misner, K. S. Thorne, and J. A. Wheeler, *Gravitation* (Freeman, San Francisco, 1973).

[46] (a) K. von Meyenn and M. Baig, 'Klein, Oskar Benjamin', in: F. L. Holmes (ed.), *Dictionary of Scientific Biography* (Scribner's Sons, New York, 1990), Vol. 17, Supplement II, pp. 480–4; (b) S. Deser, 'Oskar Klein', *Physics Today*, June 1977, p. 67.

[47] W. Pauli and J. Solomon, 'La théorie unitaire d'Einstein et Mayer et les équations de Dirac', *Journal de physique et le radium* 3 (1932), (a) Part I: p. 452, (b) Part II: p. 582. Reprinted in Ref. [23], Vol. 2, Part I: pp. 461–72, Part II: pp. 473–80.

[48] (a) J. A. Schouten and D. van Dantzig, 'Generelle Feldtheorie', *Zeitschrift für Physik* **78**, 639 (1932). (b) *Note*: In Refs [41a,b] κ is not Einstein's but Pauli's gravitational constant $\kappa_{Pauli} = c^2 \kappa_{Einstein}$ (see Eq. (409) and footnote 320 on p. 163 of Ref. [30b]). It follows that f_{ik} has the dimension of a reciprocal length, and not of a length. Note also that in Ref. [41b] Pauli writes h for $\hbar$. In the Introduction (Ref. [41b], p. 341) Pauli claims that l is a dimensionless number, whereas Eq. (7.38) shows it to be a reciprocal length. This error is corrected by Pauli in his letter [417] to Oskar Klein of 18 July 1935 (Ref. [2b], p. 423).

[49] (a) E. Schrödinger, 'Diracsches Elektron im Schwerefeld I', *Sitzungsberichte der Preussischen Akademie der Wissenschaften 1932. Phys.-Math. Klasse*, p. 105; (b) V. Bargmann, 'Bemerkungen zur allgemein-relativistischen Fassung der Quantentheorie', *Sitzungsberichte der Preussischen Akademie der Wissenschaften 1932. Phys.-Math. Klasse*, p. 346; (c) V. Bargmann, 'Über den Zusammenhang zwischen Semivektoren und Spinoren und die Reduktion der Diracgleichungen für Semivektoren', *Helvetica Physica Acta* 7, 57 (1934); (d) H. Casimir and B. L. van der Waerden, 'Algebraischer Beweis der vollständigen Reduzibilität der Darstellungen halbeinfacher Liescher Gruppen', *Mathematische Annalen* **111**, 1 (1935); (e) W. Pauli, 'Gruppentheorie und Quantenmechanik', Notes by C. P. Enz (1954, unpublished); (f) W. Pauli, 'Continuous Groups in Quantum Mechanics', CERN-Publication 56-31, 1956. Reprinted in: G. Höhler (ed., with the cooperation of S. Flügge, F. Hund, and F. Trendelenburg), *Ergebnisse der exakten Naturwissenschaften* (Springer, Berlin and Heidelberg, 1965), Vol. 37, pp. 85–104.

[50] (a) 'Tagung der Schweizerischen Physikalischen Gesellschaft', *Helvetica Physica Acta* **6**, 246 (1933); (b) W. Pauli, 'Über die Intensität der Streustrahlung bewegter freier Elektronen', *Helvetica Physica Acta* **6**, 279 (1933). Reprinted in Ref. [23], Vol. 2, pp. 622–9. (c) H. Casimir, 'Über die Intensität der Streustrahlung gebundener Elektronen', *Helvetica Physica Acta* **6**, 287 (1933).

[51] (a) S. S. Hanna, D. Kurath, and G. A. Peterson, 'David Rittenhouse Inglis', *Physics Today*, June 1997, pp. 109–10; (b) W. M. Elsasser, *Memoirs of a Physicist of the Atomic Age* (Hilger, Bristol, 1978); (c) H. Rubin and P. Olson, 'Walter M. Elsasser', *Physics Today*, February 1993, p. 99; (d) H. A. Strauss and W. Röder (eds.), *International Biographical Dictionary of Central European Emigrés, 1933-1945. Volume II. The Arts, Sciences and Literature* (Saur, Munich, 1983), Part A–K, p. 344.

8
Life with Franca

A new assistant and two weddings

Finally, he lifted his head and said, 'Who are you?' I answered, 'I am Weiss-kopf. You asked me to be your assistant.' He replied, 'Oh, yes. I really wanted Bethe, but he works on solid-state theory, which I don't like, although I started it.' [1]

This welcome by Pauli evidently was not the sort to boost Weisskopf's self-assurance. In fact, he had wondered before why Pauli had chosen him in the first place. Fortunately, Weisskopf 'had been well prepared by Peierls for events like this' (Ref. [2], p. 76). Before becoming Pauli's assistant Weisskopf had been in Cambridge on a Rockefeller Fellowship. There he met Peierls who by now 'was a refugee from Hitler's Germany' (Ref. [2], p. 72). Showing him Pauli's letter, 'Peierls told me not to worry. "Pauli is actually a very nice, almost childish person", he said. "He always says what he thinks, and that's what sometimes offends people"' (Ref. [2], p. 74). Weisskopf told Pauli, 'the Klein-Kaluza approach to general relativity—I am unable to understand that. . . . I am ready to work on everything else.' He accepted the conditions because he was already somewhat bored by the work: 'Pauli gave me some problem to study . . . and after a week he asked me what I had done about it. I showed him my solution, and he said, "I should have taken Bethe after all (*ich hätte doch den Bethe nehmen sollen*)"' (Ref. [2], pp. 75–6).

But these personal problems really pale in the face of the economic and political background in Europe: extended unemployment in the wake of the Great Depression and the rise of the National Socialists to absolute power in Germany, followed by the eviction of Jews from state positions as a consequence of the law for the re-establishment of the notion of state functionaries (*Berufsbeamtentum*) of 7 April 1933 (see page 77). This meant no less than the 'selling out of German universities'. In physics and mathematics Einstein, Schrödinger, Born, Stern, Franck, Weyl, von Mises, Courant, and many less well-known professors were forced to leave or left in protest (see Ref. [3a], p. 211, also Ref. [3b], Chapter 15). Zurich had its share of social and political difficulties, and they did not leave Weisskopf untouched.

With a 2000-year history starting as a settlement of the Helvetes, in 15 BC Zurich became a Roman customs station between northern Europe and the

Alpine passes, secured by a castle on the Lindenhof which at that time was a true island between the two rivers of the Limmat and the Sihl (see page 216). In 1220 Zurich became a Free City of the German Empire and in 1351 joined the Swiss Confederation. In 1519 Ulrich Zwingli (1484–1531), a priest at Zurich's main church, the Grossmünster on the right bank of the Limmat, started the reformation which turned Zurich into an industrious and pious city. This spirit radiated to the west through Berne to the Geneva of Jean Calvin (1509–64). Zurich witnessed a period of cultural bloom during the sixteenth and eighteenth centuries, competing in the sciences with Geneva. At the Vienna Congress of 1814–15 the Swiss Confederation was granted perpetual neutrality, and in 1848 it adopted the form of today's democracy. Zurich then underwent major expansion, overtaking Geneva as Switzerland's largest city. During World War I, from 1914 to 1917, Lenin (Wladimir Iljitsch Uljanow 1870–1924) lived in Berne and Zurich. With the influx of Jewish refugees from Germany starting in 1933, Zurich became one of the most important cultural centres in German-speaking Europe, particularly its theatre, the Schauspielhaus (see also Ref. [2], p. 76). But unemployment and quarrels between the communists and the frontists commanded by the rising German Nazis gave rise to tensions, particularly in the outer industrial quarters to the north-west of the city. Inasmuch as foreign interests were involved, neutrality demanded—at least in theory—that in these quarrels the police acted even-handedly.

In 1932 the spectre of unemployment also touched ETH. Swiss students and assistants had complained to the Fremdenpolizei (Foreigners' Police) of the canton that a number of assistant positions, for which Swiss candidates were available, were occupied by foreigners. This complaint reached the federal government in Berne which on 13 July 1932 informed President Rohn at ETH. Rohn in turn informed the persons concerned, among whom were Pauli and Scherrer (see Ref. [4], footnote 91, p. 47). Thus started a somewhat toilsome correspondence between the latter and the president. On 18 July Scherrer told Rohn the obvious: namely that, in order to give Swiss physicists a chance to occupy positions abroad, reciprocity was essential, and that foreign collaborators often bring much welcome stimulation (Ref. [4], document I.55). Pauli, who fully concurred with Scherrer's arguments, justified the appointment of Peierls by observing that no Swiss candidate of equal qualification existed (Ref. [4], I.56). Thus prepared by Pauli's careful argumentation, the president voiced no objections against either the appointment of Casimir or that of Weisskopf—at least concerning the winter term of 1933/4 and the summer term of 1934 (Ref. [4], I.66 and I.71). Starting in the latter, however, the general economic crisis also affected federal employees in that a deduction of 4% on all fixed salaries was applied (see Ref. [4], I.71), and this deduction even increased to 8% in 1936 (see Ref. [4], I.86). In the winter term of 1934/5 President Rohn again raised the foreigner question (Ref. [4], I.73). However, on mentioning explicitly that the two Swiss Heinz Schild (b. 1910), who in 1933 wrote his diploma paper under Pauli (see Ref. [4], p. 434), and Ernst Carl Gerlach Stueckelberg (1905–84) were not available (Ref. [4], I.74), Pauli had his wish granted to prolong Weisskopf's tenure for the winter term of 1934/5 (Ref. [4], I.75). Also an extension for the following summer term encountered no problems (Ref. [4], I.79).

Then, however, unexpected political complications arose. Weisskopf recounts: 'During my first year I lived at a pension or boardinghouse, presided over by a Mrs. Biske. . . . Among the other people staying at Mrs. Biske's was a man named Heinz Kurella and his girlfriend, Charlotte. He was an avowed Communist who had fled Germany and was secretly working for a weekly Communist newsletter. . . . I struck up a friendship with him and Charlotte, whom we all called "Red Countess" because of her aristocratic good looks. Although Heinz and I disagreed on many political issues, it was very interesting and stimulating to talk to him and Charlotte. We often played cards together in the evenings, and he meticulously entered our scores in a notebook. Later I was to regret this completely innocent habit of his.' Weisskopf continues: 'One day the Swiss police got wind of Kurella's Communist activities. When he saw police officers coming up the steps of the pension, he managed to get out the back door and escape with the help of Mrs. Biske. But Charlotte was caught and arrested. . . . [She] was eventually freed and expelled from Switzerland. Kurella . . . was . . . executed during the Stalin purges' (Ref. [2], pp. 76–7).

On 24 January 1935 the Swiss Minister of Internal Afairs, Bundesrat (Federal Counsellor) Philipp Etter (1891–1977) informed President Rohn that confidential information from the 'Swiss Patriotic Union' told him that a certain Victor Weisskopf, assistant in theoretical physics at ETH, 'is believed to be actively working for the IInd and the IIIrd International [the Komintern founded by Lenin in 1919]'. He wished to know whether Weisskopf was permanently employed at ETH (translated from Ref. [4], I.76). Weisskopf was called to the office of President Rohn who asked 'whether I had ever been a member of the Communist party or had any connection to the party'. The president accepted Weisskopf's assurances to the contrary, 'and he cordially issued me out of his office. Naturally I was puzzled by this encounter.' But then the Fremdenpolizei summoned him to their offices. 'On the desk I saw a large folder with my name and the ominous words *Kommunistische Umtriebe* (Communist Activities) written on it. . . . To my amazement it contained a copy of every letter I had received since I had been in Switzerland. I was shocked that in this free country my mail had been opened.' But then the police officer mentioned Kurella: '"we have proof that you were very close to him, since his notebook contains repeated references to you, using your first name". I realized that they must have found Kurella's records of our card-game scores' (Ref. [2], pp 91–92). It was thanks to President Rohn that Weisskopf was not expelled immediately. His assistantship was even extended to the summer term of 1935 and the winter term of 1935/6 (Ref. [4], I.79,84). On 2 July 1935 Weisskopf, with Pauli's knowledge, sent President Rohn a formal declaration refuting any connection to the Communist Party or any related organization and having no intention to seek such a connection in the future (Ref. [4], I.81). But he comments: 'In 1935, however, the *Fremdenpolizei* was no joking matter. Some of the officers were Nazi sympathizers' (Ref. [2], p. 93).

Victor Friedrich Weisskopf was born in Vienna in 1908, the second of two sons and a daughter of a well-to-do Jewish family of a lawyer, in which classical music was cultivated. For Victor, who played the piano himself, music became a lasting love. Developing strong socialistic leanings during high school, he entered the University of Vienna at 18, where he was introduced to theoretical physics by Hans Thirring (see pages 25f). After two years he went to Göttingen where he worked with Born and with Wigner and in April 1931 received his doctorate. After a visit to Heisenberg in Leipzig and an assistantship with Schrödinger in Berlin he spent the summer of 1932 in Leningrad, Moscow, and Kharkov with Landau, before spending one year on a Rockefeller Fellowship in Copenhagen and in Cambridge. In Copenhagen he fell in love with a girl from Bohr's circle, Ellen Tvede, whom he married during his time in Zurich and with whom he spent 57 happy years. After Zurich, Weisskopf went back to Copenhagen, then to Kiew, and finally to Rochester in the United States. From 1943 to the end of the war he was engaged on the Manhattan Project to build the atom bomb in Los Alamos, New Mexico. Since 1946 he has been a professor at MIT, the Massachusetts Institute of Technology in Cambridge in the United States. During this time he regularly visited Europe and, in particular, ETH and from 1961 to 1965 he was Director-General of CERN in Geneva. In 1956 he was awarded the Planck Medal. He played an important part in the development of quantum field theory, particularly the problems of renormalization and of the so-called Lamb shift, and he was the inventor of the optical model in nuclear theory on which he wrote an authoritative book with John M. Blatt (b. 1921). He received numerous honorary degrees.

Weisskopf describes his wedding plans as follows: 'After my first year in Zurich, Ellen and I decided to get married. I had to ask Pauli for permission to take ten days off and go to Copenhagen. In general, Pauli didn't like his assistants to leave in the middle of the term, so I was anxious as I walked to his office. When I asked for a leave, he quickly asked me the purpose of my trip. When I told him I wanted to get married, he said, to my great surprise, "Oh, I approve, because I am also getting married." And when I told him whom I was going to marry, he said, with a smile, "Oh, the Daughter of the Regiment," referring to Ellen's popularity among Bohr's disciples' (Ref. [2], p. 77). This is a nice story which allows me to switch over now to that of Pauli's marriage. However, the story demands some historical correction because Weisskopf's wedding took place on 4 or 5 September 1934, according to Refs [2] and [5a], p. 1776, respectively, while Pauli's was on 4 April of that year. Also, September was in the middle of the long summer vacation and, therefore, Weisskopf did not need any permission. According to his own account (see Ref. [5b], p. 83), Weisskopf's leave was for 10 days in January 1934, and its purpose, very likely, was to get engaged. Be that as it may, on the Wednesday after Easter, 4 April 1934 (see *Note* [5c]), Pauli married Franca Bertram in London (see page 248). This may well have been Pauli's first visit there, for, on 1 May of the previous year, he had written to Dirac 'I was never in England' (Ref. [3a], letter [308]).

Franziska (Franca) Pauli-Bertram (1901–87) was born in Munich on 16 December 1901, the only child of Hans Bertram and Maria Schwitz. Hans Bertram was born on 7 July 1874 in Kufstein, Austria [5d]. Franca lived with her divorced mother in Italy and in Cairo, Egypt, where she went to high school (Gymnasium) [5e], but during World War I she was back in Munich where in 1918 she attended a commercial school [7a]. This was the year when Pauli came to Munich to study with Sommerfeld, encountering chaos and upheaval there (see page 49). According to the City Archive (Stadtarchiv) of Zurich, in 1922 Franca lived in Zurich under her mother's name Schwitz. But starting in 1929 she is registered as a secretary (*Büroangestellte*) under the name Bertram, first at Albisstrasse 7, then at Birmensdorferstrasse 15, and in 1932 at Hadlaubstrasse 17. After the death of her mother, Franca lived with her Italian stepfather in Zurich [5e, 7a]. Unfortunately, Franca was not very communicative about her pre-Pauli period in Zurich—she had been the secretary of an important communist politician [7b]. This, possibly, was the physicist Friedrich Adler (1879–1960), a friend of Einstein's who from 1907 to 1911 was a private dozent at the University of Zurich, and in 1916 shot the Austrian Prime Minister Karl Count von Stürgkh (1859–1916). Sentenced to death but pardoned in 1918, Adler became secretary-general of the Socialist Workers' International from 1923 to 1940 [7c,d]. He was an acquaintance of the Paulis.

There are many hints concerning Franca's life in Zurich in a novel by the Zurich writer Kurt Guggenheim (see page 216). The title of this novel, *Gerufen und nicht gerufen* [7e], is a reference to the Latin inscription over the entrance to Jung's house in Küsnacht (see page 244); Jung himself occurs in the novel. Although it is of course impossible to draw conclusions for Franca's biography, this book reflects well her very practical intelligence. In Chapter IX Franca (alias Josefine Spitz), later called Jolande, is described as coming back from the United States after the war with her husband Pauli (alias Paul Mende). Waiting for a taxi in front of the railway station facing Bahnhofstrasse she reflects on her pre-Mende days and her acquaintances of that period, none of whom would be interested in knowing her now—except perhaps Guggenheim (alias Karl Dinhard), the writer. Josefine's move from Munich to Zurich and her relations are described in Chapter XXII which starts with a description of Mende and his wife at their home in Zollikon (alias Küsnacht) after the war.

The fact is that Franca and Kurt Guggenheim had been lovers. When their relationship broke up and Franca was distressed, the Guggenbühls, who were friends both of Pauli and of Guggenheim (the latter was godfather to one of the Guggenbühl children), invited Franca and Pauli for the inauguration of their new home (*Hausräuke*) at Wehrenbachhalde 22 (now Eierbrechtstrasse 72) in the north-eastern district of Eierbrecht. Adolf Guggenbühl had known Pauli since about 1929, and was fascinated by Pauli's strange personality. Pauli was a frequent weekend guest of the Guggenbühl family; they went for excursions together into the nearby mountains, to Amden, above Walensee, and to Wildhaus in Toggenburg, the Thur Valley. Pauli liked Adolf junior (b.

1923) very much, '*den Adolfi mag ich einfach gut* (little Adolf I simply like)'. In 1936, just when the Spanish Civil War broke out, the Guggenbühls and the Paulis with sister Hertha and stepmother Maria Rottler (see pages 10–11) spent their vacation at the seaside resort of Fiume, now the Croatian Rijeka, on the Adriatic. Later, in 1938–40 when the Paulis lived in their new house in Zollikon, Adolf junior often went for walks with Pauli in the neighbouring woods [7b].

Adolf Guggenbühl was the founder, together with his wife Helen (born Huber) (1899–1966) and her brother Fortunat Huber (1896–1984), of the Swiss magazine *Schweizer Spiegel*. In another, much longer novel entitled *Alles in Allem*, Kurt Guggenheim pictures the life in Zurich between the beginning of the century and the end of World War II [7f]. This story is built around the biography of Adolf Guggenbühl senior (alias Walter Abt junior) and his family and that of Kurt Guggenheim (alias Aaron Reiss). In this novel many actors appear under their real names: Albert Einstein, the Federal Counsellor Ludwig Forrer, Trudi Schoop of the dance school (see page 209), the well-known Swiss painter Ferdinand Hodler, Professor Alfred Kleiner, Einstein's thesis adviser at the university, the mayor of the city Emil Klöti, Colonel Henri Guisan, who became the general of the Swiss Army during World War II, Mussolini, Emperor Wilhelm II, and many others, but not Pauli, nor Franca.

At the party with the Guggenbühls in summer 1933 Pauli tells Miss Bertram about America. She is fascinated by him. 'He sees everything in a totally different light to her. On departure Guggenbühl asks Pauli if he would accompany Miss Bertram home. He asks where she lives: Hadlaubstrasse [17; he lives at Hadlaubstrasse 41]. "Oh then I might take you along, indeed", says Pauli. Miss Bertram is deeply offended by this. Internally she turns away from him. But they meet again. They are so totally different but yet comprehend each other so well. At last, Pauli says, "now we marry", and it is quite natural, they both are young and Miss Bertram is ingenuous and enjoying life' [7g]. Shortly after they had met at the Guggenbühls, Franca joined Pauli in his apartment at Hadlaubstrasse 41, into which he had moved with Kate [7a] (see page 210).

In mid-August Pauli and his sister Hertha spent a week as guests of Max Born in Selva Gardena (Wolkenstein) in South Tirol (see page 77) where Pauli met many friends. One of them, whom he had met before in Berlin [7a], was the famous pianist Arthur Schnabel whose concerts in Zurich later were among Pauli's frequent attendances at the Tonhalle, Zurich's concert hall. At the end of August Pauli went by car on vacation to southern France, probably with Franca (see Ref. [3a], letter [320] and p. 729). At Christmas he and Franca went to Vienna where Pauli presented her to his father (Ref. [3a], p. 245). And at Easter 1934, on the way to their wedding in London, they visited Pauli's friend Hecke in Hamburg (Ref. [3a], p. 306). On 15 March Bohr congratulated Pauli on his forthcoming marriage and invited both of them to his new grand home at the Carlsberg brewery (see page 90) for the week after Easter (Ref. [3a], letter [366], p. 309). Less than four months later Bohr and his wife lost their eldest son Christian Bohr (b. 1916) (see page 90) in a sailing accident in the Kattegat

Fig. 8.1 Franca Pauli-Bertram at wedding, 4 April 1934 (CERN)

separating Denmark from Sweden (see Ref. [3a], letters [376], [377], [380]). On Pauli and Franca's condolences Bohr answered in touching words, knowing 'that you [Pauli] are the same faithful friend in joy and in sorrow' (translated from Ref. [3a], letter [381] of 3 August 1934).

'At another party at the Guggenbühls Jung is also present. He is sitting across the table from Franca but he does not talk to her. When Franca learns of Pauli's first marriage she observes to Pauli that Jung should have known that this new binding could lead to a devastating catastrophe. "Why did Jung not talk to her?" But Pauli says that Jung knew from Pauli's dreams that the binding will be good' [7g]. These comments by Franca reflect some mistrust towards Jung: 'Pauli, the extremely rational thinker, subjected himself to total dependence on Jung's magical personality' [7g]. 'Jung has thrown Pauli into the arms of Franca' [7a]. Franca was a personality with strong opinions. Contrary to what is written on pages 247–8, she claimed that 'on Pauli's decision to marry, Jung reacts brutally, he turns completely away from Pauli. For Pauli this disengagement of the binding to Jung is catastrophic: On skiing at Zürs Pauli suddenly says to Franca "the earth is shaking". He wants to thrash someone (*verprügeln*)' [7a]. Pauli and Franca spent a skiing vacation at Zürs on the Arlberg, Austria, from 22 December 1934 to 4 or 6 January (see postcard [399] to Heisenberg on 11 December 1934 in Ref. [3a]). They were both taking skiing lessons and Wolfgang advanced faster than Franca [7a]. But the earth shaking under Pauli's feet must have been a very real experience: The desire to thrash someone is reminiscent of the insecurity expressed in Pauli's letter [30] to Jung quoted on page 249. We must remember that two months earlier Pauli had written to Jung on 26 October, perhaps under Franca's influence, saying that he wished to discontinue his visits with Jung, 'unless anything particular should happen to me' (see page 247).

While the Paulis were at Zürs, the Weisskopfs spent a skiing vacation at the small resort of Hochkrumbach, only some 15 km from Zürs, from where they paid the Paulis a short visit (see Ref. [2], p. 79). The Paulis and the Weisskopfs were on good, although at that time not close, terms. But on one occasion Pauli was angry at both his wife and his assistant. The latter had published a paper on the self-energy of the electron containing an essential error of sign (see Ref. [2], p. 80), and Pauli expressed his anger by not talking to him for a while. In his distress Weisskopf took the exceptional step of seeking solace with Franca. But Pauli was not talking to her either, because she had caused some damage to their car [7a]. Weisskopf wrote: 'Franca was a wonderful person. Pauli could not have been an easy person to live with, and Franca made his life much more pleasant. She provided him with a home where he felt at ease and could pursue his many interests in a calm atmosphere' (Ref. [2], p. 78). 'Pauli never talked about physics with Franca, "physics I prefer to talk with my assistants", but they talked about philosophy. His main concern was the unification of the rational and the irrational' [7a].

As already mentioned, Zurich in the mid-1930s was an important centre of German culture. Rudolf Humm, Scherrer's friend from Göttingen (see page 198) wrote: 'Zurich then was blessed with several literary Salons which, however, were not kept by "good" Zurich women (those had more modest interests) but by Jewish ladies or then ladies married to Jews. Such was the Salon *Rosenbaum* at Stadelhoferstrasse [a short street off Rämistrasse near Bellevue; the house is number 26, called "Baumwollhof", see Ref. [8b], p. 85 f.], a true aviary for passing foreign or native folks endowed with spirit' (translated from Ref. [8a], pp. 105–6). Wladimir Rosenbaum (1894–1984) at that time was at the height of his career as one of Zurich's best-known lawyers. But when in July 1936 General Franco started the Spanish Civil War and the Swiss Federal Council, in the name of neutrality, decreed an embargo on the supply of weapons to Spain (meaning its leftist government), Rosenbaum, who found this decree unacceptable, became actively involved in the weapons trade. Eventually he was caught, and the ensuing impeachment, aided by a wave of anti-Semitism, in the spring of 1937 led to an abrupt end to his career (see Chapters VII and VIII of Ref. [8b]).

After describing many celebrities, among them the painter Max Ernst, the writer Bernard von Brentano, the lyric Max Pulver who had turned to graphology, and the Swiss sculptors Karl Geiser and Hermann Haller (1880–1950) who was the son of Einstein's boss at the Patent Office in Berne and who in 1936–7 created a beautiful bust of Pauli (exhibited in the Pauli Room at CERN, Geneva), Humm writes: 'The lady of the house, born Ducommun (her grandfather [Elie Ducommun (1833–1906)] was recipient of a Nobel Peace Prize [1902]), of French-Swiss (*welscher*) and Bernese coinage, was a colleague who, under the pseudonym *Aline Valangin* [1889–1986], in the 1930s published two volumes of Tessinese short stories of the nicest cheerfulness and freshness' (translated from Ref. [8a], p. 107). Aline and Wladimir knew Jung well, both from analyses and from social contacts [8b].

Fig. 8.2 Bust of Pauli by Zurich sculptor Hermann Haller, 1936–7, at Pauli's home in Zollikon (photo by C. P. Enz)

Early in 1935 the Rosenbaums had invited the Viennese writer Elias Canetti (1905–94) to their house for a reading from his works. Among the guests were James Joyce, Max Pulver, Bernard von Brentano, and Pauli. Jung and Thomas Mann had been invited but did not attend. Canetti wrote that after the lecture, Pauli, for whom he had great esteem, 'held me a benevolent little discourse, sensed that my thoughts drifted off and begged me somewhat more urgently to listen to him since, after all, he had also listened to me'. The reason for Canetti's distraction was that Pauli's appearance reminded him of the writer Franz Werfel. But Pauli's manner of speaking was quite different: 'he wished to guide me on a Jungian path. . . . We parted on the best understanding' (translated from Ref. [8c], p. 191). Later in the evening Pulver told Canetti that he had got samples of Goebbels' and Göring's handwriting 'and even another one that you may guess but that is completely secret [Hitler?]' (translated from Ref. [8c], p.

193). Bernard von Brentano (1901–64), who lived in Küsnacht, was a friend of Pauli; they met almost every week [7a].

In physics, meanwhile, Zurich was busier than ever. In January 1933 Scherrer's new large lecture hall (see page 198) was inaugurated by a special demonstration lecture 'repeated three times', as Scherrer told President Rohn in his annual report, thanking him for his confidence in letting Scherrer plan the new structures himself (Ref. [4], I.68). Also mentioned in the report is that Sommerfeld from Munich and Franck from Göttingen had given talks during the year and that another *Physikalische Vortragswoche* had been organized, this time on very low temperatures and on nuclear physics. Among the 10 main conferences were those of Keesom from Leiden (see page 272), Joliot from Paris, and Patrick Maynard Stuart Blackett (1897–1974) from Cambridge who spoke on the emerging new field of cosmic ray physics (Ref. [4], p. 54). This meeting took place at ETH from 28 June to 1 July 1933. Born had come from Selva Gardena to negotiate with Blackett on an offer from Cambridge, on which occasion he probably invited Pauli to Selva Gardena (Ref. [3a], pp. 170 and 207).

In October 1934 Pauli attended an international meeting on metal physics in Geneva [8d] where he met Sommerfeld and where he had extended discussions with Bethe about neutron–proton exchange forces and, in particular, a possible connection with β-radioactivity (see letter [392] in Ref. [3a]). This meeting had been organized by Jean Weiglé (1901–68), the director of the physics department of Geneva University. There was an excursion by funicular car to the nearby Salève mountain, from which one enjoys a spectacular view of Geneva and the lake and, on the opposite side, of Mont Blanc. Salève, however, is situated in French territory. Now Sommerfeld had no French visa, so Weiglé took him illegally to the top in his two-seated open car. Weiglé's wife was the daugther of the Rabbi of New York, a beautiful and amiable woman who gave elegant invitations. Weiglé's assistant was André Mercier (1913–99), who later became professor of theoretical physics at the University of Berne. His mother, who lived in Geneva, also gave an invitation to Sommerfeld [8e].

With his three-part paper 'On the constitution of atomic nuclei' [9a], stimulated by the discovery of the neutron and written in the second half of 1932, which earned him an invited talk at the Seventh Solvay Congress of 22–29 October 1933 in Brussels, Heisenberg became one of the leaders in the field. But, in spite of Pauli's and Scherrer's urging, Heisenberg did not attend the *Vortragswoche* (Ref. [3a], Pauli's letter [312] of 16 June), nor did he follow Pauli's proposition to meet in Selva Gardena or elsewhere on 10 August (Ref. [3a], letter [321] of 25 July; see also Pauli's letter [320] of 22 July). The Nazi authorities were meddling ever more deeply with academic affairs and even with private life. Concerned about the exodus he saw at his university in Leipzig and elsewhere, Göttingen in particular, on 20 April Heisenberg had asked the 75-year-old Planck, who was considered a symbol of moral integrity, for a meeting in Berlin. A second meeting took place after Planck had an audience with Adolf Hitler on 16 May. More meetings, including one with Max von

Fig. 8.3 Pauli and Arnold Sommerfeld at the Metal Congress in Geneva, October 1934 (CERN)

Laue, followed (see Ref. [3b], Chapter 15, Refs [10a], Chapter 12 and [10b], Chapters 16 and 17). Heisenberg continued to trust the noble side of the German soul, which inevitably drew criticism from contemporaries and historians alike. Rose wrote: 'In June 1933 he [Heisenberg] informed Max Born, quite without embarrassment, that "in spite of some nasty things that have been happening here within the workings of science itself [one assumes he means the dismissals of Jewish professors], I know that among those in charge in the new political situation, there are men for whose sake it is worth sticking it out. Certainly in the course of time the splendid things will separate from the hateful." Again there is a wishful self-deluding fallacy involved here' (Ref. [10b], p. 247; see Ref. [3a], p. 168 for more details in Heisenberg's original letter to Born of 2 June 1933).

In his extensive correspondence with Heisenberg in the years 1933–4 Pauli's greetings were always warm, even tender. But this did not mean that Pauli did not apply his proverbial criticism to Heisenberg's activities. On 29 September 1933 he wrote to Heisenberg: 'Stern also told me on his passage through Zurich, that you had got bestowed the Planck Medal in Würzburg while you were staying in Copenhagen (how sharp your momentum then was determined!). Without just formally omitting a congratulation to this fact, I want to wish you, as a *complementary viewpoint*, the following from my whole heart: *May the spirit that reigns over Planck's scientific production and his personal life not overwhelm too much your publications and your life!* (In a closer analysis I indeed find in the activity in question Planck's traits, which I find in the deep—not superficial!—sense as sloppy (*schlampig*); whereas it seems to me that such traits are totally missing, e.g. with H. A. Lorentz or Bohr.)' And Pauli continues: 'In this connection I must emphasize that several turns in your review in Naturwissenschaften [11a] of Planck's book [speeches and conferences on *Ways of physical epistemology*] struck me unpleasantly. If you admit that statements on "reality of the external world" make sense at all—to wit, are hypotheses that may be true or false—you already give the devil of the "ism-philosophy" the small finger and soon he takes you the whole hand.— Maybe we will converse about it also in Brussels. Very hearty greetings from your faithful W. Pauli' (translated from letter [323] in Ref. [3a], p. 214).

I venture to explain this criticism as follows: 'Pauli had no "persona", at least not in his private life' [7g]. This means that he did not present a studied 'face' towards other people. Planck, on the other hand, cultivated a Kantian image of a humble consciousness that all knowledge is based on hypotheses, but that there is a 'real outside world'—Kant's '*Ding an sich*' (thing in itself)—and a 'moral principle' somewhere far beyond, for which we ought to strive (see Ref. [11a]). This is, I believe, what Pauli considered '*im tiefen Sinne als schlampig*' (in the deep sense as sloppy), because it allowed Planck 'to follow a straight and almost too certain path, also where unfathomable epistemological abyss menace to the right and to the left of the path' (translated from Ref. [11a]). For Pauli 'the *a prior character* of Kant's rationally formulated ideas, laid down once and for all, is thus transferred to the pre-existent images (archetypes) present and operating outside of consciousness ("in the unconscious")' (Ref. [12a], p. 126).' But these images are not 'somewhere far beyond' but may be discovered inside ourselves if we only have the honesty to plunge beyond our 'persona' into our deeper self. To Pauli's criticism Heisenberg responded: 'What you write about Planck's philosophy surely contains a grain of truth. I also have hinted to it myself in the last sentence [see the above quotation from Ref. [11a]]—to a stronger critique I considered myself not to be authorized. As to the objective side of the problem I believe, however, in agreement with Planck, that the sentence "there exists a real external world" may be used as well also for the definition of the words contained in it as any other more positivist sentence' (translated from letter [325] of 7 October 1933, Ref. [3a], p. 218).

Even Heisenberg's assistant Felix Bloch (see page 205), who had a Swiss passport, felt insecure with the Nazi law. Cassidy writes: 'Bloch had already applied for a Rockefeller stipend to work in the United States before he learned of the anti-Semitic laws. According to the rules of the Education Board, which administered the stipend, a fellow had to specify where he or she would work after the end of the grant. At Heisenberg's urging, Bloch indicated Leipzig. As the new law became public, Bloch wrote a worried letter to Bohr, then in the United States, asking Bohr to intervene with the Education Board to change his future location to Copenhagen. . . . Bohr agreed, prompting a thank-you from Werner [Heisenberg] "for . . . your efforts on behalf of our young physicists, whose well-being lies on all our hearts," and an apology "for all of that which is now happening in this country."' And Cassidy adds: 'Little remains to illuminate the extent of Heisenberg's role as conduit to Bohr and Pauli, both of whom were active in international refugee organizations. Most of the correspondence on such matters was destroyed when Pauli and Bohr themselves joined the exodus from Europe during the war' (Ref. [3b], pp. 320–1). Some evidence for the statement that Pauli had destroyed compromising correspondence may be seen in the fact that no counterparts exist of the three letters mentioned on page 8 that in November 1938 Pauli had addressed to Dirac on behalf of his cousin Felix Pauli (see also Ref. [3a], p. 376, in particular footnote 8, and the comment on pp. 303–4). Before Heisenberg took the train to Stockholm with his mother to attend the Nobel festivities, on 25–26 November he made 'a [not so] mysterious trip to Zurich [see Ref. [3a], p. 228 and Pauli's letter [333] to Heisenberg, p. 232] on behalf of refugee physicists' (Ref. [3b], p. 324).

Heisenberg received the Nobel Prize of 1932 'for the creation of quantum mechanics, the application of which has, *inter alia*, led to the discovery of the allotropic forms of hydrogen' and not, as Pauli had assumed in his congratulary letter [330] of 11 November 1933, 'for your famous and up till now unrefuted *hydrodynamic dissertation*' [12b]. In this letter Pauli continued: 'Be this as it may, I congratulate you quite heartily (with much less mixed joy than the other time for the Planck Medal) and rejoice also in general about the fact that this time there are all theorists that got the Prize' (translated from Ref. [3a], p. 225). Simultaneously, Schrödinger and Dirac were together awarded the Prize of 1933 'for the discovery of new productive forms of atomic theory'. In these citations the reference to 'the allotropic [ortho and para] forms' (Heisenberg) and to 'new productive forms' (Schrödinger and Dirac) are a compromise to conform with the statutes of the Nobel Foundation which emphasize the applied aspect of a discovery.

Among Pauli's students in 1932 was Konrad Bleuler (1912–92) who in 1935 wrote a diploma thesis on quantum electrodynamics (see Ref. [4], p. 434), discussing his problem also with assistant Weisskopf. The regular duties of Pauli's assistant were minor, as Weisskopf observes in his recollections: 'My main duties were to be ready for discussions of his work and to assist him in any new developments. However, I once actually became his accomplice. His wife had asked him to change his eating habits in an attempt at reducing his famous girth. But Pauli loved sweets of all kinds, and many afternoons he would want to continue our discussions in a nearby *Konditorei* (café) that had wonderful pastries. I had to promise never to mention these clandestine outings to Mrs Pauli (Ref. [2], p. 85).

In spring 1935 some excitement came from a quite unexpected source. Pauli wrote to Heisenberg on 15 June: 'Einstein once again has expressed himself publicly on quantum mechanics, namely in the issue of Physical Review of 15 May [12c] (in cooperation with Podolsky and Rosen—not a good company, by the way). As is well known, this is a catastrophe each time when it happens.' But later in the letter Pauli wrote: 'Quite independently from Einstein, it seems to me that, with a systematic foundation of quantum mechanics, one should *start* more from the composition and separation of systems than is the case hitherto (e.g. with Dirac). This, indeed, is—as Einstein has perceived *correctly*—a very fundamental point in quantum mechanics' (translated from letter [412], Ref. [3a], pp. 402 and 404).

The Einstein–Podolsky–Rosen (EPR) paper did not leave the main players in the quantum drama indifferent. Schrödinger wrote to Pauli: 'For me this note was cause to rethink once more the case (which, obviously, we have all essentially known long since) and others related to it' (translated from letter [413], Ref. [3a], p. 406). Heisenberg sent with his letter [414] to Pauli a long appendix with the heading 'Is a deterministic completion of quantum mechanics possible?' And Bohr wrote a response in the *Physical Review* [12d]. What EPR realized is that, unless environmental influences wash out the effects— *decoherence*—division of a quantum system does not result in independent 'elements of reality' but leaves *synchronicities* between the parts at all times after division. And this, for Einstein, was cause to question the completeness of quantum mechanics. This 'EPR-paradox' has engendered an uninterrupted discussion to this day (see e.g. Ref. [12e]).

In the winter term of 1935/6 Pauli received his first invitation to the Institute for Advanced Study in Princeton, New Jersey, 70 km south-west of New York, then called the Flexner Institute after its first director Abraham Flexner (1866–1959). This independent research centre had been founded in 1930 with an initial donation of $ 5 million by the American merchant Louis Bamberger and his sister Mrs Felix Fuld. It was first housed in Fine Hall on the university campus but in 1940 moved into a new building called Fuld Hall just outside the campus [13a,b]. On 2 July Pauli addressed a request for a leave of absence to President Rohn, at the same time proposing that assistant Weisskopf be retained for the winter term and be commissioned to take over Pauli's regular lectures on statistical mechanics and quantum theory (Ref. [4], I.80). This request was discussed and granted by the School Council on 12 July (Ref. [4], I.82). During the absence of Pauli and Franca the Weisskopfs took over their apartment at Hadlaubstrasse [7a].

Of course, the Paulis' stay in Princeton was not intended for work only, but also to do some travelling and to show Franca the country. For this purpose Pauli planned to buy a new car which he wanted to bring back to Switzerland. On 2 November 1935 he wrote to Weisskopf from Princeton: 'Chrysler in Zurich replied on inquiry that they agree to sell me an importation bonus (*Einfuhrkontingent*) of 600 Francs for a Plymouth. That is still a good deal! Now I only wait whether Dodge also will reply positively, in order to perhaps choose between the two. The 1936 models just came out, the Plymouth costs 800 and the Dodge 900 Dollars, new from the factory' (translated from Ref. [13c], letter [421d], p. 776). In the same letter Pauli tells Weisskopf that at the end of November he will drive to Baltimore where on the 29th and 30th a meeting of the American Physical Society will take place, and that in January he will probably come through Chicago, on their way back from California where Pauli and Franca stayed from the end of December to mid-January (see Ref. [13c], letters [425b,c]). At the Institute Pauli gave lectures on 'The Theory of Positrons and Related Topics' [13d].

On 8 April 1936 Pauli wrote to his assistant: 'Your letter from the last day, at which you were sitting in the Zurich institute, I have received and have gladly taken notice that you consider my influence on you not at all as so injurious.' And he informed Weisskopf about his travel plans: 'on 18 April the boat leaves New York, on 26 we arrive in [Le] Havre, and then I drive my new Plymouth car to Zurich' (translated from letter [427e], Ref. [13c], pp. 786–7). With his former car, Pauli had had 'a *genuine* Pauli effect, i.e. without my direct intervention or visible exterior causes', he had told Weisskopf on 5 September 1934 (translated from letter [381a], Ref. [13c], p. 763). And this had cost Pauli a lot of money, forcing him and Franca to come back to Zurich from the vacation earlier than planned.

When Weisskopf went back to Copenhagen, Pauli had problems finding a successor, as he told Wentzel on 24 February 1936 (letter [427], Ref. [3a]). Pauli's choice was Nicholas Kemmer, a German from Hanover of Russian origin who had done his Ph.D. work under Wentzel (Ref. [4], I.85). But President Rohn insisted that Pauli should look for a Swiss candidate and mentioned Dr Ludwig who had visited him, Rohn, twice (Ref. [4], I.86). Guido Ludwig (b. 1907) was a Swiss from Scanf in the Grisons who had a doctorate from Munich, but Pauli found Kemmer much better. Finally, a compromise was reached, by which both Kemmer and Ludwig were appointed for the summer term 1936 to a half-position with a monthly pay of 200 SFr each, from which the 8% crisis contribution was deducted (Ref. [4], I.89). Nicholas Kemmer was born in St Petersburg in 1911, but emigrated to England in 1916 with his parents and in 1921 to Germany where he studied at Göttingen. In 1933 he came to Zurich where he attended the university and in 1935 obtained his Ph.D. under Wentzel; in 1936–8 he was at Imperial College in London and from 1946 to 1953 at Cambridge University as a lecturer. In 1953 he became the successor of Max Born at the University of Edinburgh. Kemmer predicted the π^0-meson; he is a Fellow of the Royal Society and the recipient of several prizes [13e,f].

The 'Anti-Dirac Paper': charge conjugation

Practically, it may not be possible to do much with this curiosity but it gave me pleasure that I was able again to stick one (eins anhängen) to my old enemy—Dirac's theory of the spinning electron. [14a]

Why this personal animosity towards an abstract matter, one may ask, particularly since this theory was such a success in explaining the properties of the spinning electron? This is a difficult question to answer, even when considering that Pauli liked shocking language in his private statements. The letter to Heisenberg, in which the above quotation occurs, is dated 14 June 1934, one week before Heisenberg submitted his important positron theory paper [14b]. In order to understand the complicated problems that quantum electrodynamics (QED) posed in the early 1930s and which are at the heart of the problems to be discussed in this section, it is necessary to go back as far as Dirac's hole theory paper, submitted on 6 December 1929 [15], and follow the intense exchange of letters between Pauli and Heisenberg on the subject—already mentioned on page 256—which developed while the shock waves from the purge of the German Jews were felt all over Europe and across the Atlantic. A clear presentation of the physical problems of this period was given by Wentzel in Section 2 of his contribution to the Pauli memorial volume [16a], and the more technical questions were treated by him in Sections 20 and 21 of Ref. [16b]. Another informative account is by Pais in Ref. [16c].

It all started with the difficulties posed by the negative-energy states alluded to on page 253. While according to Eq. (6.32) the particle density $\rho = (\psi^*\psi)$ is non-negative by construction, the Dirac equation (6.70) for plane waves (written there for the free neutrino) admits both signs of the energy, $\pm\hbar c\sqrt{k_c^2 + \mathbf{k}^2}$. This had the effect that, even for a physical process involving only positive-energy states, virtual intermediary states of negative energy could not be ignored. It led, among other things, to the Klein paradox that electrons with positive energy traversing a potential well much higher than the electron's energy emerged on the other side with negative energy [17a,b]. In the hope of remedying this defect, Dirac then introduced the assumption that 'all the states of negative energy are occupied except perhaps a few of small velocity' [15], their stability being assured by the exclusion principle. (In this interpretation the Klein paradox corresponds to a pair annihilation.) Dirac at first identified these 'vacant states or "holes"' as protons [15], but in a second paper [18], influenced by Weyl's new book [19], he concluded that there should exist positive anti-electrons. Weyl had written: 'Indeed, according to it [Weyl's theory] the mass of a proton should be the same as the mass of an electron; furthermore, no matter how the action is chosen (so long as it is invariant under interchange of right and left), this hypothesis leads to the essential equivalence of positive and negative electricity under all circumstances' (Ref. [19], English translation p. 263). These *positrons* were soon discovered by Carl David Anderson (b. 1905) between September 1932 and March 1933 (see Refs [16c] and [20], Section 15(f)).

It is interesting to note that 'the asymmetry between the two kinds of electricity' represented by the proton and electron had already been emphasized by Pauli in his *Encyklopädie Article* of 1921 (Ref. [21], pp. 192, 202, and 205). In his *Handbuch Article* of 1933 Pauli observes that by analogy there should also be antiprotons, and he continues: 'The factual absence of such particles then is reduced to a special initial state, in which there is indeed only one kind of particles. This appears to be unsatisfactory already because of the fact that the laws of nature in this [Dirac's] theory are symmetrical with respect to electrons and anti-electrons' (translated from Refs [22a], p. 246 and [22b], p. 197). And in his letter to Heisenberg of 16 June 1933 Pauli writes: 'I do not believe in the hole theory because I would like to have asymmetry between positive and negative electricity in the laws of nature (it does not satisfy me to shove the empirically observed asymmetry into one of the initial states of the world)' (translated from letter [12], Ref. [3a], p. 169). This asymmetry, in fact, has remained a problem in cosmology to this day (see e.g. the entry 'baryon asymmetry' in the index of Ref. [23a]). But Pauli discarded these observations in the 1958 edition of his *Handbuch Article*. His vacillating view about the symmetry between positive and negative electricity is again manifest in letter [323] to Heisenberg of 29 September 1933: 'If it is true that the laws of nature are perfectly symmetric with respect to positive and negative electricity (and all observed differences are shoved upon the *initial state* of our environment), if therefore, there also would exist a negative proton—then the free neutron would have to be allowed to exist in *two* states, provided its magnetic moment is not zero' (translated from Ref. [3a], p. 213).

Then at the Seventh Solvay Congress on 22–29 October 1933 where Heisenberg gave his invited talk on nuclear physics (see the previous section) Dirac reported on a new approach to his hole theory using a generalization to two times t, t' of the density matrix introduced by von Neumann in 1927 (see Refs [22a,b], Ziff. 9),

$$(x|R|x') = \sum_r \psi_r(x) \otimes \psi_r^*(x'), \qquad (8.1)$$

here written as a 4×4 matrix and summed over occupied positive- and negative-energy states. Dirac's motivation for using the density matrix was that the singularities occur on the light cone $x - x' = 0$ where the subtraction of unphysical contributions is of the ambiguous form $\infty - \infty$ whereas, for $x \neq x'$, the calculation could be done without encountering infinite expressions. Taking the limit $x' \to x$ afterwards, it was possible to isolate and subtract the unphysical contributions. On 11 November Dirac sent Pauli (see Pauli's response [332] of 14 November, Ref. [3a], p. 229) 'an exciting letter', as Pauli told Heisenberg the same day (letter [333], Ref. [3a], p. 232). This is Dirac's letter [330a] in Ref. [13c], p. 756. Although Pauli was not happy that the density matrix 'is used in such an essential way, since the latter appears to me somewhat artificial' (translated from letter [332], Ref. [3a], p. 229), he asked Dirac for more information concerning the split

$$R = R_a + R_b, \qquad (8.2)$$

discussed in Dirac's letter [330a], where R_b is free from singularities (letter [335] of 2 December). In letter [332] Pauli also told Dirac that he himself had explicitly calculated R_a in the force-free case at the Ann Arbor summer school in 1930. Dirac answered both of Pauli's letters on 7 December (letter [335a] in Ref. [13c], p. 757). He agreed with Pauli that 'it is not very satisfactory to use the density matrix' and suggested that Pauli and Heisenberg might succeed without it, 'then your theory need not be equivalent to mine, but may be an improvement on mine'. In letter [336] of 11 December Pauli begged Heisenberg 'Could you ask Dirac when you see him, *which is his method* for the determination of the singularities of R and how his expression for R_a looks *exactly*?' (translated from Ref. [3a], p. 236).

Heisenberg, aided by Pauli, had developed a formalism (see letters [319], [326]–[331]) which Pauli urged him in letter [336] to compare with that of Dirac and, always trusting Heisenberg's intuition, to generalize it beyond the fixed-particle-number or Hartree–Fock approximation used by Dirac in Eq. (8.1). On 21 January 1934 Pauli sent with his letter [343] to Heisenberg an appendix 'On the quantum-electrodynamic formulation of the hole theory', containing his own way of understanding the theory. There, as in Dirac's Solvay report, the emphasis is on the current density j^μ but also on the energy–momentum tensor $T_{\mu\nu}$. However, in spite of the secure existence of the positron, the negative-energy electrons were still treated as if they were physically present: 'In order to obtain finite values of j_ν and $T_{\mu\nu}$ with the occupation of infinitely many states of negative energy (already in the force-free case), certain quantities must be subtracted from the usual quantum-electrodynamic expressions for j_ν, $T_{\mu\nu}$' (translated from Ref. [3a], p. 257). On 27 January Pauli informed Dirac: 'Heisenberg and myself have now found out a possible formulation of the theory of holes—which is relativistic and gauge invariant—even in the case where the interaction with the radiation field is coming into play and the Hartree–Fock-method does not apply. On the other hand the mentioned formulation seems to guide [point] to an infinite polarization of the vacuum' (letter [344b], Ref. [13c], p. 758). This creation of an infinity of virtual electron–positron pairs in an electric field turned out to be the main problem.

The quantities to be subtracted from the usual expression (6.32) for j^μ and from

$$T_{\mu\nu} = \frac{1}{2}\Re\overline{\psi}\left\{\gamma^\mu\left(\hbar c\frac{\partial}{\partial x^\nu} + ie\Phi_\nu\right) + \gamma^\nu\left(\hbar c\frac{\partial}{\partial x^\mu} + ie\Phi_\mu\right)\right\}\psi \qquad (8.3)$$

are expressed by Eq. (8.1) as

$$\delta j^\mu = -ie \lim_{x' \to x} \mathrm{Tr}(\gamma^4\gamma^\mu(x'|R|x)) \qquad (8.4)$$

and as

$$\delta T_{\mu\nu} = \frac{\hbar c}{2} \lim \Re\mathrm{Tr}(\gamma^4\gamma^\nu(x'|R_\mu|x) + \gamma^4\gamma^\mu(x'|R_\nu|x)), \qquad (8.5)$$

respectively. While in the force-free case the 4×4 matrices $R = R^0$ and $R_\mu = R^0_\mu = \partial R^0/\partial x_\mu$ are well known (see (8.13) below), they have to be determined in the general case by using the gauge invariance of j^μ and $T_{\mu\nu}$, energy–momentum conservation of $T_{\mu\nu} + S_{\mu\nu}$ where $S_{\mu\nu}$ is the Maxwell tensor, and the equations of motion. Assuming the form $R_\mu = R^0_\mu - (e/\hbar c)r_{\mu\nu}\gamma^4\gamma^\nu$, Pauli was able to determine the symmetric matrix $r_{\mu\nu}$ as a linear function of quantities $H_{\mu\nu}$—his Eq. (20)—which are determined by the electromagnetic fields (6.24) through the equation $\Box H_{\mu\nu} = -F_{\mu\nu}$.

This impressive piece of formal theory, in which Dirac's subtraction idea of Eq. (8.2) is taken over by Pauli, now became the basis for further discussions between Pauli and Heisenberg (letters [344]–[351]). Already unhappy with some of Heisenberg's remarks, Pauli writes in the last of these letters: 'This feeling of uneasiness was mightily enhanced when yesterday I got Dirac's manuscript of his long awaited paper [23b]'. This uneasiness is reflected in letter [351] by the detailed discussion of Dirac's arguments. This letter is signed 'yours (drowned in Dirac's formulae) W. Pauli' (translated from Ref. [3a], pp. 275 and 277). Dirac's paper [23b], written in his typical lucid style, contains a second density matrix R', in which the sum runs over all unoccupied states. In $R_F = R + R'$ the sum then runs over the 'full' distribution which, for free electrons and at equal times, is just the closure relation

$$(\mathbf{x}t|R_F|\mathbf{x}'t) = \sum_r u_r(\mathbf{x}) \otimes u_r^*(\mathbf{x}') = 1\delta(\mathbf{x} - \mathbf{x}'). \tag{8.6}$$

Here $r \equiv (\sigma, \lambda)$, $\sigma = \pm$ and $\lambda = \pm$ being, respectively, the sign of the spin and of the energy $\hbar\lambda\omega$, where $\omega \equiv +c\sqrt{\mathbf{k}^2 + k_c^2}$. The wavefunctions $u_r(\mathbf{x})$ may be expressed as Fourier transforms for fixed ω of the functions $u_{\sigma\lambda}(\mathbf{k})$ defined by Eq. (6.70) with κ_c replaced by $k_c = mc/\hbar$; they satisfy the Schrödinger equation

$$H_0 u_r(\mathbf{x}) = \hbar\lambda\omega u_r(\mathbf{x}); \quad H_0(\nabla) \equiv \hbar c\gamma^4(\vec{\gamma} \cdot \nabla + k_c). \tag{8.7}$$

The passage from Eq. (8.6) to times $t' \neq t$ is obtained with the help of the integral form of (8.7),

$$u_r(\mathbf{x}t) = e^{-iH_0 t/\hbar} u_r(\mathbf{x}) \tag{8.8}$$

and by writing $\delta(\mathbf{x} - \mathbf{x}') = (2\pi)^{-3} \int d^3k \exp(i\mathbf{k} \cdot (\mathbf{x} - \mathbf{x}'))$. Now the relations (6.30) imply $(H_0(i\mathbf{k})/\hbar)^2 = \omega^2$, from which it follows that

$$e^{-iH_0(i\mathbf{k})t/\hbar} = \frac{1}{\hbar\omega}\{\hbar\omega\cos\omega t - iH_0(i\mathbf{k})\sin\omega t\}. \tag{8.9}$$

Expressing here the first term in the braces as $\hbar(\partial/\partial t)\sin\omega t$ and noting that $(\partial/i\partial t) - H_0(\nabla)/\hbar = c(\gamma^\mu \partial/\partial x^\mu - k_c)\gamma^4$ one finally obtains, with $\bar{u}_r = u_r^*\gamma^4$,

$$(x|R_F|x') = \sum_r u_r(x) \otimes \bar{u}_r(x') = -iS(x - x'), \tag{8.10}$$

which is Eq. (17) in Dirac's paper [23b]. Here and below

$$S^{()}(x) \equiv \left(\gamma^\mu \frac{\partial}{\partial x^\mu} - k_c\right)\Delta^{()}(x), \tag{8.11}$$

where the upper brackets are reserved for an index, and

$$\Delta(x) \equiv -\int \frac{d^3k}{(2\pi)^3} e^{i\mathbf{k}\cdot\mathbf{x}} \frac{\sin\omega t}{\omega} = \Delta^+(x) + \Delta^-(x) \tag{8.12}$$

is the Jordan–Pauli function (6.18) generalized to the finite mass m of the electron.

Of particular interest is the density matrix $R_1 = R - R'$ because it is symmetric with respect to the sign of the charge, as Heisenberg emphasizes in his paper [14b], where $R_1/2$ is called R_S. The distinction with these density matrices reflects the separation into positive and negative frequences (energies) which Pauli had consistently emphasized since his *Handbuch Article* of 1933 (Refs [22a], p. 248 and [22b], p. 200) and up to his *Field Quantization* of 1950–1 (Ref. [24], p. 10) and which also appears in Eqs (6.13), (6.16). Thus the density matrices R and R_1 are obtained by assigning different weights to positive and negative frequencies:

$$(x|R|x') = \sum_r \delta_{\lambda,-} u_r(x) \otimes \bar{u}_r(x') = -iS^-(x - x') \qquad (8.13)$$

and

$$(x|R_1|x') = \sum_r (-\lambda) u_r(x) \otimes \bar{u}_r(x') = S^1(x - x'). \qquad (8.14)$$

These two functions, which are Dirac's Eqs (15) and (20), respectively, are defined by Eq. (8.11) with

$$\Delta^\pm(x) \equiv \mp i \int \frac{d^3k}{(2\pi)^3 2\omega} e^{i(\mathbf{k}\cdot\mathbf{x} \mp \omega t)} \qquad (8.15)$$

and with

$$\Delta^1(x) \equiv i(\Delta^+(x) - \Delta^-(x)) = \int \frac{d^3k}{(2\pi)^3} e^{i\mathbf{k}\cdot\mathbf{x}} \frac{\cos \omega t}{\omega}, \qquad (8.16)$$

the last function being called $cD'(x)$ by Wentzel (Eq. (20.56) of Ref. [16b]). The expressions (8.10) to (8.16), which all satisfy the Klein–Gordon equation (5.34), are the invariant functions which some 15 years later played a crucial role in the development of the modern scattering- or S-matrix formulation of QED (see e.g. Ref. [25]). The definitions used here are those of Eqs [3.21, 25], [5.2, 4, 5] of Pauli's *Field Quantization* [24]. As shown explicitly by Dirac, the function $\Delta(x)$ has the same δ-function singularities (6.19) on the light cone as the Jordan–Pauli function, while $\Delta^1(x)$ has pole singularities proportional to $(\mathbf{x}^2 - t^2)^{-1}$.

Introducing the electromagnetic potentials $\Phi_\mu = (\mathbf{A}, A_0)$, Dirac supposed that these singularities are unaltered and only the functions multiplying them are modified. A local gauge transformation (6.49) may then be used to generate the substitution $H_0(\nabla) \to H_0(\nabla + (ie/\hbar c)\mathbf{A})$ by setting $\chi = -\int \mathbf{A} \cdot d\mathbf{x}$ in the identity

$$e^{-i(e/\hbar c)\chi} \left(\nabla + \frac{ie}{\hbar c} \mathbf{A} \right) e^{i(e/\hbar c)\chi} = \nabla + \frac{ie}{\hbar c} (\mathbf{A} + \nabla \chi). \qquad (8.17)$$

Similarly, the substitution $i\hbar(\partial/\partial t) \to i\hbar(\partial/\partial t) + eA_0$ may be generated. Combining the two, one then derives the identity

$$e^{-i(e/\hbar c)\int \Phi_\nu dx^\nu} \gamma^\mu \frac{\partial}{\partial x^\mu} e^{i(e/\hbar c)\int \Phi_\nu dx^\nu} = \gamma^\mu \left(\frac{\partial}{\partial x^\mu} + \frac{ie}{\hbar c}\Phi_\mu\right). \quad (8.18)$$

Finally, Dirac argued that, writing R in the form (8.2) where R_a is completely fixed by the fields and contains all the singularities, the physics is contained in the non-singular term R_b—which is Heisenberg's *'wirkliche'* (true) density matrix r in Ref. [14b]. Hence, the current density and energy–momentum tensor must be identified as

$$j^\mu = -ie\text{Tr}[(x|R_b|x)\gamma^\mu] \quad (8.19)$$

and

$$T_{\mu\nu} = \frac{1}{2}\Re\text{Tr}\left[\left\{\gamma^\mu\left(\hbar c\frac{\partial}{\partial x^\nu} + ie\Phi_\nu\right) + \gamma^\nu\left(\hbar c\frac{\partial}{\partial x^\mu} + ie\Phi_\mu\right)\right\}(x|R_b|x)\right], \quad (8.20)$$

for which Heisenberg gave the complete demonstration of the conservation laws in Ref. [14b].

After several more letters Pauli, on 17 February 1934 (letter [359] of Ref. [3a]) proposed to Heisenberg that they write a 'three-man-paper' together with Weisskopf who was just completing his first semester as Pauli's assistant (see the preceding section) and had aided Pauli with several calculations. The appendix to letter [359] with the title 'Contributions to the theory of electrons and positrons', probably represents an early draft of the first two sections of such a paper. In Section 1 Pauli again points to the difficulty with Dirac's assumption that occupied negative-energy states produce no field, here for the charge fluctuation squared. Indeed, according to Heisenberg's calculations in letter [358] the latter diverges even in the absence of forces and averaged over a finite time, this expression certainly being observable. In Section 2 subtraction terms δj^μ and $\delta T_{\mu\nu}$ are again introduced, which, however, are not specified as in Eqs (8.4), (8.5) but are to be determined by the conditions mentioned earlier. This appendix, together with the one added to letter [343] discussed above, show that Pauli was much more than a 'sounding board and critic' for Heisenberg (Refs [16c], p. 100 and [20], p. 380). He also suggested that Weisskopf calculate the self-energy of the electron according to this subtraction technique [26]. Weisskopf's (corrected) result was a logarithmic divergence for both the static and the dynamic parts, while the one-electron theory yielded a linear and a quadratic divergence for the static and the dynamic parts, respectively.

But then pessimism gained the upper hand with Pauli: 'I am not very happy with the hole theory, it's really a very hideous theory', he laments, and: 'I don't believe a syllable, and the entire hole theory pleases me the less the more I think about it' (translated from letters [360] and [363] of 23 February and 2 March 1934, Ref. [3a]). He drew Weisskopf more and more into discussion of the elaborations in Heisenberg's letters and, finally, let him carry on this correspondence. Weisskopf was not backward in adopting his master's fashion of criticizing Heisenberg: 'We would be very curious to hear your opinion on these "devastating" attacks!' he wrote. And: 'By the way, I wanted to tell you on higher order that Pauli, "disgusted, turns away from such subtraction physics"!' (translated from letters [370a], [370b], and [371a] of 27 April, 1 and 16 May 1934, Ref. [3a], pp. 321, 323, and 324). On 14 June Pauli took over again: 'We [Pauli and Weisskopf] agree completely with you on all essential points but, on the other hand, the results seem to justify entirely my disgust with limes acrobatics and "subtraction physics"' (translated from letter [373], Ref. [3a], p. 327).

Pauli hinted at the deeper reason for his unhappiness six years later when, commenting on Dirac's belief that there still must exist some mathematical trick to eliminate the logarithmic singularities, he wrote to Weisskopf on 10 March 1940: 'As concerns Dirac's last sentence, I had answered him critically that it would seem evil to me if one could make a convergent quantum electrodynamics for arbitrary values of charge and restmass. For, then it would seem rather hopeless to understand the latter' (translated from Ref. [13c], letter [594]). This remarkable statement, namely that the particular value (7.2) of the fine structure constant, but also the mass values, should be singled out by leading to convergent results, has not, however, been realized to this day. Seven months later Alfred Landé (see pages 92–7), who was then in Columbus, Ohio, sent Pauli a manuscript (see the appendix to letter [608] in Ref. [2]), in which a value $\alpha^{-1} = 137.1273...$ (see Eq. (7.2)) was obtained from an eigenvalue problem. Pauli's reaction was hesitant; he answered Landé on 15 October 1940: 'Your whole approach is of such a character as to be either a) a terrible nonsense or b) a totally fundamental discovery. . . . Now it depends on whether the agreement with experience is arranged by God Almighty or by the devil' (translated from Ref. [2], letter [607]).

Letter [373] quoted above is noteworthy for two reasons. First, it coins a new technical term, namely subtraction physics, which in the ensuing correspondence between Pauli and Heisenberg became a recurrent expression until 1939 (see the entry 'Subtraktionsphysik' in the subject index of Ref. [3a]). Secondly, this letter contains the first announcement of Pauli's work on the Klein–Gordon (KG) equation (see the quotation introducing this section). In his paper [27a] with Weisskopf, which was submitted on 27 July 1934, Pauli justifies reconsideration of this equation by the argument that in a second-quantized form of the Dirac equation, Dirac's a priori arguments against the KG equation cannot be upheld. In fact, Dirac had used the hypothesis of a non-negative particle density ρ to conclude that the wave equation must be of first order in $\partial/\partial t$ (see Ref. [27a], footnote p.711), thus excluding the KG equation (5.34) as a possible alternative (see page 184). Of course, in the early 1930s there was also not much motivation for a wave equation describing particles of spin 0, since nuclei like the α-particle are composites, and the existence of Yukawa's particle, the pion or π-meson, was clarified only in 1947 (see Ref. [20], Section 18(b)). This is why Pauli called the KG equation a 'curiosity' with which 'it may not be possible to do much' [13b]. However, the problem was important enough for Pauli to lecture on it at the Institut Henri Poincaré in Paris where he was on a visit in March 1935. Besides repeating much of what is in Ref. [27a], this paper, 'Relativistic quantum theory of the particles obeying Einstein–Bose statistics' [27b], discusses an important problem not treated there (see below).

The idea to create an institute in Paris with a name devoted to mathematical physics goes back to 1926 when the famous mathematician George David Birkhoff (1884–1944) from Harvard University came to Paris on a mission from the International Education Board, the Rockefeller Foundation, to donate to the University of Paris an important sum in honour of the French school of mathematics. The usual condition of the foundation that the offer of $100 000 be complemented by local funds was met by the Baron Edmond de Rothschild while the land was offered by the university. In 1913 Birkhoff had become famous for his proof of the theorem of geometry conjectured by Henri Jules Poincaré (1854–1912) and later became a corresponding member of the Académie des Sciences. The institue was built in 1928 and inaugurated the year after, just before Louis de Broglie received the Nobel Prize (see the biography of Proca by his son in Ref. [27c], pp. A-7 and A-10, and Ref. [27d]). In 1930 Alexandre Proca (1897–1955) became editor-in-chief of the *Annales de l'Institut Henri Poincaré*, a position he kept almost to the end of his life (Ref. [27c], p. A-9).

To see Pauli's arguments more clearly it is of interest to compare the two theories in the framework of the Lagrangian formalism on page 186. Writing the Lagrangian density (6.35) for complex, and possibly multi-component, matter fields ψ, ψ^*, the four-current density, in appropriate units, is

$$j^\mu = \frac{ie}{\hbar c}\left(\frac{\partial \mathcal{L}_m}{\partial \psi_{,\mu}}\psi - \frac{\partial \mathcal{L}_m}{\partial \psi^*_{,\mu}}\psi^*\right) \qquad (8.21)$$

where $\psi_{,\mu} \equiv \partial\psi/\partial x^\mu$. With the help of gauge invariance one shows that this expression for j^μ indeed satisfies the continuity equation (6.33) (see e.g. Ref. [16b], Section 3). Now the Lagrangian density for the KG equation is, in natural units, where $|\psi|^2$ has the dimension of an inverse volume,

$$\mathcal{L}_m = -\hbar c^2(\psi^*_{,\mu}\psi_{,\mu} + k_c^2\psi^*\psi) \qquad (8.22)$$

(see e.g. Ref. [16b], Section 8). With these expressions for $\mathcal{L}_m$ and $j^4 = -ie\rho$ it then follows that for the particle density, expressed in term of the canonical momenta (6.37), $\pi = \hbar\dot{\psi}^*$ and $\pi^* = \hbar\dot{\psi}$,

$$\rho = \frac{i}{\hbar}(\psi^*\pi^* - \psi\pi) \qquad (8.23)$$

(compare Ref. [27a], Eq. (16) and Ref. [27b], Eq. (8)), which can have both signs (see below). The important point emphasized by Pauli is that, contrary to the case of the Dirac equation, the energy in the case of the KG equation is strictly non-negative. Indeed, it follows by applying Eqs (6.37) to the Lagrangian density (8.22) that the KG Hamiltonian density is

$$\mathcal{H}_m = \frac{1}{\hbar}|\pi|^2 + \hbar c^2(|\nabla\psi|^2 + k_c^2|\psi|^2) \geq 0 \qquad (8.24)$$

(compare Ref. [27a], Eqs (4), (7) and Ref. [27b], Eq. (2)).

Second quantization with Bose–Einstein (BE) statistics is obtained by imposing the canonical commutation relations (CCR) (6.39) between ψ, ψ^* and π, π^*,

$$[\pi(\mathbf{x},t),\psi(\mathbf{x}',t)] = [\pi^+(\mathbf{x},t),\psi^+(\mathbf{x}',t)] = -i\hbar\delta(\mathbf{x}-\mathbf{x}').$$ (8.25)

This becomes more explicit in a development into plane waves

$$\psi(x) = \sum_{\mathbf{k}} i\sqrt{\frac{1}{2V\omega}}(b_{\mathbf{k}}e^{-i\omega t} - \bar{b}^+_{-\mathbf{k}}e^{i\omega t})e^{i\mathbf{k}\cdot\mathbf{x}}$$

$$\pi(x) = \sum_{\mathbf{k}} \hbar\sqrt{\frac{\omega}{2V}}(\bar{b}_{\mathbf{k}}e^{-i\omega t} + b^+_{-\mathbf{k}}e^{i\omega t})e^{i\mathbf{k}\cdot\mathbf{x}},$$ (8.26)

where V is the volume of the system and $b_{\mathbf{k}}$, $\bar{b}_{\mathbf{k}}$ satisfy the commutation relations (6.14). Defining the particle number operators

$$N_{\mathbf{k}} = b^+_{\mathbf{k}}b_{\mathbf{k}}, \quad \overline{N}_{\mathbf{k}} = \bar{b}^+_{\mathbf{k}}\bar{b}_{\mathbf{k}},$$ (8.27)

it is easy to calculate the total particle number from Eq. (8.23). With the order of the operators given there one finds

$$\int_V \rho d^3x = \sum_{\mathbf{k}}(N_{\mathbf{k}} - \overline{N}_{\mathbf{k}}).$$ (8.28)

Similarly the total energy is obtained from Eq. (8.24), the result being independent of the ordering of the operators,

$$\int_V \mathcal{H}_m d^3x = \sum_{\mathbf{k}} \hbar\omega(N_{\mathbf{k}} + \overline{N}_{\mathbf{k}} + 1).$$ (8.29)

In Eqs (8.28), (8.29) $N_{\mathbf{k}}$ and $\overline{N}_{\mathbf{k}}$ may take all non-negative integer values, and it becomes evident that $\bar{b}^+_{\mathbf{k}}$ creates antiparticles whose numbers are $\overline{N}_{\mathbf{k}}$.

This is the result of Section 2 in Refs [27a,b]. In Section 3 the interaction with the electromagnetic field is introduced, and Section 4 of Ref. [27a] is devoted mainly to the problem of pair creation and annihilation, for which the cross-section in the limit $\hbar\omega \gg mc^2$ is found to be a factor $4/7$ times smaller than the one calculated according to the positron theory by Bethe and Heitler [29a]. Pauli and Weisskopf also calculate the vacuum polarization due to a Coulomb field $\Phi_0 \propto q^{-2}$ (not q^{-4}, as indicated) and find it to be of opposite charge and diverging logarithmically with the momentum cut-off. Pauli's main point in this investigation of the KG equation, however, is as follows (see Section 1 of Refs [27a,b]). Because of the possibility of pair creation and annihilation, and because of the antiparticle interpretation, the one-particle theory must be rejected. But when the Dirac case is not considered in its one-particle form but also in second quantization, then the analogy with Eqs (8.28),

(8.29) becomes perfect, provided that quantization is introduced according to FD statistics, i.e. with anticommutators, as will now be shown.

In the case of the Dirac equation, $\mathcal{L}_m$ is given by Eq. (6.43), and one finds with Eqs (6.37) and $\pi = i\hbar\psi^*$, $\pi^* = 0$,

$$\mathcal{H}_m = \hbar c\overline{\psi}(\vec{\gamma} \cdot \nabla + k_c)\psi, \tag{8.30}$$

while Eqs (8.21) applied to (6.43) yield Eqs (6.32). Note that here, because the Dirac equation is of first order in $\partial/\partial t$, ψ and ψ^* are the only fields, whereas in the KG equation, which is of second order in $\partial/\partial t$, the number of fields is doubled to ψ, ψ^*, π, π^*. The development of the Dirac spinor wavefunction is conveniently written as

$$\psi(x) = \frac{1}{\sqrt{V}} \sum_{\sigma k} (a_{\sigma k} u_{\sigma+}(\mathbf{k})e^{-i\omega t} + \overline{a}^+_{\sigma-k} u_{\sigma-}(\mathbf{k})e^{i\omega t})e^{i\mathbf{k}\cdot\mathbf{x}}$$

$$= \sum_r A_r u_r(x). \tag{8.31}$$

Here $u_{\sigma\lambda}(\mathbf{k})$ satisfies Eq. (6.70) with k_c instead of κ_c, $\sigma = \pm$ and $\lambda = \pm$ are, as above, the signs of spin and energy, respectively, and $a_{\sigma k}$, $\overline{a}_{\sigma k}$ are annihilation operators satisfying the anticommutation relation (6.11). Inserting the expression (8.31) into the particle density operator $\rho = (\psi^+\psi)$ and into the Hamiltonian density (8.30), making use of the conditions

$$(u^*_{\sigma\lambda}(\mathbf{k})u_{\sigma'\lambda'}(\mathbf{k})) = \delta_{\sigma\sigma'}\delta_{\lambda\lambda'}, \tag{8.32}$$

one has to remember that the order of the operators is again undetermined. It is therefore legitimate to anticommute all creation operators to the left of the annihilation operators, called the normal order, thus avoiding zero-point infinities. (Of course the analogous procedure could have been applied to Eq. (8.29).) One finds

$$\int_V \rho d^3x = \sum_{\sigma k} (N_{\sigma k} - \overline{N}_{\sigma k}) \tag{8.33}$$

and

$$\int_V \mathcal{H}_m d^3x = \sum_{\sigma k} (N_{\sigma k} + \overline{N}_{\sigma k}), \tag{8.34}$$

in complete analogy with Eqs (8.28), (8.29). It is interesting to note that with commutators instead of anticommutators, $\overline{N}_{\sigma k}$ in Eqs. (8.33), (8.34) would have appeared with the opposite sign, just as in the one-particle theory. Iwanenko and Sokolow [29b] used this fact as an argument in favour of the correct spin-statistics connection (see below).

From the second form of Eq. (8.31) it is also easy to write down the quantization condition for Dirac spinors. Indeed, using $\{A_r, A^+_s\} = \delta_{rs}$ one finds with (8.10)

$$\{\psi_\alpha(x), \overline{\psi}_\beta(x')\} = -iS_{\alpha\beta}(x - x'), \tag{8.35}$$

which is Eq. [3.26] of Ref. [24].

The more serious problem is to quantize the KG theory with anticommutators, which is the essence of Section 3 of Ref. [27b]. But before entering into this discussion, the following fundamental relation is established: With the aid of the expression (8.23) and the CCR (8.25) it is easy to calculate algebraically

$$-i\hbar^2[\rho(\mathbf{x},t),\rho(\mathbf{x}',t)] = (\psi^+(\mathbf{x},t)\pi^+(\mathbf{x}',t) + \psi(\mathbf{x},t)\pi(\mathbf{x}',t))\delta(\mathbf{x}-\mathbf{x}').$$
(8.36)

But since Pauli and Heisenberg had proven the Lorentz invariance of the CCR in their 1929 paper (see page 187), it follows from (8.35) that, in all generality,

$$[\rho(x),\rho(x')] = 0 \text{ if } (x-x')^2 > 0,$$
(8.37)

i.e. for any space-like separation, and this is even true with interactions. This is Eq. (9) of Ref. [27b] which, as Pauli stresses, allows one to refer to ρ as a measurable distribution. In order to prove the impossibility of quantization following the exclusion principle, Pauli then divides ψ into its positive and negative frequency parts called ψ_1 and ψ_2 and defined in Eq. (8.26). Insertion of $\psi = \psi_1 + \psi_2$ in the current density (8.21) gives rise to three groups of terms, namely those containing ψ_1 only, ψ_2 only, and mixtures. Multiplying these three groups with real coefficients c_1, c_2, and c_3, respectively, one may ask the question about how much the theory may be generalized in this way without violating Eq. (8.36). One finds $c_1 = c_2$ and $c_3 = \pm c_1$. Therefore, up to trivial factors, Eq. (8.23) is the only admissible form of ρ. If one then does the same calculation with anticommutators but maintaining (8.36), the only solution turns out to be $c_1 = c_2 = c_3 = 0$. This means that it is impossible to quantize the spin-0 KG theory according to FD statistics.

The result just reported, although without calculations, was communicated by Pauli to Heisenberg as early as 28 June 1934: 'The question whether in this [KG] theory a quantization of the matter waves with exclusion principle may be carried through, still required a more exact investigation. . . . The outcome, interestingly, is *negative: Based on the formalism established so far*, it is *not* possible to achieve, with quantization according to the exclusion principle, that simultaneously, 1. relativistic and gauge invariance hold, 2. the eigenvalues of the energy are positive (whereas with quantization according to Bose–Einstein statistics both are fulfilled automatically)' (translated from letter [375], Ref. [3a], pp. 334-5). This is the first tentative proof of the general spin-statistics theorem. But Pauli's proof described above was rejected by Wigner's student J. S. de Wet (see the introduction to Ref. [31a]). The reason was that the splitting of ψ into positive and negative frequency parts is *non-local*, as is seen from the canonical commutation relations these parts satisfy, as well as from the corresponding split of π. Therefore, the charge density $\rho(x)$ calculated with coefficients c_1, c_2, c_3 is also non-local and the condition (8.36) cannot be applied.

Returning to the density matrix, Heisenberg, in his paper [14b], went beyond Dirac, on the one hand, by giving in the fixed-particle-number or Hartree–Fock approximation the proof of the conservation laws and, on the other, by sketching, in the framework of the fully quantized theory, a programme to treat the ambitious problems of the scattering of light by light or strong fields, as discussed by Delbrück in Ref. [30a]. But as for applications, the paper only recalculates (and corrects) the charge density induced by a static potential A_0 which Dirac had calculated to first order in the coupling constant α in his Solvay paper (see Ref. [25], pp. 82–91) and which, independently, Furry and Oppenheimer had considered (see Eq (4.4) of Ref. [30b]).

This problem, surprisingly, was also taken up by Pauli during his stay at the Institute for Advanced Study in Princeton in the winter of 1935/6 (see the previous section). Pauli, together with a young research fellow, M. E. Rose, took up formula (34) in Heisenberg's paper [14b] and evaluated it in a more general and rigorous way, considering also time-varying fields [31b]. Heisenberg's result

$$\delta\rho = -\frac{\alpha}{15\pi k_c^2}\Delta\rho,$$

which is smaller than Dirac's by a factor of 4, is the special case (23) of Eqs (3), (9), and (21a) of Ref. [31b] where the expression 'subtraction physics' figures explicitly. The density ρ in the last equation is related to the applied potential A_0 in the usual way by $\Delta A_0 = -4\pi\rho$. Morris Erich Rose (b. 1911) received his Ph.D. in theoretical physics from the University of Michigan in 1935. After the Institute for Advanced Study, where he stayed during the academic year 1935–1936, Rose went to Cornell University and, after spending the war years at Princeton University, he moved to the University of Virginia in 1961. Rose wrote two books: in 1955 on *Multipole Fields* and in 1957 on *Elementary Theory of Angular Momentum* (Ref. [5a], p. 1443).

The same result (8.37) was also rederived by Weisskopf in a nice review of the problems of subtraction physics [32a]. This paper is of particular interest because Weisskopf's calculations do not make use of the density matrix at all, but determine the interaction effects directly in the physical quantities themselves by using certain criteria for identifying non-physical (and, surprisingly, at the same time also always diverging) contributions. As an example, Weisskopf determines in a quite simple way the modifications, in slowly varying fields, of the dielectric and of the permeability tensors that Heisenberg and his students Hans Euler and Bernhard Kockel had calculated with great effort following Heisenberg's programme mentioned above (see e.g. Ref. [32b]). It is also interesting that Weisskopf does the same calculation in the framework of the KG theory of Ref. [27a], obtaining very similar expressions. This all shows, first, that the density matrix approach was not essential and, secondly, that the negative-energy states of the positron theory were not the cause that made subtraction physics necessary. In fact, some 15 years later, subtraction physics was still in use but cast into the new terminology of the renormalization theory. And as regards Dirac's density matrix approach, it is just a particular example of an invariant regularization, of which a more general example is the introduction of auxiliary masses, as used in the late 1940s by Feynman, by Stueckelberg and Rivier, and by Pauli and Villars (see e.g. Ref. [33a]).

What had changed was that, instead of speaking of negative-energy states the latter were now understood as the charge conjugate states of physical electrons. Although the name charge conjugation had been introduced by Kramers in a little-known paper to the Amsterdam Academy on 27 November 1937 [33b], it was Pauli who gave this symmetry a general mathematical basis. Indeed, during his visit to the Institut Henri Poincaré in Paris, Pauli gave a second paper, 'Mathematical contributions to the theory of the Dirac matrices' [34a]. In the introduction to this paper Pauli writes that he will first give 'an elementary proof for the use of the physicists' of the fundamental theorem that between two sets of matrices γ^μ satisfying the relations (6.30) there exists a non-singular transformation. He then derives certain consequences concerning the behaviour of Dirac wavefunctions under Lorentz transformations 'as also the existence of a correspondence between the wavefunctions of positive energy and those of negative energy, a correspondence which is unique and also invariant with respect to the Lorentz transformations. The existence of this correspondence is well known and Mr. L. de Broglie . . . uses it in his works on the nature of light. We will establish this correspondence (and this is new) without resorting to a numerical specialization of the Dirac matrices' (translated from Ref. [34a], p. 110). Here and later in the paper the reference to de Broglie—who probably had invited Pauli—is very likely a sign of politeness but also of Pauli's sympathy and admiration for this nobleman whose polite French manners impressed him.

The transformation of charge conjugation appears in Eq. (32) of Ref. [34a]. In the notation of Eq. (8.31) it reads $\overline{u}_{\sigma\mp}(\mathbf{k}) = C u_{\sigma\pm}(-\mathbf{k})$ (note that in Ref. [24], Eqs [6.1], C is called C^{-1}), so that

$$\psi^c = C^{-1}\overline{\psi}; \ \overline{\psi}^c = C\psi, \qquad (8.38)$$

where the transformation matrix C is expressed in terms of another matrix B as (see Eq. (33) in Ref. [34a] where, for Hermitian matrices γ^μ, $A = \gamma^4$)

$$C = B\gamma^5. \qquad (8.39)$$

The defining property of the matrix B is that it transforms the γ^μ into their *transposed* forms, $(\overline{\gamma}^\mu)_{\alpha\beta} \equiv (\gamma^\mu)_{\beta\alpha}$:

$$\overline{\gamma}^\mu = B\gamma^\mu B^{-1}; \ \overline{B} = -B \,; B^+B = 1. \qquad (8.40)$$

From (8.39), (8.40) it then follows that (see also Eqs [6.3] in Ref. [24])

$$\overline{C} = -C; \ C^+C = 1; \ \overline{\gamma}^\mu = -C\gamma^\mu C^{-1}. \qquad (8.41)$$

The remarkable fact about this matrix B, mentioned in footnote 1 on p. 120 of Ref. [34a], is that Pauli had introduced it in a much more general context in his excursion into five dimensions in summer 1933 (see page 268). Indeed, the first of Eqs (8.40) appears as Eq. (14*) on p. 354 of Ref. [34b] where the matrices γ^μ, called α_μ, however, satisfy much more general relations (6.30), in which the $\delta_{\mu\nu}$ on the right-hand side is replaced by the metric tensor $g_{\mu\nu}$.

(Note that a matrix A appears in Ref. [34b] as Eq. (14) on p. 346.) There is yet another place where the matrix B appears, namely in the paper 'Contributions to the mathematical theory of the Dirac matrices' [34c], written in honour of Pieter Zeeman's 70th birthday on 25 May 1935 and submitted on 1 February 1935. Here B already has the physical meaning that 'it makes possible a relativistically invariant association of states of positive energy to states of negative energy' (translated from Ref. [34c], p. 42). Honouring Zeeman with a paper which was not unrelated to Pauli's 'disgust with "subtraction physics"' bears a strange symmetry to Pauli's struggle with the anomalous Zeeman effect which had made him so unhappy some years earlier (see page 92).

Finally, Pauli introduced charge conjugation as a general symmetry in his review 'Relativistic Field Theories of Elementary Particles' [35] which originally was intended for the Solvay Congress of 1939 that was cancelled because of the outbreak of World War II (see the footnote on p. 203 of Ref [35]) but which Pauli submitted only on 8 May 1941 from the Institute for Advanced Study in Princeton. In Part II of Ref. [35] Eqs (8.38) and (8.41) appear as Eqs (87) and (88a,b). But this part of Ref. [35] also gives the invariant functions (8.12) and (8.16) in Eqs (22) and (22') and rederives Eqs (8.28), (8.29) and (8.33), (8.34) in Eqs (12, 14, 20) and (98a,b), respectively. Later in this chapter more will be said on the circumstances related to this paper.

Moving into the new house at Zollikon

During these vacations between terms I had made a beautiful journey to England and Holland and had lectured at Cambridge, Leyden and Utrecht while in the meantime my wife had organized the moving into our new house at Zollikon. We like it very much there, gradually it will also be arranged inside and we hope that some time soon you will see it. [36]

These are the main events of March 1938 which Pauli reports here to Sommerfeld on 22 May, the day before the Paulis moved to Zollikon (see Ref. [4], II.19). It was Pauli's first visit to Cambridge, and Franca wished 'to send him far off (*in die Ferne*) during removal' (translated from letter [487e], Ref. [13c], p. 818). Franca was an eminently practical woman who looked at the world with realism. This was a blessing for Pauli who very much lived in his thoughts. It was natural, therefore, that Franca, who also had a keen sense for the visual arts, wished to arrange the new house according to her ideas when Pauli was away. Zollikon, which is just 4 km from Zurich's Bellevue Platz, is the first municipality adjacent to the city along the lake, and on its far side joins Küsnacht, the residence of C. G. Jung. This whole lakeside is a favourite location of Zurich's well-to-do business and intellectual communities and for this reason is known locally as the 'Gold Coast'.

The Paulis house is situated at Bergstrasse 35, a rather steep road which runs from the crossroads of Dufourplatz all the way up to the Forchstrasse at Zollikerberg, the main road from Zurich to the Forch pass between the hills of the Wassberg and Pfannenstil and down to Esslingen, the terminus of the Forchbahn, the tramway from Zurich which runs along the road. There, on the hillside, the Paulis had chosen a piece of land and asked Professor Dunkel, Pauli's colleague at the School of Architecture of ETH, to draw up the plans. The nice but rather modest house—Pauli used the diminutive '*Häuschen*' in his letter of 9 July to Uhlenbeck (Ref. [3a], p. 586)—which still exists in a slightly modified form, had on its lower level the entrance hall, a guest room, and the garage. On the upper level there was a large living room with a fireplace and a door to a small lawn. The living room facing the road, and allowing a view of the lake and the Albis mountain chain behind it, had on the windowsill the Pauli-bust by Hermann Haller (see page 288). It continued into a somewhat narrow dining area, which led to the kitchen. One floor up were the bedroom and Pauli's study.

Fig. 8.4 The Pauli house at Bergstrasse 35, Zollikon, by architect William Dunkel, 1938 (CERN)

William Dunkel (1893–1980), a Swiss citizen born in New York, grew up in Buenos Aires and attended high school in Lausanne, and then studied architecture in Dresden and Munich. He had been nominated at ETH in 1929, one year after Pauli, and there formed a whole generation of architects—the Swiss playwright and novelist Max Frisch (1911–91) was a student from 1936 to 1940. Dunkel also created many examples of interior architecture [37a]. He had therefore drawn the plan for the desk in Pauli's study which was executed in cherry-wood by a local craftsman ([37b]; the present biography is being written at this desk, which I inherited). Here Pauli could feel at home and in peace—but he paid a price. While it was just a few minutes' walk from Hadlaubstrasse to his office, this became something of an expedition: first, a few minutes' walk down to the bus-stop at Dufourplatz, then changing at Bellevue for the streetcar up to Häldeliweg, from where a short walk led him to the institute at Gloriastrasse 35.

Fig. 8.5 Charles Enz at Pauli's desk at home, 2000 (photo by C. P. Enz)

But there is also a darker side to the quotation at the outset of this section, since it tells us that Pauli was away in Cambridge when his sister Hertha came to Zurich on Monday 14 March as a refugee from the Nazis who the day before had annexed Austria (see page 18). Indeed, according to letter [494] of Ref. [3a], dated Cambridge 15 March, Pauli left Cambridge on the 20th or 21st. Austria's 'annexation' (*Anschluss*) by the German Reich also had unpleasant consequences for Pauli. On 1 April 1938 Switzerland decreed the obligation of a visa for holders of Austrian passports (Ref. [4], p. 73). The immediate consequence for Pauli was that it became doubtful whether he could attend the large international physics conference to be held in Warsaw from 30 May to 3 June. On 21 April he wrote to Heisenberg, who was himself prevented from going because he was 'harassed by Stark and the Gestapo' (translated from letter [515] from Pauli to Uhlenbeck, Ref. [3a], p. 586; see also Ref. [3b], Chapter 20): 'I myself would like to travel there but am afraid the passport- or visa-difficulties will prevent me' (translated from letter [501], Ref. [3a]). Finally Pauli decided not to go because he would have had to travel via Italy, Yugoslavia, and Hungary (letter [525], Ref. [3a]).

On the advice of President Rohn, on 24 April Pauli then wrote directly to the city mayor (Stadtpräsident) of Zurich, Emil Klöti (1877–1963) (see page 286), in order to enquire about the conditions of his naturalization: 'The intention to naturalize in Zurich where I feel at home I have already had for many years.

Now naturalization has become very urgent for me because, as an Austrian and half-Aryan, I wish to avoid, if at all possible, German nationality which is foreign to me. In this situation I dare to ask you whether the municipality of Zurich would allow me to submit now the request of naturalization for me and my wife.' And he gives as witnesses the names of 'Herrn Prof. Rohn, as well as the colleagues from the Physics Department who know me more closely, Prof. Fritz Fischer and Prof. Paul Scherrer who amiably had offered it' (translated from Ref. [4], pp. 81–2). Fritz Fischer (1898–1947), a Swiss who in 1928 had become director of Siemens and Halske in Berlin and in 1933 was nominated as professor for technical physics at ETH, became known for his development of the *Eidophor* process for large-screen projection of television images (Refs [37a] and [5a], p. 569).

Mayor Klöti answered two days later that according to existing law, naturalization in Zurich required 'an uninterrupted establishment in Switzerland of 12 years whereof the last 6 years in Zurich', which condition in Pauli's case 'will be fulfilled only in April 1940. The ten years' establishment in Switzerland whereof the last five years must fall on Zurich suffices only in so far as the candidates are adapted with respect to dialect (*mundartlich angepasst*)' (translated from Ref. [4], p. 82). Pauli rarely tried to speak Swiss German and Franca hated it when he did. Thus Pauli could not avoid acquiring a German passport, which he probably got on 29 November 1938 since, according to Ref. [13c], p. xxviii, this passport expired on 29 November 1940. In the letter of 29 May 1940 to Frank Aydelotte, the director of the Institute for Advanced Study in Princeton, quoted in Ref. [13c] on p. xxviii, Pauli describes this situation as follows: 'By the fact that Switzerland didn't make possible my naturalization in the moment of the annexation of Austria by Germany I was forced to accept a German passport. . . . The German Consulate counted me to the half-Aryans without further examination and so I got a non-Jewish (that means without J) passport. Actually I suppose I am after German law 75 per cent Jewish. This would mean that in the case of a German occupation of Switzerland I would be really menaced and treated as a Jew.' The infamous J-mark had been introduced by Germany under the decree of 5 October 1938, based on an agreement with Switzerland of 29 September 1938, in order to allow the Swiss authorities after Austria's occupation 'a gap-less control of the entrance of German emigrants', without being constrained to introduce a compulsory visa for holders of a German passport, which would have displeased Germany (see footnote 107 in Ref. [4], p. 73 and also the Chapter *Der dunkle Punkt* (The dark spot) in Ref. [38a]).

Following this exchange with Mayor Klöti, President Rohn and Pauli agreed that, for the moment, it would be unwarranted to undertake any further action on the question of Pauli's naturalization. But in November 1939 Pauli again contacted President Rohn, considering that in May 1940 the formal conditions for naturalization would be fulfilled. The latter was able to inform Pauli that, in spite of the state of war in Europe, the Federal Offices in Berne still accepted applications for naturalization and suggested that Pauli address a request to the

Department of the Interior of the Canton of Zurich. But for unknown reasons Pauli instead addressed his request to the federal authorities—who did not respond, however (see Ref. [4], II.20,21). So let us leave this subject for the moment and return to academia.

After the compromise of the double assistance from Kemmer and Ludwig during the summer term of 1936 (page 295) Pauli made a lucky choice with far-reaching consequences: namely, Markus Fierz, a Swiss who later became Pauli's most trusted discussion partner among the physicists, on philosophical, psychological and historical as well as physical questions. Markus Eduard Fierz was born in 1912 with his twin brother Heinrich Karl who became a well-known analytical psychiatrist. Their father Hans Eduard had become professor of technical chemistry at ETH in 1917 and their mother Linda, born David, had been an analytical psychologist collaborating with C. G. Jung (see e.g. Ref. [38b]). At high school in Zurich Markus Fierz was stimulated by physics, mathematics, and philosophy, which subjects he studied for three semesters at Göttingen. When the Nazis came to power he returned to Zurich where he worked for his Ph.D. with Wentzel at the university at the same time as Bargmann and Kemmer. In his 'Retrospective from Hönggerberg' (Ref. [39], pp. 218–21) Fierz gives a vivid picture of the institute where Edgar Meyer and Gregor Wentzel were the professors and Richard Bär, Kurt Zuber, and Marcel Schein were Privatdozenten. Compared with ETH this was quite a poor enterprise—Wentzel had no assistant—but the colloquium and the theoretical seminar were common events.

After his doctorate in 1936 Fierz visited Heisenberg's institute in Leipzig during the summer term, from where he went with Heisenberg and his assistant Euler to one of the well-known international conferences at Copenhagen. This was the meeting on nuclear physics which took place from 17 to 20 June 1936 (see Bohr's letter [431] to Pauli, Ref. [3a], p. 449) and where about 50 people crammed into the famous lecture hall. Fierz remembers: 'At lunch in a laboratory with a glass roof, a true hothouse, Pauli asked me whether I would like to become his assistant. I was rather frightened, especially because Pauli told me: "I know you are not as experienced as Casimir or Weisskopf but I will give it a try with you".' Fierz then ponders: 'Well, in fact, it went well, against all expectations, mainly because—according to all the stories—Pauli treated me considerably better than his former assistants' (translated from Ref. [39], p. 221; see also Ref. [3a], p. 448). It goes without saying that Fierz did extremely well; he has to his credit four important papers published together with Pauli. In proposing Fierz as assistant for the winter term of 1936/7 Pauli emphasized that Heisenberg had recommended him warmly (Ref. [4], I.93–5). Fierz stayed in this position until spring 1940 (see the decrees I.98, 101, II.2, 10, 16, and 18 in Ref. [4]). In the autumn of 1938 he became a Privatdozent at ETH (see Pauli's and Scherrer's reports II.6 and 11 in Ref. [4]) and in 1940 at the University of Basle where in 1943 he became a professor. Fierz spent the winter term of 1950/1 at the Institute for Advanced Study in Princeton and for the year 1959/60 he was director of the Theory Division of CERN in Geneva.

In 1960 ETH nominated him as the successor of Pauli; he retired in 1977. In 1979 Fierz received the Planck Medal of the German Physical Society; he is an honorary member of the Swiss Physical Society and the Zurich Physical Society (see the dust jacket of Ref. [39]).

From 30 June to 4 July 1936 there was another *Vortragswoche* at ETH, this time on nuclear physics. Among the participants were Cockcroft and Oliphant from the Cavendish Laboratory in Cambridge, who had produced the first artificial elements, and Professor Hess from Innsbruck, the discoverer of cosmic rays, as Scherrer enthusiastically reported to President Rohn (Ref. [4], I.91; see also the comment in Ref. [3a], p. 453). In the following year of 1937 the summer vacation again offered Pauli plenty of opportunities for travel. On 9 September there was a meeting in Copenhagen where he went with Franca and where he gave a talk on the 'ultrared catastrophe' (see footnote *d* in Ref. [3a], p. 531 and Pauli's letter [482] to Bohr in Ref. [3a]), on which more will be said below. After Copenhagen the Paulis paid a visit to the Kleins in Stockholm on their way to Moscow (see Pauli's letter [481a] to Klein in Ref. [3a]). In Moscow Pauli attended the Second All-Union Congress which took place from 20 to 26 September and where, in addition to his lecture on conservation laws described on page 227, Pauli gave a communication on 'Some Basic Remarks about the Theory of β-Decay' [40]. And from 18 to 21 October he participated in Luigi Galvani's 200th birthday celebrations in Bologna where O. W. Richardson presented a review on H_2 and H_2^+, the subject of Pauli's thesis work (see pages 63f), and where Pauli gave a critical analysis of the long-wavelength problem in bremsstrahlung which was variously called the 'infrared' or the 'ultrared catastrophe' [41].

In Moscow the Paulis stayed at the Hotel Metropol and were the guests of Piotr Kapitza (see page 134) [7a]. Then they visited Fritz ('Fisl') Houtermans and Charlotte ('Schnax') Houtermans in Kharkov (Refs. [7a,g])—Pauli had been best man at their wedding in Tbilisi (see page 220). Born in the Baltic region, Friedrich Georg Houtermans (1903–66) grew up with his mother in Vienna. In 1921 he studied physics at Göttingen where he became a close friend of George Gamow and where he met his future wife Charlotte. He was an extremely inventive experimental physicist and an untiring story-teller. Because of his pronounced leftist sympathies he had decided to leave the Technische Hochschule in Berlin with wife and daughter Giovanna ('Bamsi') Houtermans when the Nazis came to power and went to England in the summer of 1933 although, being one-quarter Jewish from his mother's side, he was not in immediate danger. However, he saw a bright future in the Soviet Union. So in 1935 he accepted an invitation to the Ukrainian Physico-Technical Institute in Kharkov which was then a leading centre in physics. There he became a good friend of Lev Landau (see page 208), and his son Jan Houtermans was also born there [42a]. However, 1937 was the height of the Stalinist terror. In October [42b] Houtermans was fired from his position at Kharkov. On 17 November he obtained an exit visa and planned to go to London where Blackett had arranged an invitation. But on 1 December he was seized at the customs office in Moscow [42a]. Until his extradition to Germany in April 1940 in the wake of the German–Soviet Non-Agression Pact of 23 August 1939, Houtermans was detained in prison in Moscow, Kharkov, and Kiev and suffered 14 extremely severe interrogations [42a,b]. Charlotte Houtermans and the two children managed to flee to Copenhagen via Riga where Bohr took care of them. She then moved to England and eventually emigrated to the United States.

Pauli, on 13 January 1938, told Weisskopf about these events: 'You may consider yourself lucky that you didn't go to Kiew that time. . . . In this country at present reigns a totally unexampled terror which manifests itself even in mass arrests of innocent scientists.' And he mentioned 'the Houtermans case': 'Schnax succeeded under great difficulties to travel to Copenhagen with the children where they have been since about 25 December and hope to obtain an entrance visa to England. The bad thing, however, is that Houtermans himself is in custody in Russia (under some vain pretext)' (translated from letter [487a], Ref. [3a], p. 547).

In his communication [40] to the Moscow meeting Pauli examined the possibility of using Fermi's expression (6.63) for the β-decay interaction in perturbation theory in higher than first order. Apart from the fundamental importance of this question it was also of relevance to Heisenberg's 'explosion theory' of the interaction of cosmic rays with matter [43a], as the latter explained in his letter [429] to Pauli of 26 May 1936 in Ref. [3a]. This explanation of Heisenberg's, in which 'showers' are the result of Fermi's β-decay mechanism in high-order perturbation theory, was opposed to the 'cascade theory' of Bethe and Heitler [29a] which had been initiated by Oppenheimer and Carlson and independently by Homi Jehaugir Bhabha (1909–66) in Cambridge. According to this theory the 'showers' are electromagnetic in nature, due to pair creation and bremsstrahlung (see the comment on p. 444 in Ref. [3a]). While $e^2/\hbar c$ is the pure number α, Eq. (7.2), the coupling constant g in Eq. (6.63) may be written as $g/\hbar c = l^2$ where l is a length. This difference between e and g has fundamental consequences, as Pauli shows in Ref. [40]. In modern language he shows essentially that Fermi's theory defined by the interaction (6.63) is non-renormalizable, while QED is (see e.g. Refs [43b], Chapter 15 and [43c]).

The details of the calculation underlying Ref. [40] were contained in Pauli's letter [435] to Heisenberg of 26 October 1936 in Ref. [3a]. The method for obtaining the required estimates is to discretize the problem by defining the fields on a lattice of lattice constant d. This was very much to Heisenberg's taste who already back in 1930 had postulated a world with a universal length (see the comment on p. 9 of Ref. [3a]) which in his later correspondence with Pauli was called the 'lattice world' (see the entry 'Gitterwelt' in the subject index of Ref. [3a]). The idea is to define the field $\psi(\mathbf{x})$ in a cube of volume $L^3 = (Zd)^3$ with periodic boundary conditions where Z is a large integer. Then $\psi(\mathbf{x}) = a_n/d^{3/2}$ where n designates the lattice positions and a_n satisfies anticommutation relations (6.9) (see Ref. [3a], p. 455). In addition, the interaction Hamiltonian (6.63) is simplified by writing it for one species of particles only and by setting $O_i = 1$. Then Eq. (6.63) becomes

$$\frac{H'}{\hbar c} = \frac{l^2}{d^3} \sum_n (a_n^* \gamma^4 a_n)^2, \tag{8.42}$$

while the unperturbed Hamiltonian is, for zero-mass fermions, given by the integral of (8.30) with $k_c = 0$,

$$\frac{H_0}{\hbar c} = \sum_n a_n^* \gamma^4 \vec{\gamma} \cdot \nabla a_n. \tag{8.43}$$

Note that this discretization of the theory is another example of a regularization, although a non-invariant one (see the previous section). The lattice constant d determines a high-frequency cut-off c/d as used, for example in Weisskopf's calculation of the self-energy of the electron (see page 301); the long-wave periodicity, on the other hand, is used only to discretize k-space. Equations (8.42), (8.43) show that, in the limit $d \to 0$, H_0 may be neglected relative to H' (which corresponds to a strong-coupling theory), so that from (8.42) one deduces a zero-point energy ε_0 and a one-particle energy ε_1 given, in units of $\hbar c$, by

$$\varepsilon_0 = \frac{Al^2}{d^3}, \; \varepsilon_1 = \frac{Bl^2}{d^3} \tag{8.44}$$

with some numerical factors A, B (see Ref. [3a], p. 457). These are Eqs (3) and (4) of Ref. [40], except that the former is given as the density $\varepsilon = \varepsilon_0/d^3$.

Pauli comments on this result as follows: 'If this were a general result then one would have to consider the Fermi theory as being refuted, since only the limit $d = 0$ has physical meaning because, apart from all the rest, the introduction of a finite d violates the Lorentz invariance of the theory' (translated from Ref. [6], p. 457). This negative appreciation of Fermi's theory is expressed even more strongly in Pauli's letter [454] to Heisenberg of 24 November 1936 where in Section 2 he criticizes 'the unwarranted application of perturbation theory ("Born approximation")' in Fermi's 'ingenious theory' (translated from Ref. [3a], p. 478). But then Pauli writes: 'One may choose a whole series of other relativistically invariant expressions of fourth order in the function ψ [for H']. Among those possibilities there is a definite one, by the choice of which the constants A and B in the equations [(8.44)] vanish' (translated from Ref. [6], p. 457; this choice is given in Heisenberg's letter [438] to Pauli, Ref. [3a], p. 462). In this case, $A = B = 0$, it is not possible to neglect H_0 relative to H'. On the contrary, H_0 alone determines the energies (8.44) which, according to Pauli, are now

$$\varepsilon_0 = \frac{A'}{d}, \; \varepsilon_1 = B' k^2 d. \tag{8.45}$$

Here the value of ε_0 may be understood by setting $-i\nabla \sim k = 2\pi/\lambda$ in (8.43) and remembering that $\lambda_{min} = d$. The expression for ε_1 is more dubious (see the footnote on p. 457 of Ref. [6] where B' is written B); here Pauli quotes Heisenberg's expression in letter [458]: 'Your result about a value of order d/λ^2 ... for the energy of *one* particle and the formation of showers when $\lambda \sim d$ appears very plausible to me' (translated from letter [459], Ref. [3a], pp. 487–8).

More important is Pauli's conclusion: 'Herewith the hopes that Heisenberg entertained in relation to the situation that in the theory of β-decay one encounters a new universal constant $[l^2]$ with the dimension of a length squared, are unfounded. In his theory of the cosmic showers Heisenberg has hoped to bring this universal length $[l]$ into relation to the characteristic wavelength λ_0 of the light particles in the cosmic showers.' Furthermore: 'If we cut off the wavelength spectrum at a certain minimum limit λ_{gr}, we may use the first approximation of this theory conditionally. However, one must refrain from wanting to examine the higher approximations of perturbation theory' (translated from Ref. [6], pp. 457–8). But this describes exactly the situation with a non-renormalizable theory! (see e.g. Chapter XV in Ref. [43b]). This also was the death-knell for the neutrino theories of light (see page 231) and the nuclear exchange forces (see e.g. Ref. [44]) which came out orders of magnitude too small. Of historic importance with regard to the above question is Pauli's letter [474] of 2 May 1937 to Heisenberg, in which he 'is inclined to drop the idea of the existence of an universal length'. He writes: 'I, however, call "theories *without* universal length" [i.e. renormalizable theories] those, in which at high particle energies the restmass may be neglected (put $= 0$), and in which precisely at high particle energies the interaction Hamiltonian contains no constant with the dimension of a length—so that then also the frequency of processes with formation of many particles does *not* arbitrarily increase as compared to those with few particles' (translated from Ref. [3a], p. 519).

In spring 1937 Pauli worked with his new assistant Fierz on two entirely different problems. On the one hand, he again took up the problem of Boltzmann's H-theorem in quantum mechanics that he had analysed in his contribution to Sommerfeld's 60th birthday celebration in 1928 (see Ref. [32] in Chapter 6), now replacing the random phase assumption by the notion of the macro-observer who is able to observe only a coarse-grained density in phase space [45]. On the other hand, Pauli's interest was captured by bremsstrahlung, i.e. the radiation (*Strahlung*) emitted by charged particles suffering a slow-down (*Bremsung*) or, more generally, traversing a field of force. It turned out, in fact, that in this situation the main problem of QED was not the 'ultraviolet catastrophe' at short wavelengths but the 'infrared catastrophe' at long wavelengths. But Pauli had competition from his former assistant Bloch whose manuscript, written with a collaborator, Arnold Nordsieck, was already in the hands of Heisenberg early in June. Contrary to how most scientists would have reacted, Pauli wrote to Heisenberg on 10 June 1937: 'Thereupon we, very likely, will not continue to treat the problem further, and I am quite happy that the Bloch–Nordsieck work came before we have lost all too much time', but he would appreciate having the manuscript for some time (translated from letter [476],

Ref. [3a]). Heisenberg, on the other hand, told Pauli on sending the manuscript: 'I'm ashamed that I have not found it [the solution] myself' (translated from letter [477], Ref. [3a]).

Pauli and Fierz's paper 'On the theory of the emission of long-wavelength light quanta' [41] was published in the proceedings of the bicentenary of Luigi Galvani's birth, which coincided with the 29th Meeting of the Italian Physical Society. Although the authors in Section 2 of Ref. [41] assure the reader that their work adds nothing new to the Bloch–Nordsieck paper [46], there are several important differences. In Ref. [41] the problem is 'regularized' by giving the electron an extension $a = 2\pi c/\omega_1$, described by a charge density $\rho(\mathbf{x}) = eD(\mathbf{x})$. Since this violates Lorentz invariance, the problem is treated non-relativistically, assuming the initial electron velocity $v_0 \ll c$. This procedure has the advantage that the scattering by a potential $V(\mathbf{x})$ may in principle be treated to any order of perturbation theory. Bloch and Nordsieck, on the other hand, consider a relativistic point electron, but are limited to the first order in V, i.e. to the Born approximation. In addition they are forced to neglect the energy loss of the electron in the scattering.

The point of departure in Ref. [46] is a free electron of energy $E = cp_0$ and momentum $\mathbf{p} = p_0\vec{\mu}$, for which it is convenient to introduce the projectors $P_{\pm}$ defined in Eq. (6.73), replacing the neutrino mass m_ν by m. They project the Dirac spinor ψ, satisfying $H_0\psi = E\psi$ with H_0 given in Eq. (8.7), onto the positive- and negative-energy states $P_{\pm}\psi = \psi^{\pm}$, and, in the notation of Ref. [46], is $P_{\pm} = (1 \pm \Lambda)/2$. Introducing the coupling to the electromagnetic field by the substitution $\mathbf{p} \to \mathbf{p} + (e/c)\mathbf{A}(\mathbf{x})$, the condition for the electron to have positive energy, $\psi^{-} = 0$, leads to an equation for $\psi^{+} = u$, in which $\mathbf{A}$ is developed into eigenmodes $s \equiv (\lambda, \mathbf{k})$ according to Eq. (6.13) but expressed in the canonical variables $P_s = (b(s) - b^{+}(s))/i\sqrt{2}$, $Q_s = (b(s) + b^{+}(s))/\sqrt{2}$. The total Hamiltonian, including the field term (6.16), is then quadratic in these variables, with linear terms due to $\mathbf{A}$. These linear terms are now eliminated by a canonical transformation that carries u over to u' which may be determined explicitly. This in turn determines $u(\vec{\mu}, n_s)$ where the n_s are the numbers of photons. Now the scattering potential V is introduced in the Born approximation between the initial state $u(\vec{\mu}, 0)$ and the final state $u(\vec{\nu}, n_s)$, neglecting, however, the energy loss $\hbar \sum n_s\omega_s$, where $\omega_s \equiv c|\mathbf{k}|$. If no photons are emitted, the result is a probability for the scattering of the electron into a given element of solid angle of the form

$$W \propto e^{-I}; \; I \propto \int_0^{\omega_1} \frac{d\omega}{\omega}. \tag{8.46}$$

Since I diverges logarithmically, $W = 0$. This result is unchanged if only a finite number of photons are emitted. However, if one sums over all possible numbers of photons the result is the same as 'one would have obtained by neglecting entirely the interaction with the electromagnetic field' (Ref. [46], p. 58).

Pauli and Fierz make a more in-depth analysis. Their paper [41] in a way inaugurates the method of the form factor which subsequently Pauli used in the theories of nuclear forces and much later also applied to the so-called Lee model. Therefore, and also because of the sceptical reactions this paper caused, adding some detail and precision may be justified. Pauli and Fierz use—as Bloch and Nordsieck obviously do—the dipole approximation $|\mathbf{k}| \ll 2\pi/a$ for the electromagnetic coupling of the electron, i.e.

$$\frac{e}{mc}\mathbf{A} \simeq \frac{e}{mc}\int \mathbf{A}(\mathbf{x})U(\mathbf{x})d^3x = \hbar \sum_s \omega_s \mathbf{a}_s P_s. \tag{8.47}$$

Here

$$\mathbf{a}_s \equiv \frac{e}{m\hbar\omega_s}\sqrt{\frac{\hbar}{L^3\omega_s}}\mathbf{e}(s)g(\mathbf{k}) \tag{8.48}$$

and the Q_s drop out of Eq. (8.47), provided the form factor $g(\mathbf{k})$ is assumed to be real,

$$g(\mathbf{k}) = g^*(\mathbf{k}) = g(-\mathbf{k}) = \int d^3x U(\mathbf{x})e^{i\mathbf{k}\cdot\mathbf{x}}; \tag{8.49}$$

in addition, $g(0) = 1$. The finite extension of the electron expressed by the frequency cut-off ω_1 is implemented by the condition

$$L^{-3}\sum_{\mathbf{k}} \frac{g^2(\mathbf{k})}{\mathbf{k}^2} = \int \frac{d^3k}{(2\pi)^3}\frac{g^2(\mathbf{k})}{\mathbf{k}^2} = \frac{\omega_1}{2\pi^2c} \tag{8.50}$$

which, for an isotropic form factor $g^2(\mathbf{k}) = G(ck)$, is Eq. (7) of Ref. [41].

The next step is to neglect the term $(e/c)^2\mathbf{A}^2$ in $(\mathbf{p} + (e/c)\mathbf{A})^2$, which, at first, the authors kept to rigorously (see letter [478], Ref. [3a]) but the justification of which is not so evident. First, an estimate of P_s in (8.47) may be obtained from the field Hamiltonian (6.16) which, in the canonical variables, reads

$$H_F(Q_s, P_s) = \frac{\hbar}{2}\sum_s \omega_s(Q_s^2 + P_s^2). \tag{8.51}$$

Obviously, the photon energy cannot exceed the initial particle energy, $\hbar\omega_s P_s^2 < \mathbf{p}^2/2m$. Hence, making use of (8.50) and (7.2),

$$\frac{e^2}{c^2}\mathbf{A}^2 = (\sum_s m\hbar\omega_s\mathbf{a}_s P_s)^2 \le \sum_s (m\hbar\omega_s\mathbf{a}_s P_s)^2 < L^{-3}\sum_s \frac{e^2\mathbf{p}^2}{2m\omega_s^2}g^2(\mathbf{k})$$

$$= \frac{\alpha}{2\pi^2}\frac{\hbar\omega_1}{mc^2}\mathbf{p}^2. \tag{8.52}$$

With the basic condition

$$\frac{\hbar\omega_1}{mc^2} \ll 1, \tag{8.53}$$

one therefore finds that $(e/c)^2\mathbf{A}^2 \ll \mathbf{p}^2$. Thus we are left with the Hamiltonian

$$H = \frac{\mathbf{p}^2}{2m} + \hbar \sum_s \omega_s \mathbf{a}_s \cdot \mathbf{p}P_s + V(\mathbf{x}) + H_F, \tag{8.54}$$

which corresponds to Eq. (9) of Ref. [41]. The canonical transformation of Bloch and Nordsieck eliminating the term linear in the P_s now becomes very simple:

$$\mathbf{p} = \mathbf{p}'; \ \mathbf{x} = \mathbf{x}' + \hbar \sum_s \mathbf{a}_s Q'_s; \ P_s = P'_s - \mathbf{a}_s \cdot \mathbf{p}'; \ Q_s = Q'_s. \qquad (8.55)$$

The quanta of $H_F(Q'_s, P'_s)$ may be called free photons since, as Pauli and Fierz remark, they 'may be observed in a spectrograph'. The term $-(\hbar/2) \sum_s \omega_s (\mathbf{a}_s \cdot \mathbf{p}')^2$ that appears in the transformed Hamiltonian (8.54) may be estimated similarly as in Eq. (8.52) to yield $-(\alpha/6\pi^2)(\hbar\omega_1/mc^2)(\mathbf{p}^2/m)$ which, according to (8.53) and (7.2), may be neglected. Dropping the primes on the canonical variables, the transformed Hamiltonian (8.54) therefore becomes

$$H = \frac{\mathbf{p}^2}{2m} + V(\mathbf{x} + \hbar \sum_s \mathbf{a}_s Q_s) + H_F. \qquad (8.56)$$

For $V = 0$ the eigenstates are then $|\mathbf{k}n\rangle$ where $|\mathbf{k}\rangle = L^{-3/2} \exp(i\mathbf{k} \cdot \mathbf{x})$ and $|n\rangle = \Pi_s h_{n_s}(Q_s)$, $h_N(Q)$ being an eigenfunction of H_F with N quanta and $n \equiv \{n_s\}$. In the presence of the scattering potential V the state may be written

$$|\psi\rangle = \sum_{\mathbf{k}'n'} |\mathbf{k}'n'\rangle \varphi(\mathbf{k}'n'). \qquad (8.57)$$

Operating on the Schrödinger equation $H|\psi\rangle = E|\psi\rangle$ from the left with $\langle n\mathbf{k}|$ and noting that $\langle \mathbf{k}|V(\mathbf{x})|\mathbf{k}'\rangle = L^{-3}\tilde{V}(\mathbf{k}' - \mathbf{k})$ is the Fourier component of $V(\mathbf{x})$, one obtains

$$\left\{ E - \frac{\hbar^2 \mathbf{k}^2}{2m} - \hbar \sum_s n_s \omega_s \right\} \varphi(\mathbf{k}n) = \sum_{\mathbf{k}'n'} M(\mathbf{k}'n') \varphi(\mathbf{k}'n'), \qquad (8.58)$$

where the zero-point energy $(\hbar/2) \sum_s \omega_s$ is subtracted in E and the matrix element reads

$$M(\mathbf{k}'n') \equiv \tilde{V}(\mathbf{k}' - \mathbf{k}) \langle n|e^{i\hbar(\mathbf{k}'-\mathbf{k})\cdot \sum_s \mathbf{a}_s Q_s}|n'\rangle. \qquad (8.59)$$

Taking as incoming state an electron with momentum $\hbar\mathbf{k}_0$ and zero photons, so that $E = \hbar^2 \mathbf{k}_0^2/2m$ and

$$\varphi(\mathbf{k}n) = L^3 \delta_{\mathbf{k},\mathbf{k}_0} \delta_{n,0} + \varphi_{out}(\mathbf{k}n), \qquad (8.60)$$

one obtains

$$\varphi_{out}(\mathbf{k}n) = \frac{1}{\Delta E(\mathbf{k}n)} \{ M(\mathbf{k}_0 0) + L^{-3} \sum_{\mathbf{k}'n'} M(\mathbf{k}', n') \varphi_{out}(\mathbf{k}'n') \}, \qquad (8.61)$$

where $\Delta E(\mathbf{k}n) \equiv \hbar^2(\mathbf{k}_0^2 - \mathbf{k}^2)/2m - \hbar \sum_s n_s \omega_s$. Equation (8.61) is an integral equation for φ_{out}. Writing $\varphi_{out} = f(\mathbf{k}n)/\Delta E(\mathbf{k}n)$, Eqs (8.60), (8.61) may also be expressed as

$$\varphi(\mathbf{k}n) = L^3 \delta_{\mathbf{k},\mathbf{k}_0} \delta_{n,0} + \frac{f(\mathbf{k}n)}{\Delta E(\mathbf{k}n)} \tag{8.62}$$

and

$$f(\mathbf{k}n) = M(\mathbf{k}_0 0) + L^{-3} \sum_{\mathbf{k}'n'} M(\mathbf{k}'n') \frac{f(\mathbf{k}'n')}{\Delta E(\mathbf{k}'n')}, \tag{8.63}$$

which are, respectively, Eqs (23) and (24) of Ref. [41]. Note that energy conservation in the scattering requires $\Delta E = 0$. However, the condition that the photons are outgoing (not incoming) waves is formally implemented by the replacement $\mathbf{k}^2 \to \mathbf{k}^2 - i\varepsilon$.

To first order in V one has

$$|f_1(\mathbf{k}n)|^2 = |M(\mathbf{k}_0 0)|^2 = |\tilde{V}(\mathbf{q}/\hbar)|^2 |\langle n|e^{i\mathbf{q}\cdot\sum_s \mathbf{a}_s Q_s}|0\rangle|^2, \tag{8.64}$$

where $\mathbf{q} \equiv \hbar(\mathbf{k}_0 - \mathbf{k})$.

The scattering cross-section for an electron emerging in an element of solid angle in the direction $\mathbf{k}$ is proportional to

$$S(E) \equiv \sum_n \delta(\Delta E(\mathbf{k}n)) |\langle n|e^{i\mathbf{q}\cdot\sum_s \mathbf{a}_s Q_s}|0\rangle|^2$$
$$= \sum_n \delta(\Delta E(\mathbf{k}n)) \Pi_s \frac{1}{n_s!} w_s^{n_s} e^{-w_s} \tag{8.65}$$

(see Eq. (31) of Ref. [41]), where

$$w_s \equiv \frac{1}{2}(\mathbf{q} \cdot \mathbf{a}_s)^2, \tag{8.66}$$

or using $\delta(x) = \int (d\sigma/2\pi) \exp(i\sigma x)$,

$$S(E) = \int \frac{d\sigma}{2\pi} e^{i\sigma(\mathbf{q}^2/2m)} \Pi_s e^{-w_s} \sum_{N=0}^{\infty} \frac{1}{N!} w_s^N e^{-i\sigma\hbar N\omega_s}$$
$$= \int \frac{d\sigma}{2\pi} e^{i\sigma(\mathbf{q}^2/2m)+\phi(\sigma)}. \tag{8.67}$$

Here

$$\phi(\sigma) \equiv \sum_s w_s \left(e^{-i\sigma\hbar\omega_s} - 1\right). \tag{8.68}$$

Since $e(s) \perp \mathbf{k}$, averaging over the angles of $\mathbf{k}$ is equivalent to averaging over the angles of $e(s)$. Hence, assuming isotropy, $g^2(\mathbf{k}) = G(ck)$, w_s in (8.68) may be replaced by $(1/6)\mathbf{q}^2\mathbf{a}_s^2$. With (8.48) one then finds

$$\phi(\sigma) = C \int_0^{\infty} \frac{d\omega}{\omega} G(\omega) \left(e^{-i\sigma\hbar\omega} - 1\right), \tag{8.69}$$

where

$$C \equiv \frac{\alpha\mathbf{q}^2}{6\pi^2 m^2 c^2}. \tag{8.70}$$

Equations (8.67), (8.69), (8.70) correspond to Eqs (32–34) of Ref. [41]. Neglecting the energy of the photons $\hbar \sum_s n_s \omega_s$ in the energy-conservation δ-function is equivalent to formally setting $\sigma = 0$ in the large parentheses of Eq. (8.69). This implies that the coupling term drops out, so that $S(E) = \delta(q^2/2m)$, in agreement with the above statement by Bloch and Nordsieck. On the other hand, without the form factor (frequency cut-off) $G(\omega)$ the integral in (8.69) is ill-behaved at the limit $\omega \rightarrow \infty$. The above calculation also shows that $\sum_s w_s \propto \int_0^\infty (d\omega/\omega)G(\omega)$, which is the divergent integral I in Eq. (8.46).

The first descriptions of the content of Ref. [41] were given by Pauli in letter [480] to Weisskopf on 20 July 1937 and in his talk at the Copenhagen conference in September (see the manuscript joined to letter [480] in Ref. [3a]). In October and November he had an exchange of letters with Peierls whom he told that 'In the meantime Bloch has surfaced here (he will be here the whole winter), and I take pains to convince him that his treatment of the retardation is incorrect.' And in the following letter Pauli wrote: 'Bloch pressed me tremendously hard (*hat mir ungeheuer die Hölle heiss gemacht*) because of the *limits of validity of the Born approximation*' (translated from letters [484] and [485], Ref. [3a]). Bloch finally had to give in, as he wrote to Bohr on 21 December 1937: 'In fact, I don't see any a priori reasons any more which speak against Pauli's result, and I now also believe that . . . the finiteness of the electron radius may surface yet also at other places in quantum theory, and that Pauli's result is an illustration for it' (translated from Ref. [3a], p. 540).

Meanwhile the political situation in Europe became increasingly gloomy. In Italy Jews were forced to leave the country. On 29 September 1938 the Munich conference between Hitler, Mussolini, Chamberlain, and Daladier sanctioned the cession of the German-dominated border regions of Czechoslovakia to the Reich, inviting a German thrust to the east. Then, on the night of 9 to 10 November a hideous general attack on the Jews in all of Germany was organized by the Nazis (see Ref. [3b], pp. 400–1). The day after this night—which became known as *Kristallnacht*—an exchange of diplomatic messages took place in Berlin sealing the introduction of the *J*-sign in the German passports of Jews, mentioned above (see Ref. [38a], p. 326).

In this sombre atmosphere, on 4 December, Sommerfeld's 70th birthday on the 5th was celebrated in Munich, for which Pauli had written the touching letter [537a] in Ref. [3a], quoted on pages 55–6. For this occasion a special issue of the *Annalen der Physik* was planned, which, however, was restricted to contributions from 'Aryan' authors. So Pauli and many other of Sommerfeld's pupils living abroad published in an issue of the *Physical Review* (see Ref. [3a], letter [518]); Pauli's contribution was about one of Sommerfeld's favourite subjects in optics [47a]. Heisenberg, who was Sommerfeld's choice as his successor in Munich, had been attacked on 15 July 1937 as a 'white Jew' in a vicious article of the weekly *Das Schwarze Korps*, published by Himmler's black-shirted SS (Ref. [3b], pp. 377ff.). And, although Himmler personally stopped all attacks on Heisenberg one year later (Ref. [3b], p. 393), Heisenberg's difficulties continued—it was until 1958 when he finally moved to Munich.

Heisenberg's determination to stay in Germany was an irrational act. He had visited the Paulis in their new house, and Pauli sounded him out about his attitudes [7a]. Heisenberg's stressful life contrasted strangely with the peace of Pauli's new study where sitting at his new desk was conductive to writing ever more philosophical letters. Two examples are Pauli's letters [528] and [534] dated 7 September and 20 October 1938 in Ref. [3a] to his friend from the Hamburg years, Erich Hecke (see page 158). But, of course, this tendency, to a large part, had its origin in his stimulating contacts with C. G. Jung which after his analysis (see pages 247–8) had continued uninterruptedly, Jung sending Pauli his new essays and books and Pauli reciprocating with new dream material.

The first letter from Pauli to Jung from the new study is dated 30 October 1938. It reports on the attempt of the 'anima' to express her own notion of time, based on periodic symbols, which also include the alternating bright and dark stripes of the wasp phobia mentioned on page 248. The last letter to Jung from the study before Pauli's departure to the United States is dated 3 June 1940. Pauli writes: 'External circumstances induce me to send you the enclosed dream material from the years 1937–9, in order that it is not going to be lost' (translated from Ref. [47b], letter [31], p. 32). In this letter the discussion of the notion of time is continued with the mention of another aspect of rhythms which had become apparent particularly in Pauli's 'Great Vision' of the world clock (see page 246), namely the image of the four-fold measure (*Viertakt*). Pauli, 'as a docile pupil of yours' (translated from letter [23] of 15 October 1938), then asks the question of whether there is not a parallel between the evolutionary origin of the human heart with its four ventricles and the process of individuation. With this question Pauli takes his leave of Jung.

The spin-statistics theorem. War in Europe

An important and interesting point is the fact that our [Pauli and Weisskopf's] theory can only be carried through with Bose statistics because here an unavoidable connection between spin and statistics begins to dawn. [47c]

The Pauli–Weisskopf paper rang the first alarm bell that there are incompatibilities in the choice of how to quantize a field; it is expressed for the first time in letter [375] from Pauli to Heisenberg of 28 June 1934, quoted on page 306. The above statement, addressed to Heisenberg on 7 November 1934, is Pauli's first realization that there might be a new fundamental law, namely the connection between the *spin*—integer or half-odd—of the particles and the *statistics*—Bose–Einstein or Fermi–Dirac, i.e. commutation or anticommutation relations—of the quantized field describing the particles. To convince himself of the generality of this connection Pauli embarked with his collaborators, first of all Fierz, but also Kemmer in London and a new diploma student, Josef Maria Jauch (1914–74), on a programmme of quantized fields with arbitrary spin. This line of research had been initiated simultaneously by Dirac and by Proca. In his paper 'Relativistic Wave Equations' [48a], submitted on 25 March 1936, Dirac developed a field theory of arbitrary $n \times 1/2$ spin based on van der Waerden's spinor calculus [48b] (see Ref. [29b] in Chapter 5). Proca's paper 'On the wave theory of positive and negative electrons' [48c], which was submitted on 28 May, on the other hand, is a spin-1 theory.

Now the great interest in the spin-1 case at the beginning was motivated by the nuclear forces which, since Yukawa's famous paper of 1935 [49], were explained by the exchange of 'mesotrons' or mesons, i.e. particles with masses intermediate between those of the nucleon (proton and neutron) and the electron. Already in his first paper (Ref. [9a] I) Heisenberg had postulated interaction energies $J(R)$ and $K(R)$ as functions of the distance R between the neutron and proton and between two neutrons, in analogy with the forces in the molecules H_2^+ and H_2, respectively. The first analogy Heisenberg thought to be due to a 'position exchange' (*Platzwechsel*) of a spin-less 'heavy electron', while the second had been suggested to him by Pauli (see footnote 2, Ref. [9a] I, p. 2). But Heisenberg's ideas of the properties of the neutron (Section 3 of paper Ref. [9a] II) were still very much prejudiced by Bohr (see page 228) when he wrote: 'The, although hypothetical, validity of Fermi statistics for the neutrons, as well as the failure of the energy law in the β-decay prove the unapplicability to the structure of the neutron of the hitherto valid quantum mechanics' (translated from Ref. [9a] II, p. 163).

However, to Pauli's 'arbitrarily stupid question: Why the nucleus should be built in analogy to the *liquid drop*, and not to the *crystalline* state?' (translated from letter [323] of 29 September 1933, Ref. [3a], p. 214) Heisenberg was able to give a precise answer, namely that 'From the empirical energy differences in nuclear spectra one may deduce the order of magnitude of the momenta with which neutrons and protons move in the nucleus'. $(p^2/2M \sim \Delta E)$. If one calculates from them the dimensions of the orbits according to $\Delta q \sim \hbar/p$, then Δq empirically becomes of the order of the nuclear dimension, i.e. the individual particles circulate in the whole nucleus and do not just oscillate around an equilibrium position. The deeper reason must lie in the neutron-proton force law' (translated from letter [325], Ref. [3a], pp. 217–18). Since the binding energy of the deuteron (the bound state of a proton and a neutron) was approximately known ($\Delta E = 2.23$ MeV), it was possible to estimate Δq, i.e. the size of the deuteron, from which the range of the nuclear force follows approximately as $r_0 \sim \Delta q/2 \sim 1.5 \times 10^{-13}$ cm.

With this in mind Hideki Yukawa (1907–81) in October 1934 had the idea that 'The nuclear force is effective at extremely small distances. My new insight was that this distance and the mass of the new particle are inversely related to each other' (quoted from Ref. [20], p. 430). This gives for this 'meson' a mass of $m_\mu \sim \hbar/r_0 c \sim 254 m_e$ in units of the electron mass, m_e. Starting in November 1936 the existence of such a particle was announced by Anderson, the discoverer of the positron (see page 296), and Seth Henry Neddermeyer [50a], and also by Street and Stevenson [50b]. However, it later turned out that this meson μ^+ with a mass of $207 m_e$ had spin 1/2 and hence could not be Yukawa's particle, often called the yukon. It was only in 1947 that the Bristol group of Cecil Frank Powell (1903–69) working with photographic emulsions discovered the π^+-meson which has spin 0 and a mass of $273m_e$, demystifying both the problem of the nuclear force and that of the hard component of cosmic radiation (see Ref. [20], Sections 17(g) and 18(b)).

Yukawa's theory was based on the analogy with electrostatics, $\Delta V = -4\pi e(\psi^*\psi)$. While the latter is valid for mass-less photons, the analogue for mesons is

$$(\Delta - \kappa^2)U = -4\pi g(\overline{\psi}_N\psi_P). \tag{8.71}$$

Here $\kappa \equiv m_\mu c/\hbar \sim 1/r_0$ and g is the coupling constant of the meson–neutron–proton interaction

$$H' = g\int d^3x\{(\overline{\psi}_N\psi_P)U + (\overline{\psi}_P\psi_N)U^*\}, \tag{8.72}$$

which is analogous to Eq. (6.63), ψ_N and ψ_P being the neutron and proton Dirac wavefunctions, respectively. In analogy to the electric potential V, Yukawa considered U to be the fourth component of a vector. While the Coulomb potential is $V(\mathbf{x}) = e\int d^3x'(\psi^*\psi)(\mathbf{x}')/R$, with $R \equiv |\mathbf{x} - \mathbf{x}'|$, the analogous calculation with (8.71) yields

$$U(\mathbf{x}) = g\int d^3x'(\overline{\psi}_N\psi_P)\frac{e^{-\kappa R}}{R}. \tag{8.73}$$

On the other hand, making use of the analogy with Eq. (6.13), the matrix element of the interaction Hamiltonian (8.73) between a neutron at rest at $\mathbf{x}$ and a proton plus a meson with wavevector $\mathbf{k}$ may be written as $\langle\mathbf{x}|H'|\mathbf{k}\rangle = gc\sqrt{\hbar/2L^3\omega_\mathbf{k}}\exp(i\mathbf{k}\cdot\mathbf{x})$ where $\omega_\mathbf{k} = c\sqrt{k^2 + \kappa^2}$. Then Heisenberg's neutron–proton interaction energy $J(R)$ is obtained in second-order perturbation theory (see Refs [16b], Section 7 and [51a], Section 4), or by inserting the expression for $W(R)$ in (8.72), as

$$J(R) = \sum_\mathbf{k}\frac{\langle\mathbf{x}|H'|\mathbf{k}\rangle\langle|\mathbf{k}|H'|\mathbf{x}'\rangle}{-\hbar\omega_\mathbf{k}} = -\frac{g^2}{4\pi^2R}\int_0^\infty\frac{k\sin kR}{k^2 + \kappa^2}dk = -g^2\frac{e^{-\kappa R}}{8\pi R}, \tag{8.74}$$

where again $R = |\mathbf{x} - \mathbf{x}'|$. The simplest mechanism of nuclear forces is the exchange of a neutral spin-0 meson, but a vector meson is needed if spin exchanges are considered. On the other hand, charge exchanges are described by the charged spin-0 meson of the Pauli–Weisskopf theory; they have the same probability as zero charge exchanges. This is the law of charge independence which allows for a description of the nucleons by an 'isotopic spin' or isospin as introduced by Heisenberg in Ref. [9a] I (see e.g. Ref. [16b], Section 14 and 15, also Ref. [13e]).

Pauli's comment on Yukawa's theory, contained in letter [487a] to Weisskopf on 13 January 1938, already quoted in the previous section, was: 'Of course, however, not only the problem of the theory of β-decay herewith remains completely open but also the self-energies and the magnetic moments of the particles become infinite' (translated from Ref. [3a], p. 549). Meson theory, in fact, was the only aspect of the theory of the nuclei that retained Pauli's interest. For the rest he repeatedly used depreciatory language in his correspondence: 'Bohr's "liquid drop model" for the nucleus, namely, interests me not at all', and: 'Anyhow, the *theory* of the nuclei at the moment is not worth the effort to be occupied with' (translated from letter [480] to Weisskopf of 20 July 1937, Ref. [3a], pp. 527–8). And to Bohr he writes on 31 August 1937: 'As concerns nuclear physics, "I must say" [this is an imitation of Bohr!] that, indeed, you often were able to evoke enthusiasm in me for something - but this time I still have great doubts whether you will succeed again' (translated from letter [482], Ref. [3a]).

Turning now to Proca's paper [48c], it is curious that, as the title indicates, Proca's motivation had been to search for a positron theory without negative energies but with a charge density having both signs as in the Pauli–Weisskopf theory. He concluded that the wavefunction ought to have four components like the Dirac spinor, and he ended up with the Lagrangian density of a complex vector field, which was not so objectionable since, as we saw on page 304, this theory may also be quantized with commutators. Thus Proca started with the Lagrangian density (see Ref. [16b], Section 12)

$$\mathcal{L}_P = -\frac{1}{2}(\phi^*_{\nu,\mu} - \phi^*_{\mu,\nu})(\phi_{\nu,\mu} - \phi_{\mu,\nu}) - \kappa^2 \phi^*_\mu \phi_\mu, \qquad (8.75)$$

where $x_4 = ict$, $\phi_4 = i\phi_0$, and, as before, $f_{,\nu} \equiv \partial f/\partial x^\nu$ [52]. The associated equations of motion (6.35) are then

$$-\kappa^2 \phi_\mu = (\phi_{\nu,\mu} - \phi_{\mu,\nu})_{,\nu}, \qquad (8.76)$$

so that, acting on them with $\partial/\partial x^\mu$, the right-hand side is zero because the bracket is antisymmetric. Hence, provided that $\kappa \neq 0$,

$$\Box \phi_\mu = \kappa^2 \phi_\mu; \ \frac{\partial \phi_\mu}{\partial x^\mu} = 0, \qquad (8.77)$$

the latter being identical with the Lorentz condition (6.23). Proca even wrote Eqs. (8.75), (8.76) in the presence of an electromagnetic field but did not introduce quantization, and he was rather vague concerning the spin value of his field. Anyway, this equation made Proca famous.

Alexandre Proca was a friend of Pauli's [7a]. He grew up in Bucharest in a family of intellectuals. There he became an engineer working in industry but holding an assistantship at the Bucharest Institute of Technology. In the autumn of 1923 during some riots in Bucharest he went to Paris. But his excellent Romanian diplomas were not recognized at the Sorbonne, so that he still had to pass the Licence—which he did with distinction. He eventually obtained French citizenship and married a Romanian girl. In 1925 Marie Curie offered him a job and was very pleased with his collaboration. But one year later Proca decided to devote his energy to theoretical physics. His first papers showed quite an independent mind—too independent perhaps for Louis de

Broglie, the uncontested master in Paris. During the war Proca established contact with colleagues in Portugal, and in 1943 he went to Porto where he actively participated in physics teaching at the faculty. The following year he was invited to join the French scientific mission to England. On his return to Paris after the armistice—Pauli saw him in Langevin's house in April 1946 (Ref. [13c], letter [809])—Proca found the sciences in France in bad shape. His response was the Séminaire Proca which became a famous institution at the Institut Henri Poincaré where well-known physicists from all over the world gave lectures and where a whole generation of young French theorists was formed. But Proca, somehow, had stolen the show from Louis de Broglie. And he paid for it by being passed over in the nominations both for a chair at the Sorbonne and later for a professorship at the Ecole Normale Supérieure. In both cases Proca had been the candidate of choice. Profoundly hurt but refraining from open criticism he literally lost his voice: in 1954 an ablation of the vocal cords had to be operated on. He died at 58 one year later (Ref. [27c], pp. A-4 to A-28).

The Proca equation received elegant and exhaustive treatment in the paper [51a] by Kemmer (see also Ref. [51b]) who, after his appointment in Zurich as Wentzel's pupil and as Pauli's assistant, became one of the leading experts at Imperial College in London on the meson theory of nuclear forces. By a development into plane waves he explicitly showed that the Proca field had two transverse polarizations and, in distinction to the Maxwell field, also a longitudinal one, thus exhibiting the spin-1 character of the field. But in this paper Kemmer did much more. By comparing Dirac's theory [48a] of $n \times 1/2$ spin fields for $n = 2$ with the Proca equation, Kemmer deduced four different but equivalent sets of field equations in the following combinations of symmetry types of spin-0 and spin-1 fields: $S-V$, $V-T$, $T-A$, and $A-P$, where S means scalar, V vector, T antisymmetric (or *skew*) tensor, A axial (or pseudo) vector and P pseudo scalar symmetry. These combinations resulted in four different generalized interaction Hamiltonians (8.73) with the same symmetry combinations, Kemmer's Eqs (49). When applied to the deuteron, Kemmer concluded that the $V-T$ combination is the only one that is in accord with experimental data on the S states both with parallel and antiparallel nucleon spins. Details of these findings were published by Fröhlich, Heitler, and Kemmer [53a] and by Bhabha [53b].

Working along the same lines, Pauli had a busy correspondence with Kemmer whose paper [51a] was submitted on 9 February 1938. In his letter [490] of 28 February Pauli writes 'that on 15th, 17th and 19th of March I am supposed to hold three lectures at Cambridge [see the previous section] on "Contributions to the Theory of Field Quantization". I want to say something, first, on the ultrared catastrophe and, second, on the quantization of field equations for particles of higher spin. Therewith I still want to figure out some special malignancies against Dirac. Could you perhaps come over to Cambridge?' And he explains: 'With Dirac's equations for particles with higher spin there still are all kinds of symmetry-gaps and quirks' (translated from Ref. [3a], p.

552). In the following letter [491] Pauli raises the question '*how* one has to introduce the electromagnetic potentials into the Proca theory. Unfortunately, I cannot quite reconstruct how *you* have done it' (translated from Ref. [3a], p. 554). Interestingly, at the end of letter [491] Pauli writes that on Sunday the 13th or Monday the 14th he comes to London 'where, however, I perhaps yet want to visit somebody' (translated from Ref. [3a], pp. 554 and 556). Was this 'somebody' Pauli's analyst Erna Rosenbaum (see pages 240f)?

Both Dirac [48a] and Proca [48c] had introduced these potentials Φ_μ simply by the substitution $\partial/\partial x^\mu \to \partial/\partial x^\mu + (ie/\hbar c)\Phi_\mu \equiv \Pi_\mu$. This is all right, as long as Π_μ occurs linearly as in Eqs (8.75), (8.76), but may lead to contradictions for products of the Πs because, according to Eqs (6.24), the latter do not commute, $[\Pi_\mu, \Pi_\nu] = (ie/\hbar c)F_{\mu\nu}$. Thus, if after the substitution of the Π in Eqs (8.77), one applies Π_μ to the first of these equations, one finds $\Pi_\mu \Pi_\nu^2 \phi_\mu = 0$, which is a new supplementary condition. Pauli discusses these difficulties—which Dirac had overlooked—in letter [491] and, more systematically, in a paper he published with Fierz, in which the cases of spin 2 and spin 3/2, for non-zero and for zero masses, are analysed [54a]. Pauli reported on this work at the Spring Meeting of the Swiss Physical Society in Brugg, west of Zurich, on 6 May 1939 [54b].

Fierz developed a theory of fields of arbitrary spin in the absence of interactions [55a]. For integer spin this work was a direct generalization of Kemmer's paper [51a] on Proca's vector field to general symmetric tensor fields of trace 0, while for half-odd spin Fierz used the spinor formalism, as Dirac had done. In both cases he constructed appropriate expressions for the energy–momentum tensor and for the current vector. Such expressions actually had been set up for the first time in Jauch's diploma paper of 1937, 'Closer investigation of Dirac's wave equations for particles of arbitrary spin' (see Ref. [4], p. 434). 'Appropriate' here means that for integer spin the energy is positive while for half-odd spin it has both signs, so that, in analogy to the cases of spin 0 and 1/2 quantization must be done with commutators and anticommutators, respectively—Dirac's 'energy operator' did not have these properties. Fierz then explicitly constructed the appropriate quantization relations thus giving an existence proof for arbitrary spin. It is only at the end of the paper (Section 8) that he also proved, by a technique later taken up by Pauli, that the wrong statistics do not work. In an appendix Fierz also treated separately the zero-mass cases of spin-1 photons and spin-2 gravitons, the quanta of the gravitational field.

This impressive work, 'On the relativistic theory of force-free particles with arbitrary spin' [55a], became Fierz's habilitation paper at ETH. In his report of 20 September 1938 (Ref. [4], II.6) Pauli emphasized that this work goes beyond Dirac in an essential way and that for integer spin 'the wave equations of the particles are even developed independently of Dirac according to a new method which, avoiding the spinor calculus, can do with ordinary tensors' (translated from Ref. [4], p. 84). This part of the work was reported by Fierz at the Spring Meeting of the Swiss Physical Society in Delémont on 7 May 1938

[55b]. There also Jauch reported on his diploma paper [55c]. Scherrer likewise wrote a very positive and sympathetic report (Ref. [4], II.11). For Pauli this work represented a verdict that he accepted not without reluctance. Indeed, on 14 January 1941 he wrote to his friend Oskar Klein from Princeton: 'In fact, I have tried for a long time to show, together with Fierz, that the higher spin values [> 1] are impossible in a relativistic theory, but eventually yet found the opposite' (translated from letter [620], Ref. [13c], p. 72). Pauli, in fact, had thought he could prove that the electromagnetic field always violates the subsidiary condition. But Fierz convinced him that this was not the case [55d]. The problem of higher spin values had also attracted Peierls' interest. However, as Pauli told Uhlenbeck in letter [515], the reasonings of Peierls' student Fred Hoyle (1915–2001), who later became a well-known astrophysicist, 'were totally wrong' (translated from Ref. [3a], p. 586).

In April 1939 Pauli went with Wentzel to Paris to participate in Dirac's lectures on 'subtraction tricks' at the Institut Henri Poincaré (letter [551], Ref. [3a], p. 639). It was during this visit to Paris that Pauli and Proca became friends (see above). Indeed, Pauli 'had a nice time with him in Paris in the Spring of the year 1939' (Ref. [13c], letter [679]).

Then, on 20 April Heisenberg wrote to Pauli: 'The Solvay Committee has summoned me to write a report for the congress of this year on "general questions, limits of the hitherto existing theories, notion of elementary particles". What should one write there?' (translated from letter [546], Ref. [3a]). Three days later he wrote that on thinking about the Solvay report he came to realize that 'an essential part concerns questions that you know better than I. I therefore want to ask you whether you would have time and would wish to take over this part. The Solvay Committee surely would agree since you are invited anyhow.' Heisenberg then sketched a plan in three sections (translated from letter [549], Ref. [3a]). Pauli answered on 27 April: 'As much as it goes against my laziness to write such a report, I still believe that for objective reasons I am not able principally to reject your proposition to take over Section 1. Therewith I, however, want to make the counter-proposition that in Section 1c I do *not* touch the *interaction* (that should be reserved for Section 2) and call the whole Section 1 "Relativistic wave equations of *force-free* particles and their quantization". I then also would treat the connection of spin and statistics and (if you wish) the spin 2 quanta of gravitation' (translated from letter [551], Ref. [3a], pp. 638–9).

Thus began a complicated process which was conditioned by the outbreak of World War II and eventually led to Pauli's two fundamental papers [35] and [56]. On 11 May 1939 Pauli gave Heisenberg the first disposition of his contribution (letter [562] of Ref. [3a]) and on 10 June he told Heisenberg that his part 'has grown somewhat longer than I originally had planned', and he gave in an appendix the final disposition which now had the two chapters 'General reasonings' and 'Consideration of special fields' (translated from letter [570], Ref. [3a], pp. 662 and 664). On 7 August Pauli sent a copy of his manuscript to Heisenberg (Ref. [3a], letter [577]) who from 22 June to 1 August 1939 had been in the United States where he received several offers, attending a symposium on cosmic rays in Chicago from 27 to 30 June and lecturing at Purdue University in Lafayette, Indiana, from 1 to 22 July (see Ref. [3b], pp. 411 ff). In order to meet the deadline of 1 July Heisenberg had sent two copies of his manuscript to Pauli on 8 June, asking him to forward his part together with Pauli's, one to the Solvay Committee in Brussels and one to Bohr in Copenhagen (letter [569] from Heisenberg to Pauli, Ref. [3a]). With the declaration of war by Britain and France on 3 September 1939 the Solvay Congress planned for October was cancelled, as was the *Physikalische Woche* at ETH, but the hope to publish a conference report in French persisted for some time. It prevented Pauli from publishing parts of his manuscript in *Reviews of Modern Physics* whose founder and editor John Torrence Tate (1889–1950) would have agreed to do so (see letter [594] from Pauli to Weisskopf of 10 March 1940, Ref. [13c], p. 24). Thus neither Pauli's nor Heisenberg's German manuscripts gave rise to journal articles; they were eventually published, respectively, in Ref. [13c], pp. 833–901 and Ref. [57], pp. 346–58. The chapter and section headings of Pauli's contribution there are the same as in the appendix to letter [570] in Ref. [3a], except for that of Chapter 1 which now reads 'Transformation Properties of the Field Equations and Conservation Laws'. This was the last collaboration between Pauli and Heisenberg for 18 years.

The parts of this manuscript that Pauli considered most important were the three parts of Section 3 from Chapter I. They became, together with a shortened introduction: Section 1. 'Units and Notations', the content of Ref. [56]: Section 2. 'Irreducible Tensors. Definition of Spins'; Section 3. 'Proof of the Indefinite Character of the Charge in Case of Integral and of the Energy in Case of Half-Integral Spin'; Section 4. 'Quantization of the Fields in the Absence of Interactions. Connection between Spin and Statistics'. This quite compact manuscript was sent from Zurich to the *Physical Review* and received there on 19 August 1940, while Pauli and Franca were on their way to the United States (see below). The translation of the German manuscript was probably done in England by Maurice Henry Lecorney Pryce (b. 1913) who had already translated the Fierz–Pauli paper [54a] (see letter [540] in Ref. [3a]) [55d]. The remaining long part of the manuscript was translated after Pauli's arrival in Princeton by Leonard Eisenbud (b. 1913), a collaborator of Eugene Paul Wigner (1902–95) at Princeton University (see letter [612] from Pauli to Weisskopf, dated 28 November 1940, Chicago, Ref. [13c]), and was received by *Reviews of Modern Physics* on 8 May 1941 [35]. One may see the difference in the style of translation of the two parts by comparing the identical (apart from the mentioned shortening) introduction 'Units and Notations'.

This paper [35], which has already been mentioned on page 309, does not include Section 5. 'Remarks on gravitational waves and gravitational quanta (Spin 2)' of Chapter II of the German manuscript, excerpts of which are contained in Refs [54a,b]. But it contains for the first time expressions for an anomalous magnetic moment which later came to be called Pauli terms. Indeed, Eq. (6.62) is Eq. (91) of Ref. [35] (see also Eq. (7.39)). And the contributions of such a moment to the current vector for spin 1 and spin 1/2 are given in

Eqs $\langle 49 \rangle$ and (92), respectively. Also, Pauli had added in Ref. [35] a Section 5. 'Applications' containing at the end six tables of formulae of cross-sections for various electromagnetic processes involving charged 'mesotrons'. They show that for spin 1 and/or an anomalous moment the cross-sections diverge with increasing energy, meaning according to the criterion mentioned in the last section, a non-renormalizable theory. For these calculations Pauli acknowledges discussions with the Australian physicist Dr Herbert Charles Corben (b. 1914) who in 1940 was at Berkeley on a fellowship from the Commonwealth Fund where, after studies in Melbourne and Cambridge, England, he obtained his Ph.D. under Oppenheimer. In January 1941 he came to Princeton where he collaborated with Pauli, who said of him 'I enjoy his presence very much' (Ref. [13c], letter [627] to Bhabha of 20 March 1941, p. 84). After spending five years in Australia Corben settled in the United States in 1946 (Ref. [5a], p. 371; see also Ref. [13c], letter [622] and footnote 9, p. 77).

In Ref. [35], Section 4, 'A special synthesis of the theories for spin 1 and spin 0', should be mentioned in particular because, referring to Richard J. Duffin (1909–96) as the inventor of a new algebra, it is the result of an intense exchange of letters between Pauli and Kemmer (Ref. [3a], letters [541] to [573], covering the months from April to June 1939; see also Ref. [13f]) which the authors ended up calling a 'novel with sequels' (letters [566, 573]). It concerns an equation analogous to the Dirac equation (6.31), in which the wavefunction has five components, one for spin 0 and four for spin 1,

$$\beta^\mu \frac{\partial \psi}{\partial x^\mu} + \kappa \psi = 0. \tag{8.78}$$

Here the 5×5 matrices β^μ form a hypercomplex number system often called a Petiau–Duffin–Kemmer algebra after its inventors [58a,b,c]. This algebra struck a chord with Pauli who during his Hamburg days had developed a predilection for modern algebra by not only attending Emil Artin's lectures on 'Representation theory of semi-simple systems' during the winter term of 1927/8 but also taking extensive notes which he now consulted again (see letter [544] in Ref. [3a] and page 149). The subject was fashionable. Klein was also working on it in Stockholm and gave a talk at de Broglie's seminar in Paris in Spring 1939, as he wrote to Pauli in October 1940, to which Pauli answered from Princeton in January 1941 (letters [603, 620, 621], Ref. [13c]). But for Pauli there also was the hope that this formalism might 'be generalized to include higher values of the spins' (Ref. [35], p. 227).

Let us now have a closer look at the published spin-statistics paper [56]. Following van der Waerden's book [48b], Section 2 gives the definitions of tensor and spinor fields $U(jk)$ and their complex conjugates $U^*(jk)$, characterized by integers $2j + 1$ and $2k + 1$ and transforming irreducibly under the proper (determinant $=1$) Lorentz group. The most important operation then is the reduction of products $U_1(j_1 k_1)U_2(j_2 k_2)$ into irreducible representations. Spin is introduced by considering irreducible representations of the subgroup of space rotations where $U(jk)$ factorizes into the product $U(j)U(k)$ whose irreducible parts $U(l)$, $l = |j - k|, ..., j + k$, are single or double valued, according to whether l is integer or half-odd. For non-zero mass m, spin s is then defined, following Fierz [55a] and also de Broglie [59a,b], by going to the rest frame where there are $2s + 1$ irreducible field components. In the case $m = 0$ 'gauge transformations of the second kind' (see page 191) allow, for a given four-momentum $\hbar k_i$, only two independent field components, and spin s is defined as the minimum of the angular momentum quantum number.

This classification of the irreducible representations of the Lorentz group—which is purely algebraic—is given in a most clear and compact form in the lectures 'Continuous Groups in Quantum Mechanics' [59c] Pauli gave in September 1955 at the new Theory Division of CERN in Copenhagen. As already mentioned on pages 136 and 270, these lectures were part of the *Spezialvorlesung* 'Gruppentheorie und Quantenmechanik' [59d] that Pauli gave at ETH during the summer term of 1954 (see Ref. [4], p. 434). Similar to Pauli's *Spezialvorlesung* on 'Problems of General Relativity' (see page 35), these lectures of 1954 look like a warming-up exercise for the CERN lectures of the following year. However, while the latter were limited to the 'Algebraic Part' of the infinitesimal groups or Lie algebras, the ETH lectures also comprised an important 'Analytical Part', in which finite transformations were considered and which began with a group-theoretic discussion of the hydrogen atom.

Returning to Ref. [56], Section 3 discusses the one-particle theory for arbitrary spin. Considering first only integral spin (tensors of rank $j + k$), the two classes, both j and k integers (class $+1$) and both j and k half-odd (class -1), are distinguished. Noting that the wavevector $k_l = -i\partial/\partial x_l$, as any four-vector $(1/2,1/2)$, is of class -1, any homogeneous and linear wave equation, irrespective of its order, must be of the form

$$\sum k U^{\pm} = \sum U^{\mp}, \qquad (8.79)$$

irrespective of complex conjugation and omitting even powers of k_i. These equations are invariant under the substitutions

$$k_i \to -k_i; \; U^{\pm} \to \pm U^{\pm}. \qquad (8.80)$$

In the above notation the energy–momentum tensor and the current density are tensors T and S of even and odd rank, respectively. On the other hand, T and S are bilinear forms in the $U^{\pm}$ with powers of k_i, and these forms are determined by the rank. Application of the substitutions (8.80) then yields $T \to T$ and $S \to -S$. Pauli concludes: 'it is remarkable that from the invariance of Eq. [8.79] against the proper Lorentz group alone there follows an invariance property for the change of sign of all the coordinates. In particular, the indefinite character of the current density and the total charge for even [this should read 'integral', not 'even'!] spin follows' (Ref. [56], p.719).

The argument for half-odd spin is analogous. Distinguishing the two classes j integral, k half-odd (class $+\epsilon$) and j half-odd, k integral (class $-\epsilon$), then complex conjugation and k_i change the class, so that any homogeneous and linear wave equation must have the form

$$\sum k U^{\pm\epsilon} + \sum k (U^{\mp\epsilon})^* = \sum U^{\mp\epsilon} + \sum (U^{\pm\epsilon})^*. \qquad (8.81)$$

These equations remain invariant under the substitutions

$$k_i \to -k_i; \ U^{\pm\epsilon} \to \pm i U^{\pm\epsilon}; \ (U^{\pm\epsilon})^* \to \mp i (U^{\pm\epsilon})^*. \qquad (8.82)$$

Again, the tensors T and S representing energy–momentum and current density, respectively, are bilinear forms in the $U^{\pm\epsilon}$ and $(U^{\pm\epsilon})^*$ with powers of k_i which are determined by the rank of T and S. The substitutions (8.82) then act as follows: $T \to -T$ and $S \to S$. Therefore, 'the energy density in every space-time point changes its sign as a result of which the total energy changes also its sign' (Ref. [56], p. 720).

From the impossibility of a positive particle density and a positive energy density in the case of integral and half-odd spin, respectively, it is concluded in Section 4 that the one-particle theory must be second quantized. Now the postulate is introduced that all physical quantities satisfy the micro-causality condition (8.36). Since all fields $U^{(r)}$, where r designates one of the four classes defined above, satisfy the Klein–Gordon equation (5.34), it follows 'that the bracket expressions of all quantities ... can be expressed by the function D [Δ, Eq. (8.12)] and (a finite number) of derivatives of it without using the function D_1 [Δ^1, Eq. (8.16)]' (Ref. [56], p. 721). Here 'bracket expression', denoted $[A, B]$, means a commutator or an anticommutator, $[A, B]_{\mp}$. Now $[U^{(r)}(x), U^{*(r)}(x')]$ is a tensor of even/odd rank if $U^{(r)}$ has integral/half-odd spin and, hence, is equal to an even/odd number of derivatives of $D(x - x')$. But D is an even/odd function of the space/time coordinates. Since, by symmetry, $X \equiv [U^{(r)}(x), U^{*(r)}(x')] + [U^{(r)}(x'), U^{*(r)}(x)]$ is even in $t - t'$, it therefore must equal an odd number of time derivatives. This is consistent for half-odd, but not for integral, spin. Hence, for integral spin, $X = 0$. But this is a contradiction for anticommutators which are positive for $x \sim x'$. Therefore: '*For integral spin the quantization according to the exclusion principle is not possible*' (Ref. [56], p. 722). The proof is then completed as follows. While it is formally possible to quantize half-odd spin fields according to Bose statistics, '*according to the general result of the preceding section the energy of the system would not be positive*' (Ref. [56], p. 722). Pauli concludes with the opinion that 'the connection between spin and statistics is one of the most important applications of the special relativity theory' (Ref. [56], p. 722).

In its compactness and rigour Ref. [56] might well be considered Pauli's most brilliant paper. Understandably, therefore, he was touchy with respect to less rigorous approaches. There were two Ph.D. theses written on the subject at the time, one by J. S. de Wet (see Ref. [31a]) at Princeton University under the supervision of Wigner and Howard Percy Robertson (1903–61), and one, 'Theory of Heavy Quanta' [60a], by Frederik Jozef Belinfante (1913–91) at Leiden under the direction of Kramers. As to de Wet's work, which is based on the canonical formalism, Pauli observes in footnote 7 of Ref. [56] that, because of the existence of supplementary conditions, the canonical formalism is not applicable to fields with spin greater than one. Pauli's relation to Belinfante's work is more complicated. He had in fact put his name on a paper which was essentially written by Belinfante [60b]. Footnote 1 of Ref. [56] refers to this paper with the comment: 'The relation between the present discussion of the connection between spin and statistics, and the somewhat less general one of Belinfante, based on the concept of charge invariance, has been cleared up by W. Pauli and F. J. Belinfante [Ref. [60b]].'

The genesis of Belinfante's version of the spin-statistics connection is, in his own words, as follows: 'I was a student of H. A. Kramers. In 1939 I wrote my Ph.D. thesis. . . . The second chapter under the title "The Undor Equation of the Meson Field" appeared in [Ref. [60c]]. Its fifth and last section 'Charge-invariance and statistics' pointed out that . . . all particles known at that time would automatically require Fermi-Dirac statistics for fields described by spinors of odd rank, and Einstein–Bose statistics for those described by spinors of even rank or by tensors, if one would postulate invariance under charge conjugation as defined by Kramers' [60d]. Probably 'the correspondence about my paper between Kramers and Pauli originated from an idea of Kramers that it would be useful for me to work a while under Pauli in Zurich' [60d]. 'Kramers then received immediately a letter from Pauli, in which Pauli pointed out that he could formulate theories in which this relation between spin and statistics would be broken notwithstanding charge invariance, while this relation would not be broken if, instead of charge invariance, one would postulate (a) positive-definite energy, (b) commutativity of observables at spacelike distances. Therefore, according to Pauli, the postulate of charge invariance was neither sufficient nor needed, and he asked Kramers to have me write a paper in which I would withdraw my claims' [60d].

But Belinfante thought otherwise: 'Instead, I wrote a paper in which I showed that charge-invariance alone would not be sufficient for relating spin to statistics, if one would admit Lagrangians . . . in which the field would be described by more than one single undor of rank N (and its conjugate). However, all "particles known" were described by a single undor. . . . I sent the manuscript of this paper to Pauli for his approval' [60d]. In the title 'the words "AND UNKNOWN" were for me all-important. To me this meant that Pauli was right from a mathematical point of view, but only by going outside of physics, which according to me should deal only with "known" particles. For this reason I told Pauli that the authors' names . . . should be Pauli and Belinfante, for, though the paper was written by me, the ideas discussed in it were primarily Pauli's' [60d]. Frederik Jozef Belinfante was born in the Hague. He got married in 1937 and was an assistant in theoretical physics at Leiden. In 1946 he emigrated with his family of three children to Vancouver, Canada. There he became an associate professor at the University of British Columbia. In 1948 he went to the United States where he became a professor at Purdue University in Lafayette, Indiana, and where he worked on quantum field theory and on cosmology (Ref. [5a], p. 146, see also Ref. [60e]).

Belinfante described particles with spin $N \times 1/2$ not by spinors but by Dirac wavefunctions $\psi(x; \zeta_1, ..., \zeta_N)$ which depend on N four-component variables ζ_r, on which act Dirac matrices γ_r^μ. ψ is assumed to be symmetric in all the ζ_r and to satisfy Eq. (8.78) with

$$\beta^\mu = \frac{1}{2} \sum_{r=1}^{N} \gamma_r^\mu, \tag{8.83}$$

where

$$[\gamma_r^\mu, \gamma_s^\nu] = 0; \ r \neq s, \ \mu, \nu = 1, ..., 4. \tag{8.84}$$

Belinfante called such wavefunctions *undors* [60c]. Now it follows from the mentioned symmetry of ψ that Eq. (8.78) is eqivalent to N identical Dirac equations (6.31) with $k_c = 2\kappa/N$ for each set of γ_r^μ. These equations, which correspond to Eq. (20) of Ref. [60c], are called 'supplementary conditions' in Pauli's letter [621] in Ref. [13c].

In this letter [621] of 28 January 1941 from Princeton Pauli described Belinfante's undor theory to Klein (see also letters [603, 620]). There Pauli wrote: 'At the moment Belinfante's formulation for particles with higher spin seems to me to be the most satisfactory one (in the case without forces—with external electromagnetic fields, the auxiliary conditions give difficulties which were discussed through special tricks by Fierz and me [54a]). For, this is a natural generalization of Dirac's theory and does not make use of spinor calculus' (translated from Ref. [13c], p. 74). An example of these difficulties with the auxiliary conditions in the presence of fields (Proca theory) was mentioned above. Pauli had also discussed Belinfante's undor theory [60c] in his *Spezialvorlesung* 'Gruppentherie und Quantenmechanik' mentioned above.

The idea of Belinfante's proof of the spin-statistics connection is rather simple. Consider the particle density for spin $1/2$ and spin 0, Eqs (6.32) and (8.23), respectively. Written as Hermitian operators they are

$$\rho = \psi^+ \psi; \ \rho = \frac{i}{\hbar} (\psi^+ \pi^+ - \pi\psi). \tag{8.85}$$

For spin $1/2$ charge conjugation is defined by the transformation of $\psi(x)$, Eq. (8.31), to $\psi^c(x)$ with $a_{\sigma\mathbf{k}}^c = \bar{a}_{\sigma-\mathbf{k}}$, $\bar{a}_{\sigma\mathbf{k}}^c = a_{\sigma-\mathbf{k}}$ and, according to Eqs (8.38), (8.41), $u_{\sigma\pm}^c(\mathbf{k}) = u_{\sigma\mp}^*(\mathbf{k}) = C u_{\sigma\pm}(\mathbf{k})$. For spin 0 the change of $\psi(x)$ and $\pi(x)$, Eqs. (8.26), to $\psi^c(x)$ and $\pi^c(x)$ is obtained with $b_{\mathbf{k}}^c = \bar{b}_{-\mathbf{k}}$ and $\bar{b}_{\mathbf{k}}^c = b_{-\mathbf{k}}$. In both cases one easily finds that

$$\psi^c(\mathbf{x}, t) = \psi^+(-\mathbf{x}, t); \ \pi^c(\mathbf{x}, t) = \pi^+(-\mathbf{x}, t). \tag{8.86}$$

These relations are interesting because, if in addition to charge conjugation C one also operates on ψ and π with space reflection, i.e. with parity P and with Hermitian conjugation which represents time reversal T, one can see that the fields ψ and π are invariant under the product CPT. This is the special case of the CPT-theorem (see pages 507–9) applied to free fields.

Since, by definition, charge conjugation changes the sign of the charge, one has in a reflection-invariant volume V,

$$0 = e^c + e = \int_V d^3x (\rho^c(-\mathbf{x}, t) + \rho(\mathbf{x}, t)). \qquad (8.87)$$

This, however, demands that ψ satisfies anticommutation and commutation relations for spin $1/2$ and 0, respectively. Such arguments were used by Pauli 10 years later when he came back to the spin-statistics theorem [61a] because Richard Phillips Feynman (1918–88) had observed at the end of Section 5 of his famous paper on 'The theory of positrons' [61b] and discussed at the beginning of Section 9 of the companion paper on 'Quantum Electrodynamics' [61c] that the wrong statistics also work, but that a probability larger than one, C_v in Ref. [61b], is obtained for the vacuum to be left unchanged by the action of an electromagnetic field. Pauli showed, for spin 0 and 1/2, that Feynman's treatment of the 'abnormal case' is equivalent to introducing negative probabilities and, hence, may be excluded by the condition that 'The metric in the Hilbert-space of the quantum mechanical states is positive definite' (Ref. [61a], p. 526). Notes on this paper entitled 'Abnormal Theories with "negative probabilities"' are published as an appendix to letter [1089] of 2 March 1950 to Feynman's collaborator Fritz Rohrlich (b. 1921) in Ref. [61d], pp. 47–52.

Thus, as Jost writes in the Pauli memorial volume, with Pauli's paper [56] 'the treatment of our subject comes to a certain closing. For force-free particles the connection between spin and statistics was cleared in a satisfactory manner' (translated from Ref. [62a], p. 115). The further development in the 1950s was directed mainly at the CPT-theorem and to a more rigorous (axiomatic) formulation of quantum field theory (see Ref. [62a]). As the major reference concerning the whole matter the book by Streater and Wightman [62b] must be mentioned. More recently Duck and Sudarshan have discussed the merits of the different contributions to the subject [62c].

At the end of the winter term of 1939/40 Fierz was offered a position of private dozent in Basle with the prospect of an associate professorship. On 22 January 1940 Pauli wrote to Jauch about it: 'Fierz's Basel matter is o.k. By the way, he is also engaged (to Miss [Menga] Biber, daughter of a gynaecologist in Zurich) and wants to marry in Spring. He already busily looks for an apartment in Basel' (translated from Ref. [13c], letter [588]). More details are given in Pauli's letter [594] to Weisskopf of 10 March already quoted. The wedding took place in grand style at Grossmünster on 29 March, followed by a 'lunch feast with lots of wine and subsequent dance.' And Pauli comments: 'My wife and I are officially invited to the celebration, and I already have great fear in view of the possibly occurring necessity to be obliged to wear a top-hat (!)' (translated from Ref. [13c], p. 26).

In spite of the uncertain times, not only the University of Basle but also the Universities of Geneva and Berne created chairs of theoretical physics; the former nominated Stueckelberg (see letter [553]) and the latter André Mercier (1913–99) (see Pauli's comment at the end of letter [555]). As Fierz's successor Pauli had in mind his last diploma student Jauch (see letter [568]). After his diploma Jauch had been in urgent need of a job. He got a teaching position in Trogen near St Gallen, but when he was granted an international exchange fellow~hip in the United States to work for a Ph.D. with Professor Edward L. Hill (b. 1904) at the University of Minnesota, he immediately accepted. By early 1940 he finished his thesis [63a], in time to become Pauli's assistant for the summer term of 1940 (Ref. [4], II.24). On 1 November 1939 Pauli had written to Jauch: 'It is very nice that you want to join me in the summer term 1940 in spite of war.' And he closed with the remark 'I had not believed in the war to the last moment, but now I also consider it likely that it will last long.— Hopefully, everything will remain undamaged in Switzerland' (translated from Ref. [3a], letter [584]).

On 3 February 1940 Pauli wrote to Kemmer in English: 'He [Jauch] is coming back from America this Spring—where (oh, oh!) he married an American lady. I am prepared to great complications because he has no money and Bretscher will know, that it is nearly impossible to live married with the salary of an assistant in Zurich. So Mrs. Tonia Jauch will probably be very discontent here' (Ref. [13c], letter [590], p. 11). Jauch, on 1 January 1940, had married Anna Tonetta (Tonia) Hegland, the daughter of a professor at the St Olaf College in Northfield, Minnesota (see Ref. [4], footnote 139, p. 109). Pauli asked President Rohn to pay Jauch 350 francs per month (Ref. [4], II.23). However, because of the crisis deduction Jauch had to make do with 1880 francs over six months (Ref. [4], II.24); in addition to this, Fierz had obtained a fidelity premium of 94 francs (Ref. [4], II.10). Egon Bretscher (b. 1901), who had been an assistant to Scherrer since 1928 and who in 1933 went to the Cavendish Laboratory in Cambridge, was a friend of Kemmer.

It would seem from the above judgement expressed to Jauch that Pauli at first did not take the Nazi threat too seriously, in spite of the increasing propaganda and diplomatic pressure felt in Switzerland. Deeply worried, the Swiss Federal Council had published on 9 December 1938 a statement, written by Philippe Etter (see page 283), calling for the 'intellectual defence of the country (*geistige Landesverteidigung*)' which was very much in the spirit of the magazine *Schweizer Spiegel* (see page 286). With few exceptions, the leading Swiss press, including the *Neue Zürcher Zeitung* but also the satirical weekly *Nebelspalter* and the Zurich cabaret *Cornichon* (see page 209), never hesitated to denounce Nazi misdeeds, for which they were attacked by the Nazi press and reprimanded by the German Embassy in Berne. On 28 August 1939 the Federal Council decreed the mobilization of the army to protect the frontiers, and the following day the National Assembly elected Henri Guisan (see page 286) as the supreme commander (general). On 1 September the German attack on Poland started.

But for Pauli life in Zurich became really dangerous on 10 May 1940 after the Nazi armies directed their 'lightning war' to the west. Already on

22 June the Armistice of Compiègne had put France under German control, leaving Switzerland completely surrounded by the Axis powers. A second general mobilization of the Swiss Army had been ordered on 10 May and large parts of the population left the open regions for the mountains in panic (see Ref. [4], p. 74). On 17 June 1940 Pauli asked President Rohn for leave allowing him to accept the 'urgent invitation' for a visiting professorship at the Institute for Advanced Study in Princeton during the winter term 1940/41 and 'to represent the Federal Polytechnic [ETH] at the congress of Columbia University in New York in September 1940. Since the technical possibilities of a journey to America are not previsible, this application may be valid only in the case of the realizability of this journey, I also wish to leave open the time of my departure.' Of utmost interest is Pauli's motivation: 'For my proposition the circumstance is of essential significance that my application for naturalization submitted in December 1939 to the Federal Department of Justice and Police in Berne [see the previous section] has remained without answer to this day.' Pauli closed his letter by proposing his colleague and friend Gregor Wentzel from the university to replace him in the main lectures (translated from Ref. [4], II.25).

When Pauli wrote this letter to President Rohn he was already in possession of a US visa for him and for Franca. According to the letter to Director Aydelotte of the Institute at Princeton already quoted in the previous section, Pauli hoped that, based on the confirmation letter from Director Aydelotte he had handed to the staff of Vice-Consul Strom at the US Consulate in Zurich, he would get the visa on his next visit there, on 4 June (see Ref. [13c], p. xxviii). President Rohn reacted swiftly: on 18 June he wrote to the head of the Police Division of the Federal Department of Justice and Police, Dr Heinrich Rothmund (1888–1961) (see Ref. [38a], p. 323), that Pauli had received a call from the United States but that, provided Swiss citizenship could be granted to him soon, he might be saved for ETH. 'For this reason we take the liberty to present the petition to you that you kindly accelerate, if at all possible, the handling of the request for naturalization of Prof. Dr. Pauli. We are in a position to strongly recommend Prof. Pauli for naturalization' (translated from Ref. [4], II.26). In its session of 20 June 1940 the Swiss School Council accepted Pauli's request fo leave (Ref. [4], II.27), and Pauli was informed by the presidency on 15 July (Ref. [4], II.30).

On 5 July Dr Rothmund answered Pauli in the negative (Ref. [4], II.28). To President Rohn he addressed a longer letter, the conclusion of which reads: 'If Professor Pauli has a rejecting attitude towards the political development in the Reich and wishes to get rid of the German nationality which he acquired without his deed through the annexation of Austria, this does not make him a Swiss yet. That, indeed, he cannot be considered one yet, follows from his characterization, reflected in one of the present police reports from Zurich, as given by a closer colleague which is quite sympathetic towards him [Pauli] and who had been interviewed about his fitness for naturalization' (translated from Ref. [4], II.31). So Pauli's difficulty was due to a colleague! Pauli of course had

Fig. 8.6 Passport photo for the departure to the United States, July 1940

noticed that some colleagues had reservations about him because of his being a Jew; he particularly resented not being invited to be a member of the Balling (ninepins) Club of the ETH professors [7a]. One of the colleagues interviewed was Leopold Ruzicka (1887–1976) who in 1939 had received the Nobel Prize in Chemistry. In a similar situation after the war the interviewing policeman told Ruzicka: 'you remember Professor Pauli? You then were the only one of the interviewed colleagues that recommended him' [63b]. President Rohn did not give up, though; he questioned the aptitude of the Zurich police report to give a correct characterization and he proposed to Dr Rothmund as new witnesses the mineralogist Paul Niggli (1888–1953) and Paul Scherrer, as well as Pauli's friend Adolf Guggenbühl (see page 286) (Ref. [4], II.32).

Pauli now seriously prepared for his and Franca's departure. On 15 June 1940 he wrote to John Ludwig von Neumann (1903–57) who was already in Princeton: 'On 5 June we at last received the American visa. . . . On Saturday, the 8th, we also obtained the Spanish transit visa (for which we had travelled specially to Berne), after we also had got the Portuguese visa on the 5th. We travelled directly to Locarno [in southern Switzerland] there to board the plane to Barcelona (and then take the following day the one from Barcelona to Lisbon). For the 10th (on Sundays there is no plane) the plane, however, was sold out, and our efforts to get on this plane failed. For Tuesday 11th we had reserved seats, but on Monday afternoon the Italian Declaration of War (long awaited by us) followed, and it was exactly the first day on which the plane didn't go any more. Since then this route is suspended. The route Locarno–

Rome and Rome–Lisbon exists (this letter follows it), but at present no transit visas through Italy are issued at all' (translated from Ref. [13c], p. xxix). Thus Franca and Pauli returned to Zurich—their house in Zollikon, of course, was rented out—and awaited further developments. This letter to von Neumann continued: 'We will seize any chance to get to Lisbon; whether it will be possible, very likely, would depend mainly on the military development of the Mediterranean War.' Observing that the US visa was valid until 3 October, Pauli then writes that, provided the visa is not used by then, the whole bureaucratic procedure (which lasts about three weeks) will have to be repeated. 'At the moment it looks again more peacefully here, however, while mid May the situation was critical' (translated from Ref. [13c], p. xxix).

On 31 July 1940 Pauli and Franca left Switzerland from Geneva (Ref. [13c], letter [600]) by train through southern France to Barcelona and Lisbon where they boarded a ship, arriving at New York on 24 August. They were picked up from there by von Neumann who drove them to Princeton. The next day Pauli informed President Rohn that he had arrived and that he intended to make use of his leave (Ref. [4], II.33). It must have been a complicated journey since Pauli wrote to Fierz from Princeton on 3 September: 'I do not want at all to begin with a description of the journey, otherwise this letter would come to no end at all' (translated from Ref. [13c], letter [600]). Franca was more explicit. She later said that during this journey Pauli at times entirely lost his composure, and only by her persuasion was he moved to continue the journey through Portugal (Ref. [13c], footnote 11, p. 4). About a month later Pauli's sister Hertha made the same journey from Lisbon (see page 18).

Notes and references

[1] Quoted from Ref. [2], p. 75.

[2] V. Weisskopf, *The Joy of Insight. Passions of a Physicist* (Basic Books, New York, 1991).

[3] (a) K. von Meyenn (ed., with the cooperation of A. Hermann and V. F. Weisskopf), *Wolfgang Pauli. Scientific Correspondence with Bohr, Einstein, Heisenberg, a.o., Volume II: 1930–1939* (Springer, Berlin and Heidelberg, 1985); (b) D. C. Cassidy, *Uncertainty. The Life and Science of Werner Heisenberg* (Freeman, New York, 1992).

[4] C. P. Enz, B. Glaus, and G. Oberkofler (eds), *Wolfgang Pauli und sein Wirken an der ETH Zürich* (vdf Hochschulverlag ETH, Zurich, 1997).

[5] (a) A. G. Debus (ed.), *World Who's Who in Science* (Marquis–Who's Who, Chicago, 1968); (b) V. F. Weisskopf, 'Meine Assistentenzeit bei Pauli', in Ref [6], pp. 80–8. (c) *Note*: The dates proposed by the editor of Ref. [3a] in footnote *a*, p. 283, and in footnote *a*, p. 310, are incorrect: The date of letter [355] is Monday 12 February 1934 (not the 13th) and Easter Sunday 1934 was on 1 April (not the 2nd). This follows from the dates of the early letters in Ref. [8] of Chapter 7, as discussed

in *Note* [16b] of that Chapter. This assignment of days of the week to a given date may of course also be checked with any newspaper of the period. (d) *Note*: Information on Hans Bertram contained in the birth certificate of Franziska Pauli-Bertram, according to letter of 17 May 1999 from Standesamt, Munich. Franca Pauli had told me [7a] that her father Hans Bertram had great merit as an orphans' councillor (*Waisenrat*) and that he had become an honorary citizen of the city of Munich. Unfortunately, neither of these two statements has been confirmed by information contained in a letter from Stadtarchiv München, dated 17 August 1999. (e) Notes of my conversation with Hilde Mutschler of 4 June 1996. Ms Mutschler had been the housekeeper for Franca Pauli from early 1960 until Franca's death in 1987. I thank Hilde Mutschler for her kind collaboration.

[6] C. P. Enz and K. von Meyenn (eds), *Wolfgang Pauli. Das Gewissen der Physik* (Vieweg, Braunschweig, 1988).

[7] (a) Notes of my conversation with Franca Pauli on 8 April 1984. (b) Notes of my conversation with Dr Adolf Guggenbühl-Craig (Guggenbühl Adolf junior), Zurich, on 7 May 1998. I thank Dr Guggenbühl for this information. (c) G. Rasche and H. H. Staub, 'Physik und Physiker der Universität Zürich, 1833–1948', *Vierteljahrsschrift der Naturforschenden Gesellschaft in Zürich* **124**, 205 (1979); (d) A. Pais, *'Subtle is the Lord . . . '. The Science and the Life of Albert Einstein* (Oxford University Press, Oxford, 1982), pp. 11–12; (e) K. Guggenheim, *Gerufen und nicht gerufen* (Benziger, Zurich, 1973; Buchclub Exlibris, Zurich, 1975). I thank Dr Beat Glaus for having pointed out this book to me. (f) K. Guggenheim, *Alles in Allem. Vol 1: 1900–1914, Vol. 2: 1914–1919, Vol. 3: 1920–1932, Vol. 4: 1933–1945* (Artemis Verlag, Zurich, 1952–5); (g) Notes of my conversations with Franca Pauli on 21 March, 6 April, and 6 May 1971.

[8] (a) R. J. Humm, *Bei uns im Rabenhaus* (Fretz & Wasmuth/Werner Classen, Zurich, 1963, 1975); (b) P. Kamber, *Geschichte zweier Leben – Wladimir Rosenbaum & Aline Valangin* (Limmat Verlag, Zurich, 1990); (c) E. Canetti, *Das Augenspiel. Lebensgeschichte 1931–1937* (Hansen Verlag, Munich and Vienna, 1985). I thank Professor Markus Fierz for pointing out this source to me. (d) *La théorie des électrons dans les métaux. Conférences internationales des sciences mathématiques, Genève, 15-18 octobre 1934*, in *Helvetica Physica Acta* **7** (1934). (e) Private information from the late Professor Mercier to me, dated 15 December 1986.

[9] (a) W. Heisenberg, 'Über den Bau der Atomkerne', *Zeitschrift für Physik*, *I*: **77**, 1 (1932); *II*: **78**, 156 (1932); *III*: **80**, 587 (1933). Reprinted in Ref. [9b], Part II, pp. 197–226. (b) W. Blum, H.-P. Dürr, and H. Rechenberg (eds), *Werner Heisenberg. Collected Works. Series A: Original Scientific Papers* (Springer, Berlin and Heidelberg, 1989).

[10] (a) W. Heisenberg, *Der Teil und das Ganze. Gespräche im Umkreis der Atomphysik* (Piper, Munich, 1969); English tranlation by A. J. Pomerans, *Physics and Beyond* (Allen & Unwin, London, 1971); (b) P. L. Rose, *Heisenberg and the Nazi Atomic Bomb Project. A Study in German Culture* (University of California Press, Berkeley, 1998).

[11] (a) W. Heisenberg, 'Max Planck, Wege zur physikalischen Erkenntnis. Reden und Vorträge', *Naturwissenschaften* **21**, 608 (1933). Reprinted in Ref. [11b], Part IV, p. 239. (b) W. Blum, H.-P. Dürr, and H. Rechenberg (eds), *Werner Heisenberg. Collected Works. Series C: Philosophical and Popular Writings* (Piper, Munich, 1986).

[12] (a) C. P. Enz and K. von Meyenn (eds), *Wolfgang Pauli. Writings on Physics and Philosophy* (Springer, Berlin and Heidelberg, 1994); (b) W. Heisenberg, 'Über die Stabilität und Turbulenz von Flüssigkeisströmungen', *Annalen der Physik* **74**, 577 (1924); (c) A. Einstein, B. Podolsky, and N. Rosen, 'Can Quantum Mechanical Description of Physical Reality Be Considered Complete?', *Physical Review* **47**, 777 (1935); (d) N. Bohr, 'Can Quantum Mechanical Description of Physical Reality Be Considered Complete?', *Physical Review* **48**, 696 (1935); (e) P. Lahti and P. Mittelstaedt (eds), *Symposium on the Foundation of Modern Physics. 50 years of the Einstein–Podolsky–Rosen Gedankenexperiment* (World Scientific, Singapore, 1985).

[13] (a) *The Encyclopedia Americana. International Edition* (Americana Corp., Danbury, CT, 1980); *The American Peoples Encyclopedia. A Modern Reference Work* (Grolier, New York, 1965); (b) A. Pais, *A Tale of Two Continents. A Physicist's Life in a Turbulent World* (Oxford University Press, Oxford, 1997), Chapter 15; (c) K. von Meyenn (ed.), *Wolfgang Pauli. Scientific Correspondence with Bohr, Einstein, Heisenberg a.o., Volume III: 1940–1949* (Springer, Berlin and Heidelberg, 1993); (d) W. Pauli, *The Theory of Positrons and Related Topics*, Lecture Notes by B. Hoffmann, IAS 1935–36 (APS Niels Bohr Library, 1 Physics Ellipse, College Park, Maryland); (e) H. A. Strauss and W. Röder (eds), *International Biographical Dictionary of Central European Emigrés, 1933–1945. Volume II/Part 1: A-K. The Arts, Sciences, and Literature* (Saur, Munich, 1983), p. 612; (f) N. Kemmer, 'Erinnerungen an Pauli', in Ref. [6], pp. 89–94.

[14] (a) Translated from: W. Pauli, letter [373] of 14 June 1934 to Werner Heisenberg, in Ref. [3a]; (b) W. Heisenberg, 'Bemerkungen zur Diracschen Theorie des Positrons', *Zeitschrift für Physik* **90**, 209 (1934). Reprinted in Ref. [9b], Part II, pp. 132–54.

[15] P. A. M. Dirac, 'A Theory of Electrons and Protons', *Proceedings of the Royal Society* A **126**, 360 (1930).

[16] (a) G. Wentzel, 'Quantum Theory of Fields (until 1947)', in Ref. [16d], pp. 48–77; (b) G. Wentzel, *Einführung in die Quantentheorie der Wellenfelder* (Deuticke, Vienna, 1943; English translation: Wiley Interscience,

New York, 1949); (c) A. Pais, 'On the Dirac Theory of the Electron (1930-1936)', in Ref. [9b], Part II, pp. 95–105; (d) M. Fierz and V. F. Weisskopf (eds), *Theoretical Physics in the Twentieth Century. A Memorial Volume to Wolfgang Pauli* (Wiley Interscience, New York, 1960).

[17] (a) O. Klein, 'Die Reflexion von Elektronen an einem Potentialsprung nach der relativistischen Dynamik von Dirac', *Zeitschrift für Physik* **53**, 157 (1929); (b) F. Sauter, 'Über das Verhalten eines Elektrons im homogenen elektrischen Feld nach der relativistischen Theorie Diracs', *Zeitschrift für Physik* **69**, 742 (1931); 'Zum "Kleinschen Paradoxon"', *Zeitschrift für Physik* **73**, 547 (1931).

[18] P. A. M. Dirac, 'Quantised Singularities in the Electromagnetic Field', *Proceedings of the Royal Society* A **133**, 60 (1931).

[19] H. Weyl, *Gruppentheorie und Quantenmechanik* (Hirzel, Leipzig, 1930), 2nd edn; English translation by H. P. Robertson, *The Theory of Groups and Quantum Mechanics* (Methuen, London, 1931).

[20] A. Pais, *Inward Bound. Of Matter and Forces in the Physical World* (Oxford University Press, Oxford, 1986).

[21] W. Pauli, *Theory of Relativity*, translated by G. Field. *With Supplementary Notes by the Author* (Pergamon Press, London, 1958).

[22] W. Pauli, 'Die allgemeinen Prinzipien der Wellenmechanik', (a) in: H. Geiger and K. Scheel (eds), *Handbuch der Physik. Volume 24, Part 1* (Springer, Berlin, 1933), 2nd edn, pp. 83–272; (b) in: N. Straumann (ed.) (Springer, Berlin and Heidelberg, 1990).

[23] (a) A. D. Dolgov, M. V. Sazhin, and Ya. B. Zeldovich, *Basics of Modern Cosmology* (Editions Frontières, Gif-sur-Yvette, 1990); (b) P. A. M. Dirac, 'Discussion of the infinite distribution of electrons in the theory of the positron', *Proceedings of the Cambridge Philosophical Society* **30**, 150 (1934).

[24] C. P. Enz (ed.), *Pauli Lectures in Physics: Volume 6. Selected Topics in Field Quantization*, translated by S. Margulies and H. R. Lewis (MIT Press, Cambridge, MA, 1973).

[25] J. Schwinger (ed.), *Selected Papers on Quantum Electrodynamics* (Dover, New York, 1958).

[26] V. Weisskopf, 'Über die Selbstenergie des Elektrons', *Zeitschrift für Physik* **89**, 27 (1934); correction, *Zeitschrift für Physik* **90**, 817 (1934).

[27] (a) W. Pauli and V. Weisskopf, 'Über die Quantisierung der skalaren relativistischen Wellengleichung', *Helvetica Physica Acta* **7**, 709 (1934). Reprinted in Ref. [28], Vol. 2, pp. 701–23. (b) W. Pauli, 'Théorie quantique relativiste des particules obéissant à la statistique de Einstein-Bose', *Annales de l'Institut Henri Poincaré* **6**, 137 (1936). Reprinted in Ref. [28], Vol. 2, pp. 781–96. (c) G. A. Proca (ed.), *Alexandre Proca (1897-1955). Oeuvre scientifique publié* (G. A. Proca, Paris, 1988); (d) E. Borel, 'Louis de Broglie et l'Institut Henri Poincaré', in Ref. [27e], pp. 437–43; (e)

A. George (ed.), *Louis de Broglie. Physicien et Penseur* (Albin Michel, Paris, 1953).

[28] R. Kronig and V. F. Weisskopf (eds), *Collected Scientific Papers by Wolfgang Pauli. In Two Volumes* (Wiley Interscience, New York, 1964).

[29] (a) H. Bethe and W. Heitler, 'On the Stopping of Fast Particles and on the Creation of Positive Electrons', *Proceedings of the Royal Society* A **146**, 83 (1934); (b) D. Iwanenko and A. Sokolow, 'Bemerkungen zur zweiten Quantelung der Dirac-Gleichung', *Zeitschrift der Sowjetunion* **11**, 590 (1937).

[30] (a) M. Delbrück, 'Zusatz bei der Korrektur', p. 144, in: L. Meitner and H. Kösters, 'Über die Streuung kurzwelliger γ-Strahlen', *Zeitschrift für Physik* **84**, 137 (1933); (b) W. H. Furry and J. R. Oppenheimer, 'On the Theory of the Electron and Positive', *Physical Review* **45**, 245 (1934).

[31] (a) J. S. de Wet, 'On the Connection Between the Spin and Statistics of Elementary Particles', *Physical Review* **57**, 646 (1940); (b) W. Pauli and M. E. Rose, 'Remarks on the Polarization Effects in the Positron Theory', *Physical Review* **49**, 462 (1936). Reprinted in Ref. [28], Vol. 2, pp. 749–52.

[32] (a) V. Weisskopf, 'Über die Elektrodynamik des Vakuums auf Grund der Quantentheorie des Elektrons', *Kongelige Danske Videnskabernes Selskab, Math.-Fys. Meddelelser* **XIV**, No. 6 (1936). Reprinted in Ref. [25], pp. 92–128. (b) W. Heisenberg and H. Euler, 'Folgerungen aus der Diracschen Theorie des Positrons', *Zeitschrift für Physik* **98**, 714 (1936). Reprinted in Ref. [9b], Part II, pp. 162–80.

[33] (a) F. Villars, 'Regularization and Non-Singular Interactions in Quantum Field Theory', in Ref. [16d], pp. 78–106; (b) H. A. Kramers, 'The use of charge-conjugated wave-functions in the hole-theory of the electron', *Proceedings of the Amsterdam Academy of Science* **40**, 697 (1937).

[34] W. Pauli, (a) Contributions mathématiques à la théorie des matrices de Dirac, '*Annales de l'Institut Henri Poincaré*' **6**, 109 (1936); (b) 'Über die Formulierung der Naturgesetze mit fünf homogenen Koordinaten. Teil II: Die Diracschen Gleichungen für die Materiewellen', *Annalen der Physik* **18**, 337 (1933); (c) 'Beiträge zur mathematischen Theorie der Dirac'schen Matrizen', in: J. D. van der Waals Jr (ed.), *Pieter Zeeman. 1865—25 Mei—1935. Verhandelingen* (Nijhoff, The Hague, 1935), p. 31. Reprinted in Ref. [28], Vol. 2: (a) pp. 753–80; (b) pp. 662–97; (c) pp. 724–36.

[35] W. Pauli, 'Relativistic Field Theories of Elementary Particles', *Reviews of Modern Physics* **13**, 203 (1941). Reprinted in Ref. [28], Vol. 2, pp. 923–52.

[36] Translated from: W. Pauli, letter [509] of 22 May 1938 to Sommerfeld, in Ref. [3a].

[37] (a) *Schweizer Lexikon 91* (Verlag Schweizer Lexikon, Lucerne, 1992). (b) Note of my telephone conversation with Franca Pauli on 15 January 1986.

[38] (a) W. Rings, *Schweiz im Krieg, 1933–1945. Ein Bericht* (Verlag Ex Libris, Zurich, 1974); (b) J. Cabot Reid, *Jung, my mother and I* (Daimond Verlag, Einsiedeln, 2001).

[39] M. Fierz, 'Rückblick vom Hönggerberg', in: M. Fierz, *Naturwissenschaft und Geschichte. Vorträge und Aufsätze* (Birkhäuser, Basle, 1988), pp. 215–25.

[40] W. Pauli, 'Einige grundlegende Bemerkungen über die Theorie des Beta-Zerfalls', in Ref. [6], pp. 454–8. The Russian original is reprinted in Ref. [28], Vol. 2, pp. 843–6.

[41] W. Pauli and M. Fierz, 'Zur Theorie der Emission langwelliger Lichtquanten', *Nuovo Cimento* **15**, 167 (1938). Reprinted in Ref. [28], Vol. 2, pp. 813–34.

[42] (a) I. B. Khriplovich, 'The eventful life of Fritz Houtermans', *Physics Today*, July 1992, p. 29; (b) V. Ya. Frenkel, Letter to the editor, *Physics Today*, June 1994, p. 104.

[43] (a) W. Heisenberg, 'Zur Theorie der "Schauer" in der Höhenstrahlung', *Zeitschrift für Physik* **101**, 533. Reprinted in Ref. [9b], Part II, pp. 275–82. (b) H. Umezawa, *Quantum Field Theory* (North-Holland, Amsterdam, 1956); (c) J. C. Collins, *Renormalization. An introduction to renormalization, the renormalization group, and the operator-product expansion* (Cambridge University Press, Cambridge, 1984).

[44] M. Fierz, 'Zur Fermischen Theorie des β-Zerfalls', *Zeitschrift für Physik* **104**, 553 (1937).

[45] W. Pauli and M. Fierz, 'Über das H-Theorem in der Quantenmechanik', *Zeitschrift für Physik* **106**, 572 (1937). Reprinted in Ref. [28], Vol. 2, pp. 797–812.

[46] F. Bloch and A. Nordsieck, 'Note on the Radiation Field of the Electron', *Physical Review* **52**, 54 (1937).

[47] (a) W. Pauli, 'On Asymptotic Series of Functions in the Theory of Diffraction of Light', *Physical Review* **54**, 924 (1938). Reprinted in Ref. [28], Vol. 2, pp. 835–942. (b) C. A. Meier, (ed., with the cooperation of C. P. Enz and M. Fierz), *Wolfgang Pauli und C. G. Jung. Ein Briefwechsel 1932–1958* (Springer, Berlin and Heidelberg, 1992); (c) Translated from: W. Pauli, letter [394] to Heisenberg of 7 November 1934 in Ref. [3a].

[48] (a) P. A. M. Dirac, 'Relativistic Wave Equations', *Proceedings of the Royal Society* A **155**, 447 (1936); (b) B. L. van der Waerden, *Die gruppentheoretische Methode in der Quantenmechanik* (Berlin, 1932); (c) A. Proca, 'Sur la théorie ondulatoire d'électrons positifs et négatifs', *Journal de Physique et le Radium* **7**, 347 (1936). Reprinted in Ref. [27c], pp. B.IV.3–9.

[49] H. Yukawa, 'On the Interaction of Elementary Particles. I', *Proceedings of the Physical–Mathematical Society of Japan* **17**, 48 (1935).

[50] (a) S. Neddermeyer and C. D. Anderson, 'Note on the Nature of Cosmic-Ray Particles', *Physical Review* **51**, 884 (1937); (b) C. Street and E. C. Stevenson, *Bulletin of the American Physical Society* **12**, 2, 13 (1937).

[51] (a) N. Kemmer, 'Quantum theory of Einstein-Bose particles and nuclear interaction', *Proceedings of the Royal Society* A **166**, 127 (1938); (b) E. C. G. Stueckelberg, 'Die Wechselwirkungskräfte in der Elektrodynamik und in der Feldtheorie der Kernkräfte. (Teil II und III)', *Helvetica Physica Acta* **11**, 299 (1938).

[52] *Note*: All the Lagrangian densities in Ref. [48c] have the wrong sign; note also that the case discussed in *Note* [22] of Chapter 6 was for a *real* vector field.

[53] (a) H. Fröhlich, W. Heitler, and N. Kemmer, 'On the nuclear forces and the magnetic moments of the neutron and the proton', *Proceedings of the Royal Society* A **166**, 154 (1938); (b) H. J. Bhabha, 'On the theory of heavy electrons and nuclear forces', *Proceedings of the Royal Society* A **166**, 501 (1938).

[54] (a) M. Fierz and W. Pauli, 'On relativistic wave equations for particles of arbitrary spin in an electromagnetic field', *Proceedings of the Royal Society* A **173**, 211 (1939). Reprinted in Ref. [28], Vol. 2, pp. 873–94. (b) W. Pauli and M. Fierz, 'Über relativistische Feldgleichungen von Teilchen mit beliebigem Spin im elektromagnetischen Feld', *Helvetica Physica Acta* **12**, 297 (1939). Reprinted in Ref. [28], Vol. 2, pp. 869–72.

[55] (a) M. Fierz, 'Über die relativistische Theorie kräftefreier Teilchen mit beliebigem Spin', *Helvetica Physica Acta* **12**, 3 (1939); (b) M. Fierz, 'Über die relativistische Theorie für Teilchen mit ganzzahligem Spin sowie deren Quantisierung', *Helvetica Physica Acta* **11**, 377 (1938); (c) J. M. Jauch, 'Über die Energie–Impuls–Tensoren und die Stromvektoren in der Theorie von Dirac für Teilchen mit Spin grösser als 1/2 h', *Helvetica Physica Acta* **11**, 374 (1938). (d) Letter dated Küsnacht, 29 September 1999, from Markus Fierz to Charles P. Enz. I thank Professor Fierz for this information.

[56] W. Pauli, 'The Connection between Spin and Statistics', *Physical Review* **58**, 716 (1940). Reprinted in Ref. [28], Vol. 2, pp. 911–17 and in Ref. [25], pp. 372–8.

[57] W. Blum, H.-P. Dürr, and H. Rechenberg (eds), *Werner Heisenberg. Collected Works. Series B: Scientific Review Papers, Talks, and Books* (Springer, Berlin and Heidelberg, 1984).

[58] (a) G. Petiau, 'Contribution à la théorie des équations d'ondes corpusculaires', *Mémoires de l'Académie royale de Belgique*, Classe des sciences **16**, 1 (1936); (b) R. J. Duffin, 'On the Characteristic Matrices of Covariant Systems', *Physical Review* **54**, 1114 (1938); (c) N. Kemmer, 'The

particle aspect of meson theory', *Proceedings of the Royal Society* A **173**, 91 (1939).

[59] (a) L. de Broglie, 'Sur un cas de réductibilité en Mécanique ondulatoire des particules de spin un', *Comptes Rendus* (Paris) **208**, 1697 (1939); (b) L. de Broglie, 'Sur les particules de spin quelconque', *Comptes Rendus* (Paris) **209**, 265 (1939); (c) W. Pauli, 'Continuous Groups in Quantum Mechanics', CERN-Publication 56–31, 1956. Reprinted in: G. Höhler (ed., with the cooperation of S. Flügge, F. Hund, and F. Trendelenburg), *Ergebnisse der exakten Naturwissenschaften* (Springer, Berlin and Heidelberg, 1965), Vol. 37, pp. 85–104. (d) W. Pauli, 'Gruppentheorie und Quantenmechanik', Notes by C. P. Enz (1954, unpublished).

[60] (a) F. J. Belinfante, 'Theory of Heavy Quanta', Thesis, University of Leiden, 1939; (b) W. Pauli and F. J. Belinfante, 'On the statistical behaviour of known and unknown elementary particles', *Physica* **7**, 177 (1940). Reprinted in Ref. [28], Vol. 2, pp. 895–910. (c) F. J. Belinfante, 'The Undor Equation of the Meson Field', *Physica* **6**, 870 (1939). (d) Letter dated Gresham OR 97030 USA, 27 December 1985, from Fred Belinfante to Charles P. Enz. (e) A. N. Gerritsen, S. Rodriguez, A. Tubis, and J. C. Swihart, 'Frederik Josef Belinfante', *Physics Today*, July 1992, p. 82.

[61] (a) W. Pauli, 'On the Connection between Spin and Statistics', *Progress of Theoretical Physics* (Kyoto) **5**, 526 (1950). Reprinted in Ref. [28], Vol. 2, pp. 1131–48. (b) R. P. Feynman, 'The theory of positrons', *Physical Review* **76**, 749 (1949); (c) R. P. Feynman, 'Space–Time Approach to Quantum Electrodynamics', *Physical Review* **76**, 769 (1949). Reprinted in Ref. [25], (b): pp. 225–35; (c): pp. 236–56; (d) K. von Meyenn (ed.), *Wolfgang Pauli. Scientific Correspondence with Bohr, Einstein, Heisenberg, a.o., Volume IV/Part I: 1950–1952* (Springer, Berlin and Heidelberg, 1996).

[62] (a) R. Jost, 'Das Pauli-Prinzip und die Lorentz-Gruppe', in Ref. [16d], pp. 107–36; (b) R. F. Streater and A. S. Wightman, *PCT, Spin and Statistics, and All That* (Benjamin, New York, 1964); (c) I. Duck and E. C. G. Sudarshan, *Pauli and the Spin–Statistics Theorem* (World Scientific, Singapore, 1997).

[63] (a) J. M. Jauch and E. L. Hill, 'On the Problem of Degeneracy in Quantum Mechanics', *Physical Review* **57**, 641 (1940); (b) L. Ruzicka, 'Auch Pauli ist einst gegangen . . .', *Tages-Anzeiger*, Zurich, Friday, 3 February 1967.

9

The war years in the United States

Extended stay in Princeton

*My home country (*Heimat*) indeed is a spiritual one, and as a spiritual human I evidently do not have to belong to one certain nation at all. But somehow something always seems to be absent, to flee, to be unreachable for me. What is it? Are perhaps both women as well as countries, only a more or less casual expression for something more profound that is missing? What is it? Do I perhaps rather not want to have it completely? I do not know.* [1]

Pauli's experience with his failed attempt to obtain Swiss citizenship (see previous chapter) bothered him more than he cared to admit and he began to look at it introspectively, as in the above passage from a letter to his psychiatrist friend and Jung disciple Carl Alfred Meier (1905–95) on 25 February 1942. A month earlier Pauli had written (in English) to George Uhlenbeck in Ann Arbor with whom he felt increasingly friendly: 'There is one thing, which occupies myself very much: I am getting more and more aware of it, that I don't belong to any existing nation. I feel that my standpoint has to be a supranational one based on general ideals or values of our whole civilisation. . . . This was already my attitude as I was 20 years old—then later during some time I tried really to become a Swiss but this was a great error of mine and I return with repentance to my old supranational point of view. (If you think that I am crazy, please let me know.)' (Ref. [2], p. 127).

In Princeton the Paulis settled at number 15 on Palmer Square [3], the centre of town and main shopping area off Nassau Street, which limits the university campus to the north. The dominating structure on this street is Nassau Hall, built in 1756 to house the university, and named after the King of England, William III of Orange (1650–1702), who was from Nassau in Germany [4a] (see also Ref. [4b], Chapter 15). To the south-west the campus borders the Princeton Golf Course, and still further in this direction is Fuld Hall, the new main building of the Institute for Advanced Study which is about a 2.5 km walk from Palmer Square along Mercer Street where, at number 112, the Einstein house is situated. Until 1940 the institute had been located on campus in Fine Hall [4a], the mathematics building which is adjacent to the north-east of the Palmer Physical Laboratory (see page 294).

On 15 October 1940 Pauli wrote to his colleague and friend, the mathematics professor at ETH Heinz Hopf (1894–1971), who also lived in Zollikon and with whom Pauli had walked in the woods around Zollikon and occasionally had played a game of chess: 'I like it very much here. There are again visits from outside, congresses and invitations. Our walks in Zollikon are now replaced by the Sunday walk of the Princeton Institute. . . . The Chianti [Italian wine] is now replaced by a quite drinkable Californian wine', but he laments: 'Only Dixi is replaceable by nothing' (translated from Ref. [2], letter [606]). Dixi was Pauli's dog that had kept him company in his study at Zollikon (see footnote 3, Ref. [2], p. 46). But also his collaborators in Zurich were not immediately replaceable. From Chicago he wrote to Weisskopf on 28 November 1940: 'I have so much time in Princeton as never before' (translated from Ref. [2], letter [612]). For the time being Pauli worked with Bargmann, who from 1937 to 1946 was a collaborator of Einstein at the Institute for Advanced Study, on the unitary representations of the Lorentz group (see Ref. [2], letters [607, 609]) which had become quite fashionable (see e.g. Ref. [2], letter [602]). This was an extension of Pauli's work with Fierz on fields of higher spin. He had been stimulated by a paper of Ettore Majorana (1906–38) [5a] (see Ref. [2], letters [598, 599]).

Majorana's rather enigmatic paper 'Relativistic theory of particles with arbitrary intrinsic angular momentum' [5a], written in Italian, containing no references to speak of and published in the not yet well-known Italian journal *Nuovo Cimento*, left almost no trace in the published literature [5b]. Pauli mentioned it for the first time in letter [595] to Bhabha on 12 April 1940, writing in English: 'Maybe the matter [of particles with higher spin] becomes simpler if one introduces a priori an infinite set of spin-values (compare *Majorana*, Nuovo Cimento 1932; but in this paper is no second-quantisation and no hole-theory [is] considered)' (Ref. [2], p. 29). Heisenberg had mentioned Majorana's name to Pauli the year before (see Ref. [6], letter [549]) in relation to the project on the Solvay report for 1939 (see page 329). He knew of the existence of this paper since Majorana had been at his institute in Leipzig in 1933 (see Ref. [6], p. 705).

Majorana's motivation for this paper had been similar to Proca's (see pages 326–7), i.e. to obtain for the electron a wave equation without negative energies. He writes: 'The indeterminacy of the sign of the energy may in fact be overcome by using equations of the fundamental type [see (6.28)] only if the wave function has an infinite number of components which can *not* be fractured into finite tensors or spinors' (translated from Ref. [5a], pp. 336–7). From Eq. (6.28) it follows indeed that, for *positive* energy and in the rest frame, $\psi^* \beta \psi > 0$. Then there exists a (non-unitary) transformation $\psi \to \varphi$ such that $\psi^* \beta \psi = \varphi^* \varphi$. And φ formally satisfies an equation (6.31) but with the γ of infinitely many rows, to be determined from the fact 'that the invariance of $\varphi^* \varphi$ means that we have to consider only unitary transformations' (translated from Ref. [5a], p. 337). Majorana, who was familiar with the German original (either its first or its second edition, 1928 or 1930, respectively) of Weyl's *Theory of Groups* *and Quantum Mechanics* [7a], generalized Weyl's Eqs. (15.4), (15.5) (Ref. [7a], p. 178) valid for infinitesimal three-dimensional rotations to those valid for infinitesimal Lorentz transformations which are parametrized by two real three-vectors **a** and **b**. These are Majorana's Eqs. (6) to (8); note that, with the identification **a** = **M**, **b** = −i**N**, Eqs (8) are the same as Pauli's commutators for the four-dimensional rotation group in Ref. [7b]. And, in analogy to Weyl's selection rules (3.9) (Ref. [7a], p. 200), Majorana set up his rules (9). He even wrote down one of the two Casimir operators of the Lorentz group (see page 270), his Eq. (10), namely

$$Z = \mathbf{a} \cdot \mathbf{b}. \tag{9.1}$$

These pioneering results preceded the systematic treatments (see e.g. the references given by Pauli in Ref. [7b]) by several years! Further details on Majorana's paper are given in *Note* [7c].

After his letter to Bhabha quoted above, Pauli wrote many letters to Fierz about Majorana's paper. In letter [598] of 3 July 1940, still from Zurich, he observes that Majorana's 'system of equations is considerably simpler than ours. It is essential, however, that with Majorana only in the rest system a particle with spin *s* is described by $2s + 1$ non-vanishing functions, in any other reference system it is described by infinitely many eigenfunctions', and 'for those my proofs of the Solvay report are *not* valid.' And he concludes: 'Anyway, the case of representations of the Lorentz group with an infinity of rows seems to me not to be sufficiently analyzed yet' (translated from Ref. [2], p. 33). In letter [600] of 3 September quoted on page 340 Pauli discusses the spectrum of the second Casimir operator,

$$J = \mathbf{b}^2 - \mathbf{a}^2, \tag{9.2}$$

and then goes on to develop over two pages an analytic treatment of the problem which he continues in letter [622] of 12 February 1941. Finally, on 29 March 1941, he presents the problem in letter [628] in terms of Belinfante-type wavefunctions (see page 335) $\varphi(\zeta; \mathbf{p})$; $\zeta = 1, \ldots, 2s + 1$.

These letters to Fierz are of considerable historic interest both as rare witnesses to Majorana's paper [5a], and also because Pauli's continued interest in particles of higher spin motivated him to study the representation theory of the Lorentz and the rotation groups more systematically. While Bargmann published the results of his collaboration with Pauli only in 1947 [7d], Pauli presented his vast experience on the subject in the *Spezialvorlesung* [7e] at ETH during the summer term of 1954 (see Ref. [8], p. 434), already mentioned several times (see pages 136, 270 and 332), which contained both an algebraic and an analytical part. The former, on the other hand, was the basis of Pauli's lectures [7b] he gave in September 1955 at the new Theory Division of CERN in Copenhagen. In Ref. [7b] the above vectors **a** and **b** are called **M** and **N** and the Casimir operators (9.1) and (9.2) G and $-2F$, respectively. The connection between these two operators is given in Ref. [7b] and also by Eq. (I) of letter [622], where Z is called C, namely

$$J = 1 - n^2 + \frac{Z}{n^2}. \tag{9.3}$$

Here $n = \min j \geq 0$ is either integer or half-odd and, as for the rotation group, $j(j + 1)$ is an eigenvalue of $\mathbf{a}^2$. Note that in Ref. [7d] Bargmann calls $J = -Q$ and $Z = R$ in his Eqs (2.12) and (2.12a), respectively. Six years after Pauli's first series of letters on the subject of the Majorana paper, in his letter [877] to Fierz Pauli expressed a much more negative opinion, saying: 'From this point of view [that definite values of mass and spin are associated to irreducible representations], e.g. Majorana's equations appear as completely arbitrary because they are reducible!' (translated from Ref. [2], p. 435).

Ettore Majorana was the son of a well-known Sicilian family in Catania. His father Fabio Massimo Majorana was the younger brother of Quirino Majorana, who presided over the Volta Congress of 1927 (see page 160). In 1921 the family of three sons and two daughters moved to Rome where Ettore did his schooling and went to university. Early on his phenomenal faculty of doing complicated calculations in his head impressed his friends. In July 1929 he and his friend Edoardo Amaldi (1908–89) received their doctorates from Fermi who in 1926 had been appointed associate professor at the University of Rome. In January 1933 Majorana went to Leipzig where he became a friend of Heisenberg. But otherwise, also in Copenhagen later that year, he remained very much alone. In November 1937 he was appointed professor of theoretical physics at the University of Naples where in the following January he started his lectures. But his life was increasingly wrapped in solitude. He disappeared under unknown circumstances from a boat on the way back from Palermo, probably in the waters off Naples on 26 March 1938 [9a].

In 1940 the theory of groups and their representations was only a passing activity of Pauli, however. Also he temporarily put aside the meson theory. His abundant time led him to devote his energy again to the problem of the divergences in the positron theory (see pages 297f). The reason was that Wentzel had formulated a strong-coupling meson theory, in which the development was not in terms of increasing powers of the coupling constant g but of decreasing powers (see e.g. Ref. [9b], p. 61) and Pauli hoped that this idea could be applied to QED. On 6 December 1940 he wrote to Weisskopf that positron theory was for him 'as if I had to jump into cold water. But once one is in it, then it is even quite pleasant. I am just about to jump in' (translated from Ref. [2], letter [613]). However, on 30 December he wrote to Weisskopf again: 'My physics is not in good shape, for the moment I have rather gone astray (*hineingefallen*) with my methods for the positron theory with large coupling ($e^2/\hbar c \gg 1$); they are no consistent approximations. Wentzel's method does not work neither because his inequality $g^2 \gg 1/\mu l$ [9c] (which, anyhow, I am suspicious of) is never fulfilled for photons ($\mu = 0$)' (translated from Ref. [2], letter [615], p. 58). The same argument, but with some more details, also appears in Pauli's letter [617] to Oppenheimer on 3 January 1941 (Ref. [2], p. 67; see also letter [619]).

On 15 October Pauli wrote to Bhabha: 'I would be very glad to be on [*sic*] the same place as you and to work with you' (Ref. [2], letter [605]). But

Bhabha was in Bangalore and could not travel (see Ref. [2], p. 77). Pauli had met Homi Jehaugir Bhabha (1909–66) in Zurich in April 1939 where the latter 'had reported extensively on his classical calculations of meson scattering' (translated from Ref. [6], letter [551], p. 637) and again in Geneva in July 1939—where, most probably, Pauli did not see Stueckelberg who that summer had one of his recurring attacks of cyclothymia (see letter [585], Ref. [6], p. 680 and footnote 4, Ref. [2], p. 6). Bhabha's work (see e.g. Ref. [10a]) was a transposition of Dirac's theory of a classical relativistic point electron, in which the diverging self-energy is compensated by a radiation reaction [10b]. Not surprisingly, Pauli called Bhabha's work 'classical subtraction physics' (Ref. [2], p. 28). Bhabha was born in Bombay where he also began his studies. He then went to Cambridge, England, where he obtained his Ph.D. in 1932. In 1940 he became a reader in theoretical physics at the Indian Institute of Science in Bangalore and a professor in 1942. From 1945 on he was the director of the Tata Institute for Fundamental Research in Bombay. As chairman of the Indian Atomic Energy Commission, in 1955 he presided over the International Conference on Peaceful Uses of Atomic Energy in Geneva. He received many honorary degrees and medals (Ref. [11], p. 171). Bhabha died in 1966 in a plane crash on Mont Blanc on his way to CERN in Geneva.

Academic life in the autumn of 1940 offered Pauli plenty of opportunities for travelling and for making new acquaintances. In the second half of October Pauli and Franca accepted the invitation to visit the Weisskopfs in Rochester in up-state New York (Ref. [2], letter [605]), and on their way back they stopped off at Ithaca where Pauli had been invited to visit Cornell University (see Ref. [2], end of letter [613]). Towards the end of November the Paulis were in Ann Arbor where Uhlenbeck had invited Pauli to lecture on 'Spin and Statistics' and where they also 'wanted a special seminar talk on Duffin-Kemmer numbers. It was a very learned algebra, all improvised, by-the-way, and afterwards people were extremely satisfied. Strange!' (translated from Ref. [2], letter [612]). From Ann Arbor the three Sommerfeld pupils (see Ref. [6], pp. 705 ff.), Otto Laporte, Arthur Rosenthal, and Pauli, sent a postcard to Paul Epstein in California, a fourth one (Ref. [6], postcard [611]). Leaving Ann Arbor, Pauli and Franca drove further west to Chicago where on 22–23 November a meeting of the American Physical Society (APS) took place and where Pauli had 'really a wonderful time'. On Saturday 30 November the Paulis started on their way back to Princeton (Ref. [2], letter [612]). Finally, from 26 to 28 December there was an APS meeting in Philadelphia where Pauli was made a Fellow of the APS and where Max Delbrück also participated (see Ref. [2], letter [613] and footnote 10, p. 56).

After this meeting Pauli wrote to Weisskopf about Delbrück: 'I still feel a strong friendly relationship with him, as often happens with two problematic characters. Also he seems to be on the way to be stabilized and to be fairly decided to stay in this country.—But it turned out that in relation to politics there are paths where I, with the best intention, am not able to follow him. Surely, you hitherto hardly have assumed that the question, whether it were desirable that Germany loses the war, were a very difficult question which requires longer reflection or even continuous brooding. But in the depth of a Prussian soul this still seems to be somewhat different (in spite of opposition to Nazi ideology and undiminished sympathy for Jews). Furthermore, problems are "deeply" pondered, if not a German world domination yet would be very interesting, although probably it will not yet be realized.' And Pauli tries to explain that 'a certain fear of his pending Americanization concurs' (translated from Ref. [2], letter [615], p. 59). Such views about the future may well have been more common among German intellectuals. Indeed, Heisenberg, on a visit to German-occupied Holland in 1943, told his Dutch physicist friends 'and so, perhaps, a Europe under German leadership might be the lesser evil [than the onslaught of Eastern hordes]' (Ref. [12a], p. 208; see also Ref. [12b], p. 473).

Max Delbrück (1906–81) belonged to an old Protestant Prussian family of noted civil servants and intellectuals. His father Hans Delbrück was a history professor at the University of Berlin and his mother was the granddaughter of the famous chemist Justus Liebig (1803–73). The youngest of seven children, Max grew up in the distinguished Berlin suburb of Grunewald. After World War I he had a great time in high school with Greek, Latin, and mathematics. Kepler became his idol, and he started to study astronomy at Tübingen at age 18. But in Göttingen he changed to quantum theory in 1928 and wrote a thesis about the lithium molecule in 1930, but failed the oral exam at first. In 1931 Delbrück went with a Rockefeller stipend to Bohr's institute at Copenhagen, and the following winter visited Pauli in Zurich. Both visits gave rise to lifelong friendships. Delbrück returned to Copenhagen the following summer where on 15 August 1932 Bohr gave the famous lecture on 'Light and Life'. This lecture and particularly Bohr's idea of complementarity influenced Delbrück's further career, although no direct consequences emerged from it. He then accepted a position offered by Lise Meitner in Berlin where his discussions with the geneticist Timofeeff-Ressovski and the physicist K. G. Zimmer (see page 259) led to his first publications on gene mutations. This work earned Delbrück a second Rockefeller Fellowship on which he went in 1937 to the United States for two years.

In Pasadena Delbrück worked in the group of Thomas Hunt Morgan (1866–1945) who in 1933 had won the Nobel Prize in Medicine. When his fellowship ended, Delbrück got the job in January 1940 as physics instructor at Vanderbilt University in Nashville, Tennessee. In 1941 he married Mary (Manny) Bruce, whom he had met the year before and with whom he had four children. In December 1940 he had been at the APS meeting in Philadelphia where he met not only Pauli (see above) but also his most important future collaborator, Salvador Edward Luria (b. 1912). Delbrück and Luria, together with Alfred D. Hershey (b. 1908) won the Nobel Prize in Medicine of 1969. Applying now also for naturalization, Delbrück's American life began in earnest in 1945. In 1949 he became a member of the US National Academy, and in 1953, the year of the discovery of the double helix by James D. Watson (b. 1928) and Francis H. C. Crick (b. 1916) which earned them the Nobel Prize of 1962, Delbrück turned to a new adventure, the fungus *Phycomyces* and its reactions with the environment. He died of cancer in Pasadena on 10 March 1981 [12c].

Starting in 1940, many US physicists, mainly experimental ones, disappeared from academic life to engage in classified defence work at the Radiation Laboratory of MIT, the Massachusetts Institute of Technology, in Cambridge, where the main task was to develop the military application of RADAR (radio detection and ranging). In this task there was strong collaboration with British scientists who had made an important advance. The Radiation Laboratory started with the nomination in mid-October 1940 of its director, Lee Alvin DuBridge (b. 1901); I. I. Rabi became associate director. Work began in early November with about 30 people. By the end of the war the workforce had grown to 3000 employees [13]. This was a matter of concern to Pauli, as he wrote to Weisskopf in the already quoted letter [612] from Chicago: 'The destruction process of experimental physics which in this country arises from the running away of the best people to MIT in Cambridge worries me yet more than you. (Although in Ann Arbor one does not notice very much of it and here nothing at all.) I don't know whether in Rochester one has not been too optimistic about this' (translated from Ref. [2], p. 54). And in the following letter he mentioned 'a very informative conversation' about the problem of national defence at MIT and physics with Rabi. But this attitude of Pauli was also motivated by his fear of becoming scientifically isolated. Indeed, in the already quoted letter [615] to Weisskopf he wrote: 'If only my physics were in better shape, then I would not care about the whole MIT' (translated from Ref. [2], p. 58). As we shall see, this fear was justified.

Isidor Isaac Rabi's parents had come to the United States from Rymanow, Galicia, in the Austro-Hungarian Empire, where Rabi was born in 1898. They settled in Lower East-side Manhattan where Rabi grew up to be a good Jew. He studied chemistry at Cornell University with a scholarship and later went to graduate school at Columbia University in New York. There one of his fellow students was Ralph Kronig who after his Ph.D. in January 1925 left for Europe (see page 110). Rabi obtained his doctor's degree in 1927 and in July of that year went to Europe with his wife Helen Rabi Newmark whom he had married the year before. After a visit to England and to Copenhagen Rabi went to Hamburg where he intended to work with Pauli but ended up doing his first molecular beam experiment in Stern's laboratory (see page 148). After a visit to Heisenberg in Leipzig where he met Oppenheimer and Teller, in March he visited Pauli in Zurich where he met Oppenheimer again, as well as many others (see page 192).

In August 1929 Rabi went back to Columbia University where he spent the next 39 years as a professor, first in theoretical physics. But in 1931 he built up the molecular beam laboratory where his new magnetic resonance method made him famous and earned him the Nobel Prize of 1944. However, during the war, Rabi spent all his time with organizing and consulting on defence work, first at the Radiation Laboratory and later as advisor to Oppenheimer who from 1943 to 1945 was director of the Manhattan Project for the development of the atomic bomb (see page 207). After the war Rabi was one of the main initiators of the Brookhaven National Laboratory on Long Island near New York in 1946, and also of CERN at Geneva in 1952 and of the International Conference on the Peaceful Uses of Atomic Energy at Geneva in 1955 (see above). He also became a member of the US President's Scientific Advisory Committe in 1957, just after the Russians had launched the Sputnik. Rabi received a large number of honours [14a,b].

When the year 1940 came to an end, Pauli began to look at an extended stay in the United States with more realism. On New Year's Day he wrote to his assistant Jauch in Zurich: 'As things now stand, I consider as practically excluded that I may travel back as long as the war lasts. No ship's company would accept me over the Atlantic as passenger in the direction America → Europe, and with a journey via Japan—Siberia I surely would be stuck in Germany. I probably will have to wait here until the war is over' (translated from Ref. [2], letter [616], p. 63). And he asked Jauch to send him his lecture notes (*Kolleghefte*) since, sooner or later, he might yet use them. Pauli's situation at the institute was indeed rather shaky, as may be seen from his letter [674] (in English) to Rabi, written from Pauli's favoured summer resort of Saranac Lake in up-state New York on 10 July 1943 (Ref. [2], p. 186): 'In December 1942, however, [director] Aydelotte told me "to look for a job" [at] latest in June 1944, the Institute not being in a position to replace my stipend.' The Rockefeller Foundation had paid Pauli a yearly scholarship of $5000 from September 1940, when he had arrived, until June 1942. But after this date the Rockefeller Foundation was willing to pay only $3000 for each of the two following years, while the institute agreed to pay a supplement of $1000. Therefore Pauli started to look for a teaching job. However, his contacts with Johns Hopkins University in Baltimore, Maryland, where David Inglis (see pages 271–2) was to be replaced, broke up when Pauli mentioned that he was receiving $5000 per year. Based on this experience Pauli concluded that 'the best solution I can see, is to do defence work, despite the difficulties of not yet being naturalized. I know that it takes many months to get a clearing for defence work and this is one of the reasons why I write you so early' (Ref. [2], letter [674]). Pauli had given this letter an official touch by having it typed on institute paper (see Ref. [2], footnote 1, p. 187). Rabi, however, answered: 'I feel very strongly that you have nothing to worry about' (Ref. [2], letter [676]).

Pauli had discussed the possibility that he himself might go into defence work earlier with Weisskopf who spoke about it to Oppenheimer on a visit to Los Alamos. Oppenheimer wrote to Pauli on 20 May 1943; this is one of only a few preserved letters from Oppenheimer to Pauli, and its content is of utmost interest. Oppenheimer wrote that his 'feeling is that at the present time it would be a waste and an error for you to do that [research directly connected with the war]. You are just about the only physicist in the country who can help to keep those principles of science alive which do not seem immediately relevant to the war, and that is eminently worth doing.' Besides, Pauli's example would be decisive for people 'who because of legal complications cannot work on military problems'. But Oppenheimer had another idea: since the fact that 'none of the people in our field are publishing work in the Physical Review' might be noticed by the enemy, 'We have often wondered whether your great talents for physics and for burlesque could not appropriately be put to use by your publishing some work in the names of a few of the men who are now engaged on things that they cannot publish.' And Oppenheimer added: 'I know that Bethe, Teller, Serber and I would be delighted to grant you that and I have no doubt that there would be many others. Do not dismiss this thought too lightly' (Ref. [2], letter [671]).

Pauli answered a month later that 'the non-scientists who give me the money ... are becoming more and more, let's say, reluctant with it. ... Although I would be glad to be helpful in the suggested way, I am afraid I should publish the few things which I have at present to say with my own name to prove to the quoted money-givers that after all I am working [on] something for their money, fearing their sense for burlesque to be rather underdeveloped.' But, in addition, Pauli believed that 'the whole Don-Quichotery would be in vain' (Ref. [2], letter [672]). It appears, in fact, that Pauli, but also Oppenheimer, viewed war in general with pessimism. The latter, apparently, had expressed the opinion that this is the 'normal state' in history, to which Pauli answered in a postscript to letter [619], dated 13 January 1941, 'that the idea that the world is now in its normal state is only a part of the truth namely in comparison with the extremely peaceful and tolerant time from 1870–1914'. And he added an even more intriguing thought, namely that 'we are now in the beginning of a deep religious crisis connected with the decline of the official church religion—a crisis, which has some similarity with the end of the antique world (I am influenced here by the Swiss historian [of civilizations] Jacob Burckhardt [1818–1897]).' Pauli, in fact, thought that neither materialism nor any other 'isms' are the motivation, but 'that the occident—distinctly different from India and China—has a 2000 year religious crisis period (a kind of puberty)' and that 'the immense technical development of the means of destruction is in no way accidental'. And Pauli added the Jungian thought that 'A remedy is probably (among others) the development of Eros in connection with the religious crisis' (Ref. [2], letter [619]).

In the real world the period between the last-quoted letter [619] and letter [672] quoted earlier witnessed historic events: On 7 December 1941 the Japanese destroyed an important part of the US Navy at Pearl Harbor; the following day the United States and Britain declared war on Japan. And on 11 December Germany and Italy, fulfilling their obligation towards Japan, declared war on the United States. This ended a period of ambiguity in the attitude of the United States which was torn between interventionism and isolationism. After the re-election in 1940 of the Democrat Franklin Delano Roosevelt (1882–1945) as president, this split even sharpened. However, by the autumn of 1941 the United States was already engaged in an undeclared war with Germany. In August 1941 Roosevelt and Winston Churchill (1874–1965) signed aboard ship the Atlantic Charter to destroy Nazi tyranny, and US destroyers began to patrol the North Atlantic. Then, on 6 April 1942, the first B-18 bomber equipped with microwave radar developed at the Radiation Laboratory spotted and sank a German submarine (Ref. [14a], p. 142).

For the moment Pauli enjoyed a secure and socially active life. At the end of January 1941 Weisskopf came to Princeton for a fortnight (Ref. [2], letter [618]), and at Easter Bethe and his wife visited [631]. And from 1 to 3 May Pauli attended the Washington meeting of the APS where he learned the latest facts on cosmic rays from Schein and his group in Chicago. Marcel Schein (1902–60) had come to the United States from the University of Zurich where he had been a Privatdozent (see page 313) [633, 635]. On 15 May Pauli was invited to a seminar on 'Wave equations for particles with higher spin' at Columbia University in New York where both Fermi and Bethe were staying [623]. During the first half of June he and Franca went south 'in our new car' on a swimming vacation to Myrtle Beach in South Carolina, a sandy shore on the Atlantic [630, 636]. Towards the end of June they drove in their new car—again a Plymouth (see page 294)—to Ann Arbor where Pauli's lectures started on 30 June. They did not stay to the end of the summer school, however, and Pauli gave his last lecture on 8 August, after which he and Franca went to California where they visited Oppenheimer at Berkeley and Bloch at Stanford. From 15 to 20 September they were back in Chicago where Pauli attended lectures on cosmic rays, and then they had 'a nice trip of 3 days' back to Princeton (Ref. [2], letters [625, 629, 633, 637, 640, 641]).

From 29 to 31 December the American Physical Society held a meeting at Princeton where Pauli gave a talk on 'Problems of the quantum theory of fields' and where he met the '*Vollbasler*' Gregory Hugh Wannier (1911–83), meaning that the latter spoke an extreme Basle dialect (Ref. [2], letters [646,647]). Wannier (see Ref. [15a]) had just published, together with Pauli's friend Kramers, 'a very good work about ferromagnetism' (translated from letter [646]), namely the proof showing that the Ising model (see page 221) in two dimensions must have a phase transition [15b]. This proof was made quantitative three years later in the famous paper [15c] by Lars Onsager (b. 1903) which Pauli mentions in letter [757] (Ref. [2], p. 297).

In the last-mentioned letter of 30 December 1941 Pauli wrote to his friend Wentzel, who was giving Pauli's lectures at ETH, that he had received 'an official letter [Ref. [8], II.49] from Rohn', in which the president expressed in kind terms, 'that I had to decide to return at the very latest on 1st of October 1942, since otherwise the chair for theoretical physics at the Polytechnic [ETH] could no longer be held vacant for me. ... It is very strange how in all letters from Zurich to me the main effect is always neglected, namely the practical impossibility of the journey for me' (translated from Ref. [2], letter [646], p. 116). On 20 June Pauli had confirmed from Myrtle Beach receipt of the president's letter of 23 May 1941 (see Ref. [8], footnote 133, p. 101), which informed him that his leave of absence had been extended to the winter term of 1941/2 (Ref. [8], II.47). In letter II.49, dated 21 October 1941, President Rohn told Pauli that 'because of your absence the physics instruction at ETH

... is somewhat disorganized. Surely, Herr Prof. Wentzel could be won for the maintenance of one or the other lecture. . . . On the other hand, at present the influence of our theoretician of the physics domain on the research in our Physics Institute is wanting' (translated from Ref. [8], II.49). This letter had been instigated by Pauli's colleague Scherrer who had written to President Rohn that research, 'because of the missing of Prof. Pauli, is considerably reduced. Particularly now that the cyclotron soon will start operation, we absolutely should have an experienced proper theoretician (*Haustheoretiker*) to whom we could address all the difficult questions of nuclear physics.' Also, 'Herr Dr. Jauch, who, without Pauli's help and severe critique, works with quite bad efficiency, would again be brought into the proper current (*Fahrwasser*) by Pauli's return' (translated from Ref. [8], II. 48).

Pauli answered on 14 January 1942, thanking the president 'for the kind evaluation of my activity at the ETH' and assuring him of his 'undiminished interest in the physics teaching and the works at the Physics Institute of ETH', regretting very much 'that, according to your opinion, the latter are unfavourably influenced'. Pauli then gives the detailed arguments why he cannot travel and closes by writing: 'In this connection I also wish to point to the fact that in the summer of 1940 I decided to engage in this leave of absence granted by you only with your agreement, after the refusal of my request of naturalization by the Swiss naturalization authority was known to you and that, therefore, you had knowledge of the fact that I continue to be legally placed under a foreign military power with all its implications' (translated from Ref. [8], II.51). President Rohn, who had received Pauli's letter only on 3 March, answered on 17 March 1942: 'I understand your situation very well and, therefore, will examine with the Federal Department of the Interior as well as with the Federal Political Department whether the possibility would exist to send you as professor of our Institute a voucher with which you could engage in the return trip', and he informed Pauli that the Swiss School Council had extended his leave of absence to the summer term of 1942 (translated from Ref. [8], II.53). But since nothing happended concerning this voucher, Pauli commented to his friend C. A. Meier on 26 May: 'It seems to me that the "Dame Helvetia" [Switzerland] consists of offices working intentionally against each other, which cannot stand each other. . . . Rohn wants me to come back, but in Berne one reflects, "just let the foreigner go" (1940) and "stay away" (1942), respectively' (translated from Ref. [2], comment p. 125).

Meanwhile Jauch, on Scherrer's proposing, was confirmed as assistant in theoretical physics for the winter term of 1941/2 and the summer term of 1942 (Ref. [8], II.50, 55). However, Jauch and his wife decided to take advantage of the offer to US citizens and their close relatives of favourable transportation by ship back to the United States. On 11 May he wrote a somewhat bitter letter to President Rohn, saying that, with his earnings at ETH he would never be able to pay back to ETH the debt he had incurred with his studies, and that he had hoped to be given a teaching assignment 'to fill the gap now that Prof. Wentzel was no longer able to replace Prof. Pauli. I am surprised and shocked that this teaching assignment was not given to me but Herrn Dr. [Paul] Huber [1910–71] who is not even a theoretical physicist' (translated from Ref. [8], II.57). On request Scherrer gave President Rohn a quite impartial evaluation of Jauch (II.59). Jauch's appointment as assistant at ETH ended on 31 May 1942 (Ref. [8], footnote 142, p. 110). His successor, proposed by Scherrer, was André Houriet (b. 1919) of St Imier in the Bernese Jura, as aid in the problem sessions for the winter term of 1942/3 (II.82) and, after his diploma with Wentzel at ETH, as assistant for the summer term of 1943 (II.86).

On 4 May 1942 the professors of Section IX, the Mathematics and Physics Faculty of ETH, convened to discuss the matter of Professors Pauli and Pólya who both said they were unable to return from the United States. The Hungarian-born Georg Pólya (1887–1985) was a mathematics professor at ETH from 1914, and in 1940 went to Brown University in the United States. Although he was Swiss, he chose to stay there. From 1942 to his retirement in 1953 Pólya was professor at Stanford University. In this meeting mentioned above a new argument was introduced into the discussion by the mathematics professor Michel Plancherel (1885–1967), who said: 'Suppose that our colleagues return to Zurich once the war is terminated and all danger is discarded. What, at that moment, would be their moral position, mainly facing the students?' (translated from Ref. [8], II.56). Psychologically this was a dangerous argument, for it encouraged colleagues with anti-Semitic or jealous feelings against Pauli or Pólya to express more openly their opposition to the former's return. That is in essence what was going to happen.

Writing to the president on 12 May, Scherrer proposed Wentzel as possible successor to Pauli and simultaneously Fierz as the successor to Wentzel at the university (II.58). The rector of ETH, the mathematics professor Walter Saxer (1896–1974), in a letter to President Rohn on 28 May invoking fidelity and perseverance, concluded: 'Anyhow, I am also of the opinion that further compromising (*Entgegenkommen*) is not warranted any more' (translated from Ref. [8], II.60). Under these circumstances President Rohn felt obliged to send Pauli and Pólya the following telegram, dated 3 June 1942: 'Swiss School Council will 22 June probably decide that your leave of absence cannot be extended over [beyond] the summer term of this year' (II.61). While Pólya cabled back with thanks that a return in time was impossible, Pauli sent the following telegram on 9 June: 'should my return for winter term 1942 be prevented by circumstances beyond my control i shall defend my rights deriving from my employment contract with the swiss federal polytechnicum [ETH] by legal procedure' (II.62). In this situation the president sought the advice of a lawyer, who concluded that in case of an impossibility to travel, Pauli's legal rights did not cease to be valid but that this would be the case at the end of the legal period, i.e. in 1948 (II.63).

Then the students also started to complain to President Rohn about the difficult situation in theoretical physics (II.68). On 29 August 1942 Bundespräsident Philipp Etter (see page 283), who now held the cyclic position of President of the Federal Council, informed President Rohn that it was not possible to provide travelling papers to Pauli. On the other hand, he confirmed that Pauli's legal rights were valid until 31 March 1948, so that an extension was warranted, since otherwise a costly early pensioning would occur (II.70). Thus Pauli's leave was extended to the winter term of 1942/3 (II.74). However, on 30 September 1942 the President of the Swiss School Council wrote to Bundespräsident Etter 'that, by moral reasons, Prof. Pauli should under no circumstances take up his teaching at ETH again after the end of the war only—unless the end of the war happens next winter; his esteem would be totally sapped and his teaching success yet only petty' (translated from Ref. [8], II.76, p. 131). Bundespräsident Etter replied on 9 December that he agreed with the proposed measure of no re-election after March 1948 because Pauli's leave for the United States was considered by the students as a kind of defection: 'if he came back at the end of the war and even if he came back now he would not have left any moral authority' (translated from Ref. [8], II.79).

After discussions in the Swiss School Council (II.81) this measure was communicated to Pauli by letter on 3 February 1943 (II.83). In a detailed answer Pauli, on 14 April, refused to accept the measure: 'About my moral authority as University teacher *after* the war you now evidently may have only an opinion: But you elevate this opinion to a fact and draw very far reaching consequences out of this anticipation. Against this attempt to raise without justification a severe moral blame against me I must protest in all firmness.' And he proposed a disciplinary measure 'at a moment of time when all accusations against me are put forward in full responsibility and when I am in a position to give account, also account for my moral duties during the war, and to conserve my reputation in the scientific world' (translated from Ref. [8], II.84, p. 139). On 10 July 1943 the crucial meeting of the Swiss School Council took place in the presence of Bundespräsident Etter. After a long discussion Councillor Ernst Bärtschi (1882–1976), the mayor of the city of Berne, made the proposition of immediate discharge or pensioning of Pauli, while the president proposed a continuation of the status quo. The result of the vote was three to three, so that the president's casting vote decided for the latter—which involved, besides Wentzel's standing in for Pauli, teaching assignments to Fierz and to Stueckelberg (II.90).

Thus, thanks to President Rohn, nothing was broken. But the problem continued to bother Rohn. On 17 May 1945 he received the following telegram from Pauli: 'have contacted swiss consulate concerning my travel arrangements for zuerich stop copy of application following = professor wolfgang pauli' (Ref. [8], II.107). Rohn convened the Swiss School Council on 2 June. The discussions of this meeting were summed up by the president as follows: 'It would indeed be very unfavourable if Prof. Pauli would resume his teaching. Therefore the Federal Council should be asked that his wages be paid till the end of March 1948, even if he does not lecture. Presumably, there will be a legal procedure anyway' (translated from Ref. [8], II.108). In response to the report of President Rohn, Federal Councillor Etter insisted that there was no motive preventing Pauli from teaching on his return if he so wished (II.117).

This situation changed, however, when in November 1945 the news of Pauli's Nobel Prize reached Zurich (see Ref. [8], II.122).

Lonely leader in peaceful research

(in connection with your remark about Stumpen) I now smoke the pipe!! *([Otto] Stern is highly pleased with it).* [16a]

This afterthought in the already quoted letter [646] to the *Stumpen* (Swiss-cigar)-smoking Wentzel in Zurich marks an important event in Pauli's biography. The occupational quality of cleaning, filling, and lighting the pipe favoured the reflective attitude necessary in Pauli's upcoming scientific solitude. On 26 May 1942 he wrote to his friend C. A. Meier in Zurich: 'Next fall I will (for reasons I cannot write because of military censorship) perhaps sit around here and have nobody to talk with about physics' (translated from Ref. [2], comment, p. 126). And, indeed, the following year he wrote to Bhabha on 23 September: 'I am rather isolated and life is becoming more dull' (Ref. [2], letter [682]).

In this contemplative period Pauli's thoughts often took a poetic turn. In May 1942 he wrote a fantasy entitled 'The contest of the genders. A philosophical comedy', in which Pauli's personal antinomy between thinking and feeling, alluded to on page 246, took the form of a dialogue. In it Aphrodite is arguing that emotions are at the root of all thoughts and even of language, while Immanuel defends the supremacy of thought and the existence of objective truths, such as the Ten Commandments. But A. claims that emotions have a proper life and that there is a secret connection between all emotions in the world—an obvious allusion to Jung's collective unconscious—whose goal is the increase of the consciousness. However, I., in a typically Paulian drive for unification of opposites, concludes that what was really before both emotion and thought, is something, in which both participate, from which emerges life and where birth and death are one and the same. On this unifying vision both A. and I. finally agree. Almost 10 years later Pauli sent this fantasy to his friend Marie-Louise von Franz (enclosed with letter [1335] in Ref. [16b]). During this period Pauli also wrote several poems which he later sent to Jung's secretary Aniela Jaffé (enclosed with letter [1166] in Ref. [16b]).

The autumn term of 1941 was still lively. Pauli lectured on the 'Theory of cosmic rays' at the Princeton Institute. His research activity now turned seriously to the physics of mesons. It was Oppenheimer who had stimulated him in 1941 to work in this field, as Pauli later told Rosenfeld and Casimir (Ref. [2], letters [757] and [780]). And it was also from Oppenheimer's institute that Pauli's first collaborator in meson physics, Sidney Michael Dancoff (1913–51), had arrived at the beginning of October 1941, just after Pauli's and Franca's return from Chicago (see the last section) [641]. Pauli was back in Chicago for the Thanksgiving meeting of the APS which took place on 21–22 November 1941 [642].

The following year, from 1 February to 1 May, Pauli again lectured at the institute, this time on 'Strong coupling theories', of which mimeographed notes were supposed to be prepared [647]. Then he accepted an invitation by the head of the physics department of Purdue University, Professor Karl Lark-Horovitz (1892–1958)—a Viennese who had come to the US in 1925 (Ref. [11], p. 1000)—for a visiting professorship at the summer school in Lafayette, Indiana, for the period from 15 May to 1 July 1942 (Ref. [2], letters [647, 657]); Pauli's lectures began on 18 May [655]. At Purdue Heisenberg's visit of 1939 (see pages 329–30) still was remembered by many [661]. Pauli had an office next to that of Julian Seymour Schwinger (1918–94) 'who has gone over to war physics almost entirely, however, but still had enough time for talking shop (*Fachsimpeln*)' (translated from Ref. [2], letter [661], p. 149). Meanwhile Franca was so bored that she attended a talk by Schwinger and was so enraptured by his presentation that she found him the most marvellous lecturer (Ref. [16c], footnote on p. 98).

Fig. 9.1 Celebration of the 45th birthday at the Princeton Institute, 1945 (CERN)

Schwinger had joined Oppenheimer's group in Berkeley in the autumn of 1939 as a fresh Ph.D. at age 21 and two years later came to Purdue University, where he bought a large, black Cadillac car (see Ref. [16c], pp. 54–6 and 90). There he and several other theoretical physicists were engaged by Hans Bethe in a collaboration on the theory of microwave propagation. Bethe, who had been asked in 1941 by the directorate of the Radiation Laboratory (see the previous section) to work on this project, became its coordinator. Purdue University was favoured because Lark-Horowitz himself was directing a programme of semiconductor research for radar detection (see Ref. [16d], pp. 293–4). From

Lafayette Pauli and Franca drove east to the northern-most part of the state of New York where they spent the summer vacation until mid-September 1942 at Lake Clear, situated on a small lake in the Adirondack mountains [657, 662].

In Lafayette Pauli had received a letter from Jauch, written in Lisbon on 7 June 1942 [658]; and another one, written on 8 July on board the ship *Drottingholm* docked in New York harbour [660]. Jauch was now looking for a job, which, because of the defence work of most US physicists, was not too difficult. He wished to collaborate with Pauli. However, the Princeton Institute had no money, as Pauli told Jauch on 26 June. But Pauli suggested that Jauch contact the head of the physics department of Princeton University, Professor Henry DeWolf Smyth [658]. On 22 July Jauch was able to inform Pauli at Lake Clear that Professor Smyth had answered favourably [662]. Thus Jauch could continue his assistantship with Pauli—nominally only for one more year, however, since on 15 July 1943 Pauli was able to send Jauch cordial congratulations for his nomination as assistant professor at Princeton University [675].

Josef Maria Jauch (1914–74) grew up in Lucerne, Switzerland, and had a lifelong love for mathematics and music. He had learned to play the violin, first with his father, then with a professional teacher and gave his first public concert in Lucerne at age 12. However, he became an orphan when he was only 15 years old. Generous friends made his entrance to ETH in 1933 possible by lending him the necessary funds. Among his teachers were Pauli and the mathematician Heinz Hopf (see the previous section). From Hopf he learned complex functions, Galois theory, and topology which remained favourite topics. But he did his diploma thesis with Pauli and Pauli's assistant Fierz, after which he went to the United States and obtained his Ph.D. under Edward Hill at the University of Minnesota (see page 337). After the war Jauch spent one year at the Bell Telephone Laboratory at Murray Hill near Princeton, and in the autumn of 1946 he joined the University of Iowa as a professor of theoretical Physics, where his closest collaborators were Fritz Coester (b. 1921) and Fritz Rohrlich (b. 1921). In 1953 the latter joined Jauch in writing the well-known book on QED, *The Theory of Photons and Electrons* [17a], published two years later. This was the first treatise that developed QED in the covariant formalism of Tomonaga, Schwinger, Feynman, and Dyson and cast it consequently into the form of Heisenberg's S-matrix (see later).

In 1958 Jauch joined the London branch of the US Office of Naval Research as a scientific liaison officer with the task of reviewing research and its institutions in Europe. And in 1959 he returned to his native land when the University of Geneva offered him the directorship of the physics department, the Ecole de Physique, which, making use of his experience gained in the United States and in London, he built up anew. Within four years he had brought the key people together, after which he withdrew from administrative work to devote his time to theoretical physics, mainly to the foundations of quantum mechanics, on which he published a book in 1968 [17b]. He also became a defender of Galileo Galilei (1564–1642) against the latter's condemnation in 1633 and, together with prominent Catholic personalities, obtained a formal Papal acquittal of Galilei. Jauch produced a poetic synthesis of these last two loves in his last book *Are Quanta Real? A Galilean Dialogue* [17c]. Jauch unexpectedly died of a heart attack on 30 August 1974 at Geneva (see Ref. [17d]).

At the APS meeting in Princeton in December 1941 mentioned in the last section, Dancoff and Pauli reported their first results on 'Strong Coupling Mesotron Theory of Nuclear Interactions', namely new values of the excitation energies of the nucleon, commonly called isobaric states, depending 'quadratically on spin and charge excesses' [18a]. These excitations occur when the coupling to the meson field is sufficiently strong. The final version of this work, 'The Pseudoscalar Meson Field with Strong Coupling' [18b], was submitted on 23 June 1942. This long paper, which yields its secrets only reluctantly, gives the most systematic treatment of Wentzel's strong-coupling theory [20a] (see the previous section), as well as a quantum mechanical evaluation of the mentioned spin and charge excesses. Previous calculations by Oppenheimer and Schwinger [20b] treated the spin and isospin dependences only classically. Among Pauli's papers I have studied in detail, this one is probably the most difficult to read. But it is not even certain that, apart from filling his plentiful spare time, Pauli's impressive calculations were worth their while since, indeed, this and the following papers were completely ignored by the post-war treatises [20c].

In this paper [18b] a single extended nucleon at rest at $\mathbf{x} = 0$ is considered, which is coupled to a pseudoscalar, isovector meson field described by Hermitian operators $\varphi_\alpha(\mathbf{x})$, $\alpha = 1, 2, 3$. This terminology becomes evident by considering the interaction Hamiltonian

$$H' = -\frac{g}{\kappa}\sqrt{2\pi\hbar c}\int d^3x U(r)\sum_\alpha \tau_\alpha(\vec{\sigma}\cdot\nabla)\varphi_\alpha, \qquad (9.4)$$

where $r = |\mathbf{x}|$, g is the (dimensionless) coupling constant and, as before, $\kappa = m_\mu c/\hbar$ [20d]. In distinction to Eq. (8.73) the nucleon here is decribed just by the real spherically symmetric shape function U, called K in Ref. [18b], or by the corresponding form factor (8.49). U is normalized according to

$$\int U(r)d^3x = 1 \qquad (9.5)$$

and defines the *radius* a of the nucleon by (see Ref. [20b])

$$a^{-1} = \int\int d^3x d^3x' U(r)\frac{1}{|\mathbf{x}-\mathbf{x}'|}U(r'), \qquad (9.6)$$

where $r' = |\mathbf{x}'|$. τ_α and σ_k are Pauli matrices (5.37) describing the isospin and spin, respectively, of the nucleon. From the invariance of H' the isovector and pseudoscalar nature of the meson field φ_α follows, the second property being due to the fact that the gradient ∇ changes sign under space reflection (parity) $\mathbf{x} \to -\mathbf{x}$. $\psi^+ = 2^{-1/2}(\varphi_1 - i\varphi_2)$ and $\psi^- = (\psi^+)^+$ create a π^+- and a π^--

Fig. 9.2 At the Institute for Advanced Study, Princeton, 1940–6

meson, respectively, and φ_3 a neutral π^0. Note that in order to describe the nuclear forces a second nucleon would have to be included in the interaction (9.4), as is evident from Eq. (8.74).

The free meson fields φ_α and the conjugate Hermitian momenta π_α are defined by the Hamiltonian

$$H_0 = \frac{1}{2} \sum_\alpha \int d^3x \{ (\hbar c)^2 \pi_\alpha^2 + (\nabla \varphi_\alpha)^2 + \kappa^2 \varphi_\alpha^2 \} \qquad (9.7)$$

and satisfy the canonical commutation relations (6.39),

$$i[\pi_\alpha(\mathbf{x}), \varphi_\beta(\mathbf{x}')] = \delta_{\alpha\beta} \delta(\mathbf{x} - \mathbf{x}'). \qquad (9.8)$$

This symmetrical pseudoscalar theory which predicted the π^0-meson (see page 295) was first formulated by Kemmer [20e]. Without the φ_3-field it becomes the charged pseudoscalar theory, which is discussed in Section 8 of Ref. [18b]. The free field Hamiltonian density of the latter is, in different units, Eq. (8.24) of the Pauli–Weisskopf theory.

Strong coupling, defined by the condition $g \gg \kappa a$, now means that the interaction (9.4) is the dominant term. Therefore, the part of the meson field which is coupled to the shape function U is large, while the part φ'_α outside the nucleon may be considered as describing the perturbing free meson field. Defining

$$\varphi^0_{\alpha k} = -\sqrt{4\pi} \int d^3x\, U(r) \frac{\partial \varphi_\alpha}{\partial x_k} = \sqrt{4\pi} \int d^3x \frac{\partial U}{\partial x_k} \varphi_\alpha, \qquad (9.9)$$

the interaction (9.4) may be expressed in terms of this projection alone as

$$H' = \frac{g}{\kappa} \sqrt{\frac{\hbar c}{2}} \sum_{\alpha k} \tau_\alpha \varphi^0_{\alpha k} \sigma_k, \qquad (9.10)$$

while the remainder in φ_α, φ'_α, satisfies the orthogonality condition

$$\int d^3x\, \varphi'_\alpha \nabla U = 0. \qquad (9.11)$$

The large, *static*, part of the meson field then has the form

$$\varphi_\alpha - \varphi'_\alpha = \frac{1}{\sqrt{4\pi}} \sum_k \varphi^0_{\alpha k} f_k(\mathbf{x})$$

which, multiplied by $\sqrt{4\pi}\partial U/\partial x_i$ and then integrated yields, according to Eqs (9.9) and (9.11),

$$\varphi^0_{\alpha i} = \sum_k \varphi^0_{\alpha k} \int d^3x \frac{\partial U}{\partial x_i} f_k.$$

Here the orthogonality in i, k of the integral is achieved by setting $f_k = \partial X/\partial x_k$ with a spherically symmetric function $X = X(r)$, since then, using $\xi \equiv X/I$,

$$\int d^3x \frac{\partial \xi}{\partial x_i} \frac{\partial U}{\partial x_k} = \delta_{ik}; \quad I \equiv \frac{1}{3} \int d^3x \nabla X \cdot \nabla U, \qquad (9.12)$$

or

$$\varphi_\alpha - \varphi'_\alpha = \frac{1}{\sqrt{4\pi}} \sum_k \varphi^0_{\alpha k} \frac{\partial \xi}{\partial x_k}. \qquad (9.13)$$

The idea to split the meson field into a large static part located at the nucleon and a smaller dynamic part outside is due to Wentzel [20a], who had used a lattice-space description (see page 315) instead of a shape function. As the nucleon is located at the lattice point $s = 1$, the meson field ψ_1 at this point corresponds to Pauli's $\varphi^0_{\alpha k}$, while ψ_s with $s \neq 1$ corresponds to φ'_α. On 20 July 1942 Pauli wrote to Wentzel from Lake Clear: 'Surely, you will soon see that the "lattice space" technically makes things only more complicated and that with the introduction of a finite extension ... of the heavy particle instead of the lattice space everything becomes much simpler to calculate.' And he assured him that 'your merit of having introduced "strong coupling" first, remains unimpaired' (translated from letter [661], Ref. [2], p. 150).

In order to express the potential energy part of H_0 in the new variables we choose X as the potential whose source is $W \propto (\overline{\psi}_N \psi_P)$ in (8.71). Since W is spherically symmetric, the same is then true of X. Hence, according to Eq. (8.73), $X(r) = \int d^3x' W(r') \exp(-\kappa R)/R$, where $R = |\mathbf{x} - \mathbf{x}'|$. Inserting expression (9.13) into (9.7), it then follows from the orthogonality condition (9.11) and Eq. (8.71) with source W that the cross-terms between $\varphi^0_{\alpha k}$ and φ'_α vanish, and the potential energy becomes, on inserting (9.10),

$$
\begin{aligned}
E_{pot} &= \frac{1}{2} \sum_\alpha \int d^3x \varphi_\alpha (-\Delta + \kappa^2) \varphi_\alpha + H' \\
&= E^0_{pot} + \frac{1}{2} \sum_\alpha \int d^3x \varphi'_\alpha (-\Delta + \kappa^2) \varphi'_\alpha,
\end{aligned}
\tag{9.14}
$$

where

$$
E^0_{pot} = \frac{1}{2I} \sum_{\alpha k} (\varphi^0_{\alpha k})^2 + \frac{g}{\kappa} \sqrt{\frac{\hbar c}{2}} \sum_{\alpha k} \tau_\alpha \varphi^0_{\alpha k} \sigma_k.
\tag{9.15}
$$

The projection $\pi^0_{\alpha i}$ of $\pi_\alpha(\mathbf{x})$ analogous to Eq. (9.9) may be chosen such that $\pi^0_{\alpha i}$ and $\varphi^0_{\beta k}$ satisfy the canonical commutation relations

$$
i[\pi^0_{\alpha i}, \varphi^0_{\beta k}] = \delta_{\alpha\beta} \delta_{ik}.
\tag{9.16}
$$

In view of (9.12) this implies that

$$
\pi^0_{\alpha i} = \frac{1}{\sqrt{4\pi}} \int d^3x \frac{\partial \xi}{\partial x_i} \pi_\alpha
\tag{9.17}
$$

and that the remainder in π_α, π'_α, satisfies the orthogonality condition

$$
\int d^3x \pi'_\alpha \nabla \xi = 0;
\tag{9.18}
$$

thus

$$
\pi_\alpha - \pi'_\alpha = \sqrt{4\pi} \sum_i \pi^0_{\alpha i} \frac{\partial U}{\partial x_i}.
\tag{9.19}
$$

One also shows that the π'_α, φ'_β commute with the $\pi^0_{\alpha i}$, $\varphi^0_{\beta k}$ but, of course, not among themselves. The kinetic energy, however, has cross terms; with

$$
4\pi \int d^3x \frac{\partial U}{\partial x_i} \frac{\partial U}{\partial x_k} = N\delta_{ik}; \quad N = \frac{4\pi}{3} \int (\nabla U)^2 d^3x,
\tag{9.20}
$$

one finds

$$
\begin{aligned}
E_{kin} = \ & \frac{(\hbar c)^2}{2} \sum_\alpha \pi^2_\alpha = \frac{(\hbar c)^2}{2} \sum_\alpha \Big\{ N \sum_k (\pi^0_{\alpha k})^2 \\
& + 2\sqrt{4\pi} \sum_k \pi^0_{\alpha k} \int d^3x \pi'_\alpha \frac{\partial U}{\partial x_k} + \int d^3x (\pi'_\alpha)^2 \Big\}.
\end{aligned}
\tag{9.21}
$$

In Eq. (9.21) the term linear in π'_α may be transformed away by a Bloch–Nordsieck shift (8.55), $\psi'_\alpha \to \psi''_\alpha + \gamma_\alpha$. Note that Pauli introduces this shift

(Eq. (75) of Ref. [18b]) only at a later stage of the calculation, which makes it much less transparent. Since the shift $\gamma_\alpha(\mathbf{x})$ must also satisfy the orthogonality condition (9.18), it is found to be $\gamma_\alpha = -\sqrt{4\pi}\sum_k \pi^0_{\alpha k}(\partial U/\partial x_k + (4\pi)^{-1}\vec{\lambda}_k \cdot \nabla\xi)$. Here $\vec{\lambda}_k$ is determined by the orthogonality (9.18) of γ_α to be $\lambda_{ki} = \delta_{ki}\lambda$. Inserting this shift in Eq. (9.21) one finds, making use of (9.12), (9.20),

$$E_{kin} = \frac{(\hbar c)^2}{2}\left\{\lambda\sum_{\alpha i}(\pi^0_{\alpha i})^2 + \sum_\alpha \int d^3x(\pi''_\alpha)^2\right\}; \quad \lambda \equiv \frac{12\pi}{\int d^3x(\nabla\xi)^2} \cdot \quad (9.22)$$

This simple result corresponds to the much more complicated result (76, 77) of Ref. [18b]. Note that to make this shift canonical one should make sure that the commutator $[\pi'_\alpha(\mathbf{x}), \varphi'_\beta(\mathbf{x}')]$ is preserved in the transformation; Pauli does this in Eq. (78) of Ref. [18b]. However, since this commutator is irrelevant for the present purpose I do not include this complication.

Introducing the charged fields $\psi^\pm$ and $\pi^\pm = 2^{-1/2}(\pi_1 \pm i\pi_2)$ (see Eqs. (110) of Ref. [18b]), the proton density is given by Eq. (8.23) or, in the units adopted here, by $(\varphi_1\pi_2 - \varphi_2\pi_1)$, so that the total charge becomes, in units of the elementary charge e,

$$\epsilon = \int d^3x(\varphi_1\pi_2 - \varphi_2\pi_1) + \frac{1}{2}(1 + \tau_3). \quad (9.23)$$

On the other hand, the mesonic isospin is given by the rotation operators in the three-dimensional isospin space, $\int d^3x(\varphi_\alpha\pi_\beta - \varphi_\beta\pi_\alpha)$, so that the total isospin is

$$T_{\alpha\beta} = \int d^3x(\varphi_\alpha\pi_\beta - \varphi_\beta\pi_\alpha) + \frac{1}{2}\tau_{\alpha\beta}, \quad (9.24)$$

where $\tau_{12} = -\tau_{21} = \tau_3$, etc. For the total angular momentum $\mathbf{L}$, Eq. (10') of Ref. [18b] is a good guess (note that the quoted equation, written with opposite sign and φ_α and π_α interchanged, is equivalent), i.e.

$$L_{ik} = \sum_\alpha \int d^3x\varphi_\alpha l_{ik}\pi_\alpha + \frac{1}{2}\sigma_{ik}; \quad l_{ik} \equiv x_i\frac{\partial}{\partial x_k} - x_k\frac{\partial}{\partial x_i}. \quad (9.25)$$

Indeed, inserting here the expressions (9.13), (9.19), the mesonic contribution to the nucleon spin $\mathbf{L}^0$ becomes

$$L^0_{ik} = \sum_{\alpha rs}\varphi^0_{\alpha r}\pi^0_{\alpha s}J_{rs;ik}; \quad J_{rs;ik} \equiv \int d^3x\frac{\partial\xi}{\partial x_r}l_{ik}\frac{\partial U}{\partial x_s}.$$

Rotational symmetry of ξ and U implies that $J_{rs;ik} \neq 0$ only if $r, s = i, k$ or $r, s = k, i$, and that special axes may be chosen such that $i = 1$, $k = 2$. In polar coordinates $x_1 = r\sin\vartheta\cos\varphi$, $x_2 = r\sin\vartheta\sin\varphi$ one has $l_{ik} = \partial/\partial\varphi$, so that, in view of (9.12), $J_{12;12} = -J_{21;12} = 1$. Hence

$$L^0_{ik} = \sum_\alpha(\varphi^0_{\alpha i}\pi^0_{\alpha k} - \varphi^0_{\alpha k}\pi^0_{\alpha i}), \quad (9.26)$$

as one would expect in view of the commutator (9.16).

The first task in the strong-coupling approximation is to diagonalize the interaction (9.10) which, in vector notation, may be written as $H' \propto \vec{\tau}\varphi\vec{\sigma}$ where φ is the real matrix with elements $(\varphi)_{\alpha k} = \varphi_{\alpha k}^0$. The idea is to bring $\vec{\tau}$ and $\vec{\sigma}$ into the same coordinate system by diagonalizing φ. Considering the real bilinear form $\mathbf{y}\varphi\mathbf{x}$ and the real symmetric matrices $C = \overline{\varphi}\varphi = \overline{C}$ and $C' = \varphi\overline{\varphi} = \overline{C'}$, where a bar means transposition, one can see that the non-negative quadratic forms $(\varphi\mathbf{x})^2 = \mathbf{x}C\mathbf{x}$ and $(\mathbf{y}\varphi)^2 = \mathbf{y}C'\mathbf{y}$ may be diagonalized by real rotation matrices A and B, respectively, such that $\mathbf{x} = A\mathbf{x}'$ and $\mathbf{y} = \mathbf{y}'B$. Then

$$\overline{A}CA = Q^2 \tag{9.27}$$

is diagonal and non-negative, so that the eigenvalues Q_r of Q are all real. And since $A\overline{A} = 1$ and $\det A = 1$,

$$\sum_{\alpha i}(\varphi_{\alpha i}^0)^2 = \mathrm{Tr}C = \sum_{r=1}^{3} Q_r^2, \tag{9.28}$$

and $\det \varphi = Q_1 Q_2 Q_3$, respectively. One may now choose B such that

$$B\varphi A = Q = \overline{Q} \tag{9.29}$$

(see Eq. (32) of Ref. [18b]). Then one also has $BC'\overline{B} = Q^2$ and

$$\vec{\tau}\varphi\vec{\sigma} = \vec{\tau}'Q\vec{\sigma}' = \sum_r \varepsilon_r Q_r, \tag{9.30}$$

where $\varepsilon_r \equiv \tau_r'\sigma_r'$. Since the τ_r' commute with the σ_s', it follows that $\varepsilon_r^2 = 1$ and $\varepsilon_1\varepsilon_2\varepsilon_3 = -1$. Therefore, Eq. (9.30) has 'singlet' and 'triplet' solutions

$$\begin{aligned} \varepsilon_1 = \varepsilon_2 = \varepsilon_3 = -1; \quad \vec{\tau}\varphi\vec{\sigma} = -(Q_1 + Q_2 + Q_3) \\ \varepsilon_r = -1;\ \varepsilon_s = \varepsilon_t = 1;\ \vec{\tau}\varphi\vec{\sigma} = -Q_r + Q_s + Q_t, \end{aligned} \tag{9.31}$$

where $r, s, t = 1, 2, 3; 2, 3, 1; 3, 1, 2$. Hence, the leading term (9.15) of the potential energy is lowest in the first, the 'singlet', case, where it becomes, with (9.28), (9.30),

$$E_{pot}^0 = \frac{1}{2I}\mathrm{Tr}(Q^2 - 2DQ); \quad D \equiv \frac{gI}{\kappa}\sqrt{\frac{\hbar c}{2}}, \tag{9.32}$$

which has the minimum

$$E_{min}^0 = -\frac{3D^2}{2I^2}\hbar c \tag{9.33}$$

at $Q = D$. With Eq. (9.29) we may define new variables $q_{\alpha\beta}$ by the symmetric matrix

$$q \equiv \varphi\overline{e} = \overline{B}QB = \overline{q}; \quad \overline{e} \equiv AB. \tag{9.34}$$

Since from (9.34) it follows that $\mathrm{Tr}q^n = \mathrm{Tr}Q^n$ for $n = 1, 2, ...$, Eq. (9.32) may be expressed in the variables

$$q' = q - D = \overline{q'}, \tag{9.35}$$

as

$$E_{pot}^0 - E_{min}^0 = \frac{1}{2I}\text{Tr}(q')^2 = \frac{1}{2I}\sum_{\alpha\beta}(q'_{\alpha\beta})^2. \tag{9.36}$$

The $q'_{\alpha\beta}$ describe the neighbourhood of the minimum. It is then to be expected that the symmetric matrix

$$p = \frac{1}{2}(e\pi + \overline{e\pi}) = \overline{p}, \tag{9.37}$$

where π is the matrix with elements $(\pi)_{k\alpha} = \pi_{\alpha k}^0$, will define variables $p_{\alpha\beta}$ conjugate to the $q_{\alpha\beta}$ and, hence, also to the $q'_{\alpha\beta}$. This is Eq. (59) of Ref. [18b] which is arrived at there after a lengthy calculation. It is clear from (9.34) that e and π do not commute. To solve for π we multiply Eq. (9.37) from the left by $\overline{e}$, the result of which may be written as $\pi = \overline{e}(2p - \overline{e\pi})$. Adding to this the identity $\pi = \overline{e}e\pi$ we obtain

$$\pi = \overline{e}\left\{p + \frac{1}{2}(e\pi - \overline{e\pi})\right\}. \tag{9.38}$$

At the minimum $q' = 0$, and it follows from (9.34), (9.35) that $\varphi = De$, so that Eq. (9.38) may be expressed in terms of the angular momentum (9.26) by defining

$$L_0^{\alpha\beta} \equiv e_{\alpha i}e_{\beta k}L_{ik}^0. \tag{9.39}$$

Indeed, one finds $L_0^{\alpha\beta} = D(e_{\beta k}\pi_{\alpha k}^0 - e_{\alpha k}\pi_{\beta k}^0)$, so that, writing (9.38) in components,

$$\pi_{\alpha i}^0 = e_{\beta i}\left(p_{\alpha\beta} + \frac{1}{2D}L_0^{\alpha\beta}\right), \tag{9.40}$$

which is Eq. (72) of Ref. [18b].

The next step is to show that Eqs (9.34), (9.37) define conjugate variables. To this end we need commutation relations between $L_0^{\alpha\beta}$ and $e_{\mu i}$. Since A describes rotations of $\mathbf{x}$ and L_{ik}^0 is a rotation operator in $\mathbf{x}$-space while B acts only on the isospin space $\mathbf{y}$, Eq. (38) of Ref. [18b] holds, i.e.

$$i[L_{ik}^0, A_{\nu r}] = \delta_{\nu k}A_{ir} - \delta_{\nu i}A_{kr}; \quad [L_{ik}^0, B_{r\alpha}] = 0. \tag{9.41}$$

Since according to (9.34) $e = \overline{AB}$ one deduces from this that

$$i[L_{ik}^0, e_{\mu\nu}] = \delta_{\nu k}e_{\mu i} - \delta_{\nu i}e_{\mu k}, \tag{9.42}$$

and from (9.39),

$$i[L_0^{\alpha\beta}, e_{\mu\nu}] = \delta_{\alpha\mu}e_{\beta\nu} - \delta_{\beta\mu}e_{\alpha\nu}. \tag{9.43}$$

Inserting $\varphi_{\beta k}^0 = q_{\beta\nu}e_{\nu k}$ from (9.34) and $\pi_{\alpha i}^0$ from (9.40), Eq. (9.16) becomes

$$\delta_{\alpha\beta}\delta_{ik} = e_{\mu i}e_{\nu k}i[p_{\alpha\mu}, q_{\beta\nu}] + \frac{1}{2D}e_{\mu i}q_{\beta\nu}i[L_0^{\alpha\mu}, e_{\nu k}].$$

Inserting here $q_{\beta\nu} \simeq D\delta_{\beta\nu}$ from (9.35) and making use of (9.43) one finds Eq. (57) of Ref. [18b], i.e.

$$i[p_{\alpha\mu}, q_{\beta\nu}] = \frac{1}{2}(\delta_{\alpha\beta}\delta_{\mu\nu} + \delta_{\alpha\nu}\delta_{\mu\beta}). \tag{9.44}$$

Now in expressing the main term proportional to $\sum(\pi^0_{\alpha k})^2$ of the kinetic energy (9.21) in the new variables one runs into the complication that the $p_{\alpha\beta}$ are not Hermitian while the $\pi^0_{\alpha i}$ are. Writing for any operator O the Hermitian average as $\langle O \rangle_H = (O + O^+)/2$, one finds with (9.26), (9.39), (9.42) $\langle e_{\beta k}L^{\alpha\beta}_0 \rangle_H = (1/2)(e_{\alpha i}L^0_{ik} + L^0_{ik}e_{\alpha i})$ and, after much algebra using Eqs (9.40), (9.42),

$$\sum_{\alpha i}(\pi^0_{\alpha i})^2 = \sum_{\alpha\beta}\langle p_{\alpha\beta}\rangle^2_H + \frac{1}{2D^2}\left((L^0)^2 + \frac{3}{2}\right), \tag{9.45}$$

which corresponds to Eqs (72a) and (42) of Ref. [18b], including a real additive constant—which Pauli suppresses. Inserting this result in Eq. (9.22) and combining with (9.14), (9.15), (9.33) one finally arrives at the result

$$H = \frac{(\hbar c)^2}{2}\left\{\lambda\sum_{\alpha\beta}\langle p_{\alpha\beta}\rangle^2_H + \frac{\lambda}{2D^2}\left((L^0)^2 + \frac{3}{2}\right) + \sum_\alpha \int d^3x(\pi''_\alpha)^2\right\}$$
$$+ E^0_{min} + \frac{1}{2I}\sum_{\alpha\beta}(q'_{\alpha\beta})^2 + \frac{1}{2}\sum_\alpha \int d^3x\varphi'_\alpha(-\Delta + \kappa^2)\varphi'_\alpha, \tag{9.46}$$

which corresponds to Eqs (76), (77) of Ref. [18b]. In the latter the coefficient of the first term, however, is N instead of λ and there is still a linear term in π''_α left over, which is due to the fact that in Ref. [18b] the Bloch–Nordsieck transformation was made much later in the calculation. For the estimation of the isobar energies, and also the magnetic moment, these terms do not matter, however.

Finally, the diagonalization transformation (9.30) must be applied to the whole Hamiltonian. To this end it is convenient to keep the isospin y-space fixed and apply the transformation only to x-space. This means setting $B = 1$ and $A = \bar{e}$. On $\vec{\sigma}$ this transformation may be expressed by a unitary operator S defined by Eqs (65a,b) of Ref. [18b], i.e. by

$$\vec{\sigma}' = \overline{A}\vec{\sigma} = S\vec{\sigma}S^{-1}; \quad \vec{\tau} = S\vec{\tau}S^{-1}. \tag{9.47}$$

When applied to the Hamiltonian (9.46) this means that we have to calculate $S(L^0)^2S^{-1}$. In Ref. [18b] this is done in the two appendices—which probably were the main contribution of Dancoff. The problem is to find the transformation $x \to x' = \overline{A}x$ which brings the x-coordinate system into coincidence with the y-coordinate system, $x' = y$, A being parametrized by Euler angles φ, ϑ, ψ. Since the calculation is complicated I will treat only the simpler case where $x_3 = y_3$. Then the only Euler angle is φ, and the transformation (9.47) is given by $S = \exp(i\sigma_3\varphi/2)$. Since $L^0_3 = \partial/i\partial\varphi$ one finds $SL^0_3S^{-1} = L^0_3 - \sigma_3/2$. In the general case of three Euler angles this result is true for all three components of $\mathbf{L}^0$, which is Eq. (68) of Ref. [18b] or

$$\mathbf{L}' = S\mathbf{L}^0 S^{-1} = \mathbf{L}^0 - \frac{1}{2}\vec{\sigma}.$$ (9.48)

This shows that $\mathbf{L}'$ takes half-odd values j and $\mathbf{L}'^2 = j(j+1)$, the ground state having the value $j = 1/2$, and the charge value (9.23) is confined to $-j \leq \epsilon \leq j$ (Eq. (74) of Ref. [18b]). Now all the ingredients are assembled to write down the isobar excitation energy. It is the part proportional to $\Delta(\mathbf{L}'^2)$ in the Hamiltonian SHS^{-1} where H is given by Eq. (9.46). Hence

$$\Delta E = \left(\frac{\hbar c}{2}\right)^2 \frac{\lambda}{D^2}\left(j(j+1) - \frac{3}{4}\right),$$ (9.49)

which is Eq. (80) of Ref. [18b]. The prefactor in the result (9.49) is, up to a number, given by $\kappa a/g^2$ for a small extension of the source, $\kappa a \ll 1$, and by $(\kappa a)^5/g^2$ for $\kappa a \gg 1$.

While ΔE is determined by the terms proportional to g^{-2} in the Hamiltonian (9.46), meson scattering is obtained from the terms proportional to g^0. In appropriate variables this part turns out to be a Hamiltonian describing a free meson field. That, in spite of this, scattering occurs, has its origin in the orthogonality relations (9.11), (9.18) applied to the new variables (Eqs (87a,b) of Ref. [18b]), the former being incorporated into the equations of motion (91) with the help of Lagrangian multipliers. The result is a scattering amplitude A, introduced in Eq. (95) which, in accord with the underlying part of the Hamiltonian, is independent of the coupling constant g. The same is then true of the total (elastic plus inelastic) cross-section, given in Eqs (103, 104) of Ref. [18b]. This is a typical feature of the strong-coupling approach which contrasts with the result obtained by perturbation theory, where cross-sections are found to be proportional to g^2. The latter had been calculated for Pauli in the case of charge-exchange scattering by F. Adler and are quoted in the Introduction, Ref. [18b], p. 86.

In his letter [642] to Uhlenbeck of 4 October 1941 Pauli wrote: 'Here is Dr. Adler, which (*sic!*) I have to occupy in some reasonable way. If you think that there is some reasonable problem in this field (cosmic rays, etc.) please let me know, I think he could be very useful' (Ref. [2], p. 111). And on 22 February 1942 Pauli told Oppenheimer: 'My last collaborator here is F. Adler, a younger theoretical physicist and *Swiss* citizen. He has a finite amount of private money left but is searching for some job (teaching or research). . . . I can recommend him with good conscience' (Ref. [2], letter [650], p. 132). Felix T. Adler (b. 1911) had come to the Institute for Advanced Study early in 1941 and stayed there until summer 1942. He had obtained his doctorate under Wentzel at the University of Zurich in 1938 and was working mainly in neutron physics.

On 12 November 1941 Pauli wrote to Serber: 'Dancoff and I made great progress with the charged pseudoscalar theory, both with the isobar separation and with the [nuclear] forces' (Ref. [2], letter [643]). And to Oppenheimer he wrote: 'I have worked out many details myself and I think that I have now a good knowledge of the whole formalism' (letter [644]). The result (9.49) for the isobar separation is quoted in several of Pauli's letters, for the first time to Oppenheimer on 12 December 1941 (Ref. [2], letter [645]). In the earlier letter [638] to Fierz of 22 July 1941 Pauli still quotes the classical result proportional to $(j^2 - 1/4)$. Equation (9.49), as well as the calculated total scattering cross-sections, may be fitted to experimental values. However, the prediction in Ref. [18b] that the magnetic moments of the proton and neutron are equal, but have opposite sign, is in contradiction with measurements. Pauli emphasized this again in his communication to the New York meeting of the APS on 22–23 January 1943 [21a], where he now advocated a theory with a classical point source and use of the so-called λ-limiting process of Wentzel [21b] and Dirac [21c] to be described in the following section. This and other reasons were such that Pauli's grand construction of a systematic strong-coupling theory did not receive the attention it would have deserved.

Two more applications of this theory were to follow, however, both devoted to the calculation of the nuclear forces. The first, Ref. [22a], by Serber and Dancoff was submitted on 23 December 1942 from the University of Illinois in Urbana, to where Dancoff returned after his collaboration with Pauli and brought with him the detailed knowledge of Ref. [18b]—the authors of Ref. [22a] thanked Pauli for permission to use his formalism. Sidney Dancoff received his Ph.D. from the University of California at Berkeley in 1939. He came to the University of Illinois as instructor the following year and then went on a National Research Fellowship to collaborate with Pauli in 1941–2. The period of 1943–5 he spent as senior physicist at the 'Metallurgical Laboratory' in Chicago, another institution that had been set up for war research and where Enrico Fermi built the first nuclear reactor. Dancoff returned to Urbana in 1945. He spent the year 1950/1 on sabbatical leave with a Guggenheim Fellowship, again at the Princeton Institute. But he died of cancer shortly after, 37 years old, leaving behind a wife and two daughters. Dancoff had been considered one of the most brilliant theoretical physicists of his generation [22b].

The second paper making use of the formalism of Ref. [22a] was by Pauli and his new collaborator Kusaka and was submitted on 24 March 1943. While Serber and Dancoff examined the 'freezing' of the exchange force, meaning that there exists a minimum of the distance R between the nucleons, at which the isospin dependence of the force disappears, the authors only considered spin-0 fields. Pauli and Kusaka, on the other hand, studied the singularity at $R \to 0$ of the force. While the Yukawa potential (8.74) has the same, integrable $1/R$ singularity as the Coulomb potential, the generalization of the integral (9.12) to two nucleons involves two derivatives of $1/R$ and leads to a non-integrable $1/R^3$ singularity. A method to overcome this difficulty was proposed by Møller and Rosenfeld in 1940, namely to admix to the pseudoscalar field a vector field with the same coupling constant, by which the bad singularity is made to disappear [24a]. Before addressing this theory himself, Pauli used to call it somewhat derisively the Møller–Rosenfeld '*Patentmixtur*' (letters [661] and

[662], Ref. [2], pp. 153 and 157). Later, in letter [672] to Oppenheimer already quoted in the last section, he was vehemently negative, saying that 'it is not satisfactory that in the meson theory one needs a vector meson, which is not derived in the theory itself, besides the pseudoscalar meson to explain the nuclear forces' (Ref. [2], p. 183).

Dancoff had left Princeton on 8 February 1942 for Urbana (letter [649]), and Pauli missed him 'because I was getting accustomed to speak with him about the subject' (letter [650], Ref. [2], p. 131). Dancoff's successor as Pauli's collaborator was Shuichi Kusaka (1915–47) whom Oppenheimer had proposed to send him. However, as Pauli told the latter on 9 February 1942, 'here is no money available from the Institute'. And he asked about Kusaka's 'scientific qualities' (Ref. [2], letter [649]). Then on 9 April Pauli told Oppenheimer 'Kusaka's arrival here was very dramatical. I think he is well' [652] and, two weeks later, 'Kusaka is working well here' [654]. 'Fortunately I have two collaborators, Jauch and Kusaka', Pauli wrote to Bhabha on 16 March 1943, since 'there are very few people left with whom I can talk physics' (Ref. [2], letter [670]).

Kusaka's collaboration led to Pauli's second publication on strong-coupling meson theory [23]. Apart from the nuclear force problem already mentioned, this paper contains a detailed spectroscopic analysis of the deuteron in the symmetrical pseudoscalar theory, as well as a calculation of the deuteron magnetic moment. The result for the latter is deceiving: only a small 'orbital' contribution is obtained. Thus, again, strong coupling gave no satisfactory results. Kusaka reported this work at the New York meeting of the APS in January 1943 [24b], after Pauli's communication [21a]. When Kusaka left Pauli later that year his financial problems continued to bother him. Director Aydelotte tried hard to find a teaching position in a small university for him, as Kusaka reported to Oppenheimer on 13 April 1943 (Ref. [2], comment, pp. 127–8). On 29 June Pauli wrote to Jauch from Saranak Lake near Lake Clear in the Adirondacks where he and Franca spent the summer vacation: 'Kusaka has written me. . . . He now goes to North Carolina and in the fall will perhaps assume a position at the Smith-College' (translated from Ref. [2], letter [673]).

Shuichi Kusaka was 5 years old when his parents emigrated from Osaka to British Columbia where he later attended the university in Vancouver. After more studies at MIT he went to Berkeley to work with Oppenheimer. Following his collaboration with Pauli he obtained a teaching position at Smith College, Northampton, Massachusetts. However, after the declaration of war against Japan, Japanese nationals in the United States were considered hostile aliens. So Kusaka's application for an extension of his stay in the United States was first refused. Finally, he was asked to do defence work at the Aberdeen Proving Grounds, New Jersey. After the war he became assistant professor at Princeton University and resumed his research work. However, on 31 August 1947 he drowned while bathing at Beach Haven, New York (Ref. [2], comment, pp. 127–8). Pauli had commented on Kusaka's difficulties with the US authorities in his letter [677] to Jauch from Saranac, New York (not Princeton) on 3 August

1943: 'This method of the bureaucracy of this country (*hiesige*) has reminded me very much of the [Swiss] Federal foreign police' (translated from Ref. [2], p. 190).

From Zurich Pauli had news from his father who had gone there in 1938 (see pages 10–11), and from Scherrer, who 'writes to me very sentimentally, that he misses me, that the Institute needs me' (Ref. [2], letter [668]). Surprisingly, the exchange of information between Pauli and Wentzel remained unimpaired by censorship. On 16 October 1942 Pauli wrote to Wentzel: 'For the interesting communication about Rohn and about H[eisenberg] many thanks. Please give the latter my greetings when he comes to Zurich in December' (translated from Ref. [2], letter [666], p. 167; see also letter [668]). This information to Pauli about the (first) invitation of Heisenberg to Zurich during the war that Scherrer had occasioned [25a] gave rise to an adventurous plan concocted by Weisskopf who recalls: 'This overpowering fear of the Nazis beating us in the race for the atom bomb caused me to react in a bizarre fashion to the news that Heisenberg would visit Switzerland in spring [17 to 25 November, not spring!] 1942. I wrote a letter to Oppenheimer suggesting that Heisenberg be kidnapped during his Swiss visit, and I even offered my services in carrying out this scheme. Oppenheimer took it seriously enough to forward the suggestion to the military authorities. Fortunately this ill-conceived endeavor never took place' (Ref. [25b], p. 119). Weisskopf's letter to Oppenheimer of 18 October 1942 is quoted on p. 166 of Ref. [2].

Heisenberg's second visit to Zurich in December 1944 is mentioned in Pauli's letter [727] to Delbrück of 27 April 1945. Cassidy gives a detailed account of it in Chapter 25 of his Heisenberg biography. He writes: 'about 20 people, including [Morris] Berg [who in late 1944 was assigned to the Office of Strategic Services at the US Embassy in Berne which was directed by Allan Welsh Dulles, the brother of John Foster Dulles, the Secretary of State in the Eisenhower Administration, and an acquaintance of Scherrer's] and several pro-German Swiss scientists, attended the lecture in the Polytechnic's [ETH] physics institute. Although Heisenberg was now practiced in treading warily before a lecture audience, he apparently managed to get himself into trouble at a private dinner party in the Scherrer home. Berg sat next to Heisenberg with open ears and a loaded pistol but was disappointed: Heisenberg's main indiscretion, later reported all the way up to Roosevelt, was a defeatist remark about Germany's failing war fortunes. . . . "How fine it would have been if we had won this war."' (Ref. [12b], p. 492).

In the 'actual world' the war ends

Personally, however, I have now the greatest doubts whether the idea of the introduction of an hypothetical world with its indirect connection with physics is really the correct way to further progress. It seems to me more likely that it will not be possible to formulate a satisfactory quantum electrodynamics, which contains an arbitrary value of $e^2/\hbar c$. [26]

On 19 June 1941 Dirac had the honour of presenting the famous Bakerian Lecture; these lectures of the Royal Society of London had been founded in 1774. One may wonder how many in the audience were able to follow Dirac's presentation, since even Pauli, when he saw the proceedings of it early in May the following year [27], warned Jauch, 'don't be frightened: it is very difficult to understand because it was not entirely clear to the author himself. I am in correspondence with Dirac about it' (translated from Ref. [2], letter [660]). On 29 July 1942 Pauli again wrote to Jauch, who had found Dirac's paper 'mystical', saying that Dirac's 'formalism may be right or wrong, but it is in no case mystical. However, it took me about two months until I understood it' (translated from Ref. [2], letter [662]). And three days later he told Bhabha: 'With great efforts I have at least partially understood Dirac's idea and have exchanged two letters about it with him. I intend to work out some examples' (Ref. [2], letter [663], p. 160).

At first, however, Pauli was 'very enthusiastic about it', as he told Dirac on 6 May 1942 (Ref. [2], letter [656], p. 142). Dirac had introduced a 'hypothetical world', in which he formulated unphysical rules governing quantum electrodynamics (QED), and these rules he connected by a physical interpretation to the 'actual world'. By this artefice Dirac hoped to get rid of the most serious divergences of the theory, noting that in QED divergent integrals are of the form

$$\int_0^\infty f(\omega)d\omega; \quad \lim_{\omega \to \infty} f(\omega) \propto \omega^n, \tag{9.50}$$

where the main values of n are $1, 0, -1$. And he also noted that at least divergences with odd n—an example is given in Eq. (8.69)—are eliminated in a QED which treats positive and negative energies $\pm\hbar\omega$ symmetrically. Pauli called them 'typical quantum effect[s]' (Ref. [2], letter [763]).

Considering the Pauli–Weisskopf theory of spin-0 fields as a guide, Dirac noted that, in a one-particle formalism, the particle density ρ given by Eq. (8.23) may be used to define, for zero-mass particles, a scalar product by

$$\langle a|b \rangle \equiv i \int d^3x \psi_a^*(x) \Upsilon_{x_o} \psi_b(x); \quad \Box \psi_{a,b} = 0, \tag{9.51}$$

where Υ_t stands for '$\partial/\partial t$ acting to the right, minus $\partial/\partial t$ acting to the left'. It then follows that, expressed in Fourier components $\varphi(\mathbf{k})$,

$$\int d^3x \rho(\mathbf{x}) = \langle | \rangle_{x_0=0} = V^{-1} \sum_{\mathbf{k}} |\varphi(\mathbf{k})|^2 (k_0' - k_0); \quad k_0^2 = k_0'^2 = \mathbf{k}^2. \tag{9.52}$$

Since ρ represents the one-particle probability density, Eq. (9.52) clearly introduces negative probabilities, together with negative energies. As mentioned, the merit of this extension of the theory to negative energies is that it gives rise to the passage from expression (9.50) to $\int_\infty^\infty f(\omega)d\omega$ and thereby gets rid of the singularities with odd n. Singularities with even n, on the other hand, may be taken care of by the λ-limiting process to be described subsequently. For spin-1/2 particles, on the other hand, the particle density given in Eq. (6.32) is never negative, $\rho = \psi^* \psi \geq 0$, and hence only positive probabilities occur in this case.

Dirac then introduced the electromagnetic potentials and the vacuum state in such a way that positive- and negative-energy photons refer, in the actual world, to the emission and absorption of photons, respectively. In the last of three very formal appendices giving the details of this theory, the probability that there are n photons in the initial state is found to be $P_n = 2(-1)^n$. Now, according to Einstein's theory of black-body radiation (see e.g. Ref. [28], Section 23), the coefficients of (induced + spontaneous) emission and of absorption are related to the expressions $a \equiv \sum_{n=0}^{\infty}(n + 1)P_n$ and $b \equiv \sum_{n=0}^{\infty} nP_n$, respectively. Writing $P_n = 2(\epsilon - 1)^n$, ϵ being a small positive number, one readily finds in the limit $\epsilon \to 0$ that $a = +1/2$ and $b = -1/2$, so that in the 'actual world' both emission and absorption coefficients are indeed positive. To this Pauli made the comment to Dirac that he finds it 'very objectionable, that in your new form of quantum-electrodynamics Einstein's laws for radiation processes are *not derived* but are an additional assumption' (Ref. [2], letter [659] of 27 June 1942). Pauli later withdrew this criticism (see letter [665], Ref. [2], p. 161); it was indeed unjustified!

As mentioned in the last section, the λ-limiting process was introduced by Wentzel [21b] and by Dirac [21c]. It is a relativistically invariant regularization procedure (see page 307) applied to the electromagnetic interactions. The latter have singularities on the mutual light cones $z_i - z_j = 0$ of the charges with position four-vectors z_i which are of the form $\delta((z_i - z_j)^2)$. These singularities may also be expressed in terms of the Jordan–Pauli function (6.18) since, as follows from Eq. (6.19),

$$D(x) = -\frac{\text{sgn } x_0}{2\pi c}\delta(x^2). \qquad (9.53)$$

In the variational principle developed by Dirac in Appendix I of Ref. [27] one may assume that in the initial and final states all charges lie outside each other's light cones, which in the metric $x^2 \equiv \mathbf{x}^2 - x_0^2$ means $(z_i - z_j)^2 > 0$. If $\pm\lambda$ is now a small time-like four-vector, $\lambda^2 < 0$, which when added to $z_i - z_j$ does not alter the latter's space-like character, $(z_i - z_j \pm \lambda)^2 > 0$, but displaces the singularities away from the light cones, then, as Dirac showed in Ref. [21c], the divergences of the form (9.50) with even n may be eliminated.

It was said above that Pauli 'work[ed] out some examples' of Dirac's theory, which he later published [29]. This paper also mentions in a footnote that Pauli had lectured on it at the Purdue summer school (see the previous section). And many 'listeners urgently asked me to write a paper about it in the Review[s] of Modern Physics, now I have strong inhibitions to write this paper, because I don't think that all principal points are really clear enough' he wrote on 20 November 1942 in a letter for Max Born's 60th birthday, which begins in old Paulian wit as follows: 'I have in my mind the clear idea not of an ordinary birthday-cake with 60 candles but of a large "*Lebkuchen*" [gingerbread] in the form of a mathematical symbol, let us say B, with 60 different indices' (Ref. [2], letter [668]). The principal merit of this paper [29] is to have generalized Dirac's scalar product (9.51) into a systematic treatment of an indefinite metric in Hilbert space to handle negative probabilities. In the early 1950s this formalism became very important indeed, first in the Bleuler–Gupta formulation of QED (see Ref. [30], Section 15c), then in Pauli's response to Feynman's criticism of the spin-statistics theorem mentioned on page 336 and finally in Pauli's treatment of the Lee model [31].

Pauli generalized Dirac's scalar product (9.51) by introducing an Hermitian metric tensor $\eta = \eta^+$ such that, for any operator A,

$$\langle a|\eta A|b\rangle^* = \langle b|\eta A^*|a\rangle. \tag{9.54}$$

Here $A^* \equiv \eta^{-1}A^+\eta$ is the adjoint of the operator A and is different from the Hermitian conjugate A^+. Observables therefore must be self-adjoint instead of Hermitian, $A = A^* \neq A^+$. On the other hand, since the metric operator η is Hermitian, it can be diagonalized, and in a discrete basis $|n\rangle$, by definition, has the eigenvalues ± 1 which then are the probabilities of these states. Pauli then showed in Section 4 of Ref. [29] in the model of two self-adjoint harmonic oscillators, one of which is Hermitian and the other anti-Hermitian, that there are two equivalent 'worlds' characterized by numbers of quanta N_+, N_- and N_a, N_b, respectively, of the two oscillators, such that $N_+ - N_- = N_a - N_b$. He then calculated the transformation function $(N_aN_b|S|N_+N_-)$ between the two worlds and found for the probability that there are $N_a = N_b = n$ quanta in the '(a,b)-world' when there are $N_+ = N_- = 0$ quanta in the '$(+-)$-world', Eq. (68) of Ref. [29], namely

$$P_n = \eta_n|(nn|S|00)|^2 = 2(-1)^n, \tag{9.55}$$

which is Dirac's result mentioned above.

In generalizing to fields, Pauli refers to Part II of his earlier review [32]. Now, however, positive and negative frequencies, $k_0 = \pm\omega$ (see Eqs (8.26)), are associated with Hermitian and anti-Hermitian oscillators, respectively. In QED this leads, as in Dirac's theory [27], to two electromagnetic potentials, Eqs (97) of Ref. [29],

$$A_\nu(x) = \frac{1}{\sqrt{2}}(U_\nu(x) + U_\nu^*(x)); \quad B_\nu(x) = \frac{1}{\sqrt{2}}(U_\nu(x) - U_\nu^*(x)), \tag{9.56}$$

which are expressed in terms of the auxiliary fields in the '$(+-)$-world',

$$U_\nu(x) = \sum_\mathbf{k} \frac{1}{\sqrt{2V\omega}}(U_{\nu+}(\mathbf{k})e^{-i\omega t} + U_{\nu-}(-\mathbf{k})e^{i\omega t})e^{i\mathbf{k}\cdot\mathbf{x}}. \tag{9.57}$$

The parenthesis in (9.57) is called $\mathbf{R_k}$ in Ref. [27]. But Pauli replaces Dirac's definition (15) of the vacuum state Ψ by Fermi's expression (6.42), $\partial U_\nu(x)/\partial x^\nu$ $\Psi = 0$, his Eq. (98).

Next Pauli introduces in Ref. [29] the λ-limiting process by the following modification of the commutator (6.17):

$$[A_\mu(x), A_\nu(x')] = \frac{i\hbar c^2}{2}\delta_{\mu\nu}(D(x - x' + \lambda) + D(x - x' - \lambda)) \qquad (9.58)$$

where, as before, λ is a small time-like four-vector. Now it follows from Eq. (6.18) that

$$\frac{1}{2}(D(x + \lambda) + D(x - \lambda)) = -\int \frac{d^3x}{(2\pi)^3}e^{i\mathbf{k}\cdot\mathbf{x}}\frac{\sin \omega_\mathbf{k}t}{\omega_\mathbf{k}}\cos(k\lambda)$$

which, according to Eq. (6.13), gives rise to a weight factor in the commutator (6.14),

$$i[b(\sigma\mathbf{k}), b^+(\sigma'\mathbf{k}')] = \cos(k_0\lambda_0 - \mathbf{k}\cdot\vec{\lambda})\delta_{\sigma\sigma'}\delta_{\mathbf{k}\mathbf{k}'}, \qquad (9.59)$$

where $k_0 \equiv \sqrt{k^2 + \kappa^2}$ and $k \equiv |\mathbf{k}|$. This is Eq. (147) of Ref. [29], which Pauli then applies to Weisskopf's calculation of the self-energy of the electron (see page 301). However, in spite of having succeeded in casting Dirac's theory [27] into a rigorous form fit to do real calculations, his conclusion about the problem of interpretation remains entirely negative: 'The arbitrariness of the rules for the translation of results concerning the hypothetical world into results concerning the actual world and the lack of uniqueness of these rules seems to indicate that new ideas and more radical changes of the present formalism will be necessary in order to get a really satisfactory quantum theory of the electromagnetic field' (Ref. [29], p. 207).

Pauli now applied the λ-limiting process to meson theory, as he had announced in his communication to the New York meeting of the APS on 22–23 January 1943 (see the last section). There he had observed that 'Such a weak coupling theory with point sources avoids all difficulties of the strong coupling theory with extended source' [21a]. The ensuing paper [33], submitted on 14 October 1943, starts with the simplest case of a neutral pseudoscalar meson field defined by the interaction (9.4), but without the index α and with τ_α replaced by one. To be suitable for the application of the λ-limiting process the formalism, to start with, has to be supplemented with some unphysical features. The latter are best introduced in the Fourier representation of the shape function $U(\mathbf{x})$ and that of the fields,

$$\varphi(\mathbf{x}) = \int \frac{d^3k}{(2\pi)^{3/2}}q(\mathbf{k})e^{i\mathbf{k}\cdot\mathbf{x}}; \quad \pi(\mathbf{x}) = \int \frac{d^3k}{(2\pi)^{3/2}}p(-\mathbf{k})e^{i\mathbf{k}\cdot\mathbf{x}}. \qquad (9.60)$$

Instead of the reality conditions for the fields and for the form factor g in (8.49), called v in Ref. [33], one requires the weaker conditions

$$g(\mathbf{k})\xi(-\mathbf{k}) = (g(-\mathbf{k})\xi(\mathbf{k}))^*; \quad \xi(\mathbf{k}) = q(\mathbf{k}),\ p(-\mathbf{k}),\ g(\mathbf{k}) \qquad (9.61)$$

and

$$q(\mathbf{k})q(-\mathbf{k}) = (q(\mathbf{k})q(-\mathbf{k}))^*; \quad p(\mathbf{k})p(-\mathbf{k}) = (p(\mathbf{k})p(-\mathbf{k}))^*. \qquad (9.62)$$

The latter guarantee the reality (Hermiticity) of the Hamiltonian, which follows from Eqs. (9.4), (9.7) with the mentioned modifications,

$$H = \frac{1}{2} \int d^3k \{ p(\mathbf{k})p(-\mathbf{k}) + k_0^2 q(\mathbf{k})q(-\mathbf{k}) \} + \frac{if}{\pi\sqrt{2}} \int d^3k g(-\mathbf{k})\vec{\sigma} \cdot \mathbf{k}q(\mathbf{k}),$$
(9.63)

where $f = g\sqrt{2\pi}/\kappa$ and units are chosen such that $\hbar = c = 1$. The commutator (9.8) becomes

$$i[p(\mathbf{k}), q(\mathbf{k}')] = \delta(\mathbf{k} - \mathbf{k}').$$
(9.64)

Note that in Ref. [33] this commutator is written as Poisson bracket (5.29), and hence without the factor i.

Pauli's trick to introduce the λ-limiting process now is as follows. Introduce new variables

$$\tilde{q}(\mathbf{k}) = g(-\mathbf{k})q(\mathbf{k}); \quad \tilde{p}(\mathbf{k}) = g(\mathbf{k})p(\mathbf{k})$$
(9.65)

which, according to Eqs (9.61), are real (Hermitian) and satisfy the weighted commutation relation

$$i[\tilde{p}(\mathbf{k}), \tilde{q}(\mathbf{k}')] = G(\mathbf{k})\delta(\mathbf{k} - \mathbf{k}'),$$
(9.66)

where, according to (9.61),

$$G(\mathbf{k}) \equiv g(\mathbf{k})g(-\mathbf{k})$$
(9.67)

is also real. Comparing with Eq. (9.59) it is evident that the λ-limiting process may now be introduced by the identification

$$G(k) = \cos(k_0\lambda_0 - \mathbf{k} \cdot \vec{\lambda}),$$
(9.68)

which depends only on $k = |\mathbf{k}|$. Equation (9.68) is possible since, contrary to Eq. (8.49) where $g(-\mathbf{k}) = g^*(\mathbf{k})$ and hence $G(\mathbf{k}) \geq 0$, here $G(k)$ may also assume negative values.

In order to see the action of the λ-limiting process, let us calculate the nucleon radius and the static nucleon self-energy. Inserting (8.49) into Eq. (9.6) one finds

$$a^{-1} = \frac{1}{2\pi^2} \int \frac{d^3k}{k^2} G(k) = \frac{2}{\pi} \int_0^\infty G(k)dk.$$
(9.69)

Inserting (9.68) here and setting $\vec{\lambda} = 0$ one finds

$$a^{-1} = \frac{2}{\pi} \int_0^\infty \cos(k_0\lambda_0)dk = \kappa J_1(\kappa\lambda_0) = \kappa \left\{ \frac{\kappa\lambda_0}{2} - \frac{(\kappa\lambda_0)^3}{16} + - \cdots \right\},$$
(9.70)

where J_1 is a Bessel function. Hence, $\lim_{\lambda_0 \to 0} a^{-1} = 0$, instead of $a^{-1} = \infty$ in the limit of a point source. From the Hamiltonian (9.63) one derives the equation of motion

$$\frac{\partial^2 \tilde{q}(\mathbf{k})}{\partial t^2} + k_0^2 \tilde{q}(\mathbf{k}) = \frac{if}{\pi\sqrt{2}} G(k)\vec{\sigma} \cdot \mathbf{k}, \tag{9.71}$$

from which the self-energy of the static nucleon is obtained by setting $\vec{\sigma} = (0, 0, 1)$. Hence $\tilde{q}(\mathbf{k}) = (if/\pi\sqrt{2}k_0^2)G(k)k_3$, which expression, when inserted in the last term of Eq. (9.63), yields

$$E_0 = -\frac{2f^2}{3\pi} \int_0^\infty G(k)\frac{k^4 dk}{k^2 + \kappa^2} \tag{9.72}$$

(which differs by a factor 2 from Eq. (29a) of Ref. [33]). Writing $k^4/(k^2+\kappa^2) = k^2 - \kappa^2 + \kappa^4/(k^2 + \kappa^2)$ and setting $G(k) = 1$ in all convergent integrals, one obtains with

$$N \equiv \frac{2}{\pi} \int_0^\infty G(k)k^2 dk; \quad \int_0^\infty G(k)\frac{dk}{k^2 + \kappa^2} = \int_0^\infty \frac{dk}{k^2 + \kappa^2} = \frac{\pi}{2} \tag{9.73}$$

the result

$$E_0 = -\frac{f^2}{3}(N - \kappa^2 a^{-1} + \kappa^4) \tag{9.74}$$

(which differs from Eq. (33a) of Ref. [33]). Applying now the λ-limiting process also to N one finds by acting with $(\partial/\partial\lambda_0)^2$ on Eq. (9.70)

$$N + \kappa^2 a^{-1} = -\left(\frac{\partial}{\partial\lambda_0}\right)^2 \kappa J_1(\kappa\lambda_0) \tag{9.75}$$

and, hence, also $\lim_{\lambda_0 \to 0} N = 0$. Therefore, the λ-limiting process renders the static nucleon self-energy finite, $E_0 = -(f^2/3)\kappa^4$.

Jauch used this procedure together with Dirac's negative-energy oscillators to calculate the nucleon magnetic moment in the pseudoscalar theory and also found a convergent result—which he reported at the New York meeting of the APS in January 1943 [34a]. One year later Pauli and Jauch reported at the New York meeting on 14–15 January 1944 calculations in which Dirac's formalism was applied to the problem of the emission of low-energy photons (see pages 317–8). In particular, they used the rule that in the 'actual world' only positive-energy photons are emitted and treated the form factor $g(\mathbf{k})$ by the method described above [34b]. Their result was not satisfactory, however, and they concluded that 'It does not seem that the difficulty can be removed without arbitrary changes of the rules applied for the physical interpretation of Dirac's formalism' [34b]. Pauli came back to this problem once more after returning to Zurich in the spring of 1946, where he presented a communication on 'Dirac's Field Quantization and Emission of Photons of Small Frequency' at the meeting of the Swiss Physical Society in Aarau on 4 May 1946 [34c]. In this note Pauli concluded more explicitly that Dirac's prescriptions for the 'actual world' mentioned above are equivalent to replacing the exponential factor (8.46) by one, and that this leads back to the situation of perturbation theory with all its shortcomings.

At the New York meeting in January 1944 Pauli met Delbrück again.'It is not yet decided whether I return to Princeton for the night or whether I spend the night with my sister (in New York)', he had written Delbrück on 4 January and that 'in a touch of thirst for knowledge' he had bought Huxley's book *Evolution*, 'and often already I would have liked to misuse you for the explanation of some technical terms' in it. At the end of the letter Pauli mentions 'a very old letter from Dorette Wagner . . . from Switzerland' which had reached him on 28 December. And he closes with 'More secrets by word of mouth' (translated from Ref. [2], letter [687]). Is this the 'Dora' on page 241? In May 1944 Pauli was invited to give a 'physics Colloquium' at Harvard University [693]. And on 8 December he was in Pittsburgh for the celebration of Otto Stern's Nobel Prize [708].

Kusaka's successor as Pauli's collaborator was Ning Hu (b. 1915) from China who arrived in Princeton in November 1943, sent by Paul Sophus Epstein (1883–1966) from the California Institute of Technology in Pasadena (see page 352) where Hu obtained his Ph.D. in 1944. Epstein had asked Pauli whether Hu could work with him. Pauli discussed the case with his colleague and friend Hermann Weyl who as permanent member of the Princeton Institute then wrote directly to Epstein on 27 October 'that Dr. Hu will be welcome in Princeton if he cares to come and work under Pauli and will be granted admission to the Institute. There is no tuition fee' (Ref. [2], comment, p. 229). The following year a Brazilian graduate student, José Leite Lopes, arrived. Another graduate student, John Markus Blatt (b. 1921) was working with Jauch at Cornell University where the latter was engaged as an instructor (see Ref. [2], letters [658, 680]). All were working on the meson theory of nuclear forces. Hu succeeded in showing that treating the nucleon relativistically by taking into account its recoil, and also the '*Patentmixtur*' of Møller and Rosenfeld [24a] (see the previous section), gave rise to non-integrable singularities of the type R^{-2} and R^{-3} in the distance R between the nucleons [35a]. Lopes submitted his thesis to Princeton University, thanking both Pauli and Jauch 'for the suggestion of this problem and many valuable discussions' [35b]. Blatt, after his Ph.D. under Jauch at Cornell University [35c], came to Princeton to collaborate with Pauli and Jauch [35d].

As in the previous years Pauli and Franca spent the summer of 1944 in the Adirondacks, not, however, at Lake Clear (see the previous section) but, as the year before, at Saranac Lake, situated on another of the many lakes in this region, about 10 km south-east of Lake Clear. The Paulis used to spend these vacations in the lake region of the Adirondacks together with the mathematician Carl Ludwig Siegel (b. 1896) with whom Pauli went swimming. During the war Siegel also had been at the Institute for Advanced Study and afterwards was professor at Göttingen University [3].

Apparently Pauli had a touch of bad temper with Hu, when he wrote to Jauch on 17 August 'that, as I had predicted in June, Hu is only useful for numerical calculations. I very much wish to throw him out, but I don't know very well to where' (translated from Ref. [2], letter [697]). However, such remarks notwithstanding, Pauli later invited Hu to collaborate with him on a systematic investigation of meson pair theories in strong coupling [36]. Since in the case of a pair theory the Hamiltonian is a quadratic form in the meson field, it may be diagonalized exactly—at least for spin-independent forces—by a very elegant method devised by Wentzel [37]. Making use of the division of the meson field into a part at the nucleon and one outside it, as introduced in the Pauli–Dancoff paper [18b] (see the previous section), this diagonalization is performed for the outside part of the field in Section 4 of Ref. [36] and yields the exact dependence of the nuclear potential on the distance R between the nucleons. The singularity found in this way was R^{-7} which is to be compared with Hu's results [35a] mentioned above. In Section 5 the influence of the spin-dependent part of the interaction on the potential is estimated and found to be small, provided the coupling is strong.

Hu stayed with Pauli till 1945. Later he worked with Heitler at the Institute for Advanced Studies in Dublin and spent some time in Bohr's institute in Copenhagen. Eventually he went back to China where he became associated with the National Peking University. For Pauli the paper [36] with Hu rounded off his activity in meson theory which he found 'to be a bit exhausted now' (Ref. [2], letter [729]). This activity had its peak in an invitation from Rabi to give a series of lectures at MIT in the autumn of 1944. In the letter already quoted Pauli told Jauch that he had received 'an invitation from Rabi to hold a number of evening lectures on "recent developments in physics" for the war physicists of M.I.T. in Cambridge, Mass.' (translated from Ref. [2], letter [697]). To Rabi he suggested starting the lectures at the end of September or in October and proposed as topics 'the present state of the meson theory and nuclear forces' and 'about quantum electrodynamics' (Ref. [2], letter [698]). On 11 September he told Rabi of his agreement for six lectures on Friday or Saturday afternoon. He hoped that Bloch, who worked at the Radiation Laboratory, would be there. 'But the main thing is that you and Uhlenbeck and, of course, Schwinger . . . will really be in Cambridge' (Ref. [2], letter [704]).

Finally, Pauli gave six lectures every fortnight on Saturday evening, starting on 7 October and ending on 16 December 1944 (see Ref. [2], letter [710]). Rigden, in his biography of Rabi writes: 'In order to help the Radiation Laboratory physicists get ready for the resumption of their academic research, Rabi started a series of biweekly seminars given by Pauli, who would survey recent developments in physics. The seminars started with a crowded room, but attendance quickly dwindled' (Ref. [14], p. 246). 'Edward [Mills] Purcell [b. 1912] who also was among the participants reported that Rabi saved the situation by inviting Schwinger to lecture every other week which had much more success' (translated from Ref. [2], comment, p. 246). Pauli was hesitant to have his notes published as a book (see Ref. [2], letter [778]). But finally these lectures appeared as six chapters in two editions [38]. In the 'Preface to the First Edition' Pauli writes: 'Despite the imperfections of my lectures, the original notes written by Dr. J. F. Carlson and Dr. A. J. F. Siegert have been amended only slightly, to preserve the informal character of the lectures and to emphasize the very provisional state of the problems in question.'

After an introductory Chapter I explaining the properties of the mesons and the different types of theories, Chapter II treats the theory of the extended source as developed by Pauli in Ref. [33] and Chapter III is devoted to meson scattering and to the magnetic moments of the proton and neutron. While Chapter V and VI discuss the theory of neutron–proton scattering and of strong coupling of the two-nucleon system, respectively, without, however, touching on pair theories, Chapter IV commands the greatest interest. This is because it is devoted to the most recent advances in general scattering theory represented by Heitler's theory of radiation damping [39a] as described in more recent editions of Heitler's famous book [39b] and Heisenberg's theory of the S-matrix [39c] which became the main tool of the relativistic QED in the late 1940s (see Ref. [30]). Since in this chapter Pauli contributes an essential element of proof and since the S-matrix will turn up again in later chapters, I give here some details on this part of Pauli's lectures.

Both Heitler and Heisenberg consider scattering states with an outgoing scattered wave proportional to $\exp[i(kr - (\omega + i\epsilon)t)]$ which vanishes in the remote past, ϵ being a positive infinitesimal. In momentum space the latter then has the form

$$|\Phi_k^+\rangle \propto \delta_+(E_0 - E_k), \tag{9.76}$$

where $(H_0 - E_0)|0\rangle = 0$ defines a free-particle reference state and

$$\delta_\pm(x) \equiv \frac{1}{2}\delta(x) \mp \frac{1}{2\pi i}\frac{1}{x}. \tag{9.77}$$

From the Schrödinger equation $(H_0 + H' - E_k)\Phi_k = 0$ it then follows that

$$(\Phi_k|0)(E_0 - E_k) = (\Phi_k|H'|0), \tag{9.78}$$

so that we may write quite generally

$$(\Phi_k|0) = (A_k|0)\delta(E_0 - E_k) + 2\pi i(k|f|0)\delta_+(E_0 - E_k). \tag{9.79}$$

Here $|A_k\rangle$ is the projection of the incident wave on the mass shell, $E_k = E_0$, which in the case of damping is different from a plane wave $|k\rangle$ because of the existence of a line width. $(k|f|0)$ is obtained by multiplying Eq. (9.79) by $(E_0 - E_k)$ and equating to (9.78). With the definition (9.77) this leads to

$$-(k|f|0) = (\Phi_k|H'|0). \tag{9.80}$$

Substituting here on the right $|0\rangle = \sum_l |\Phi_l\rangle(\Phi_l|0)$ and using Eq. (9.79) for $(\Phi_l|0)$, iteration on the mass shell leads to Heitler's integral equation (see Ref. [39b], Section 15, Eq. (33)). As Pauli observed in his letter [757] to Rosenfeld, this 'integral-equation method circumvents the divergence difficulties' (Ref. [2], p. 296).

Since the aim here is not radiation damping but the S-matrix, we may replace the first term in (9.79) by an incoming plane wave $(k|0)$. Pauli introduces the S-matrix by defining the scattering amplitude as the outgoing wave part $(\Phi_k^+|0)$ of Eq. (9.79). Since according to (9.77), $\delta(x) = \delta_+(x) + \delta_-(x)$,

$$(\Phi_k^+|0) = \{(k|0) + 2\pi i(k|f|0)\}\delta_+(E_0 - E_k). \qquad (9.81)$$

The S-matrix is then the restriction of the braces in (9.81) to the mass shell,

$$(A|S|0) = (A|1|0) + 2\pi i(A|f|0), \qquad (9.82)$$

or written as operators,

$$S = 1 + 2\pi i f. \qquad (9.83)$$

But this matrix only makes sense if the integrated scattering probability sums to one, i.e. if for any state $|A)$

$$\int_B |(A|S|B)|^2 = \int_B (A|S|B)(B|S^+|A) = 1. \qquad (9.84)$$

A sufficient condition therefore is the unitarity

$$SS^+ = 1, \qquad (9.85)$$

or, according to (9.83),

$$f - f^+ = 2\pi i f f^+. \qquad (9.86)$$

Concerning the proof of Eq. (9.86) Pauli comments: 'The proof to be given is one that was developed by Pauli and Bargmann. It is particularly adapted to this case and does not make use of Heitler's integral equation nor of different times as used in the proof given by Heisenberg' (Ref. [38], second edition, p. 48). It goes like this: write

$$|\Phi_k) = |k) + F^+|k); \qquad (9.87)$$

then it follows from Eq. (9.80) that

$$-(k|f|0) = (k|H'|0) + (k|FH'|0), \qquad (9.88)$$

or written as operators and their Hermitian conjugates

$$-f = H' + FH'; \quad -f^+ = H' + H'F^+. \qquad (9.89)$$

In order to eliminate H', multiply the first equation by F^+ from the right and the second by F from the left. Their difference is then

$$Ff^+ - fF^+ = H'F^+ - -FH' = f - f^+, \qquad (9.90)$$

where on the right Eqs (9.89) have been used. Equation (9.90) may also be written in the form

$$(k|f - f^+|0) = \sum_l \{(k|F|l)(l|f^+|0) - (k|f|l)(l|F^+|0)\}. \qquad (9.91)$$

Now, according to (9.79), (9.87)

$$(k|F|0) = 2\pi i(k|f|0)\delta_+(E_0 - E_k), \qquad (9.92)$$

and hence, restricting Eqs (9.91), (9.92) to the mass shell, so that $(A_l|F|0) = i\pi(A_l|f|0)$, one finds

$$(A|f - f^+|0) = 2\pi i(A|ff^+|0), \qquad (9.93)$$

which is Eq. (9.86).

As mentioned above, Pauli's lectures at MIT were proposed by Rabi 'In order to help the Radiation Laboratory physicists get ready for the resumption of their academic research.' The end of the war was expected in a not too distant future. In the United States Roosevelt, although ill and worn, was re-elected for a fourth term in 1944, and Harry S Truman became vice-president. However, a few months after his inauguration Roosevelt died and Truman took over. On the fronts, by 1944 German defeat had become obvious everywhere. It had already been signalled at the end of 1942 by the Battle of Stalingrad and became realistic with the landing of the Allied Forces in northern France on 6 June 1944. On 8 May 1945 German capitulation occurred. Four days previously, Pauli had written to Bohr after reading in the newspapers 'that the Germans in Denmark have surrendered . . . all my good wishes and congratulations are with you in this joyful event' (Ref. [2], letter [728]).

But the war in the Pacific continued. There the end was hastened by the dropping of the two atom bombs on Hiroshima and Nagasaki on 6 and 9 August. Weisskopf comments: 'Oppenheimer, of course, was very interested in all of these questions of the effect of the bomb on the political future of the world, and he had many discussions with Bohr [when the latter was at Los Alamos]. . . . Oppie [Oppenheimer] was our hero and mentor. He so strongly influenced our thinking that we did not discuss alternatives for the use of the bomb other than the destruction of Japanese cities' (Ref. [41], pp. 147–8). On 2 September 1945 the Japanese capitulated.

The Nobel Prize

Over the following years you helped to pave the way for my relations with the Institute for Advanced Study which had become the permanent seat of your own activity. It was also to become, during the Second World War, a peaceful asylum for me, at a time when Switzerland was not yet my legal home country. After your 60th birthday had been spent in a somewhat restless mood, these difficult war years ended with that feast in Princeton at which Einstein had deeply impressed us and where you also had found such friendly words for me. [42a]

These are Pauli's friendly words pronounced by him in his allocution for Weyl's 70th birthday on 9 November 1955 in Zurich, already quoted several times. They recall the years that Pauli and Franca spent together with Hermann Weyl and his family at the Institute for Advanced Study during the war (see page 39) and allude to the two main social events after the end of the war, just 10 years earlier: Weyl's 60th birthday on 9 November 1945 and Pauli's Nobel feast on 10 December. Niels Bohr's 60th birthday had occurred still one month earlier, on 7 October. For that occasion Pauli had organized and edited a double issue of *Reviews of Modern Physics* which, in 25 contributions, covered many domains of physics and on 250 pages honoured the founder of the quantum theory of the atom [42b]. Among the authors were Einstein and Lise Meitner, and the great absentees were Heisenberg and a vanquished Germany. Pauli's own contribution was his paper [36] with Hu on meson pair theories. But in addition he wrote a biographical eulogy [42c] which contains a concise presentation of the early history of quantum theory and is an excellent example of Pauli's masterly way of describing the Copenhagen Interpretation (see pages 249–51).

At Pauli's fifth lecture at MIT on the 9 December 1944 the decision had been taken that Pauli should be editor of a special issue of *Reviews of Modern Physics* in honour of Bohr's 60th birthday (Ref. [2], letter [710]). Also in December Pauli sent out a typewritten circular letter of invitation to former collaborators and friends of Bohr (see e.g. letter [711] in Ref. [2]). The content of the papers was to be strictly scientific. At the business meeting of the American Physical Society held during the New York meeting on 19–20 January 1945 Pauli was nominated to the Board of Editors of the APS for a three-year term [42d]. On 27 August 1945 Pauli wrote to Rosenfeld that 26 papers had been received (Ref. [2], letter [763]). One was by Rosenfeld himself. However, since it was an essay on Bohr, and not a scientific paper, the acting editor J. William Buchta notified Rosenfeld that his contribution could not be accepted (see letter [757] in Ref. [2]). The Bohr issue 'finally appeared on October 26th. It was technically not possible to speed up the print much more than we actually did and we hope that Bohr himself will also enjoy the journal when he obtains it a bit later than on his birthday. . . . Here the issue is a success and many students will buy it for themselves.' Pauli wrote to Oskar Klein on 2 November (Ref. [2], letter [783]). His editorial efforts were acknowledged by the acting editors of the journal in an 'Editor's Note: The editors of the REVIEWS OF MODERN PHYSICS are greatly indebted to Professor W. Pauli who has collected manuscripts for this issue and who has done much of the editorial work. His collaboration has been invaluable' (Ref. [42b], Table of contents).

At the Institute for Advanced Study social contacts among the institute members were cultivated at lunch in the cafeteria on the top floor or at afternoon tea served every weekday in the Common Room on the main floor and, of course, also privately at dinner parties and other gatherings in Princeton or nearby New York. Pauli's closer contacts among the Faculty members were, apart from Hermann Weyl, the brilliant mathematician John von Neumann (1903–57) and the art historian Erwin Panofsky (1892–1968) whose knowledge of Renaissance art was legendary, and among the other members the mathematicians Kurt Gödel (1906–78) and Carl Ludwig Siegel, then considered the most important mathematician after Weyl (see Ref. [4b], Sections 16.3, 4). Siegel, who counted among Pauli's friends, used to join Pauli and Franca on their summer vacation in the Adirondacks [3].

A wider circle of contacts also included Eugene Wigner (1902–95) at the university. Of the latter, who was known for his extreme politeness, Pauli wrote to Rabi: 'Yesterday I saw Wigner. He said "I am not so polite any longer as I have been before." I said "I am much more polite now than I have been before." Then we shook hands' (Ref. [2], letter [776]). Pauli, apparently, had come to pay more attention to his 'persona' (see page 292). That he thought this to be a serious matter is apparent from letter [759] to Jauch whom he lectured: 'You must consider: persona there must also be in the manners between different people' (translated from Ref. [2], p. 300). Einstein, of course, belonged to Pauli's discussion partners, too, both at the institute and at Einstein's house. Franca, on the other hand, was on friendly relations with Einstein's sister Maja Winteler—they read Dante together—and with stepdaughter Margot Einstein, both of whom lived in the Einstein house at 112 Mercer Street [3].

Out of Pauli's discussions with Einstein sprang a common paper [43a]. It took up a problem that Einstein had investigated a few years earlier, namely that in the absence of matter, $T_{ik} = 0$, Einstein's field equations (2.3) under conditions to be specified should have the Euclidean metric as the only solution. Using Greek instead of Roman indices as on page 30, where the $\Gamma^\alpha{}_{\mu\nu}$ are defined (for the definition of $R_{\mu\nu}$ see Eq. (94) of Ref. [43b]), an equivalent form of these equations reads:

$$R_{\mu\nu} \equiv \frac{\partial \Gamma^\alpha{}_{\mu\alpha}}{\partial x^\nu} - \frac{\partial \Gamma^\alpha{}_{\mu\nu}}{\partial x^\alpha} + \Gamma^\beta{}_{\mu\alpha}\Gamma^\alpha{}_{\nu\beta} - \Gamma^\alpha{}_{\mu\nu}\Gamma^\beta{}_{\alpha\beta} = 0. \qquad (9.94)$$

In Pauli's *Encyklopädie Article* [43b] of 1921 this problem is mentioned on p. 182 where in footnote 365 it is said that this theorem 'has not yet been proved for the general case'. And another footnote refers to the *Supplementary Note 18* where the Einstein–Pauli paper [43a] is mentioned. What Einstein had proven is that there exists no solution of Eqs (9.94) satisfying the following conditions: 1) stationarity $\partial g_{ik}/\partial x^4 = 0$, 2) g_{ik} is singularity free, 3) asymptotically, $g_{44} = -1 + \mu/r$ for $r \to \infty$, r being the distance from the origin of the spatial coordinate system.

Einstein's and probably very strongly also Pauli's motivation for reconsidering the problem was the formal resemblance of the stationarity condition 1) with the cylinder condition (7.3) of the five-dimensional theory of Kaluza and Klein. The authors write: 'When one tries to find a unified theory of the gravitational and electromagnetic fields, he cannot help feeling that there is some truth in *Kaluza's* five-dimensional theory' (Ref. [43a], p. 131). Since in the Kaluza–Klein theory the fifth dimension is assumed to be space-like one faces the problem of a four-dimensional space. On this fact the authors make the following most interesting speculation: 'Assume that in such a theory the fields corresponding to non-singular solutions are not point-like but linearly extended in a four-dimensional space. The geometrical configuration of several coexisting fields of this character would, then, more or less resemble the configuration of the objects of a three-dimensional space' (Ref. [43a], p. 131). But this is just the modern idea of strings to describe elementary particles (see e.g. Ref. [43c]). In Ref. [43a] Einstein and Pauli consider a space of $n \geq 4$ dimensions and show that under the above conditions 1) and 2) and for large r the leading terms of the metric tensor satisfying Eqs (9.94) have the form $g_{ik} = \delta_{ik}(1 + m/r)$, $g_{i\rho} = 0$, $g_{\rho\sigma} = \pm\delta_{\rho\sigma} + m_{\rho\sigma}/r$, where $i, k = 1, 2, 3$ and $\rho, \sigma = 4...n$. They then show that $m = m_{\rho\sigma} = 0$. Hence, the deviations of the metric tensor from the Euclidean one must decrease more rapidly than $1/r$.

Einstein and Pauli's paper [43a] was submitted on 4 January 1943. More than two and a half years later a letter from a young man in Strasburg, France, with the name André Léon Lichnérowicz (b. 1915) [11], reached Pauli, probably still on his summer vacation—which in 1945 was Lake Clear again (Ref. [2], letter [767]). Lichnérowicz wrote to Pauli that he had at last got hold of Einstein and Pauli's paper. And he was keen to inform Pauli that he himself had published a quite similar theorem in 1938 [44a] and in his Ph.D. thesis of 1939 [44b], using, however, a quite different method. Back in Princeton Pauli answered on 21 September. He found Lichnérowicz's assumption that, for $r \to \infty$, g_{44} must approach the value -1 from one side 'very unnatural'. In turn, Lichnérowicz, in his second letter [777], found Einstein's asymptotic form $g_{44} = -1 + \mu/r$ 'not satisfactory' (Ref. [2], p. 317). Pauli answered again on 15 November, and on 15 December—in time to congratulate Pauli on his Nobel Prize—Lichnérowicz wrote a third letter containing many details.

In the year following this exchange of letters, Lichnérowicz succeeded in proving the general theorem conjectured on p. 182 of Ref. [43b] that under the above assumptions 1) and 2) the metric tensor satisfying Eqs (9.94) and going over to the Euclidean metric for $r \to \infty$ is Euclidean everywhere. In this communication, which was presented to the Paris Academy by Elie Cartan, Lichnérowicz mentions his exchange of letters with Pauli, saying that the latter had pointed out to him "the interest it would have to dispose of any auxiliary hypotheses' (translated from Ref. [44c]). The fascinating aspect of this story now is that Pauli gave a detailed account of Lichnérowicz's theorem in his lectures at ETH in 1953, of which I had prepared the notes [45a] (see pages 35 and 261). Therefore, rather than entering into the details of Lichnérowicz's proof it is of more interest here to follow Pauli's own presentation given in Ref. [45a], and which is only mentioned in passing in the *Supplementary Notes 18* and *23* of Ref. [43b]. It also shows that the essential elements of the proof are already contained in the communication [44c].

In his lectures [45a] Pauli based the proof of Lichnérowicz's theorem on the Kaluza–Klein formalism (see pages 261f), reducing the dimensionality from 5 to 4. He had taken this idea from the Ph.D. thesis of Yves-René Thiry done under Lichnérowicz in Paris, which Pauli received with a dedication (Ref. [45b], Chapter III). The point of departure therefore is the line element (7.8) transposed to four dimensions,

$$ds^2 = g_{\mu\nu}dx^\mu dx^\nu = \overline{g}_{ik}dx^i dx^k - V^2(dx^4 + \psi_r dx^r)^2, \qquad (9.95)$$

where $i, k = 1, 2, 3$. The ψ_i are the three-dimensional analogues of the electromagnetic potentials Φ_μ, and the analogues of the fields $F_{\mu\nu}$ defined by Eq. (6.24) are $h_{ik} = \partial\psi_k/\partial x^i - \partial\psi_i/\partial x^k$. Since $ds^2 < 0$ if all $dx^i = 0$, the sign of the second term in (9.95) must be negative, in distinction from the second term in Eq. (7.8). The regularity condition (2) means that the determinant $\| g_{\mu\nu} \| = -q^2 V^2 \neq 0$ where $q^2 \equiv \| \overline{g}_{ik} \|$. Furthermore, since $ds^2 > 0$ if $dx^4 = 0$, it follows that $\overline{g}_{ik}dx^i dx^k > V^2(\psi_r dx^r)^2 \geq 0$ or $q^2 > 0$.

In Chapter 7 of Ref. [45a] Pauli had calculated for the Kaluza–Klein theory the complicated expression of the five-dimensional analogue of the tensor (9.94) and in Chapter 8 he translated it to Eqs (9.94) with stationarity condition 1). The result for the latter is

$$R_{44} = -\frac{1}{q}(q\overline{g}^{ab}V(V)_{,a})_{,b} + \overline{g}^{ab}V_{,a}V_{,b} - \frac{1}{4}V^2 h^{ab}h_{ab} = 0,$$

$$R_{i4} = \frac{1}{2Vq}\overline{g}_{ib}(qV^3h^{ba})_{,a} = 0,$$

$$R_{ik} = \overline{R}_{ik} - \frac{V^2}{2}\overline{g}^{ab}h_{ia}h_{kb} + \frac{1}{2V^2}(V^2)_{;i;k} - \frac{1}{4V^4}(V^2)_{,i}(V^2)_{,k} = 0,$$

$$(9.96)$$

where $h^{ik} \equiv h_{ab}\overline{g}^{ia}\overline{g}^{kb}$, $f_{,i} \equiv \partial f/\partial x^i$, and the covariant derivative $a^i_{;k}$ is as defined on page 30.

From these equations the following deductions are made. First, from $\psi_k \overline{g}^{ki}$ $R_{i4} = 0$ it follows that

$$(qV^3\psi_k h_{ka})_{,a} = qV^3 h_{ka}\psi_{k,a} = -\frac{1}{2}qV^3 h_{ka}h_{ka}. \qquad (9.97)$$

When integrated over a sphere S of radius r in three-space the left-hand side of (9.97) vanishes for $r \to \infty$, provided that $g_{\mu\nu}$ becomes Euclidean sufficiently rapidly ($\psi_i \to 0$ more rapidly than $r^{-1/2}$). In bringing $\overline{g}_{ik}$ locally to diagonal form, $\overline{g}_{ik} = g_i\delta_{ik}$, $g_i > 0$, one finds $h^{ka}h_{ka} = (h_{ka})^2/g_k g_a \geq 0$, so that integration of (9.97) over S yields $h_{ka} \equiv 0$, since h_{ka} is a tensor. Hence, the third term of R_{44} in (9.96) vanishes and integration of qR_{44} over S yields, under the mentioned assumption, $V_{,a} \equiv 0$, or $V = \text{const}$.

Finally, Eqs (9.96) yield $0 = R_{ik} = \overline{R}_{ik}$. But $\overline{R}_{ik}$ has the same number, six, of independent components as has the Riemann tensor $\overline{R}_{hijk}$ in three dimensions. But from a count of the number of components of this tensor Riemann's theorem follows, which states that in any dimension the vanishing of this tensor is necessary and sufficient for Euclidean geometry to hold. This proves that $g_{\mu\nu}$ is Euclidean. Now $h_{ka} \equiv 0$ still leaves for the ψ_i the freedom of gauge, $\psi_i = f_{,i}$. When inserted in (9.95) this means replacing dx^4 by dx'^4 where $x'^4 = x^4 + f(x^1 x^2 x^3)$. But this is just the transformation (7.5) leaving the cylinder condition (7.3) invariant, translated to the four-dimensional case with the stationarity condition 1). This demonstration shows that the result of Einstein and Pauli in Ref. [43a] had become obsolete; Pauli did not mention it in his lectures [45a]. In turn he mentioned the analogy of the above proof to that of potential theory where one shows that the solution of $\Delta\varphi = 0$ under the condition that $\varphi = 0$ in the limit $r \to \infty$ is $\varphi \equiv 0$. Considering the integral over S of $\varphi\Delta\varphi$ one concludes that $\nabla\varphi = 0$.

It is likely that in this collaboration Einstein became impressed by Pauli and that this became the motivation for Einstein's telegram to the Nobel Committee in Stockholm on 19 January 1945 proposing Pauli for the forthcoming Prize (see page 125). It is equally likely that this led Einstein, who retired from the institute Faculty in 1946, to propose Pauli as his successor. In a memorandum Einstein, together with Weyl, recommended Pauli over Oppenheimer for a professorship at the Institute for Advanced Study (see Ref. [45c], p. 135). On 17 June 1945 Director Aydelotte wrote to Pauli: 'It gives me great pleasure to extend to you on the recommendation of the Faculty, approved by the Board of Trustees, an invitation to become a member of the Faculty of the School of Mathematics of the Institute. I am authorized to offer you a salary of $10000 per year plus a payment of 1500 per year by the Institute toward a retiring allowance on condition that you make a similar payment from your own salary. ... We all of us feel that in the five years you have spent at the Institute you have made a place for yourself which no one else could fill. ... I sympathize with your feeling of obligation to the Technische Hochschule [ETH] in Zurich, and can readily understand that you must secure a release from them before you can accept a permanent appointment here ... but I must say that on the basis of such advice as I have been able to secure, I doubt whether such a return would be feasible until you and Mrs. Pauli have qualified for American citizenship' (*Pauli Archive*, CERN, PLC Bi 47, also Ref. [2], comment, pp. 278–9).

On Thursday 14 June (there is an inconsistency of dates here) Pauli relates this news to Rabi, writing: 'Last Saturday I obtained an official offer of a permanent position at the Institute here' (Ref. [2], letter [738]). And on 12 July he writes to Bhabha: 'The Princeton Institute for Advanced Study offered me a professorship in the occasion of Einstein's retirement; I have not decided anything yet' [750]. Director Aydelotte's letter signalled the end of the lonely period. 'By the fall of 1946 the main decisions will probably be made', Pauli wrote to Rabi on 1 July [744]. As a first step he had sent on 16 May the telegram to President Rohn of ETH mentioned on page 360, announcing his travel plans for Zurich. The further decisions also concerned an invitation by Rabi to Pauli and to Oppenheimer to join the Columbia Faculty (Ref. [2], letters [744, 764] and comment, p. 285). The official letter of invitation to Pauli for a professorship at Columbia University is dated 14 January 1946 and signed by the Acting President Dr. Frank D. Fackenthal (*Pauli Archive*, CERN, PLC Bi 3). The return to peace physics had induced great movement in the appointment questions among US universities. John Hasbrouk van Vleck from Harvard had asked Pauli for his advice 'comparing Schwinger and Bethe' and concluded that 'it is probable that Schwinger will ultimately be a member of our staff'

(Ref. [2], letters [742]; see also [746]), which he was.

However, concerning the future of the world at large, Pauli was more pessimistic. He was very glad to have kept out of war physics altogether [731]. To Bohr who was striving for an 'open world' (Ref. [2], comment, p. 314) Pauli wrote in his birthday letter: 'What I wish you is that your attitude toward the general problems of the world, which seems to be more active and hopeful than my own, may be justified by the future developments' [772]. On 1 October 1945 he wrote to his friend Carl Alfred Meier in Zurich: 'The Atom Bomb is a very evil thing, also for physics, I think. The politicians, of course, are at a complete loss and talk in a demagogic way of a *secret* which, evidently, does not exist (the true secret is the nature of the nuclear forces). Although most people say that I see ghosts, I am afraid that physics gets more or less subdued by military censorship and that free research, in principle, is gone' (translated from Ref. [2], comment, p. 314).

On 1 September 1945 Pauli, still at Lake Clear, received a telegram from Scherrer in New York, and 'the same day had a very friendly longer telephone conversation with him. His well-known method of improvization in the last moment, however, has not worked: I did not want to drive down to New York, since time was short and I had not made a room reservation, and he could not arrange to come up here any more. He planned to travel back to Europe already at the end of the week' (translated from Ref. [2], letter [768] to Jauch). According to a letter that Pauli wrote to his friend Meier in Zurich, Scherrer had been in the United States for a longer visit. '(What I heard first was a rumour that somebody had seen him in a bar in Washington)' (translated from Ref. [2], comment, p. 312). This was Scherrer's privileged tour to visit US industrial installations mentioned on page 199 (see also footnote 159 in Ref. [8], p. 161).

Reflecting on his time spent in the United States, Pauli wrote to Casimir: 'The past years have been rather lonesome, particularly '42 and '43. Last year I saw Uhlenbeck regularly at M.I.T. during my course there. I had a good contact with him, both in scientific and in human matters. I am often thinking on Kramer's old statement about me, that "my heart is better than my mind"' (Ref. [2], letter [780], p. 322). But then a hint in letter [781] by Oskar Klein of 16 October 1945 brought Pauli back to the present and to what lay in store for him: 'I think we have some reason to hope that we shall see you and Mrs. Pauli here this winter which would be very pleasant indeed.' In his reply Pauli observed: 'I wonder how the mysterious remarks of yours about Mrs. Pauli and me will be clarified. (I made the constatation, however, that some Swedish newpapermen seem to be "after me".)' [783]. The first cheers came from Bohr who on 16 November telegraphed: 'Heartiest congratulations from the whole Copenhagen Institute' [786].

Pauli answered Bohr on Sunday, 25 November: 'It was a great exciting surprise that the Nobelprize was awarded to me this year, although I had thought already a week earlier, when the congratulation telegram of you and your wife arrived, that it was a good omen' (Ref. [2], letter [788]). But then Pauli faced the problem 'whether or not I should go to Stockholm on December

10th'. The US authorities offered Pauli an exit and re-entrant permit for the trip [788], and the Statehouse of New Jersey in Trenton, according to Franca [3], was opened specially to establish it. Issued on 21 November 1945, the visa was valid for two months, starting on 1 December, and gave as 'Nature of business *To accept the Nobel Prize*' (*Pauli Archive*, CERN, PLC Bi 52; see also Ref. [2], footnote 2, p. 330). But Pauli 'finally decided to postpone my participation in the ceremony in Stockholm to next year, after having heard that Stern and Rabi are doing the same. I have asked the Swedish Academy whether they could agree to this proposal' (letter [788]; see also [789]). Otto Stern had been awarded the Prize for 1943 in 1944 'for his contributions to the development of the molecular ray method and his discovery of the magnetic moment of the proton'. And Rabi had received the Prize of 1944 'for his resonance method for recording the magnetic properties of atomic nuclei'. Both had been prevented from going to Stockholm by the war.

Pauli gives as reason for not going to Stockholm in 1945 'particularly the possibility of a delay by such a trip of my getting naturalized' (Ref. [2], letter [788]). For him the more urgent task was to get his relations with the ETH in Zurich normalized, and without a US passport he was unable to travel. Indeed, he wrote in letter [780] to Casimir already quoted: 'I am still Professor in Zürich, where my position was kept open for me. Therefore I feel to have some obligation toward my colleagues in Zürich and I decided, first to make a trip to Zürich as soon as possible and to consider the farer future later' (Ref. [2], p. 322). Pauli's renunciation settled a complicated and delicate problem.

Fig. 9.3 Pauli and John Archibald Wheeler at Pauli's Nobel Prize celebration at the Princeton Institute, November 1945 (CERN)

It now was up to the director of the Institute for Advanced Study to substitute for the Nobel festivities, and he did so in grand style, all the more so since Pauli's was the first Nobel Prize awarded to a resident member. He sent out official invitations to a large number of scientists and officials as follows: 'The Director, Trustees and Faculty of the Institute of Advanced Study cordially invite you to a dinner in honor of Professor and Mrs. Wolfgang Pauli celebrating the award to him of the Nobel Prize. Fuld Hall, Monday evening, December 10, 1945 at seven o'clock. Black tie' (Ref. [2], comment, p. 329). Among the guests, accompanied by their spouses, were representatives of the Swedish Embassy in Washington and of several universities. I. I. Rabi, J. F. Carlson, S. A. Goudsmit, and A. J. F. Siegert had come from MIT, W. F. Colby, O. Laporte, and D. M. Dennison from the University of Michigan, and J. H. van Vleck from Harvard. The Carnegie Institute of Technology in Pittsburgh was represented by I. Estermann (Otto Stern apparently was unable to come). From Princeton University had come Dean L. P. Eisenhart and Professors J. M. Jauch, R. Ladenburg, H. N. Russell, and J. A. Wheeler. Among the members of the Princeton Institute were Einstein and his collaborators W. Mayer and V. Bargmann, and Pauli's collaborator N. Hu, then the mathematicians H. Weyl, J. von Neumann, K. Gödel, and O. Veblen and the art historian E. Panofsky (see Ref. [2], comment, p. 329).

In his welcome address Director Aydelotte declared 'The awarding of a Nobel Prize to a newly appointed member of our Faculty is just the kind of endorsement of our choice that we value and that we ought to expect.' And later on he said: 'We hope that in a few weeks we shall have the pleasure of welcoming Dr. and Mrs. Pauli to American citizenship' [46a]. The next speaker was Hermann Weyl who, in his *encomium* already quoted on pages 39 and 43, observed that 'Pauli for many years has been the conscience and criterion of truth for a large part of the community of theoretical physicists.' And he concluded with the observation: 'Pauli has all his life been deeply interested in philosophy. The wisdom of the Chinese sages seems to have a special appeal for him. No wonder that his sympathies are with those who are not willing to sacrifice the spiritual for the secular, and who are not willing to accept efficiency as the ultimate criterion' [46b].

After Weyl, quite unexpectedly and unannounced, Einstein rose to give an impromptu address. 'He was like an old Greek', Franca later recalled [3]. Pauli himself described this event almost 10 years later in a letter to Max Born, dated 24 April 1955, shortly after Einstein had died:

Einstein's death has also reached me personally. Now such an affection-ate, fatherly friend is no more. Never will I forget the discourse that he has pronounced about me and for me in 1945 in Princeton after I had received the Nobel Prize. It was like a king who abdicates and who installs me as sort of a 'son of choice' [Wahl-Sohn] as the successor. Unfortunately, no records of this discourse of Einstein exist (it was improvised and a manuscript does not exist neither). (Translated from *Pauli Archive*, CERN, PLC Bi 1; see also Pauli's letter [1020] to Einstein in Ref. [2]).

Pauli recalled this impressive event in his letter for Einstein's 70th birthday: 'Your 70th birthday on 14 March of this year [1949] is for me a welcome occasion to tell you, together with my hearty congratulations, how strongly I had felt the personal sympathy that you presented me in Princeton, and how unforgettable your speech at the Institute celebration in December 1945 has remained for me' (translated from Ref. [2], letter [1020]).

The last eulogy was by Panofsky who had become a close friend of Pauli. He observed that the Prize citation 'inexcusably omitted ... the celebrated "Pauli Effect"'. And he gave his own example. As a matter of fact, Pauli and Panofsky had already met in 1928 in Hamburg, when a common friend had invited both of them to an outdoor restaurant where he introduced them. It so happened that when they rose in the afternoon, Panofski and the friend both 'had been sitting in whipped cream for about three hours—but not Pauli'. But then Panofsky introduced the notion of 'Pauli Effect in Reverse' which 'affects minds and souls, and affects them beneficially, stimulating them, making them more conscious of themselves, charging them with an induced electric current, as it were—even if they happen to be minds and souls of *humanists*'. Later in his speech Panofsky recalled: 'One of my fondest recollections in connection with our Laureate is the picture of his wistful face when he once mentioned, in the woods behind the Institute, how much impressed he had been by a reading of *Kepler*.' Then Panofsky spoke at length about Kepler, thereby signalling the great interest Pauli would take in this personality in the years to come [46d].

Born in Hanover (Germany) in 1892, Erwin Panofsky studied art history in Freiburg (Germany), Munich, and Berlin, where he obtained his Ph.D. in 1914. He then went to Hamburg, where he became an assistant at the university and a disciple of the famous art historian and private scholar Abraham (Abi) Moritz Warburg (1866–1929). In 1916 Panofsky married Dorothea Moss (1885–1965) from the influential Jewish family of publishers. The couple had two sons; the younger of them, Wolfgang K. H. Panofsky (b. 1919), became a well-known US physicist. In 1920 Panofsky became a private dozent and in 1926 a professor of art history at the University of Hamburg. In 1931–3 he also taught at the Institute of Fine Arts of New York University. In 1934 the family emigrated to the United States, where Panofsky first taught at Princeton University. From 1935 to his retirement in 1963 he was a professor of art history at the Institute for Advanced Study, after which he taught again at New York University. Panofsky died in 1968. His fame is based mainly on his many works dedicated to Albrecht Dürer [46e].

After Panofsky's eulogy Pauli rose to give the principal address on a personal history of the exclusion principle [46c]. He began with a sketch of his scientific development since the days in Munich and Copenhagen, where the anomalous Zeeman effect had made him so unhappy (see pages 91–2) and ended with a moving tribute to the hospitality and the freedom offered to him during the war by the Institute for Advanced Study and a glance into the future: 'We know, however, that this further development can take place only in the same atmosphere of free investigation and unhampered exchange of scientific results between nations that existed at the time of the disclosure of the exclusion principle. I am therefore very glad to be able to give this short historical survey here in Princeton's Institute for Advanced Study, which in the difficult years of the war, by support of pure and free research irrespective of applications, made it possible for me and others to continue our scientific work' [46c].

Fig. 9.4 Pauli with, from left to right, Ning Hu, José Leite Lopez, and Josef Maria Jauch, November 1945 (CERN)

In one of the many letters of acknowledgement for the received congratulations Pauli wrote to Klein: 'I have now every reason to be confident in my future' (Ref. [2], letter [791]). For the moment he was awaiting his naturalization in order to make the planned trip to Europe and put in order his relations with ETH. On 24 to 26 January 1946 Pauli participated in the New York meeting of the APS 'where [Marcel] Schein talked on the artificial generation of mesons by the 100 MeV γ-rays of the Betatron of the General Electric in Schenectady [New York].' And: 'I suddenly found Lise Meitner sitting just in front of me during Schein's lecture, she had arrived here the day before.' Pauli wrote to Lamek Hulthén in Lund, Sweden (Ref. [2], letter [800]). In this letter Pauli told Hulthén that 'The preparations for my trip to Europe are going well according to plan and I shall leave in about 4 weeks.—My first station will probably be Dublin.' Then he asked Hulthén whether the Nobel Lecture 'is supposed to be a technical one'. In his reply Hulthén wrote 'Probably you will have more to tell us about the results [of Schein] and the theoretical conclusions when you come to Sweden' [804]. So Pauli planned to give his Nobel Lecture in Stockholm in early spring. Indeed, on 12 February 1946 a second travel permit had been issued to Pauli, giving as 'Countries to be visited *Sweden, Denmark, England, Ireland, Switzerland*' and as 'Nature of business *Nobel Prize lectures en route*' (*Pauli Archive*, CERN, PLC Bi 38). But Pauli's US passport was issued on 19 January 1946 and was valid for four years (see Pauli's letter to the US Consulate-General in Zurich of 18 August 1949, *Pauli Archive*, CERN, PLC Bi 91).

This was not a time of quiet research. In January Pauli was solicited by Karl K. Darrow, the long-time secretary of the APS, on the question of organizing a large conference in New York on the physics of elementary particles similar to the Solvay Congresses. The two met in New York, together with Duncan MacInnes from the Rockefeller Institute, which was to be the sponsor. John Archibald Wheeler from Princeton was considered for the organization (Ref. [2], letters [796, 799]). However, in February Pauli and Wheeler came up with another idea. Bohr had almost simultaneously proposed such a conference to be held in Copenhagen, so it would be more efficient, they suggested, to spend the money on travelling subsidies for scientists from the United States to Copenhagen. But MacInnes did not like the idea; he was thinking of a more modest conference with fewer and mainly young participants in some country inn. This idea eventually came to be realized in the Shelter Island and Pocono conferences of spring 1947 and 1948, respectively, which were of historic importance (see Ref. [16d], Chapter 4). Meanwhile, however, the first international physics conference after the war was held in Cambridge, England, in July 1946.

At the end of February 1946 Pauli and Franca boarded the ship *Gripsholm* for the Old Continent (see Scherrer, in Ref. [8], II.125). They first visited Dublin, Ireland, where Pauli held seminars with Schrödinger, Heitler, and Heitler's collaborator H. W. Peng at the Institute for Advanced Studies (Ref. [2], letters [803, 807]). Then Pauli and Franca visited England and Paris where they were invited to Langevin's house, meeting Proca and Mrs. Solomon there [809] (see pages 326–7 and 265). Finally they reached Zurich at the beginning of April. In all evidence the trip to Stockholm had been postponed.

Notes and references

[1] Translated from: W. Pauli, letter of 25 February 1942 to C. A. Meier, quoted in Ref. [2], p. 125.

[2] K. von Meyenn (ed.), *Wolfgang Pauli. Scientific Correspondence with Bohr, Einstein, Heisenberg a.o., Volume III: 1940–1949* (Springer, Berlin and Heidleberg, 1993).

[3] Notes of my conversation with Franca Pauli on 8 April 1984.

[4] (a) *The American Peoples Encyclopedia. A Modern Reference Work* (Grolier, New York, 1965); (b) A. Pais, *A Tale of Two Continents. A Physicist's Life in a Turbulent World* (Oxford University Press, Oxford, 1997).

[5] (a) E. Majorana, 'Teoria relativistica di particelle con momento intrinseco arbitrario', *Nuovo Cimento* **9**, 335 (1932); (b) D. M. Fradkin, 'Comments on a Paper by Majorana Concerning Elementary Particles', *American Journal of Physics* **34**, 314 (1966).

[6] K. von Meyenn (ed., with the cooperation of A. Hermann and V. F. Weiss-kopf) *Wolfgang Pauli. Scientific Correspondence with Bohr, Einstein, Heisenberg a.o., Volume II: 1930–1939* (Springer, Berlin and Heidelberg, 1985).

[7] (a) H. Weyl, *The Theory of Groups and Quantum Mechanics*, trans-lated from the second (revised) edition by H. P. Robertson (Dover, New York, 1931); (b) W. Pauli, 'Continuous Groups in Quantum Mechan-ics', CERN-Publication 56-31, 1956. Reprinted in: G. Höhler (ed., with the cooperation of S. Flügge, F. Hund, and F. Trendelenburg), *Ergeb-nisse der exakten Naturwissenschaften* (Springer, Berlin and Heidelberg, 1965), Vol. 37, pp. 85–104. (c) *Note*: In order to derive Eqs (13) of Ref. [5a] one must remember that the γ_μ transform like the coordinates x_μ and, therefore, have the same commutators with **a** and **b** as the x_μ, the latter being easy to calculate. Indeed, Pauli's infinitesimal operators of the four-dimensional rotation group may be written

$$J_{\lambda\mu} = x_\lambda p_\mu - x_\mu p_\lambda, \tag{A}$$

where $p_\mu = \partial/i\partial x_\mu$ and $[p_\mu, x_\nu] = -i\delta_{\mu\nu}$. One finds

$$[x_\nu, J_{\lambda\mu}] = i(x_\lambda \delta_{\mu\nu} - x_\mu \delta_{\lambda\nu}). \tag{B}$$

Substitution for x_μ in (B) by γ_μ with $x_4 = ix_0$ and $\gamma_4 = i\gamma_0$, and $J_{23} = a_1$, $J_{31} = a_2$, $J_{12} = a_3$ and $J_{41} = ib_1$, $J_{42} = ib_2$, $J_{43} = ib_3$ one readily arrives at Eqs (13) of Ref. [5a]. Note that these substitutions include the passage from the four-dimensional rotation group to the Lorentz group.

With (13) one easily verifies that γ_0^2 commutes with all the γ_μ, as well as with **a** and **b**. Therefore, $\gamma_0^2 \propto \mathbf{1}$ and, hence, the eigenvalues of γ_0 are all proportional to ± 1. But since in the rest frame, $W(\varphi^* \gamma_0 \varphi) = mc^2(\varphi^* \varphi) > 0$, it follows that $\gamma_0 \propto \mathbf{1}$ and one may put $\gamma_0 = j + 1/2$. From $\psi^* \psi = \varphi^* \gamma_0 \varphi$, it then follows that $\psi = \sqrt{\gamma_0}\varphi$ or Eq. (15) of Ref. [5a].

Introducing electromagnetic fields Majorana finds for the electron a mag-netic moment of $+\mu_B/2$, instead of $-\mu_B$, which forces him to include Pauli terms (see pages 218, 269–70, and 330) in his theory. This anomaly must be related to the auxiliary conditions discussed on page 335. (d) V. Bargmann, 'Irreducible representations of the Lorentz group', *Annals of Mathematics* **48**, 568 (1947); (e) W. Pauli, 'Gruppentheorie und Quan-tenmechanik', *Spezialvorlesung* ETH, Sommersemester 1954, notes by C. P. Enz (unpublished).

[8] C. P. Enz, B. Glaus, and G. Oberkofler (eds), *Wolfgang Pauli und sein Wirken an der ETH Zürich* (vdf Hochschulverlag ETH, Zurich, 1997).

[9] (a) E. Amaldi, 'Ettore Majorana, man and scientist', in: A. Zichichi (ed.), *Strong and weak interactions—Present problems* (Academic Press, New York, 1966), pp. 10–77; (b) G. Wentzel, *Einführung in die Quantentheorie der Wellenfelder* (Deuticke, Vienna, 1943; English translation, Wiley Interscience, New York, 1949). (c) Note that here Pauli uses units $\hbar = c = 1$. With the actual values of $\hbar$ and c Pauli's inequality, which is dimensionless, reads $g^2/\hbar c \gg \hbar/\mu lc$, where l is a typical length. Here g is the same as in Eqs (8.71)–(8.74).

[10] (a) H. J. Bhabha, 'Classical Theory of Mesons', *Proceedings of the Royal Society* A **172**, 384 (1939); (b) P. A. M. Dirac, 'Classical Theory of Radiating Electrons', *Proceedings of the Royal Society* A **167**, 148 (1938).

[11] A. D. Debus (ed.), *World Who's Who in Science* (Marquis–Who's Who, Chicago, 1968).

[12] (a) H. B. G. Casimir, *Haphazard Reality. Half a Century of Science* (Harper & Row, New York, 1983); (b) D. C. Cassidy, *Uncertainty. The Life and Science of Werner Heisenberg* (Freeman, New York, 1992); (c) P. Fischer, *Licht und Leben. Ein Bericht über Max Delbrück, den Wegbereiter der Molekularbiologie* (Universitätsverlag Konstanz, Constance, 1985).

[13] L. A. DuBridge, 'History and Activities of the Radiation Laboratory of the Massachusetts Institute of Technology', *Review of Scientific Instruments* **17**, 1 (1946).

[14] (a) J. S. Rigden, *Rabi. Scientist and citizen* (Basic Books, New York, 1987); (b) N. F. Ramsey, 'I. I. Rabi' (Obituary), *Physics Today*, October 1988, p. 82.

[15] (a) C. P. Enz, 'Gregory H. Wannier (1911–1983)', *Helvetica Physica Acta* **57**, 136 (1984); (b) H. A. Kramers and G. H. Wannier, 'Statistics of the Two-Dimensional Ferromagnet. Part I, II', *Physical Review* **60**, 252, 263 (1941); (c) L. Onsager, 'Crystal statistics I. A two-dimensional model with an order-disorder transition', *Physical Review* **65**, 117 (1944).

[16] (a) Translated from: W. Pauli, letter [646] of 30 December 1941 to G. Wentzel, in Ref. [2], p. 119; (b) K. von Meyenn (ed.), *Wolfgang Pauli. Scientific Correspondence with Bohr, Einstein, Heisenberg a.o., Volume IV/Part I: 1950–1952* (Springer, Berlin and Heidelberg, 1996); (c) J. Mehra and K. A. Milton, *Climbing the Mountain. The Scientific Biography of Julian Schwinger* (Oxford University Press, Oxford, 2000); (d) S. S. Schweber, *QED and the Men Who Made It: Dyson, Feynman, Schwinger, and Tomonaga* (Princeton University Press, Princeton, NJ, 1994).

[17] (a) J. M. Jauch and F. Rohrlich, *The Theory of Photons and Electrons* (Addison-Wesley, Cambridge, MA, 1955); (b) J. M. Jauch, *Foundations of Quantum Mechanics* (Addison-Wesley, Reading, MA, 1968); (c) J. M. Jauch, *'Are Quanta Real?' A Galilean Dialogue* (Indiana University

Press, Bloomington, 1973); (d) C. P. Enz and J. Mehra, 'A Personal Introduction', in: C. P. Enz and J. Mehra (eds), *Physical Reality and Mathematical Description* (Reidel, Dordrecht, 1974).

[18] (a) S. M. Dancoff and W. Pauli, 'Strong Coupling Mesotron Theory of Nuclear Interactions' (Abstract, Proceedings APS), *Physical Review* **61**, 387 (1942); (b) W. Pauli and S. M. Dancoff, 'The Pseudoscalar Meson Field with Strong Coupling', *Physical Review* **62**, 85 (1942). Reprinted in Ref. [19], Vol. 2, pp. 953–76.

[19] R. Kronig and V. F. Weisskopf (eds), *Collected Scientific Papers by Wolfgang Pauli. In Two Volumes* (Wiley Interscience, New York, 1964).

[20] (a) G. Wentzel, 'Zum Problem des statischen Mesonfeldes', *Helvetica Physica Acta* **13**, 269 (1940); 'Zum Problem des statischen Mesonfeldes. Nachtrag', *Helvetica Physica Acta* **14**, 633 (1941); (b) J. R. Oppenheimer and J. Schwinger, 'On the Interaction of Mesotrons and Nuclei', *Physical Review* **60**, 150 (1941); (c) J. M. Blatt and V. F. Weisskopf, *Theoretical Nuclear Physics* (Springer, New York, 1952, 1979). (d) Note that Pauli has chosen units such that $\hbar = c = 1$ (see footnote 13 of Ref. [18b]). (e) N. Kemmer, 'The Charge-Dependence of Nuclear Forces', *Proceedings of the Cambridge Philosophical Society* **34**, 354 (1938).

[21] (a) W. Pauli, 'On Strong-Coupling and Weak-Coupling Theories of the Meson Field' (Abstract, Proceedings APS), *Physical Review* **63**, 221 (1943); (b) G. Wentzel, 'Über die Eigenkräfte der Elementarteilchen. I., II., III.', *Zeitschrift für Physik* **86**, 479, 635 (1933); **87**, 726 (1934); (c) P. A. M. Dirac, 'La théorie de l'électron et du champ électromagnétique', *Annales Institut Henri Poincaré* **9**, 13 (1939).

[22] (a) R. Serber and S. M. Dancoff, 'Strong Coupling Mesotron Theory of Nuclear Forces', *Physical Review* **63**, 143 (1943). (b) Information received from the University Archives, Urbana, Illinois, through Professor Hans Frauenfelder, Department of Physics, University of Illinois (now at Los Alamos National Laboratory). I thank Professor Frauenfelder for his precious help.

[23] W. Pauli and S. Kusaka, 'On the Theory of a Mixed Pseudosacalar and a Vector Meson Field', *Physical Review* **63**, 400 (1943). Reprinted in Ref. [19], Vol. 2, pp. 977–93.

[24] (a) C. Møller and L. Rosenfeld, 'On the field theory of nuclear forces', *Kongelige Danske Videnskabernes Selskab, Mat.-Fys. Meddelelser* **17**, Nr. 8, 1 (1940); (b) S. Kusaka and W. Pauli, 'On the Theory of a Mixed Pseudo-Scalar and a Vector Meson Field' (Abstract, Proceedings APS), *Physiscal Review* **63**, 221 (1943).

[25] (a) In Ref. [12b], p. 491, this visit is mentioned without a date in relation to Heisenberg's second invitation in December 1944. (b) V. F. Weisskopf, *The Joy of Insight. Passions of a Physicist* (Basic Books, New York, 1991).

[26] Quoted from: W. Pauli, letter [692] of 13 April 1944 to P. A. M. Dirac, Ref. [2], p. 222.

[27] P. A. M. Dirac, 'BAKERIAN LECTURE. The physical interpretation of quantum mechanics', *Proceedings of the Royal Society* A **180**, 1 (1942).

[28] C. P. Enz (ed.), *Pauli Lectures on Physics: Volume 4. Statistical Mechanics*, translated by H. R. Lewis and S. Margulies (MIT Press, Cambridge, MA, 1973).

[29] W. Pauli, 'On Dirac's New Method of Field Quantization', *Reviews of Modern Physics* **15**, 175 (1943). Reprinted in Ref. [19], Vol. 2, pp. 1001–33.

[30] C. P. Enz (ed.), *Pauli Lectures on Physics: Volume 6. Selected Topics on Field Quantization*, translated by H. R. Lewis and S. Margulies (MIT Press, Cambridge, MA, 1973).

[31] G. Källén and W. Pauli, 'On the Mathematical Structure of T. D. Lee's Model of a Renormalizable Field Theory', *Kongelige Danske Videnskabernes Selskab, Mat.-Fys. Meddelelser* **30**, 3 (1955). Reprinted in Ref. [19], Vol. 2, pp. 1261–81.

[32] W. Pauli, 'Relativistic Field Theories of Elementary Particles', *Reviews of Modern Physics* **13**, 203 (1941). Reprinted in Ref. [19], Vol. 2, pp. 923–52.

[33] W. Pauli, 'On Applications of the λ-Limiting Process to the Theory of the Meson Field', *Physical Review* **64**, 332 (1943). Reprinted in Ref. [19], Vol. 2, pp. 1034–46.

[34] (a) J. M. Jauch, 'On the Meson Field Theory of the Magnetic Moment of Proton and Neutron' (Abstract, Proceedings APS), *Physical Review* **63**, 222 (1943); (b) W. Pauli and J. M. Jauch, 'On the Application of Dirac's Method of Field-Quantization to the Problem of Emission of Low Frequency Photons' (Abstract, Proceedings APS), *Physical Review* **65**, 332 (1944). Reprinted in Ref. [19], Vol. 2, p. 1047. (c) W. Pauli, 'Diracs Feldquantisierung und Emission von Photonen kleiner Frequenz', *Helvetica Physica Acta* **19**, 234 (1946). Reprinted in Ref. [19], Vol. 2, pp. 1076–9.

[35] (a) N. Hu, 'The Relativistic Correction in the Meson Theory of Nuclear Force', *Physiscal Review* **67**, 339 (1945); (b) J. L. Lopes, 'High Energy Neutron-Proton Scattering and the Meson Theory of Nuclear Forces with Strong Coupling', *Physical Review* **70**, 5 (1946); (c) J. M. Blatt, 'On the Meson Charge Cloud Around a Proton', *Physical Review* **67**, 205 (1945); (d) J. M. Blatt, 'On the Heavy-Electron Pair Theory in the Limit of Strong Coupling', *Physical Review* **69**, 285 (1946).

[36] W. Pauli and N. Hu, 'On the Strong Coupling Case for Spin-Dependent Interactions in Scalar- and Vector-Pair Theories', *Reviews of Modern Physics* **17**, 267 (1945). Reprinted in Ref. [19], Vol. 2, pp. 1053–72.

[37] G. Wentzel, 'Zur Paartheorie der Kernkräfte', *Helvetica Physica Acta* **15**, 111 (1942).

[38] W. Pauli, *Meson Theory of Nuclear Forces* (Wiley Interscience, New York, 1946, 1948). Second edition and Preface to the First Edition reprinted in Ref. [19], Vol. 1, pp. 939–1011.

[39] (a) W. Heitler and H. W. Peng, 'The influence of radiation damping on the scattering of mesons', *Proceedings of the Cambridge Philosophical Society* **38**, 296 (1942); (b) W. Heitler, *The quantum theory of radiation* (Oxford University Press, Oxford, 1936, 1944, 1954); (c) W. Heisenberg, 'Die "beobachtbaren Grössen" in der Theorie der Elementarteilchen. I, II, III', *Zeitschrift für Physik* **120**, 513, 673 (1943); **123**, 93 (1944). Reprinted in Ref. [40], Part II, pp. 611–86.

[40] W. Blum, H.-P. Dürr, and H. Rechenberg (eds), *Werner Heisenberg. Collected Works. Series A: Original Scientific Papers* (Springer, Berlin and Heidelberg, 1989).

[41] V. Weisskopf, *The Joy of Insight. Passions of a Physicist* (Basic Books, New York, 1991).

[42] (a) Translated from: W. Pauli, Allocution at the occasion of the 70th birthday of Hermann Weyl on 9 November 1955, item [2183] in Ref. [42e]. (b) 'In Commemoration of the Sixtieth Birthday of Niels Bohr. October Seventh 1945', *Reviews of Modern Physics* **17**, Nos 2 and 3, April–July, 1945; (c) W. Pauli, 'Niels Bohr on His 60th Birthday', *Reviews of Modern Physics* **17**, 97 (1945). Reprinted in Ref. [19], Vol. 2, pp. 1048–52. (d) *Proceedings of the American Physical Society, Physical Review* **67**, 199 (1945); (e) K. von Meyenn (ed.), *Wolfgang Pauli. Scientific Correspondence with Bohr, Einstein, Heisenberg, a.o., Volume IV, Part III: 1955-1956* (Springer, Berlin and Heidelberg, 2001).

[43] (a) A. Einstein and W. Pauli, 'On the Non-Existence of Regular Stationary Solutions of Relativistic Field Equations', *Annals of Mathematics* **44**, 131 (1943). Reprinted in Ref. [19], Vol. 2, pp. 994–1000. (b) W. Pauli, *Theory of Relativity*, translated by G. Field. *With Supplementary Notes by the Author* (Pergamon Press, London, 1958). (c) D. Bailin and A. Love, *Supersymmetric Gauge Field Theory and String Theory* (Institute of Physics Publishing, Bristol, 1994).

[44] A. Lichnérowicz, (a) 'Espaces-temps extérieurs réguliers partout', *Cmpts. Rendus* (Paris) **206**, 313 (1938); (b) 'Sur certains problèmes globaux relatifs au système des équations d'Einstein', Thèse, Paris, 1939; (c) 'Sur le caractère euclidien d'espaces-temps extérieurs statiques partout réguliers', *Comptes Rendus* (Paris) **222**, 432 (1946).

[45] (a) 'Probleme der Allgemeinen Relativitätstheorie, Vorlesung gehalten an der Eidg. Techn. Hochschule, Zürich, 1953, von Prof. W. Pauli. Ausgearbeitet von Ch. Enz' (handwritten, 111 pp.), *Pauli Archive*, CERN; (b) Y. R. Thiry, *Thèse. Etude mathématique des équations d'une théorie unitaire à quinze variables de champ* (Gauthier-Villars, Paris, 1951); (c) E. Regis, *Who Got Einstein's Office?* (Addison-Wesley, Reading, MA,

1987). I thank Professor J. Mehra for having drawn my attention to this reference.

[46] (a) F. Aydelotte, 'Introductory Remarks'; (b) H. Weyl, 'Encomium'; (c) W. Pauli, 'Remarks on the History of the Exclusion Principle', in: *Science* **103**, (a) p. 215; (b) p. 216; (c) p. 213 (1946); (d) E. Panofsky, Speech for Pauli's Nobel Prize, *Pauli Archive* CERN, PLC Bi 264; (e) H.A. Strauss and W. Röder (eds), *International Biographical Dictionary of Central European Emigrés, 1933-1945, Vol. II/Part 2: L-Z* (Saur, Munich, 1983).

10
The famous professor returns

And Zurich becomes home

I found Zurich not at all to be an alpine village but a very international place. Among my students and collaborators are not only Swiss but also Swedes and Belgians. The political mood of 'Dame Helvetia' [Switzerland] who is well-known for her caprice, at present is favourable and particularly very friendly towards England (The Swiss hotels now are all replete with Englishmen). [1]

'Franca (alias Josefine Spitz), later called Jolande, is . . . coming back from the United States after the war with her husband Pauli (alias Paul Mende). Waiting for a taxi in front of the railway station facing Bahnhofstrasse she reflects on her pre-Mende days'. This is how on page 285 Pauli's and Franca's return to Zurich after six years in America was described following the novel *Gerufen und nicht gerufen* by Kurt Guggenheim. They arrived 'in the first days of April' (Ref. [2], letter [810]). But since their house in Zollikon was occupied by a tenant they took up residence at Hotel Bellerive on Utoquai in Zurich (Ref. [3], II.132, 136).

At ETH President Rohn was impatient: 'It is indeed very urgent that at last we know exactly what Herr Professor Pauli is up to. His behaviour towards our Institute is utterly strange', he wrote to Scherrer on 16 March 1946 (translated from Ref. [3], II.126). Then, on 25 April, Rohn at last could report on a meeting with Pauli, after which he was able to decree that Pauli would take up his teaching during the summer term (II.128). As before the war, Pauli announced a regular course of four hours weekly and a special one of two hours, the first being 'Thermodynamics' and the latter 'Introduction to the principles of wave mechanics' (II.131). On 30 April President Rohn made this programme official (II.133). At the same time he annulled, in agreement with Wentzel, the latter's teaching commission for the summer term of 1946 (II.135). Then he convened the School Council on 4 of May for a discussion and subsequent resolution concerning Pauli's resumption of teaching and his naturalization.

In his introductory statement Rohn observed that Pauli had been able to travel to Europe only after having become a US citizen. Then he said: 'I was surprised about the change in his character and his mind. While before and also during his stay in the US he was often arrogant, now he was singularly mild and modest in his appearance, so that our conversation took a quite different turn than I had supposed on the basis of our correspondence of recent years' (translated from Ref. [3], II.134). Rohn also mentioned that since his Nobel Prize, Pauli had received three offers, two in the United States and a tentative one in England. 'In spite of this, Professor Pauli wishes to resume his activity in Zurich and to continue it durably. In this sense he also begged me to forget all misunderstandings that had occurred, and said that he had always suffered under the pressure of the refusal of his application for naturalization, because the statement saying that he was not apt for assimilation is, in fact, the same as a declaration of his unworthiness to become a Swiss and to teach at a Swiss University' (translated from Ref. [3], II.134, p. 167).

On 26 April President Rohn had been in Berne to discuss Pauli's naturalization with Bundesrat Etter and the persons responsible for questions of naturalization, the upshot of which was 'that both the Federal Department of the Interior as well as the Federal Foreigner's Police would continue to examine the question of Pauli's naturalization with much goodwill' (translated from Ref. [3], II.134, p. 168). At the meeting of the School Council, Rohn invited the president of the city of Zurich, Dr. Adolf Lüchinger, himself a member of the School Council (see Ref. [3], p. 5), to present his assessment of Pauli's chances in the city of Zurich which, however, were quite unfavourable: Pauli could scarcely hope to be a citizen before 1958! However, he said, in Zollikon the conditions would be less strict and, if all worked together, including the federal authorities, a more favourable solution might be found. Many of the councillors, however, doubted whether Pauli's change of mind reported by President Rohn would be durable.

Following the recommendation of President Rohn, Pauli and Franca moved to Zollikon on 25 July, where they stayed at the inn *Zum Rössli*, situated at Alte Landstrasse 86 (II.152, 156), the road through the village of Zollikon connecting Zurich with Küsnacht. Then on 21 November 1946 Pauli wrote to Schrödinger: 'On 2 October we have again moved into our house in Zollikon, everything went smoothly, the tenant left punctually. It was a strange experience to be in this house again after 6 years' (translated from Ref. [2], letter [850]). In June 1946 Pauli's friend Adolf Guggenbühl (see pages 285–6) had also contacted President Rohn in order to see what he could do on the question of Pauli's naturalization (Ref. 3, II.140). On 11 July President Rohn confirmed to Pauli the settlement of his financial claims during his leave in the United States, as well as the salary increase they had agreed upon, the total salary now amounting to 21390 francs per annum, all extras included (II.145). On 23 July Rohn told Bundesrat Etter that Pauli had authorized him to resume the procedure of naturalization in Pauli's name. (II.149) This he did on 12 August by letter to Dr A. von Reding, the head of the service of citizenship at the Federal Department of Justice and Police in Berne (II.153). But in spite of all these measures Pauli's naturalization still took three more years.

On 12 August 1946 Pauli wrote to Rabi at Columbia University in New York: 'I know you will not be pleased with my final decision: to keep my Professorship in Zurich. Believe me, that this decision, which is due to various circumstances, was awfully hard for me' (Ref. [2], letter [828]). On the same day he wrote two official letters of renunciation to Director Aydelotte at the Institute for Advanced Study and to President Fackenthal of Columbia University (see the last section) (*Pauli Archive*, CERN, PLC Bi 48 and 3). He then informed his friends, one after the other, about his decision. On 19 September Pauli related the news to Einstein, accompanied by his appreciation of Zurich and of 'Dame Helvetia' given at the outset of this section. He considered it 'still good that some physicists remain in Europe'. While he judged working conditions in the United States to be favourable for the near future, 'For the farther future (in about 5 years, say) I, however, see there the big danger of an interference of the military with physics (with or without the detour over a committee of non-physicists in civilian clothes).' But, in this letter to Einstein, Pauli concluded: 'It is questionable whether rational considerations alone may lead to decisions at all. It seems to me more and more that critical reason alone can only lead to Hamlet-like inaction' (translated from Ref. [2], letter [835] to Einstein). One of Pauli's and Franca's reasons to stay in Zurich was undoubtedly the nice house they had built in Zollikon in 1938 (see pages 309–10).

In physics, the latest subject was Heisenberg's scattering or S-matrix. As we saw on pages 384–6, Pauli had discussed this concept in Chapter IV of his MIT lectures in the autumn of 1944, where he had given a formal proof of the unitarity of S. In late summer 1945 he had begun corresponding with Christian Møller (1904–80) in Copenhagen who was studying the analytical properties of S (Ref. [2], letters [737] to [816]). And early in 1946 Heitler's collaborator Shih-Tsun Ma (b. 1913) came to the Princeton Institute for three terms to work with Pauli on the S-matrix. After Pauli's departure this collaboration continued by correspondence and lasted until the end of 1946 (Ref. [2], letters [811] to [856]). Pauli also used Ma for other duties like forwarding Pauli's mail and sending out reprints. Most importantly, Pauli asked Ma on 26 April 1946 to send him his personal copy of the *Handbuch Article* of 1933 (see pages 249f) which Pauli had left at the Institute with other items and which he now needed for the announced *Spezialvorlesung* (Ref. [2], letter [811]).

Heisenberg's motivation in the first of the four papers on 'The observable quantities in the theory of the elementary particles' [4a,b,c,d] was his belief that the divergences occurring in the theory, the self-energy of the electron and the vacuum polarization, 'probably must be understood as the expression of the fact that with the phenomena in question a new universal constant of the dimension of a length plays a decisive part' (translated from Ref. [4a], p. 513). This fundamental length would have to be of the order of the range of the nuclear forces, $\sim 10^{-13}$ cm (see pages 324–5). Heisenberg thought that 'also the future theory, of course, should in the first place contain relations between "observable quantities"' (translated from Ref. [4a], p. 514). Most important among those depending on the fundamental length would be the energies of stationary states and the cross-sections of scattering, emission, and absorption processes, i.e. the asymptotic wavefunctions at infinity. And the latter should be described by a new quantity, the *characteristic* or S-*matrix* which has the general form (9.83), where the one characterizes the incoming (undisturbed) wave and f the scattered wave. It is interesting to note that it was Fermi who for the first time had treated the case of the emission and absorption of particles in this way in his theory of β-decay (see pages 228–30).

Foremost among the properties of S is the unitarity (9.85). While on pages 385–6 a general operator proof was given, in Ref. [4a] Heisenberg uses asymptotic states at $t \to +\infty$ for this purpose. Of great practical importance is the problem of the connection between S and the Hamiltonian. In the second paper [4b] Heisenberg considers models in which $S = \exp(i\eta)$ is directly related to the interaction part of the Hamiltonian, in the form

$$\eta = \varepsilon \lambda^{n-4} \int d^4x \varphi^n, \tag{10.1}$$

where λ has the dimension of length and $n \geq 4$, where $n > 4$ describes the emission or absorption of particles. Note that, to lowest order in the interaction H', it follows from Eqs (9.83), (9.89) that $\eta/2\pi \simeq f \simeq -H'$, which is the meaning of Eq. (10.1). In the late 1940s this perturbative treatment became the point of departure of the covariant formulation of QED. Examples of a spherically symmetric perturbation $H' = V(r)$ are discussed in the fourth paper [4d].

Heisenberg's third paper [4c] gives in Section I.a) the following important argument for the determination of the stationary states from the S-matrix: Writing the asymptotic form of Eq. (9.87) for the case $H' = V(r)$ as

$$\Phi_k(r) = e^{-ikr} - S(k)e^{ikr}, \tag{10.2}$$

the stationary states are determined by the Schrödinger equation (9.78) with $E_k \propto k^2$ for $k = -i\kappa$. Now, since for such a value of k the second term in (10.2) diverges in the limit $r \to \infty$, it follows that $S(-i\kappa) = 0$. This implies that $S(k)$ must be an analytic function in the lower half k-plane and has zeros on the negative imaginary axis which determine the stationary states. Heisenberg had got this idea from his friend Kramers on his visit to Leiden in the autumn of 1943 mentioned on page 353 (see Ref. [6], p. 606).

Pauli's first contact with Heisenberg after the war was in a letter the latter had written end of April 1946. At the end of the war Heisenberg had been interned together with other German atomic scientists, first near Versailles in France and then at Farm Hall near Cambridge in England. It was at Farm Hall that they learned about the American atom bombs used over Japan. Heisenberg returned to Germany in January 1946 (see Ref. [7], Chapters 26 and 27). Pauli answered Heisenberg's letter on 26 May: 'Your papers on the S-matrix I have read with great interest, but they remain a mere programme as long as no method is indicated to determine the S-matrix theoretically. I hope that sooner or later a possibility will present itself where we may again discuss about physics' (translated from Ref. [2], letter [818]). And he closes with congratulations to Otto Hahn who had received the Chemistry Nobel Prize of 1944 in 1945 (see page 217). On 9 September Pauli informed Heisenberg about

Ma's result that the S-matrix may have 'wrong' zeros which do not correspond to stationary states. This discovery [8] is mentioned for the first time in Pauli's first letter to Ma from Zurich on 26 April: 'I discussed with several persons your redundant zeros of the S-matrix in the case of a potential $Ae^{-\mu r}$ and we found it interesting' (Ref. [2], letter [811]).

On 20 May Pauli had written to Møller in Copenhagen: 'I have finally decided to stay without an assistant for this term. If I shall stay here longer a possible candidate is also the Swiss Dr Jost who is now in Copenhagen. I shall ask you later about his qualities' (Ref. [2], letter [816]). Møller answered on 25 August: 'Dr Jost is very much looking forward to working in your institute' [830]. In Copenhagen Jost had collaborated with Møller and with Dirk ter Haar on the S-matrix [9a,b]. In particular he had examined the 'wrong' zeros of $S(k)$ which had been discovered by Pauli's collaborator Ma in Princeton [8]. On 22 October Jost reported his results to Heisenberg in an Appendix (Ref. [2], pp. 390–4) to Pauli's letter [841], in which Pauli wrote to Heisenberg that 'Herr Jost has written the very learned paper, on the results of which he reports here, with Møller in Copenhagen' (translated from Ref. [2], letter [841]) [9c].

Jost writes in his memoirs 'Recollections: Things read and things experienced' [10a] that he had left Zurich by bus in early May 1946, after having completed his Ph.D. under Wentzel at the university on the charge dependence of nuclear forces [10b]. In Copenhagen he met Abraham Pais from Holland who had arrived in January. Both born in 1918, they became lifelong friends. Res Jost grew up in the city of Berne where his father was a physics teacher at the municipal Gymnasium. His studies for a diploma in mathematics, physics, and chemistry teaching at his home university during World War II suffered frequent interruptions by military service. Towards the end of the war he spent one semester in Zurich where the lectures by Heinz Hopf in mathematics and by Gregor Wentzel in theoretical physics left a lasting impression on him. During a longer sojourn in Zurich Jost was fascinated by Heisenberg who had come for a seminar on the S-matrix. This led Jost to study this new approach, of which, however, he later became a severe critic (Ref. [10a], pp. 13–14; see also Ref. [10c], pp. 1–2).

On his selection as Pauli's assistant Jost recalls: 'In the autumn of the year 1946, quite unexpectedly, I was summoned (*aufgefordert*) by Wolfgang Pauli to become his assistant. It was impossible to reject this offer, and hence I cut short my stay in Copenhagen after less than half a year, boarded an airplane for the first time and arrived in Zurich at the beginning of October.' And he comments: 'That I returned to Zurich was obvious. What, however, could have motivated Pauli in the year 1946, against the rapid stream of emigrants from Europe, to decline the tempting offers from the United States and to return to the old continent, the raped one? On this serious question I have only frivolous answers' (translated from Ref. [10a], pp. 15–16). On 11 October President Rohn announced the appointment of Res Jost from Wynigen (Canton of Berne) as assistant in theoretical physics for the winter term 1946/7 with an overall salary of 3000 francs (Ref. [3], II.160).

Jost's 'learned paper' [9c] already carries the trace of his predilection for mathematical rigour which marks his entire career. It begins by introducing, for any non-singular value of k, two particular solutions $f(k, r)$ and $f(-k, r)$ of the Schrödinger equation (9.78) with $H' = V(r)$ which have the asymptotic form $f(\pm k, r) \sim \exp(\mp ikr)$. Then $\Phi_k(r)$ is a linear combination of the two with the property that $\Phi_k(0) = 0$. If we write $f(k) \equiv f(k, 0)$ the solution therefore may be written as

$$\Phi_k(r) = f(k, r) - \frac{f(k)}{f(-k)} f(-k, r). \tag{10.3}$$

Comparison of the asymptotic form of (10.3) with Eq. (10.2) then yields Jost's basic relation

$$S(k) = \frac{f(k)}{f(-k)}. \tag{10.4}$$

In this form the distinction between true and wrong zeros becomes simple: true zeros are zeros of the numerator and wrong zeros are divergences of the denominator. It turns out that these wrong zeros do not occur for potentials which are exactly zero above a certain value r_0 of r because in this case the asymptotic formulae are then exact. [9c] But it was only in the late 1950s that the wrong zeros were recognized as remnants of singularities in a crossed channel of the analytical S-matrix which is then a function of two complex variables, the Mandelstam variables s and t, the crossed channels of a four-particle process $ab \to cd$ being $ac \to bd$ and $ad \to bc$ (see Ref. [6], p. 607 and the references 16 there).

Pauli remained sceptical, particularly also towards the idea of a fundamental length, because he could see no connection with the fine structure constant (7.2) (Ref. [2], letter [833]). Later he wrote to Heisenberg: 'To put it drastically, I would not be surprised if your "universal" length were to reveal itself as a mere craze' (translated from Ref. [2], letter [899]). He kept his conviction that without a theory that determines the fine structure constant 'no real progress is possible anymore' (translated from Ref. [2], letter [854]). Not without astonishment one reads later in this letter to Heisenberg of 25 December 1946: 'Your assumption that I would be interested in "German Atom Bombs" (i.e. in papers about this or related subjects) has surprised me very much' (translated from letter [854]). This casts an interesting light on Heisenberg's attempt to justify his war-time activities to his friend. On 11 July in the following year Pauli's views about the S-matrix had become even more negative: 'I personally consider the idea of an analytic continuation of the S-matrix to be a complete flop,' he wrote to Heisenberg (translated from Ref. [2], letter [899], p. 462; see also Ref. [6], p. 607).

At the beginning of the summer vacation of 1946 Pauli spent some days in the mountains at Crans in Valais and attended the first international conference after the war in Cambridge, England, which lasted from 22 to 27 July (see Ref. [2], letters [826, 827]) and where he gave a paper on the difficulties in field theory [11a]. Reviewing first the methods of the λ-limiting process and of Dirac's negative probabilities he then discussed the S-matrix along the lines of Chapter IV of his MIT lectures (see pages 383f), showing that from a perturbation theory point of view the form

$$S = \frac{1 - i\pi K}{1 + i\pi K} \tag{10.5}$$

with a Hermitian matrix K seems to be more appropriate than Heisenberg's $S = \exp(i\eta)$. He concluded by mentioning the two limits of static fields and the emission of many low-frequency photons as possible test cases for a future theory.

At the end of August Pauli and Franca went on holiday to Morcote on Lake Lugano in southern Switzerland, from where he described to Jauch the quite favourable situation at Zurich. 'Among the younger Swiss I found two who largely surpass the average: Thellung and Schafroth. But both are still beginners. Houriet went to Geneva and so I was without assistant during the summer term (Villars is house-theoretician with Scherrer). But the coming term Jost will have this post who has done his Ph.D. with Wentzel and is now in Copenhagen' (translated from Ref. [2], letter [832]). All these names will become more familiar in the course of this narrative.

On 29 October 1946 Pauli wrote to Bohr: 'My wife and I have now finally decided to go to Stockholm on the occasion of this year's Nobel celebration and I shall give [t]here my lecture at about December 12th. It would be a nice occasion for us to see you and your family in Copenhagen afterwards, that means in the week between December 15th and 21st' (Ref. [2], letter [842]). Also present with Pauli at the Nobel festivities were Otto Stern and Otto Hahn (Ref. [2], letters [845, 854]) and, of course, the winner of the Prize for 1946, Percy Williams Bridgman (1882–1961) from Harvard, who received it 'for the invention of an apparatus to produce extremely high pressures and for the discoveries he made in the field of high pressure physics'. Rabi, however, did not attend; he never gave his Nobel Lecture. Stern's lecture was on 12 December and Pauli's on the 13th. In the already quoted letter [854] to Heisenberg of 25 December Pauli wrote: 'The festival days in Stockholm were very tiring but beautiful, in Copenhagen it was nice to reside again with Bohr who now has returned to physics' (translated from Ref. [2], letter [854]).

In Stockholm Pauli also saw his friend Oskar Klein again. There also Lise Meitner met Hahn 'again after a long time. I don't know whether they came to terms completely about everything', Pauli wrote to Schrödinger (translated from Ref. [2], letter [863]; see page 212). As for Pauli's Nobel Lecture, according to Bengt Nagel, the long-time member of the Nobel Committee for Physics who as a young man sat in the audience, it was not brilliant [11b]. The written version [11c], however, is an impressive historical document (see page 125). In his conclusion Pauli repeats his views about 'a correct theory': no infinite zero-point energies, no infinite zero charges, no 'mathematical tricks to subtract infinities', no 'hypothetical world'.

Since Pauli's 10-year appointment at ETH was coming to an end in March 1948, the School Council examined an extension for another 10-year period, as usual one year in advance, on 7–8 February 1947. Councillor Ernst Bärtschi, the president of the city of Berne, summed up the situation as follows: 'About Pauli we have already heard so many disagreeable things before that his present attachment to the ETH appears to me like a transmutation. In serious scientific circles he is considered an authority. I am glad of the fact that he has decided to remain at ETH' (translated from Ref. [3], II.167). Thereupon Pauli's extension was accepted by the School Council, and the Bundesrat confirmed it on 14 March 1947. The only remaining problem, however, was Pauli's naturalization. On this question the Zurich authorities informed President Rohn on 20 February 1947 that Pauli's case could not be treated before February 1948 (II.170).

As I experienced myself as a fresh student, after the war ETH was a very lively place, full of hardy Scandinavian students, which caused much excitement among the girls of Zurich. According to the quotation at the start of this section, Pauli also had some Swedish students in his group. Professor Torsten Gustafson of the University of Lund sent two of his pupils, Carl-Erik Fröberg in spring 1946 and Bertil Nilsson in the autumn (Ref. [2], letters [814 and 836]). Many visitors passed through Zurich. In September 1946 Pauli had a conversation with the famous mathematician and philosopher Bertrand Russell (1872–1970) who had come for three lectures (Ref. [2], letter [835]). Among Pauli's colleagues Rozental and von Laue called in (letters [851] and [863]), and Pauli's friend Otto Stern came for three weeks in the winter of 1946/7 [863], which was the beginning of Stern's yearly commuting between Zurich and Berkeley. In the new year of 1947 Pauli wrote to Casimir that 'in Zurich life is quite agreeable, I am also very satisfied with my assistant Jost' (translated from Ref. [2], letter [858]).

Another dear friend from Hamburg must be mentioned here: Erich Hecke, who, as dean of the Faculty in 1924, had welcomed Pauli as a Privatdozent (see page 158). Although Hecke was Pauli's senior by 13 years, this friendship meant very much to Pauli. After the war Hecke was in poor health, and since food still was scarce in Germany, his Copenhagen mathematics friends, in particular Bohr's brother Harald, invited him to Copenhagen in the late autumn of 1946 (Ref. [2], letter [845]). There Pauli saw him for the last time on his journey to Stockholm in December [868]; Hecke died there two months later on 13 February. In the project of letter [868] Pauli wrote to Hecke's widow: 'It was a human relationship which went far beyond the common spiritual and scientific interest.' And: '"At an advanced hour" (as he used to say) which we often spent in the company of Mosel wine and music, sometimes walking until sun rise, then things discussed concerned intimate matters that even touched the religious sphere' (translated from Ref. [2], letter (project) [868]). The only witnesses to this relationship in the Pauli correspondence are the two letters [528] and [534] in Ref. [11d]. Erich Hecke studied mathematics in Breslau, Berlin, and Göttingen where he wrote a thesis on number theory in 1910 and where he became a Privatdozent in 1912. After two years as a professor at Basle and one year at Göttingen he went to Hamburg in 1919, when the university was founded (see page 89). Hecke became well known for his contributions to special analytic functions and operator theory [13].

In physics the puzzle of the mesons continued to confuse the issues of both meson theory and nuclear forces and the cosmic radiation. While, because of their Coulomb repulsion, positive mesons were thought to decay, rather than being absorbed, the absorption of negative mesons indicated that the interaction was up to 10^{12} times weaker than the theoretical meson–nucleon coupling of Eqs (8.72) or (9.4). In the years 1942–3 Japanese researchers had already proposed the existence of two kinds of mesons. Such a proposal was independently discussed at the conference on Shelter Island on the eastern tip of Long Island, east of New York, which took place at the beginning of June 1947. The first preliminary experimental evidence came at the end of May 1947 from the Bristol group of Cecil Frank Powell (1903–69) working with photographic emulsions, and this result was confirmed in October [14a]. Thus, finally, there were two mesons, the π-meson or pion with a mass of $m_\pi = 277 m_e$ which is responsible for the strong interaction of Eqs (8.72), (9.4) but which decays by weak interaction into a μ-meson or muon with a mass of $m_\mu = 207 m_e$ which itself decays into an electron (mass m_e), both decays being accompanied by the emission of a neutrino (see Ref. [14b], Section 18b).

The Shelter Island conference, which was organized by Oppenheimer and sponsored by the US National Academy, turned out to be a decisive event also with respect to QED (see Refs [15a] and [15b], Chapter 4). Under the directorship of Rabi, Columbia University had become the leading laboratory producing spectacular new experimental results. One of them, reported at Shelter Island, was a split of $\sim 10^9 \mathrm{s}^{-1}$ between the $n = 2$, $^2S_{1/2}$, and $^2P_{1/2}$ levels of the hydrogen atom which, according to Dirac's theory, should be degenerate. This experiment had become possible only by the development of the new microwave technology during the war at the Radiation Laboratory of MIT (see page 354). Indeed, the authors Willis Eugene Lamb (b. 1913) and Robert Curtis Retherford (1912–81) wrote: 'The great wartime advances in microwave techniques in the vicinity of three centimeters wave-length make possible the use of new physical tools for a study of the $n = 2$ fine structure states of the hydrogen atom' [15a]. Weisskopf recalls: 'A sensational development on the opening day catapulted quantum electrodynamics into being the center of interest at the conference. At the first session, Willis Lamb announced that he and his collaborator, Robert C. Retherford, had reliably measured a small shift of a hydrogen level [$^2S_{1/2}$] away from the value [$^2P_{1/2}$] it should have had according to the fundamental equations of quantum mechanics' (Ref. [17], p. 168). Later on Weisskopf writes: 'I felt guilty that I had not tried harder to calculate that shift before its final discovery. In principle, I could have done it in 1936 and certainly later by following some of the suggestions that Kramers had given me in 1940. He had argued that the difficulty of the extremely large mass (or energy) of the electron might be circumvented by comparing the energy of the free electron to that of the electron bound in the hydrogen atom' (Ref. [17],

p. 169).

Shortly after Shelter Island this mass renormalization was indeed used by Bethe in an approximate non-relativistic calculation of the shift of the $^2S_{1/2}$ level. Using a plausible cut-off of mc^2, Bethe did obtain the required value $1.040 \times 10^9 \text{s}^{-1}$ (see Ref. [14b], Section 18(c)1). At the end Bethe acknowledged: 'This paper grew out of extensive discussions at the Theoretical Physics Conference on Shelter Island, June 2 to 4, 1947' [18a]. Bethe's paper stimulated a host of more sophisticated calculations by Richard Phillips Feynman (1918–88), [18b] by Norman M. Kroll and Lamb [18c], by Schwinger [18d], and by Bruce French (b. 1921) and Weisskopf [18e]. In this last reference both mass and charge renormalization are performed explicitly. On 9 December 1947 Oppenheimer wrote to Pauli about 'our meeting last Spring at Ram Island', Ram Head being the eastern tip of Shelter Island (Ref. [2], letter [919]). He reported three main results of QED corrections, namely the 'Lamb shift' just mentioned, Schwinger's correction $-(\alpha/2\pi)\mu_B$ to the magnetic moment of the electron [19], and Schwinger's finite radiative corrections to the electron scattering [20]. Ten days later Rabi reported on the experimental results from his laboratory concerning the correction δg to the g-value (4.11), obtained by measurements of the $^2S_{1/2}$ state of sodium and of the $^2P_{1/2}$ state of gallium (Ref. [2], letter [921]), the result being $\delta g = 0.00244 \pm 0.00006$ [21]. All this shows that the United States had now vigorously taken the lead in particle physics.

But Pauli had been occupied with other things. On 5 May 1947 he wrote to Bohr 'concerning the stopping of mesons in dense substances' that 'I don't wish to be occupied too much with complicated problems of this kind which do not seem to be of any fundamental significance. Your arguments are certainly qualitatively correct but whether they are sufficient to explain the whole missing factor 10^{12}, can only be decided by more quantitative computations. I am sure that there will be many younger theoreticists, who will soon be able to decide this question' (Ref. [2], letter [880]). On 17 September Pauli wrote to Fierz from Uppsala 'that in Bristol the number of photographs, on which the creation of a light meson out of a heavier one is to be seen ... has now grown so much ... that I am inclined to consider this result as experimentally certified' (translated from Ref. [2], letter [912]). Pauli had attended the Copenhagen conference between 15 and 20 September and at the beginning of October he was invited to Lund in Sweden, where he stayed for 10 days, until the beginning of the winter term at ETH (Ref. [2], letter [909]). On the way back he hoped to see Lise Meitner in Stockholm [913]. In Zurich, meanwhile, Gregor Wentzel, after a highly successful career of 20 years of teaching and research, resigned from his professorship at the University to take up a call from the University of Chicago. He left Zurich in March 1948 (see the letter [894] to Stern in Ref. [2]).

In letter [880] Pauli told Bohr that 'Recently a philosophical periodical (called *Dialectica*) was founded in Zurich by Prof. Gonseth and a group of mathematicians in which the epistemological fundaments of mathematics and of different parts of natural sciences, including physics, are being discussed.

I gave my consent to consult the redaction about articles regarding the foundations of physics. There is a particular interest for your point of view of complementarity and I was asked by the redaction whether I am willing to organize a particular issue of the *Dialectica* on complementarity and foundations of quantum mechanics which should appear in about one year' (Ref. [2], letter [880]). After Bohr had reacted positively on 16 May 1947 to the invitation for a contribution, Pauli sent out a circular letter on 9 July in German [897] which was later followed by another one in English [898], that mentioned the contributions already received or promised and in which a deadline of May 1948 was indicated, so that the issue could appear at the beginning of July of that year.

The foundation of *Dialectica* in 1947 as an international review of philosophy of knowledge was due to Pauli's mathematics colleagues at ETH, Ferdinand Gonseth (1890–1975) and Paul Isaak Bernays (1888–1977), and to the French philosopher Gaston Bachelard (1884–1962) (see Ref. [2], comment p. 438). The French-speaking Gonseth from Sonvilier (Canton of Berne), who became the long-time director of *Dialectica*, had studied mathematics at ETH where he obtained his Ph.D. in 1916 and where he became a Privatdozent one year later. In 1919 he became an associate professor in applied mathematics at the University of Zurich and in 1920 a full professor at Berne. Starting in 1929 Gonseth taught higher mathematics in French at ETH and when in 1947 a chair of philosophy of science was created at ETH, it was Gonseth who was nominated for it. On his retirement in 1960 Gonseth settled in Lausanne. He had published in French on the foundations and philosophy of mathematics. In 1961 he became an officer of the French Legion of Honour [22a,b]. Bernays was born in London into a well-known German Jewish family of scholars and merchants; his great-grandfather, Isaac Bernays, had been Rabbi of Hamburg. Bernays went to high school in Cologne, and then studied in Berlin and Göttingen where he obtained his doctor's degree in 1912. He spent the following five years as a Privatdozent at the University of Zurich, after which he returned to Göttingen as assistant professor in Hilbert's institute. Evading the Nazi threat, Bernays joined ETH in 1934 as a lecturer and from 1945 until his retirement in 1959 was an associate professor. He was famous for his work on the foundation of mathematics (see also Ref. [2], pp. 458-9) [22a,b].

Pauli wrote a detailed editorial for the special issue of *Dialectica* on Complementarity, in which he placed the different contributions into the general context [23]. Concerning his own opinion, he emphasized that only by renouncing the possibility of drawing general conclusions from results gained on individual atomic systems in particular states is it possible to keep the usual notion of space and time and of a 'closed system'. He concluded that 'It is in this sense that I consider the quantum mechanical description to be *complete*' (Ref. [23], p. 309). In addressing the application of non-Aristotelian logic made in many of the more mathematical contributions to the issue, Pauli wrote: 'the physicists (among them myself) have great resistance against the acceptance of new axioms of logic' (Ref. [23], p. 309), by which he reiterated his conviction that experimental facts have precedence over mathematical formalism. He closed his editorial with the observation concerning the "'elementary" particles in nature', that 'We are here only in the very beginning of a new development of physics, which will certainly lead to still further generalizing revisions of the ideals underlying the particular description of nature which we today call the classical one' (Ref. [23], p. 311).

Complementarity was again the subject when Pauli was invited for a second time to give a lecture to the Zurich Philosophical Society (see page 256) in February 1949 [24]. The above renunciation necessary in quantum theory led Pauli to the following observation in his second lecture: 'As a result of the development of atomistics and quantum theory since 1910 physics has gradually been compelled to abandon its proud claim that it can, in principle, understand the whole universe. All physicists who accept the development . . . of the mathematical formalism of wave mechanics, must admit that while at present we have exact sciences, we no longer have a scientific picture of the universe (*Weltbild*).' But Pauli saw in this renunciation 'the germ of progress towards a unified total world-picture, of which the exact sciences are only a part. In this I would like to see the more general significance of the idea of complementarity' (Ref. [24], pp. 35–6). This desire for unification became more and more the guiding idea of Pauli's thinking.

In the second lecture to the Zurich Philosophical Society two new ideas appear for the first time, which frequently recur in later essays in Ref. [25]. The first is the opposite of the detached observer of classical physics who views the automatic evolution of the physical system passively from the outside. Pauli writes: 'The observer or instruments of observation which modern microphysics has to consider thus differs essentially from the detached observer of classical physics' (Ref. [24], p. 40). The second idea is the notion of transformation (*Wandlung*) which Pauli repeatedly uses in its alchemistic meaning to express what technically is called the 'reduction of the wave packet': 'The course of the events, taking place according to predetermined rules, is interrupted by this observation, and a transformation (Wandlung) is evoked with an unpredictable result' (Ref. [24], p. 40). Here the influence of Jung's intense occupation with alchemy is undeniable.

Kepler, background physics, and synchronicity

The encounter of your research with alchemy is for me a serious symptom of the fact that the development tends toward a closer merging of psychology with the scientific experience of the processes in the world of material bodies. This probably means a rather long path, of which we experience only the beginning and which will be connected in particular with a continuing relativising critique of the notion of space-time. [26a]

In this first letter from Pauli to Jung after the former's return to Zurich, dated 23 December 1947, Pauli begins by giving Jung his consent to put his (Pauli's) name on the act of foundation, dated 24 April 1948, of the C. G. Jung Institute in Küsnacht. But then Pauli engages in the above reflection on space and time, which continues with the following truly Paulian remark: 'As is well known [26b], space and time have been put by Newton so-to-speak to the right hand of God (remarkably (*pikanterweise*), on the place that had been left vacant by the Son of God who was expelled by him [Newton] from there)' (translated from Ref. [27], letter [33]). This letter, however, was not the first contact between Pauli and Jung after the war. Indeed, in letter [832] of 29 August 1946 quoted in the last section Pauli told Jauch, who knew Meier from his assistant days in Zurich: 'I see C. A. Meier very often. Once he took me along to Jung at Bollingen on the upper lake of Zurich [see page 244]. The latter is writing a book on the *chymische Hochzeit* [wedding] or *conjunctio sol [et] luna* of the alchemists. He seems to be completely absorbed by this. May not this binding be too strong?' (translated from Ref [2], letter [832]). The book Pauli refers to here is Jung's last great work *Mysterium Coniunctionis* [28a] (see page 245) which, according to Jung's preface of October 1954, he had started more than 10 years before (Ref. [28a], Vol. 1, p. 11).

Carl Alfred Meier (1905–95), a disciple of Jung, often played the role of messenger between Pauli and Jung. Among his friends Meier was called Fredi, which led Pauli to name him $C + A = F$ in his correspondence. Meier grew up in Switzerland's northern-most city of Schafhausen near the Rhine falls. After studies in Paris, Vienna, and Zurich, where he obtained his medical doctorate in 1931, he, like Jung, became an assistant at the Psychiatric Clinic Burghölzli of the University of Zurich. Later he was Jung's assistant in the latter's lectures and seminars at ETH. In 1936 he established himself as an analyst with a private practice. In 1949 he became a Privatdozent in psychology at ETH. When asked by President Rohn, Pauli gave unofficially his appreciation of Meier's habilitation paper, whose goal was to point out parallels in modern psychoanalytic practice to the ancient rituals of incubation (see Ref. [3], document III.20). In 1959 Meier was nominated a professor at ETH. In 1948, at the foundation of the C. G. Jung Institute, a postgraduate teaching institution for analytical psychology, Meier became its first president, which position he kept until 1958. In 1954 and 1959 he lectured in the United States. Meier published many important works on Jungian psychology, in particular a course on Jung's complex psychology in four volumes. [28b] Like Jung, Meier, during his long career, collected an impressive private library of rare and precious books, which after his death in 1995 was transferred to the main library of ETH. Meier kept the classical heritage alive by spending his summer vacations, often in company with Pauli, in southern Italy or Sicily—the ancient Greater Greece—and in winter visiting Rome [22a, 28c].

In letter [35] of 7 November 1948 Pauli sent Jung two dreams (Ref. [27], number [32]; see Note [29]). In the first, of 25 October 1946, two figures make their appearance, which later frequently recur: the "Blond", a tall man somewhat younger than Pauli, and the dark "'Persian" . . . who is not admitted to study at the Institute of Technology (opposition to the prevailing scientific collective opinion). The "Blond" and the "Persian" are perhaps dual aspects of the same figure (both never occur at the same time)' (translated from Ref. [27], p. 34). In the second dream, of 28 October 1946, the "Blond" stands beside Pauli, who reads 'in an old book about the lawsuits of the Inquisition against the followers of the Copernican doctrine (Galilei, Giordano Bruno), as also about *Kepler's image of the Trinity.*' The dream continues: 'Then the Blond says *"The men whose wives have objectivated the rotation are accused".*' Later in the dream 'the Blond says sadly to me (apparently referring to the book): "The judges don't know what rotation or turning means, therefore they are unable to understand the men." With the urgent sound of a teacher he continues: *"But you do know what rotation is!"* "Of course", I say at once, "the turning and the circulation of the light, this evidently belongs to the basic knowledge' (translated from Ref. [27], p. 35).

This 'basic knowledge' consisted in Pauli's earlier dreams on rotation, to begin with, in the 'Great Vision' of the 'World Clock' described on pages 246–7, but in particular also those numbered 26 and 37 in Ref. [30]. In dream 26 a shooting star begins a left-handed circulation (Ref. [30], Section 227) and in dream 37 a dark centre is circumscribed by light (Ref. [30], Section 258). Jung explains the latter as follows: 'The light always points to the consciousness and hence first runs along the periphery [of the dark self]' (Ref. [30], Section 259). As concerns the enigmatic statement about 'the men whose wives have objectivated the rotation' Pauli writes to Jung: 'The "accusation" also refers to the resistance from the side of the collective opinion (see above, remark to the previous dream). From the superior point of view of the process of becoming conscious (*Bewusstwerdung*) the accusation refers to the fact that the men did not know where their wife (= anima) is and what *her* role was in the process of perception' (translated from Ref. [27], p. 35).

This, however, is not the whole story. Indeed, directly after describing the second dream Pauli writes: 'Thereupon I woke up very shaken. The dream was an experience of *numinous* [unspeakable, mysterious, frightening] character which influenced my conscious attitude in an essential way. *It then motivated me to resume the work on Kepler. Evidently then (17th century) a projection of the mandala and rotation symbolism towards the exterior had occurred*' (translated from Ref. [27], p. 35, my italics). Now, in Panofsky's eulogy for Pauli's Nobel Prize in Princeton we have seen 'how much impressed he [Pauli] had been by a reading of *Kepler*' (see page 395). This shows that the beginning of Pauli's work on Kepler goes as far back as the early 1940s in Princeton where Kepler's works were available to Pauli in the library of the Institute for Advanced Study. There also, and throughout this work, Panofsky gave Pauli his advice concerning the history and the sources of ideas; he is the main recipient of Pauli's acknowledgements in the published work (Ref. [31a], p. 167; Ref. [31b], p. 220).

In Pauli's publication 'The Influence of Archetypal Ideas on the Scientific Theories of Kepler' of 1952 [31a,b], Kepler's work of youth, *Mysterium Cosmographicum* is quoted as follows: '*The image of the triune God is in the spherical surface, that is to say, the Father's in the centre, the Son's in the outer surface, and the Holy Ghost's in the equality of relation between point and circumference.'* (Ref. [31b], p. 225). Further on Pauli writes: 'The symbolical images and archetypal conceptions are what cause him [Kepler] to seek natural laws. For this reason we also regard Kepler's view of the correspondence between the sun with its surrounding planets and his abstract spherical picture of the Trinity as primary: *because he looks at the s un and the planets with this archetypal image in the background he believes with religious fervour in the heliocentric system*—by no means the other way around, as a rationalistic view might cause one erroneously to assume' (Ref. [31b], p. 232).

Thus, perhaps, the 'objectivation of the rotation' by the wives, i.e. the anima—namely the personification of a female nature in the unconscious of a man—may be seen as Eve's apple, or as a dance, or as Kepler's archetypal sphere—or as space in general. This last interpretation is indeed the one that Pauli gives in letter [35] to Jung mentioned above: 'Since you have told me that you now are occupied particularly with the rotation symbolism of mandalas [see pages 245–7] I take the liberty of sending you the precise text of one of my dreams which occurred about 2 years ago and where the rotation—*and therewith the notion of space*—occupied the central position' (translated from Ref. [27], p. 38, my italics). This quotation refers to the second dream described above; it also shows the fascinating give and take of ideas between Pauli and Jung.

According to Jung (Ref. [30], Section 313), Pauli's vision of the 'World Clock' represents the essence of space–time; it is a picture of the medieval macrocosm consisting of the revolving concentric spheres of the fixed stars, of the sun, of the five known planets, and of the moon. This mythical picture of space–time became more quantitative, mainly during the Renaissance, by distance measurments among the heavenly bodies. Thus, Kepler, in his first book *Mysterium Cosmographicum*, published at age 25 in Graz, Austria, could use the relative distances given by Nicolaus Copernicus (1473–1543) for his purely geometrical ordering of the spheres of Saturn, Jupiter, Mars, Earth, Venus, and Mercury between the Platonic bodies of the cube, the tetrahedron, the dodecahedron, the icosahedron, and the octahedron, respectively (see Ref. [33], pp. 249–51). Thirteen years later Kepler published his principal work *On the motion of the star Mars*, in which the first two of his famous three laws are formulated, namely that (1) the orbits of the planets are not circles but ellipses with the sun in one of the foci, and (2) the motion of the planets is not uniform but such that the line joining the planet to the sun sweeps equal areas in equal times. Pauli desribes this spectacular evolution as follows: 'My attention was therefore directed especially to the seventeenth century, when, as the fruit of a great intellectual effort, a truly scientific way of thinking, quite new at the time, grew out of the nourishing soil of a magical–animistic conception of nature. For the purpose of illustrating the relationship between archetypal ideas and scientific theories of nature Johannes Kepler (1571–1630) seemed to me especially suitable, since his ideas represent a remarkable intermediary stage between the earlier, magical–symbolical and the modern, quantitative–mathematical descriptions of nature' (Ref. [31b], p. 222).

In Kepler's spherical picture symbolizing the divine Trinity, as also in his geometrical arrangement of the planet spheres described above, time is conspicuously absent. Pauli comments on this absence as follows: 'Perhaps the lack of a symbolism of time in Kepler's spherical picture is related to the lack of any suggestion of quaternity' (Ref. [31b], p. 234). And he finds this 'all the more significant since Kepler had an excellent knowledge of the Pythagorean numerical speculations' (Ref. [31b], pp. 233–4). Concerning this picture, Pauli writes in letter [929] to Fierz on 7 January 1948: 'Indeed—as I gather from conversations with C. A. Meier—Kepler's spherical symbol of the Trinity gives the psychologists some headache. It acts like a mandala, but isn't one in so far as it contains neither 4 cardinal points nor any other hint of the number four' (translated from Ref. [2], p. 497). Pauli gave two lectures on the subject at the Psychological Club of Zurich on 28 February and on 6 March 1948, for which he had written an extended abstract (Ref. [27], pp. 199–204 and Ref. [34], pp. 509–14). Apparently, these lectures were discussed intensely among the participants.

More generally, in his Kepler article Pauli expresses the view 'that intuition and the direction of attention play a considerable role in the development of the concepts and ideas'. And he concludes that, in order to establish the connection between sense perception and concepts, 'It seems most satisfactory to introduce at this point the postulate of a cosmic order independent of our choice and distinct from the world of phenomena' (Ref. [31b], p. 220). This order manifests itself in 'a "matching" [Zur-Deckung-Kommen] of inner images pre-existent in the human psyche with external objects and their behavior. . . . These primary images . . . are called by Kepler archetypal. . . . Their agreement with the "primordial images" or archetypes introduced into modern psychology by C. G. Jung and functioning as "instincts of imagination" is very extensive' (Ref. [31b], p. 221). In the comment to the two dreams [32] mentioned in Ref. [27] Pauli writes to Jung: 'As you know, I then hit upon that strange fellow R. Fludd . . . whose anima had not objectivated the rotation for him since, indeed, the former could still find her expression in the Rosicrucian mysteries' (translated from Ref. [27], p. 35). As we shall see, in the course of his work on Kepler, Pauli's discovery of Robert Fludd (1574–1637) made for an important psychological counter-balance.

In Ref. [31b] we read on p. 244: 'Kepler's views on cosmic harmony, essentially based on quantitative, mathematically demonstrable premises, were incompatible with the point of view of an archaic–magical description of nature as represented by the chief work of a respected physician and Rosicrucian, Robert Fludd of Oxford.' In his Harmonices Mundi of 1619 Kepler violently criticized Fludd's work, whereupon Fludd responded with a detailed polemic against Kepler, to which Kepler 'replied with an Apologia . . . that was followed by a Replicatio . . . from Fludd' (Ref. [31b], p. 244). Further on Pauli writes: 'The intellectual "counter world" with which Kepler here clashed is an archaic–magical description of nature culminating in a mystery of transmutation (*Wandlungsmysterium*). It is the familiar alchemical process that by means of various chemical procedures releases from the *prima materia* the world-soul dormant in it and in so doing both redeems matter and transforms the adept. Fludd, unlike Kepler, had no original ideas of his own to proclaim; even his alchemical notions are formulated in a very primitive form' (Ref. [31b], p. 244).

In letter [929] to Fierz, Pauli writes that Fludd plays 'the role of an author who is untroubled by any specific knowledge and who reproaches the other [Kepler] that he has a point of view diverging from his own archaic one—and that this, somehow, were bad. But "somehow" (Fludd, of course, knows nothing about it) it is indeed bad, i.e. defective. Sometimes Fludd plays the role of a fool who tells the truth' (translated from Ref. [2], p. 497). This defect, in fact, is a retrieval of the projection. Indeed, in the comment to the two dreams [32] in Ref. [27] Pauli writes: 'But *Fludd knew where the anima was* with Kepler and the other scientists: she had wandered from matter into the knowing subject, which evoked the greatest mistrust in Fludd, since then—outside of the Rosicrucian mysteries—she escaped the control by the consciousness' (translated from Ref. [27], p. 35). And in the extended abstract of his two lectures Pauli writes: 'Generally, one has the impression that Fludd is always wrong where he lets himself be carried away into an astronomical or physical discussion. Nevertheless, the polemic between Fludd and Kepler seems still to be of relevance also to modern man. Indeed, an important hint is contained in Fludd's reproach raised against Kepler "you force me to defend the dignity of the *quaternary*" [see also letter [929], Ref. [2], p. 497]. This is for modern man a symbol for the *completeness of experience* (*Erleben*) which is not possible within the mode of consideration of the natural sciences. The former is superior in the archaic point of view, which also tries to express with its symbolic pictures the emotions and the valuations of the soul through feeling' (translated from Ref. [27], p. 203). Pauli grants Fludd the whole of Appendix II to let him praise 'the dignity of the number four'.

On 9 March 1948 Pauli wrote to Fierz that 'in relation to my work on the Kepler lectures, about 4-6 weeks ago, suddenly and immediately (*brühwarm*) Scotus Eriugena [9th century] occurred to me' (translated from letter [941], Ref. [2], p. 513). He then describes how he started to study this Platonist author. The result of these studies is Appendix III of Refs [31a,b], the opening remark of which reads: 'The controversy between Kepler and Fludd is connected, from the point of view of the history of ideas, with the existence in the Middle Ages of two different philosophical trends which I may designate briefly as the *Platonic* and the *alchemistic* (or hermetic)' (Ref. [31b], p. 277). Further on he writes: 'The Platonists, as we have seen in Kepler's case also, favoured in general a trinitarian attitude in which the soul occupies an intermediary position between mind and body. It may be of considerable interest to know, however, that in the earliest Platonic thinker of the Middle Ages, Scotus Eriugena, the idea of quaternity can also be found' (Ref. [31b], p. 278). Pauli then discusses the latter's model of the four-fold division of nature, observing that it did not fit into the dogma of the Church which is based on the trinity of God (see also the appendix to letter [1255] to Jung's secretary Aniela Jaffé on 16 June 1951 in Ref. [35]).

Pauli continued to discover new aspects of Fludd's work and personality well into the year 1950, so that his occupation with the Kepler article extended to the end of that year (see letter [1178] to Panofsky on 11 December 1950 in Ref. [35]). And he became more and more personally involved with it; he believed 'that I carry "Kepler" as well as "Fludd" in myself and that it is for me a necessity to arrive at a synthesis of this pair of opposites, as best I can' (translated from letter [1188] to Fierz on 25 December 1950, Ref. [35], p. 226). During his stay in Princeton from the end of November 1949 to the end of April 1950, Pauli also gave a lecture on his Kepler article at the Institute for Advanced Study, as he reports in his letter [1080] of 14 February 1950 (Ref. [35], p. 21) to Jung's collaborator Marie-Louise von Franz (1915–98). She had translated Fludd's Latin texts for Pauli since 1947 when, on 27 October, he addressed her in the first letter saying: 'About your "being late" I am not at all "angry". My motto is *"Everything comes to him who knows how to wait"* [in English, italics added]. (This I once telegraphed from the United States to C. A. Meier, whereupon he wrote back that he was glad but that, unfortunately, he did not understand at all what I meant with it, after all.—In your case I have reasons, however, to assume that your reaction to my motto will not be the same.)' (translated from letter Hs. 176:4, ETH, Wissenschaftshistorische Sammlung, WHS). From Princeton Pauli sent Aldous Huxley (1894–1963) comments on his Kepler study and told him that he would send him a copy of Ref. [32b] as soon as he was back in Zurich (letter [2269] in Ref. [36a]).

Towards the end of the Kepler article Pauli draws the conclusion with respect to modernity: he sees 'in the fact that the "quaternary" attitude of Fludd corresponds, in contrast to Kepler's "trinitarian" attitude, from a psychological point of view, to a greater *completeness of experience (Erleben)*' (Ref. [31b], p. 258). However: 'It is obviously out of the question for modern man to revert to the archaistic point of view that paid the price of its unity and completeness by a naive ignorance of nature. His strong desire for a greater unification of his world view, however, impels him to recognize the significance of the pre-scientific stage of knowledge for the development of scientific ideas ... by supplementing the investigation of this knowledge (*Erkenntnis nach aussen*), directed inward (*Erkenntnis nach innen*)' (Ref. [31b], p. 259). Pauli closes the Kepler article by repeating his ideas about the detached observer (see the previous section) and about the transformation or transmutation (*Wandlung*) in the process of observation in microphysics mentioned at the end of the last section.

With respect to the first of these two ideas Pauli wrote in letter [33] to Jung on 23 December 1947, quoted at the outset of this section: 'Modern microphysics puts the observer again in his microcosm as a little master of the creation, with the capacity for a (at least partial) free choice and of, in principle, uncontrollable effects on the observed.' And he asked the intriguing question: 'If, however, these phenomena depend on the manner, how (with what experimental device) they are observed, do not then perhaps *also* exist phenomena (extra corpus) which depend on *who* observes them (i.e. on the constitution of the psyche of the observer)? ... should there not then be sufficient space for all sorts of curiosities for which, in the end, the distinction of "physical" and "psychical" loses its meaning?' (translated from Ref. [27], p. 36). In the following letter [34], dated 16 June 1948, Pauli speaks of 'projection of psychical facts onto properties of matter' (translated from Ref. [27], p. 38).

At this time Pauli made a systematic study of 'physical terms as archetypal symbols' in an essay of June 1948 which he called 'Modern examples of background physics' [36b], and which begins as follows: 'Under "background physics" I understand the occurrence of quantitative notions and conceptions of physics in spontaneous fantasies in a qualitative, transposed, hence symbolic sense. The existence of this phenomenon is known to me for about 12 to 13 years from personal dreams which evolve completely uninfluenced by other persons' (translated from Ref. [36b], p. 176). At the beginning, these dreams, in which physical notions such as 'wave, electric dipole, thermo-electricity, magnetism, atom, electron shells, atomic nucleus, radioactivity' occurred as symbols (translated from Ref. [36b], p. 176), appeared shocking to Pauli's rational mind, like an *abuse* of scientific terminology. But later he recognized the objective and independent character of these dreams or fantasies. 'In addition', he writes, '*the purely psychological interpretation* seems to me to comprehend *only one half of the facts. The other half is the disclosure of the archetypal foundation* of the notions actually applied *in today's physics*' (translated from Ref. [36b], p. 177).

On 12 August 1948 Pauli wrote to Fierz about the origin of this essay [36b]: 'In the allocution that C. G. Jung gave at the opening of the institute named after him, this subject [of the symbol] occupied a relatively broad space. With his formulation, that "quaternity ... is also inherent in the essence of the observed micro-physical process" I, however, did not and do not agree; I am almost certain that within physics in the restricted sense the quaternity of Jungian psychology finds no application.—I therefore have decided to formulate with the aid of examples, in an as positive and synthetic way as possible, what *I myself*, in fact, understand with respect to this subject, as a physicist and also as a single individual (whereby I endeavoured to make clear also the differences of my ideas from those of C. G. Jung).' And Pauli adds: 'The result was a small essay which, after discussion with C. A. Meier, I sent C. G. Jung in June' (translated from Ref. [2], letter [971], p. 560).

As to these differences, Pauli saw a real danger to his reputation as a physicist in the possibility that Jung could quote him wrongly; he later wrote to Fierz 'that I *dream* of physics the same way as Herr Jung (and other non-physicists) *think* about physics. The danger of this situation lies in the fact that Herr Jung could publish nonsense about physics and thereby could even refer to me. This it is essential to prevent, and to turn the matter towards the positive. I can *not* simply subtract myself (*mich drücken*)! But every time I have spoken with Herrn Jung (about the "synchronistic" phenomenon [see below] and similar things), a certain spiritual fructification takes place' which, with Pauli, manifested itself in dreams (translated from letter [1058] to Fierz on 26 November 1949, Ref. [2], p. 708).

In Section 2 of Ref. [36b] Pauli discusses as examples dreams in which a *doubling* (spectral lines) or a *separation* (isotopes) occurs, and he writes: 'In order to identify the "second sense" of this group of dreams, their assertion, to start with, must be translated into a *neutral* language with respect to the distinction of the physical and the psychological. Such a translation always contains hypothetical elements, also the physical assertions should not be taken too seriously' (translated from Ref. [36b], p. 179). He then gives a tentative 'dictionary': "'Frequency", on the one hand, determines a specific energetic state and, on the other, considered in time, a regular repetition [see Eqs (4.1) and (5.30)].' Later on Pauli observes that this connection appears 'from the point of view of classical physics as totally unexpected, even as irrational (I still remember exactly the strong intellectual shock which this state of affairs and its consequences dealt me as a student' (translated from Ref. [36b], p. 180). Both frequency and energy are well defined only in an oscillation and a measurement, respectively, that last for ever. Energy, however, cannot be destroyed (conserved), i.e. in alchemist language, it is '*increatum*' (non-created), as was matter, i.e. '*prima materia*', in alchemy. Thus energy—and momentum—, on the one hand, and motion in time—and space—, on the other, are '*two opposite (complementary) aspects of reality*' (translated from Ref. [36b], p. 181).

In Section 3 this pair of opposites is translated into psychology as 'time-less objective psychical' and 'ego-consciousness in the time' (see the diagram on p. 187 of Ref. [36b]). As we shall see, for Jung such pairs of opposites were a favoured means of understanding. Seeking still other examples of separation, Pauli mentions the act of giving birth, which in the psychological understanding may be seen as 'detachement of the spiritual body to autonomous existence' in the state of meditation (translated from Ref. [36b], p. 183). In this connection the picture of the alternating clear and dark stripes appears again, which in the early 1930s had given rise to Pauli's wasp phobia (see page 248), but here is seen as a symbol of multiplication which 'often appears tied with the figure designated as "anima" in psychology. . . . It seems to represent a succession of complementary-opposite states, of which one is imagined as spiritual-timeless (aion), the other as material-temporal (chronos)' (translated from Ref. [36b], pp. 183–4).

In Section 4 Pauli expresses the view that 'to the totality consisting of physics and psychology a quaternity is associated' (translated from Ref. [36b], p. 188) in the sense of the diagram mentioned above, namely of two pairs of opposites, a physical one and a psychical one, which may be arranged at the tips of a cross, called a *quaternio* by Jung (see below). Pauli associates this picture with the second of two dreams belonging together, which both occurred around mid-March 1948 and to which Pauli has added the following note: 'According to my experience, with me dreams, in which the quaternity symbol and in particular the birth of something new plays an essential role, occur preferentially in the season of the equinoxes, i.e. end of March or end of September. The two dreams mentioned here are typical "equinoctial dreams". With me the two equinoxes are times of a relative psychic instability which manifests itself as well negatively as also positively (creatively)' (translated from Ref. [36b], footnote 10, p. 189).

This second dream is the remarkable 'egg dream', which consists of seven successive pictures—recall the importance to Pauli of the number seven, which acquired relevance for the first time in the fact that his sister Hertha was born in Pauli's seventh year (see page 12). In the first picture a woman comes with a bird and the bird lays a big egg. In picture 2 the egg divides itself into two. In picture 3 Pauli approaches and notices that he holds an additional egg with a blue shell in his hand. In picture 4 Pauli divides this third egg into two, so that there now are two blue eggs. In picture 5 the four eggs transform into two lines of the formulae $\cos \delta/2$ and $\sin \delta/2$. In picture 6 these four formulae arrange into

$$\frac{\cos \delta/2 + i \sin \delta/2}{\cos \delta/2 - i \sin \delta/2}.$$

And in picture 7 Pauli says 'But the whole result, in fact, is $e^{i\delta}$, and this is a circle.' Then the formula disappears and a circle appears (Ref. [36b], pp. 189–90).

Later on Pauli observes: 'The imaginary unit $i = \sqrt{-1}$ is a typical symbol. ... In this dream it has the irrational function to unite the pairs of opposites and, hence, to constitute the one-ness.' However, Pauli continues: 'In no way can I pretend to be able to give an "interpretation" of the two mentioned dreams; it even seems to me as if such an interpretation would require the further development of *all* the sciences. The decisive role that mathematical symbols play for the realization of the "One" in dream II, anyhow, seems to me to speak in favour of the fact that today the uniting force of mathematical symbolism is far from being exhausted; I even consider it probable that the latter reaches farther than physics' (translated from Ref. [36b], pp. 191, 192).

The essay closes with an evocation of the need for a unification of our world view: the role of physics is to describe the *lawful* in nature, i.e. the *reproducible*. 'As a consequence of this restriction inherent in the essence of physics, not only everything feeling-like (*Gefühlsmässige*), valuing (*Wertende*) and emotional stays outside of it on the psychological opposite side, but out of this root also springs the *statistical character* of its assertions which, particularly with atomic processes, must (apart from special cases) renounce in principle to grasp the individual case. This, however, does not mean an incompleteness of quantum theory within physics ... but an incompleteness of physics within the whole of life' (translated from Ref. [36b], p. 192). It is this incompleteness then that leaves room for the description of meaningful coincidences between psychical and physical happenings which Jung came to call *synchronistic*.

In letter [35] to Jung on 7 November 1948 Pauli writes: 'Our conversation of yesterday about the "synchronicity" . . . of dreams and external events . . . was a great help to me, and I wish to thank you again very much for it.' And in parentheses he asks: 'Do you use this term "synchronous" also when dream and external event are separated by about 2–3 months?' (translated from Ref. [27], p. 38). This is the first mention of the term 'synchronicity' in the correspondence between Pauli and Jung (see footnote *a* in Ref. [27], p. 39). Concerning the origin of this notion Jung recalls in a letter of 25 February 1953 that 'Prof. Einstein was then several times my evening guest for dinner. . . . It was Einstein who gave me the first impulse to think of a possible relativity of time as well as space and their psychic dependence. More than 30 years later grew out of this suggestion my relation with the physicist Prof. W. Pauli and developed into my thesis of the psychic synchronicity ..' (translated from Ref. [27], p. 3). These encounters of Jung and Einstein must have occurred between January and June 1919 when Einstein was in Zurich to give a series of lectures at the university (Ref. [37], p. 525).

In the systematic description of these psychic dependences in space and time, Jung hesitated between an explanation as *transcendental meaning* and as magic causality. He gave preference to the former, in order not to enter into conflict with the empirical notion of causality, arguing as follows: 'The great difficulty is that we have absolutely no scientific means of proving the existence of an *objective* meaning which is not just a psychic product' (Ref. [38b], Section 915). But, 'Although meaning is an anthropomorphic interpretation it nevertheless forms the indispensable criterion of synchronicity' (Ref. [38b], Section 916).

With letter [36] of 22 June 1949 Jung sent Pauli a first draft of his synchronicity article [38a,b] for a critical reading, hoping 'that I do not unduly burden your precious time with this intention (*Ansinnen*). However, your judgement in this matter is so important to me that I have put aside all hesitations in this respect' (translated from Ref. [27], p. 40). Pauli answered less than a week later with a long letter, in which he makes many important comments that merit consideration in detail. First he observes in relation to Rhine's experiments on the statistics of guessing cards at a distance [40] that, with the latter, 'I cannot see any archetypal foundation (or am I wrong here?). But to me the latter is essential for the understanding of the phenomena in question.' With respect to Jung's proposed 'statistical experiment about the horoscopes of married and single persons' Pauli observes: 'Personally, I have a much stronger relation to such happenings, in which an external event coincides with a dream, than to the behaviour of statistical series' (translated from letter [37], Ref. [27], pp. 40, 41). Towards Fierz, Pauli's opinion was more openly negative: 'This whole kind of experiment, in which all irrational factors are excluded and the unconscious has no possibility to act, obviously could not proceed differently (it's a curious thought that it is us physicists who have to call the attention of the psychologists of the unconscious *to this*!)! . . . For, here the reproducible is concerned, and not the unique' (translated from letter [1091] of 20 March

1950, Ref. [35], p. 55).

Pauli's next comment on Jung's first draft on synchronicity concerns 'the idea of a *sense in chance*—i.e. of events that coincide in time but are not causally connected', in relation to which he refers to Schopenhauer's essay 'On the Apparent Design in the Fate of the Individual' [41]. Quoting Schopenhauer, Pauli writes: 'He there postulates a *"last unit of necessity and of chance"* which to us appears as "power that connects all things—also those which the causal chain leaves untouched by any mutual relation—in such a way that they meet at the appropriate moment." Thereby he compares the causal chains with the meridians directed along the time and the simultaneous events with parallels of latitude—in exact correspondence with your "meaningful cross-connections".' Pauli, the great Schopenhauer lover, observes later on that 'This essay of Schopenhauer has exercised a lasting and fascinating effect on me, and he seemed to me to anticipate a future turn in the natural sciences. But while Sch. still wished to adhere unconditionally to rigorous determinism in the sense of the classical physics of his time, we now have recognized that in the atomic realm the physical events cannot be followed along causal chains through space and time' (translated from letter [37], Ref. [27], p. 41). In the published version Jung discussed this work of Schopenhauer in detail (see Ref. [38a], pp. 10–12, Ref. [38b], pp. 16–20).

In the second half of letter [37] to Jung, Pauli gives a detailed and apparently very particular answer to Jung's questions 'about the possibility to relate with the synchronicity hypothesis some of the physical facts mentioned by you'. Admitting that the answer is very difficult because it is connected with his 'background physics', Pauli proposes to consider radioactivity as an example, because 'The physical phenomenon of radioactivity consists in the transition of an unstable initial state of the atomic nuclei of the active substance into its stable final state . . . whereby the radioactivity eventually ceases. By analogy, the synchronistic phenomenon, based on an archetypal foundation, accompanies the transition from an unstable state of consciousness into a new stable state which is in equilibrium with the unconscious, and in which the synchronistic border phenomenon has again disappeared' (translated from letter [37], Ref. [27], pp. 44–5).

Again the notion of time here causes difficulties, because 'The moments of time of the decay of the *single* atoms are strictly not determined by natural law but, according to modern view, *they even do not exist independently from their observation* [i.e. observation of their energy level].' Since in this case it is unknown whether the *single* atom is in its initial or its final state, Pauli concludes that this 'corresponds to the association of the single individual with the collective unconscious through an archetypal content *which is unconscious to him*. The observation of the state of consciousness of the individual, which lifts him out of this collective unconscious and makes the synchronous phenomenon disappear, corresponds to the determination of the energy level of the *single* atom in a particular experiment' (translated from letter [37], Ref. [27], p. 45).

After this letter [37], Pauli and C. A. Meier visited Jung at his retreat in

Bollingen on 14 July 1949 to discuss further all the questions raised and also other examples. During the nights, Pauli had two dreams; on 29 June 1949 he dreamt of a landing aircraft, from which foreign people emerged. The voice of the 'stranger' (*der Fremde*) says: 'You should not exaggerate your difficulties with the notion of time. The dark girl has only to do a small journey, in order to determine the time!' (translated from Ref. [35], p. 258). In letter [37] quoted above, this 'stranger' is characterized as 'the archetypal background constellated by the system of the scientific notions of our time' and has been known to Pauli since about 1934 (translated from Ref. [27], p. 43). On 2 July 1949 Pauli had a dream in which a blond woman with two dark boys and a light girl are present. After a while, only the girl remains who now reads a book (Ref. [35], pp. 258–9). 'Suddenly she says to me "but there comes my teacher!" and already the tall light-dark man enters.' This is another recurring dream figure who now explains to Pauli that 'The ordinary mathematics is still too difficult for her. Therefore I have to give her popular mathematics books for the moment' (translated from Ref. [35], p. 259).

Pauli sent these two dreams with comments to Jung with letter [38] of 4 June 1950, 'As a consequence of our conversation of yesterday' (translated from Ref. [27], p. 46). Jung answered on 20 June with his own interpretations of the dreams: 'The "stranger" wants to induce the anima, i.e. the female sentimental and sensitive side of the personality, to study mathematics, and this via "archetypal" mathematics where the integer numbers still are (qualitative) *archetypes of order*. Indeed, with their help the synchronicity phenomenon may be captured (mantic methods!), and a more unified world view may be produced' (translated from letter [39], Ref. [27], p. 49). And Jung draws the following *quaternio*:

$$\begin{array}{c} \text{space} \\ \text{causality} \;-\!\!|\!-\; \text{correspondentia} \\ \text{time} \end{array} \qquad (10.6)$$

The exchanges between Pauli and Jung on synchronicity continued well into the year 1951. Jung sent Pauli a new version of his synchronicity article, to which Pauli responded on 24 November 1950 with letter [45]: 'After the turn that your Chap. II "The astrological argument" now has taken, a further convergence of our points of view seems to have taken place.' Referring to the criticism of applying statistical methods to irreproducible events expressed, for example in letter [1091] to Fierz quoted above, Pauli writes to Jung: 'Now, however, the unconscious has entered again in the form of the "vivid interest of the trial person [*Versuchsperson*, V.P.] on the one hand, and of the psychic state of the astrologer on the other" ... whereby your statement "of the ruinous influence of the statistical method on the quantitative comprehension of synchronicity" ... appears as the essential result of your whole statistical investigation' (translated from letter [45], Ref. [27], p. 56). In the published version [38b] of Chapter II Jung writes: 'Compared with Rhine's work the great disadvantage of my astrological statistics lies in the fact that the entire experiment was carried out on [by] only one subject, myself. . . . I was thus in the position of a subject who is at first enthusiastic, but afterwards cools off on becoming habituated to [as in] the ESP [extrasensory perception] experiment' (Ref. [38b], p. 90).

Although this admission raises doubts about the value of Jung's considerable effort invested in the research of Chapter II of Ref. [38a,b], Pauli's opinion on this question was not as clear cut. Indeed, in letter [1057] to Fierz on 6–7 November 1949 Pauli voiced a somewhat more subtle view, which he got after having talked shortly to Jung the day before during the break in the latter's talk at the Psychological Club. The new thought was 'that with the application of the notion "meaningful connection" (*Sinnzusammenhang*) I do not see such a sharp separation between the reproducible and the single case as we had assumed it before. Is it really so much different than in quantum mechanics when Jung assumes a meaningful connection between the psychic situation (state of the consciousness) of the observer and the external happening? Let's leave the horoscopes aside, but also the "I Ging" presupposes the reproducibility of a situation (otherwise no symbols would in fact be possible)' (translated from Ref. [2], p. 707).

More constructive was Pauli's influence in the elaboration of a quaternio, although he could not accept the above form (10.6). In letter [45] he writes: 'I, therefore, wish to offer for discussion here the following *proposition of compromise* for a quaternary scheme which avoids the opposition of time and space and perhaps unites the advantages of your scheme and my former scheme of 1948 [unknown],' (translated from letter [45], Ref. [27], p. 60):

$$\text{causality} \quad -|- \quad \text{synchronicity} \qquad (10.7)$$

energy (conservation)

space–time continuum

Jung responded to Pauli's letter [45] six days later from Bollingen, expressing his own understanding of synchronicity as follows: 'Inasmuch as synchronicity for me represents first of all a mere being, I am inclined to subsume under the notion of synchronicity all cases which refer to a causally not thinkable to-be-so (*So-Sein*)' (translated from letter [46], Ref. [27], p. 63). And he proposed a slight variation to the quaternio (10.7) which he also takes over into the published version of Chapter IV (see Ref. [38a], p. 102, Ref. [38b], p. 137). There Jung comments on quaternio (10.7) as follows: 'This schema satisfies on the one hand the postulates of modern physics, and on the other hand those of psychology' (Ref. [38b], p. 137).

Jung continued to appreciate greatly Pauli's aid in his endeavour, inviting him to Küsnacht for dinner on 23 December 1950 (Ref. [27], letter [48]) and writing on 13 January 1951 from Bollingen: 'I owe you special thanks for having given me courage' (translated from letter [49], Ref. [27], p. 70). On 20 January 1951 and on 3 February 1951 Jung at last gave two talks on the subject to the Psychological Club of Zurich (see Ref. [27], footnote a, p. 49). Apart from the problems already discussed, Jung's synchronicity article contains in Chapter III a masterful historical account of the forerunners of the synchronicity idea, from Lao-Tse's *Tao Te King* to Leibniz's *Prestabilized Harmony*.

Chapter IV, which Jung calls 'Zusammenfassung' (summary), appeared to Pauli as being more than that, namely 'as C. G. Jung's spiritual testament which presses away from the particular "analytical psychology" towards natural philosophy in general and the psycho-physical problem in particular'. Pauli now called the general case of reproducible acausal arrangements, including those of quantum mechanics, 'meaningful correspondences', and tried to view the rare psycho-physical happenings as particular cases (translated from Ref. [35], letter [1417] of 3 June 1952 to Fierz). Jung's article appeared in 1952 under the same cover as Pauli's Kepler article [32a], for which Pauli proposed the title 'Naturerklärung und Psyche' (Ref. [35], letter [1267] of 25 July 1951 to von Franz).

His Majesty Julian Schwinger and regularization

I believe that in a future theory neither the functions, which are singular at the light cone, nor the three dimensional δ-function in the commutators of the canonical formalism will occur and that the present theory can only describe averages over small but finite space time regions approximately correct[ly]. The dimensions of these critical regions do not seem to be a 'universal length', but must depend on the masses of the particles involved. As a substitute for concepts not yet precisely known appear now auxiliary masses and it seems to me only a formal preliminary measure to let them finally tend to infinity. [42]

With this vision of a future theory Pauli concluded his long letter of 24 January 1949 to Schwinger, in which he responded to the second of Schwinger's fundamental papers on QED, Refs. [43a,b] and [20], with a detailed discussion of the problem of regularization (see pages 307 and 318), to which, in Pauli's view, Schwinger had not paid enough attention in Ref. [43b]. As was said on page 362, Pauli had closer contact with Schwinger in 1942 at Purdue University. In Spring 1947 Schwinger was among the many physicists who were greatly stimulated by the meeting at Shelter Island mentioned on pages 412–3. Thus, when a second such meeting from 30 March to 2 April 1948 was organized at Pocono Manor situated in the Pocono mountains, some 130 km north of Philadelphia, 'the main event was Schwinger's marathon performance on the first day', in which he presented his covariant formalism (Ref. [14b], Section 18(c)1; see also Ref. [15b], Chapter 4). In Part I of Ref. [43a], of this long and elegant series of papers, Schwinger recasts the canonical formalism of Heisenberg and Pauli (see pages 186f) into a form which is manifestly invariant under both the Lorentz and the gauge groups. He also adopts Pauli's charge conjugation (8.38) in Eq. (1.3) of Ref. [43a] and introduces the Jordan–Pauli quantization (6.17) in his Eq. (2.28), the analogue (2.29) for the electrons being discussed below.

He evidently considers all this to be common knowledge which requires no particular references. Indeed, the only publications prior to 1939 quoted in Ref. [43a] are Dirac's density matrix regularization paper, Ref. [23b] of Chapter 8, and Heisenberg's generalization, Ref. [14b] of Chapter 8, which are both rejected as fruitless 'subtraction physics'.

Of particular importance for the formulation of a systematic renormalization theory is the interaction representation. The truly astonishing fact is that Schwinger had introduced this representation back in 1934 in an unpublished paper written at the age of 16, when he was a student at the City College of New York (see Ref. [44a], p. 14). No wonder, then, that he did not care to give detailed references in Part I, Ref. [43a]. Julian Seymour Schwinger (1918–94) grew up in Manhattan, New York, in a middle-class Jewish family. His parents had come from different parts of Poland. Schwinger did his Ph.D. at Columbia University in 1937 with a paper co-authored with Edward Teller. In 1945 Schwinger was appointed to the chair at Harvard University for which Pauli had written the letter of recommendation [717] in Ref. [2]. Schwinger stayed at Harvard until 1971 when he moved to the University of California at Los Angeles. In 1947 he married Clarice Carrol whom he had met during his stay at the Radiation Laboratory [44a]. Schwinger, together with Tomonaga and Feynman, received the Nobel Prize of 1965 'for their fundamental work in quantum electrodynamics, with deep-ploughing consequences for the physics of elementary particles'.

If Φ is the constant state vector of the Heisenberg representation, then Schwinger defines the interaction representation on a space-like surface $\sigma(x)$ containing the point x as

$$\Psi[\sigma] = U[\sigma]\Phi, \tag{10.8}$$

where the unitary operator U satisfies the equation of motion (2.5) of Ref. [43a],

$$i\hbar c\frac{\delta U[\sigma]}{\delta\sigma(x)} = \mathcal{H}'(x)U[\sigma]. \tag{10.9}$$

Note that it is the condition to render this time derivative covariant that makes the surfaces σ necessary. See Pauli's lectures *Feldquantisierung*, Ref. [44b], Section 21, where the interaction representation is introduced referring to Tomonaga [44c] (also mentioned by Schwinger, [43a]) and to Dirac, Fock and Podolsky [44d].

In Eq. (10.9) the interaction Hamiltonian density $\mathcal{H}'$, according to Eq. (6.37), is just the negative of the interaction Lagrangian density obtained from Eq. (6.43) by inserting the current density (6.32), i.e.

$$\mathcal{H}'(x) = -j^\mu(x)\Phi_\mu(x), \tag{10.10}$$

which is Eq. (2.7) of Ref. [43a] and Eq. [19.2] of Ref. [44b]. The main purpose of the interaction representation is to define the S-matrix, which Schwinger calls the *invariant collision operator*:

$$S = U[\infty, -\infty]. \tag{10.11}$$

$U[\sigma_2, \sigma_1]$ transforms $\Psi[\sigma_1]$, defined in (10.8), into $\Psi[\sigma_2]$ and satisfies the integral equation (4.5) of Ref. [43a],

$$U[\sigma_2, \sigma_1] = 1 - \frac{i}{\hbar c} \int_{\sigma_1}^{\sigma_2} \mathcal{H}'(x) U[\sigma, \sigma_1] d^4 x, \qquad (10.12)$$

where $d^4 x = d^3 x dx_0$ and $x_0 = ct$. To lowest order in $\mathcal{H}'$ this has the form (10.1) considered by Heisenberg. In Eq. (4.16) of Ref. [43a] Schwinger also formally derives the expression (10.5) preferred by Pauli.

Part II, Ref. [43b], of Schwinger's series of papers begins with an invariant definition of the vacuum state Ψ_0. According to Eqs (6.5), (6.12) Ψ_0 satisfies, for bosons and fermions, the conditions $b_r \Psi_0 = a_r \Psi_0 = 0$, where $b_r = b(\lambda \mathbf{k})$ and $a_r = a_{\sigma,+}, a_{\sigma,-}^+$ are the annihilation operators of photons and of electrons ($\lambda = +$) and positrons ($\lambda = -$), respectively. Now, as in Eq. (6.15), the annihilation part of a free field operator is characterized by a time dependence proportional to $\exp(-i\omega t)$ and may therefore be obtained by the projection

$$\left(\frac{1}{2\pi i} \int_{C_+} \frac{d\tau}{\tau} e^{\pm i \omega \tau} \right) e^{\mp i \omega t} = \frac{1}{2\pi i} \int_{C_+} \frac{dz}{z + t} e^{\pm \omega z} = \begin{cases} e^{-i\omega t} \\ 0 \end{cases} \qquad (10.13)$$

where the path C_+ is along the real axis, avoiding the singularity by a semi-circle in the lower half plane. The result (10.13) follows by closing the path with a large semi-circle in the upper/lower half z-plane. Schwinger casts this projection into covariant form in Eqs (1.1a) and (1.47) of Ref. [43b]; in Pauli's *Feldquantisierung* [44b] it is given by Eqs [4.3]. Writing for electrons and positrons

$$\psi(x) = \sum_r a_r u_r(x), \qquad (10.14)$$

it follows from the closure relation (8.10) and the anticommutation relations (6.11) that

$$\{\psi(x), \overline{\psi}(x')\} = -iS(x - x'). \qquad (10.15)$$

This is Eq. (2.29) of Ref. [43a] mentioned above and Eq. [3.26] of Ref. [44b].

Schwinger's next task in Ref. [43b] is the calculation of *vacuum polarization*, to which Pauli's letter to Schwinger quoted above refers, and which is also the content of Chapter 2 of the *Feldquantisierung* [44b]. Defining the vacuum state by the initial condition (2.3) of Ref. [43b],

$$\Psi_0 = \lim_{\sigma \to -\infty} \Psi[\sigma], \qquad (10.16)$$

the induced current density may be written as

$$\langle j^\mu(x) \rangle = \langle U^{-1}[\sigma, -\infty] j^\mu(x) U[\sigma, -\infty] \rangle_0, \qquad (10.17)$$

where $\langle \ldots \rangle_0$ designates a vacuum expectation value. Since one must require that $\langle j^\mu(x) \rangle_0 = 0$, the lowest-order contribution to (10.12) making use of (10.10) yields the linear response

$$\langle j^\mu(x)\rangle = \frac{i}{\hbar c}\int_{-\infty}^{\sigma} d^4x' \langle [j^\mu(x), j^\nu(x')]\rangle_0 \Phi_\nu(x'). \tag{10.18}$$

Insertion of the current density (6.32) into the commutator leads to an expression of the form

$$[AB, CD] \equiv A\{BC\}D - AC\{BD\} + \{AC\}DB - C\{AD\}B,$$

so that with Eq. (10.15)

$$\begin{aligned}[j^\mu(x), j^\nu(x')] = \frac{ie^2}{2}\{&[\psi_\alpha(x), \overline{\psi}_\beta(x')](\gamma^\nu S(x'-x)\gamma^\mu)_{\beta\alpha}\\
&-[\psi_\alpha(x'), \overline{\psi}_\beta(x)](\gamma^\mu S(x-x')\gamma^\nu)_{\beta\alpha}\},\end{aligned} \tag{10.19}$$

where in addition an antisymmetrization of expressions $\overline{\psi}_\beta(x)\psi_\alpha(x')$ was applied. This means that, in order to guarantee that $\langle j^\mu(x)\rangle_0 = 0$, the current density is written as $j^\mu(x) = -(ie/2)[\psi_\alpha(x), \overline{\psi}_\beta(x)]\gamma^\mu_{\alpha\beta}$. This prescription was introduced by Pauli quite generally in Equation (96) of his review paper of 1941 [44e] (see also Ref. [44b], p. 22). Eq. (10.19) agrees with the expression (2.9) of Ref. [43b].

Making use of the conditions on Ψ_0 discussed above, one obtains with (10.14)

$$\begin{aligned}\langle [\psi_\alpha(x), \overline{\psi}_\beta(x')]\rangle_0 &= \sum_{rs} u_{r\alpha}(x)\overline{u}_{s\beta}(x')\langle [a_r, a_s^+]\rangle_0 = \sum_r \lambda u_{r\alpha}(x)\overline{u}_{r\beta}(x')\\
&= -S^1_{\alpha\beta}(x-x'),\end{aligned} \tag{10.20}$$

where the last equality follows from (8.14). This is Eq. (1.68) of Ref. [43b] and Eq. [4.14] of Ref. [44b]. Thus, finally,

$$\begin{aligned}\langle [j^\mu(x), j^\nu(x')]\rangle_0 = \frac{ie^2}{2}\mathrm{Tr}\{&S^1(x'-x)\gamma^\mu S(x-x')\gamma^\nu\\
&-S^1(x-x')\gamma^\nu S(x'-x)\gamma^\mu\},\end{aligned} \tag{10.21}$$

which is Eq. (2.10) of Ref. [43b] and the second Eq. [7.1] of Ref. [44b]. Here the Dirac operator in the general definition (8.11) of the functions $S^{(\,)}$ may be evaluated explicitly by observing that the trace of an odd number of γ-matrices vanishes and $\mathrm{Tr}\gamma^\mu\gamma^\nu = 4\delta_{\mu\nu}$, $\mathrm{Tr}\gamma^\kappa\gamma^\mu\gamma^\lambda\gamma^\nu = 4(\delta_{\kappa\mu}\delta_{\lambda\nu} - \delta_{\kappa\lambda}\delta_{\mu\nu} + \delta_{\kappa\nu}\delta_{\lambda\mu})$. Making use of the properties

$$\Delta(-x) = -\Delta(x); \quad \Delta^1(-x) = \Delta^1(x) \tag{10.22}$$

that follow from Eqs (8.12), (8.16), Eq. (10.21) may be written in the form

$$\frac{1}{2}\langle [j^\mu(x), j^\nu(x')]\rangle_0 = -ie^2 K_{\mu\nu}(x-x') \tag{10.23}$$

with

$$\frac{1}{2}K_{\mu\nu}(x) \equiv \frac{\partial\Delta^1}{\partial x^\mu}\frac{\partial\Delta}{\partial x^\nu} + \frac{\partial\Delta^1}{\partial x^\nu}\frac{\partial\Delta}{\partial x^\mu} - \delta_{\mu\nu}\left(\frac{\partial\Delta^1}{\partial x^\alpha}\frac{\partial\Delta}{\partial x^\alpha} + k_c^2\Delta^1\Delta\right). \tag{10.24}$$

This agrees with Eq. (2.15) of Ref. [43b] and with Eqs [7.5], [7.6] of the *Feldquantisierung* [44b]. In the latter this result is derived by a direct calculation of $\langle \psi_\alpha(x)\overline{\psi}_\beta(x')\rangle_0$ and $\langle \overline{\psi}_\beta(x')\psi_\alpha(x)\rangle_0$.

Introducing the invariant sign $\varepsilon(x) = \pm 1$ for $\pm x_0 > 0$, the causal projection $t' < t$ occurring in Eq. (10.18) may be expressed by the projector $(1 + \varepsilon(x - x'))/2$,

$$\langle j^\mu(x) \rangle = \frac{i}{2\hbar c} \int_{-\infty}^{+\infty} d^4 x' \langle [j^\mu(x), j^\nu(x')] \rangle_0 (1 + \varepsilon(x - x')) \Phi_\nu(x'). \quad (10.25)$$

This is Eq. (2.18) of Ref. [43b] and Eq. [8.3] of Ref. [44b]. The important special case of vacuum polarization is obtained by the condition that the expression (10.18) vanishes in the limit $\sigma \to +\infty$, i.e. no real pairs have been created. Hence

$$\langle j^\mu(x) \rangle = \frac{1}{2\hbar c} \int_{-\infty}^{+\infty} d^4 x' \langle [j^\mu(x), j^\nu(x')] \rangle_0 \varepsilon(x - x') \Phi_\nu(x'), \quad (10.26)$$

which is Eq. (2.19) of Ref. [43b]. Here a new function

$$\overline{K}_{\mu\nu}(x) \equiv -\varepsilon(x) K_{\mu\nu}(x) \quad (10.27)$$

occurs which may be evaluated by introducing the new invariant function

$$\overline{\Delta}(x) = -\frac{1}{2} \Delta(x)\varepsilon(x) \quad (10.28)$$

defined by Eq. (2.21) in Ref. [43b] and by Eq. [5.7] in Ref. [44b]. Since according to (8.12), $\Delta(\mathbf{x}, 0) = 0$, so that $\Delta(x)\partial\varepsilon(x)/\partial x^\mu = (1/ic)\delta_{\mu 4}\Delta(x)\delta(t) = 0$, one obtains by partial integration of (10.24)

$$\frac{1}{4}\overline{K}_{\mu\nu}(x) = \frac{\partial\Delta^1}{\partial x^\mu}\frac{\partial\overline{\Delta}}{\partial x^\nu} + \frac{\partial\Delta^1}{\partial x^\nu}\frac{\partial\overline{\Delta}}{\partial x^\mu} - \delta_{\mu\nu}\left(\frac{\partial\Delta^1}{\partial x^\alpha}\frac{\partial\overline{\Delta}}{\partial x^\alpha} + k_c^2 \Delta^1 \overline{\Delta}\right). \quad (10.29)$$

This is Eq. [7.15] of Ref. [44b] and corresponds to Eq. (2.20) of Ref. [43b]. It is also Eq. (1) of Pauli's letter [1001] to Schwinger quoted at the beginning of this section.

Equation (10.26) gives the vacuum polarization linear in the external field Φ_ν, but, of course, higher powers of Φ_ν also contribute. This problem had been suggested by Peierls when Pauli visited him in Birmingham in spring 1949 after delivering the Rouse Ball Lecture in Cambridge (Ref. [2], letter [1009]). Pauli gave this problem to his new collaborator from Lund, Gunnar Källén (1926–68), whom Professor Gustafson (see page 411) had asked Pauli on 17 February to send him for the coming summer term (Ref. [2], letter [1010]). By 10 July Pauli could write to Peierls that Källén had solved the problem [45a], and it was a total success (Ref. [2], p. 678). Meanwhile the problem of the magnetic moment of electrons and nucleons was investigated by another of Pauli's collaborators, Quin Luttinger [45b]. Joaquin Mazdak Luttinger (1923–97) had obtained his Ph.D. from MIT in 1947 and was in Zurich on a Swiss–American exchange fellowship [45c] (see also Ref. [15b], Section 10.2).

Pauli had received a preprint of Schwinger's paper [43b] at Christmas 1948, as he told Wentzel in letter [1000] of 22 January 1949. In this letter, in which much of the discussion of the letter to Schwinger is repeated, Pauli also mentions that he is working with Scherrer's assistant Felix Marc Hermann Villars (b. 1921) on an extension of the regularization procedure to the problem of the magnetic moment of the electron (see also letter [995] to Bohr). Villars had done his doctoral thesis at Scherrer's institute and Pauli, on his return from the United States, on 10 May 1946, had been the second referee. This, in fact, was Pauli's first official act back at ETH (see Ref. [3], II.137). On 10 May 1949 Pauli and Villars submitted a paper, in which all these problems were discussed in detail. This paper [46a] appeared in the issue of the *Reviews of Modern Physics* dedicated to Einstein's 70th birthday on 14 March 1949 and for which Abraham Pais had collected the manuscripts. Pais had asked Pauli to write an introduction to the volume, but Pauli felt that he 'actually had done his duty towards Einstein' with his contribution in the 'Einstein Festschrift' [46b]. 'Also I feel I am not in as close personal contact as with Bohr, since the pre-history in both cases is quite different for me. As personal an introduction as in the Bohr issue [see page 387] I cannot write about Einstein' (translated from letter [993] of 26 December 1948 to Pais, Ref. [2], p. 586).

According to Eq. (10.29) the difficulties are due to the singularities of the functions $\Delta^1(x)$ and $\overline{\Delta}(x)$. It is necessary therefore to exhibit these singularities explicitly. This is done by starting from appropriate integral representations for these functions deduced in *Note* [47]:

$$\Delta^1(x) = -\frac{k_c^2}{2\pi^2 c} \int_0^\infty d\alpha \sin\left(\frac{1}{4\alpha} + \alpha\xi\right)$$

(10.30)

and

$$\overline{\Delta}(x) = \frac{k_c^2}{4\pi^2 c} \int_0^\infty d\alpha \cos\left(\frac{1}{4\alpha} + \alpha\xi\right),$$

(10.31)

where $\xi \equiv -k_c^2 x^2 \equiv k_c^2 \lambda$. These are Eqs (1a,b) of Ref. [46a], where $\overline{\Delta}$ is designated as a bold Δ. Making use of a further parametrization introduced by Schwinger in the appendix to the second paper [43b], it is possible to extract the following singularity structure, which again is derived in *Note* [47]:

$$\Delta^1(x) = \frac{1}{4\pi^2 c}\left\{-\frac{2}{\lambda} + \frac{1}{2}k_c^2 \log \lambda + \ldots\right\}$$

(10.32)

and

$$\overline{\Delta}(x) = \frac{1}{4\pi c}\left\{\delta(\lambda) - \left(\frac{k_c^2}{4} + \ldots\right)\theta^+(\lambda) + \ldots\right\},$$

(10.33)

where $\theta^+(\lambda) = 1$ or 0, if $\lambda > $ or < 0. These are, essentially, Eqs (4) and (5) of Ref. [46a]. It must be emphasized that this derivation rests essentially on the appendix to Ref. [43b] which is an impressive example of Schwinger's virtuosity.

Now since $k_c = mc/\hbar$, introducing auxiliary masses M_i and defining the regularized functions Δ^1 and $\overline{\Delta}$ by Eqs (3) of Ref. [46a],

$$\Delta_R^1(x) = \sum_i c_i \Delta^1(x; M_i) \; ; \; \overline{\Delta}_R(x) = \sum_i c_i \overline{\Delta}(x; M_i),$$

(10.34)

it follows immediately from Eqs (10.32), (10.33) that all the singularities of Δ^1 and $\overline{\Delta}$ and their derivatives are removed from the regularized functions Δ^1_R and $\overline{\Delta}_R$ by the two conditions

$$\sum_i c_i = 0 \; ; \; \sum_i c_i M_i^2 = 0, \tag{10.35}$$

which are Eqs (I) and (Ia) of Ref. [46a]. If one now sets $M_0 = m$ and $c_0 = 1$ but lets $M_i \to \infty$ for $i \neq 0$, it follows from Eqs (8.12), (8.16) that the regularized expressions (10.34) reduce to their normal form. This formal device was first introduced independently by Feynman [18b] and by Stueckelberg and Rivier [48a] and most explicitly in the doctoral thesis [48b] submitted to the University of Lausanne by Dominique Rivier (1918–98).

This general idea, which Pauli called 'formalistic' in Ref. [46a], in contrast to the 'realistic' point of view of also describing the auxiliary particles by a Hamiltonian, may be modified in several ways. First, instead of a discrete mass spectrum as in Eqs (10.34) a continuous mass spectrum may be introduced which is indicated particularly in Fourier transformation. Secondly, instead of regularizing Δ^1 and $\overline{\Delta}$ separately one may regularize bilinear expressions as occur, for example, in Eq. (10.29). As Pauli wrote in his letter to Schwinger (Ref. [2], p. 612), this possibility had been suggested to him by his collaborator from Poland, Jerzy Rayski (b. 1917). In Section 4 of Ref. [46a] it is indeed shown that in order to guarantee gauge invariance $\partial \langle j^\mu (x) \rangle / \partial x^\mu = 0$, i.e. $\partial \overline{K}_{\mu\nu} / \partial x^\mu = 0$—which also implies a vanishing photon self-energy—regularization must be applied to the bilinear expression (10.29) as a whole (note that in Ref. [46a] $\overline{K}_{\mu\nu}$ is called $K_{\mu\nu}$). In the same Section 4 it is also shown that the electron self-energy giving rise to a mass renormalization Δm requires a different regularization prescription—which actually confirms Schwinger's result (3.97) of Ref. [43b]. In letter [1000] to Wentzel, Pauli comments on this last finding as follows: 'Since the time I have eliminated Schwinger's notation, I am *much more* convinced of the physical correctness of his results' (translated from Ref. [2], p. 608).

This Pauli–Villars regularization was too technical for the *Feldquantisierung* lectures; it is only mentioned in Sections 7 and 29 of Ref. [44b]. But, as Villars observes in his review of the subject in the Pauli memorial volume, later developments showed that 'Pauli was rightly concerned with finding a consistent prescription for dealing with otherwise undefined mathematical expressions, as arose in various problems through confluence of singularities in products of Green's functions' (Ref. [49a], p. 84). Pauli himself commented in his letter to Schwinger as follows: 'It seemed to me necessary to come into the open now because people are already beginning to publish contradictory statements. Therefore I first repeated some calculations of Schwinger's Part II "under cover of the regulator". ... This technical device enables me to go around safely among people, who contradict each other, like one walks in the rain with an umbrella' (letter [1002] to Bethe, Ref. [2], p. 621). The letter to Schwinger had been typed, and only the formulae were in Pauli's hand; it simplified the production of copies which Pauli sent to Bethe, Bohr, Heitler, and Ma.

Pauli, of course, was waiting impatiently for Schwinger's reply—which, however, did not come. In fact, in one of his conversations with Jagdish Mehra in 1988 Schwinger observed: 'I was neither a great letter writer nor a telephonist' (Ref. [44a], p. 92). Instead, one of Schwinger's students, Bryce Seligman DeWitt, to whom Schwinger had shown the letter, wrote to Pauli. Although this letter is lost (see the comments in Ref. [2], pp. 592–3 and p. 625; see also Ref. [15b], pp. 347–52, Ref. [44a], pp. 264–5, and letter [1039] to Peierls, Ref. [2], p. 697), we know from Pauli's reply [1038a] (Pauli calls him Mr Seligman) that the letter was dated 2 May. In fact, DeWitt had already written to Pauli on 1 April about his own work with Schwinger (see Pauli's reply [1025a] in Ref. [2]). After receiving DeWitt's second letter, i.e. in May (not February) 1949, Pauli reported this non-event to Oppenheimer: 'My discussion with Schwinger, in which he never participated himself, makes me think on "His Majesty's" (Julian's) psychology. . . . His Majesty permitted one of his pupils (B[ryce] Seligman [DeWitt]) to break the "blockade" of the E.T.H., by Harvard and to write me a letter, but he *refused to read this letter* himself! The content of this diplomatic note (it was a very long one) is only this, that His Majesty had a kind of revelation on some Mt. Sinai, to put always $\partial\Delta^1/\partial x_\nu = 0$ for $x = 0$ (in contrast to $\partial\delta(x)/\partial x_\nu$, which has the same symmetry properties), wherever it occurs. We are calling this equation "the revelation" here, but it did not help our understanding' (Ref. [2], letter [1006]). And Pauli continued that DeWitt and another Schwinger pupil, Roy Glauber, wanted to visit Zurich but were unable to obtain a letter of recommendation from "His Majesty". In his first letter to Pauli, DeWitt, in fact, had expressed the wish to come to Zurich, but Pauli had told him in his reply [1025a] that 'a recommendation from Professor Schwinger is all the more indispensable as you do not give any other reference'. On p. 265 of Ref. [44a] we are informed, however, that Schwinger wrote 'strong letters of recommendation' for both DeWitt and Glauber.

On the morning of 31 August 1949 Isidor Rabi entered Pauli's office at ETH quoting Friedrich Schiller's play *Wilhelm Tell* in exclaiming: 'Make way, the bailiff cometh' (*Platz, der Landvogt kommt*). He thus announced the arrival of His Majesty in Zurich on Friday afternoon 2 September by plane from Paris. 'Since Rabi had been in Paris together with Mr and Ms Schwinger, their presence in Paris is sufficiently proved. In addition, Rabi proposed to hold on Saturday afternoon (before noon, he thought, would be an unfair advantage for the people of Zurich, since His Majesty used to sleep till noon) a "peace conference", with himself [Rabi] as chairman, between His Majesty and a small group of theorists at Gloriastrasse 35 [the physics institute]' (translated from letter [1051] of 1 September 1949 to Fierz, Ref. [2], p. 696). Apparently this "peace conference" proceeded to the satisfaction of both sides. Indeed, on 14 December Pauli wrote to Paul T. Matthews in Cambridge that he was glad of having had a long discussion with Schwinger in Zurich (Ref. [2], p. 720). In his interviews with Mehra in 1988, Schwinger recalls: 'They must have confronted me with it [the Pauli–Villars regularization]. I have no memory, but I must have said I didn't like it. But it was very amicable' (Ref. [44a], p. 306, also Ref. [15b], p. 352).

On Monday 5 September the party moved to Basle where an international conference on 'experimental methods (counters, accelerators) of [nuclear] physics, and field theories' took place till Friday the 9th (letter [1043] to Bhabha; see also [1047] to Fierz) and Schwinger gave an invited talk (see letter [1087] in Ref. [35]). There were also invited talks by Stueckelberg [50a] and by Dyson [50b]. It was Dyson's first encounter with the physicists of the continent (Ref. [15b], p. 552). In particular, he met Pauli and Stueckelberg for the first time. For Stueckelberg this was an important event. Ernst Carl Gerlach Stueckelberg von Breidenbach (1905–84) grew up in Basle where he obtained his Ph.D. in 1927. After staying with Sommerfeld in Munich he became a research associate and then assistant professor at Princeton University. Returning to Switzerland in 1933 he became a Privatdozent at the University of Zurich and in 1935 professor at the University of Geneva. From 1942 to 1957 he was also a professor at Lausanne. In 1941 he viewed positrons as electrons travelling backward in time. After introducing, with Rivier, the causal propagator, in 1951 he developed with another student, André Petermann, the idea of the renormalization group. In 1976 Stueckelberg was presented with the Planck Medal. As 'Freiherr von Breidenbach zu Breidenstein und Melsbach' and as an officer in the Swiss Army he recognized both his German and his Swiss roots, and he also identified deeply with his Lutheran faith [50c].

In the following week, Pauli and many others moved from the northern tip of Switzerland to the southern tip where, across the border, in Como, the continuation of the international conference, this time on cosmic rays, was taking place from 12 to 19 September (see letter [1043]) and where Schwinger gave another lecture. Although proceedings of both conferences exist, the one at Basle, organized by the Swiss Physical Society [50d], and the one at Como, organized by the Italian Physical Society [50e], neither of Schwinger's talks is reproduced. The Basle lecture, however, is mentioned in the preface by the editors Paul Huber and Markus Fierz. This was the first trip to Europe by Schwinger and his wife Clarice Schwinger who became very fond of Pauli and firm friends with Franca. From Como the Schwingers continued their tour of Europe to Florence and Nice, returning home on the *Queen Mary* in time for the beginning of the autumn term (Ref. [44a], pp. 305–7).

Between the Basle and Como conferences, on Sunday 11 September, Pauli's father, who lived at Cäcilienstrasse 8 in the neighbourhood of Zurich University, celebrated his 80th birthday; it was an occasion for numerous friends to gather in Zurich (see page 11). A note on this event is appended to the minutes of the conversation between Pauli and the new president of the School Council (Ref. [3], III.25). At the end of 1948 President Rohn had gone into retirement and the agrochemist Professor Hans Pallmann (1903–65) became his successor. Through this conversation, which took place on 10 August, the new president wished to gain more personal knowledge of the famous member of his school.

They spoke about many things, but for Pauli the most important topic was his invitation to the Institute for Advanced Study for the coming winter term of 1949/50. Pauli also mentioned that after the departure of his assistant Jost on 1 October, who had been with Pauli for three years, he wished to appoint Robert Schafroth to this position. In Pauli's absence the new assistant would have to take over Pauli's regular course on 'Optics' (Ref. [3], III.25).

Max Robert Schafroth (1923–59) had just completed his thesis with Pauli on 'The higher radiation–theoretic approximations to the Klein–Nishina formula' [51a,b]. It was a long and complicated work, in which mass and charge renormalization as well as regularization were applied to higher-order electron–photon scattering processes and which required the evaluation of a large number of complicated integrals. Schafroth asked me to do this latter task as a pre-diploma exercise together with my co-student and friend from St Gallen, Hans Iklé (1906-96), a cousin of the later undersecretary of defence in the Reagan administration, Fred Iklé (see Ref. [17], p. 276). Schafroth acknowledged our work at the end of his second paper [51b], and he reported his results at the Spring Meeting of the Swiss Physical Society at Brienz near Interlaken on 7 May 1949 [51c].

An additional complication in Schafroth's work, already addressed by Jost [52a], was the simultaneous emission of long-wavelength photons as studied by Pauli and Fierz (see pages 317–22). It should be mentioned also that before Schafroth, Pauli's Italian collaborator Ernesto Corinaldesi (b. 1923) from Rome (see the comment in Ref. [35], p. 7) had already worked on the problem of the Compton effect of spin-0 particles; he presented a communication together with Res Jost at the Spring Meeting of the Swiss Physical Society at Yverdon on Lake Neuchâtel on 8 May 1948 [52b]. There Luttinger and Jost presented a communication on the subject of radiation–theoretic corrections [52c]. In his recollections Jost writes about this period: 'The most fascinating development during my time of three years as assistant was the renormalization theory tied with the researchers S. Tomonaga, J. Schwinger, R. Feynman and F. J. Dyson. (The great E. C. G. Stückelberg-von Breidenbach we did not understand.)' (translated from Ref. [10a], p. 16). Pauli wrote a very favourable report on Schafroth's thesis for the School Council (Ref. [3], III.2). His request to appoint him under the same conditions (category A) as Jost (Ref. [3], III.26), however, ran into difficulties because of Schafroth's lack of experience as assistant (III.31). But when Schafroth accepted the less favourable conditions of 5000 francs per year instead of 6000, plus inflation supplement (III.33), the appointment was confirmed by President Pallmann on 13 October 1949 (III.34). Pauli reported these results on the Compton effect to the Eighth Solvay Conference which was held in Brussels from 27 September to 2 October 1948 (Ref. [12], Vol. 2, pp. 1127–9; see also Ref. [52e], p. 259).

Pauli's subsequent invitation to the Institute for Advanced Study was important to him for several reasons. First, he had not been in the United States since over three years and, secondly, in 1947 his friend Robert Oppenheimer had succeeded Director Aydelotte. In addition, Res Jost, with whom Pauli had developed close, almost fatherly relations, as well as Felix Villars and Quin Luttinger also came to the Princeton Institute in the autumn. Pauli called these three 'my children' in letter [1064] to Rabi, written in Princeton on 19 December 1949. It was Abraham Pais who had urged Oppenheimer to invite some of Pauli's collaborators (Ref. [10c], p. 3); Pais had joined the Institute in 1946. While Luttinger and Villars were invited for the academic year 1949/50, Jost stayed there until 1955, mainly at the personal invitation of Oppenheimer (Ref. [10a], p. 16). Jost came to Princeton with his wife Hilde Jost Fleischer (b. 1922), a native of Vienna and herself a doctor in physics whom Res had married in Berne earlier in 1949 (Ref. [10c], p. 3). After Princeton Luttinger held various positions and in 1960 he went to Columbia University as a professor, working mainly and very successfully in solid-state theory [45c]. Villars went from Princeton to MIT, starting as a research associate and becoming a professor in 1960 [52d].

Since Pauli was requested to answer the Princeton invitation by early September, President Pallmann rushed to give his written agreement (Ref. [3], III.30) and informed the School Council only afterwards, the latter confirming it on 8 October (II.32). Pauli planned to go to Princeton in November and be back at the beginning of May 1950 at the latest (Ref. [2], letter [1043]). However, new difficulties related to his nationality appeared because Pauli's US passport expired on 19 January 1950 (see page 397). So Pauli on 18 August contacted the American Consulate General in Zurich in order to obtain 'advice concerning the technical arrangements of my (and Mrs Pauli's) imminent trip to the United States' [53a]. The Consulate requested a confirmation of employment from ETH (Ref. [3], III.27). This apparently solved the problem. Pauli had not told the Consulate that on 25 July 1949 he had at last obtained Swiss citizenship from the municipality of Zollikon [53b] (not in the second half of August as stated in document III.32 of Ref. [3]). Thus everything seemed to be clear for Pauli's and Franca's departure. They had reservations for the *Ile de France* from Le Havre on 10 November. Villars left on 7 November from Liverpool, Luttinger was already in the United States and Jost followed later (Ref. [2], letter [1052] to Fierz). However, Pauli's and Franca's papers were not ready in time, so they finally left Zurich on 22 November 1949 (Ref. [3], III.36) and boarded the ship *Mauretania* (Ref. [2], letter [1058]).

Lecturing on the new field theory

Only on the occasion of your last visit to Zurich did I notice that the business of the re-appointment to the chair of theoretical physics at the Zurich University meant much more to you than I had assumed on the grounds of the few of your previous remarks about this matter. From your letter I see, unfortunately, that now you are even left in a considerable depression. As is well known, in such a state one does usually not see and judge things correctly, but I am quite willing to consider the fact that you have written me this letter as a sign of your continuing personal confidence in me. Had you shown me this

*confidence earlier already, you would have given me the opportunity therewith
for a discussion about the motives of my point of view in this whole question.*
[54]

As was said on page 413, Wentzel had left Zurich in March 1948, following
a call from the University of Chicago. In order to bridge the gap left in the
teaching of theoretical physics at the university until a successor was appointed,
the Minister of Education of the Canton of Zurich, Robert Briner, contacted
President Rohn on 22 May 1948 proposing that the university students be
authorized to follow Pauli's courses at ETH (Ref. [3], III.14), to which the
president had no objections (III.15). For the succession the preferred candidate
was Walter Heitler (1904–81) at the Dublin Institute for Advanced Studies. As
the above quotation shows, this posed a problem to Fierz and even caused him
serious depression. Fierz of course would have loved to move to Zurich, and
he rightly considered himself a valid candidate.

Although in the quoted letter [989] of 5 December 1948 Pauli emphasizes
that 'in the now effectuated call of Heitler I have not given the swaying motion,
but I indeed approved of the decision by the Faculty', his authority was of
course overwhelming. Pauli's rationale was 'that Wentzel's departure requires
a more essential compensation than a simple transposition of your activity
from Basel to Zurich. In this case an important place in the seminar [of Swiss
theoretical physics] would stay vacant which would weigh all the more as the
theoretical physics in Bern [Mercier] and Geneva [Stueckelberg] now is what it
indeed is.' But in this letter to Fierz Pauli went even further saying 'That I have
welcomed you as "psycho-physical" partner should not only put our spiritual
relation on a broader basis, but it was also a welcome opportunity for me to give
you, independently from the issue of the re-appointment of the professorship
in Zurich, a sign of my personal confidence' (translated from Ref. [2], p. 583).

Heitler, who in 1927 had created at this same university his theory of the
covalent chemical bond, began his new activity in Zurich in the winter term
of 1949/50, just when Pauli left Zurich on 22 November (Ref. [3], III.36) to
go to Princeton. In the conversation of 10 August 1949 between Pauli and
President Pallmann mentioned in the last section, Pauli had expressed himself
as against a substitution in his teaching by Heitler and in favour of Schafroth
because of the former's unfamiliarity with the place (Ref. [3], III.25). Heitler
took over from Wentzel's assistant Konrad Bleuler (1912-92) who had been
in this position with Wentzel since 1943 (Ref. [3], III.132). Pauli arrived in
Princeton on 29 November and found that "It is fun to be in a town where
everybody knows me in the shops (telling me that I became fat)' (letter [1061]
to Rabi of 13 December 1949 from Princeton, not Lund, as indicated in Ref.
[2]). In February Bohr also came to the institute on a visit (Ref. [35], letters
[1087, 1091]). There Pauli met his old friends, particularly Erwin Panofsky,
in whom the Kepler manuscript 'produced great enthusiasm', as Pauli wrote
to C. A. Meier on 26 February. In addition he mentioned in this letter that
'About the time of the [spring] equinox [see page 424] I am supposed to repeat
here the Kepler talk (in German) before a small circle consisting of historically

interested people as well as of a German [technical] *physicist* with the name of [Max] Knoll and his wife who is interested in Jungian psychology' (translated from letter [1085], Ref. [35], p. 36; see also letter [1080]).

But Pauli's main preoccupation was the new field theory. In letter [1061] just quoted he continues: 'Physics is as difficult here than in Zurich and my talent to get new good ideas is decreasing (sign of age?). But the picture which I have in mind both on the achievements and the limitations of this *renormalization-*[business] is becoming more and more round.' And observing 'This is a funny idea of Oppe[nheimer] to make this "Pauli-show" at the February physics meeting in New York', he asks Rabi: 'Would the show really [be] good for something?' To Pauli's hesitation Rabi answered on 16 December: 'I therefore think it is somewhat of a duty for an old *Quantengreis* to let himself be seen and heard as a part of the community of physics' (Ref. [35], note 1, p. 8). The 'show' was an invited paper at the New York meeting of the American Physical Society held from 2 to 4 February 1950 [55]. Pauli's main concern now was the question of which of the field theories were renormalizable, a problem he had discussed with Heisenberg back in 1937 (see page 317) and for which he had given quantitative criteria in the tables of cross-sections contained in the last section of his review paper of 1941 [44e] (see page 330).

Pauli expressed this concern explicitly in letters [1059] of 1 December to Jauch and [1062] of 14 December 1949 to Paul Taunton Matthews (1919–87) [56]. To Jauch he wrote: 'What interests me very much at the moment is the circumstance that the principle of "renormalization" of charge and mass suffices only with particular interactions to render the results convergent and unique. It certainly does *not* work with charged particles of spin 1 and electromagnetic interaction with the radiation field and with many meson theories, perhaps it also works with electromagnetic interaction of spin-0 particles (this is the most important still undecided question) and it certainly works in the positron theory (i.e. spin-1/2 particles). That it works at all sometimes looks at the present state of the theory like "more luck than reason"' (translated from Ref. [2], p. 711). And to Matthews Pauli wrote: 'Your results seem to indicate that this division of interactions into two classes coincides exactly with another one proposed by Heisenberg [57a,b]: at that time we called "interactions of the first kind" those for which the cross section for the possible processes (today one would say the elements of the "S-matrix") computed by perturbation theory, do not increase with the energy of the particle involved (more precisely: do not increase stronger than logarithmically with the energies), whilst for "interactions of the second kind" they do increase essentially with energy. . . . In 1938 I pointed out to Heisenberg that his earlier dimensional considerations are not sufficient to decide whether a given interaction is of the first or second kind and I particularly emphasized that the interaction of charged spin-1 particles with the electromagnetic field is of the *second* kind. . . . This general principle to discuss the properties of different kinds of interactions you will find again in the last section of my article in the Reviews of Modern Physics 1941 [44e]." (Ref. [2], p. 718).

The spin-0 case was found soon after to be renormalizable by Fritz Rohrlich (b. 1921), a collaborator of Feynman at Cornell University and a student of Schwinger at Harvard, as Pauli wrote to Dyson and to Matthews on 16 January 1950 (Ref. [35], letters [1072, 1073]), coming back from a small meeting in Rochester. However, Dyson had observed that compensating terms $\sim \phi^{*2}\phi^2$ are essential in this connection. Now, as Pauli related to Dyson on 9 February (letter [1079]), the same problem was addressed again by David Feldman (b. 1921) who from the beginning of 1949 to summer 1950 was at the Princeton Institute, collaborating with Chen Ning Yang (b. 1922) who came to the Institute in the autumn of 1949. 'Feldman is trying to obtain a general proof that the higher approximations of electrodynamics of charged scalar mesons can all be made convergent with renormalizations plus such additional terms $\sim \phi^{*2}\phi^2$. ... The general proof, however, is not easy' (letters [1079, 1089]).

By now Dyson's name was firmly established through his two fundamental papers [58a,b] on QED, the first showing the equivalence of the Feynman and Schwinger theories, [59a,b] and [43a,b], respectively, and the second giving the demonstration that QED is renormalizable. It was the first such theory. Dyson grew up in Berkshire, England, where his father taught music at a boys' school. After college, in 1941, he started studies mainly of mathematics at Cambridge University, and in 1943 joined war research at Bomber Command. In the autumn of 1947 he went on a Commonwealth Fellowship and strong recommendations from Kemmer at Cambridge to Cornell University in the United States for graduate studies. There he interacted a great deal with Bethe and Feynman. In September 1948, after attending the Ann Arbor summer school, he came to the Institute in Princeton. It is there that the Feynman–Schwinger connection became clear to him and where he wrote the two papers [58a,b]. On his return from the Solvay Conference in Brussels, Oppenheimer, who had become the Institute's director the year before, arranged a series of seminars by Dyson. Dyson returned to the Institute in 1950 and became a professor there in 1953 [15b].

The magic formula which allowed Dyson to connect Feynman's intuitive pictures of particle lines in space–time to Schwinger's formalism was the result of a formal resolution of Schwinger's Eq. (10.12) written for the S-matrix (10.11). By iteration one obtains

$$S = \sum_{n=0}^{\infty} S_n \qquad (10.36)$$

with

$$S_n = \left(-\frac{i}{\hbar c}\right)^n \int_{-\infty}^{\infty} d^4x_1 \mathcal{H}^I(x_1) \int_{-\infty}^{\sigma_1} d^4x_2 \mathcal{H}^I(x_2) \ldots \int_{-\infty}^{\sigma_{n-1}} d^4x_n \mathcal{H}^I(x_n). \qquad (10.37)$$

Here $\mathcal{H}^I$ is the interaction energy density (10.10), to which is added a term

$$\delta\mathcal{H}'(x) = -\delta m \overline{\psi}(x)\psi(x) \qquad (10.38)$$

compensating the self-energy effects in S and, depending on the physical process considered, also an external field $\Phi^e_\mu(x)$ is added to $\Phi_\mu(x)$. Now the n-fold space–time integral in (10.37) is the $(n!)^{-1}$ fraction of the integral extended over all n-fold space–time. Therefore, symmetrizing over all $n!$ permutations of $(1 \ldots n)$, one may write

$$
S_n = \frac{1}{n!} \prod_{r=1}^{n} \left(-\frac{i}{\hbar c} \int_{-\infty}^{\infty} d^4x_r \right) P(\mathcal{H}^I(x_1) \ldots \mathcal{H}^I(x_n)), \qquad (10.39)
$$

where for any (bosonic) operators $F_1(x_1) \ldots F_n(x_n)$ *Dyson's time-ordered or P-product* is defined in Eq. (29) of Ref. [58a] as

$$
P(F_1(x_1) \ldots F_n(x_n)) = F_{p_1}(x_{p_1}) \ldots F_{p_n}(x_{p_n}) \; ; \; \text{if } t_{p_1} > t_{p_2} > \cdots > t_{p_n}, \qquad (10.40)
$$

$p_1 \ldots p_n$ being a permutation of $1 \ldots n$. With this notation the S-matrix (10.36) may formally be written in the compact form of Eqs [22.2], [22.4] of Pauli's *Feldquantisierung* [44b],

$$
S = P \exp\left(-\frac{i}{\hbar c} \int_{-\infty}^{\infty} d^4x \mathcal{H}^I(x) \right). \qquad (10.41)
$$

Dyson in Ref. [58a] then goes on to show that a typical product in S_n,

$$
Q_n = \overline{\psi}(x_{i_1}) \otimes \psi(x_{i_1}) \ldots \overline{\psi}(x_{i_n}) \otimes \psi(x_{i_n}) \Phi(x_{j_1}) \ldots \Phi(x_{j_m}), \qquad (10.42)
$$

may be resolved into unpaired operators determined by the initial and final states and a sum of products, Eq. (40) of Ref. [58a],

$$
\begin{aligned}
Q'_n = \; & \epsilon P(\overline{\psi}(x_1) \otimes \psi(x_{r_1})) \ldots P(\overline{\psi}(x_n) \otimes \psi(x_{r_n})) \\
& \times P(\Phi(x_{s_1}) \otimes \Phi(x_{t_1})) \ldots P(\Phi(x_{s_h}) \otimes \Phi(x_{t_h})), \qquad (10.43)
\end{aligned}
$$

where ϵ is a sign depending on whether the permutation of the $\overline{\psi}$ and ψ from (10.42) to (10.43) is even or odd. In the matrix element of the process considered only $\langle Q'_n \rangle_0$ occurs, which gives rise to the new invariant functions

$$
\langle P(\Phi_\mu(x)\Phi_\nu(x'))\rangle_0 = \frac{1}{2}\hbar c^2 \delta_{\mu\nu} D^c(x - x') \qquad (10.44)
$$

and

$$
\langle P(\psi_\alpha(x)\overline{\psi}_\beta(x'))\rangle_0 = -\frac{1}{2}\epsilon(x - x') S^c_{\alpha\beta}(x - x'). \qquad (10.45)
$$

Dyson calls the functions D^c and S^c in Eqs (41) and (43) of Ref. [58a] D_F and S_F in honour of Feynman. The above notation is that of Pauli's *Feldquantisierung* [44b] which was Stueckelberg's choice (see e.g. Ref. [48b]), the upper index c standing for causal.

For the evaluation of these functions the P-product of both bosonic and fermionic operators is needed, which in terms of the step function θ^+ introduced in Eq. (10.33) is

$$P(A(x)B(x')) = A(x)B(x')\theta^+(x-x') \pm B(x')A(x)\theta^+(x'-x). \quad (10.46)$$

Expressing θ^+ here by the sign function $\varepsilon(x)$ as $\theta^+(\pm x) = (1 \pm \varepsilon(x))/2$ and using Eq. (10.28) and the analogues of (6.17) and (10.44), i.e.

$$\langle P(A(x)B(x'))\rangle_0 = \frac{1}{2}\hbar c^2 \Delta^c(x-x'), \quad (10.47)$$

one finds that

$$\Delta^c(x) = \Delta^1(x) - 2i\overline{\Delta}(x). \quad (10.48)$$

In Ref. [44b] this function is given in Eqs [5.13] and [13.9]. Similarly one finds for Dirac spinors, with Eqs (10.15), (10.20), (10.45),

$$S^c(x) = \varepsilon(x)S^1(x) + iS(x). \quad (10.49)$$

Finally, making use of the integral representations (10.30), (10.31) in Eq. (10.48) one obtains Dyson's Eq (45) of Ref. [58a], or

$$\Delta^c(x) = -i\frac{k_c^2}{2\pi^2 c}\int_0^\infty d\alpha e^{-i\left(\frac{1}{4\alpha}+\alpha\xi\right)}, \quad (10.50)$$

where $\xi \equiv -k_c^2 x^2$. $\overline{\Delta}$ and, according to (10.48), therefore also Δ^c, are *Green's functions*, i.e. solutions of the inhomogeneous equation, Eq. [5.6] of Ref. [44b],

$$(\Box - k_c^2)\overline{\Delta}(x) = -\frac{1}{c}\delta^4(x). \quad (10.51)$$

Dyson's magic formula (10.39), together with the factorization (10.43), now give rise to the celebrated Feynman graphs that Feynman introduced in Ref. [59b] and for which Dyson in Ref. [58b] gives the following general rules: each factor (10.45) represents a directed electron/positron line, each factor (10.44) represents an undirected photon line, and each factor $\mathcal{H}^I$ in (10.39) gives rise to a vertex in which two equally directed electron/positron lines and one photon line meet. This graphical understanding of field theory was not to Pauli's liking; his mind was not visual, as his poor performance at the blackboard showed. Indeed, Thellung recalls: 'Pauli didn't draw well; his figures on the blackboard were rather rudimentary.' And further on he says: 'When someone drew Feynman graphs on the blackboard he said: "This is sentimental painting" (*Stimmungsmalerei*)' (translated from Ref. [60a], pp. 100 and 101).

A quite different problem on which Pauli and Dyson exchanged letters in February 1950 (Ref. [35], letters [1079, 1081, 1083]) concerned the difficulty with the 'longitudinal photons'. This problem, and in particular Pauli's appendix to letter [1079], are described in detail in Section 15 of the *Feldquantisierung* [44b], where the commutation relations (c.r.) (6.17) are called 'strong'.

In fact, they violate the Lorentz condition (6.23) since, according to (6.18), (6.20), $\partial D(x)/\partial x^\mu \neq 0$. One therefore has to use either the 'weak' c.r. (6.22) or the supplementary condition (6.42). It is easy to see that with the polarization vectors of Eq. (6.13) the supplementary condition (6.42) reduces to the two conditions $(b(3\mathbf{k}) + ib(4\mathbf{k}))\Psi = 0$ and $(b^+(3\mathbf{k}) + ib^+(4\mathbf{k}))\Psi = 0$ which in the notation of [15.10] are Eqs [15.14] of Ref. [44b]. This problem was not, however, Pauli's main concern, as he writes on 26 February to Wentzel who had invited him to Chicago for a talk, 'but in all principal questions I am "stuck" and before me is an impenetrable wall. (That I can laugh about mistakes which in part younger people, in part such famous gentlemen as Heisenberg [footnote: "He has sent me a paper. It is shocking!" See Pauli's letters [1088, 1098] to Heisenberg], make therewith is only a feeble consolation.) I know increasingly better why this or that, here or there cannot work. But would this be a pleasant colloquium talk in Chicago?' (translated from Ref. [35], letter [1086]).

Pauli and Franca left Princeton on 12 April 1950 (letter [1098]) with an invitation to Pauli to become a permanent member of the Institute for Advanced Study (see Ref. [35], letter [1403]). They sailed from New York to Le Havre on board the *Ile de France*. From there they travelled to Paris where Pauli attended the conference on elementary particles, organized by the Centre National de Recherche Scientifique (CNRS) under the direction of Alexandre Proca (see pages 303 and 323–8) from 24 to 29 April and taking place at the Institut Henri Poincaré. From the United States Feynman also attended the conference and, like Pauli, gave one of the invited talks. After the conference Feynman visited Zurich on Pauli's invitation and addressed the joint ETH–University colloquium (Ref. [60b], p. 332). Pauli had prepared his Paris talk during the voyage (see the comment on pp. 92–3 of Ref. [35]). At the conference, on 25 April, Pauli celebrated his 50th birthday. For this occasion Robert Oppenheimer and his wife Kitty, who had become a close friend of Franca, sent Pauli a telegram (Ref. [35], letter [1107]). Pauli missed Scherrer's 60th birthday on 3 February, for which a special issue of *Helvetica Physica Acta* appeared and Jean Weiglé from Geneva had written a congratulary address [60c].

Pauli's talk 'Present State of the Quantum Theory of Fields. The Renormalization.' [61a,b] opens with the following summary: 'The idea of "renormalization" of charges and masses ... is based on the hope, that only non observable quantities like self-charges and self-masses are divergent or ambiguous but that all observable quantities ... are unique. ... This hope however turned out only [to] be justified for the particular kind of interaction occurring in quantum electrodynamics (or the very similar theories of *neutral* scalar or vector mesons with vector coupling). For all other kind[s] of interactions genuine or "primitive" divergences remain even after renormalization' (Ref. [61b], p. 83). In Ref. [61a] the particularities of all these cases are summarized in Table I (missing in [61b]). In Section 2 technical details of the Schwinger–Dyson theory are summed up in Table II (Slide 1 in [61b]) which are followed by two important remarks, the first: 'The essential feature[s] of the theory are always a) the commutation rules for the unperturbed fields b) the vacuum-expectation values of quantities bilinear in the field components. As was shown by Bohr and Rosenfeld [62a], these quantities can be considered as measurable in principle, *after averaging them over finite space-time regions.*' The second remark concerns the interaction representation used in this theory: 'it seems to me of great methodical interest, that *C. N. Yang* succeeded to define all quantities of physical importance (as the S-matrix) also with the Heisenberg representation (time independent state vectors) at once [62b]. The latter has the advantage not to use any curved surfaces [σ] and their normals, thus showing more directly the relativistic invariance of the defined quantities' (Ref. [61b], p. 85).

This form of QED is sketched on p. 69 of Ref. [61a] and on p. 86 of Ref. [61b]. It is also described in detail in Section 24 of the *Feldquantisierung* [44b]. In the Heisenberg representation the time evolution from incoming free fields $F^{in} = \Phi_\mu^{in}, \psi^{in}$ is governed by Eqs [24.1] of Ref. [44b],

$$\Phi_\mu(x) = \Phi_\mu^{in}(x) + ie \int d^4x' D^{ret}(x - x')j^\mu(x')$$

$$\psi(x) = \psi^{in}(x) - ie \int d^4x' S^{ret}(x - x')\gamma^\mu \Phi_\mu(x')\psi(x'). \quad (10.52)$$

Here the retarded Green's functions are defined as

$$D^{ret} = \overline{D} - \frac{1}{2}D \; ; \; S^{ret} = \overline{S} - \frac{1}{2}S. \quad (10.53)$$

For $t \to +\infty$ the fields again approach free fields $F^{out} = \Phi_\mu^{out}, \psi^{out}$, and the S-matrix now defines the *canonical transformation* [24.5] of Ref. [44b],

$$F^{out}(x) = S^{-1}F^{in}(x)S. \quad (10.54)$$

In Section 3 of Refs. [61a,b] the procedures of renormalization and regularization are described and summarized in Tables III and IV (Slides 2 and 3 in [61b]). And Pauli closes with the belief: 'A further progress can only be reached by quite new ideas sufficient to determine theoretically the masses of the particles occurring in nature and presumably also the value of the fine structure constant, as the lower bound of the space–time regions occurring in the discussed averaging can only be discussed in connection with these mass values and is therefore a problem beyond the range of the present ideas' (Ref. [61b], p. 89). In Ref. [61b] some additional comments on Divergences are added.

Back in Zurich on 1 May 1950 (letters [1098, 1114] in Ref. [35]), Pauli had to start teaching immediately, the summer term having already started. Replete with all the new knowledge on field theory he also began his *Spezialvorlesung* 'Ausgewählte Kapitel aus der Feldquantisierung', which has already been mentioned many times. Pauli continued this course during the following winter term of 1950/1 (see Ref. [3], p. 433). These published lectures [44b] are based on notes edited by Urs Hochstrasser (b. 1926) and Pauli's assistant Schafroth. Hochstrasser had written his diploma paper under Pauli in 1948 (Ref. [3], p. 434) and then went into applied mathematics. After holding various professorships in the United States, he became in 1958–61 the first scientific counsellor to the Swiss embassy in Washington. Back in Switzerland he held several positions connected with science policy.

Sections 1 to 12 of the *Feldquantisierung* [44b] were given by Pauli during the short summer term and Sections 13 to 30 during the winter term. The two lecture periods were separated by the long summer vacation, during which Pauli went with Franca to the Engadin in the Grisons and from there to Italy, where Pauli participated with Schafroth and Roy Glauber, visiting from Harvard, at the meeting of the Italian Physical Society in Bologna (Ref. [35], letter [1151] and footnote 1, p. 162). The vacations had the effect that the invariant functions were introduced twice by Pauli, first in Sections 3 and 5 and again in Sections 13 and 18 (see the editorial note [A-4] in Ref. [44b]). In the German original which, like the cycle of the five basic courses, was published as a paperback volume, these two periods were marked as *Teil I* and *Teil II*. There was no division to the seven chapters of the English translation, however. The course, in fact, was not a systematic introduction into the subject but rather a discussion of the salient points of ongoing research, and the tools were introduced as needed. Among the participants to Pauli's course were his diploma student Philippe Choquard (b. 1929) and probably Igal Talmi (b. 1925) who in November 1951 submitted his Ph.D. thesis on 'Nuclear Spectroscopy with Harmonic Oscillator Wave Functions', on which Pauli wrote a very favourable report (see Ref. [3], III.73, 77). At ETH on a fellowship from Utrecht Laurens Jansen (b. 1923) also took part (see footnote 4 in Ref. [63a], p. 90). Armin Thellung (b. 1924) had left Zurich to become Kronig's assistant in Delft where he worked for his Ph.D. with Pauli. I missed the entire course because of illness.

The last chapter of Ref. [44b], which in the original *Feldquantisierung* figures as *Anhang* (appendix) stands out not only because of its unusual content but also because of its history, which has been recounted in an article by Choquard and Steiner [63b]. This is *Feynman's path-integral approach* to quantum mechanics and to QED, Refs [64a] and [64b], respectively. During the winter of 1949/50 that Pauli spent in Princeton, this problem was the subject of research and discussion at the Institute for Advanced Study between Cécile Morette (b. 1922) from Paris who spent the years 1948–50 at the institute and of Léon Charles van Hove (1924–90) from Brussels who was there in 1949–50. Morette and van Hove also presented their work in a discussion to Pauli who then wrote some research notes on it (see Ref. [63b]). In his lectures at ETH on the subject that took place towards spring 1951 Pauli developed his own version of Feynman's theory. This part of the lectures caught the interest of Choquard who, after his diploma with Pauli in 1951 (Ref. [3], p. 435), decided to write a Ph.D. thesis on the subject, in which he extended Pauli's presentation in several respects [65].

Pauli wrote a very favourable report on Choquard's thesis (Ref. [3], III.124), the first paragraph of which is an excellent introduction to the subject; it reads: 'Taking up older attempts by Dirac, Feynman has brought into focus the discussion of special solutions $K(q, t; q', t')$ of the Schrödinger equation which depend on a pair of points (q, q') and which for $\tau = t - t' = 0$ go over into the

Fig. 10.1 Pauli on an excursion of the ETH Physics Institute, 1952

(singular) Dirac function $\delta^{(n)}(q - q')$ [not α]. It is indeed possible in this way to formulate in a new, independent manner, to start with in the ordinary non-relativistic wave mechanics, the relation between wave mechanics and classical mechanics. While in the usual presentation of the theory the transition from the latter to the former is contained in the formal prescription for the substitution of the classical Hamiltonian by the Hamilton operator, according to Feynman this takes place in a prescription, based on classical mechanics, for the formation of the kernel $K_c(q, t; q', t')$ which for sufficiently small τ approximates the exact quantum mechanical kernel $K(q, t; q', t')$. Once the latter is found for small τ it can, due to group-theoretical properties, in principle be found by iteration with the help of an appropriate passage to the limit. Feynman's rules to form K_c soon turned out to be identical with the old Wentzel-Kramers-Brillouin (WKB) method for time-dependent solutions' (translated from Ref. [3], III.124).

The kernel $K(q, t; q', t')$ is the solution $\psi(q, t)$ of the Schrödinger equation

$$\frac{\hbar}{i} \frac{\partial \psi}{\partial t} + H\psi = 0 \qquad (10.55)$$

for arbitrary initial values $\psi(q', t')$,

$$\psi(q, t) = \int d^n q' K(q, t; q', t') \psi(q', t'), \qquad (10.56)$$

so that

$$K(q, t; q', t) = \delta^{(n)}(q - q'). \qquad (10.57)$$

If the Hamiltonian H is time independent, one may write

$$K(q, t; q', t') = K(q, q'; t - t').$$ (10.58)

This kernel has the group property

$$\int d^n q' K(q, q'; \tau) K(q', q''; \tau') = K(q, q''; \tau + \tau').$$ (10.59)

Classically, the canonical variables (p', q') at time $t' = t - \tau$ and (p, q) at time t are connected by a canonical transformation $(p', q') \rightarrow (p, q)$ generated by a function $S(q, q')$ as

$$p_k = \frac{\partial S}{\partial q_k} \; ; \; p'_k = -\frac{\partial S}{\partial q'_k}.$$ (10.60)

Quantum-theoretically this transformation is described by $K(q, q'; \tau)$ and, according to Dirac [66], it corresponds to $\exp(iS/\hbar)$, where

$$S(q, t; q', t') = \int_{t'}^{t} L(q, q'; t'') dt''$$ (10.61)

is the classical action, the classical path being determined by the minimum of S. The discussions in Princeton in the winter of 1949/50 concerned the problem of how this correspondence can be made into an equation. It turned out [63b] that one has to multiply $\exp(iS/\hbar)$ by $(2\pi i\hbar)^{-n/2}\sqrt{D}$ where D is the functional determinant

$$D = \frac{\partial(p'_1 \ldots p'_n)}{\partial(q_1 \ldots q_n)} = \left\| -\frac{\partial^2 S}{\partial q_i \partial q'_k} \right\|.$$ (10.62)

This is, up to a sign, van Hove's result (see Ref. [63b], p. 654).

In the *Feldquantisierung* [44b] Pauli shows that, for a Hamiltonian

$$H(p, q) = \sum_k \frac{1}{2m_k} (p_k + A_k(q))^2 + V(q)$$ (10.63)

and a Lagrangian $L = \sum p_k \dot{q}_k - H$, i.e.

$$L(\dot{q}, q) = \sum_k \left(\frac{m_k}{2} \dot{q}_k^2 - \dot{q}_k A_k(q) \right) - V(q)$$ (10.64)

(note the misplaced bracket in L, Eqs [30.15] of Ref. [44b]), the classical kernel [30.9] so constructed,

$$K_c(q, q'; \tau) = (2\pi i\hbar)^{-n/2}\sqrt{D} e^{\frac{i}{\hbar} S(q, q'; \tau)}$$ (10.65)

gives rise to a Schrödinger-like equation

$$\frac{\hbar}{i} \frac{\partial K_c}{\partial \tau} + H K_c = \left\{ Q_0 + \frac{\hbar}{i} Q_1 + \left(\frac{\hbar}{i} \right)^2 Q_2 \right\} K_c.$$ (10.66)

Since for any logarithmic variation of the expression (10.65)

$$\delta K_c = \left(\frac{i}{\hbar}\delta S + \frac{1}{2D}\delta D\right)K_c \tag{10.67}$$

one finds

$$Q_0 = \frac{\partial S}{\partial \tau} + \sum_k \frac{1}{2m_k}\left(\frac{\partial S}{\partial q_k} + A_k\right)^2 + V, \tag{10.68}$$

so that $Q_0 = 0$ is the Hamilton–Jacobi equation, while

$$Q_1 = \frac{1}{2D}\left\{\frac{\partial D}{\partial \tau} + \sum_k \frac{1}{m_k}\frac{\partial}{\partial q_k}\left(\left(\frac{\partial S}{\partial q_k} + A_k\right)D\right)\right\} \tag{10.69}$$

and

$$Q_2 = \sum_k \frac{1}{2m_k\sqrt{D}}\frac{\partial^2\sqrt{D}}{\partial q_k^2}. \tag{10.70}$$

Following Eq. [30.17] in Ref. [44b] it is shown that $Q_1 = 0$ (see *Note* [67]). Therefore, the only non-Schrödinger term left is Q_2, which is called 'false term' in Ref. [44b]. Note that Feynman in Section 6 of Ref. [64a] considers the one-dimensional example $H = p^2/2m + V(x)$ and verifies the Schrödinger equation by calculating Eq. (10.56) for $t - t' = \epsilon$ and with K given by Eq. (10.65) with $D = 1$.

The basic requirements on K_c are Eqs (10.57) and

$$\lim_{\tau\to 0}\frac{K - K_c}{\tau} = 0. \tag{10.71}$$

Dividing the time interval τ into N equal parts $\epsilon = \tau/N$ such that, within ϵ, K_c is a sufficient approximation to K in the sense (10.71), it is possible, by making use of the group property (10.59), to express K by Eq. [30.21] of Ref. [44b],

$$K(q^{(\infty)}, q^{(0)}; \tau) = \lim_{N\to\infty}\int d^n q^{(1)}\ldots d^n q^{(N-1)}\prod_{\alpha=0}^{N-1}K_c(q^{(\alpha+1)}, q^{(\alpha)}; \epsilon). \tag{10.72}$$

Here the multiple integral may be interpreted as an integral over all classical paths. This is Feynman's fundamental path-integral expression for the wavefunction, Eq. (12) of Ref. [64a], based on the classical action S. Note that Feynman does not explicitly calculate the prefactor in Eq. (10.65)—which he calls $1/A$—except in the example mentioned above (where he sets $D = 1$; see also footnote 15 where $D = \rho$). In Ref. [64b] this path-integral treatment is applied to QED, describing the interaction of radiation with matter by forced harmonic oscillators. This is the last subject mentioned in Pauli's *Feldquantisierung*.

Richard Phillips Feynman (1918–88) spent his childhood on Long Island, just east of Brooklyn, New York, where he had his own laboratory. His father had come from a Jewish family in Minsk, Byelorussia; his mother's family was from Germany. Feynman attended college at MIT and in 1939 went to graduate school at Princeton where he was assigned to the young John Archibald Wheeler (b. 1911). While Wheeler and Feynman developed an action-at-a-distance formulation of electrodynamics, in May 1942 Feynman submitted his doctor's dissertation on 'the principle of least action in quantum mechanics' Since December 1941 he had worked in Princeton on the atomic bomb project and early in 1943 he was one of the first to move to Los Alamos. After the war Feynman joined Hans Bethe in Cornell University and in 1950 he moved to Caltech at Pasadena [15b, 60b].

For the spring vacation, early in April 1951, Pauli and Franca went to Sicily for about three weeks , Pauli's first visit to this 'strange country mixed of Greece, Albania, Italy with wonderful byzantine Art, particularly mosaics (in which Franca is most interested)' (Ref. [35], letter [1220] to Pais). On his return Pauli participated in the 'International Forum Zurich' organized by his colleague Gonseth (letter [1224]). On 30 April he wrote the following letter: 'Dear Bryce DeWitt and Cécile Seligman [DeWitt]! . . . I am sending to both of you my most affectionate congratulations to your marriage, also in the name of my wife' (letter [1230]). DeWitt had spent the year 1949/50 at the Institute for Advanced Study where he met both Cécile Morette and Pauli. In the autumn of 1950 Morette and de Witt visited the Paulis in Zurich as fiancés (Ref. [35], letter [1153] and footnote 1, p. 295). In letter [1230] Pauli continued: 'I shall be in Copenhagen and I am also planning to visit your summer school at least for a few days. I intend to talk this over with Jost, as soon as he arrives in Zurich.' In July Pauli attended the Copenhagen conference (letter [1232]) and from 15 to 22 August he visited the first of the summer schools at Les Houches on the north slope of Mont Blanc which had been initiated by Cécile DeWitt-Morette [68]. From Zurich Heitler and Schafroth also attended, and also Jost, who spent the summer vacation from Princeton in Switzerland (Ref. [35], postcard [1272]). Pauli was back in Les Houches the following year 1952 (see letter [1444] in Ref. [35]).

The quoted letter [1230] in Ref. [35] also mentions Sommerfeld's fatal accident: 'He was literally running into a truck (it is true that he has been deaf recently, but otherwise of good health).' Pauli wrote the obituary in Ref. [25], pp. 69–71. The quoted letter [1230] ends with the parenthesis: 'Cécile Morette may be interested in the way I treated the Feynman-action principle in my mimeographed lecture. It is a kind of generalization of the WKB-method to time-dependent solutions.'

At ETH meanwhile, on 6 July 1950 Pauli was unanimously elected Abteilungsvorstand, i.e. head of the mathematics and physics section, for two years (Ref. [3], III.43). His first official action in this position was to send a letter to President Pallmann dated 30 October 1950 concerning the duration of the diploma papers and the associated taxes (Ref. [3], III.47). And his last action was to send a letter to the president on 16 July 1952 proposing the creation of an institute for the history and philosophy of science by Ferdinand Gonseth

(Ref. [3], III.99). During this period, and more precisely from 1948 to 1953, Pauli's and Scherrer's secretary at the physics department was Käthi Gemperle (1923–59) who on 1 August 1952 married Pauli's assistant Schafroth (Ref. [3], p. 197).

Also during this period, due to pressure from Scherrer's expanding enterprise, the Institute for Theoretical Physics had to move from the first floor, the c-level, of the physics building at Gloriastrasse to the third floor, the e-level. Thus Pauli had to give up his office 4c, which was adjacent to the lecture room 6c, in favour of 3e. This was not a bad deal though, since from 3e, which was situated directly over the entrance, one had a commanding view of the lake and the chain of the Alps. But in addition, Pauli gained the back-rooms 14e to 17e opposite his new office that were previously occupied by the meteorology department which moved to a new building (Ref. [3], p. 189). Room 14e–15e became a discussion room, 16e the assistant's office, and 17e later became Jost's office.

Max Robert Schafroth (1923–59) grew up in the Bernese town of Burgdorf at the centre of the Emmental. His father was a career colonel in the Swiss Army. Schafroth finished Gymnasium in Berne where in 1941 he began to study theoretical physics. These studies continued with Pauli at ETH where he obtained his diploma in 1948. After his Ph.D. one and a half years later he was Pauli's assistant till March 1953. In Zurich he became interested in superconductivity, in which field he built a distinguished career, competing— to the amusement of Pauli—with John Bardeen (b. 1908) and his collaborators Leon N. Cooper (b. 1930) and John R. Schrieffer (b. 1931). This orientation led him to accept a research fellowship in 1953 with Herbert Fröhlich (b. 1905) in Liverpool. After one year he and his wife Käthi went to Sydney, Australia, where Schafroth became a lecturer and in 1957 a reader at the university. In 1955 he spent three months at the Institute for Advanced Study in Princeton where he met Leon Cooper who was there during the year 1954/5 and with whom he discussed the idea of electron pairs as carriers of superconductivity. In Sydney Schafroth extended his ideas on the subject with J. M. Blatt (see page 382) and S. T. Butler. In contrast to the overlapping pairs of the Bardeen–Cooper– Schrieffer theory, the pairs in Schafroth's theory are dilute and correspond much better to the situation in the newer high-temperature superconductors (see e.g. Sec. 20 in Ref. [69]).

In July 1958 Schafroth was nominated to the chair of theoretical physics at Geneva University and he asked me to join him as associate professor. He visited Geneva during the Annual International Conference on High Energy Physics at CERN from 30 June to 5 July where he also met N. N. Bogoliubov, another expert on superconductivity. On a farewell tour in Australia with his wife before taking up his new position in Geneva on 1 September, the small postal plane on which they were travelling crashed on 29 May 1959 about a thousand miles (1600 km) north of Brisbane. Robi and Käthi Schafroth left behind their son Markus Konradin Schafroth (b. 1955) who was then looked after by Schafroth's parents. Robi Schafroth was a tall, elegant, and good-natured man who spoke several languages fluently; he had had a very pleasant relationship with Pauli and with the students.

Notes and references

[1] Translated from: W. Pauli, letter [835] of 19 September 1946 to Einstein, in Ref. [2].

[2] K. von Meyenn (ed.), *Wolfgang Pauli. Scientific Correspondence with Bohr, Einstein, Heisenberg, a.o., Volume III: 1940–1949* (Springer, Berlin and Heidelberg, 1993).

[3] C. P. Enz, B. Glaus, and G. Oberkofler (eds), *Wolfgang Pauli und sein Wirken an der ETH Zürich* (vdf Hochschulverlag ETH, Zurich, 1997).

[4] W. Heisenberg, 'Die beobachtbaren Grössen in der Theorie der Elementarteilchen'. (a): I, *Zeitschrift für Physik* **120**, 513 (1943); (b): II, *Zeitschrift für Physik* **120**, 673 (1943); (c): III, *Zeitschrift für Physik* **123**, 93 (1944); (d): unpublished. Reprinted in Ref. [5], Part II, (a): pp. 611–36; (b): pp. 637–66; (c): pp. 667–86; (d): pp. 687–98.

[5] W. Blum, H.-P. Dürr, and H. Rechenberg (eds), *Werner Heisenberg. Collected Works. Series A. Original Scientific Papers* (Springer, Berlin and Heidelberg, 1989).

[6] R. Oehme, 'Theory of the Scattering Matrix (1942–1946). An Annotation', in Ref. [5], Part II, pp. 605–10.

[7] D. Cassidy, Uncertainty. The Life and Science of Werner Heisenberg (Freeman, New York, 1992).

[8] S. T. Ma, 'Redundant zeros in the discrete energy spectra in Heisenberg's theory of the characteristic matrix' (Letter), *Physical Review* **69**, 668 (1946).

[9] (a) D. ter Haar, 'On the redundant zeros in the theory of the Heisenberg matrix', *Physica* **12**, 501 (1946); (b) R. Jost, 'Bemerkungen zur vorstehenden Arbeit', *Physica* **12**, 509 (1946); (c) R. Jost, 'Über die falschen Nullstellen der Eigenwerte der S-Matrix', *Helvetica Physica Acta* **20**, 256 (1947).

[10] (a) R. Jost, 'Erinnerungen: Erlesenes und Erlebtes', in Ref. [10d], pp. 11–19; (b) R. Jost, 'Zur Ladungsabhängigkeit der Kernkräfte in der Vektormesonentheorie ohne neutrale Mesonen', *Helvetica Physica Acta* **19**, 113 (1946); (c) A. Pais, 'Res Jost. January 10, 1918—October 3, 1990', in Ref. [10d], pp. 1–9; (d) K. Hepp, W. Hunziker and W. Kohn (eds), *Res Jost. Das Märchen vom Elfenbeinernen Turm. Reden und Aufsätze* (Springer, Berlin and Heidelberg, 1995).

[11] (a) W. Pauli, 'Difficulties of Field Theories and of Field Quantization', *Proceedings of the International Conference on Fundamental Particles and Low Temperatures* (Cambridge, 1947), pp. 5–10. Reprinted in Ref. [12], Vol. 2, pp. 1097–102. (b) Private communication from Professor

Bengt Nagel, Royal Institute of Technology, Stockholm, 8 April 1986. I thank Professor Nagel for this information. (c) W. Pauli, 'Exclusion Principle and Quantum Mechanics' in: *Les Prix Nobel en 1946* (Norstedt & Söner, Stockholm, 1948), pp. 131–47. Reprinted in Ref. [12], Vol. 2, pp. 1080–96 and in Ref. [25], pp. 165–81. (d) K. von Meyenn (ed.), *Wolfgang Pauli. Correspondence with Bohr, Einstein, Heisenberg, a.o., Volume II: 1930–1939* (Springer, Berlin and Heidelberg, 1985).

[12] R. Kronig and V. F. Weisskopf (eds), *Collected Scientific Papers by Wolfgang Pauli. In Two Volumes* (Wiley Interscience, New York, 1964).

[13] W. Killy and R. Vierhaus (eds), *Deutsche Biographische Enzyklopädie (DBE)* (Saur, Munich, 1996).

[14] (a) C. M. G. Lattes, G. P. S. Occhialini, and C. F. Powell, 'Observations on the tracks of slow mesons in photographic emulsions', *Nature* **160**, 453, 486 (1948); (b) A. Pais, *Inward Bound. Of Matter and Forces in the Physical World* (Oxford University Press, Oxford, 1986).

[15] (a) W. E. Lamb, Jr and R. C. Retherford, 'Fine Structure of the Hydrogen Atom by a Microwave Method', *Physical Review* **72**, 241 (1947). Reprinted in Ref. [16], pp. 136–8. (b) S. S. Schweber, *QED and the Men Who Made It: Dyson, Feynman, Schwinger, and Tomonaga* (Princeton University Press, Princeton, NJ, 1994).

[16] J. Schwinger (ed.), *Selected Papers on Quantum Electrodynamics* (Dover, New York, 1958).

[17] V. F. Weisskopf, *The Joy of Insight. Passions of a Physicist* (Basic Books, New York, 1971).

[18] (a) H. Bethe, 'The Electromagnetic Shift of Energy Levels', *Physical Review* **72**, 339 (1947). Reprinted in Ref. [16], pp. 139–41. (b) R. P. Feynman, 'Relativistic Cut-Off for Quantum Electrodynamics', *Physical Review* **74**, 1430 (1948); (c) N. M. Kroll and W. E. Lamb, Jr, 'On the Self-Energy of a Bound Electron', *Physical Review* **75**, 388 (1949). Reprinted in Ref. [16], pp. 414–24; (d) J. Schwinger, 'On Radiative Corrections to Electron Scattering', *Physical Review* **75**, 898 (1949). Reprinted in Ref. [16], pp. 143–4; (e) J. B. French and V. F. Weisskopf, 'The Electronic Shift of Energy Levels', *Physical Review* **75**, 1240 (1949).

[19] J. Schwinger, 'On Quantum-Electrodynamics and the Magnetic Moment of the Electron', *Physical Review* **73**, 416 (1948). Reprinted in Ref. [16], p. 142. Note that in this reference the correction to the magnetic moment is erroneously given as $(\pi/2)\alpha$.

[20] J. Schwinger, 'Quantum Electrodynamics, III: The Electromagnetic Properties of the Electron–Radiative Corrections to Scattering', *Physical Review* **76**, 790 (1949). Reprinted in Ref. [16], pp. 169–96.

[21] H. M. Foley and P. Kusch, 'On the Intrinsic Moment of the Electron', *Physical Review* **73**, 412 (1948). Reprinted in Ref. [16], p. 135.

[22] (a) *Who's Who in Switzerland 1968/69* (Nagel Publishers, Geneva, 1969); (b) *Schweizer Lexikon 91* (Verlag Schweizer Lexikon, Lucerne, 1992).

[23] W. Pauli, 'Editorial', *Dialectica* **2**, 307 (1948). Reprinted in Ref. [12], Vol. 2, pp. 1107–111.

[24] W. Pauli, 'Die philosophische Bedeutung der Idee der Komplementarität', *Experientia* **6**, pp. 72–75 (1950); English translation, 'The Philosophical Significance of the Idea of Complementarity', in Ref. [25], pp. 35–42.

[25] C. P. Enz and K. von Meyenn (eds), *Wolfgang Pauli. Writings on Physics and Philosophy* (Springer, Berlin and Heidelberg, 1994).

[26] (a) Translated from: W. Pauli, letter [33] of 23 December 1947 to C. G. Jung, in Ref. [27]. (b) *Note*: Pauli here refers to Fierz's lecture of 21 December 1942 'Isaac Newton, sein Charakter und seine Weltansicht', *Vierteljahrsschrift der Naturforschenden Gesellschaft in Zürich* **83**, 198 (1943). See also Fierz's masterful essay 'On the origin and the significance of Isaac Newton's doctrine of the absolute space' in Ref. [26c]. See also letter [926] in Ref. [2]. (c) M. Fierz, 'Über den Ursprung und die Bedeutung der Lehre Isaac Newtons vom absoluten Raum', *Gesnerus* (Schweizerische Gesellschaft für Geschichte der Medizin und der Naturwissenschaften) **11**, pp. 62–120 (1954).

[27] C. A. Meier (ed., with the cooperation of C. P. Enz and M. Fierz), *Wolfgang Pauli und C. G. Jung. Ein Briefwechsel 1932–1958* (Springer, Berlin and Heidelberg, 1992).

[28] (a) C. G. Jung, *Mysterium Coniunctionis. Untersuchungen über die Trennung und Zusammensetzung der seelischen Gegensätze in der Alchemie, Gesammelte Werke. 14* (Walter-Verlag, Olten, 5. Auflage 1990); (b) C. A. Meier, *Lehrbuch der komplexen Psychologie C. G. Jungs* (Walter-Verlag, Olten, 1977), 4 vols; (c) B. Glaus, 'Carl Alfred Meier (1905–1995) und seine Bibliothek', LIBRARIUM. Zeitschrift der Schweizerischen Bibliophilen Gesellschaft **40**, 113 (1997).

[29] *Note*: As said explicitly in letter [35] of Ref. [27], at least the second of these two dreams was joined to this letter.

[30] C. G. Jung, *Psychology and Alchemy, Collected Works, 12* (Princeton University Press, Princeton, NJ, 1968 2nd edn).

[31] W. Pauli, (a) 'Der Einfluss archetypischer Vorstellungen auf die Bildung naturwissenschaftlicher Theorien bei Kepler', in Ref. [32a], pp. 111–94; (b) 'The Influence of Archetypal Ideas on the Scientific Theories of Kepler'. English translation by Priscilla Sitz, in Ref. [32b], pp. 147–240. Reprinted in Ref. [25], pp. 219–79. Page numbers of quotations refer to the reprint.

[32] C. G. Jung and W. Pauli, (a) *Naturerklärung und Psyche* (Rascher Verlag, Zurich, 1952); (b) *The Interpretation of Nature and the Psyche* (Bollingen Series LI, Pantheon Books, New York, 1955).

[33] A. Koestler, *The Sleepwalkers. A history of man's changing vision of the Universe* (Hutchinson, London, 1959).

[34] C. P. Enz and K. von Meyenn (eds), *Wolfgang Pauli. Das Gewissen der Physik* (Vieweg, Braunschweig, 1988).

[35] K. von Meyenn (ed.), *Wolfgang Pauli. Scientific Correspondence with Bohr, Einstein, Heisenberg, a.o., Volume IV/Part I: 1950–1952* (Springer, Berlin and Heidelberg, 1996).

[36] (a) K. von Meyenn (ed.), *Wolfgang Pauli. Scientific Correspondence with Bohr, Einstein, Heisenberg, a.o., Volume IV, Part III: 1955–1956* (Springer, Berlin and Heidelberg, 2001); (b) W. Pauli, 'Moderne Beispiele zur "Hintergrundsphysik"', in Ref. [27], Appendix 3, pp. 176–92.

[37] A. Pais, *'Subtle is the Lord . . . ' The Science and the Life of Albert Einstein* (Oxford University Press, Oxford, 1982).

[38] C. G. Jung, (a) 'Synchronizität als ein Prinzip akausaler Zusammenhänge', in Ref. [32a], pp. 1–107. Reprinted in Ref. [39a], pp. 457–553. (b) 'Synchronicity: An Acausal Connecting Principle', in Ref. [32b], pp. 5–146. Reprinted in Ref. [39b], pp. 417–531.

[39] C. G. Jung, (a) *Gesammelte Werke, 8* (Walter-Verlag, Olten, 3. Auflage 1987); (b) *The Collected Works, 8* (Bollingen Series XX, New York, 1960).

[40] J. B. Rhine, *New Frontiers of the Mind* (New York, 1937); *The Reach of the Mind* (London, 1948).

[41] A. Schopenhauer, 'Über die anscheinende Absichtlichkeit im Schicksale des Einzelnen', in: A. Hübscher (ed.), *Arthur Schopenhauer. Sämtliche Werke* (Brockhaus, Wiesbaden, 1972). *Parerga und Paralipomena* Vol. 1/1, pp. 213–237.

[42] W. Pauli, letter [1001] of 24 January 1949 to Schwinger, in Ref. [2], pp. 618–19.

[43] J. Schwinger, 'Quantum Electrodynamics'. (a) 'I. A Covariant Formulation', *Physical Review* **74**, 1439 (1948); (b) 'II. Vacuum Polarization and Self-Energy', *Physical Review* **75**, 651 (1949).

[44] (a) J. Mehra and K. A. Milton, *Climbing the Mountain. The Scientific Biography of Julian Schwinger* (Oxford University Press, Oxford, 2000); (b) C. P. Enz (ed.), *Pauli Lectures on Physics: Volume 6. Selected Topics in Field Quantization*, translated by S Margulies and H. R. Lewis from: U. Hochstrasser and M. R. Schafroth (eds), *Ausgewählte Kapitel aus der Feldquantisierung*, reprinted by Boringhieri, Turin, 1962 (MIT Press, Cambridge, MA, 1973). (c) S. Tomonaga, 'On a relativistically invariant formulation of the quantum theory of wave fields', *Progress of Theoretical Physics* **1**, 27 (1946). Reprinted in Ref. [16], pp. 156–68; (d) P. A. M. Dirac, V. A. Fock, and B. Podolsky, 'On Quantum Electrodynamics', *Physikalische Zeitschrift der Sowjetunion* **2**, 468 (1932). Reprinted in Ref. [16], pp. 29–40; (e) W. Pauli, 'Relativistic Field Theories of Elementary Particles', *Reviews of Modern Physics* **13**, 203 (1941). Reprinted in Ref. [12], Vol. 2, pp. 923–52.

[45] (a) G. Källén, 'Higher approximations in the external field for the problem of vacuum polarization', *Helvetica Physica Acta* **22**, 637 (1949); (b) J. M. Luttinger, 'A note on the magnetic moment of the electron', *Physical Review* **74**, 893 (1948); 'On the magnetic moments of neutron and proton', *Helvetica Physica Acta* **21**, 483 (1948); 'On the magnetic moment of the nucleons in meson theory' (letter), *Physical Review* **75**, 1277 (1949); (c) P. W. Anderson, R. M. Friedberg, and W. Kohn, 'Joaquin M. Luttinger' (Obituary), *Physics Today*, December 1997, p. 89.

[46] (a) W. Pauli and F. Villars, 'On the Invariant Regularization in Relativistic Quantum Theory', *Reviews of Modern Physics* **21**, 434 (1949). Reprinted in Ref. [12], Vol. 2, pp. 1116–26 and in Ref. [16], pp. 198–208. (b) W. Pauli, 'Einstein's contribution to quantum theory', in: P. A. Schilpp (ed.), *Albert Einstein. Philospher scientist* (The Library of Living Philosophers, London, 1949); German version, *Einsteins Beitrag zur Quantentheorie* (Kohlhammer Verlag, Stuttgart, 1955). Reprinted in Ref. [12], Vol. 1, pp. 1013–22. English translation in Ref. [25], pp. 85–94.

[47] *Note*: Here Schwinger's Appendix in Ref. [43b] is followed in essence.
Using $\delta(k^2 + k_c^2) = (c/2\omega)(\delta((\omega/c) - k_0) + \delta((\omega/c) + k_0))$ Eq. (8.16) becomes [5.5] of Ref. [44b],

$$\Delta^1(x) = \frac{1}{c} \int \frac{d^4k}{(2\pi)^3} e^{ikx} \delta(k^2 + k_c^2) \, . \tag{A}$$

From (10.28) it follows by integration in a small four-cylinder closed at $x_0 = \pm\epsilon$

$$\int d^4k(\Box - k_c^2)\overline{\Delta} = - \int d^3x \frac{\partial\overline{\Delta}}{\partial x_0}\Big|_{-\epsilon}^{+\epsilon} = -1,$$

where in the last step use was made of $(\partial\Delta/\partial t)_{t=0} = -\delta^3(\mathbf{x})$ which follows from (8.12). Hence, $(\Box - k_c^2)\overline{\Delta} = -(1/c)\delta^4(x)$, the Fourier transform of which yields

$$\overline{\Delta}(x) = \frac{1}{c}P \int \frac{d^4k}{(2\pi)^4} \frac{e^{ikx}}{k^2 + k_c^2} \, , \tag{B}$$

where the principal value P is defined by

$$\frac{P}{u} \mp i\pi\delta(u) = \frac{1}{u \pm i\epsilon} = \mp i \int_0^\infty da e^{\pm iau}.$$

This allows (A) and (B) to be written

$$\Delta^1(x) = \frac{1}{c} \int_{-\infty}^{+\infty} da \int \frac{d^4k}{(2\pi)^4} e^{ikx} e^{ia(k^2 + k_c^2)} \tag{C}$$

and

$$\overline{\Delta}(x) = -\frac{i}{2c} \int_{-\infty}^{+\infty} da \frac{a}{|a|} \int \frac{d^4 k}{(2\pi)^4} e^{ikx} e^{ia(k^2 + k_c^2)}. \tag{D}$$

Writing $ak^2 + kx = a(k + x/2a)^2 - x^2/4a$ and making the shift $k \rightarrow k - x/2a$ the k-integrations are easily done since

$$I_{\pm} \equiv \int_{-\infty}^{+\infty} dk e^{(\pm ia - \epsilon)k^2} = \sqrt{\frac{\pi}{2a}}(1 \pm i); \; a > 0.$$

Indeed, $I_{\pm}^2$ may be evaluated in polar coordinates. Thus

$$\int d^4 k e^{ikx} e^{\pm ia(k^2 + k_c^2)} = \pm i \frac{\pi^2}{a^2} e^{\pm i(ak_c^2 - x^2/4a)}.$$

Hence, (C) and (D) become

$$\Delta^1(x) = -\frac{1}{8\pi^2 c} \int_0^\infty \frac{da}{a^2} \sin\left(ak_c^2 - \frac{x^2}{4a}\right) \tag{E}$$

and

$$\overline{\Delta}(x) = \frac{1}{16\pi^2 c} \int_0^\infty \frac{da}{a^2} \cos\left(ak_c^2 - \frac{x^2}{4a}\right). \tag{F}$$

With the substitution $\alpha = 1/4k_c^2 a$ Eqs. (10.30) and (10.31) readily follow. Next consider the substitution $\alpha = (1/2\sqrt{|\xi|}) \exp \theta$ which yields

$$\frac{1}{4\alpha} \pm \alpha \xi = \pm \sqrt{|\xi|} \begin{cases} \cosh \theta \\ \sinh \theta. \end{cases}$$

Thus, for $\xi > 0$,

$$\Delta^1(x) = -\frac{k_c^2}{4\pi^2 c \sqrt{\xi}} \int_{-\infty}^{+\infty} d\theta e^\theta \sin(\sqrt{\xi} \cosh \theta) = \frac{k_c^2}{4\pi c \sqrt{\xi}} N_1(\sqrt{\xi}).$$

Here N_1 is a Bessel function of the second kind whose singularities are well known. With $\xi = k_c^2 \lambda$ one finds Eq. (10.32). Now write $\overline{\Delta}(x) = (k_c^2/4\pi^2 c) F'(\xi)$ with

$$F(\xi) = \int_0^\infty \frac{d\alpha}{\alpha} \sin\left(\frac{1}{4\alpha} + \alpha \xi\right)$$

$$= \int_{-\infty}^{+\infty} d\theta \begin{cases} \sin(\sqrt{\xi} \cosh \theta) \\ - \sin(\sqrt{\xi} \sinh \theta) \end{cases} = \begin{cases} \pi J_0(\sqrt{\xi}) \\ 0. \end{cases}$$

Thus, in terms of the step function θ^+ introduced in (10.33),

$$\overline{\Delta}(x) = \frac{k_c^2}{4\pi c} \left\{ \delta(\xi) J_0(0) + \theta^+(\xi) J_0'(\sqrt{\xi}) \frac{1}{2\sqrt{\xi}} \right\},$$

which has the singularities of Eq. (10.33).

[48] (a) D. Rivier and E. C. G. Stueckelberg, 'A Convergent Expression for the Magnetic Moment of the Neutron', *Physical Review* **74**, 218 (1948); (b) D. Rivier, 'Une méthode d'élimination des infinités en théorie des champs quantifiés. Application au moment magnétique du neutron', *Helvetica Physica Acta* **22**, 265 (1949).

[49] (a) F. Villars, 'Regularization and Non-Singular Interactions in Quantum Field Theory', in Ref. [49b], pp. 78–106; (b) M. Fierz and V. F. Weisskopf (eds), *Theoretical Physics in the Twentieth Century. A Memorial Volume to Wolfgang Pauli* (Wiley Interscience, New York, 1960).

[50] (a) E. C. G. Stueckelberg and D. Rivier, 'A propos des divergences en théorie des champs quantifiés', in Ref. [50d], p. 236; (b) F. J. Dyson, 'The Radiation Theory of Feynman', in Ref. [50d], p. 240; (c) C. P. Enz, 'Ernst Stueckelberg' (Obituary), *Physics Today*, March 1986, p. 119; (d) 'Internationaler Kongress über Kernphysik und Quantenelektrodynamik', *Helvetica Physica Acta* **23** *Supplementum III* (1950); (e) 'Congresso internazionale di fisica dei raggi cosmici', *Nuovo Cimento* **6**, *Supplemento* (1949).

[51] M. R. Schafroth, Höhere strahlungstheoretische Näherungen zur Klein-Nishina-Formel, *Helvetica Physica Acta*, (a) *I*: **22**, 501 (1949); (b) *II*: **23**, 542 (1950); (c): **22**, 392 (1949).

[52] (a) R. Jost, 'Compton Scattering and the Emission of Low Frequency Photons', *Physical Review* **72**, 815 (1947); (b) E. Corinaldesi and R. Jost, 'Die höheren strahlungstheoretischen Näherungen zum Compton-Effekt', *Helvetica Physica Acta 21*, 183 (1948); (c) R. Jost and J. M. Luttinger, 'Strahlungstheoretische Korrekturen zur Paarerzeugung und zur Bremsstrahlung', *Helvetica Physica Acta* **21**, 391 (1948); (d) *American Men and Women of Science* (Bowker, New York, 1989); (e) J. Mehra, *The Solvay Conferences on Physics. Aspects of the Development of Physics since 1911* (Reidel, Dordrecht, 1975).

[53] (a) W. Pauli, typed letter dated 18 August 1949 to the US Consulate General, Zurich, *Pauli Archive*, CERN, PLC $\overline{\text{Bi } 91}$; (b) My telephone conversation of 12 February 1991 with the Zivilstandsamt Zollikon.

[54] Translated from: W. Pauli, letter [989] of 5 December 1948 to Fierz, in Ref. [2].

[55] W. Pauli, 'Recent developments in quantized field theories' (title only), *Physical Review* **78**, 315 (1950).

[56] A. Salam, 'Paul Matthews' (Obituary), *Physics Today*, October 1987, p. 142.

[57] W. Heisenberg, (a) 'Zur Theorie der "Schauer" in der Höhenstrahlung', *Zeitschrift für Physik* **101**, 533 (1936); (b) 'Über die in der Theorie der Elementarteilchen auftretende universelle Länge', *Annalen der Physik* **32**, 20 (1938). Reprinted in Ref. [5], Part II, (a): pp. 275–82; (b): pp. 301–14.

[58] F. J. Dyson, (a) 'The Radiation Theories of Tomonaga, Schwinger, and Feynman', *Physical Review* **75**, 486 (1949); (b) 'The S Matrix in Quantum Electrodynamics', *Physical Review* **75**, 1736 (1949). Reprinted in Ref. [16], (a): pp. 275–91, (b): pp. 292–311.

[59] R. P. Feynman, (a) 'The Theory of Positrons', *Physical Review* **76**, 749 (1949); (b) 'Space–Time Approach to Quantum Electrodynamics', *Physical Review* **76**, 769 (1949). Reprinted in Ref. [16], (a): pp. 225–35, (b): pp. 236–56.

[60] (a) A. Thellung, 'Pauli als Lehrer', in Ref. [34], pp. 95–104; (b) J. Mehra, *The Beat of a Different Drum. The life and science of Richard Feynman* (Oxford University Press, Oxford, 1994); (c) J. Weiglé, 'Eulogy for the 60th birthday of Paul Scherrer', *Helvetica Physica Acta* **23**, 4 (1950).

[61] W. Pauli, (a) 'Etat actuel de la théorie quantique des champs. La renormalisation', in: P. Auger and A. Proca (organizers), *Particules fondamentales et noyaux*. Paris, 24–29 Avril 1950, *Colloques internationaux du Centre National de Recherche Scientifique* **38** (Editions CNRS, Paris, 1953), pp. 67–77. Reprinted in Ref. [12], Vol. 2, pp. 1165–75; (b) 'The idea of renormalization of charges and masses', Appendix to letter [1106], Ref. [35], pp. 83–92.

[62] (a) N. Bohr and L. Rosenfeld, 'Field and Charge Measurements in Quantum Electrodynamics', *Physical Review* **78**, 794 (1950); (b) C. N. Yang and D. Feldman, 'The S Matrix in the Heisenberg Representation', *Physical Review* **79**, 972 (1950).

[63] (a) K. von Meyenn (ed.), *Wolfgang Pauli. Scientific Correspondence with Bohr, Einstein, Heisenberg, a.o., Volume IV, Part II: 1953–1955* (Springer, Berlin and Heidelberg, 1999); (b) Ph. Choquard and F. Steiner, 'The Story of van Vleck's and Morette-van Hove's Determinants', *Helvetica Physica Acta* **69**, 636 (1996).

[64] R. P. Feynman, (a) 'Space–Time Approach to Non-Relativistic Quantum Mechanics', *Reviews of Modern Physics* **20**, 267 (1948); (b) 'Mathematical Formulation of the Quantum Theory of Electromagnetic Interaction', *Physical Review* **80**, 440 (1950). Reprinted in Ref. [16], (a): pp. 321–41, (b): pp. 257–74.

[65] Ph. Choquard, 'Traîtement semi-classique des forces générales dans la représentation de Feynman', *Helvetica Physica Acta* **28**, 89 (1955).

[66] P. A. M. Dirac, 'The Lagrangian in Quantum Mechanics', *Physikalische Zeitschrift der Sowjetunion* **3**, 64 (1933). Reprinted in Ref. [16], pp. 312–20.

[67] *Note*: Successively applying $\partial/\partial q_i'$ and $\partial/\partial q_j$ to Eq. (10.68) and calling $\varphi_{ji} = \partial^2 S/\partial q_j \partial q_i'$ one obtains Eq. [30.17] of Ref. [44b],

$$\frac{\partial \varphi_{ji}}{\partial \tau} + \sum_k \frac{1}{m_k}\left(\frac{\partial}{\partial q_j}\left(\frac{\partial S}{\partial q_k} + A_k\right)\right)\varphi_{ki} + \sum_k \frac{1}{m_k}\left(\frac{\partial S}{\partial q_k} + A_k\right)\frac{\partial \varphi_{ji}}{\partial q_k} = 0$$

(A)

If now Φ_{ji} is the subdeterminant of φ_{ji}, so that

$$\sum_i \Phi_{li}\varphi_{ji} = D\delta_{lj},$$ (B)

and, for any variation,

$$\delta D = \sum_{ij} \Phi_{ji}\delta\varphi_{ji},$$ (C)

multiplication of (A) by Φ_{ji} and summation over i and j are seen to yield $Q_1 = 0$. I thank Professor Ph. Choquard for an illuminating discussion on this point.

[68] B. S. DeWitt, 'Theoretical Physics in the Alps', *Physics Today*, December 1951, p. 22.

[69] C. P. Enz, *A Course on Many-Body Theory Applied to Solid-State Physics* (World Scientific, Singapore, 1992).

11
The one world: physis and psyche

The 'piano lesson': The 'Lecture to the Foreign People'

As in Austria during the First World War, in this year [1945] in the U.S.A. I suddenly had the feeling that I was placed in a 'criminal' atmosphere—and this at the time when those 'A-bombs' were dropped. ... My anima became very irritable and occasionally produced eruptions of anger until I had departed from the U.S.A. (in February). [1]

Pauli describes here a certain *feeling complex* of his which he designates as the 'pacific' one. Another one, as Pauli adds in a footnote, has to do with his Jewishness. And, although in Switzerland the anima does not bother him, he realizes that 'Now, however, something perhaps more evil has occurred: although my (geographical) neighbourhood now is *not* involved in the "collective blame", my profession is: *Physics.*' His conscious attitude is 'that it is quite unreasonable, really, to give up physics as a consequence of these politico-military events', because of the usual argument that scientists are not responsible for the applications which may be useful or noxious. 'Not so the unconscious, the anima. Perhaps, in this connection she develops resistances against physics and ensures that nothing intelligent (or only little) comes to my mind. Will the new event of 1945 now induce me to a migration in the spiritual domain? Perhaps such a step is even already constellated? Or should this "pacific" complex, as belonging to the shadow, be separated from the anima, so that the latter will then react "reasonably" again?' And Pauli concludes: 'I know, it is a question of fate which ultimately will not be decided by the Ego' (translated from Ref. [2], letter [1239]).

Pauli addressed these fateful questions to Marie-Louise von Franz on 17 May 1951 with the comment 'I would not know anybody that would be more competent for it [the judgement of the whole complex]' (translated from letter [1239]). This shows how close Pauli had come to Marie-Louise von Franz in the years since he addressed her in the first letter in October 1947 (see page 421). Psychologically, a transmission had occurred between the two. They now had a steady exchange of letters and saw each other regularly, taking walks together or going for an excursion on one of the regular boat trips on the lake, discussing Pauli's dreams and interpretations as well as Jung's ideas. How far von Franz had come to understand Pauli's thinking is most eloquently expressed in the draft [1257] of the only existing letter from her to Pauli, dated 17 June 1951. Pauli considered her as his 'femme inspiratrice', and became irritated when he had the impression that she did not play this role well enough (Ref. [2], letter [1300]).

About this time there was much talk among people near Jung about the latter's latest work, 'Answer to Job' [3a]. Pauli was also aware of it, as letters [1228], [1270], and [1277] to von Franz indicate. And, although Pauli read Jung's 'Answer to Job' only in September 1952 (see below), the above reflections express the same anxiety, for, indeed, Jung writes in 'Answer to Job': 'Since the Apocalypse we now know again that God is not only to be loved, but also to be feared. He fills us with evil as well as with good, otherwise he would not need to be feared; and because he wants to become man, the uniting of his antinomy must take place in man. This involves man in a new responsibility. He can no longer wriggle out of it on the plea of his littleness and nothingness, for the dark God has slipped the atom bomb and the chemical weapons into his hands and given him the power to empty out the apocalyptic vials of wrath on his fellow creatures. Since he has been granted an almost godlike power, he can no longer remain blind and unconscious. He must know something of God's nature and of metaphysical processes if he is to understand himself and thereby achieve gnosis of the Divine' (Ref. [3a], Section 747).

Pauli's dissatisfaction with the science enterprise—which in letter [58] to Jung is expressed as 'the physicist who must compensate the one-sidedness' (translated from Ref. [4], p. 86)—also manifested itself in his dreams. On 15 March 1947 Pauli dreamt about two young physicists whom he took to be Americans, one of whom says disappointingly: 'Always this unobliging (*unverbindliche*) archetypal thinking; the physics treatises are full of it' (translated from Ref. [2], pp. 255–6). Thus Pauli looked out for a more relevant physics, in which physis and psyche were considered as having equal importance. In letter [58] to Jung, Pauli writes later on: 'Indeed, it seems to me that in the *complementarity of physics* with its overcoming of the pair of opposites "wave–particle" kind of a *model or example of that other, more comprehensive coniunctio* is present. . . . The smaller "coniunctio" in the framework of physics, namely the quantum- or wave mechanics construed by physicists, indeed exhibits, quite independently from its inventors, certain features that seem also to be realizable for the overcoming of the other pairs of opposites mentioned on p. 3 [p. 87: Fludd–Kepler, psychology–physics, intuitive feeling–thinking according to the natural sciences, Holland–Italy, mystics–natural science]' (translated from Ref. [4], pp. 92-3). Significant in this development is Pauli's dream figure of the "stranger" (see Ref. [5a]), who in the dream of 24 November 1948 'emerges from the river' and has the 'traits of the "Persian", of the "Blond" and of the *shadow*' (translated from Ref. [2], p. 258; see pages 416–17).

Pauli gives a detailed characterization of the "stranger" in the (second) letter [44] of 16 November 1950 to Mrs Emma Jung in Ref. [4], concerning her work on the *Graal legend* [6]. According to this characterization the "stranger" is a 'double-layered' dream figure, 'on the one hand, a spiritual figure of light of superior knowledge, on the other hand a chtonic [earthen] spirit of Nature. ... His acts are always telling, his words definitive, if often unintelligible. Women and children like to follow him, and he often tries to instruct them. Above all, he considers his entire environment (particularly me) as totally ignorant and uneducated as compared to himself. ... he is in a certain sense an "anti-scientist", where under "science" here the methods of the natural sciences have to be understood in particular, above all those that today are taught at Institutes of Technology and Universities. These latter he perceives ... as the place and the symbol of his oppression, to which (in my dreams) he sometimes also sets fire. When he is paid too little attention, he manifests himself by all means, e.g. through synchronistic phenomena (which he, however, calls "radioactivity" [see page 426])' But Pauli emphasizes that 'the relation of the "stranger" to the natural sciences ultimately is not a destructive one'. He longs for redemption, but his liberation will come only in a form of culture 'which will validly express the quaternity' (translated from Ref. [4], p. 54).

As Pauli wrote to Jung on 23 December 1953, he sees the problem of establishing 'lines of development of a future enlarged physics' in the relation of physics to other sciences and to psychology, even to life as a whole. But 'Since today the natural sciences draw their dynamics from the archetype of the quaternity, hereby also the ethical problem of the evil is constellated which, indeed, has become manifest particularly through the atom bomb' (translated from letter [66], Ref. [4], p. 130). This letter [66], which is reprinted as number [1694] in Ref. [7], is noteworthy for its four appendices deposited at ETH (Wissenschaftshistorische Sammlung, WHS, Hs. 1056: 30880–30883), of which only no. 30881 is reproduced in Ref. [4] (Appendix 7) and in Ref. [7] (Appendix to letter [1694]). More interesting, however, is no. 30880, since this is Pauli's *horoscope*, the origin of which is unknown to me. Considering Pauli's 'very rejecting conscious attitude towards horoscopes and astrology' (translated from letter [38], Ref. [4], p. 46), it is unlikely that he attached any significance to this document. This rejection was quite irrational, however (see letter [1672], Ref. [7], p. 351).

Still, from a biographical point of view this horoscope is of interest [8a]. In particular, one may hope to gain some understanding of the fact that, as Pauli says, 'With me the two equinoxes are periods of a relative psychic instability which may manifest itself as well negatively as also positively (creatively)' (translated from footnote 10 of the *Hintergrundsphysik*, Ref. [4], p. 189). In Pauli's horoscope the spring equinox (the boundary between the zodiacal signs of Pisces and Aries) is situated at the boundary between the seventh house which might be called the 'house of the conjunction', and the 8th house, or 'house of the unconscious'. And the autumn equinox lies at the boundary between the first house, or 'house of the ego', and the second house, or 'house of the material things' [8a]. Thus Pauli's instabilities may reflect instabilities of these boundaries between 'horoscope houses'. A more obvious explanation, however, is that the equinoxes mark times of subtle changes in the world surrounding those of us living in the moderate zones of the earth (see, however, *Note* [8b]).

On 25 September Pauli sent Jung's secretary Aniela Jaffé (1903–91) a letter from Brussels where, from 25 to 29 September, the Ninth Solvay Conference devoted to solid-state physics took place. This letter [1284] has the title 'Equinoctial Thoughts'. Having read Plotin and Jung's *Aion* [9] and observing that things in the world are favourable for thinking about the evil, Pauli wrote 'As well the good as also the evil are only *absence of the void*.' And: 'The void is not "nothing" but indeed something extremely efficacious. But it, in fact, is the void because it escapes the rendering evident (*Veranschaulichung*) through images and also through words. The void also is synonymous with a deeper *unity* of physical and psychic happenings ("neutral language!"). This unity seems to occur as a consequence of a process of conjugation, i.e. it acts in the sense of ordering and not as "cause" in the narrow sense. The total absence of conscious purposes puts it beyond the human good and evil. The ethical point of view is necessary for man but on ethical grounds alone the ultimate reality of the "void" (= the "one") seems to me not to be attainable' (translated from Ref. [2], pp. 370-1). This and many other letters show the significant exchange of ideas that Pauli also had with Aniela Jaffé. She had fled Nazi Germany and in 1937 had met Jung who, at the occasion of the foundation of the C. G. Jung Institute in 1947, appointed her as his secretary (see the comment in Ref. [2], p. 135). The quoted text shows the influence that Jung's book *Aion* had on Pauli.

Pauli acknowledges this influence in letter [58] of 27 February 1952, where he writes to Jung 'about various considerations and amplifications that your book "Aion" [9] has produced with me. Also apart from astrology, on which we seem not to be of the same opinion, there still remains very much that has captured my interest.' And he refers especially to Chapter V, in which the Augustinian notion of *privatio boni* (i.e. of the evil being mere absence of the good) is discussed, and to Chapters XIII and XIV, which consider the unconscious state of God (Section 299) and God as paired opposites (Section 400). These ideas considerably stimulated Pauli's thinking. In the just quoted letter he writes: 'As indeed you may know that in religious and philosophical regard my roots are in Lao-tse and Schoppenhauer. While from this basis your analytical psychology—and, as I believe, also your personal spiritual attitude in general—always appeared well accessible to me, I must confess that the specially Christian religiosity and its notion of God in particular, has remained to this day quite inaccessible to me, as well sentimentally as intellectually. (Although I have *no sentimental resistances* against the idea of a capricious tyrant like Jahwe, the excessive arbitrariness in the cosmos which this idea implies, seems to me, however, to be an anthropomorphism which is untenable in natural philosophy).' And, commenting on Schopenhauer's rejection of such a God 'because the evil would have to fall back on him', Pauli writes: 'However, critically I myself wish to say to this, *that, what is rejected here is only the idea of a human-like conscience of God*' (translated from Ref. [4], pp. 76, 77).

It must be emphasized that in this view Pauli was in complete accord with Jung's writings, as well as in *Aion* and also in 'Answer to Job'. Indeed, in commenting on Job 9: 32 Jung writes in footnote 13, Section 600 of Ref. [3a]:

'The naïve assumption that the creator of the world is a conscious being must be regarded as a disastrous prejudice which later gave rise to the most incredible dislocations of logic. For example, the nonsensical doctrine of the *privatio boni* would never have been necessary had one not had to assume in advance that it is impossible for the consciousness of a good God to produce evil deeds. Divine unconsciousness and lack of reflection, on the other hand, enable us to form a conception of God which puts his actions beyond moral judgement and allows no conflict to arise between goodness and beastliness.' But, as we saw, Pauli could not identify with Jung's Christian and particularly Protestant religiosity. In his letter [1492] of 30 October 1952 to von Franz he writes: 'Besides, I also have written, as a particular task of diligence, a letter to [Gershom] Scholem, in which I compare myself to a horse, which—standing on the ground between the two chairs of orthodoxy and of rationalism—likes to kick in all directions' (translated from Ref. [2], p. 791).

Pauli, as we saw, considered himself 'incredulous' as concerns the specific Christian faith. It is in this attitude that 'after overcoming some resistances' he chose the autumn equinox, 19 September 1952, to begin reading 'Answer to Job', 'quite uncritically' and 'as an agreeable entertainment reading' (translated from Ref. [4], letter [58], p. 87). This letter [58] of 27 February 1953 to Jung, whose declared themes were *considerations of a non-believer on psychology, religion and your Answer to Job*', grew into a meditation of over 12 pages, spiced with several dreams. The night following his reading Pauli dreamt that he rode on a local train with Bohr. On changing trains he noticed the 'dark girl'—the dark anima (see Ref. [5a]), and he comments: 'The "dark girl" for me was always the counter-pole to protestantism, that *"man's religion* which allows no metaphysical representation of woman" [Ref. [3a], Section 753]. The pair of opposites catholicism—protestantism bothered me in dreams for long periods.' Remember that Pauli's parents had converted to Protestantism (see page 10). Pauli continues: 'It is a conflict between an attitude which does not accept the reason at all or too little and another attitude which does not accept the anima. This pair of opposites later appeared time and again in many forms. ... It is a pair of opposites which apparently demands overcoming through a coniunctio' (translated from Ref. [4], pp. 86 and 87).

Later in letter [58] Pauli describes a dream he had during the night from 27 to 28 September 1952, in which the 'dark girl' assumes 'Chinese' aspects, i.e. aspects of totality (*ganzheitliche*): The 'Chinese girl' [see Ref. [5a]] descends a staircase dancing like in a ballet. 'I follow her and see that the stairs lead into a *lecture room*. In there "the foreign people" wait for me. The "Chinese girl" signifies to me that I should ascend the podium and speak to the people, apparently to give them a lecture. While I still wait, she "dances" all the time rhythmically from below up the stairs, through the open door into the open and then down again, always holding the index finger of the left hand with the left arm upwards and the right arm and the index finger of the right hand downwards. The repeated application of this rhythm now has a strong effect in that, gradually, a rotation movement (circulation of the light) results. ... When

I mount the podium of the lecture room I wake up' (translated from Ref. [4], p. 90; see also Ref. [5a]). A somewhat more spontaneous version of this and the previous dream is contained in the appendix to Pauli's letter [1472] of 12 October 1952 to Marie-Louise von Franz in Ref. [2].

Further on in letter [58] to Jung, Pauli makes the following comment: 'The motive of the not yet assumed professorship seems to me very important, for it shows the resistances of the conscience against the "professorship". *The unconscious expresses a blame against me, that I have withheld the public something specific, something like a confession, that I have not followed my "call" because of conventional resistances.'* (italics added). And observing 'that the mathematical natural sciences for me, as for anybody else who practises them, are an extremely strong binding on a [typically Western] tradition', Pauli then gives the following details: 'In the sense of this tradition and of my conscious attitude, everything which belongs to the counter-position of the natural sciences, because related with the feeling, was a private affair. On the other hand, the people in the lecture room expect a professor who teaches as well the natural sciences as also their feelingwise—intuitive counter-position, perhaps even including ethical problems. The people in the lecture room are, against my resistances, of the opinion that also this extended subject of the "lecture", while being personal, yet were interesting for the public' (translated from Ref. [4], p. 91). Here, I believe, Pauli has expressed the fundamental antinomy of his later life, an antinomy which stayed with him to his death. For, the 'confession' of talking openly on this 'extended subject' might have met the lack of judgement of an ignorant public, but it might also have endangered the considerable prestige Pauli enjoyed among the community of physicists.

While in the quoted letter [58] to Jung no specific content of Pauli's 'Lecture to the Foreign People' is given, this is not so in the 'Active Fantasy about the Unconscious' with the title 'Die Klavierstunde' (the piano lesson) [10], which Pauli had written at the end of October 1953 and 'dedicated to Miss Marie-Louise v. Franz in friendship'. In this document the content of the 'Lecture to the Foreign People' expresses Pauli's views on biology. This looks like a diversion from the mentioned 'extended subject'. However, Pauli's purpose is to bring elements of complementarity and of synchronicity into the understanding of biology. This means that, besides causal connections, relations of meaning or purpose are also considered important. Thus Pauli wrote to Heisenberg on 29 December 1953 that meaning, in fact, should characterize something unique. In particular, in the natural sciences meaning should designate *'hereditary modifications in direction to adaptation (to the physical milieu)'* (translated from Ref. [7], letter [1698], p. 405).

As Pauli writes to Delbrück from Princeton on 4 February 1954, he 'became a bit more interested [in biology] than in earlier times since last autumn. It started with some remarks of Heisenberg . . . who found a rather old Lamarckian book by A. Pauly [11a] still (be) worth to be read today. . . . Then I talked with O. Klein about the matter, who told me about his friend Runnström . . . in Stockholm, who always attacks Darwinism, after he had a couple of drinks and is then re-attacked by others who had less drinks than he. (Which seems to me rather characteristic.) Then Bohr's letter, which you saw, arrived.' This is followed by an impressive list of recent biological publications that Pauli had read, and he observes that 'Here [in Princeton] I have so much more time than in Zurich' (Ref. [7], letter [1712], pp. 451, 452). Bohr's letter, most probably, is letter [1700] in Ref. [7], which shows how deeply Bohr himself was engaged in this kind of biological thinking.

In the *'Klavierstunde'* [10], which precedes these letters, the 'Lecture to the Foreign People' is the insert from Section 32 to Section 43 that is introduced by the voice of the "master", namely the "stranger" who 'in the meantime has become very intimate to me' (translated from Ref. [4], letter [69], p. 138). He says: 'Younger brother!', meaning Max Delbrück, 'who now stands by the window and smiles kindly towards me.' Pauli stands at the window and speaks to the foreign people outside who applaud and call Pauli's name. Central to the lecture is the notion of chance, first of all in quantum mechanics. Considering the latter as the liberation from the detached observer (see page 415), Pauli believes, 'that, starting from this insight, only an advance is possible and that this leads directly to the *phenomena of life*.' In the latter, chance occurs in the form of mutations, of which Pauli distinguishes two types, the spontaneous and the induced ones. And in the theory of evolution, according to Pauli, neo-Darwinism postulates 'purpose-less or blind chance', while in Lamarckism 'purpose-ful or directed chance' is assumed. This Pauli comments on by saying 'Chance always fluctuates, but sometimes it fluctuates systematically' (translated from Ref. [10], Sections 28-38).

Thus Pauli believes that synchronicity brings in a third way of functioning of the phenomena of life: 'One therefore has the impression that *the external physical circumstances on the one hand, and inherited changes of the genes (mutations) adapted to the former on the other hand, although they are not connected causally-reproducibly, but still have occurred once meaningfully and purposefully as an indivisible whole, together with the external circumstances—correcting the "blind", accidental fluctuations of the happening mutations'*. And he concludes: 'According to this hypothesis, which distinguishes itself as well from the Darwinian view as from the Lamarckian view, we here encounter in fact the sought-for *third type* of natural laws which *consists in a correction of the fluctuations of the chance by meaningful or purposeful coincidences of causally not connected events.*' A further hypothesis advanced by Pauli is that *'this holistic occurrence of meaningful coincidences points to a psychologic factor in the biological evolution going hand in hand with it and appearing on a higher level as emotionality or excitement.'* Here Pauli refers to Rhine's extrasensory perception (ESP) experiments [11b] (translated from Ref. [10], Section 40 to Section 42. See Ref. [11c], Section 8).

Not surprisingly, Pauli's 'younger brother' Delbrück vigorously rejected this way of understanding biology, although he had been seduced into it by Bohr's inaugural lecture 'Light and Life' [11d], given at the Second International Congress on Light Therapy in Copenhagen on 15 August 1932 (see page

353). Indeed, in this lecture Bohr had proposed to see in the psycho-physical parallelism an example of complementarity between the freedom of the will and organic functions. But, as Müller-Herold writes: 'As is well known, Delbrück's search for complementarity in biology has ended differently: in a Newtonian campaign of submission of epochal extent. Delbrück became the very founder of molecular biology' (translated from Ref. [11c], p. 164). Delbrück himself has painted a fascinating picture of the development of his own understanding of biology in his Nobel Lecture of December 1969 [11e].

Pauli's letter [1712] to Delbrück quoted above closes with a parody due to John von Neumann of what Pauli calls the 'orthodox view', according to which biological evolution is governed 'by taking *only chance* (so-called: "natural") selection and *random* mutations into account'. He writes: 'So you see, that I am partly critical and "heretic", partly *open*-minded. I am on the side of Bohr only as far [as] he tries to keep the situation in biology *open* and the orthodox ones try to "*close* the subject". But I am glad to belong to a science, where one is considered to be "almost normal" if one does *not* say, that the buildings (which I see outside the window of this room) *are all the result of the Brownian motion of stone particles* (and that they are the fittest buildings which have survived so many others and have therefore attracted men to live in them)' (Ref. [7], pp. 451 and 452; see also letters [1715] and [1716]).

Pauli, who spent the first four months of 1954 at the Institute for Advanced Study in Princeton, met Delbrück on 10 March in New York, 'which was very amusing. He became very emotional, i.e. vexed to angry, when the discussion turned to doubts at the empirical correctness of the time scale of biological evolution which, theoretically, should result from "random mutations" plus "selection". He spoke of a "conspiracy of unemployed physicists against biology" ... "this is simply quite stupid" and—"this is not interesting at all"' (translated from letter [1744] of 15 March to Heisenberg, Ref. [7], p. 524). Pauli again took up the question of the time scale of biological evolution in the second part of his essay [12a] dedicated to Jung's 80th birthday on 26 July 1955. There he writes: '*It would have to be shown that on the basis of the assumed model anything fit for a purpose (Zweckmässiges) which is in fact present had a sufficient chance of arising within the empirically known time-scale. A consideration of this sort has however nowhere been attempted.*' (Ref. [12b], p. 162). More recently, however, such a quantitative determination of a realistic evolution has been performed by successive minimization of the duration in an ensemble of model pathways [12c].

To Weisskopf, Pauli writes on 8 February 1954: 'If one disregards the *time scale* one may invent all sorts of possibilities.' But further on in this letter [1716] he ponders: 'In fact, I am glad that Bohr's letter burst this whole "bubble" physics-biology! And I know well that also Max Delbrück personally, as an immigrant ["*Zugereister*", a typically Viennese expression] to biology from physics in this matter, is psychologically in a somewhat difficult position— which is why he also has assigned to me the role of a "peace-angel" between him and Bohr. However, I fear that he will soon be of the opinion that therewith he made the deer the gamekeeper! But perhaps it will yet improve' (translated from Ref. [7], pp. 465 and 466). An 'improvement' of sorts—it signalled the beginning of biotechnology (see Ref. [11c], Section 3)—came with the fundamental paper [14] of Watson and Crick on deoxyribonucleic acid, DNA, which Pauli mentions for the first time in letter [1718] to his assistant Thellung.

The 'Lecture to the Foreign People' is the relatively small centre-piece of Pauli's 'Klavierstunde' [10], the beginning and the end of which having a much more personal touch. The fantasy opens with the words: 'It was a foggy day and for some time already I had a serious grief. Indeed, here were *two* schools: in the older one only words but not the meaning were understood, in the other more recent one only the meaning was understood but not the words. I could not bring the two schools together.' Meant here are theoretical physics with its mathematical language and psychology with its interpretations. 'Then I thought, the only thing still left to me were to visit a girl that lives in Küsnacht. It was at Hornweg 2. . . . But there was so much about which the girl could not talk, I much liked that, for thus I always could imagine that it was not really that different from my grief and that therefore she certainly would understand me' (translated from Ref. [10], Sections 1 and 2). Hornweg 2 on the shore of Lake Zurich at Küsnacht was indeed Marie-Louise von Franz's address. It may also be mentioned that the C. G. Jung Institute (see pages 416 and 465) is situated on the same Hornweg at No. 28.

Then the voice of the 'master' sounds, 'like that of a sea-captain. It says "*time reversal*"' and paper cones appear 'with the summit downward'. These are light cones, and their reversal signals a blending of different epochs and, indeed, we are suddenly in Vienna: 'I was a pupil who held a folder of music notes in his hand. I knew exactly, we are in the year 1913. . . . As before, a large piano, a grand piano, was in the room with old furniture. Against the grand piano leaned a *lady* with dark hair who was like a familiar old friend. She was a very distinguished lady and I had to talk to her very respectfully. When I approached her at the piano she extended her hand saying: "You haven't played the piano for a long time. I want to give you a piano lesson".' This lady is unmistakably Pauli's grandmother Bertha (see page 4). Saying that he had looked forward to this lesson, Pauli tells the lady of his grief and of the girl whose grief was that her mother had destroyed her femininity. 'However, then I thought that this could not be. Since otherwise, how could something that is destroyed excite my feeling?' (translated from Ref. [10], Sections 3 and 4).

After a pause Pauli hears the 'master's' voice saying: 'captain', and he narrates: 'Here in Vienna lives a captain who has a sick daughter, a sick soul.' This captain is Marie-Louise's father, who now blends in with the captain (centurion) of Capernaum (Matthew 8:5) but also with the father of Pauli's grandmother (see page 4) and with the captain of Köpenick of Carl Zuckmayer's play of this name. Pauli says: 'A captain of Köpenick once told the people that the black keys of the piano were only holes where the white is missing.' In associating 'white' with 'good' and 'black' with 'bad' this statement is seen to be a parody on the *privatio boni* mentioned earlier in this section, and the theologians who preached it are identified as 'captains of Köpenick'. Note that the historic captain of Köpenick was the cobbler Wilhelm Voigt who in 1906 donned a captain's uniform, arrested the mayor of Köpenick and confiscated the city's cash-box. Switching to communist ideologues Pauli observes: 'In the East there is a new form of the captain of Köpenick, namely a sect with virulent theologians. These are the red slaves, about whom the alchemists already have written. They are dangerous because they have guns and cannons whereas the old black theologians have none left and had also to renounce the stakes' (translated from Ref. [10], Sections 6–12).

Pauli and the lady talk about many things and from time to time play an accord on the piano. Eventually, after Pauli had given the lecture to the foreign people, he concludes: 'This instruction by the lady made a very great impression on me. Grown modest I say to her: "The hour has been long already, now I must leave for my men's world, mixing with the people. But I will come again."' One last time the "master's" voice resounds, and 'the lady takes a ring from her finger ... "You surely know the ring from your school of mathematics. It is the <ring i>".' In his long letter [60] of 31 March 1953 to Jung in Ref. [4] Pauli refers to the three rings in Gotthold Ephraim Lessing's play *Nathan der Weise*, of which he says in footnote 17 that the *Homousia*—being of like substance—of these rings and people 'is the *one* and true ring which encompasses the void' (translated from Ref. [4], p. 111). In the 'Klavierstunde' Pauli then says: 'The i makes the Void and the One into a pair. At the same time it is the operation of rotation by one quarter of the whole ring.' This is the complex plane (recall the egg dream with the mathematical formula on page 424). But Pauli also refers here to the description of quantum mechanics by complex wavefunctions where the i connects particles and waves, 'it is the atom, the in-dividuum in Latin. ... It makes the time into a static picture', since from $t \to i\tau$, $\exp(-i\omega t) \to \exp(-\omega\tau)$ follows . The fantasy ends with the 'master's' voice speaking, transformed, out of the centre of the ring to the lady: 'Stay merciful' (translated from Ref. [10], Sections 54–70; see Ref. [5a]).

This sounds like an invocation of Pauli's anima. But since the latter was often projected on Marie-Louise von Franz, the recipient of the 'Klavierstunde' (see e.g. letter [1741], Ref. [7], p. 519), the wish could also be specifically addressed to her (see, however, Ref. [7], footnote 24, p. 341). For her reactions to Pauli sometimes took a wrathful turn, as may be seen from Pauli's letter [1624] of 16 August 1953 to her in Ref. [7], in which Pauli, referring to his own 'stupidity' (*Verblödung*) writes: 'If in this you could produce only just a fraction of the vitality which in the last months appeared in your eruptions of wrath, this already would be a considerable help to me' (translated from Ref. [7], p. 250). Marie-Louise von Franz (1915–98), Jung's closest collaborator, was born in Munich, the second daughter of a professional imperial ('*k. und k.*') captain. During World War I her mother and the two daughters lived most of the time in their holiday home in Berchtesgaden near Salzburg, while the father was on the Italian front. When Marie-Louise was 12 years old the family moved to Zurich where she attended the Freie Gymnasium. This period was marked by an intense religious search, as she was dissatisfied with traditional Christianity. She devoted her interest to the science of the living but felt confined by the demands of her mother who did not understand her.

At age 18 Marie-Louise met C. G. Jung who had invited her and a group of young people to Bollingen. After finishing Gymnasium in 1933 she was hesitant about whether to study mathematics or ancient languages. It was a dream that finally made her turn to the languages and attend the University of Zurich. During her studies she helped Jung in the analysis of alchemist texts with translations and the organization of literature. In this endeavour Jung came across a text called *Aurora consurgens* which became a complement to his last work *Mysterium coniunctionis* (see pages 245 and 416) and which he left his pupil to achieve. Von Franz worked on it for 15 years [15a]. In 1952 she published an essay on *The dream of Descartes*, [15b] on which Pauli made extended comments (see the appendix to letter [1326] in Ref. [2]). She had hoped to publish this work together with Jung's essay on synchronicity and Pauli's Kepler article and was disappointed when this did not materialize. For a while this disappointment perturbed her relationship with Pauli (see Pauli's letter [1270] of 8 August 1951 to her in Ref. [2]).

In 1940 Marie-Louise obtained her doctorate, after which she earned her living by tutoring. On Jung's suggestion she cohabited with one of Jung's pupils, the Scot Barbara Hannah. This friendship lasted for 40 years until the latter's death in 1985. After Jung's death in 1961 von Franz carried on his unfinished work, in particular the problem concerning man and his symbols. She worked on this subject until 1968, undertaking intensive studies of mathematics, in which she endeavoured to show the archetypal nature of numbers (see also letter [1383] in Ref. [2] that Pauli addressed to her on 9 March 1952). The result was her well-known book *Number and Time* [15c], the content of which she presented on lecture tours in the United States, Germany, and France. In her last years Marie-Louise von Franz suffered from Parkinson's disease and it became difficult to communicate orally with her. But her intensity of work did not diminish. The last problem she studied was death, and she even thought that her in-depth drilling had made something of Hades stick to her and manifested itself in her disease [16].

Pauli's close relationship with Marie-Louise von Franz, but also his exchanges with Jung and Fierz and some others, had now become an essential part of his life. Thus, on 27 April 1952 he answered Robert Oppenheimer, who again had invited him to join the Institute for Advanced Study (Ref. [2], letter [1369]): 'with increasing age I get still more hesitant to make a drastic change in the entire form of my life'. In addition he wrote: 'I am more urgently needed and more indispensable here than I ever could be in the States, particularly in connection with the new plans for a European Research Institute of Physics' (Ref. [2], letter [1403]). The discussion concerning the creation of a European Laboratory for Nuclear Research (Conseil Européen pour la Recherche Nucléaire, CERN)) came to a conclusion in February 1952 and in October of that year a site near Geneva was selected [17]. Pauli considered the project 'not entirely stupid ... since there also jobs for theoreticians will be available' (translated from letter [1454] of 25 August 1952 to Stern). In October 1954 Felix Bloch became the first director-general (see Pauli's letter [1888] of 11 October 1954 to Schafroth, Ref. [7], p. 784).

On 4 November 1952 Pauli and Franca left Zurich, following a long-standing invitation by Bhabha to visit India. Pauli comments on this visit to his friend Panofsky as follows: 'India is a strange country of the most extreme contrasts, from the most beautiful temples (on the island of Elephanta they are magnificent) to mass misery, from airplanes to serpents and vultures, from palms with coco-nuts to bacterias. My wife got sick by it (now she slowly recovers) and I was enormously stimulated and "cranked-up" (*angekurbelt*). Thus India has produced a pair of opposites with my wife and myself.' And later he comments: 'And, anyway, I must reject the whole tendency of spiritualization which is hostile towards matter' (translated from letter [1570] of 7 May 1953, Ref. [7], pp. 146 and 147).

The retrospection into his own life contained in the 'Klavierstunde' is taken up again by Pauli in letter [1669] of 6 November 1953 to von Franz. It begins with the observation that 'with me the shadow earlier was projected entirely on the father, and only in the course of time has the former disconnected himself from the projection'. Having dreamt of his blond stepmother (see pages 10–11), Pauli then concludes that one aspect of his anima had been 'invested' in her while another aspect was projected on Marie-Louise von Franz and still another one also on his wife Franca. But now the anima was to be withdrawn from all these projections. And Pauli begins to understand many things: 'I made 2 sketches of six-pronged stars which I attach for you, one of the phase of youth of my life (till 1928 when my mother died and I was called to Zurich), one of the phase after the second world war till recently. It seems to me that the second follows from the first by a fairly simple transformation [a clockwise rotation by 60°], the pendulum, in fact, had swung back to the other side. Much lies in store between the two phases: my father married for the second time soon after the death of my mother (a woman of *my* age!), then great crises (a personal one with me around 1930 which ended positively with my marriage; then the general crisis, the passage through the Hitler period with the second world war as conclusion). Such is the *rhythm* of life, *it* determines how the different phases of life and their orientations emerge from each other' (translated from Ref. [7], pp. 342 and 343).

In these pictures that Pauli joined to letter [1669] the relational problem is always between pairs of opposites situated on opposite prongs of the stars: 'light anima—shadow, ego—mother and spiritual self—dark anima'. Pauli's comment on the first star reads: 'Religion: rationalistic enlightenment. The evil I have strongly experienced in the first world war, however, I thought that it were possible to eliminate it by reforms!' The comment on the second star reads: 'The relation to physics cools. A positive relation to the Kabbala (and the Chassidim) arises (see also Fludd).—Thence: *Kepler article*—Good and evil appear relativized and do *not* coincide with light (spiritual) and dark (chthonic).—*Religion*: is searched—Values of the phase of youth threaten to be lost. At the height of this phase a compensational stagnation occurs with anima-projections on real women' (translated from Ref. [7], pp. 345, 346).

However, Pauli then says in this letter [1669] that 'Now with me a large transformation is in the making, as you already have seen correctly: the drawing with the six-pronged star does not correspond any more to reality at all. For, the light (spiritual) and the dark (chthonic) aspect of the figures now have approached each other so far that it is quite clear to me that there exists only *one* anima who is as well the Light one as the Dark one as, in addition, the real woman: The chthonic and its relation to the spiritual—as you have said—with me now seems to be ripe for the becoming conscious (*Bewusstwerdung*).' This, Pauli believes, is the synthesis of the youth phase and the later phase. 'Thus, out of 2 × 6 first arises the 12 which corresponds to the *zodiac* and its natural script, namely its old partition into *four* trigones.' And he concludes: 'That is the goal which I have in mind, *there* I wish to arrive! ... *My path proceeds via* 2 × 6 *to the zodiac, for the still older is always the new.*' Thus Pauli's retrospection has turned into foresight; it made him indulgent even towards Christianity: 'Meanwhile I look forward full of confidence towards everything that further happens to me' (translated from Ref. [7], pp. 343,

A vision of gauge field theory

Dear παις!
I worked out what Pauli said in Leiden and I found that he is not so much more clever than μξ, after all (a paradox!).

...

The whole note for you is of course written in order to drive you further into the real virgin-country ('Neuland'); therefore I tried to write it so, that an amplification of the group (I) is suggested. And this amplification has to be dealt [with] together with the field-quantization.

Yours old μξ [18]

This letter [1614] from Pauli to Abraham Pais on 25 July 1953 is remarkable for several reasons. First, there are the Greek names παις (pais, the child) and μξ (Mooksy) that were used by Pauli in many letters to Pais for Pais and for himself, respectively. It is perhaps significant that children often occurred in Pauli's dreams (see e.g. Pauli's letters [38] to Jung of 4 June 1950 and [69] of 23 October 1956 in Ref. [4], pp. 46 and 148, respectively, and the dream of 2 July 1949 in Ref. [2], pp. 258–9; See also Ref. [2], footnote 1, p. 114). On the other hand, according to Pauli's widow Franca, the name 'Mooksy', which is taken from James Joyce's novel *Finnegans Wake*, had been given to Pauli by his friend and colleague at the Institute for Advanced Study, Carl Ludwig Siegel (see page 388; see also footnote 15 in Ref. [7], p. 191). In addition, the date of the letter is supplemented with the notice 'Ad "usum Delphini"' only', which means 'for use of the dauphin (dolphin)', the King of France's eldest son, a formula which during the reign of Louis XIV was used in the education of the former (see Ref. [7], footnote 1, p. 228). This suggests that Pauli considered Pais as his spritual heir. Then Pauli, in the first line of the letter, is jokingly talking of Pauli as if he is somebody other than $\mu\xi$. The core of the above quotation, however, is an invitation to Pais for a ride into 'the real virgin-country', of which Pauli had had a vision in Leiden. What he saw was entirely new physics.

Fig. 11.1 Smoking a pipe, on the occasion of his election as Foreign Member of the Royal Society, 23 April 1953

From 22 to 27 June 1953 the Lorentz and Kammerling Onnes Centennial Conference was held in Leiden (see the comment in Ref. [7], pp. 205–7), at which Pauli participated and Pais gave a paper on 'Isotopic Spin and Mass Quantization' [19a,b]. At that time new heavy particles called V_1^0 and V_1^+ which decayed into a nucleon and a π-meson had been detected. In his paper [19a] Pais proposed a formula for the mass spectrum of the heavy particles which he deduced from a Dirac equation for the nucleons, containing an extra term depending on isotopic spin, his Eq. (5),

$$\left(\gamma_\mu \frac{\partial}{\partial x\mu^\mu} + \frac{\vec{\tau} \cdot \mathbf{K}}{\Lambda} + M\right)\psi = 0. \tag{11.1}$$

Here $\hbar M/c$ is the nucleon mass, $\vec{\tau}$ are the Pauli matrices (5.37) in isospin space, called ω-space, $\mathbf{K}$ is an orbital angular momentum vector in ω-space and Λ is a parameter with the dimension of a length which is adjusted to yield the masses of the V-particles, $\Lambda \sim 0.9\ M^{-1}$.

In the discussion after Pais' communication Pauli asked the question 'whether the transformation group with constant phases can be amplified in a way analogous to the gauge group for the electromagnetic potentials in such a way that the meson-nucleon interaction is connected with the amplified group. The main problem hereby seems to me the proper incorporation of the coupling constant into the group' (Ref. [19a], p. 887). As shown on page 43, 'the gauge group for the electromagnetic potentials' is obtained by making the phase f in Eq. (2.12) dependent on x. Similarly Pauli wanted to determine the new field—which, he hoped, would describe pions—*by assigning an isospin group to every space–time point* x, an enlargement that in modern mathematics is called a fibre bundle, but which Pauli here and in letter [1614] quoted at the outset calls an amplification. This, of course, is much more ambitious than the mentioned prescription for the phase f in electrodynamics; it means investigating the structure of the enlarged group.

It is fascinating to note that in 1934 Pauli had already had similar ideas concerning the gauge group of electrodynamics. Indeed, in letter [373] of 14 June 1934 to Heisenberg, which was extensively quoted in Chapter 8, Pauli had made a number of intriguing remarks, one being 'that in our formalism of quantum electrodynamics the gauge invariance does not come out very nicely. . . . —Shouldn't there, in the future theory, yet be something like a q-number gauge invariance?' In Pauli's opinion already at that time '*Also in a future theory, to the conservation law of the charge should correspond a transformation group of the Hamiltonian, and indeed one which may be considered as a generalization of to-day's theory.*' (translated from Ref. [20a], p. 330). These ideas Pauli took up again in his letter [1595] of 3 July 1953, in which he wrote Pais 'Dear Pais (soon not $\pi\alpha\iota\varsigma$ anymore)!', implying that the latter would soon solve the problem of the 'amplification' of the ω-space. This problem much excited Pauli who continued: 'and I gradually begin to understand what I had said in the discussion after your talk [see the quotation at the outset].' (translated from Ref. [7], p. 189). In this letter [1595] Pauli took as the guiding model Klein's group of transformations (7.5), which was an 'amplification' to five dimensions. But for the mathematics he was guided by Schrödinger's work [20b] mentioned on page 268, which he presented at about the same time in his lectures [20c] of the summer term 1953 (see Ref. [21], p. 434). Pauli first considered it natural to set, in analogy to Eq. (7.28),

$$\tau_\alpha \tau_\beta + \tau_\beta \tau_\alpha = 2g_{\alpha\beta}(x) \qquad (11.2)$$

(see Pauli's letters [1594] to Fierz and [1595] to Pais), but he gave up this equation again. Then on 21 July 1953 he wrote Pais: 'Have made further progress in the understanding of my discussion remark in Leiden. This kind of differential geometry is amusing—but the true "virgin country" is second quantization with a generalized group.' (translated from Ref. [7], postcard [1607]).

Pauli had attached to letter [1614] quoted at the outset of this section an extended appendix with the heading 'Written down July 22 till 25 in order to see how it is looking. Meson–nucleon interaction and differential geometry' (Ref. [7], pp. 229–32; reprinted in Ref. [22a], pp. 171–5). A second appendix he attached to letter [1682] of 6 December 1953 to Pais (Ref. [7], pp. 369–9; translated in Ref. [22a], pp. 175-81). The central equation of these two appendices, of which a handwritten copy by Freeman Dyson exists in the *Pauli Archive* at CERN, is the transformation equation of the fibre bundle of the rotation group in ω-space,

$$\Omega'^A = C_B^A(x)\Omega^B, \tag{11.3}$$

where

$$g_{AB}(x)\Omega^A\Omega^B = 1, \tag{11.4}$$

$A, B = 1, 2, 3$. Equation (11.3) is Pauli's *Assumption I*.

Pauli then considers covariant vectors f_A in ω-space and f_i in space–time transforming, for fixed $x = x'$, as follows:

$$f'_i = f_i + \frac{\partial \Omega^B}{\partial x^i} f_B; \ f'_A = \frac{\partial \Omega^B}{\partial \Omega'^A} f_B. \tag{11.5}$$

The tensor components g_{AB}, g_{Ai}, and g_{ik} transform like the corresponding products of vectors, in particular

$$g'_{Ai} = \frac{\partial \Omega^B}{\partial \Omega'^A}\left(g_{Bi} + \frac{\partial \Omega^C}{\partial x^i} g_{BC}\right). \tag{11.6}$$

The transformation of contravariant vectors is, for $x'^i = x^i$,

$$f'^i = f^i; \ f'^A = \frac{\partial \Omega'^A}{\partial \Omega^B} f^B + \frac{\partial \Omega'^A}{\partial x^k} f^k \tag{11.7}$$

The rules for f_i and for f^A are now cast into a form analogous to those for f^i and f_A, respectively, by defining vectors with underlined lower index i and underlined upper index A as follows:

$$f_{\underline{i}} \equiv f_i - g_i^A f_A; \ f^{\underline{A}} \equiv f^A - g_k^A f^k. \tag{11.8}$$

Note that the sign in Eq. (6) of Ref. [7], p. 230, which corresponds to the second Eq. (11.8), is wrong, as will become evident below. Eqs (11.8) have the effect that

$$f'_i = f_{\underline{i}}; \ f'^{\underline{A}} = \frac{\partial \Omega'^A}{\partial \Omega^B} f^{\underline{B}}. \tag{11.9}$$

In order to verify the laws (11.9) we need the transformation of g_i^A. Defining $g^{\underline{AB}}$ by its transforming like a product of two $f^{\underline{A}}$, it follows from the second Equation (11.9) that it is the reciprocal of g_{AB} and hence may be used to raise indices. With (11.6) one then finds (note that here the underlining of upper indices is superfluous)

$$g'^B_i \equiv g'_{Ai}g'^{AB} = \frac{\partial \Omega'^B}{\partial \Omega^C}g^C_i + \frac{\partial \Omega'^B}{\partial x^i}. \tag{11.10}$$

Making use now of Eqs. (11.5), (11.8), (11.10) for the primed quantities one verifies Eqs (11.9) which indeed confirms the sign in the second equation (11.8).

It is easy to see from the transformation of the g-tensor that one may choose g_{ik}, g_{Ai}, and g_{AB} to be homogeneous functions in the Ω^A of degree $n + 2$, $n + 1$, and n, respectively. Choosing $n = 0$ Pauli's *Assumption II* then reads

$$g_{Ai} = f_{AB,i}(x)\Omega^B. \tag{11.11}$$

With the help of (11.6) the transformation of $f_{AB,i}$ becomes, in the notation $\lambda^A_B \equiv \partial \Omega^A / \partial \Omega'^B$,

$$f'_{AB,i} = \frac{\partial g'_{Ai}}{\partial \Omega'^B} = \lambda^C_A \left(\lambda^D_B f_{CD,i} + \frac{\partial \lambda^D_B}{\partial x^i} g_{CD} \right). \tag{11.12}$$

This is analogous to the gauge transformation $A'_i = A_i + \partial \lambda / \partial x^i$ in electrodynamics. Thus the $f_{AB,i}(x)$ play the role of potentials. Using g^{ik} and g_{AB} for raising and lowering the respective indices, application of the definitions (11.8) yields

$$g_{A\underline{i}} \equiv g_{Ai} - g^B_i g_{AB} = 0; g^{A\underline{i}} \equiv g_{A\lambda i} - g^A_k g^{ik} = 0. \tag{11.13}$$

The physics now comes in by applying the group described above to generalize the Dirac equation of Eqs (6.30), (6.31). In so doing Pauli emphasizes that the following has strong analogies with the Klein–Kaluza theory described on pages 261f. Defining γ_i and γ_A in analogy to Eq. (7.28) by

$$\gamma_\rho \gamma_\sigma + \gamma_\sigma \gamma_\rho = 2g_{\rho\sigma}; \ \rho = A, i; \ \sigma = B, k \tag{11.14}$$

and similarly with upper indices, and using the definitions (11.8) to write

$$\gamma_{\underline{i}} \equiv \gamma_i - g^A_i \gamma_A; \gamma^{\underline{A}} \equiv \gamma^A - g^A_k \gamma^k, \tag{11.15}$$

one concludes from Eqs (11.13) that $\gamma_{\underline{i}}$, γ_A and γ^i, $\gamma^{\underline{A}}$ are anticommuting pairs of matrices. Introducing covariant derivatives as defined in Eq. (7.33), a Dirac equation of the form of Pais' Eq. (11.1) is

$$\gamma^k \psi_{;k} + \gamma^A \frac{\partial \psi}{\partial \Omega^A} + M\psi = 0. \tag{11.16}$$

Making use of the homogeneity relation (11.11), in which the index A is raised with the help of $g^{\underline{AB}}$, and using the second Eq. (11.15), this Dirac equation becomes Pauli's Eq. (7) (note that the underlining of the upper A in f is superfluous),

$$\gamma^k \psi_{;k} + \left(\gamma^{\underline{A}} + \gamma^k f^A_{B,k}(x) \Omega^B \right) \frac{\partial \psi}{\partial \Omega^A} + M\psi = 0. \tag{11.17}$$

In analogy to Pais' Eq. (11.1) one may put here $\gamma^{\underline{A}} = \gamma^5 \tau^A$. This is a highly interesting result, except for the fact that $f^A_{B,k}(x)$ is a vector field and hence does not describe pions! It is possible, in principle, to form a pseudoscalar from a tensor, although this is quite artificial. Using the covariant derivatives defined on page 30, $\partial a_i / \partial x^k - \partial a_k / \partial x^i = a_{i;k} - a_{k;i}$ is a tensor. Therefore one may construct a tensor $F^A_{ik} = -F^A_{ki}$ as follows:

$$F^A_{ik} \equiv f^A_{B,ik} \Omega^B \equiv \frac{\partial g^A_k}{\partial x^i} - \frac{\partial g^A_i}{\partial x^k} + \frac{\partial g^A_i}{\partial \Omega^B} g^B_k - \frac{\partial g^A_k}{\partial \Omega^B} g^B_i, \tag{11.18}$$

which is Pauli's Eq. (8). Since this equation is homogeneous of first degree in the Ω^A and, according to (11.11), $g^A_i = f^A_{B,i} \Omega^B$, it follows that

$$f^A_{B,ik} = \frac{\partial f^A_{B,k}}{\partial x^i} - \frac{\partial f^A_{B,i}}{\partial x^k} + f^A_{C,i} f^C_{B,k} - f^A_{C,k} f^C_{B,i}, \tag{11.19}$$

which is Pauli's Eq. (8a). Thus the $f^A_{B,i}$ play the role of potentials and the $f^A_{B,ik}$ the role of field strengths. And, in analogy to electrodynamics, the vanishing of the field strengths everywhere in space–time implies that, by a transformation (11.3), the potentials, i.e. the g^A_i, may be made to vanish. This is what Pauli calls his main result. Note that in the fibre bundle with the metric tensor $g_{\rho\sigma}$, $f^A_{B,ik}$ has a form analogous to the curvature tensor $R^\mu_{\nu\rho\sigma}$.

The mass of Pauli's field $f^A_{B,i}(x)$ also was a problem, however. Being a gauge field analogous to the photon and because no length had been introduced, a mass zero suggested itself. Pauli's letter [1682] of 6 December 1953 to Pais, which accompanied the second appendix discussed above, therefore turned out 'to be somewhat gloomy'. First, Pauli thought that 'new features should only occur in the *quantized* field'. Secondly, he observed that the rotations in ω-space 'can *not* be an exact law of nature, as it only holds as long as electromagnetism is neglected'. His third objection concerned the mentioned vector character and vanishing mass (Ref. [7], pp. 367–8). But he also complained about Pais to his assistant Thellung on 28 January 1954 from Princeton, that he 'has nothing new on the ω-space, is very restless but has no ideas' (translated from Ref. [7], letter [1707], p. 436). From January to April 1954 Pauli was at the Institute for Advanced Study (see Ref. [21], III.117, 122, 123, 125). In Princeton he gave a seminar on his work on 10 February, as Pais reports in his autobiography (Ref. [22b], p. 321).

Abraham Pais (1918–2000) grew up, together with a younger sister, in Amsterdam where his father was the headmaster of the Sephardic, i.e. Portuguese Jewish, School. He obtained his Master's degree under George Uhlenbeck at Utrecht in 1940. When the latter left for the United States that summer, Pais became the student of Léon Rosenfeld (1904–74), under whom he obtained his Ph.D. in 1941, in the middle of the German razzias against Jews. In March 1943 Pais went into hiding (his sister had been deported) and towards the end of the war he spent some weeks in prison. In January 1946 he went to Copenhagen where he met Pauli in Bohr's home. In September of that year he went to the United States where he settled at the Institute for Advanced Study, becoming a professor in 1950 and a US citizen in 1954. He got married in 1956 and in 1958 his son was born. After his divorce he joined the Rockefeller Institute in New York in 1963 and when it became a University he was appointed professor, in 1981. The following year Pais, after three years of work, published his famous biography of Einstein 'Subtle is the Lord . . .' [22c] and in 1985 he published his second important book, Ref. [22d]. In his later years he spent the winter in New York and the summer in Copenhagen where he married for the third time in 1990 and where he died in 2000. Pais is known for the idea of associated production of strange particles and also the Pais–Piccioni effect of oscillations of the neutral K-meson system ($K^0, \overline{K}^0$) (see Refs [22b] and [22d], Section 20(b)).

According to Eq. (11.4), g_{AB} and Ω^A are real, and therefore, according to (11.11), so is $f^A_{B,i}(x)$. However, with two such fields one may form a complex vector field. Such a field had been proposed by Oskar Klein back in 1938 to describe charged intermediate bosons, analogous to Yukawa's meson, as the mechanism of β-decay [22e]. Then in late 1957, after experiments had confirmed that the β-decay interaction (6.63) proceeds by a $V–A$ coupling, several physicists again proposed such bosons (see Ref. [22d], Section 21(e)). The latter attained permanent status in the electroweak theory of Salam and Weinberg [23a,b]. These gauge bosons, today called $W^\pm$ and Z^0, had been detected by Rubbia and his group at CERN [23c]. Their masses in units of the proton mass m_p are $M_W \simeq 85.6$ and $M_Z \simeq 97.2$ [23d] (see also Ref. [17], Vol. 3, Section 10.3). Abdus Salam and Steven Weinberg, together with Sheldon L. Glashow, were awarded the Nobel Prize of 1979 'for their contribution to the theory of the unified weak and electromagnetic interaction between elementary particles, including *inter alia* the prediction of the weak neutral current'. Carlo Rubbia and Simon Van der Meer received the Nobel Prize of 1984 'for their decisive contributions to the large project, which led to the discovery of the field particles W and Z, communicators of weak interaction'.

It must be added (see Ref. [22a]) that Chen Ning (Frank) Yang (b. 1922), who in 1949–54 was a member and in 1955–66 a professor of the Institute for Advanced Study, had also had the idea of generalized gauge fields and worked it out together with a graduate student, Robert Laurence Mills (b. 1927) [24a]. They presented the results at the Washington meeting of the APS in April 1954 [24b], but Yang had already given a seminar on the subject at the Princeton Institute on 23 February when Pauli was in the audience. Yang reports: 'Soon after my seminar began, when I had written down on the blackboard

$$(\partial_\mu - i\epsilon B_\mu)\psi = 0,$$

Pauli asked "What is the mass of this field B_μ?" I said we did not know. Then I resumed my presentation, but soon Pauli asked the same question again. I said something to the effect that that was a very complicated problem, we had worked on it and had come to no definitive conclusions. I still remember his repartee: "That is not sufficient excuse." I was so taken aback that I decided, after a few moments' hesitation, to sit down. There was general embarassment. Finally Oppenheimer said "We should let Frank proceed." I then resumed and Pauli did not ask any more questions during the seminar.' The next day Yang found a message by Pauli saying 'I regret that you made it almost impossible for me to talk with you after the seminar' (see Ref. [7], note [1726] of 24 February 1954). Yang continues: 'I went to talk to Pauli. He said I should look up a paper by E. Schrödinger [20b], in which there were similar mathematics. . . . But it was many years later when I understood that these were all different cases of the mathematical theory of connections on fibre bundles'. And Yang concludes: 'Pauli was the first physicist who showed intense interest in our paper. That is not surprising, since he was familiar with Schrödinger's work and had himself tried to relate interaction with geometry' (Ref. [24c], pp. 20 and 21).

Probably in response to this exchange, Pauli gave Yang the note [1727] of Ref. [7], in which he comments on his equations (11.11), (11.19), and (11.17) for Yang's special case, referring explicitly to Schrödinger's paper [20b]. On 26 February 1954 he asked Thellung from Princeton to tell his American collaborator William A. Barker the following: 'In the special case that the spinor ψ does not depend on ω, Yang and Mills (at present at Brookhaven) have derived equations [24b] that become identical with mine, which last autumn he [Barker] had treated by the sweat of his brow together with Gulmanelli' (translated from letter [1729], Ref. [7], p. 501). William Alfred Barker (b. 1919) had come from the University of St. Louis, Missouri, in 1953 and stayed with Pauli until 1955 [25a] (see also footnote 28 in Ref. [7], p. 438).

Paolo Gulmanelli (b. 1928) from Pavia [25b] had probably met Pauli at the conference on non-local field theories which took place in Turin, Italy, from 9 to 14 March 1953 and where Pauli gave an invited talk. Both Pauli and Gulmanelli had worked on this problem [26a,b]. Pauli had become interested in the subject from discussions he had had at the physics conference in Copenhagen in June 1952. He had taken up the problem again after his return from India in January 1953. But by the time of the Turin conference his pessimism had again gained the upper hand. Pauli was back in Italy in September on holiday with Franca to Forte dei Marmi on the Ligurian coast south of La Spezia. From there they went to Calgary on the south coast of Sardinia where, from 23 to 27 September 1953 a meeting of the Italian Physical Society took place and from where Pauli, Konrad Bleuler, and Gulmanelli sent postcards to assistant Thellung and to the Kronigs (Ref. [7], [1636] and [1637]).

When Gulmanelli came to Zurich in September 1953 to work with Pauli at ETH until November 1954 (see footnote 12 in Ref. [7], p. 119), he also switched to Pais' ω-space for a while; in November Pauli spoke twice on the subject in the

Zurich seminar (see letters [1659, 1676] in Ref. [7]). Back in Italy, Gulmanelli also gave a series of seminars at the Institute for Theoretical Physics of the University of Parma [25b], which he then edited in the form of Notes [27a]. Gulmanelli later commented on Pauli's endeavour as follows: 'Pauli was at the beginning very amused about his generalization of Pais' ω-space and, if I may add, he appeared very fond of his trick of index underlining. In so far as I know, there have been no lectures by Pauli. He gave me just the basic equations, that you find in Section II [of Ref. [27a]] and some hints how to proceed with the Dirac equation in ω-space. Also the equations of Section IV [the generalized Dirac equation in Ref. [27a]] have been checked by Pauli. Unfortunately (but who might blame [him] for it at that time?), as soon as he got convinced that the "gauge" field *had* to be vectorial (plus a scalar or pseudoscalar component at the price of very artificial hypotheses) and that mass terms . . . are *bound* to be identically zero, then he gave it all up and everything was thrown into the basket' [27b].

On his 53rd birthday, on 25 April 1953, Pauli received news of his election as Foreign Member of the Royal Society, on which Max Born had sent him congratulations (see letters [1563] and [1571] in Ref. [7]). In his response [1571] Pauli mentioned his lecture on Probability and Physics he had given at the meeting of the *Swiss Society of Natural Sciences* in Berne in 1952 [28a] (see also the comment on pp. 148–9 in Ref. [7]) and expressed the hope that Einstein would not be offended by what Pauli had written. Pauli had criticized Einstein's statement: 'There is such a thing as the real state of a physical system, which exists objectively, independently of any observation or measurement, and can in principle be described by the modes of expression used in physics' (Ref. [13b], p. 47), writing: 'However, these formulations of Einstein's are only a paraphrase of the ideal of a special form of physics, namely the "classical" form. This ideal, so pertinently characterized by Einstein, I would call that of the *detached observer*. In point of fact "existent" and "non-existent", or "real" and "unreal", are not unique characterizations of complementary qualities, which can be checked only by statistical sequences of experiments using different arrangements, freely chosen, which may in some cases be mutually exclusive' (Ref. [13b], p. 47). As in his lecture on the 'The Philosophical Significance of the Idea of Complementarity' (pages 414–15) and in his 'Piano Lesson' (pages 462 and 467f) Pauli here again sums up his criticism in the notion of the detached observer.

As Pauli wrote to Pais on 21 April 1953 (Ref. [7], letter [1558]; see also [1567]), on 2 May 1953 the Swiss Physical Society convened in Geneva at the new physics building of the university. This building had been inaugurated on 23 October 1952 with the bestowal of an honorary degree to Paul Scherrer who had contributed to its planning (see Ref. [2], comment on pp. 783–4). The director of the institute, Richard Charles Extermann (b. 1911) was supervising a search for a professor of theoretical physics. Felix Villars and Gregory Wannier were candidates (see Ref. [7], footnote 4, p. 118), but it was only in 1960 that a stable solution emerged with the nomination of Josef Jauch. At the Geneva meeting of 2 May 1953 the Swiss Physical Society elected Pauli vice-president and voted on a resolution 'expressing the satisfaction of the Society to see the European Centre of Nuclear Research [CERN] being established at Geneva and its recognition towards the scientists and politicians who contribute to the realization of this grand project' (translated from Ref. [29a], p. 378). At the meeting of the Swiss Physical Society in Vevey on the shore of Lake Geneva, on 7–8 May 1955, Pauli was elected president for a two-year term and the acting president Professor Paul Huber of Basle University, read a commemorative address for Einstein (written by Markus Fierz) who had passed away in April of that year. On Saturday afternoon, 8 May, the participants to the meeting were guests of Professor and Mrs Auguste Piccard (1884–1962), the famous explorer of the stratosphere and the ocean depths, at their mansion in Chexbres above Vevey, overlooking Lake Geneva (see Refs [29b], pp. 297-8).

Before CERN was established at Geneva by the ratification of the Convention on 29 September 1954 (see Ref. [17], Vol. 1, p. 226), Bohr supervised a group of younger theoretical physicists in Copenhagen (see page 332). One of them was Kurt Alder (b. 1927) who in 1951 had won a prize with his diploma under Pauli (Ref. [21], footnote 349, p. 435) and to whom Pauli, in summer 1952, asked President Pallmann to pay a grant permitting him to work in Copenhagen (Ref. [21], III.100). During the winter term of 1951/2 Gunnar Källén (1926–68), a brillant student of Professor Torsten Gustafson in Lund, and Walter Thirring (b. 1927), the son of Hans Thirring (see pages 25–7), were working at ETH, as Pauli tells Thellung, who is back in Zurich from Delft for the Christmas vacation. In this letter [1327] of 20 December 1951 in Ref. [2] Pauli also suggests that Thellung joins him at the traditional Christmas Party of the whole Physics Institute of ETH that took place at the 'Morgensonne' restaurant near Zurich Zoo on Zürichberg. Pauli was always present at these gatherings (see page 199).

For the summer term of 1953 Pauli agreed to welcome as a guest Bernard Jouvet (1927-78) from Paris (Ref. [21], III.105). But his foremost concern was to create the position of an associate professorship of theoretical physics 'for the exoneration of the full professorship' (translated from Ref. [21], III.111, 115). Pauli's main candidate for this position was Res Jost, whom he proposed to invite as guest lecturer for the summer term of 1954 (Ref. [21], III. 112, 116, 120). After discussion in the Swiss School Council (Ref. [21], III.121, 126), on 7 April 1954 President Pallmann was able to inform Pauli that the Swiss Federal Council had agreed to the creation of an associate professorship of theoretical physics. In the session of 19 June 1954 the Swiss School Council examined the candidacies of Schafroth, Jauch, Gunnar Källén from Sweden, Villars, Jost, Bleuler, and Maurice Jean proposed by Louis de Broglie (Ref. [21], III. 132). The decision, taken three weeks later, was no surprise: Jost was elected for a three-year period beginning 1 April 1955, with a yearly pay of 17050 francs plus an age supplement of 2200 francs (III.134), and he accepted (III.135; see also the comment in Ref. [7], pp. 560–1). Pauli wrote to Jost on 6 October 1954: 'First of all, once again the joy of all colleagues in Zurich that you have accepted! Your election will surely appear soon in the newspaper. I already rejoice in your coming, for your amusement there will be provision, e.g. by the relativity congress in June, the jubilee of the Polytechnic in the autumn of 1955 and, of course, I now will lecture during the summer term . . . on "Problems of Quantum Statistics" [see Ref. [21], p. 434]' (translated from Ref. [7], letter [1885], p. 778).

Fig. 11.2 Pauli on an excursion of the ETH Physics Institute to the Säntis mountain, 28 July 1953 (photo by C. P. Enz)

From 8 to 13 August 1953 Pauli lectured again at the summer School in Les Houches (Ref. [7], p. 186, see page 451). This time his topic was the old problem of the H-theorem (see pages 200f) which, since his paper with Markus Fierz in spring 1937 (see page 317), had remained one of the favourite subjects of discussion between Fierz and him. Encouraged by Pauli, Fierz formulated his views in his letters [1577] of 21 May and [1613] of 25–27 July 1953 in Ref. [7]. Two days before his departure Pauli again asked Fierz's opinion [1618] and, fresh from his discussions in Les Houches, he reported his views to Oskar Klein who also was thinking about the problem [1623].

During Pauli's absence in Princeton in the first quarter of 1954, his assistant Thellung had to teach the main course on 'Optics and the Theory of Electrons' (see Ref. [21], III.125 and p. 432). Thellung had succeeded Schafroth on 1

April 1953 as Pauli's assistant for one year with a yearly pay of 9300 francs plus 4 % inflation supplement (see Ref. [21], III.110). Being a music lover and playing the piano himself, he used to organize visits to Zurich's Concert Hall, the Tonhalle, for Pauli and his collaborators, in which also Konrad Bleuler, from the university, often participated. After a concert the group would go to Kronenhalle, a traditional restaurant near Bellevue Square, where Pauli would order Swiss white wine and Viennese sausages with horseradish (see page 197). This 'Music Club' became quite a tradition, as letters [2247] and [2287] between Pauli and Thellung from the year 1956 in Ref. [30a] show.

Fig. 11.3 Pauli with Armin Thellung and Charles Enz, on the Säntis mountain, 28 July 1953.

Armin Eugen Albert Thellung (b. 1924) grew up in Zurich where his father taught botany at the university. While he was assistant of Kronig in Delft from 1949 to 1952, Thellung worked on his Ph.D. thesis with Pauli on the magnetic moment of the proton in meson theory [30b], which he submitted to ETH in late 1951; Pauli wrote a very favourable report (see Ref. [21], III.78). With Kronig, Thellung worked on the quantization of the hydrodynamics of ideal fluids (see the comment in Ref. [7], p. 434). After his assistantship with Pauli, Thellung went to Birmingham in 1956 to become a lecturer in Peierls' institute. In 1958, the year he got married, he was nominated to a second chair of theoretical physics beside Walter Heitler at the University of Zurich. He held visiting professorships in 1961 at Perth, Western Australia, and 1964–5 at Cornell University in the United States.

Later in the year 1954 Pauli wrote an article in honor of Jung's 80th birthday on 26 July 1955 with the title 'Ideas of the Unconscious from the Standpoint of

Fig. 11.4 Pauli with, from left to right, Ernst Heer, Margaret Schmid, Armin Thellung, and Konrad Bleuler, on the Säntis mountain, 28 July 1953 (photo by C. P. Enz)

Natural Science and Epistemology' [12a,b]. In the first section, 'The Problem of Observation', Pauli reviews the notion of the unconscious in its evolution throughout Jung's work. The second section, 'Applications of the Ideas of the Unconscious in Quantitative Sciences', considers aspects of the arrangement of our conceptions (*Angeordnetsein unserer Vorstellungen*) and ends with a discussion of the notion of chance in evolution very similar to the one in the 'Piano Lesson' (see pages 462f). Of particular interest, however, is Pauli's description in the first section of the ideas of a relatively unknown Swiss psychiatrist with the name Charles de Montet who had sent Pauli his book *Evolution vers l'essentiel* [31a] with a dedication addressed to the 'physicien de l'atome et psychologue', but whom Pauli never met. This book is an essay on the notion of complementarity in physics, epistemology, physiology, sexuality, psychology, ethics and religion. Charles de Montet (1880–1951) became Privatdozent of psychotherapy and medical psychology at the Faculty of Medicine of Lausanne University in 1911 and director of the clinic at Mont Pélerin above Vevey until its closure after World War I. He published several books on psychology and related subjects [31b].

In his letter [1848] of 23 July 1954 to Marie-Louise von Franz Pauli writes on de Montet's book [31a] that it emphasizes a particular aspect of Heisenberg's uncertainty relations, 'namely the situation "the sacrifice—and the choice".
. . . Since then I am convinced that the *archetypal situation of the sacrifice* plays an essential role with the notion of measurement in quantum theory (every physicist will see the connection at once). In view of this I now have read more carefully the passages on the psychology of the sacrifice in the "Roots of the Consciousness" [32].' And he continues: 'Going beyond *de*

Montet I then realized at once an essential incompleteness of the quantum-physical "sacrifice": with the physical measurement the observer—after he has interfered with the external world by the choice of the experimental setup—immediately absconds again without "wetting his fur". As a consequence *he himself does not transmute*, and what transmutes is only the state of the observed system' (translated from Ref. [7], pp. 717–18). In his birthday essay for Jung this situation is also described—although with somewhat more restraint—and de Montet's book is quoted.

The summer of 1954 was again quite eventful. On 1–4 July, still during the summer term, Pauli was in Lund for a conference on 'Rydberg and the Periodic System of Elements' [33a]. On 29 May the university there had bestowed an honorary doctorate on Pauli (see Ref. [7], letters [1756] and [1779], also the comment on pp. 700–1). In the framework of the Second International Congress on the Philosophy of Science which was organized in Zurich under the presidency of Ferdinand Gonseth from 23 to 28 August (see Ref. [7], comment, pp. 722–3), on 24 August Pauli was chairing a symposium on 'Phenomenon and Physical Reality' [33b]. Heisenberg had first agreed to participate [1797], but later withdrew [1861], so that Pauli looked forward to the event with some anxiety, as he told Fierz on 10 August. [1864] In the introduction to his essay, Pauli characteristically says: 'My general tendency is rather to hold a middle course between extreme directions.' And speaking in Part 2, on 'Logical Structure of Physical Theories', he goes so far as to state about the observer: 'Indeed I myself even conjecture that the observer in present-day physics is still too completely detached, and that physics will depart still further from the classical example' (Ref. [33b], p. 132). This view, very likely, was motivated by the fact that still no theory was in sight explaining the value (7.2) of the fine-structure constant.

The day after the symposium Pauli and Franca went on holiday to Cervia on the Adriatic, south of Ravenna (Ref. [7], letter [1872]), until the Solvay Congress on the theory of metals took place from 13 to 17 September and at which superconductivity was the main theme. Finally, a large international congress of historians and philosophers was planned in Mainz on the Rhine in October 1954, but had to be postponed to March 1955 (Ref. [7], letter [1822] and comment pp. 629–30). Pauli had prepared a manuscript on 'Science and Western Thought' [33c], on which he commented to Erwin Panofsky in his letter [2004] of 6 February 1955 in Ref. [34] with the following statement which also figures in the introductory passages of the published version (Ref. [33c], p. 139): 'The main theme of my report is the relational problem: knowledge of salvation—scientifc knowledge.' In this text which—at least in the German original—may well be considered the most beautiful of his essays, Pauli writes towards the end: 'Yet I believe that there is no other course for anyone for whom narrow rationalism has lost its force of conviction, and for whom also the magic of a mystical attitude, experiencing the external world in its crowding multiplicity as illusory, is not effective enough, than to expose himself in one way or another to these accentuated contrasts and their conflicts. It is precisely by this means that the scientist can more or less consciously tread a path of inner salvation. Slowly then develop inner images, fantasies or ideas, compensatory to the external situation, which indicate the possibility of a mutual approach of poles in the pairs of opposites' (Ref. [33c], p. 147). To Marie-Louise von Franz, whom Pauli now addressed by the German 'Du', he wrote on 23 January 1955 that for the preparation of this talk 'your copy of *The Tibetan Book of the Great Liberation* [35a]—which still rests on my desk—was very useful to me' (translated from Ref. [34], letter [1988]). Pauli was very pleased with the effect of his conference, as he wrote to the organizer of the Mainz congress, Professor Hermann Göring (Ref. [34], letter [2056]).

The neutrino manifests itself at last

Proof of the existence of the neutrino was slow to come. The first news of its detection reached Zurich late in 1953. To celebrate the event Pauli and a group of faithfuls climbed Uetliberg above Zurich. On the way down late that evening, recounts William Barker, 'Pauli was a little wobbly from the red wine we had at dinner. (He had graciously responded to many individual toasts.) [Konrad] Bleuler said to me: "Take his left arm—I'll take his right arm, we can't afford to lose him now." Later when we were about midway, Pauli turned to me with a comment I shall always treasure, "Remember, Barker," he said, "all good things come to the man who is patient".' [35b]

On 6 July 1953 Pauli mentioned in his letter [1596] to Fierz that, according to a recently published note [36a], the absorption of free neutrinos was 'close to the experimental detection' (translated from Ref. [7], p. 192). Six months later the same authors, Frederick Reines (1918–98) and Clyde Lorrain Cowan (1919–74), announced that they had indeed detected free neutrinos [36b]. It was this news that gave rise to the celebration on Uetliberg mentioned above. Reines and Cowan's first idea had been to use a nuclear explosion as a source of antineutrinos, but they then settled on the idea of using a reactor. In their first experiment they worked at the Hanford reactor at Hanford in the state of Washington, but for their improved measurements [37a,b] they went to the Savannah River reactor in South Carolina.

Reines and Cowan used as a source the antineutrinos $\bar{\nu}$ produced by the neutrons n emerging from a nuclear reactor and decaying with a mean lifetime of 14.8 minutes (Ref. [23d], p. 93) according to the reaction

$$n \to p + e^- + \bar{\nu}. \tag{11.20}$$

They detected the $\bar{\nu}$ by the reaction

$$\bar{\nu} + p \to n + e^+, \tag{11.21}$$

where the protons p were supplied by the water filling a large tank which also contained cadmium atoms. The positrons e^+ and neutrons n produced by the reaction (11.21) were detected by the γ-rays emerging from the annihilation of the e^+ by electrons e^- and from the delayed absorption of the n by cadmium nuclei, which have a large absorption cross-section for neutrons. These γ-rays produced photons in an organic scintillation liquid which was monitored by a large array of photomultipliers. In order to suppress background counts, Reines and Cowan introduced delayed coincidences between the annihilation γ-rays and the absorption γ-rays. This experiment is described in detail in Pauli's essay on the history of the neutrino (see Ref. [38a,b], Section 4, in particular Fig. 5).

On 14 June 1956 Reines and Cowan sent Pauli the following telegram from Los Alamos where they were stationed: 'We are happy to inform you that we have definitely detected neutrinos from fission fragments by observing inverse beta decay of protons. Observed cross-section agrees well with expected six times ten to minus forty-four square centimeters. Frederick Reines, Clyde Cowan.' The telegram, which is reproduced in Fig. 8 of Ref. [39a], was addressed to Zurich University but, apparently, it was sent directly to ETH. By the next day Pauli sent the following message: 'Frederick REINES and Clyde COWAN. Box 1663, LOS ALAMOS, New Mexico. *Thanks for message. Everything comes to him who knows how to wait. Pauli*' (see page 421). The original sheet of paper with the quoted text in Pauli's hand was destined for the secretary, Margaret Schmid (b. 1923). It is in the *Pauli Archive* at CERN (PLC Bi) and contains the additional note in the hand of the secretary: 'erl. 15.6.56/15.35 h als night letter', meaning 'executed (*erledigt*) 15 June 1956 at 3.35 p.m. as night letter.' Surprisingly, Frederick Reines wrote to me on 11 April 1986 that he had never received this night letter, and he attached a copy of his and Cowan's telegram [39b] (I sent him a copy of Pauli's night letter on 22 September 1986).

From 20 to 23 June 1956 Pauli was in Geneva, attending the CERN symposium on pion physics, where he announced the detection of the neutrino [39c]. Reines erroneously writes in Ref. [39a], p. 15, that his telegram 'was forwarded to him [Pauli] at CERN where he interrupted the meeting . . . to read the telegram' and, quite inexplicably, that 'Pauli and some friends consumed a case of champagne in celebration.' There exists nothing to support this claim, however. From Geneva Pauli went directly to Lindau on Lake Constance until 27 June to participate at the meeting of the winners of Physics Nobel Prizes. During this time his lectures on 'Thermodynamics and the Kinetic Theory of Gases' (see Ref. [21], p. 432) were cancelled, while the associated problem session was left to his new assistant Enz to supervise (Ref. [21], III.165, 166). Long after Cowan's death, Reines, then at the University of California at Irvine, shared the Nobel Prize of 1995 with Martin L. Perl from Stanford University, Reines 'for the detection of the neutrino' and Perl 'for the discovery of the tau lepton'.

Parallel to the experiments of Reines and Cowan a different effort to detect neutrinos was undertaken by Raymond Davis Jr at Brookhaven National Laboratory on Long Island, east of New York. Between 1952 and 1954 Davis exposed a drum of carbon tetrachloride (CCl_4) cleaning fluid outside the shield of the Brookhaven graphite reactor, hoping to detect argon-37 atoms produced after a few weeks according to the reaction

$$Cl^{37} + \bar{\nu} \to Ar^{37} + e^-. \tag{11.22}$$

[40a,b] This reaction had originally been proposed by B. Pontecorvo and by L. W. Alvarez (see note 3 in Ref. [39a]). According to (11.20) the reaction (11.22) should not occur at all with antineutrinos—provided that ν and $\bar{\nu}$ are not identical or, more generally, that the number of light particles or leptons is conserved, as is in fact the number of heavy particles or hadrons.

In 1956 Davis exposed 3800 litres of the same cleaning fluid to the Savannah River reactor, [40c] (see Davis's letter [2380] of 31 October 1956 to Pauli in Ref. [34]), the same time as Reines and Cowan were measuring there. But, as Pauli notes on p. 210 of Ref. [38b], the result of this experiment was still inconclusive, giving only an upper limit for the reaction cross-section of 0.9×10^{-45} cm^2/atom. [40c] It was only in the early 1990s that Davis got his first result. He still worked with the same reaction (11.22) but now, in order to shield from cosmic rays, 1.5 km underground at the Homestake gold mine in Lead, South Dakota, and his neutrino source now was the sun. Using a 400 000 litre tank filled with tetrachloroethylene (C_2Cl_4) and extracting on average 15 argon-37 atoms after every month, he obtained the first evidence of a neutrino flux from the sun. During the period of strong solar flares in June 1991 Davis collected up to 90 argon-37 atoms per month [40a].

More recently, a different technique has been used by a Japanese group under Masatoshi Koshiba at the Kamiokande and, still more recently, at the Super-Kamiokande facilities in Japan. These experiments detect Cerenkov radiation given off by electrons, positrons, or muons produced by incoming neutrinos scattering on nucleons in a huge underground tank of highly purified water. In this way it is possible to measure the arrival time, energy, and direction of incoming neutrinos. Quite unexpectedly, the Kamiokande experiments allowed detection of a neutrino burst coming from the supernova 1987A explosion observed on 23 February 1987 in the Large Magellanic Cloud, some 170 000 light years from the earth [40d]. Raymond Davis Jr and Masatoshi Koshiba received the prestigious Wolf Prize of the year 2000 for their achievements [40e].

The embarrassing fact, however, is, that the measured neutrino flux from the sun is only some 30–50% of the flux predicted by the standard model of the sun [40a] (see also Ref. [40f], p. 98). An even more serious problem posed by the neutrino is its identity. First, neutrino ν and antineutrino $\bar{\nu}$ could be identical, a possibility that had been considered by Majorana (see pages 349f) [40g]. This would imply that the double β-decay of neodymium ^{150}Nd ($Z = 60$) into samarium ^{150}Sm ($Z = 62$) could proceed without emission of Majorana neutrinos (see Ref. [37b], p. 449). On the other hand, today we know of three charged, weakly interacting particles or leptons, the electron e^-, the muon μ^-, and the tau τ^- discovered by Perl (see above), forming with their associated antiparticles and neutrinos the three lepton families. If these neutrinos have non-vanishing masses, they may transform into each other, which manifests itself in oscillations. Such oscillations are the subject of much discussion and research, as is the problem of the neutrino mass [40h]. Homestake, Kamiokande, and similar underground facilities (see e.g. Ref. [40f], p. 98) are inaugurating a new way of doing neutrino research which may be called neutrino astronomy. For more detail on all these questions see Refs [40i and 40j].

However, for Pauli and his collaborators in Zurich, at that time, the neutrino was not an object of research. On 7 April 1954 President Pallmann reconfirmed Armin Thellung as Pauli's assistant for another year starting 1 April 1954 (Ref. [21], III.127) and on 22 April 1955 for a third year (III.144). And, as reported in the previous section, Res Jost commenced his work as associate professor in the summer term of 1955, giving a three-hour lecture on 'Selected Chapters from Analytical Mechanics and Quantum Theory' (see Ref. [21], p. 434). Pauli also got a new collaborator, David Speiser (b. 1926), a student of Fierz and a nephew of the famous professor of mathematics, particularly of group theory, Andreas Speiser, in Basle. On 26 October 1954 Pauli had written to Fierz that he completely agreed that Speiser should come to Zurich; he had even expected him that day (Ref. [7], letter [1902], p. 824). Speiser stayed with Pauli until December 1957, when he became assistant to André Mercier in Berne. He was later named professor of theoretical physics at the Université Libre of Louvain, Belgium, where he made his name in the history of physics and mathematics, particularly with the edition of the works of the Basle mathematicians Leonhard Euler and the Bernoullis (see the comment in Ref. [7], p. 807).

In Zurich David Speiser met his future wife Ruth Bär, the second daughter of Ellen Weyl from her first marriage to the physicist Richard Bär (see page 39). Also Ellen's first daughter Marianne Olsen, after having studied herself physics at ETH, married a physicist, the Dane Jorgen Lykke Olsen (b. 1923), who in 1961 became professor of low-temperature physics at ETH. The home of the Bär family at Bergstrasse on the slope of Zürichberg was a centre of much social and cultural activity, in which Pauli and Franca regularly participated. Res and Hilde Jost also belonged to this circle. Pauli's and Weyl's close relations, as recalled in Pauli's allocution for Weyl's 70th birthday on 9 November 1955, in fact, have been mentioned several times in this biography (see Chapters 2, 3, 6, and 9).

During Speiser's time in Zurich a multiple Pauli effect happened, as Thellung recounts: 'In commemoration of the 50th anniversary of the Special Theory of Relativity, on 26 May 1955 in the evening, Pauli gave a talk on Einstein to the Zurich Physical Society [see Ref. [41a]]. Before, Kronig (who was in Zurich on his yearly visit), Jost, David Speiser and I met for dinner at the 'teetotal restaurant Zürichberg', near the tram terminus near the Zoo. On the way from the restaurant to Pauli's talk the following happened: Speiser, discovering that the gasoline tank of his Lambretta [scooter] was empty, went to a filling station. There the Lambretta suddenly caught fire! It was extinguished with the water from a jug but was not usable any more, so that Speiser had to walk. I found my bike with flat tyres and, hence, also had to walk. Kronig, finally, went by tram—a stretch he had travelled many times already—but he forgot to get out at Gloriastrasse, and noticed it only many stops later. However, we all arrived in time for Pauli's talk! Scherrer also was with the party. Pauli appeared highly amused by our adventures and apparently considered them as typical Pauli effects.' Indeed, a defining feature of a Pauli effect was that Pauli himself never experienced any harm.

Thellung's story continues: 'That evening we all missed Konrad Bleuler. He had asked me before the lecture to excuse him to Pauli because he was exhausted. Reason: He had abundantly quarrelled with [Hans] Staub, the full professor of experimental physics at the university. When Pauli heard that, he spontaneously said: "Let's send Bleuler a postcard!" And he wrote: "Thellung says, you are exhausted. I am very sorry for that. The experimental physicist of the ETH has such effects only on women!" And the latter added: "Read and approved! Scherrer"' (translated from Ref. [21], pp. 199–200). Hans Heinrich Staub (1908–80) had been professor of experimental physics at the University of Zurich since 1949. He had studied physics at ETH and, after his Ph.D., was assistant there from 1931 to 1937, when he went to Caltech in Pasadena. During the war, from 1942 to 1946 he worked on the Manhattan atomic bomb project at Los Alamos and then became a professor at Stanford University.

In the early 1950s Pauli sat for several oil-paintings by the Zurich artist Ernst Morgenthaler (1887–1962), whose wife Sacha was a friend of Franca. Sacha Morgenthaler was well known in Zurich for the beautiful dolls she manufactured. One of Ernst Morgenthaler's portraits of Pauli is exhibited in the seminar room of the theoretical physics institute at the Hönggerberg campus of ETH. Also exhibited in the same building, in the physics library upstairs, is a bust of Pauli by the Zurich sculptor Hermann Hubacher (1885–1976). Pauli, however, found that in Morgenthaler's portrait he looked like a master butcher during times of bad business, and Franca considered it a caricature [41b]. But Ernst Morgenthaler made a very nice pen drawing of Pauli which he published in his charming little book *A painter recounts* (Ref. [41c], p. 180). The German poet and novelist Hermann Hesse (1877–1962), who since 1919 had lived in Tessin, the Italian part of Switzerland, was a good friend of Morgenthaler and had written a beautiful foreword to his book. In this book there is a letter to Hesse, in which Morgenthaler writes about Pauli: 'He also is a great admirer of yours and asked me recently to tell you occasionally that he has nothing to do with the atom bomb, that he abhors it and is sad about the misuse of the results of research.—Pauli does not read newspapers. When I told him he may well be right, we the others are reading every morning who fell off a bike, who knocked his head and who, somewhere, on the way home was run over . . . to which Pauli observed, he indeed would be more interested to read about the people who came home safely' (translated from Ref. [41c], p. 181).

From 13 to 18 June 1955 Pauli participated at the Pisa conference on elementary particles. Since it took place during term time, Pauli's course on 'Electrodynamics' and his special lectures on 'Problems of Quantum Statistics' (see

Fig. 11.5 Sketch of Pauli by Zurich painter Ernst Morgenthaler (from: E. Morgenthaler, *Ein Maler erzählt*, Zurich, 1957)

Ref. [21], pp. 432 and 434) were cancelled, while the problem session for the former was to be held by his assistant Thellung (Ref. [21], III.147, 148). In the following month the International Congress 'Fünfzig Jahre Relativitätstheorie' took place in Berne from 11 to 17 July 1955; Pauli was president and André Mercier secretary [42a,b]. This congress had been planned before Einstein's death which occurred at Princeton Hospital, early in the morning of 18 April

1955 (Ref. [22c], p. 477). The following day Pauli wrote to Léon Rosenfeld: 'The sad news about Einstein yesterday affected me very much, I feel that with him a certain chapter of physics is leaving us. I could not well refuse to write a short article on him for the *Neue Zürcher Zeitung*' (Ref. [34], letter [2070]). This is the article mentioned above, which appeared on 22 April 1955 [41a].

In his opening address to the relativity congress Pauli mentioned Einstein's attachment to Berne where in 1905 he had created relativity theory, saying that 'at his home in Princeton the visitor could see only a single diploma: it was an honorary membership of the "Naturforschende Gesellschaft in Bern"' and that Einstein 'expressed his joy and gratitude about our report on the [preparation of] the Jubilee conference, which appeared to him to be promising'. Pauli closed the address with the wish 'that Einstein's wisdom, which is for me always clearly visible even in what I believe to be his errors, will be a good guide to the reports and discussions of this conference and will show us the right way towards the future' (Ref. [42a], p. 27).

Reviewing in his closing speech (which he gave in German) the main contributions to the congress, Pauli said: 'I believe the most important we have heard in addition [to the experimental verification and the cosmological problem] was the report by Lichnérowicz [see page 389] on the Cauchy initial-value problem in the nonlinear field equations of general relativity. I associate a very large value to the study of these problems, because I definitely think that also with the field quantization . . . it will play an essential role.' He then turned to the classical notion of field which he was inclined 'to consider unsatisfactory not only heuristically but also in a deeper manner. Herewith several things have come to light that bothered me already in my youth when I wrote the Encyklopädie article on relativity theory and which since, as Mr. Born has said, in fact did not receive a sufficient clarification.' This was the problem of measuring a field that Pauli had addressed at age 19, criticizing Weyl's notion of field in the interior of the electron (see page 36). He then said 'The question rather is that, in proceeding to measure a field one needs a test body, and it causes me great uneasiness if, with the same breath, the test body is again considered as a field.' In particular, Pauli was perturbed that 'for instance, it is logically possible to write down the Maxwell equations in vacuum without introducing a charge'. In conclusion, considering the quantization of the gravitational field, Pauli hoped that the uncertainty of the light cone 'could yet perhaps contribute to overcome the divergence problems of the field quantization' (translated from Ref. [42b], pp. 263, 265, 266, and 267).

In his own contribution [42c] to the congress Pauli reviewed Einstein's G-field (the g_{ik} on page 29) 'which, according to Einstein, is just the ether in a new form' but which 'retains its conceptual independence over against matter. Einstein had repeatedly stated that he would find it more satisfactory if the G-field were to vanish identically in the absence of matter. He called this fundamental principle "Mach's principle" in honour of Ernst Mach who paved the way for later thought on the general theory of relativity by his critique of absolute space. Nevertheless it can be said that Mach's principle does not follow from the equations of the general theory of relativity alone without special assumptions, which are hard to justify. The existence of a non-vanishing G-field in a space-time universe free from matter remains logically possible, according to these equations. And in so far as the G-field exists, space and time are not empty' (Ref. [42d], pp. 109–10).

The relativity congress in Berne was a big event [43], which allowed the younger participants like Thellung and myself from ETH and Walter Thirring and Michel Kervaire from Berne to see many illustrious personalities, specially during the boat trip on Lake Thun. Of particular interest there were the Russian participants, the mathematician Paul Alexandroff and the physicist Vladimir A. Fock, accompanied by a quite jovial party secretary. In this year of 1955 Walter Thirring became a Privatdozent in Berne, after having been assistant to Mercier in 1953–4. Thirring had become a visiting professor in the United States in 1956 and an associate professor in Berne in 1958. In 1959 he was elected full professor in his native Vienna (see the comment in Ref. [7], pp. 775–6).

André Mercier (1913–99) grew up in Geneva where he attended the prestigious Collège Calvin and where in 1922 he obtained his Ph.D. at the university. He continued his studies under Elie Cartan in Paris at the Institut Henri Poincaré and at the Collège de France. After a stay at Bohr's institute in Copenhagen and an assistantship in mechanics at ETH he became Weiglé's assistant in Geneva (see page 290), then a Privatdozent, and in 1939 a professor in Berne. Mercier was a highly educated man with considerable social activities. At Berne he held the position of Dean of the Science Faculty for a year, and in 1967–8 he was rector of the university. He was a member of many national and international learned societies and received many honours [44a].

But the dominant and colourful personality in physics in Berne at that time was Friedrich Georg Houtermans (1903–66), an old friend of Pauli's (see pages 220 and 314–5). Houtermans became a professor in Berne in 1952. He inherited an institute which he described to Casimir as follows: 'If you want to see an authentic early-twentieth-century laboratory, come and visit me. ... But you have to come soon, for I am going to change all that.' Here is Casimir's description: 'A little tower with a winding staircase was part of the physics laboratory. The heavy central column bore on its top the zeropoint of the Swiss geodesic survey. It was also supposed to provide a vibrationless support but, according to Houtermans, the stability left something to be desired. Hence his remark: "When I stand in my institute and push I can wiggle the whole of Switzerland"' (Ref. [44b], p. 223).

Continuing the story from page 315, Houtermans, alias Fisl, after his liberation in Germany, divorced Charlotte, alias Schnax, and married another woman, with whom he also had children. But in Berne he got divorced again and remarried Schnax who had returned from the United States; Pauli again was the Best Man. Pauli called this situation 'the Houtermans statistics: the uneven women are identical'. However, after some time Schnax realized that they had lived too long apart and returned to the United States. Then Fisl married wife number 4 [44c,d]. Houtermans was a chain-smoking, hard-drinking story-teller and a brilliant scientist who initiated geophysics and satellite-related research at the institute in Berne.

About the same time that Pauli's interest in the possible detection of the neutrino had been aroused by Reines and Cowan's announcement [36a], Houtermans and Thirring in Berne set out to estimate the mean free path of the neutrino by calculating the cross-section σ for the scattering of antineutrinos $\bar{\nu}$ on electrons e^-. They first noted that a more efficient scattering process than (11.21) is the reaction

$$\bar{\nu} + e^- \to \bar{\nu}_\mu + \mu^-, \tag{11.23}$$

applied to second order. However, since this calculation diverges when the Fermi interaction (6.63) is used, the authors replaced it by a square-well potential with a width given by the Compton wavelength of the muon and a depth proportional to g^2. In ordinary matter of mass density $\sim 1\text{g}/\text{cm}^3$ the result was a mean free path of $\sim 10^{23}$cm. But the authors observed that by assigning to the neutrino a magnetic moment with the reasonable value $10^{-10}\mu_B$, the electromagnetic interaction shortened the mean free path to 10^{20}cm [44e].

Later in 1955, from 17 to 23 October, ETH celebrated its 100th anniversary with a multitude of events (see Ref. [21], p. 194). The official ceremony was held on the 21st; it began with a welcome address by Rector Karl Schmid, professor of German literature (the husband of the Zurich actress Elsie Attenhofer of the *Cabaret Cornichon*, see pages 209 and 337) and was followed by a speech by the President of the Swiss Government, Bundesrat Max Petitpierre. In the academic ceremony on 22 October, 22 honorary doctor diplomas were distributed. Among the recipients were Bundesrat Philipp Etter (see pages 283 and 337), Henri Cartan (Paris), Theodore von Kármán (Caltech), the three Chemistry Nobel Prize winners Peter Debye (Cornell University), Paul Karrer (Zurich University, who was host to Pauli's father during the war, see pages 10–11) and Hermann Staudinger (University of Freiburg, Germany), and the architect Frank Lloyd Wright (USA). On the 20th Pauli and Scherrer also gave talks in the framework of the academic training courses organized by the Society of Former Polytechnicians (see Ref. [21], III.154 and p. 194).

Pauli spoke on 'Problems of today's physics'. One of the participants, Wilhelm Frank, a former student of Pauli's, recalls: 'Contrary to his usual habit Pauli, probably in order to conform to the special occasion, had prepared a manuscript. But, obviously, he had difficulties adhering to the text. The tension in the overfilled big lecture hall at Gloriastrasse rose with Pauli's pains to follow the manuscript. Suddenly he threw away the manuscript and began to talk freely, not always grammatically perfect and stylistically elegant but with great impressiveness. The core of his considerations was the so-called fine-structure constant. He illustrated the task, to accept the so-designated coupling constant ... not as mere quantity of measurement but to understand it as one of the actual main problems of theoretical physics.—Pauli's performance was honoured by an abundant applause' (translated from Ref. [21], p. 194). For this celebration a memorial volume describing the institutes and personalities of ETH during its 100 years of existence was produced [44f]. And the *Neue Zürcher Zeitung* honoured the event in its edition of 21 October with a presentation of the eight recipients of Nobel Prizes of whom Pauli was the last.

In October 1955 Pauli was elected a member of the Swedish Academy of Sciences as successor to Einstein. For this event President Pallmann, on 18 October, sent Pauli his congratulations (Ref. [21], III.153). And on 9 November the president sent Pauli a condolatory message on the death of his father who had died in Zurich on the 4th (see page 11 and Ref. [21], III.155). From 29 November to 1 December Pauli spent three days in Hamburg where he had not been for a long time and where on 30 November he repeated his Mainz talk. It was the memorable visit described on page 146, which gave rise to the romantic encounter with the woman he had not seen for 30 years. In the long letter [69] of 23 October 1956 to Jung with the heading 'Statements of the psyche' Pauli muses on this event: 'But as with a tale by E. Th. A. Hoffmann, it seemed to me that in parallel an inner, fairy, archetypal action was playing. I particularly reflected on the "return of the soul" [which is possible only if the feeling is not impaired by the thinking; see Ref. [45], Section 488, also Section 503]; by the way, on 29th November there was a full moon' (translated from Ref. [4], p. 150). This visit to Hamburg is described in great detail in Pauli's letters [2214] of 13 December 1955 to Paul Rosbaud and [2253] of 2 March 1956 to Fierz in Ref. [34] (see also the comment referring to this visit in Ref. [34], pp. 428–9).

Notes and References

[1] Translated from: W. Pauli, letter [1239] to M.-L. von Franz in Ref [2], p. 306.

[2] K. von Meyenn (ed.), *Wolfgang Pauli. Scientific Correspondence with Bohr, Einstein, Heisenberg, a.o., Volume IV, Part I: 1950–1952* (Springer, Berlin and Heidelberg, 1996).

[3] (a) C. G. Jung, 'Antwort auf Hiob', in Ref. [3b], pp. 361–471. English translation, 'Answer to Job', in Ref. [3c], pp. 355–470 ; (b) C. G. Jung, *Zur Psychologie westlicher und östlicher Religion, Gesammelte Werke, 11* (Walter-Verlag, Olten, 5. vollständig revidierte Auflage 1988). (c) English translation: *Psychology and Religion: West and East, Collected Works, 11* (Princeton University Press, Princeton, NJ, 1969), 2nd edn.

[4] C. A. Meier (ed., with the cooperation of C. P. Enz and M. Fierz), *Wolfgang Pauli und C. G. Jung. Ein Briefwechsel 1932–1958* (Springer, Berlin and Heidelberg, 1992).

[5] (a) H. van Erkelens, 'Pauli und Jungs "Antwort auf Hiob"', in Ref. [5b], pp. 67–88; (b) H. Atmanspacher, H. Primas, and E. Wertenschlag-Birkhäuser (eds), *Der Pauli-Jung-Dialog und seine Bedeutung für die moderne Wissenschaft* (Springer, Berlin and Heidelberg, 1995).

[6] E. Jung and M.-L. von Franz, *Die Graalslegende in psychologischer Sicht* (Rascher Verlag, Zurich, 1960).

[7] K. von Meyenn (ed.), *Wolfgang Pauli. Scientific Correspondence with Bohr, Einstein, Heisenberg a.o., Volume IV, Part II: 1953–1954* (Springer, Berlin and Heidelberg, 1999).

[8] (a) *Note*: Pauli's horoscope has been interpreted for me in an extensive report by Ms Theres Lötscher. In its essence this report confirms the features of Pauli's life and character as described in this biography. I thank Ms Lötscher for her kind collaboration and correspondence. I also thank her and her husband Bernhard for innumerable discussions on Pauli and Jung, (b) *Note*: A better astrological explanation of Pauli's sensitivity to the equinoxes might be the moon-lilith opposition on the equinoxial axis in his horoscope, "Lilith" being the outer focus of the moon's elliptic orbit.

[9] C. G. Jung, *Aion. Beiträge zur Symbolik des Selbst, Gesammelte Werke, 9,II* (Walter Verlag, Olten, 8. Auflage 1992); English translation, *Aion. Researches into the Phenomenology of the Self, Collected Works, 9,II* (Princeton University Press, Princeton, NJ, 1968) 2nd edn.

[10] W. Pauli, 'Die Klavierstunde. Eine aktive Phantasie über das Unbewusste. Frl. Dr. Marie-Louise v. Franz in Freundschaft gewidmet', Ref. [5b], pp. 317–30. Reprinted in Ref. [7], pp. 330–40.

[11] (a) A. Pauly, *Darwinismus und Lamarckismus. Entwurf einer psychophysischen Teleologie* (Reinhardt, Munich, 1905); (b) J. B. Rhine, *Extra-Sensory Perception* (Bruce Humphries, Boston, 1934); (c) U. Müller-Herold, 'Vom Sinn im Zufall: Überlegungen zu Wolfgang Paulis "Vorlesung an die fremden Leute"', in Ref. [5b], pp. 159–77; (d) N. Bohr, 'Licht und Leben', *Naturwissenschaften* **21**, 245 (1933). English translation, 'Light and Life', *Nature* **131**, 421 (1933); (e) M. Delbrück, 'A Physisicst's Renewed Look at Biology: Twenty Years Later', *Science* **168**, 1312 (1970).

[12] (a) W. Pauli, 'Naturwissenschaftliche und erkenntnistheoretische Aspekte der Ideen vom Unbewussten', *Dialectica* **8**, 283 (1954). Reprinted in Ref. [13a], pp. 113–28 and in Ref. [13c], Vol. 2, pp. 1212–30. (b) English translation, 'Ideas of the Unconscious from the Standpoint of Natural Science and Epistemology', in Ref. [13b], pp. 149–64; (c) H. Kuhn, 'Selforganization of Nucleic Acids and the Evolution of the Genetic Apparatus', in: H. Haken (ed.), *Synergetics* (Teubner, Stuttgart, 1973), p. 157–77. Reprinted in: W. Hoppe, W. Lohmann, H. Markl, and H. Ziegler (eds), *Biologie—Ein Lehrbuch* (Springer, Heidelberg, 1977), p. 662.

[13] (a) W. Pauli, *Aufsätze und Vorträge über Physik und Erkenntnistheorie* (Vieweg, Braunschweig, 1961). Reprinted as '*Physik und Erkenntnistheorie*. Mit einleitenden Bemerkungen von Karl von Meyenn' (Vieweg, Braunschweig, 1984), (b) English translation: C. P. Enz and K. von

Meyenn (eds), *Wolfgang Pauli. Writings on Physics and Philosophy* (Springer, Berlin and Heidelberg, 1994); (c) R. Kronig and V. F. Weisskopf (eds), *Collected Scientific Papers by Wolfgang Pauli. In Two Volumes* (Wiley Interscience, New York, 1964).

[14] J. D. Watson and F. H. C. Crick, 'A structure of deoxyribose nucleic acid' *Nature* **171**, 737 (1953).

[15] (a) C. G. Jung, *'Mysterium coniunctionis.* Dritter Teil: *Aurora consurgens.* Ein dem Thomas von Aquin zugeschriebenes Dokument der alchemistischen Gegensatzproblematik, von Dr. M.-L. von Franz' (Rascher, Zurich, 1957); (b) M.-L. von Franz, *'Der Traum des Descartes.* Zeitlose Dokumente der Seele' (Studien a.d. C. G. Jung Institut, Zurich, 1952); (c) M.-L. von Franz, *Zahl und Zeit* (Klett, Stuttgart, 1970; Suhrkamp Taschenbuch, 1980).

[16] M. Schoch, 'Auf der Suche nach dem grossen Einen'. Marie-Louise von Franz—Porträt einer Psychologin, *Neue Zürcher Zeitung*, 5 February 1996, Nr. 29.

[17] A. Hermann, J. Krige, U. Mersits, and D. Pestre, *History of CERN*, Volume I: *Launching the European Organization for Nuclear Research* (North-Holland, Amsterdam, 1987).

[18] W. Pauli, letter [1614] of 25 July 1953 to Pais, in Ref. [7].

[19] (a) A. Pais, 'Isotopic Spin and Mass Quantization', *Physica* **19**, 869 (1953). (b) *Note*: The issue of *Physica*, in which Pais' paper appeared contained, inserted facing page 744, a photograph of 52 of the participants identified by name.

[20] (a) K. von Meyenn (ed., with the cooperation of A. Hermann and V. F. Weisskopf), *Wolfgang Pauli. Scientific Correspondence with Bohr, Einstein, Heisenberg, a.o., Volume II: 1930–1939* (Springer, Berlin and Heidelberg, 1985); (b) E. Schrödinger, 'Diracsches Elektron im Schwerefeld I', *Sitzungsberichte der Preussischen Akademie der Wissenschaften 1932. Phys.-Math. Klasse*, p. 105; (c) 'Probleme der Allgemeinen Relativitätstheorie. Vorlesung, gehalten an der Eidg. Techn. Hochschule, Zürich, 1953, von Prof. Dr. W. Pauli. Ausgearbeitet von Ch. Enz' (handwritten, 111 pp.), *Pauli Archive*, Section 10.

[21] C. P. Enz, B. Glaus, and G. Oberkofler (eds), *Wolfgang Pauli und sein Wirken an der ETH Zürich* (vdf Hochschulverlag ETH, Zurich, 1997).

[22] (a) L. O'Raifeartaigh, *The Dawning of Gauge Theory* (Princeton University Press, Princeton, NJ, 1997); see also: L. O'Raifeartaigh and N. Straumann, 'Gauge theory: Historical origins and some modern developments', *Reviews of Modern Physics* **72**, 1 (2000); (b) A. Pais, *A Tale of Two Continents. A Physicist's Life in a Turbulent World* (Princeton University Press, Princeton, NJ, 1997); (c) A. Pais, *'Subtle is the Lord . . .' The Science and the Life of Albert Einstein* (Oxford University Press, Oxford, 1982); (d) A. Pais, *Inward Bound. Of Matter and Forces in the Physical World* (Oxford University Press, New York, 1986); (e)

O. Klein, 'Sur la théorie des champs associés à des particules chargés', in: *Les nouvelles théories de la physique. Varsovie, 30 mai–3 juin 1938* (Institut international de coopération intellectuelle, Warsaw, 1938), pp. 81–98.

[23] (a) S. Weinberg, 'A Model of Leptons', *Physical Review Letters* **19**, 1264 (1967); (b) A. Salam, 'Elementary Particle Physics', in: N. Svartholm (ed.), *Proc. 8th Nobel Symp.* (Almqvist and Wiksell, Stockholm, 1968), p. 367. Reprinted in: C. H. Lai (ed.), *Gauge theory of weak and electromagnetic interaction* (World Scientifc, Singapore, 1981), (a): p. 185; (b): p. 188; (c) C. Rubbia, 'Physics results of the UA1 Collaboration at the CERN proton-antiproton collider', in: K. Kleinknecht (ed.), *Proc. Int. Conf. on Neutrino Physics and Astrophysics* (Dortmund, 1984), p. 1; (d) Particle Data Group, *Particle Physics Booklet* (American Institute of Physics, New York, 1996), pp. 6–7.

[24] (a) C. N. Yang and R. L. Mills, 'Conservation of Isotopic Spin and Isotopic Gauge Invariance', *Physical Review* **96**, 191 (1954); (b) C. N. Yang and R. L. Mills. 'Isotopic Spin Conservation and a Generalized Gauge Invariance', *Physical Review* **95**, 631 (1954); (c) C. N. Yang, *Selected Papers, 1945–1980 with Commentary* (Freeman, San Francisco, 1983).

[25] (a) *American Men & Women of Science. 1989–90. 17th Edition* (Bowker, New York, 1989); (b) *Who's Who in Atoms* (Harrap Research Publications, London, 1965).

[26] (a) W. Pauli, 'On the Hamiltonian Structure of Non-Local Field Theories', *Nuovo Cimento* **10**, 648 (1953). Reprinted in Ref. [13c], Vol. 2, pp. 1176–95; (b) P. Gulmanelli, 'On a theorem in non-local field theories', *Nuovo Cimento* **10**, 1582 (1953).

[27] P. Gulmanelli, (a) *Su una teoria dello spin isotopico* (Casa editrice Pleion, Milano, no date); (b) letter to C. P. Enz, dated 29 November 1984, Pavia. I thank Professor Gulmanelli for this correspondence.

[28] (a) W. Pauli, 'Wahrscheinlichkeit und Physik', *Dialectica* **8**, 112 (1954). Reprinted in Ref. [13a], pp. 18–23 and in Ref. [13c], Vol. 2, pp. 1199–211. (b) English translation, 'Probability and Physics', in Ref. [13b], pp. 43–8.

[29] 'Minutes of the Assembly of the Swiss Physical Society', *Helvetica Physica Acta*: (a) **26**, 377 (1953); (b) **28**, 297 (1955).

[30] (a) K. von Meyenn (ed.), *Wolfgang Pauli. Scientific Correspondence with Bohr, Einstein, Heisenberg, a.o., Volume IV, Part III: 1955–1956* (Springer, Berlin and Heidelberg, 2001); (b) A. Thellung, 'Höhere mesontheoretische Näherungen zum magnetischen Moment des Protons', *Helvetica Physica Acta* **25**, 307 (1952).

[31] (a) C. de Montet, *Evolution vers l'essentiel* (Rouge, Lausanne, 1950); (b) 'Dr. Charles de Montet' (Necrology), *Gazette de Lausanne*, 24 July 1951.

[32] C. G. Jung, *Von den Wurzeln des Bewusstseins* (Rascher, Zurich, 1954); English translation, *Transformation Symbolism in the Mass*. Reprinted in Ref. [3b], pp. 217–310; English translation, Ref. [3c], pp. 296–448.

[33] W. Pauli, (a) 'Rydberg and the Periodic System of Elements', *Proc. Rydberg Centennial Conf. on Atomic Spectroscopy* (Universitetes Årsskrift, Lund, 1955), Vol. 50, no. 21. Reprinted in Ref. [13b], pp. 73–7 7 and in Ref. [13c], Vol. 2, pp. 1231–5. German translation, 'Rydberg und das periodische System der Elemente', in Ref. [13a], pp. 43–7; (b) 'Phänomen und physikalische Realität', *Dialectica* **11**, 35 (1957). Reprinted in Ref. [13a], pp. 93–101 and in Ref. [13c], Vol. 2, pp. 1350–61. English translation, *Phenomenon and Physical Reality*, in Ref. [13b], pp. 127–35 ; (c) 'Die Wissenschaft und das abendländische Denken', in: M. Göhring (ed.), *Europa—Erbe und Auftrag*. Internationaler Gelehrtenkongress, Mainz 1955 (Steiner Verlag, Wiesbaden, 1956), pp. 71–9. Reprinted in Ref. [13a], pp. 102–12 and in Ref. [13c], Vol. 2, pp. 1290–8. English translation, 'Science and Western Thought', in Ref. [13b], pp. 137–48.

[34] K. von Meyenn (ed.), 'Wolfgang Pauli. Scientific Correspondence with Bohr, Einstein, Heisenberg, a.o., Volume IV, Part III: 1955–1956' (Springer, Berlin and Heidelberg, 2001).

[35] (a) W. Y. Evans-Wentz, *The Tibetan Book of the Great Liberation* (Oxford University Press, Oxford, 1955); (b) C. P. Enz, 'Wolfgang Pauli (1900–1958). A Biographical Introduction', in Ref. [13b], pp. 13–26, 19. The quotation is taken from Ref. [35c]. (c) W. Barker, Letter to *Physics Today*, February 1979, p. 11.

[36] F. Reines and C. L. Cowan, Jr, (a) 'A Proposed Experiment to Detect the Free Neutrino', *Physical Review* (letter) **90**, 492 (1953); (b) Detection of the Free Neutrino', *Physical Review* (letter) **92**, 830 (1953).

[37] (a) C. L. Cowan, Jr, F. Reines, F. B. Harrison, H. W. Kruse, and A. D. McGuire, 'Detection of the Free Neutrino: a Confirmation', *Science* **124**, 103 (1956); (b) F. Reines and C. L. Cowan, jun., 'The Neutrino', *Nature* **178**, 446 (1956).

[38] W. Pauli, (a) 'Zur älteren und neueren Geschichte des Neutrinos', Ref. [13a], pp. 156–80; (b) English translation, 'On the Earlier and More Recent History of the Neutrino', Ref. [13b], pp. 193–217.

[39] (a) F. Reines, 'The Early Days of Experimental Neutrino Physics', *Science* **203**, 5 January 1979, p. 11; (b) Frederick Reines, letter to C. P. Enz, dated Irvine, California 92717, 11 April 1986; (c) W. Pauli, 'Announcement', *CERN Symp. on High Energy Accelerators and Pion Physics, Geneva, 11–23 June 1956 (CERN Proceedings*, 1956), Vol. 2, p. 259. Reprinted in Ref. [13c], Vol. 2, p. 1312 and in Ref. [39d], p. 480. (d) C. P. Enz and K. von Meyenn (eds), *Wolfgang Pauli. Das Gewissen der Physik* (Vieweg, Braunschweig, 1988).

[40] (a) R. Zimmerman, 'The Shadow Boxer', *The Sciences*, January/February 1996, pp. 16–19; (b) R. Davis, Jr, 'Attempt to Detect the Antineutrinos

from a Nuclear Reactor by the $Cl^{37}(\bar{\nu}, e^-)A^{37}$ Reaction', *Physical Review* **97**, 766 (1955); (c) R. Davis, Jr, 'An Attempt to Detect the Neutrinos from a Nuclear Reactor by the $Cl^{37}(\bar{\nu}, e^-)A^{37}$ Reaction', *Bulletin of the American Physical Society* (1956), UA5, p. 219; (d) M. M. Waldrop, 'Supernova 1987A: Notes from All Over', *Science* **236**, 522 (1987); (e) 'We hear that', *Physics Today*, March 2000, p. 91; (f) T. Ypsilantis 'The Hellaz Neutrino Detector', *Europhysics News* **27**, 97 (1996); (g) E. Majorana, 'Teoria simmetrica dell'elettrone e del positrone', *Nuovo Cimento* **14**, 171 (1937); (h) M. van der Heijden and M. de Jong, 'Neutrino oscillations NOW', *CERN Courier*, November 1998, p. 17; (i) K. Winter (ed.), *Neutrino Physics* (Cambridge University Press, Cambridge, 2000); (j) P. Ramond, *Journeys Beyond the Standard Model* (Perseus Books, Cambridge, Mass., 1999), Chapter 8.

[41] (a) W. Pauli, 'Impressionen über Albert Einstein', *Neue Zürcher Zeitung*, 22 April 1955. Reprinted in Ref. [13a], pp. 81–4 and in Ref. [13c], Vol. 2, pp. 1237–8; English translation, 'Impressions of Albert Einstein', in Ref. [13b], pp. 113–16. (b) Notes of my telephone conversation with Hilde Mutschler on 13 January 1988. (c) E. Morgenthaler, *Ein Maler erzählt* (Diogenes Verlag, Zurich, 1957).

[42] W. Pauli, (a) 'Opening Talk', Ref. [43], p. 27; (b) 'Schlusswort durch den Präsidenten der Konferenz', Ref. [43], pp. 261–7; (c) 'Relativitätstheorie und Wissenschaft', Ref. [43], pp. 282–6. Reprinted in Ref. [13c], Vol. 2, (a): p. 1299; (b): pp. 1300–6; (c): pp. 1307–11. (b) is also reprinted in Ref. [39d], pp. 153–9; (c) is also reprinted in Ref. [13a], pp. 76–80. (d) English translation of (c), 'The Theory of Relativity and Science', in Ref. [13b], pp. 107–11.

[43] A. Mercier and M. Kervaire (eds), *Fünfzig Jahre Relativitätstheorie*, *Helvetica Physica Acta*, Supplement IV (1956).

[44] (a) J. Lacki, 'André Mercier (1913–1999)', *Archives des Sciences* (Geneva) **53**, 69 (2000); (b) H. B. G. Casimir, *Haphazard Reality. Half a Century of Science* (Harper & Row, New York, 1983); (c) I. B. Kriplovich, 'The eventful life of Fritz Houtermans', *Physics Today*, July 1992, p. 29; (d) Notes of my conversation with Franca Pauli on 6 April 1971; (e) F. G. Houtermans and W. Thirring, 'Zur freien Weglänge von Neutrinos', *Helvetica Physica Acta* **27**, 81 (1954); (f) *Eidgenössische Technische Hochschule—Ecole Polytechnique Fédérale, 1855–1955* (Buchverlag der Neuen Zürcher Zeitung, Zurich, 1955).

[45] C. G. Jung, 'The Psychology of the Transference' (1946), in: C. G. Jung, *Collected Works* (Princeton University Press, Princeton, NJ, revised and augmented edn, 1966), Vol. 16.

12
The symmetry breaks

<hr>

Dreams of symmetry, the *CPT*-theorem and 'ghosts'

In a dream in the Spring of 1951 the word 'Automorphism' (taken from mathematics) came flying towards me. This is a word for the picturing of a system on itself, a reflection of the system into itself, for a process, that is, in which the inner symmetry, the connexional richness (relations) of a system reveals itself. In abstract algebra there also are 'elements engendering the automorphism' . . . and to those, very likely, correspond in the analogy the 'archetypes' as arranging factors, as you have defined and conceived them in 1946. [1]

In the above quotation from letter [55] of 27 February 1952 to Jung, Pauli has in mind Jung's definition of the notion of archetype, as given in Jung's *Psychology of the Transference* of 1946 [4a], e.g. in Section 354. And referring to Chapter XIV of Jung's *Aion* [4b], he comments on this quotation as follows: 'I have understood the dream of that time (it was a real exam, with the "stranger" [see page 463] as examiner, in which the word "automorphism" has acted like a "mantra" [i.e. a magic word or spell, see letter [1219] in Ref. [3]]) in such a way that a *concept on a higher level [Oberbegriff]* was searched which should comprehend as well your notion of archetype as also that of the physical law of nature' (translated from Ref. [2], p. 81, also Ref. [3], p. 560). In Chapter XIV of *Aion* [4b] Jung, in Section 410, draws a square with sides A, B, C, D, whose corners are replaced by little squares with corners a, b, c, d and a_n, b_n, c_n, d_n, $n = 1, 2, 3$, a figure permitting a multitude of automorphisms. Jung gives the following comment on this figure: 'The formula reproduces exactly the essential features of the symbolic process of transformation. . . . What the formula can only hint at, however, is the higher plane that is reached through the process of transformation and integration' (Ref. [4a], Section 410). An automorphism may indeed be viewed as such a transformation (see letter [1285] to von Franz in Ref. [3]).

The symmetry operation of reflection, in particular, often occurred in Pauli's dreams, e.g. in the egg dream of March 1948 (see page 424), picture 4, on which Pauli comments as follows: 'This change consists in the appearance of the fourfoldness which occurs by reflection of the twofoldness' (translated from Ref. [2], p. 191). The same idea is expressed in the interpretation of the dream described in Pauli's letter [60] of 31 March 1953 to Jung: 'Through "reflection" of the consciousness from this [the "same" and the "other"] arises the quaternity' (translated from Ref. [2], p. 109). Reflection also occurs with the leaf of the trefoil (see the picture on p. 148 in Ref. [2]) in the dream of 20 May 1955, which in his interpretation appears to Pauli *as lower reflected image of the three popes* of the earlier dream (20 July 1954)' (translated from Ref. [2], p. 148). And on 27 November 1954 Pauli dreamt of a 'dark woman' who was with him in a room where experiments were performed, which produced 'reflexes'. While the other people in the room consider these reflexes as 'real objects', the 'dark woman' and Pauli know that they are only 'reflections'. This creates a secret which separates Pauli and the 'dark woman' from the other people. But the secret fills them with fear (Ref. [2], p. 161).

For the period from 4 January to 18 April 1954 Pauli asked again for leave to go to Princeton (see Ref. [5], III.117, 122, 123). At the Princeton Institute Pauli encountered a new problem, as he reports to Oskar Klein on 25 December 1954: 'Beginning of April of this year (shortly before my trip back to Zurich and Källén's trip back to Lund), (when) the young, very gifted Chinese T. D. Lee there held a lecture on his example for renormalizable fields [6] which we all immediately realized to be a very instructive one (It is organized such that what elsewhere appears as "Tamm–Dancoff approximation", in this model is *exact*. In return he, however, then has to renounce the relativistic invariance.)... In the discussion after Lee's talk *Källén* contradicted him very youthful-vehemently. But *Lee* seemed to me to answer him well (very Chinese-calmly). My sympathy was with *both* of them and I immediately had the impression that here there yet was something to be cleared up. Based on Källén's argument my impression immediately went in the direction that, probably, in Lee's model the S-matrix is not generally unitary and that he had fallen into the trap of an indefinite metric ... without noticing it' (translated from Ref. [7], letter [1954], p. 942). This is the Lee model [6], on which Pauli and Källén exchanged many letters in the autumn of 1954 (see also the comment on pp. 913–14 in Ref. [7]), and about which they wrote the paper [8], dedicating it to Bohr on his 70th birthday.

In Ref. [8] the model is defined by a Hamiltonian H_0 describing heavy fermions V and N and light bosons θ interacting via H_{int},

$$H_0 = \sum_{\mathbf{p}} \{E_V(\mathbf{p})\psi_V^+(\mathbf{p})\psi_V(\mathbf{p}) + E_N(\mathbf{p})\psi_N^+(\mathbf{p})\psi_N(\mathbf{p})\} + \sum_{\mathbf{k}} \omega a_{\mathbf{k}}^+ a_{\mathbf{k}},$$

$$(12.1)$$

$$H_{int} = -g_0 \sum_{\mathbf{pk}} \frac{f(\omega)}{\sqrt{2V\omega}} \{\psi_V^+(\mathbf{p})\psi_N(\mathbf{p}-\mathbf{k})a_{\mathbf{k}} + a_{\mathbf{k}}^+\psi_N^+(\mathbf{p}-\mathbf{k})\psi_V(\mathbf{p})\};$$

$$(12.2)$$

to this a mass renormalization term

$$\delta H = -\delta m \sum_{\mathbf{p}} \psi_V^+(\mathbf{p})\psi_V(\mathbf{p})$$

$$(12.3)$$

is added (see also the appendix to letter [1900] in Ref. [7]). While $E_V(\mathbf{p})$, $E_N(\mathbf{p})$, and $\omega(\mathbf{k})$ may, in principle, be any functions, they are chosen to have the values m, m, and $\sqrt{k^2 + \mu^2}$, respectively, in order to avoid inessential complications. The cut-off $f(\omega)$ in Eq (12.2) is introduced to keep sums finite. Quantization is introduced by the usual commutation and anticommutation relations .

The reactions $V \leftrightarrow N + \theta$ described by H_{int} imply that the total particle numbers $N_I = \sum_{\mathbf{p}} n_I(\mathbf{p})$, $I = V, N$ and $N_\theta = \sum_{\mathbf{k}} n_{\mathbf{k}}$, where $n_I(\mathbf{p}) = \psi_I^+(\mathbf{p})\psi_I(\mathbf{p})$ and $n_{\mathbf{k}} = a_{\mathbf{k}}^+ a_{\mathbf{k}}$, give rise to the conserved quantities

$$Q_1 \equiv N_V + N_N; \quad Q_2 \equiv N_N - N_\theta, \tag{12.4}$$

so that

$$[H, Q_i] = 0; \quad i = 1, 2. \tag{12.5}$$

Eigenstates of H_0 are of the form $|n_V n_N n_{\mathbf{k}}\rangle$, while eigenstates of H are characterized by definite values of the operators (12.4). Thus $Q_1 = Q_2 = 0$ defines the physical vacuum $|0\rangle$, and $Q_1 = 1$, $Q_2 = 0$ corresponds to the two types of states $|1_V, 0, 0\rangle$ and $|0, 1_N, 1_{\mathbf{k}}\rangle$. Hence, the physical one–V-particle state has the form

$$|V(\mathbf{p})\rangle = C\{|1_V(\mathbf{p}), 0, 0\rangle + \sum_{\mathbf{k}} \phi_{\mathbf{k}} |0, 1_N(\mathbf{p} - \mathbf{k}), 1_{\mathbf{k}}\rangle\}; \tag{12.6}$$

it is an eigenstate of H,

$$H|V\rangle = (m + \omega_0)|V\rangle. \tag{12.7}$$

Inserting Eqs (12.1)–(12.3) and (12.6) into (12.7) one finds the two conditions

$$\delta m + \omega_0 = -g_0 \sum_{\mathbf{k}} \frac{\phi_{\mathbf{k}} f(\omega)}{\sqrt{2V\omega}} \tag{12.8}$$

and

$$(\omega - \omega_0)\phi_{\mathbf{k}} = g_0 \frac{f(\omega)}{\sqrt{2V\omega}}. \tag{12.9}$$

Eliminating ϕ from these two equations yields

$$\delta m + \omega_0 = -\frac{g_0^2}{2V} \sum_{\mathbf{k}} \frac{f^2(\omega)}{\omega(\omega - \omega_0)}. \tag{12.10}$$

In particular, $\omega_0 = 0$ for $\mathbf{p} = 0$, and hence

$$\delta m = -\frac{g_0^2}{2V} \sum_{\mathbf{k}} \frac{f^2(\omega)}{\omega^2}. \tag{12.11}$$

Inserting in (12.6) $\phi_{\mathbf{k}}$ from (12.9) the normalization factor is found to be given by

$$C^{-2} = 1 + \frac{g_0^2}{2V} \sum_{\mathbf{k}} \frac{f^2(\omega)}{\omega^3}, \tag{12.12}$$

which also determines the scaling factor

$$N \equiv \langle 0|\psi_V(\mathbf{p})|V(\mathbf{p})\rangle = C\langle 0|0\rangle = C. \tag{12.13}$$

Introducing a rescaled coupling constant

$$g = Ng_0, \tag{12.14}$$

the combination of the last three relations yields

$$N^2 = 1 - \frac{g^2}{g_{crit}^2}, \tag{12.15}$$

where

$$g_{crit}^{-2} \equiv \frac{1}{2V} \sum_{\mathbf{k}} \frac{f^2(\omega)}{\omega^3}, \tag{12.16}$$

or with (12.14)

$$\frac{1}{g^2} = \frac{1}{g_0^2} + \frac{1}{g_{crit}^2}, \tag{12.17}$$

so that $g^2 < g_{crit}^2$, provided that g_0 is real.

Eliminating δm from Eqs (12.10), (12.11) the result may be written

$$\frac{1}{g^2} h(\omega_0) \equiv \omega_0 \left\{ \frac{1}{g_0^2} + \frac{1}{2V} \sum_{\mathbf{k}} \frac{f^2(\omega)}{\omega^2(\omega - \omega_0)} \right\} = 0. \tag{12.18}$$

Källén and Pauli transform this expression further using Eqs (12.17), (12.16), which leads to their Eq. (36), namely

$$h(\omega_0) \equiv \omega_0 \left\{ 1 + \frac{g^2}{2V} \sum_{\mathbf{k}} \frac{f^2(\omega)\omega_0}{\omega^3(\omega - \omega_0)} \right\} = 0. \tag{12.19}$$

However, the form (12.18), in which g_0^2 is the parameter, is more appropriate for a discussion of the zeros of $h(\omega_0)$ since it is g_0^2 that is considered to allow for negative values. In this case, $g_0^2 < 0$, it is easy to see from Eq. (12.18) that there is an extra zero $\omega_0 < \mu$, which for large enough $|g_0^2|$ is $-\lambda < 0$, and, as Källén and Pauli note, 'The corresponding eigenstate is not a scattering state, but will represent another state of the V-particle' (Ref. [8], p. 11). The important feature of this state $|V_{-\lambda}\rangle$ is that its norm is negative, so that an *indefinite metric* (see page 378) has to be introduced (see Ref. [8], p. 12).

It was in June 1954 that Pauli turned his attention to the discrete symmetries of reflections which, as mentioned at the outset, had been anticipated in his dreams. After having discussed time reflection in his *Les Houches* lectures of 1952 (see page 451), Pauli had sent a copy to Gerhard Lüders (1920–95), a pupil of Heisenberg's in Göttingen (see Lüders' letter [1776] of 26 April 1954 in Ref. [7] and the appendix; see also the comment in Ref. [7], pp. 581–4). During his stay at the Theory Division of CERN in Copenhagen and also in Geneva in 1953, Lüders had written the paper [10a], showing that in a relativistic field theory with an interaction consisting of completely symmetrized products of Bose fields and completely antisymmetrized products of Fermi fields the operations of time reversal T and charge conjugation C are equivalent. In his letter [1832] of 16 June 1954 Pauli considers Lüders' basic idea to be correct. But he insists that time reversal should be applied simultaneously with space reflection or parity P. And he emphasizes the close connection of the problem with his spin-statistics paper [10b] of 1940 (see pages 331–4). On 30 June Lüders wrote to Pauli: 'With great interest I have learned that in any field theory which is covariant under the proper Lorentz group a four-dimensional reflection including charge reversal may be defined' (translated from letter [1838], Ref. [7]). This is the CPT-theorem.

Stimulated by his 1940 paper and by Lüders' work, Pauli now wrote a paper himself and sent a copy to Lüders with his letter [1914] of 10 November 1954. This paper, 'Exclusion Principle, Lorentz Group and Reflection of Space-Time and Charge' [11], Pauli again dedicated to Bohr on his 70th birthday. But the *'Dedication'* in Ref. [11] has a more personal touch; it recalls the 'long and still continuing common pilgrimage since the year 1922, in which so many stations are involved'. In particular Pauli hints at the struggle of arriving at a satisfactory interpretation of the new formalism of quantum mechanics: 'After a brief period of spiritual and human confusion, caused by a provisional restriction to "Anschaulichkeit", a general agreement was reached following the substitution of abstract mathematical symbols, as for instance psi, for concrete pictures' (Ref. [11], p. 30). And in a footnote he refers 'to Bohr's favourite verses of Schiller' from *Sprüche des Konfuzius* to justify the 'richer "fulness" of plus and minus signs'—which is naturally associated with a study of reflections—by the gain of 'an increasing "clarity"' (Ref. [11], p. 31).

In this paper [11] Pauli considers three types of reflections, C, PT, and CPT, which he calls *particle–antiparticle conjugation AC, weak reflection WR*, and *strong reflection SR*, respectively. And he notes that Lüders in his Copenhagen paper [10a] showed the equivalence of C and PT and hinted at a more general validity of CPT, adding 'His proof, however, is not easy to understand, because it uses unnecessary assumptions' (Ref. [11], p. 33). Under weak reflection PT, a charged scalar field ϕ and a spinor field ψ transform according to

$$\phi'(x) = \phi^+(-x); \ \phi^{+\prime}(x) = \phi(-x) \tag{12.20}$$

and

$$\psi'(x) = B^{-1}\overline{\psi}(-x); \ \overline{\psi}'(x) = \psi(-x)B, \tag{12.21}$$

respectively, where B in Eq. (12.21), which Pauli calls Ω, is the matrix in Eq. (8.40). Applied to the respective current densities (8.21), (8.22), and (6.32) one sees that there is no sign change. This confirms the fact that charge does not change sign under PT.

As Pauli emphasizes in Ref. [11], the more important case, however, is that of strong reflection CPT, which is already contained in his 1940 paper [10b]. Indeed, introducing the four classes ± 1 and $\pm \epsilon$ of page 332 and calling $2j = m$, $2k = n$, the classes $+1$ and -1 are *bosons* with m and n both even and both odd, respectively, while the classes $+\epsilon$ and $-\epsilon$ are *fermions* with m even, n odd and m odd, n even, respectively. For the boson and fermion cases, strong reflection CPT is governed by Eqs (8.80) and (8.82), respectively. In the paper [11] this is expressed, respectively, by the second and the first equation (T) on p. 35, which is also the same as the respective Eqs (31) on p. 49, namely

$$
\begin{aligned}
u'(n,m) &= -i(-1)^n u(n,m) = i(-1)^m u(n,m); \quad n+m \text{ odd} \\
u'(n,m) &= (-1)^n u(n,m) = (-1)^m u(n,m); \quad n+m \text{ even.}
\end{aligned}
\tag{12.22}
$$

Note that here $u(n,m)$ is an object with $(n+1)(m+1)$ components; thus $u(0,0)$ is a scalar, $u(1,0)$ and $v(0,1)$ are two-component spinors, and $u(1,1)$ is a vector. In the representation where γ^5 is diagonal (see the footnote in Ref. [11], p. 48) the first line of Eqs (12.22) applied to a Dirac spinor written as two spinors $u(1,0)$ and $v(0,1)$ then yields Eqs (27a) on p. 49,

$$
\psi'(x) = i\gamma^5 \psi(-x), \quad \overline{\psi}'(x) = i\overline{\psi}(-x)\gamma^5.
\tag{12.23}
$$

This result agrees with the relation Schwinger gives in footnote 8 of Ref. [12a].

In the application of the general result (12.22) to an interaction Hamiltonian containing a product of N fermion fields, a factor $(-i)^N(-1)^n$ appears, where $n = \sum_{k=1}^{N} n_k$. This gives rise to an extra factor $(-1)^\nu$ for $N = 2\nu$ and for $N = 2\nu + 1$. The conclusion to be drawn from this result is formulated by Pauli as follows: 'To dispose of this extra factor we have now to apply the *second quantization with anticommutators for Fermions. . . , commutators for Bosons. This enables us first to assume that all products of field quantities have to be antisymmetrized for all permutations of the Fermion fields and symmetrized for all possible positions of the Boson fields*' (Ref. [11], p. 50). Now, due to Schwinger's definition of the time-reversal operation adopted by Pauli, the strong reflection CPT in addition requires an overall inversion, i.e. a reading of a T-transformed formula from right to left. It turns out, however, that the mentioned sign $(-1)^\nu$ is just cancelled by this inversion. One is then led to the CPT-*theorem* which states that *with the above ordering prescription the general transformation* (12.22) '*is sufficient to guarantee the existence of an SR [strong reflection] for all L_4 invariant local field theories in which the ordering of the products is properly taken into account*' (Ref. [11], p. 50, italics added).

In the acknowledgements of Ref. [11] Pauli thanks R. Jost and G. Lüders 'for interesting discussions on the generality of the "strong reflection"', which had taken place 'in Les Houches during the summer of 1952'. Two years later Jost observed that if the proper Lorentz group L_4 is embedded in the complex orthogonal group, then the operation PT can be reached from the identity by a continuous path in the parameter space. The CPT-theorem then may be understood simply as an analytic continuation [12b].

But Pauli did not waste any time with this problem. In letter [1914] mentioned above he writes to Lüders: 'I myself now have once more switched to another field of work' (translated from Ref. [7], letter [1914]). Indeed, Pauli was back again at the Lee model which, however, began to irritate him. On 13 November 1954 he writes to Källén: 'At night I begin to sleep badly (with me this is seldom), namely because the "Lee model", i.e. the zero-cone, begins to excite me' (translated from Ref. [7], letter [1917]). In his previous letter [1913] Pauli had characterized this zero-cone as having—as a consequence of 'the wrong sign of N^2' in Eq. (12.15)—a hyperbolic basis containing the origin. Pauli eventually put his collaborator from England, Kurt Metzer, on the problem who, however, had to return to England in early Spring 1955 for health reasons (see letter [2092] in Ref. [13a]). Another collaborator came in April 1955, namely Werner Theis from Hamburg. He worked on meson theory, but he liked to discuss also the other problems treated in Pauli's group. He stayed in Zurich until October 1956 and we became good friends.

On Saint Nicholas's Day, 6 December 1954, the heading of Pauli's letter [1939] to Fierz reads: 'Sanniklaus. Have looked a bit into the glass. The onions send greetings!' Here the first word betrays Pauli's defective knowledge of Basle German; it should read 'Santiclaus', meaning Saint Nicholas. But this, as well as the 'onions' may be excused by Pauli's admission of having drunk 'a bit'. The following letter is written in the form of a dialogue with the title 'The Exam'. Its interest stems from the fact that it is the origin of the notion of ghost that Pauli has introduced into field theory to designate 'something that is left to exist only in the "renormalization" of the fields'. More precisely, it is 'An energy value which, with the aid of a *negative probability*, revenges itself for the injustice inflicted upon Nature and Mathematics by renormalization' (translated from Ref. [7], letter [1939]). And three days later Pauli writes Källén: 'I now consider it almost certain that the "ghosts" soon will be found in all mathematical papers in which theoretical physicists are "renormalizing"' (translated from Ref. [7], letter [1942]). Indeed, "ghosts" became quite ubiquitous in the analysis of such models.

At the Pisa conference on the elementary particles, which took place from 12 to 18 June 1955, Pauli gave a review of the different models [13b]. These were: the Lee model [6], the non-linear spinor equation of Heisenberg, namely the Dirac equation (6.31) with k_c replaced by $l^2\bar{\psi}\psi$, l being a length [14a], the photon pair theory, which is a QED with fixed extended charges, in which the linear field term is eliminated by a shift [14b], and the charged scalar meson pair theory [14c]. Pisa 'was very funny', Pauli writes. 'Heisenberg wanted to convince me by all means of his theory with foul mathematics' (translated from letter [2111] of 20 June 1955 to Fierz in Ref. [13a]).

It was at Pisa that Pauli officially introduced the notion of '*ghost*' for the 'new discrete stationary states whose contribution to the conserved sum of "probabilities" is negative' (Ref. [13b], p. 704). The two Hilbert spaces in Heisenberg's model, according to Pauli, were spanned by pairs of states characterized by positive and negative probabilities and for which, therefore, he found the terminology '*dipole ghost*' to be appropriate. In the photon pair theory, on the other hand, it was the eigenfrequency squared, Ω_0^2, which for $g_0 < 0$ acquired a discrete negative value (see Fig. 1 in Ref. [14b]). For this situation Pauli proposed in his Pisa paper [13b] the terminology '*ghost of the second kind*'. And he observed that: 'This negative square of an eigenfrequency also appears as a complex pole of the scattering S-matrix' (Ref. [13b], p. 708). This fact is implicit in Eqs (5.49), (5.42'), and (2.57), (2.58) of Ref. [14b].

In his Pisa paper [13b] Pauli quoted my work [14b] in advance as 'doctoral thesis in Zürich'. And in parallel he mentioned an earlier, more restricted treatment of non-relativistic QED by N. G. van Kampen [14d], a student of Kramers. Van Kampen already had defined the mass renormalization by the mentioned shift to eliminate the linear field term (Eqs (9) of Ref. [14d] and (2.59') of Ref. [14b]) and had given an expression that could be interpreted as self-energy (Eq. (16) of Ref. [14d], to be compared with Eq. (4.13) of Ref. [14b]). But what pleased Pauli in my work was that I too had found a 'ghost'. I submitted my thesis in November, as may be seen from Pauli's report to the ETH School Council, dated 25 November 1955 (Ref. [5], III.157). In this report Pauli writes: 'Already in the case of a single "electron" Enz shows here the impossibility of the transition to the limit of the local interaction. With a given finite value of the renormalized mass a "breakdown of the photon vacuum" occurs here already at a finite maximum value of the cut-off momentum (instead of the negative probabilities with the Lee model): The Hamiltonian does retain its Hermitian character, but it loses its positive definite one' (translated from Ref. [5], III.157). This work had been written in addition to my regular activity as theoretical collaborator in the Institute for Semiconductor Physics of Professor Georg Busch (1908–2000).

At the University, Heitler who knew of my work from the talk I had given in the common Monday-afternoon seminar, found my renormalization prescriptions interesting. In a short note devoted to these questions [14e], he observed that: 'if correct [they] should be derivable from the relativistic theory, at least in certain cases' (Ref. [14e], p. 303). This is what Heitler did in this reference. Walter Heitler was born in Karlsruhe and studied in Berlin and Munich. He spent 1926–7 in Copenhagen and Zurich on a Rockefeller Fellowship, after which he was assistant in Göttingen until 1929 and Privatdozent until 1933. From 1933 to 1941 Heitler was a research fellow at the University of Bristol, and then became a professor at the Dublin Institute for Advanced Studies. From 1946 to 1949 he was director of this institute when he received the call to the University of Zurich. Apart from his fundamental work on the homeopolar bond, in 1936 he wrote the well-known book on *The Quantum Theory of Radiation* [14f] which had several editions and translations into other languages. In his later years Heitler wrote books on epistemological and religious themes. He was a Fellow of the Royal Society of London, a member of the Royal Irish Academy, and he had an honorary doctorate from the University of Dublin.

Fig. 12.1 Pauli and his secretary Margaret Schmid

On 6 March Pauli wrote to President Pallmann proposing me as his assistant, to succeed Thellung on 1 April 1956 (Ref. [5], III.162, 163). This proposal was written by the secretary Margaret Schmid (b. 1923), but signed by Pauli in Princeton where he was staying from the beginning of January to mid-April (Ref. [5], III.160). He and Franca came back by ship, as he told me on 9 April in his first letter to his new assistant (see letter [2266] and also letter [2263] to Thellung in Ref. [13a]), in which he also mentioned that W. Krolikowski from Warsaw wished to join his group. In the first summer of my assistantship with Pauli, he and I stayed with our wives at the lovely Villa Monastero in Varenna on the shore of Lake Como, where from 15 July to 4 August, a *summer school on magnetism* [16a] took place. Pauli loved to attend the summer events of the Italian Physical Society where he was usually honoured by special attention. So it was this time: the president of the Society, Professor Giovanni Polvani (1892–1970), the director of the periodical *Il Nuovo Cimento* (see footnote 1 to letter [2343] in Ref. [13a]), publicly saluted the 'stimatissimo professore' (see Ref. [16a], p. 811). But while I gave a paper on the magnetic susceptibility of crystals [16b], Pauli, who, as we saw, was only selectively interested in the solid state, was just enjoying the setting. On a walk in the hills above Varenna with Pauli and Heini Gränicher (b. 1924), who later became professor at ETH, I took some photographs. On another excursion Pauli and Franca and my wife Ilse Enz and I took a taxi to the slopes above the lake, from where one could enjoy a spectacular view. But the taxi stalled and the driver looked at the motor. Timidly I said: 'Pauli effect!' But Pauli apparently did not appreciate my remark. It is only now, writing this biography, that I understand why: in a true Pauli effect, Pauli never experienced any inconvenience himself! From Varenna Pauli and Franca went on holiday to Forte dei Marmi in the Gulf of Genova, from where he sent me a picture postcard on 29 August. On 23 September Pauli attended the autumn meeting of the Swiss Physical Society in Basle, where he spent a pleasant time with Fierz (see letter [2342] in Ref. [13a]).

Fig. 12.2 Walking at Esino, with Heini Gränicher, during the Varenna Summer School, July 1956 (photo by C. P. Enz)

Then a new academic year began. As reported on page 485, Thellung had introduced the tradition of organizing visits to concerts for Pauli. One day in the first year of my assistantship, at the beginning of the concert season, I received in my institute mail a picture postcard from the Schifflände-Bar (see page 197) in downtown Zurich, written by Pauli, dated 4 October 1956 and addressed to 'Herrn Dr. Ch. Enz (vormals Dr. A. Thellung), Physikgebäude ETH, Zürich, Gloriastr. 35'. Here the parentheses 'formerly Dr. A. Thellung' were already a veiled critique! Pauli wrote in German: "Not official. Dear Mr Enz, of course you haven't told me *anything* about the fact that today there is a *concert* with Isaak Stern as violinist and orchestra (Beethoven + Schubert). Now, of course, it's sold out! In the hope for a better future, yours W. Pauli' (translated). This postcard is reproduced in Ref. [16c], p. 106 (see below and item [2347] in Ref. [13a]). That, of course, was a bitter experience to start with. But, fortunately, Pauli never bore a grudge. And being a music lover myself, I was pleased to continue Thellung's tradition.

As with my predecessors, my relationship with Pauli turned out to be most pleasant. He would come in around 11 a.m., taking the bus from Zollikon to Bellevue Square and from there the tram to Gloriastrasse. After looking through

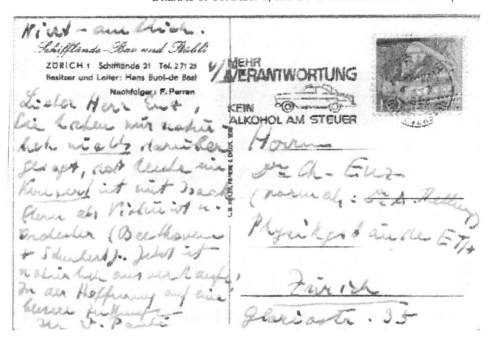

Fig. 12.3 A postcard from Pauli

his mail, he would read or write letters, then come to my office to discuss the news in physics or hand me a new preprint. For lunch we usually went to the cafeteria in the main building of ETH, a 10-minute walk from the physics institute; often visitors joined us. But Pauli hated to be alone at the institute. If for some reason I were not in, he would call my home to find out why. But it was not that he urgently needed his assistant's help—the latter's only task, in fact, was to keep Pauli's collections of reprints and preprints in order. As mentioned on page 452, Pauli's assistant occupied the office 16e. It had no view, except into an inner courtyard with Scherrer's laboratories on the c-level, where Pauli often went for afternoon tea. And the adjacent office 17e was Jost's. From the beginning, Res Jost regularly came over to my office in the morning for a chat, and he deeply impressed me with his erudition.

At the beginning Jost's relationship with Pauli was very friendly, even intimate. They called each other by their first names and by the German 'Du'; Jost was the spiritual son [16d]. However, as Fierz later observed to me, Jost had become too close to Pauli. Fierz, on the other hand, always kept a respectful distance, addressing him as 'Herr Pauli'. After some time Jost, in his daily visits to my office, began to complain about Pauli. The difference had grown out of a competition between independent projects by Pauli and Jost to visit the Princeton Institute during the winter term of 1957/8; both were friends of Oppenheimer. But they could not leave at the same time because of teaching obligations. Pauli let Jost know his plan on 28 October 1956; he thought he had

Fig. 12.4 Pauli's first and last assistants, Ralph Kronig and Charles Enz, at Kronig's home in Küsnacht, 1981

priority. On 4 November Jost offered Pauli his excuses out of fear of a permanent rupture. But finally, Pauli asked for leave only on 28 October 1957, and this for the period from 15 January to 1 June 1958. This time he planned to visit the University of California at Berkeley (Ref. [5], IV.19). Jost's request for leave to visit the Princeton Institute during the first half of the winter term of 1957/8, on the other hand, had been written on 14 May 1957 and had Pauli's reserved consent (Ref. [5], IV.3). Pauli's reservation signalled the coming rupture which, somehow, was written in the stars, and it also included Franca. Res Jost and his wife Hilde suffered because of it. Res continued to come to my office, but the leisurely chat often gave way to complaints. Pauli, on the other hand, voiced his criticism to me about Jost's scientific performance. As Franca later observed to me, the Pauli–Jost relationship had one and a half good years and one and a half years of breach [16d].

In the autumn of 1956 Pauli continued to think about the possible neutrino processes implied by Eqs (11.20) and (11.21), in particular the double β-decay and the Davis reaction (11.22). His main concern now was how the conservation of the number of leptons (see page 490) or the lepton charge, as he called it, could be verified. The extreme case of non-conservation would be the *Majorana theory*, in which $\psi^c = \psi$ (only $\psi + \psi^c$ interacts but not $\psi - \psi^c$). On 15 February 1957 Pauli wrote to Fierz: 'The immediate association to Majorana was of course "Aha, particles and antiparticles should not exist anymore, these one wants to take away from me (as one takes away a symbol from somebody)!"

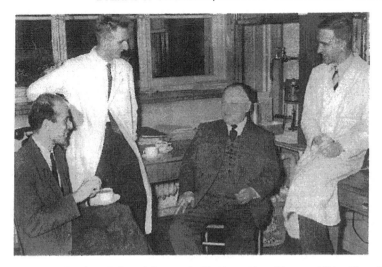

Fig. 12.5 Tea with, from left to right, Armin Thellung, Ernst Heer, and Fritz Gimmi at Scherrer's institute, 1955

This inspires *fear* in me. I also know that already since autumn the conservation of the lepton charge is enormously important in physics—considered rationally, perhaps *too important*. I am afraid it could turn out to be false' (*Pauli Archive*, translated).

Pauli noted that matrix elements depend only little on the neutrino mass—because of its smallness—and that, in the limit of a vanishing neutrino mass, $m_\nu = 0$, a considerable liberty of description occurs, since in this case there exist two canonical transformations of the neutrino field, namely

$$\psi' = a\psi + b\gamma^5 C^{-1}\overline{\psi}; \quad \overline{\psi}' = a^*\overline{\psi} + b^*\psi C\gamma^5; \quad |a|^2 + |b|^2 = 1, \qquad (12.24)$$

where C is the matrix of Eqs (8.38)–(8.41) and γ^5 assures the correct sign, and

$$\psi' = e^{i\alpha\gamma^5}\psi; \quad \overline{\psi}' = \overline{\psi}e^{i\alpha\gamma^5}. \qquad (12.25)$$

'Canonical' here means that in both cases, (12.24) and (12.25), ψ' and $\overline{\psi}'$ satisfy the same anticommutation relations (8.35) as ψ and $\overline{\psi}$. This liberty of the mass-less neutrino field may be expressed in terms of the substitution of ψ_ν in Eq. (6.63) by the general linear combination (compare Eq. (1) in Ref. [17a])

$$\phi \equiv (g_{Ii} - f_{Ii}\gamma^5)\psi_\nu^c - (g_{IIi} + f_{IIi}\gamma^5)\psi_\nu \qquad (12.26)$$

which, however, violates both parity and lepton conservation.

In the notation of Eqs (8.38) and with

$$\Psi \equiv \begin{pmatrix} \psi_\nu \\ \psi_\nu^c \end{pmatrix}; \quad U \equiv \begin{pmatrix} a & b\gamma^5 \\ -b^*\gamma^5 & a^* \end{pmatrix}; \quad U^+U = 1 \qquad (12.27)$$

this *Pauli group* (12.24), (12.25) reads

$$\Psi' = U\Psi; \quad \Psi' = e^{i\alpha\gamma^5}\Psi \tag{12.28}$$

and the expression ϕ given by (12.26), may be written

$$\phi = \Gamma_i\Psi; \quad \Gamma_i \equiv (-(g_{IIi} + f_{IIi}\gamma^5),\ g_{Ii} - f_{Ii}\gamma^5). \tag{12.29}$$

Dropping the index i and splitting Γ into right (R) and left (L) parts,

$$\Gamma = \Gamma^R \frac{1 - \gamma^5}{2} + \Gamma^L \frac{1 + \gamma^5}{2}, \tag{12.30}$$

one finds, in the notation of Eqs (5) of Ref. [17a],

$$\begin{aligned}
\Gamma^R &= (-g_{II} + f_{II},\ g_I + f_I) \equiv (G_2, G_1); \\
\Gamma^L &= (-g_{II} - f_{II},\ g_I - f_I) \equiv (-F_2, F_1).
\end{aligned} \tag{12.31}$$

Defining Γ' by $\Gamma'\Psi' = \Gamma\Psi$, the first transformation (12.28) has the effect

$$\Gamma' = \Gamma U^+ = \Gamma^R \frac{1 - \gamma^5}{2} U^{+R} + \Gamma^L \frac{1 + \gamma^5}{2} U^{+L} = \Gamma'^R + \Gamma'^L, \tag{12.32}$$

or explicitly,

$$\begin{aligned}
(G_2', G_1') &= (G_2, G_1) \begin{pmatrix} a^* & b \\ -b^* & a \end{pmatrix}, \\
(-F_2', F_1') &= (-F_2, F_1) \begin{pmatrix} a^* & -b \\ b^* & a \end{pmatrix},
\end{aligned} \tag{12.33}$$

which is equivalent to Eqs (10) of Ref. [17a] with $\alpha = 0$. From (12.33) one sees that

$$K + L \equiv G_2 G_2^* + G_1 G_1^*; \quad K - L \equiv F_2 F_2^* + F_1 F_1^* \tag{12.34}$$

are *invariants*. Inserting Eqs (12.31) here and again attaching indices i and j to the first and second factors, respectively, one finds the Hermitian invariants K_{ij} and L_{ij} of Eqs (13) and (14), respectively, of Ref. [17a]. Equations (12.33) show, however, that there is the additional invariant

$$I \equiv \frac{1}{2}(G_2 F_1 - G_1 F_2). \tag{12.35}$$

Inserting Eqs (12.31) here and attaching indices i and j as before, the expression separates into a symmetric part $I_{ij} = I_{ji}$ and an antisymmetric part $J_{ij} = -J_{ji}$, both of which must be separately invariant. These are the invariants of Eqs (15) and (16), respectively, of Ref. [17a].

The physical significance of this transformation group is that S-matrix elements are also invariant and hence can only depend on the invariants (12.34),

(12.35). Parity conservation is then described by the invariant conditions $L_{ij} = 0$ and lepton conservation by $I_{ij} = 0$ and $J_{ij} = 0$. It was then my task to calculate the general matrix elements for the mentioned double processes [17b]. In October 1956 Pauli exchanged letters with Fierz about the transformations leaving the β-interaction invariant (see letters [2364]–[2366] in Ref. [13a]). However, by the end of 1956 neither of the references [17a,b] had been written up. Pauli's paper [17a] and my note [17b] were submitted together only on 14 May 1957. On 17 December 1956 the Swiss School Council and on 29 January 1957 the Federal Council reconfirmed Pauli for a fourth period of 10 years as professor of theoretical physics at ETH (Ref. [5], III.169 and 172). Then in January came the news of parity violation, the very day Pauli gave his evening conference on the neutrino (see page 520).

Pauli began to think seriously about writing up his ideas on parity and the lepton charge during the spring vacation of 1957. On Tuesday, 19 March, he wrote me a long letter from his office on Gloriastrasse, asking me to be present on Friday, and on the 20th he wrote an addendum. In this letter the general form (12.26) of the neutrino field in the β-decay interaction (6.63) and the transformation (12.25) are discussed, mainly in relation to the Davis reaction (11.22). Also the invariants K_{ii} and I_{ii} already appear. Later in the month Pauli went to England (see Ref. [13a], footnote 1 to letter [2352]), from where he sent me a postcard on the 31st, together with Thellung. Then on 9 April he wrote me a letter from the CERN Theory Division in Copenhagen. In this letter the transformation (12.24) appears for the first time and Pauli observes that nobody knows it. He therefore considered it necessary to publish a letter to the editor. On 15 April Pauli wrote again, this time from his office and to the holiday home of my family in western Switzerland, saying that during Easter he wanted to try to write a letter to the editor. This led to Refs [17a,b]. In addition to these references I later calculated the cross-section for the production and reabsorption of neutrinos in terms of the above invariants [17c].

Parity violation and the spinor model

> *Now, where shall I start? It is good that I did not make a bet. It would have resulted in a heavy loss of money (which I cannot afford); I did make a fool of myself, however (which I think I can afford to do)—incidentally, only in letters or orally and not in anything that was printed. But the others now have the right to laugh at me.* [18a]

At the end of June 1956 Pauli received a preprint from T. D. Lee and C. N. Yang with the title 'Is parity conserved in weak interactions?' The authors concluded that the experimental evidence of parity conservation was almost non-existent, and they proposed specific experiments to check it [18b]. The reason for this doubt in parity conservation had been the then widely discussed 'θ–τ puzzle', namely the fact that the K^+- meson has, among others, the two decay modes $K^+ \to 2\pi$, the θ-mode, and $K^+ \to 3\pi$, the τ-mode [18b]. Since the pions have negative parity, i.e. are described by pseudoscalar fields (see pages 364–5), the indicated two decay modes indeed have different parity. But Pauli did not take seriously Lee and Yang's doubts of parity conservation. He came to my room with the preprint and left it with me. However, immediately several experiments were started which all confirmed parity violation. These results left the community of particle physicists puzzled and won Lee and Yang the Nobel Prize in 1957 'for their penetrating investigation of the so-called parity laws, which has led to important discoveries regarding the elementary particles'.

Fig. 12.6 Pauli's 'Cockchafer Speech' at the gathering of the Physics Nobel Prize winners at Lindau, June 1956 (CERN)

In the letter to Weisskopf of 27 January 1957 quoted at the outset of this section, Pauli describes his reaction to the news of parity violation: 'Now the first shock is over and I begin to collect myself [*mich zusammenzuklauben*] again (as one says in Munich).' And he continues: 'Yes, it was dramatic. On Monday 21st at 8:15 p.m. I was supposed to give a talk about "past and recent history of the neutrino". At 5 p.m. the mail brought me three experimental papers: C. S. Wu [19a], Lederman [19b] and Telegdi [19c]; the latter was so kind to send them to me. The same morning I received two theoretical papers, one by Yang, Lee and Oehme [20a], the second by Yang and Lee [20b] about the two-component spinor theory. The latter was essentially identical with the paper by Salam [20c], which I received as a preprint already six to eight weeks ago and to which I referred in my last short letter to you. . . . (At the same time a letter came from Geneva by Villars with the New York Times article.)' [18a]. Villars wrote again from CERN in Geneva on 25 January, reporting on the attempts of the theory group there concerning the θ–τ puzzle, but Pauli asked me to answer. In the experiment of Ms Wu of Columbia University, New York, the asymmetry in the angular distribution of the β-decay electrons emitted by polarized Co^{60} nuclei was measured [19a]. In the Lederman experiment μ^+-mesons resulting from decaying π^+ in the cyclotron at Columbia University were stopped in carbon, and the asymmetry of the emerging e^+ measured. [19b] The same asymmetry was detected in nuclear emulsions by Friedman and Telegdi [19c].

On Wednesday, 16 January, 1957, the *New York Times* printed the news about parity violation on its front page with the title 'Basic Concept in Physics Is Reported Upset in Tests'. Around the spring equinox of 1957 Pauli wrote to Jung: 'At the moment physics is occupied with reflections, my dreams were so already before, namely parallel to mathematical works which now have gained actuality. That, however, I first have to digest spiritually' (translated from Ref. [2], letter [72]). On 5 August Pauli was ready to write to Jung in detail 'a somewhat odd mixture of physics and psychology' (translated from Ref. [2], letter [76], p. 159), attaching to his letter the report from the *New York Times*, which is reproduced in Appendix 10 of Ref. [2]. The first part of this letter [76] is on physics and the second part on psychology. In conclusion of the physics part Pauli writes: 'Thus it is now certain that "God *still* is weakly left-handed"— as I like to express it', adding 'Of such a possibility I had not thought in the slightest before January of this year.' In the psychological part of letter [76] Pauli observes that: 'After these happenings of January which have delivered to me and other physicists (e.g. also Fierz) a severe shock, Mr. Fierz asked me the question why, in fact, in 1954–1955 I had come to occupy myself with the mathematics of the reflections, surely psychological background phenomena must have played a role there' (translated from Ref. [2], p. 161). That was a synchronistic phenomenon.

This letter [76] also reports two dreams that occured on 15 March and on 15 May 1957. In the first a younger dark man who is surrounded by a faint halo of light presents a manuscript to Pauli, whereupon Pauli shouts at the man: 'How come you dare ask me to read this work?' And Pauli wakes up. Pauli interprets this dream as his conventional resistance to certain ideas (e.g. parity violation?). In the second dream the same man gets into Pauli's car. Appearing now as a policeman he commands sharply: 'You come with me!' They approach a desk where an unknown dark lady sits. The policeman now asks her, equally sharply: 'Director Spiegler, please!' On the word 'Spiegler' ('reflector') Pauli wakes up in terror (translated from Ref. [2], p. 163). Pauli eventually gave this calamity a more humorous turn by writing the following condolatory notice: 'It is our sad duty to make known that our dear friend of many years, PARITY, gently passed away on 19 January 1957 after a short suffering under further experimental interference. In the name of the survivors, $e \, \mu \, \nu$' (translated from Ref. [21a], p. 192; this was an addendum to the letter of 29 January 1957 to Fierz).

During the winter term of 1956/7 Pauli's special course was again on 'Problems of General Relativity' (see Ref. [5], p. 434). Subjects of particular interest discussed in these lectures were the conformal invariance of the Dirac equation for mass 0, i.e. for the neutrino, cosmological questions as discussed in Jordan's recent book [21b], and, most interesting, the problem of Mach's principle which recently had been brought up again by Sciama [21c]. On the evening of 21 January 1957 Pauli, by invitation of the Zurich Society of Natural Science, gave in Scherrer's large auditorium 22c the celebrated lecture 'On the Earlier and More Recent History of the Neutrino' [22] mentioned above. Pauli's preparatory notes for this important historical account cover eight pages, between letters [2349] and [2350] in Ref. [13a], and are dated 24 December 1956. These notes also include Pauli's discussion remarks after Heisenberg's report at the Solvay Congress of 1933 concerning the proposed neutral particle (see page 226). The most detailed and 'correct' neutrino 'story', however, Pauli told in letters [2350], [2356], [2379], and [2410] to Franco Rasetti (b. 1901) of 6, 8, and 30 October and 10 December 1956 in Ref. [13a]. On 5 December 1956 Lise Meitner sent Pauli with her letter [2408] in Ref. [13a] a copy of his famous letter to the 'Radioactive Ladies and Gentlemen' (see page 215).

Also during the winter term Wojciech Królikowski, a student of Professor Jan Rzewuski in Warsaw, was visiting Pauli. In the summer term of 1957 Pauli gave a course on 'Selected Chapters from Wavemechanics' (see Ref. [5], p. 434). After introducing charge conjugation and space reflection for spinors in the first part, he devoted the second part to analytical properties of vacuum expectation functions following Heisenberg's collaborator Harry Lehmann (1924–98) in Göttingen [24] and derived the CPT-theorem along the lines of Jost [12b]. Of this course I prepared typed lecture notes for Pauli. On 1 April my assistantship with Pauli was extended for a second year (Ref. [5], III.173 and IV.1). During the summer term I gave two seminars on 'The Light Particles and Parity Violation', on 11 June at the University of Neuchâtel at the invitation of Jean Rossel (b. 1918) and on 6 July at the University of Berne at the invitation of Houtermans. At the seminar in Berne everybody suffered from the heat, so Houtermans proposed after the seminar that everybody went swimming in the Aar river.

In Neuchâtel Konrad Bleuler (1912–92) had become professor of theoretical physics in the autumn of 1956; he stayed there until Spring 1960 when he accepted a call to the University of Bonn in Germany. During his time in Neuchâtel Bleuler got to know quite well the famous Swiss playwright and novelist Friedrich Dürrenmatt (1921–90). Both of them lived half-way up the Chaumont mountain outside Neuchâtel. Dürrenmatt then was interested in meeting physicists whom he would serve a whisky or, better yet, one of those old Bordeaux wines, for which his cellar was famous. He, in fact, considered only the latter as wines, all the rest he called beverages. And he would ask the physicists whether they were sure that in continuing with their research they would not end up with a contradiction. Bleuler was happy to introduce to Dürrenmatt the colleagues he invited for seminars. One of them was Pauli whom, however, Bleuler invited privately. Out of these acquaintances grew Dürrenmatt's well-known play *The Physicists* which takes place at a psychiatric clinic on the shore of Lake Neuchâtel and which had its première on 21 February 1962 at the Schauspielhaus in Zurich.

Fig. 12.7 At lunch with C. N. Yang and assistant Enz at a garden restaurant near the Physics Institute, about July 1957 (photo by C. P. Enz)

In July 1957 Chen Ning Yang (b. 1922), who since 1955 had held a professorship at the Institute for Advanced Study, was visiting with Pauli. He 'gave a good report', Pauli told Oppenheimer in a letter written while on holiday in Ronchi, Italy(?), and all three of us had lunch at an open-air restaurant on the slope of Zürichberg where I took some photographs. The summer vacation of 1957 allowed Pauli again to make some pleasant trips abroad. From 9 to 16 September he participated at the International Congress on Nuclear Structure in Rehovot, Israel. For this journey he had received from President Pallmann 2000 francs of travelling money (Ref. [5], IV.2, 4). Pauli reported to the president on this congress on 4 October (Ref. [5], IV.15). In Rehovot Pauli was particularly happy to see Ms Chien-Shiung Wu (1912-97) who, as we saw, had done one of the first experiments on parity violation, and to whom Pauli felt very much attracted. From 22 to 28 September the large and elegant International Conference on Mesons and Recently Discovered Particles took place, the first part in Padua and the second part in Venice. Both Pauli and I attended but without giving papers. I had obtained a travel grant of 350 francs from President Pallmann (Ref. [5], IV.12, 13), for which I thanked him with a report on the conference (Ref. [5], IV.16). The end of summer also ended Pauli's exchange of letters with C. G. Jung and Marie-Louise von Franz. A year earlier Pauli had withdrawn from the curatorship of the C. G. Jung Institute, as he told C. A. Meier on 22 August 1956 in his letter [2331] in Ref. [13a] (see also the two letters in Ref. [2], Appendix 9). This and the following letter [2355] contain a severe personal criticism of C. A. Meier.

Fig. 12.8 Pauli with, from left to right, Charles Enz, Valentin Telegdi, Philippe Choquard, and Aloisio Janner at the International Conference on Elementary Particles, Padua–Venice, September 1957

Then, during the first half of the winter term of 1957/8, Pauli lectured on the 'Theory of the Weak Interactions' (see Ref. [5], p. 434). It is in these lectures that Pauli, following T. D. Lee, introduced the notation of right- and left-handed neutrinos, in accord with Eq. (12.30), namely (see Eqs (7.1))

$$\psi_\nu^R \equiv \frac{1 - \gamma^5}{2}\psi_\nu; \ \psi_\nu^L \equiv \frac{1 + \gamma^5}{2}\psi_\nu. \tag{12.36}$$

Since γ^5 anticommutes with the γ^μ it follows that the Dirac equation (6.31) mixes ψ^R and ψ^L, unless $m_\nu = 0$. This is best seen in the representation in which γ^5 is diagonal,

$$\gamma^4 = \begin{pmatrix} 0 & 1 \\ 1 & 0 \end{pmatrix}; \; \gamma^5 = \begin{pmatrix} -1 & 0 \\ 0 & 1 \end{pmatrix}; \; \vec{\alpha} = \begin{pmatrix} \vec{\sigma} & 0 \\ 0 & -\vec{\sigma} \end{pmatrix}, \tag{12.37}$$

where $\vec{\alpha} \equiv i\gamma^4\vec{\gamma}$, and ψ_ν is split into the two-component spinors φ^R and φ^L. For the Fourier components this looks as follows:

$$\begin{aligned}
\left(\vec{\sigma} \cdot \mathbf{k} - \frac{\omega}{c} \right) \varphi^R &= -k_c \varphi^L \\
\left(-\vec{\sigma} \cdot \mathbf{k} - \frac{\omega}{c} \right) \varphi^L &= -k_c \varphi^R.
\end{aligned} \tag{12.38}$$

Thus, for $m_\nu = 0$, the right- and left-handed neutrino is polarized parallel and antiparallel to its momentum, respectively (see fig. 3 in Ref. [22]). It turns out that only φ^R couples in the weak interactions, so that one may set $\varphi^L = 0$. This leads to a 'two-component theory'. Pauli also discussed the canonical transformation (12.24) in a form equivalent to Eqs (12.33).

Late in the autumn of 1957 Heisenberg came to see Pauli in Zurich for mathematical advice on the problem of his non-linear spinor model (see the previous section) which he defined by the Lagrangian density

$$\mathcal{L}' = \overline{\psi}\gamma^\mu \frac{\partial\psi}{\partial x^\mu} + \frac{l^2}{2}(\overline{\psi}\psi)^2. \tag{12.39}$$

For over a year Pauli had steadfastly resisted Heisenberg's invitation to collaborate in this endeavour. But now the latter claimed to have found a modified Lagrangian which in addition to the trivial gauge transformation also had Eqs (12.25) as invariance group. Pauli was now interested in seeing whether Heisenberg's Lagrangian could be modified in a way so as to be invariant also under his (Pauli's) group (12.24). The latter group had gained considerable appeal since, as Pauli and Heisenberg wrote in their common 'preliminary report'—which, although unpublished, is reprinted in Heisenberg's *Collected Works* [25]: 'As [12.24] is isomorphic to the group of unitary transformations of two complex variables, with the determinant 1, and hence with the 3-dimensional rotation group, it can be connected with the rotation group of the isospin, while [12.25] can be connected with the baryonic number' (Ref. [25], p. 338). This had been shown in Ref. [26a] by Feza Gürsey (b. 1921), who was in the United States on leave from Istanbul.

The final form of Eq. (11) in Ref. [25] of Heisenberg's and Pauli's Lagrangian density,

$$\mathcal{L} = \overline{\psi}\gamma^\mu \frac{\partial\psi}{\partial x^\mu} \pm \frac{l^2}{2} \sum_\mu (\overline{\psi}\gamma^\mu\gamma^5\psi)^2, \tag{12.40}$$

although invariant under the group (12.25), is not invariant under the group (12.24)—unless ψ and $\overline{\psi}$ are replaced by Ψ and $\overline{\Psi}$, respectively, as defined

Fig. 12.9 Pauli with Werner Heisenberg (Max-Planck-Gesellschaft, Berlin).

in Eqs (12.27). In fact, one shows for both groups that, with an arbitray γ^5-dependence,

$$\overline{\Psi}'\gamma^\mu f(\gamma^5)\Psi' = \overline{\Psi}\gamma^\mu f(\gamma^5)\Psi. \tag{12.41}$$

Apart from this group structure the model contained other revolutionary features, one of which the authors describe as follows: 'In constructing this quantum theory it will be possible to give a precise mathematical meaning to the three approximations, which are usually distinguished by the terms strong interaction, electromagnetic interaction and weak interaction' (Ref. [25], p. 339). Exploiting the nonlinearity of the model, Heisenberg previously had calculated a numerical value for the fine-structure constant α of Eq. (7.2) (see e.g. Heisenberg's letter [2089] of 11 May 1955 to Pauli in Ref. [13a]). Another novelty occurring in this 'preliminary report' is introduced by these words: 'For the spinor-model, here presented, it is fundamental that the vacuum is degenerate. We shall assume here at most a four-fold degeneration' (Ref. [25], p. 342). Heisenberg's idea from the start had been that, due to its nonlinearity, Eq. (12.39) or (12.40) should describe all elementary particles as composites, starting from the neutrino. This composition of the elementary particles is exhibited in the table on p. 347 of Ref. [25]. Towards the end of 1957 Pauli's enthusiasm with the magic of this group structure took on unreal proportions. Phone calls, telegrams, letters were exchanged and Heisenberg who, according to Franca, was much more temperate than Pauli [16d], made his appearance in Zollikon or at Gloriastrasse. On 1 December 1957 Pauli wrote to Heisenberg: 'In the mean time I had supper. Now I have a strong sentiment of security. Dear Heisenberg: *It cannot, in fact, be different!* But—now what? Help to get further! In the mean time I also continue to think about it' (translated). On

13 May 1958 Pauli, however, wrote to Fierz about Heisenberg: 'He believes that when he publishes with me, then it is also again 1930! I already became embarrassed, how he ran after me!' (*Pauli Archive*, translated).

Fig. 12.10 Pauli with assistant Enz at the Christmas party of the ETH Physics Institute, 1957

As always Pauli was present at the Christmas party of the Physics Institute which this year was held there. On 7 January 1958, on the occasion of the dedication to the ETH Physics Institute of a bust of Albert Einstein created by the Zurich sculptor Hermann Hubacher (1885–1976) and commissioned by Einstein's biographer Carl Seelig (1894–1962), Pauli had agreed to give the inaugural address [26b] outside Scherrer's large auditorium (see Ref. [5], IV.11 and footnote 301, p. 296). This address is reprinted in the collection of essays [26b] by Pauli. After the ceremony President Pallmann invited Pauli and Franca for dinner at the restaurant of the Saffran Guild on Limmatquai (Ref. [5], IV.29). On 17 January 1958 Pauli left Zurich for Berkeley in California, after having asked President Pallmann on 28 October for leave of absence from 15 January to 1 June 1958 (Ref. [5], IV.19, 22). On 18 January Pauli wrote to Heisenberg from the Physics Institute in Milan, saying that the people from *Nuovo Cimento* could print a paper like theirs within 14 days. On 20 and 21 January he sent me postcards from Naples, on board an Italian ship heading for New York, telling me: 'the "preliminary report" [Ref. [25]] which Göttingen now sends out should not yet be printed in *this* form. Surely it still contains many mistakes in the detail' (translated). From the ship Pauli sent Heisenberg a telegram and also wrote to him.

Fig. 12.11 Pauli and Franca, January 1958 (CERN)

Before his departure Pauli had written to Ms Wu at Columbia University, offering to give a private seminar on arrival about his work with Heisenberg. Ms Wu first arranged a lunch at the Faculty Club, at which about 15 people were present, including Yang, Lee, and Pais. 'Everybody was very polite to Pauli, but nobody was enthusiastic about his ideas. (That is in fact an understatement.)', Yang wrote to me on 14 October 1985 [27a]. Since this was the time of the New York meeting of the APS, Pauli had a large audience. In the ensuing discussion a younger, quite critical generation manifested itself. But on Pauli's observation that this theory was perhaps somewhat crazy, Bohr, who happended to be there, made the famous remark that, perhaps, it was not crazy enough [27a]. On 1 and 2 February 1958 Pauli wrote to Heisenberg from New York, mentioning the discussions he had with Norman Myles Kroll (b. 1922) and with Yang, most probably after his seminar, and, he insisted, not to precipitate publication.

Chien-Shiung Wu (1912–97) was born in Liuho, a small town near Shanghai. In 1934 she obtained a B.Sc. degree in physics at the National Central University in Nanjing and then worked at the National Academy of Sciences in Shanghai. After moving to the United States in 1936 she received her Ph.D.

at Berkeley in 1940 for studies on the fission products of uranium, joining the Manhattan Project at Columbia University in 1944. From 1946 to her retirement in 1981 she was a member of the Physics Faculty of Columbia University where in 1958 she became a full professor. During this period she devoted her main research effort to β-decay, publishing the first result on parity violation in January 1957 [19a]. In 1975 Ms Wu served as president of the APS, and in the same year she received the National Medal of Science. In 1978 the prestigious Wolf Prize was bestowed upon her [27b].

In Berkeley Pauli, pondering the spinor model, on 11 March 1958, wrote Amos de-Shalit (1926–69), who in 1951 had obtained his Ph.D. with Scherrer and Pauli at the ETH (Ref. [5], III.75): 'My mood with this whole spinor-model is very fluctuating. (Regarding Heisenberg I have the feeling, that the situation is slowly growing over his head; certainly he needs vacations. I was myself very much overworked at X-mas and in January. Now I feel normal again.) In the moment my opinion on all this is more negative. (Perhaps it will change again.)' (*Pauli Archive*). But Pauli came to the conclusion that he had to withdraw. He wrote on 7 April to Heisenberg the following *'final decision*: I must totally drop the plan to publish with you a work "On the isospin group in the theory of elementary particles"'. On 8 April 1958 he sent the following circular letter to all his colleagues who had received the preprint: 'As essential parts of the preprint with the above title [25] don't agree any longer with my opinion, I am forced to give up the plan to publish a common paper with Heisenberg on the subject in question.' And he continued: 'Particularly I am now convinced that the degeneration of the vacuum should not be used in order to explain the possibility of a half-integer difference between ordinary spin and isospin for some strange particles. The idea of a unification of the spinor field seems to fail here and I believe that one should try to introduce, besides spinors with isospin 1/2, either other spinors with isospin 0, or at least one scalar field with isospin 1/2 ("Goldhaber model"), in order to reach an interpretation of the elementary particles' (*Pauli Archive*).

While Pauli was at Berkeley I had to take over the main course on 'Optics and the Theory of Electrons' during the second half of the winter term and to start the course on 'Thermodynamics' in the first half of the summer term (see Ref. [5], IV.19). From Berkeley Pauli sent me 11 letters. In the first one of 25 February he thanked me for the lecture notes on 'Selected Chapters from Wavemechanics' mentioned above. I had sent him a copy which, he said, served him well, particularly the second part on the CPT-theorem. He also reminded me of my reappointment for another year as his assistant, which I should ask the secretary Margaret Schmid to prepare for his signature (see Ref. [5], letter of 7 March 1958 and documents IV.35 and IV.37). As for his preprint with Heisenberg, Pauli wrote that it was not being sent out by Heisenberg's institute in Göttingen because he, Pauli, was not yet satisfied with it. 'I am not at all convinced that the spinor model works (in spite of Heisenberg's optimism), but I am also not convinced that it does *not*. The bad thing is that no reasonable methods exist for the treatment of the eigenvalue problem.

(The Tamm–Dancoff method I wish to leave exclusively with Heisenberg and his collaborators)' (translated). In the second letter, dated 26 February, Pauli observed that the degenerate vacuum was Heisenberg's invention in order to make strange particles.

Meanwhile in Göttingen Heisenberg was receiving some publicity. The Hamburg daily *Die Welt*, for instance, announced in its edition of 5 March 1958 that Heisenberg, on the occasion of the celebration of Max Planck's 100th birthday, would give in Berlin on 25 April an account of his 'formula', which would be understandable to a broad audience. The newspaper even reproduced the 'formula', namely the two spinor equations following from the Lagrangian densities (12.39) and (12.40). In his letter of 4 March Pauli asked me: 'Have you in the mean time heard of Heisenberg's radio- and newspaper-advertisement, with him in the principal role as super-Einstein, super-Faust and super-human? His passion for publicity seems insatiable (what does he want to compensate with it?)' (translated). And on 14 March he wrote that the measure of Heisenberg's advertisement is at least one macro-scherrer, alluding to Scherrer's liking of publicity. 'But at the same time H. writes me letters of lamentation, what annoyance he is having with the press' (translated).

In Berkeley Pauli regularly attended the solid state seminars of Professor Charles Kittel (b. 1916), in which superconductivity was the main topic. It was also in Berkeley Pauli learnt that at the 100th anniversary celebration for Max Planck in Berlin the Planck Medal was to be bestowed upon him. On 2 May President Pallmann sent his congratulations (Ref. [5], IV.40). On Friday, 25 April 1958 the president of the Association of German Physical Societies, Professor F. Trendelenburg, in his opening speech of the Berlin meeting had announced that Pauli was the winner of the Planck Medal of 1958 and read the laudation, but he admitted that, unfortunately, he was unable to present the medal because Pauli was in the United States (see Ref. [27c], pp. 45-7).

This letter to me, dated 14 March, contains also Pauli's first reference to the plans for a new physics institute, the following letter of 22 March announcing Pauli's ideas concerning the theory group. And he attached a sketch of the requirements of the theory group to his letter of 11 April. The question of a new institute building was raised for the first time in a conversation between Scherrer and President Pallmann on 27 July 1957 (see Ref. [5], IV.14, point III). In parallel with this planned new building, Pauli and Scherrer, in a letter to the president on 23 October 1957, had also suggested the nomination of two or three new professors (Ref. [5], IV.18). In his letter of 19 April 1958 to me, Pauli expresses his satisfaction that Pierre E. Marmier (1922–73), who had become associate professor of nuclear physics in 1957 (Ref. [5], IV.8), is taking over the coordination of the new physics building. In the same letter Pauli voices his displeasure that Jost, after his promising start with his CPT-paper, has not brought any other stimulating ideas in physics to Zurich. Towards the end of this letter Pauli, who for the first time had to sign forms to get money from the Committee for Atomic Sciences of the Swiss National Fund, presided by Scherrer, rails against 'these bureaucrats' who apparently do not care about deadlines. (This letter is reprinted as document IV.38 in Ref. [5].)

The next two letters from Berkeley to me, dated 21 and 29 April 1958, still address 'these bureaucrats'. The second one, however, also gives Pauli's plan for his return, which comprises a visit to the Brookhaven National Laboratory before boarding a flight from New York to Zurich on 1 June. This second letter closes with the following hexameters addressed to the physicists at the seminar on field theory held in Oberwolfach in the Black Forest, Germany (translated):

> *Sweet as honey flows from the mouth of the experts the humbug*
> *Far from the foundations of physics appears it as blue-coloured vapour.*

On 12 May the Swiss National Fund confirmed Pauli's request (Ref. [5], IV.32), granting him 30 000 francs over one year for the appointment of advanced collaborators (Ref. [5], IV.41). In his last letter from Berkeley, dated 17 May, Pauli, after thanking me for my congratulations on his Planck Medal and after discussing possible candidates to be invited during the winter term of 1958/9, closes with remarks about the 'LSZ-Club' of Heisenberg's collaborators Harry Lehmann (1924–98), Kurt Symanzik (1923–83), and Wolfhart Zimmermann (b. 1928). Pauli thinks 'that the masses [of the particles] should *not* be supposed to be given and that these and the *S*-matrices should follow from the same overall system of equations. (One must construct *models* [inserted by pencil].) And probably there even does not exist such a system of equations that is without contradiction. In the quantum field theory we hardly will succeed as cheaply as the LSZ-Club and Heisenberg believe' (translated from the letter of 17 May 1958). Pauli particularly appreciated Lehmann. On 9 July 1958 he wrote to Fierz: 'I spent a nice evening with *Lehmann* and again had an excellent human and spiritual understanding with him' (*Pauli Archive*, translated).

From 9 to 14 June 1958 Pauli participated in the 11th Solvay Conference on 'Astrophysics, Gravitation, and Structure of the Universe' (see Ref. [28a], pp. 381–7), for which he had obtained leave of absence from President Pallmann and I had to teach his course (Ref. [5], IV.43, 45). Between 30 June and 5 July Pauli and I were in Geneva for the Annual International Conference on High Energy Physics at CERN, for which permission had been granted by President Pallmann on 9 June (Ref. [5], IV.46), Pauli also received a subsidy (Ref. [5], IV.52, 55). At this conference, on Tuesday, 1 July, a session on *Fundamental theoretical ideas* was held in the large auditorium of the Physics Department of Geneva University. Pauli was chairman, and Yukawa was the first speaker; Heisenberg was the second (Ref. [28b], pp. 117 and 119). The third speaker was Pauli whose talk was on 'The Indefinite Metric with Complex Roots' [28c]. In order to prepare for the session Pauli had exchanged several letters with Vladimir Glaser (1924–84) who was then at the Theory Division of CERN. In an earlier letter, dated 24 April 1958, Pauli had asked Glaser to convey Pauli's greetings to the meeting at Oberwolfach, mentioning the above 'dedication' he had asked Walter Thirring to transmit [28d].

As chairman, Pauli opened the session with the following remarks: 'This session is called "fundamental ideas" in field theory, but you will soon find out or have already found out that there are no new fundamental ideas. So what you shall hear are substitutes for fundamental ideas, and it works in the same way as I am the substitute for a rapporteur. So, you will also see that there are two kinds of ignorance; the rigorous ignorance and the more clumsy ignorance. You will also hear that many speakers will want to form new credits for the future. I am personally not very willing to give such credits but it is everybody's own choice, what he wants to do in this respect' (Ref. [28b], p. 116, see also Ref. [16c], p. 94). During Heisenberg's talk Pauli made some critical remarks which the former accepted graciously. Later in the year Pauli acted polemically against Heisenberg, drawing an empty rectangle and writing 'This is to show the world, that I can paint like Titian; only technical details are missing. W. Pauli.'

Fig. 12.12 Pauli and Franca at a congress of the Italian Physical Society, September 1958 (CERN)

From 21 July to 9 August 1958 Pauli again participated in a Varenna Summer School (see the previous section), this time on 'Mathematical Problems of the Quantum Theory of Particles and Fields.' Heisenberg, who was also there, recalls that this was the last meeting between the two friends and that Pauli asked him to continue with the problem of the spinor model. He, Pauli, did not have the forces to do it any more (Ref. [28e], pp. 319–20). Then Pauli went on holiday with Franca to Forte dei Marmi, from where he again sent me a picture postcard on 3 September, telling me that 'in Varenna it was merry but scientifically unsatisfactory' (translated), and that he would be at his office on

9 September, but that, if I still wished to have some more days' holiday, he would not mind. On 9 September he wrote a letter from his office to our holiday place, saying that he was somewhat anxious about my health and 'I am very much looking forward to see you again, but as far as I *myself* am concerned, you could very well stay longer' (translated). On 10 September Pauli had a conversation with President Pallmann at his home in Zollikon, in which he told the president that I should stay with him for another year, after which I should go to another research centre. He also said that he would be very glad if my pay could be raised by 5000 francs a year and that at present I was getting 1150 francs per month (Ref. [5], IV.66). This astonishing statement was the result of a conversation between Franca and my wife Ilse, who, on being invited to the Pauli home happened to tell Franca that I was teaching at an evening school. Franca was scandalized, finding it unworthy for a Pauli assistant to do that. And she immediately asked her husband to see to it that I earn enough—which he did (see Ref. [5], IV.68).

On 5 March 1958 Jost had an audience with President Pallmann, in which the latter learned of Jost's differences with Pauli. The president, however, viewed it rather as a conflict between generations (Ref. [5], IV.34). On 26 March the president extended my appointment as assistant for another year (Ref. [5], IV.35, 37). But his main problems with physics were the new building, for which a site on the Hönggerberg north of the city was considered, and the new professors who should occupy it. On 26 November 1956 President Pallmann and other notables responsible for science policy in Switzerland had received a circular letter about the problems of physics in the country, written by two of Scherrer's former assistants, Hans Frauenfelder (b. 1922) and Peter Stähelin (b. 1924) in Urbana, Illinois (see Ref. [5], III.168). They maintained that the excellent state of research and teaching in Switzerland was in danger of being left behind if contact with the 'leading countries'—evidently, the United States—should be lost. On 10 July the president had an exhaustive discussion with Pauli about the Hönggerberg project and the planned nominations. In this conversation Pauli worried how without a car, he would be able to reach this place which is situated at the opposite end of Zurich from Zollikon (Ref. [5], IV.56).

On 13 May 1958 Pauli wrote to Fierz: 'I now expect that I will have difficulties of adaptation, with people (e.g. with Jost things are fairly bad), as well as with physics. As a matter of fact, I am not seventy yet, i.e. not old enough yet to retire quietly and entirely from physics (e.g. into the history of physics) but, on the other hand, I have not the force any more to work out myself *those* problems that interest me. Other colleagues at my age develop a drive for power and beguile their boredom with *organizing* and *administering*. But this to me is even more boring than anything else! However, I feel that I *burden* the younger generation *too much* with a claim that they should work out something that interests me! I do not see clearly myself and *therefore* I do not know *how* I should "dissipate the boredom". This then is a difficult problem for me' (*Pauli Archive*, translated).

For the winter term of 1958/9 Pauli had invited Henry Stapp (b. 1928) from Berkeley with an overall pay of $5000 (Ref. [5], IV.58) which President Pallmann exceptionally accepted, although it surpassed the norm (Ref. [5], IV.87). Pauli's two courses during the winter term were 'Wavemechanics' and 'Many-Body Theory' (Ref. [5], pp. 432 and 434). He had been invited by the Association of German Physical Societies to give the gala speech at its convention in Essen on 3 October 1958. The Planck Medal was supposed to be presented to him at this convention. However, owing to the controversy with Heisenberg, Pauli had cancelled this invitation on short notice, so that Professor Trendelenburg was again obliged to confess that the medal could not be presented because Pauli had told him that he had to be in Zurich for compelling reasons (Ref. [27c], pp. 91–6). Thus the medal remained in the hands of Professor Trendelenburg. Judging from the astonishingly negative comments concerning Planck, both the scientist and the person, that Pauli had voiced to Heisenberg when the latter received the Planck Medal in 1933 (see pages 291–2), Pauli apparently did not view the bestowal of this medal to him as an honour.

From 20 to 22 November Pauli spent three days in Hamburg where on the 21st the university bestowed an honorary doctorate upon him. He had asked his editor-friend Paul Rosbaud (1896–1963) and his former student Hans Frauenfelder who happened to be in Zurich to accompany him. A third colleague and friend joining him on this trip was Hans Daniel Jensen (1907–73) from Heidelberg [29a]. Pauli apparently felt insecure; according to Franca, in the summer of 1957 he had often complained about stomach pains after lunch [16d]. At the CERN conference of June/July 1958 Pauli had also complained about stomach pains to Piet Cornelis Gugelot (b. 1918), one of Scherrer's pupils and professor in Amsterdam [29b].

During his stay at Berkeley Pauli had been invited to become a Faculty member. This was an interesting offer since, besides the high quality of research, Pauli had many friends there, in the first place Max Delbrück and his collaborator Jean Weiglé (1901–68). Weiglé had been director of the Physics Institute of Geneva University until 1948 when he resigned to become a research associate in Delbrück's group, much to the regret of his students in Geneva. However, he came back regularly and kept a research group in Geneva, which included Eduard Kellenberger (b. 1920), who obtained his Ph.D. in 1953, and Werner Arber (b. 1929) who obtained his doctorate under Weiglé and Kellenberger in 1958 and who won the Nobel Prize in Medicine in 1978.

Pauli, however, had many reasons to like his life in Zurich. He let his decision depend on how ETH would treat him in the future. On 4 October 1958 the Swiss School Council, deliberating on the question of an exceptional salary increase, noted that Pauli's yearly income lagged behind Scherrer's by more than 8000 francs. And it was decided, first, to ask the Federal Council through the Federal Department of the Interior to raise Pauli's yearly salary by 1200 francs and, secondly, to add to his pay a sum of 4000 francs from private funds (Ref. [5], IV.71). President Pallmann informed Pauli about this on 10 October

(IV.75, 84). Jost, on 7 October, assured the president that 'I have no plans for irreversible acts in the proper sense' (translated) but that he wished to ask for unpaid leave of absence during the summer term of 1959 (IV.85).

On 6 October 1958, from his office on Gloriastrasse, Pauli wrote an important retrospective letter of 12 pages to his friend Max Delbrück (see pages 352–4). Seeing all the details before him he wrote: 'Now the problem is to bring some order into the somewhat heterogeneous matter of this letter, which runs the risk of becoming much too long. What is common to everything, is the motive of pieces of the past which come back because they are again of actual significance' (translated from p. 1). In this letter Pauli first writes about Lise Meitner, who 'on 7 November of this year has her 80th birthday', and about the neutrino. Mentioning the biologist Timofeeff-Ressovsky (see pages 259 and 353) then brings Pauli to the 'Physics/Biology' problem and to Delbrück's 'quasi-political dream', which the latter recounted 'on that memorable evening in April [1958] in your house in Pasadena where we 3 [Delbrück, his wife Manny Delbrück, and Pauli] came close to one-another in a human sense. In particular, however, I cannot forget the dear manner, which goes back to old times, how on the following day you took leave of me. I have the impression that here something has been renewed between us and that it is important' (translated from letter of 6 October 1958, p. 1, *Pauli Archive*). This was to be a leave for ever.

The journey ends in hospital room 137

I was so afraid of meeting you, afraid that with you I yet would lose my composure. And then I was so glad and reassured when you sat next to me in the crematorium. At the Pauli's we so often talked about you and I know how much both of them loved you. Still in Hamburg we spoke together about you. I have not yet recovered from the shock. Pauli has much enriched my life, and upon this wealth I will live to my end. [30]

The following account is based on notes I took at the time. On Friday afternoon 5 December from 2 to 4 p.m., Pauli as usual held his *Spezialvorlesung*, which this semester was on 'Many-Body Theory' (see Ref. [5], p. 434). Quite exceptionally I was not present because I had been invited to talk at the Physics Colloquium at the University of Berne on 'Forces between macroscopic bodies' (the Casimir effect) in the late afternoon. During his lecture Pauli suffered from stomach pains. He felt abandoned and had to order a taxi himself, the journey by tram and bus was too painful. The following day, Saturday the 6th, Pauli was taken to the Red Cross Hospital in Zurich, on Gloriastrasse near the Physics Institute. On Sunday morning Franca called me and I then called Pauli. On Monday evening I went to the hospital where I had a conversation with Franca, before Pauli returned from a medical examination. He asked me whether I had seen the room number. When I answered in the negative, he said: '137!' (see Eq. (7.2)). I said that this must mean good luck, but he was not convinced. Over the following days I had to teach Pauli's main course on 'Wavemechanics' (see Ref. [5], p. 432). On the 10th I informed President Pallmann of Pauli's illness and told him that Pauli had asked me to take over his main course (Ref. [5], IV.90). I went to the hospital several times; Pauli's mood varied. On one occasion he said: 'Soon Heisenberg will present his theory to Nasser [1918–70, Egyptian President].'

On Saturday the 13th Pauli had an operation. The report on the operation by Privatdozent Dr med. Franz Deucher mentions an enormous tumour in the pancreas which is unoperable. In the novel *Gerufen und nicht Gerufen* by the Zurich writer Kurt Guggenheim mentioned on pages 285 and 404, this episode is described in Chapter XXVI. There Pauli is called Paul Mende and the surgeon is Dr Däubler . Einstein, alias Onkel Chaim Mendeles and Heisenberg, alias Kempf, in Göttingen occur also. (Note that in September 1958 Heisenberg moved from Göttingen to Munich.) On Monday morning of the 15th, around 8.30 a.m., Ms Pauli called to say that Pauli had died. On this day Professor F. Trendelenburg, the president of the Association of German Physical Societies, had intended to deliver at last the Planck Medal to Pauli (see the previous section). So again this was not possible. Pauli's last reading at the hospital was 'The Life of Albertus Magnus' [31a]. According to Jung's secretary Aniela Jaffé, Pauli still wished to speak to one single person: C. G. Jung [31b]. I went to Zollikon to help Ms Pauli organize the funeral and then spent a long evening at the Fraumünster Post Office near the main churches of Zurich, sending out telegrams using the list of addresses she and I had prepared. On 16 December the rector of ETH, Professor Albert Frey-Wyssling (1900–88), sent out an official condolatory notice in the name of the authorities, teachers, and students of ETH (Ref. [5], IV.91). And on 19 December President Pallmann sent Franca an expression of condolence from ETH (Ref. [5], IV.96). In the evening of 17 December I had a conversation with Res Jost. He mentioned that during the times of his friendship with Pauli, the latter had told him of a dream which frightened Jost because it contained a rejection and a presentiment of death. Jost refused to be influenced by this dream, he could not follow Pauli on the path of the latter's dreams.

On Saturday 20 December, before noon, Pauli was cremated in the presence of family members and friends. As mentioned at the outset, in the crematorium I sat next to Paul Rosbaud (1896–1963), Pauli's publisher and friend (see the preface in Ref. [21a]). On a walk with Rosbaud after the ceremony he told me that Pauli's enthusiasm during his collaboration with Heisenberg a year earlier (see the previous section) had been inspired by the following dream. A boy and a girl are in Pauli's room the moment he enters. He calls: 'Franca, here are two children!' Pauli interpreted the two children as the new ideas of his and Heisenberg's theory (see also the children in Pauli's dream on the 'World Clock' on pages 246, 323, and 417). Rosbaud also mentioned a letter of 14 pages on questions of faith that Pauli had written to him.

At 3 p.m. the official ceremony was held at the Fraumünster Church which, like St Peter's and Grossmünster, dates back to the Carolingian period and is known today for its stained-glass windows by Marc Chagall and Augusto Giacometti. The non-religious ceremony consisted of addresses by Scherrer, Weisskopf (see Ref. [23], p. XXV), Bohr, Fierz, and Pauli's friend Adolf Guggenbühl (see pge 285). At 5 p.m. Franca gave a reception at the guild house 'Zur Meise' across the street from the church on the Limmat river, where just over two months earlier, on 10 October, she and Pauli had attended the wedding party of David Speiser and Ruth Bär (see page 491), a splendid black-tie dinner with dancing at which my wife and I had also participated. Professor Trendelenburg had been invited as the bearer of the Planck Medal. He handed the medal to Franca at the church ceremony and she gave it to me at the end, since she had to shake so many hands. Being very busy with the organization, I put the medal on a church bench. At the reception she asked me where the medal was—and I suddenly realized that I had left it at the church. I hurried back and, fortunately, found the medal where I had put it. The medal now rests in safety in the *Pauli Archive* at CERN, together with all of Pauli's diplomas.

Franca's task to arrange things after Pauli's death was daunting. I helped her to sort out manuscripts according to subject and put them into boxes. These boxes still exist in the *Pauli Archive*. Franca kept the urn containing Pauli's ashes at home in a wardrobe for several years, but finally decided to have them interred at the cemetery in Zollikon, where they rest next to the remains of his colleague and friend, the mathematics professor at ETH Heinz Hopf (see page 349). At the Physics Institute on Gloriastrasse Pauli's office 3e had to be cleared—a delicate enterprise since Jost, who still was only associate professor, was now in command, at least for practical matters. Freeman Dyson wrote to President Pallmann on 15 January 1959, urging him to promote Res Jost to be Pauli's successor, because otherwise ETH would very likely lose him (Ref. [5], IV.117). Fierz in Basle also wrote in the same sense on 19 January (IV.118). Jost, on the other hand, discussing the future of theoretical physics at ETH with me, thought it was unavoidable that Fierz would become Pauli's successor (IV.151)—or at least that he would share the succession with Jost (IV.164). The latter's immediate goal, however, was to become a full professor (IV.121). Concerning the clearing of Pauli's office, I was in a somewhat ambiguous situation with Franca Pauli and Res Jost. For the former my proximity to Jost at the institute made me appear untrustworthy. For Jost, however, I was a remainder of the Pauli era which for him was definitively over. But in spite of this, Jost in fairness renewed my assistant position for another year until the end of March 1960 (Ref. [5], IV.122, 133).

On 10 February 1959 President Pallmann wrote to Franca, enquiring about the possibility of ETH acquiring Pauli's personal library and, hopefully, also his scientific assets, including the considerable reprint collection; and he offered the services of the ETH library (Ref. [5], IV.137). Franca answered a week later, proposing to talk about the matter and emphasizing that her main concern was to preserve for posterity the image of the physicist Wolfgang Pauli (IV.138). On 2 March the director of the ETH library, Dr Paul Scherrer (no relation to the physicist) established an estimate of the content of Pauli's office, for which he proposed a purchase value of 4300 francs (IV.139), which price the president communicated to Franca on 6 March (IV. 141). President Pallmann also proposed the creation of a 'Pauli Room', an idea Franca took up, expressing the wish that the whole scientific estate, which she wished to donate to ETH, could be placed in such a 'Pauli Room' (IV.142). The president responded on 17 March, thanking Franca for her offer and proposing a 'Pauli Room' in the future institute at Hönggerberg.

This project, however, eventually failed. On 6 May Jost informed the president about the serious shortage of rooms in theoretical physics, and he proposed to use Pauli's office 3e for the teaching staff (Ref. [5], IV.155). The president, however, informed Jost a week later that he could not put Pauli's room at the disposal of theoretical physics as long as the question of the donation of the Pauli estate was not settled (IV.156). Then on 31 August Fierz had a phone conversation with President Pallmann (IV.188). Fierz informed the president that he had learned from Ms Pauli and also from conversations at CERN, where during the year 1959/60 he had been director of the Theory Division, that Franca had the intention to donate the estate of her late husband to CERN, and this after she had expressed in spring her intention to donate it to ETH! Even CERN's Director-General C.J. Bakker (1904–60), however, agreed with Fierz that the Pauli estate should go to ETH (IV.188).

It is obvious that this surprising idea of donating the Pauli estate to CERN had arisen, as a consequence of the rift between Franca and Jost, in her conversations with Weisskopf, a close friend since the days of his assistantship with Pauli in the mid-1930s (see pages 281f). Since his sabbatical year 1957/8 that he had spent at CERN, Weisskopf had kept in close contact with this laboratory and in December 1960 returned there to become its director-general on 1 July 1961 (see Ref. [32], pp. 218–20). In the evening of 31 August 1959 President Pallmann had tried to talk to Franca on the phone but she said she was expecting visitors and would call back (IV.188). Then on 9 November he received a long letter from her, reviewing the whole negotiation and ending with the information that she had disposed of her late husband's estate differently (IV.200). Thus the establishment of the Pauli estate at CERN, indirectly, was the result of the breach between Pauli and Jost.

More complications were to come, as I reported in Ref. [33a]. On 15 July 1957 Pauli had signed a contract with Interscience Publishers, Inc., New York, to publish the English translation of his essays by Robert Schlapp now contained in Ref. [21a]. When Pauli died, 'Franca very courageously took command of all editorial projects concerning her late husband. These projects now also included an agreement with Vieweg, Braunschweig, that Pauli himself had signed on 29 October 1958 to publish a German edition of the essays with the title "Aufsätze über Physik und Erkenntnistheorie" [33b] and a plan suggested by Rosbaud to publish the Collected Papers with Interscience [33c]. The "Aufsätze" were published in 1961 with the addition of "und Vorträge" in the title and of an obituary speech (*Trauerrede*) by V. F. Weisskopf. . . . On the other hand, the Collected Papers appeared in 1964 [9]. While none of

THE JOURNEY ENDS IN HOSPITAL ROOM 137 | **537**

these publications acknowledges the important rôle played by Paul Rosbaud, a memorial volume which originally had been planned to celebrate Pauli's 60th birthday on 25 April 1960 does mention Rosbaud in the preface [34]' (taken from Ref. [33a], p. 4). Pauli's relations with Paul Rosbaud were very friendly. Pauli called him 'Steinklopferhansl', who is a somewhat equivocal figure from a comedy with this name by Ludwig Anzengruber (see Ref. [7], comment, pp. 731–2). This 'Steinklopferhansl' is a poor stone mason living at the edge of a village community but who, by his harmless remarks, directs the things, unnoticed by the other members—a good and witted chap. How much Pauli knew about Rosbaud's behind-the-scenes activity is not evident.

Born in Graz, Austria, on 18 November 1896 Paul Rosbaud, the brother of the well-known conductor and expert in modern music, Hans Rosbaud, obtained his doctorate in X-ray physics in Berlin where he started his editorial career. With his Austrian charm and his enthusiasm he made friends among the scientific elite all over Europe. In the early 1930s Dr Ferdinand Springer offered him the position of scientific adviser with the prestigious publishing house run by him and his brother, Julius Springer, in Berlin. Rosbaud became increasingly hostile to the Nazi regime and during World War II was the most valuable scientific informant of the British government in Germany. After the war Rosbaud resumed his editing career in London, and in 1951 he became, together with Robert Maxwell, the founder, co-owner, and then the scientific director of a new venture called Pergamon Press, which, however, he left in 1956 in profound bitterness because of serious differences with Maxwell, to take up consulting functions with the leading scientific publishing houses. He was honoured in 1962 with the Tate Medal for scientific editing awarded by the American Institute of Physics. Paul Rosbaud died from leukaemia in 1963 (taken from Ref. [21a], pp. 2–4; see also Ref. [35]).

Pauli had asked his English-speaking colleagues M. H. L. Pryce (b. 1913), Kemmer, and Rosenfeld to check the translation of the more philosophical essays. Having received somewhat ambiguous comments on this matter after Pauli's death, Franca lost confidence in Rosbaud. A letter she wrote to E. Proskauer, the editor-in-chief of Interscience, on 27 January 1960 contained incorrect and offensive information concerning Rosbaud [33c] who, deeply hurt, sent me several documents on 4 April 1960 with the aim of setting things straight. 'Thus the Schlapp translations went into total oblivion, and I was convinced that they were lost. So my surprise was immense when Professor Kemmer, having secured the single carbon copy of the whole translation in Schlapp's home in Edinburgh, contacted Springer Publishers in December 1988 and, after examining the legal situation with them, sent the whole package to the Pauli Estate at CERN for reconsideration' (Ref. [33a], p. 4). Asked to act as the editor, I found the task too considerable to do by myself and I proposed to share the work with Karl von Meyenn who had already edited the second printing of the German edition of Pauli's essays [23].

Franca also got involved with the project of publishing the Pauli-Jung correspondence within the *Collected Works* of C. G. Jung. In its edition of 7

May 1963 the *Neue Zürcher Zeitung* ran a notice by the editor of the Jung correspondence, Dr Gerhard Adler in London, asking recipients of Jung letters to send them to Jung's daughter Ms Marianne Niehus-Jung. On 7 August 1963 Dr Adler contacted Franca in a personal letter. But Franca's reaction was strongly negative. On 22 August she wrote to Dr Adler: 'As heir and legal successor of my late husband, Prof. Pauli, I must decidedly reject the idea that the letters *of Prof. Pauli to* Prof. Jung are used in one way or another, independently to what extent, in which manner and to what purpose' (*Pauli Archive*, translated). Finally, Dr Adler thanked Franca on 10 June 1966 for her permission to publish Jung's letter [8] of 29 October 1934 to Pauli in Ref. [2], but leaving out the last passage, in which Jung thanks Pauli for the news about his personal well-being.

In 1954 Pauli had two dreams which in retrospect might be interpreted as a premonition of his death in December 1958, of which he was not aware, however. In the dream of 20 July 1954 Niels Bohr officially informs Pauli that three Popes, of whom one has the name Johannes, donate a new house to Pauli (Ref. [2], p. 135; for a slightly different version see Ref. [13a], appendix to letter [2209], also letter [2367]). In the second dream, of 28 August 1954, Pauli walks from the tram stop on a winding uphill path to a new building of ETH, where he finds his office. On a table in the office he finds two letters, the first one, which is signed by President Pallmann, contains an order for Pauli to pay 568 francs. The contents of the second letter, whose sender is a 'philosophical choir (*Gesangsverein*)', are nice red cherries, which he eats (Ref. [2], p. 138; for a slightly different version see Ref. [7], appendix to letter [1883], and Ref. [13a], appendix to letter [2209]). In his comment on this dream Pauli observes that he had occasionally used the expression 'philosophical choir' because 'the philosophy of the *contemporary* professional philosophers seems to me not really to be done with and for the intellect but appears to me as an implicit (*verklausulierte*) feeling attitude' (translated from Ref. [2], p. 138).

As for the Popes, the following reflection was communicated to me by Markus Fierz in two letters dated 11 and 21 August 1992. Concerning the figures of the Popes in general, in his first letter to me Fierz observes that Bohr may be called the 'Pope of the Quanta'. And as regards the number 568 he asks: Who was Pope in the year 568? Answer: Johannes III, from 561 to 574. Now one finds in history books that the most important event during the papacy of Johannes III was the intrusion into Italy of the Langobards under King Alboin in the year 568! Fierz then observes that the election of Cardinal Roncalli as Pope Johannes XXIII took place on 28 October 1958. 48 days later Pauli died. Therefore Popes, and particularly those named Johannes, were connected to Pauli's fate. And the strange number 568 enhances the strength of this connection (see also Fierz's letter [2213] in Ref. [13a]).

As for the 'new house', its situation described in the dream of 28 August 1954 (Ref. [2], p. 138) resembles surprisingly that of the future physics building on Hönggerberg (see the previous section). Pauli interprets it as a 'reformed ETH' where besides physics and mathematics, other, new subjects are also to be taught by Pauli. It is a place of 'reunion of the *opposites*' (Ref. [2], p. 138), as are churches (p. 149) or cathedrals which should be built to revere the 'isomorphy', as said by Pauli's mathematics teacher Emil Artin (1898–1962) from Hamburg in Pauli's dream of 6 September 1954 (p. 139). To Pauli isomorphy stands for the demand to reconcile old insights with a new form (p. 141). In a certain sense complementarity also has this reconciliatory meaning. Indeed, Pauli writes: 'It seems to me that *psychologically*, Bohr's complementarity is a model or a prefiguration of the *unification of opposites*' (translated from letter [2345] of Ref. [13a]). In the dream of 12 April 1955 the 'new house' appears as a laboratory where experiments 'with two neutrinos' are done, which Pauli interprets as a place where a synthesis of analytical psychology, physics, and biology occurs (Ref. [2], p. 146). And in the dream of 12 August 1955 it is a house in a place called 'Enzdorf' or 'Lenzdorf', given to Pauli and his wife (p. 148). Here the names 'Lenz' and 'Enz' symbolize the span of Pauli's scientific life from Munich to Hamburg to Zurich and to his death. But 'Always the new house is *the place where a unification of pairs of opposites, a coniunctio, occurs*' (translated from Ref. [2], p. 136).

In looking back from the threshold of the twenty-first century, Pauli appears as a towering figure of the twentieth, as a man of uncompromising scientific honesty, whose superb intellect and memory let him feel at home not only in science but also in the history of civilization and in the realm of the soul. Pauli's refusal to compromise did not favour speculation—except in the Jungian sense of the amplification of dream figures—and his deep insight often let him see the weakness of an idea before elaborating on it. As a consequence, his fundamental contributions to physics were all the result of analytical hard work. This is true particularly for the exclusion principle (pages 119f), for the idea of the neutrino (pages 213 and 488f) and for the spin-statistics relation (pages 323f). But this method of research did not enhance Pauli's popularity. In a survey by the British journal *Physics World* of the year 2000 Pauli does not figure among the 10 most important physicists ever. Neither had it helped to overcome the stagnation of quantum field theory which in mid-century, quite contrary to Pauli's hopes, ran into the rigidity of axiomatics. In fact, the former gave way to a more intuitive and much freer manner of doing physics only towards the end of the century (see e.g. Ref. [36a]), paralleled by the spectacular advances in superconductivity, which served as a model for spontaneous symmetry breaking and made use of a degenerate vacuum (see e.g. Ref. [36b]).

But Pauli's idea 'that the masses [of the particles] should *not* be supposed to be given', expressed in his letter to me of 17 May 1958 (see the last section), has been taken care of in the so-called Higgs mechanism, by which masses are generated through the coupling to a special scalar field (see e.g. Ref. [36a]). And the unifying efforts in the theory of elementary particles all culminated in today's so-called standard model (see e.g. Ref. [36a]). Besides the unification of the weak and electromagnetic interactions into electroweak theory, which had been honoured by the Nobel Prize of 1979 to Sheldon L. Glashow, Abdus Salam and Steven Weinberg mentioned on page 480, the standard model is based on quantum chromodynamics, which is a gauge field theory with fermionic matter (see pages 474f). And it was 'for having placed particle physics theory on a firmer mathematical foundation' that Gerhard 't Hooft and Martinus J. G. Veltman were awarded the Nobel Prize of 1999 [37].

So much for the developments in physics. But what was the future of Pauli's 'background physics'? Well, it represented his personal dream world during a period of his life, and therefore it died with him. However, Pauli developed ideas about the future of religion. In one of the rare documents addressed to his sister Hertha after his visit to Israel in the summer of 1957 (see Ref. [7], p. 962; see also the last section) he writes: 'I do not believe in the possible future of mysticism in the old form. However, I do believe that the natural sciences will out of themselves bring forth a counter pole in their adherents, which connects to the old mystic elements. On this subject I have tried to find general formulations in my paper on Kepler [Ref. [21a], paper 20], as also in a lecture on "Science and Western Thought" [Ref. [21a], paper 16]' (translated from Ref. [7], p. xxxv). At the same time Pauli writes on 11 October 1957 to the science historian Shmuel Sambursky whom he had met on his trip to Israel (see Ref. [7], p. 964): 'In opposition to the monotheist religions—but in unison with the mysticism of all peoples, including the Jewish mysticism—I believe that the ultimate reality is *not* personal.' And enumerating the different religious figures, Pauli concludes: 'It is the task of man to *irrealize* (teaching of Yoga) these figures by personal relations with them. Now, in this sense Jahwe for me is a local demon, who develops his action particularly in Israel. How then has he behaved with me? He was relatively mild, he only beat me gently on the left ear' (translated from Ref. [7], p. xxxv).

Notes and References

[1] Translated from: W. Pauli, letter [55] of 27 February 1952 to C. G. Jung, Ref. [2], p. 81. Reprinted in Ref. [3], letter [1373], p. 560.

[2] C. A. Meier (ed., with the cooperation of C. P. Enz and M. Fierz), *Wolfgang Pauli und C. G. Jung. Ein Briefwechsel 1932–1958* (Springer, Berlin and Heidelberg, 1992).

[3] K. von Meyenn (ed.), *Wolfgang Pauli. Scientific Correspondence with Bohr, Einstein, Heisenberg, a.o., Volume IV, Part I: 1950–1952* (Springer, Berlin and Heidelberg, 1996).

[4] C. G. Jung, (a) *Psychology of the Transference*, in Ref. [4c], Vol. 16, Part Two, Section III; (b) *Aion*, Ref. [4c], Vol. 9, Part II; (c) *Collected Works* (Princeton University Press, Princeton, NJ, 1967–78).

[5] C. P. Enz, B. Glaus, and G. Oberkofler (eds), *Wolfgang Pauli und sein Wirken an der ETH Zürich* (vdf Hochschulverlag ETH, Zurich, 1997).

[6] T. D. Lee, 'Some Special Examples in Renormalizable Field Theory', *Physical Review* **95**, 1329 (1954).

[7] K. von Meyenn (ed.), *Wolfgang Pauli. Scientific Correspondence with Bohr, Einstein, Heisenberg, a.o., Volume IV, Part II: 1953–1954* (Springer, Berlin and Heidelberg, 1999).

[8] G. Källén and W. Pauli, 'On the Mathematical Structure of T. D. Lee's Model of a Renormalizable Field Theory', *Kongelige Danske Vidensk-abernes Selskab, Math.-Fys. Meddelelser* **30**, No. 7 (1955). (Dedicated to Professor Niels Bohr on the occasion of his 70th birthday.) Reprinted in Ref. [9], Vol. 2, pp. 1261–81.

[9] R. Kronig and V. F. Weisskopf (eds), *Collected Scientific Papers by Wolfgang Pauli. In Two Volumes* (Wiley Interscience, New York, 1964).

[10] (a) G. Lüders, 'On the Equivalence of Invariance under Time Reversal and under Particle–Antiparticle Conjugation for Relativistic Field Theories', *Kongelige Danske Videnskabernes Selskab, Math.-Fys. Meddelelser* **28**, No. 5 (1954); (b) W. Pauli, 'The Connection between Spin and Statistics', *Physical Review* **58**, 716 (1940). Reprinted in Ref. [9], Vol. 2, pp. 911–17.

[11] W. Pauli, 'Exclusion Principle, Lorentz Group and Reflection of Space–Time and Charge', in: W. Pauli (ed., with the assistance of L. Rosenfeld and V. Weisskopf), *Niels Bohr and the Development of Physics* (Pergamon, London, 1955), pp. 30–51. Reprinted in Ref. [9], Vol. 2, pp. 1239–60.

[12] (a) J. Schwinger, 'The Theory of Quantized Fields. I', *Physical Review* **82**, 914 (1951); (b) R. Jost, 'Eine Bemerkung zum CPT-Theorem', *Helvetica Physica Acta* **30**, 409 (1957).

[13] (a) K. von Meyenn (ed.), *Wolfgang Pauli. Scientific Correspondence with Bohr, Einstein, Heisenberg, a.o., Volume IV, Part III: 1955–1956* (Springer, Berlin and Heidelberg, 2001); (b) W. Pauli, 'Remarks on Problems Connected with the Renormalization of Quantized Fields', *Nuovo Cimento* **4**, Supplement, 703 (1956). Reprinted in Ref. [9], Vol. 2, pp. 1282–9.

[14] (a) W. Heisenberg, F. Kortel, and H. Mitter, 'Zur Quantentheorie nicht-linearer Wellengleichungen, III', *Zeitschrift für Naturforschung* **10a**, 425 (1955). Reprinted in Ref. [15], Part III, pp. 214–35. (b) C. P. Enz, 'Wechselwirkungskräfte und Renormalisation in der Photonpaar-Theorie', *Nuovo Cimento* **3**, Supplement, 363 (1956); (c) W. Thirring, 'Renormalisation der Meson-Paartheorie', *Helvetica Physica Acta* **28**, 344 (1955); (d) N. G. van Kampen, 'Contribution to the Quantum Theory of Light Scattering', *Kongelige Danske Videnskabernes Selskab, Math.-Fys. Meddelelser* **26**, No. 15 (1951); (e) W. Heitler, 'Renormalization in Non-Relativistic Field Theories', *Nuovo Cimento* **5**, 302 (1957); (f) W. Heitler, *The Quantum Theory of Radiation* (Clarendon, Oxford, 1954, 3rd edn).

[15] W. Blum, H.-P. Dürr, and H Rechenberg (eds), *Werner Heisenberg. Collected Works. Series A: Original Scientific Papers* (Springer, Berlin and Heidelberg, 1989).

[16] (a) L. Giulotto (ed.), *International School of Physics 'Enrico Fermi'*, *Course IV*, Varenna, Italy, July 15–August 4, 1956, *Nuovo Cimento* **6**, Supplement (1957), pp. 808–1238; (b) C. P. Enz, 'Magnetic Susceptibility of Electrons in Periodic Fields', in Ref. [16a], p. 1224; (c) C. P. Enz and K. von Meyenn (eds), *Wolfgang Pauli. Das Gewissen der Physik* (Vieweg, Braunschweig, 1988); (d) Notes of my conversation with Franca Pauli of 8 April 1984.

[17] (a) W. Pauli, 'On the Conservation of the Lepton Charge', *Nuovo Cimento* **6**, 204 (1957). Reprinted in Ref. [9], Vol. 2, pp. 1338–49. (b) C. P. Enz, 'Fermi Interaction with Non-Conservation of "Lepton Charge" and of Parity', *Nuovo Cimento* **6**, 254 (1957); (c) C. P. Enz, 'The Production and Absorption of Neutrinos in Beta-Decay Theory', *Helvetica Physica Acta* **31**, 69 (1958).

[18] (a) W. Pauli, letter to V. Weisskopf of 27 January 1957, reproduced in facsimile in Ref. [9], Vol. 1, pp. xiii–xvi; English translation ibid. pp. xvii–xviii; (b) T. D. Lee and C. N. Yang, 'Question of Parity Conservation in Weak Interactions', *Physical Review* **104**, 254 (1956).

[19] (a) C. S. Wu, E. Ambler, R. W. Hayward, D. D. Hoppes, and R. P. Hudson, 'Experimental Test of Parity Conservation in Beta Decay' (letter), *Physical Review* **105**, 1413 (1957); (b) R. L. Garwin, L. M. Lederman, and M. Weinrich, 'Observations of the Failure of Conservation of Parity and Charge Conjugation in Meson Decays: the Magnetic Moment of the Free Muon' (letter), *Physical Review* **105**, 1415 (1957); (c) J. I. Friedman and V. L. Telegdi, 'Nuclear Emulsion Evidence for Parity Nonconservation in the Decay Chain' (letter), *Physical Review* **105**, 1681 (1957).

[20] (a) T. D. Lee, R. Oehme and C. N. Yang, 'Remarks on Possible Noninvariance under Time Reversal and Charge Conjugation', *Physical Review* **106**, 340 (1957); (b) T. D. Lee and C. N. Yang, 'Parity Nonconservation and a Two-Component Theory of the Neutrino', *Physical Review* **105**, 1671 (1957); (c) A. Salam, 'On Parity Conservation and Neutrino Mass', *Nuovo Cimento* **5**, 229 (1957).

[21] (a) C. P. Enz and K. von Meyenn (eds), *Wolfgang Pauli. Writings on Physics and Philosophy* (Springer, Berlin and Heidelberg, 1994); (b) P. Jordan, *Schwerkraft und Weltall* (Vieweg, Braunschweig, 1955, 2nd edn); (c) D. W. Sciama, 'On the Origin of Inertia', *Monthly Notices of the Royal Astronomical Society* **113**, 34 (1953).

[22] W. Pauli, 'Zur älteren und neueren Geschichte des Neutrinos', in Ref. [23], pp. 156–80. Reprinted in Ref. [9], Vol. 2, pp. 1313–1337. English translation, 'On the Earlier and More Recent History of the Neutrino', in Ref. [21a], pp. 193–217.

[23] W. Pauli, *Aufsätze und Vorträge über Physik und Erkenntnistheorie* (Vieweg, Braunschweig, 1961). Reprinted as *Physik und Erkenntnistheorie*. Mit einleitenden Bemerkungen von Karl von Meyenn (Vieweg, Braunschweig, 1984).

[24] H. Lehmann, 'Über Eigenschaften von Ausbreitungsfunktionen und Renormierungskonstanten quantisierter Felder', *Nuovo Cimento* **11**, 342 (1954).

[25] W. Heisenberg and W. Pauli, 'On the Isospin Group in the Theory of the Elementary Particles' (unpublished). Reprinted in Ref. [15], Vol. III, pp. 337–51.

[26] (a) F. Gürsey, 'Relation of Charge Independence and Baryon Conservation to Pauli's Transformation', *Nuovo Cimento* (letter) **7**, 411 (1958); (b) W. Pauli, 'Albert Einstein in der Entwicklung der Physik', in Ref. [23], pp. 85–90; English translation, 'Albert Einstein and the Development of Physics', in Ref. [21a], pp. 117–23.

[27] (a) Letter from C. N. Yang to C. P. Enz, dated Stony Brook, 14 October 1985. I thank Professor Yang for this information. (b) R. L. Garwin and T. D. Lee, 'Chien-Shiung Wu', *Physics Today*, October 1997, p. 120; (c) *Physikalische Verhandlungen. Verbandsausgabe*, Vol. 9 (1958).

[28] (a) J. Mehra (ed.), *The Solvay Conferences on Physics. Aspects of the Development of Physics since 1911* (Reidel, Dordrecht, 1975); (b) B. Ferretti (ed.), *1958 Annual International Conference on High Energy Physics at CERN* (CERN, Geneva, 1958); (c) W. Pauli, 'The Indefinite Metric with Complex Roots', Ref. [28b], p. 127. Reprinted in Ref. [9], Vol. 2, pp. 1381–2. (d) I thank Dr André Martin from the Theory Division of CERN for copies of these letters; (e) W. Heisenberg, *Der Teil und das Ganze. Gespräche im Umkreis der Atomphysik* (Piper, Munich, 1969); English translation by A. J. Pomerans, *Physics and Beyond* (Allen & Unwin, London, 1971).

[29] (a) E-mail from H. Frauenfelder of 20 October 2000, Urbana, IL. I thank Professor Frauenfelder for this information; (b) Notes of my conversation with P. C. Gugelot, Zurich, 1 March 1997. I thank Prof. Gugelot for this information.

[30] Translated from: P. Rosbaud, letter to Charles P. Enz, London, 23 December 1958.

[31] (a) R. Baumgardt, *Der Magier. Das Leben des Albertus Magnus* (Funk, Munich, 1949); (b) H. van Erkelens, *Wolfgang Paulis Begegnung mit dem Geist der Materie*, *Jungiana*, Series A, Vol. 4 (1992).

[32] V. Weisskopf, *The Joy of Insight. Passions of a Physicist* (Basic Books, New York, 1991).

[33] (a) C. P. Enz, 'The History of this Translation: Paul Rosbaud, Friend and Publisher of Wolfgang Pauli', Preface in Ref. [21a], pp. 1–5; (b) *Vereinbarung* signed on 16 October 1958 by Friedr. Vieweg & Sohn. Copy dated by Pauli, in *Pauli Archive*, CERN, Geneva. (c) Correspondence between Franca Pauli and Eric S. Proskauer, Editor-in-Chief, 'Interscience Publishers, Inc.', in *Pauli Archive*, CERN, Geneva.

[34] M. Fierz and V. F. Weisskopf (eds), *Theoretical Physics in the Twentieth Century. A Memorial Volume to Wolfgang Pauli* (Wiley Interscience, New York, 1960).

[35] A. Kramish, *The Griffin* (Houghton Mifflin, Boston, 1986).

[36] (a) P. Ramond, *Journeys Beyond the Standard Model* (Perseus Books, Cambridge, MA, 1999); K. Huang, *Quarks, Leptons and Gauge Fields* (World Scientific, Singapore, 1992), 2nd edn; (b) P. W. Anderson, *Basic Notions of Condensed Matter Physics* (Benjamin/Cummings, Menlo Park, CA, 1984).

[37] G. B. Lubkin, 'Nobel Prize to 't Hooft and Veltman for Putting Electroweak Theory on Firmer Foundation', *Physics Today*, December 1999, p. 17.

Name index

Subject index

16543740R00333

Made in the USA
Middletown, DE
17 December 2014